W9-ATX-684

Processes and Design for Manufacturing

SECOND EDITION

Sherif D. El Wakil
Chancellor Professor and Chairman
Department of Mechanical Engineering
University of Massachusetts Dartmouth

PWS PUBLISHING COMPANY
I(T)P An International Thomson Publishing Company

Boston • Albany • Bonn • Cincinnati • London • Melbourne • Mexico City
New York • Paris • San Francisco • Singapore • Tokyo • Toronto • Washington

To the memory of Mamdouh El-Wakil, M.D., Ph.D.

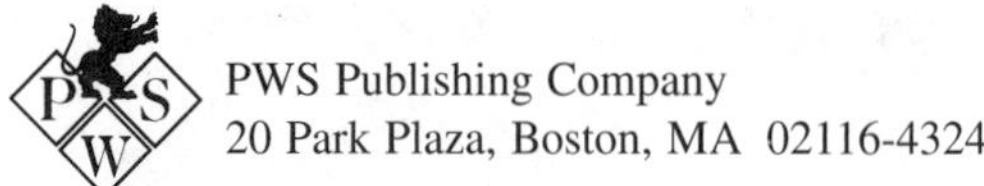

PWS Publishing Company
20 Park Plaza, Boston, MA 02116-4324

Printed in the United States of America.
98 99 00 01 02 — 10 9 8 7 6 5 4 3 2 1

Sponsoring Editor: *Bill Barter*
Developmental Editor: *Mary Thomas Stone*
Production Editor: *Andrea Goldman*
Manufacturing Coordinator: *Andrew Christensen*
Marketing Manager: *Nathan Wilbur*
Copyeditor: *Jean Peck*
Proofreader: *Helen Walden*
Composition: *The Clarinda Company*
Cover Design: *Jean Hammond*
Interior Design: *Julia Gecha*
Text Printer/Binder: *R R. Donnelley & Sons/Crawfordsville*
Cover Printer: *Phoenix Color Corp.*

Library of Congress Cataloging-in-Publication-Data

El Wakil, Sherif D.
Design for manufacturing processes / Sherif D. El Wakil. -- 2nd ed.
p. cm.
Rev. ed. of: Processes and design for manufacturing. Englewood Cliffs, N.J.: Prentice Hall, 1989.
Includes bibliographical references and index.
ISBN 0–534–95165–1
1. Manufacturing processes. 2. Metal-work. 3. Computer integrated manufacturing systems. I. El Wakil, Sherif, Processes and design for manufacturing II. Title.
TS183.E47 1998 97-31812
670.42--dc21 CIP

I(T)P®
International Thomson Publishing
The trademark ITP is used under license.

For more information, contact:
PWS Publishing Company
20 Park Plaza
Boston, MA 02116

International Thomson Publishing Europe
Berkshire House
168-173 High Holborn
London WC1V 7AA
England

Thomas Nelson Australia
102 Dodds Street
South Melbourne, 3205
Victoria, Australia

Nelson Canada
1120 Birchmont Road
Scarborough, Ontario
Canada M1K 5G4

International Thomson Editores
Campos Eliseos 385, Piso 7
Col. Polanco
11560 Mexico D.F., Mexico

International Thomson Publishing GmbH
Königswinterer Strasse 418
53227 Bonn, Germany

International Thomson Publishing Asia
221 Henderson Road
#05-10 Henderson Building
Singapore 0315

International Thomson Publishing Japan
Hirakawacho Kyowa Building, 31
2-2-1 Hirakawacho
Chiyoda-ku, Tokyo 102
Japan

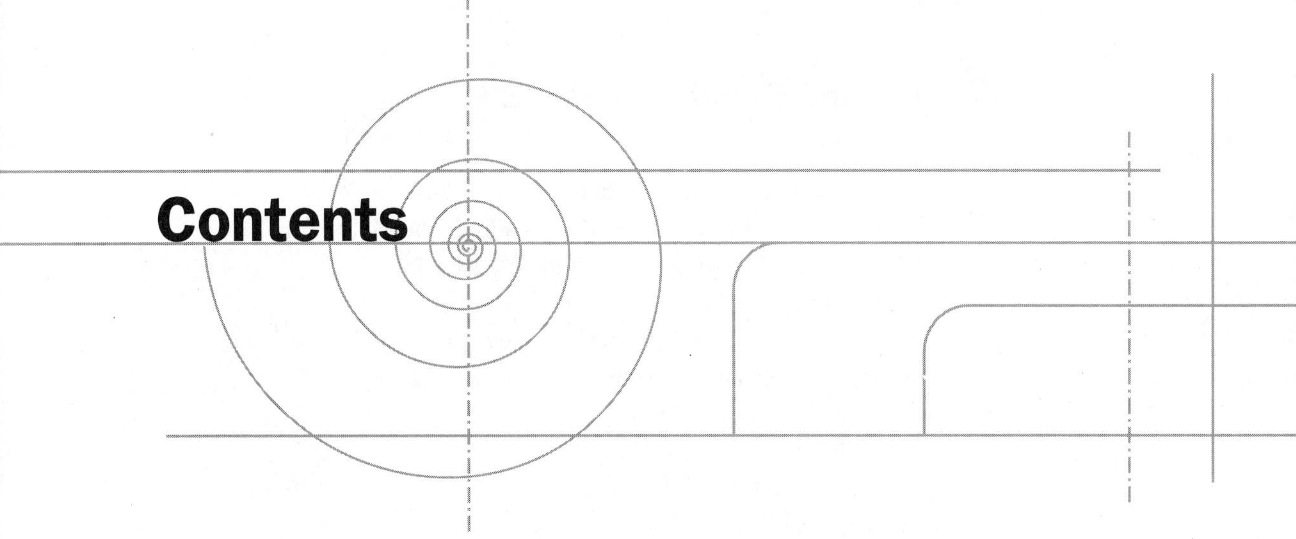

Contents

Chapter 3 Casting and Foundry Work 33

Chapter 4 Joining of Metals 84

Chapter 5 Metal Forming 139

Chapter 9 Physics of Metal Cutting 331

Chapter 10 Machining of Metals 361

Chapter 11 Product Cost Estimation 415

Preface

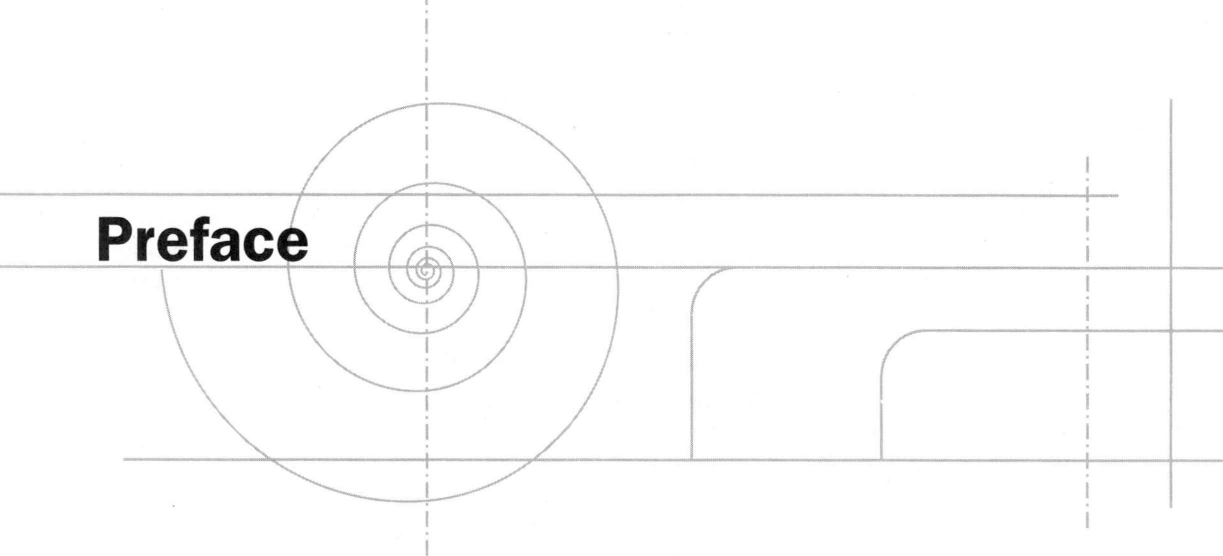

At the time the author's first book on processes and design for manufacturing was published, the main concern of the manufacturing/engineering academic community was the erroneous picture of manufacturing as involving little more than manual training (i.e., manual skills acquired by on-site training in machine shops and the like). Unfortunately, this distorted view of manufacturing was created and fueled by the shallow, descriptive, and qualitative manner in which the vast majority of books then covered the subject. Now, design for manufacturing is a "hot topic," and engineers in all disciplines are beginning to realize its strategic importance. Many government programs are aimed at enhancing the efficiency of product development and design. The present text serves to provide engineering students with the knowledge and skills required for them to become good product designers.

The design component in this book has been strengthened by adding four new chapters:

- Chapter 2, Concurrent Engineering
- Chapter 11, Product Cost Estimation
- Chapter 12, Design for Assembly
- Chapter 13, Environmentally Conscious Design and Manufacturing

Also, whenever applicable, chapters have been supplemented by design examples illustrating the interaction between design and manufacturing and showing how products can be designed for producibility, taking factors like the lot size into consideration. In addition, some design projects, which were previously assigned at the University of Massachusetts Dartmouth, have been given at the end of several chapters. Students are encouraged to use computational tools like spreadsheets and other software for modeling and analysis.

The text has also been supplemented with an appendix that covers the fundamentals of materials engineering. It provides a basis for understanding manufacturing processes, as well as for selecting materials during the product design process. It is aimed at engineering students who have not taken materials science as a prerequisite for a course on manufacturing processes but is not meant as a substitute for any materials science textbook.

The author wishes to acknowledge the contributions of the many corporations and individuals who supplied various figures and photographs or provided software to aid in producing this book, chief among them Silverscreen. Thanks are also extended to reviewers of the manuscript:

Mary C. Kocak, Pellissippi State Technical Community College

Zhongming (Wilson) Liang, Purdue University—Fort Wayne

Wen F. Lu, University of Missouri—Rolla

Antonio Minardi, University of Central Florida

Charles Mosier, Clarkson University

Masud Salimian, Morgan State University

Richard D. Sisson, Jr., Worcester Polytechnic Institute

Joel W. Troxler, Montana State University

David C. Zenger, Worcester Polytechnic Institute

Yuming Zhang, University of Kentucky

A note of gratitude also goes to Ana Gonzalez for her hard work in typing the manuscript. The author wishes to thank the staff at PWS Publishing, especially Andrea Goldman and Jean Peck, for their encouragement and support. Finally, the author must express his profound gratitude to his wife and children for their patience as the huge task of completing this second edition unfolded. God knows the sacrifice they gave.

Sherif D. El Wakil
North Dartmouth, Massachusetts

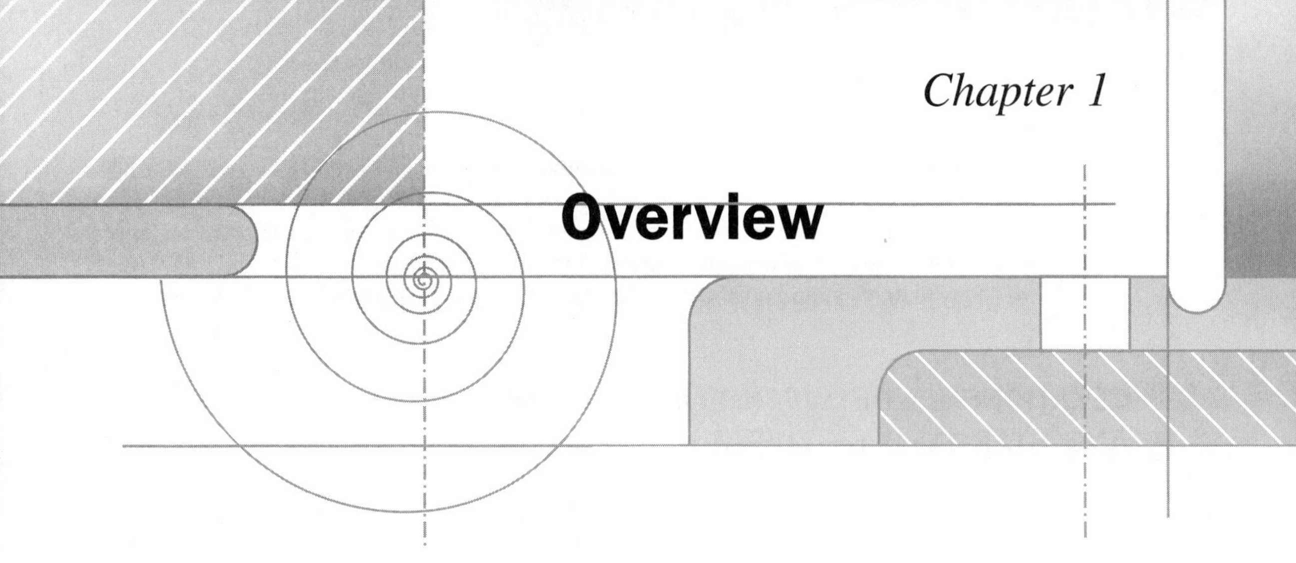

Chapter 1

Overview

INTRODUCTION

Before learning about various manufacturing processes and the concept of design for manufacturing, we first must become familiar with some technical terms that are used frequently during the planning for and operation of industrial manufacturing plants. We also must understand thoroughly the meaning of each of these terms, as well as their significance to manufacturing engineers. The explanation of the word *manufacturing* and its impact on the life-style of the people of industrialized nations should logically come at the beginning. In fact, this chapter will cover all these issues and also provide a better understanding of the design process, as well as the different stages involved in it. Finally, the concept of *design for manufacturing* and why it is needed will be explained.

1.1 DEFINITION OF MANUFACTURING

Manufacturing can be defined as the transformation of raw materials into useful products through the use of the easiest and least expensive methods. It is not enough, therefore, to process some raw materials and obtain the desired product. It is, in fact, of major importance to achieve this goal by employing the easiest, fastest, and most efficient methods. If less efficient techniques are used, the production cost of the manufactured part will be high, and the part will not be as competitive as similar parts produced by other manufacturers. Also, the production time should be as short as possible in order to capture a larger market share.

The function of a manufacturing engineer is, therefore, to determine and define the equipment, tools, and processes required to convert the design of the desired product into reality in an efficient manner. In other words, it is the engineer's task to find out the most appropriate, optimal combination of machinery, materials, and methods needed to achieve economical and trouble-free production. Thus, a manufacturing engineer must have a strong background in materials and up-to-date machinery, as well as the ability to develop analytical solutions and alternatives for the open-ended problems experienced in manufacturing. An engineer must also have a sound knowledge of the theoretical and practical aspects of the various manufacturing methods.

1.2 RELATIONSHIP BETWEEN MANUFACTURING AND STANDARD OF LIVING

The standard of living in any nation is reflected in the products and services available to its people. In a nation with a high standard of living, a middle-class family usually owns an automobile, a refrigerator, an electric stove, a dishwasher, a washing machine, a vacuum cleaner, a stereo, and, of course, a television set. Such a family also enjoys health care that involves modern equipment and facilities. All these goods, appliances, and equipment are actually raw materials that have been converted into manufactured products. Therefore, the more active in manufacturing raw materials the people of a nation are, the more plentiful those goods and services become; as a consequence, the standard of living of the people in that nation attains a high level. On the other hand, nations that have raw materials but do not fully exploit their resources by manufacturing those materials are usually poor and are considered to be underdeveloped. It is, therefore, the know-how and capability of converting raw materials into useful products, not just the availability of minerals or resources within its territorial land, that basically determines the standard of living of a nation. In fact, many industrial nations, such as Japan and Switzerland, import most of the raw materials that they manufacture and yet still maintain a high standard of living.

1.3 OVERVIEW OF THE MANUFACTURING PROCESSES

The final desired shape of a manufactured component can be achieved through one or more of the following four approaches:

1. Changing the shape of the raw stock without adding material to it or taking material away from it. Such change in shape is achieved through plastic deformation, and the manufacturing processes that are based on this approach are referred to as *metal forming processes*. These processes include bulk forming processes like rolling, extrusion, forging, and drawing, as well as sheet metal forming operations like bending, deep drawing, and embossing. Bulk forming operations are covered in Chapter 5, and the working of sheet metal is covered in Chapter 6.
2. Obtaining the required shape by adding metal or joining two metallic parts together, as in welding, brazing, or metal deposition. These operations are covered in Chapter 4.

3. Molding molten or particulate metal into a cavity that has the same shape as the final desired product, as in casting and powder metallurgy. These processes are covered in Chapters 3 and 7, respectively.
4. Removing portions from the stock material to obtain the final desired shape. A cutting tool that is harder than the stock material and possesses certain geometric characteristics is employed in removing the undesired material in the form of chips. Several chip-making (machining) operations belong to this group. They are exemplified by turning, milling, and drilling operations and are covered in Chapter 10. The physics of the process of chip removal is covered in Chapter 9.

1.4 TYPES OF PRODUCTION

Modern industries can be classified in different ways. These classifications may be by process, or by product, or based on production volume and the diversity of products. Classification by process is exemplified by casting industries, stamping industries, and the like. Classification by product indicates that industries may belong to the automotive, aerospace, and electronics groups. Classification based on production volume identifies three distinct types of production: mass, job shop, and moderate. Let us briefly discuss the features and characteristics of each type. We will also discuss the subjects in greater depth later in the text.

Mass Production

Mass production is characterized by the high production volume of the same (or very similar) parts for a prolonged period of time. An annual production volume of less than 50,000 pieces cannot usually be considered as mass production. As you may expect, the production volume is based on an established or anticipated sales volume and is not directly affected by the daily or monthly orders. The typical example of mass-produced goods is automobiles. Because that type attained its modern status in Detroit, it is sometimes referred to as the *Detroit type.*

Job Shop Production

Job shop production is based on sales orders for a variety of small lots. Each lot may consist of up to 200 or more similar parts, depending upon the customer's needs. It is obvious that this type of production is most suitable for subcontractors who produce varying components to supply various industries. The machines employed must be flexible to handle frequent variations in the configuration of the ordered components. Also, the personnel employed must be highly skilled in order to handle a variety of tasks that differ for the different parts that are manufactured.

Moderate Production

Moderate production is an intermediate phase between the job shop and the mass production types. The production volume ranges from 10,000 to 20,000 parts, and the machines employed are flexible and multipurpose. This type of production is gaining popularity in industry because of an increasing market demand for customized products.

1.5 FUNDAMENTALS OF MANUFACTURING ACCURACY

Modern manufacturing is based on flow-type "mass" assembly of components into machines, units, or equipment without the need for any fitting operations performed on those components. That was not the case in the early days of the Industrial Revolution, when machines or goods were individually made and assembled and there was always the need for the "fitter" with his or her file to make final adjustments before assembling the components. The concepts of mass production and interchangeability came into being in 1798, when the American inventor Eli Whitney (born in Westboro, Massachusetts) contracted with the U.S. government to make 10,000 muskets. Whitney started by designing a new gun and the machine tools to make it. The components of each gun were manufactured separately by different workers. Each worker was assigned the task of manufacturing a large number of the same component. Meanwhile, the dimensions of those components were kept within certain limits so that they could replace each other if necessary and fit their mating counterparts. In other words, each part would fit any of the guns he made. The final step was merely to assemble the interchangeable parts. By doing so, Eli Whitney established two very important concepts on which modern mass production is based—namely, interchangeability and fits. Let us now discuss the different concepts associated with the manufacturing accuracy required for modern mass production technologies.

Tolerances

A very important fact of the manufacturing science is that it is almost impossible to obtain the desired nominal dimension when processing a workpiece. This is caused by the inevitable, though very slight, inaccuracies inherent in the machine tool, as well as by various complicated factors like the elastic deformation and recovery of the workpiece and/or the fixture, temperature effects during processing, and sometimes the skill of the operator. Because it is difficult to analyze and completely eliminate the effects of these factors, it is more feasible to establish a permissible degree of inaccuracy or a permissible deviation from the nominal dimension that would not affect the proper functioning of the manufactured part in any detrimental way. According to the International Standardization Organization (ISO) system, the nominal dimension is referred to as the *basic size* of the part.

The deviations from the basic size to each side (in fact, both can also be on the same side) determine the high and the low limits, respectively, and the difference between these two limits of size is called the *tolerance.* It is an absolute value without a sign and can also be obtained by adding the absolute values of the deviations. As you may expect, the magnitude of the tolerance is dependent upon the basic size and is des-

FIGURE 1.1 The relationship between tolerance and production cost

ignated by an alphanumeric symbol called the *grade.* There are 18 standard grades of tolerance in the ISO system, and the tolerances can be obtained from the formulas or the tables published by the ISO.

Smaller tolerances, of course, require the use of high-precision machine tools in manufacturing the parts and, therefore, increase production cost. Figure 1.1 indicates the relationship between the tolerance and the production cost. As can be seen, very small tolerances necessitate very high production cost. Therefore, small tolerances should not be specified when designing a component unless they serve a certain purpose in that design.

Fits

Before two components are assembled together, the relationship between the dimensions of the mating surfaces must be specified. In other words, the location of the zero line (i.e., the line indicating the basic size) to which deviations are referred must be established for each of the two mating surfaces. As can be seen in Figure 1.2a, this determines the degree of tightness or freedom for relative motion between the mating surfaces. Figure 1.2a also shows that there are basically three types of fits: clearance, transition, and interference.

In all cases of *clearance fit,* the upper limit of the shaft is always smaller than the lower limit of the mating hole. This is not the case in *interference fit,* where the lower limit of the shaft is always larger than the upper limit of the hole. The *transition fit,* as the name suggests, is an intermediate fit. According to the ISO, the internal enveloped part is always

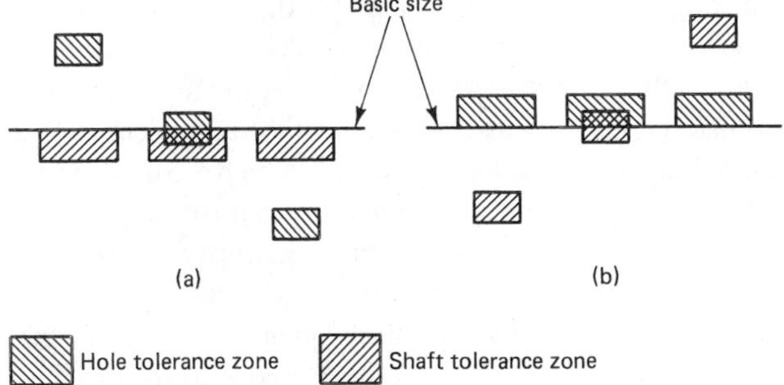

FIGURE 1.2 The two systems of fit according to the ISO: (a) shaft-basis system; (b) hole-basis system

referred to as the *shaft,* whereas the surrounding surface is referred to as the *hole.* Accordingly, from the fits point of view, a key is the shaft and the keyway is the hole.

It is clear from Figures 1.2a and b that there are two ways for specifying and expressing the various types of fits: the *shaft-basis* and the *hole-basis* systems. The location of the tolerance zone with respect to the zero line is indicated by a letter, which is always capitalized for holes and lowercased for shafts, whereas the tolerance grade is indicated by a number, as previously explained. Therefore, a fit designation can be H7/h6, F6/g5, or any other similar form.

Interchangeability

When the service life of an electric bulb is over, all you do is buy a new one and replace the bulb. This easy operation, which does not need a fitter or a technician, would not be possible without the two main concepts of interchangeability and standardization. *Interchangeability* means that identical parts must be able to replace each other, whether during assembly or subsequent maintenance work, without the need for any fitting operations. Interchangeability is achieved by establishing a permissible tolerance, beyond which any further deviation from the nominal dimension of the part is not allowed. *Standardization,* on the other hand, involves limiting the diversity and total number of varieties to a definite range of standard dimensions. An example is the standard gauge system for wires and sheets. Instead of having a very large number of sheet thicknesses in steps of 0.001 inch, the number of thicknesses produced is limited to only 45 (in U.S. standards). As you can see from this example, standardization has far-reaching economic implications and also promotes interchangeability. Obviously, the engineering standards differ for different countries and reflect the quality of technology and the industrial production in each case. Germany established the DIN (Deutsche Ingenieure Normen), standards that are finding some popularity worldwide. The former Soviet Union adopted the GOST, standards that were suitable for the period of industrialization of that country.

1.6 THE PRODUCTION TURN

In almost all cases, the main goal of a manufacturing project is to make a profit, the exception being projects that have to do with the national security or prestige. Let us establish a simplified model that illustrates the cash flow through the different activities associated with manufacturing so that we can see how to maximize the profit. As shown in Figure 1.3, the project starts by borrowing money from a bank to purchase machines and raw materials and to pay the salaries of the engineers and other employees. Next, the raw materials are converted into products, which are the output of the manufacturing domain. Obviously, those products must be sold (through the marketing department) in order to get cash. This cash is, in turn, used to cover running costs, as well as required payment to the bank; any surplus money left is the profit.

We can see in this model that the sequence of events forms a continuous cycle (i.e., a closed circuit). This cycle is usually referred to as the *production turn.* We can also realize the importance of marketing, which ensures the continuity of the cycle. If the prod-

FIGURE 1.3
The production turn

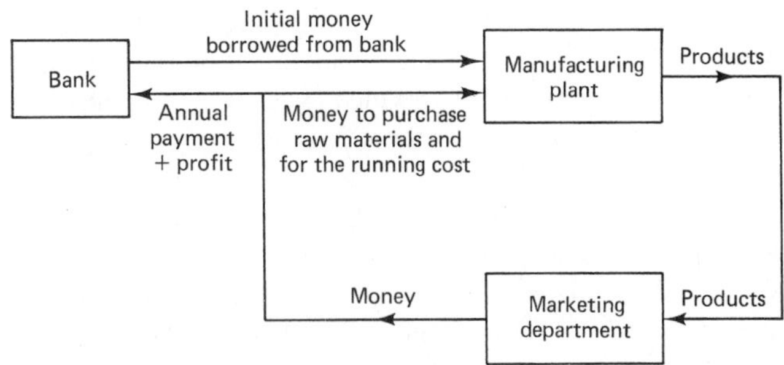

ucts are not sold, the cycle is obviously interrupted. Moreover, we can see that maximum profit is obtained through maximizing the profit per turn and/or increasing the number of turns per year (i.e., running the cycle faster). Evidently, these two conditions are fulfilled when products are manufactured in the easiest and least expensive way.

1.7 PRODUCT LIFE CYCLE

It has been observed that all products, from the sales viewpoint, go through the same *product life cycle,* no matter how diverse or dissimilar they are. Whether the product is a new-model airplane or a coffeemaker, its sales follow a certain pattern or sequence from the time it is introduced in the market to the time it is no longer sold. The main difference between the cycles of these two products is the span or duration of the cycle, which always depends upon the nature and uses of the particular product. As we will see later when discussing concurrent engineering in Chapter 2, it is very important for the designer and the manufacturing engineer to fully understand that cycle in order to maximize the profits of the production plant.

It is clear from Figure 1.4 that the sales, as well as the rate of increase in sales, are initially low during the *introduction* stage of the product life cycle. The reason is that the consumer is not aware of the performance and the unique characteristics of the product.

FIGURE 1.4
The product life cycle

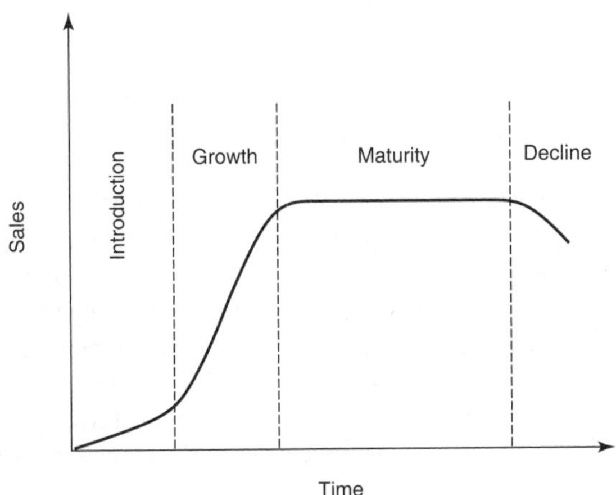

Through television and newspaper advertisements and word-of-mouth communication, a growing number of consumers learn about the product and its capabilities. Meanwhile, the management works on improving the performance and eliminating the shortcomings through minor design modifications. It is also the time for some custom tailoring of the product for slightly different customer needs, in order to serve a wider variety of consumers. As a result, the customer acceptance is enhanced, and the sales accordingly increase at a remarkable rate during this stage, which is known as the *growth* stage. However, this trend does not continue forever, and, at a certain point, the sales level out. This is, in fact, the *maturity* stage of the life cycle. During this stage, the product is usually faced with fierce competition, but the sales will continue to be stable if the management succeeds in reducing the cost of the product and/or developing new applications for it. The more successful the management is in achieving this goal, the longer the duration of the maturity stage will be. Finally, the *decline* stage begins, the sales fall at a noticeable rate, and the product is, at some point, completely abandoned. The decrease in the sales is usually due to newer and better products that are pumped into the market by competing manufacturers to serve some customer need. It can also be caused by diminishing need for the uses and applications of such a product. A clever management would start developing and marketing a new product (B) during the maturity stage of the previous one (A) so as to keep sales continuously high, as shown in Figure 1.5.

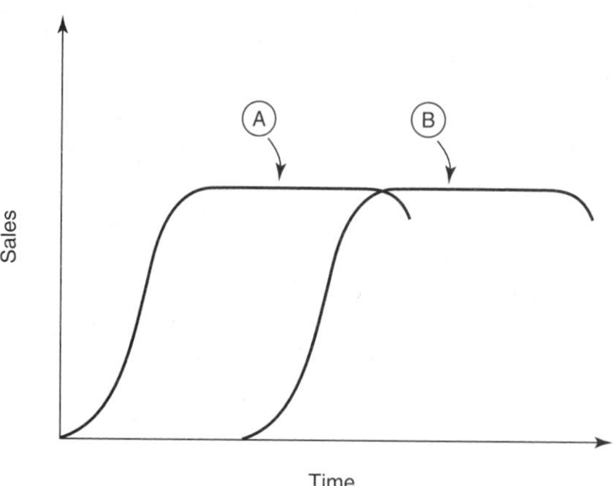

FIGURE 1.5
The proper overlap of products' life cycles

1.8 TECHNOLOGY DEVELOPMENT CYCLE

Every now and then, a new technology emerges as result of active research and development (R & D) and is then employed in the design and manufacture of several different products. It can, therefore, be stated that *technology* is concerned with the industrial and everyday applications of the results of the theoretical and experimental studies that are referred to as *engineering*. Examples of modern technologies include transistor, microchip, and fiber optics.

The relationship between the effectiveness or performance of a certain technology and the effort spent to date to achieve such performance is shown graphically in Figure 1.6. This graphical representation is known as the *technology development cycle.* It is also sometimes referred to as the *S curve* because of its shape. As can be seen in Figure 1.6, a lot of effort is required to produce a sensible level of performance at the early stage. Evidently, there is a lack of experimental experience since the techniques used are new. Next, the rate of improvement in performance becomes exponential, a trend that is observed with almost all kinds of human knowledge. At some point, however, the rate of progress becomes linear because most ideas are in place; any further improvement comes as a result of refining the existing ideas rather than adding new ones. Again, as time passes, the technology begins to be "exhausted," and performance levels out. A "ceiling" is reached, above which the performance of the existing current technology cannot go because of social and/or technological considerations.

An enlightened management of a manufacturing facility would allocate resources and devote effort to an active R & D program to come up with a new technology (B) as soon as it realized that the technology on which the products are based (A) was beginning to mature. The production activities would then be transferred to another S curve, with a higher ceiling for performance and greater possibilities, as shown in Figure 1.7. Any delay in investing in R & D for developing new technology may result in creating a gap between the two curves (instead of continuity with the overlap shown in Figure 1.7), with the final outcome being to lose the market to competing companies that possess newer technology. In fact, the United States dominated the market of commercial airliners because companies like Boeing and McDonnell Douglas knew exactly when to switch from propeller-driven airplanes to jet-propulsion commercial airliners. This is contrary to what some major computer companies did when they continued to develop and produce mainframe computers and did not recognize when to make the switch to personal computers. Current examples of technological discontinuity include the change from conventional telecommunications cables to fiber optics for communication and information transfer.

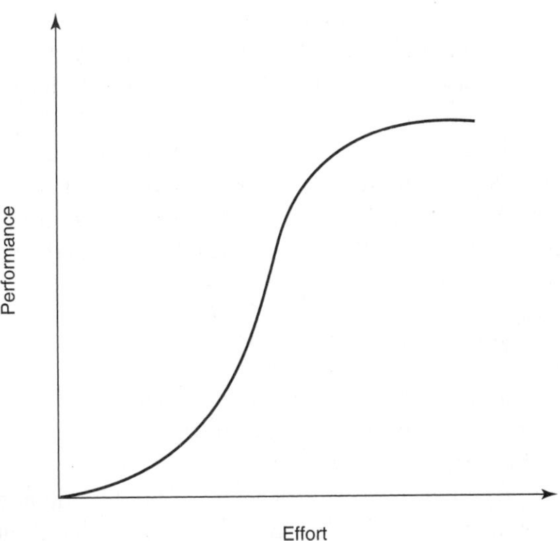

FIGURE 1.6
The technology development cycle (or S curve)

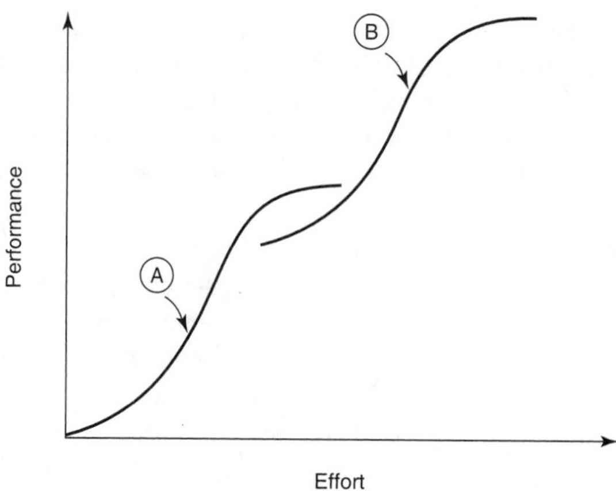

FIGURE 1.7 Transfer from one S curve to another

1.9 THE DESIGN PROCESS

An engineer is a problem solver who employs his or her scientific and empirical knowledge together with inventiveness and expert judgment to obtain solutions for problems arising from societal needs. These needs are usually satisfied by some physical device, structure, or process. The creative process by which one or more of the fruits of the engineer's effort are obtained is referred to as *design.* It is, indeed, the core of engineering that provides the professional engineer with the chance of creating original designs and watching them become realities. The satisfaction that the engineer feels following the implementation of his or her design is the most rewarding experience in the engineering profession. Because design is created to satisfy a societal need, there can be more than one way to achieve that goal. In other words, several designs can address the same problem. Which one is the best and most efficient design? Only time will tell because it is actually the one that would be favored by the customers and/or the society as a whole.

Although there is no single standard sequence of steps to create a workable design, E. V. Krick has outlined the procedure involved in the design process, and his work has gained widespread acceptance. Following is a discussion of the stages of the design process according to Krick. (See the references at the back of the book for more detailed information.)

Problem Formulation

As illustrated in Figure 1.8, problem formulation is the first stage of the design process. This phase comes as a result of recognizing a problem and involves defining that problem in a broad perspective without getting deep into the details. It is also at this stage that the engineer decides whether or not the problem at hand is worth solving. In other words, this stage basically constitutes a feasibility study of the problem arising from a recognized need. The designer should, therefore, realize the importance

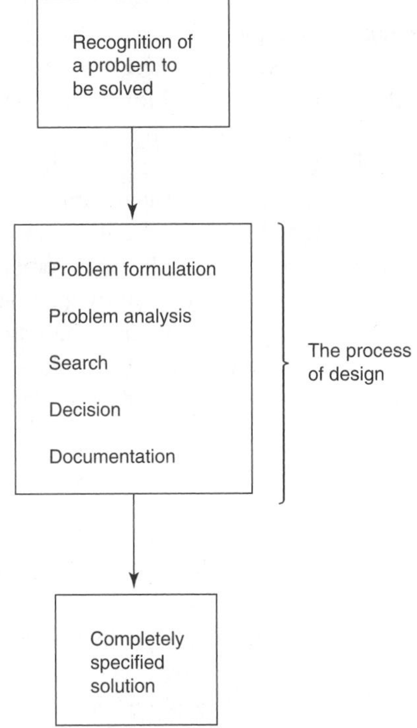

FIGURE 1.8
The design process (*Adapted from Krick,* An Introduction to Engineering and Engineering Design, *2nd ed. New York: John Wiley, 1969)*

of this stage. Neglecting it may result in wasting money in an effort to solve a problem that is not worth solving or wasting time on details that make it extremely difficult to get a broad view of the problem so as to select the appropriate path for solving it. The formulation of a problem can take any form that is convenient to the designer, although diagrammatic sketching (in particular, the black-box method) has proven to be a valuable tool.

Problem Analysis

The second stage involves much information gathering and processing in order to come up with a detailed definition of the problem. Such information may come from handbooks, from manufacturers' catalogs, leaflets, and brochures, as well as from personal contacts. You are strongly advised to seek information wherever you can find it; workers at all levels of a company may have some key information that you can use. The end product should be a detailed analysis of the qualitative and quantitative characteristics of the input and output variables and constraints, as well as the criteria that will be used in selecting the best design.

Search for Alternative Solutions

In the third stage, the designer actively seeks alternative solutions. A good practice is to make a neat sketch for each preliminary design with some notes about its pros and cons. All sketches should be kept even after a different final design is selected so that if that final design is abandoned for some reason, a designer does not have to start from the beginning again. It is also important to remember not to end the search for alternative solutions prematurely, before it is necessary or desirable to do so. Sometimes, a designer gets so involved with details of what he or she thinks is a good idea or solution that he or she will become preoccupied with these details, spending time on them instead of searching for other good solutions. Therefore, you are strongly advised to postpone working out the details until you have an appropriate number of viable solutions.

It is, indeed, highly recommended to employ collaborative methods for enabling the mind to penetrate into domains that might otherwise remain unexplored. A typical example is the technique of brainstorming, where a few or several people assemble to produce a solution for a problem by creating an atmosphere that encourages everyone to contribute with whatever comes to mind. After the problem is explained, each member comes up with an idea that is, in turn, recorded on a blackboard, thus making all ideas evident to all team members.

Decision Making

The fourth stage involves the thorough weighing and judging of the different solutions with the aim of being able to choose the most appropriate one. That is, *trade-offs* have to be made during this stage. They can be achieved by establishing a decision matrix, as shown in Figure 1.9.

As can be seen in Figure 1.9, each of the major design objectives is in a column, and each solution is allocated a row. Each solution is evaluated with regard to how it fulfills each of the design objectives and is, therefore, given a grade (on a scale of 1 to 10) in each column. Because the design objectives do not have the same weight, each grade must be multiplied by a factor representing the weight of the design function for which it was given. The total of all the products of multiplication is the score of that particular solution and can be considered as a true indication of how that solution fulfills the design objectives. As you can see, this technique provides a mechanism for rating the various solutions, thus eliminating most and giving further consideration to only a few.

The chosen design is next subjected to a thorough analysis in order to optimize and refine it. Detailed calculations, whether manual or computational, are involved at this point. Both analytical and experimental modeling are also extensively employed as tools in refining the design. It is important, therefore, to now discuss modeling and simulation. A *model* can be defined as a simplified representation of a real-life situation that aids in the analysis of an associated problem. There are many ways for classifying and identifying models. For example, models can be descriptive, illustrating a real-world counterpart, or prescriptive, helping to predict the performance of the actual system. They can also be deterministic or probabilistic (used when making decisions under uncertainty). A simple example of a model is the free-body diagram used to determine the internal tensile force acting in a wire with a weight attached to its end. There are many computer tools (software) that are employed by designers to create models easily and quickly. Examples include geometric modeling and finite element analysis software packages. On the other hand, *simulation* can be defined as the process of experimenting with a model by subjecting it to various values of input pa-

FIGURE 1.9
A simple decision matrix

Design Objectives	Production Cost			Reliability			Durability			
Alternatives	**Level of satisfaction**	**Weight factor**	**Weight**	**Level of satisfaction**	**Weight factor**	**Weight**	**Level of satisfaction**	**Weight factor**	**Weight**	**Overall Score**
Design 1	9.0	0.5	4.5	7.0	0.3	2.1	8.0	0.2	1.6	8.2
Design 2	8.0	0.5	4.0	8.0	0.3	2.4	8.0	0.2	1.6	8.0
Design 3	9.0	0.5	4.5	9.0	0.3	2.7	9.0	0.2	1.8	9.0

rameters and observing the output, which can be taken as an indication of the behavior of the real-world system under the tested conditions. As you can see, simulation can save a lot of time and effort that could be spent on experimental models and prototypes. This saving is particularly evident when computer simulation is employed. Still, simulation would not eliminate design iterations but rather would minimize their number. You are, therefore, urged to make use of these tools whenever possible.

Documentation

In the fifth and last stage of the design process, the designer organizes the material obtained in the previous stage and puts it in shape for presentation to his or her superiors. The output of this stage should include the attributes and performance characteristics of the refined design, given in sufficient detail. Accordingly, the designer must communicate all information in the form of clear and easy-to-understand documents. Documentation consists of carefully prepared, detailed, and dimensioned engineering drawings (i.e., assembly drawings and workshop drawings or blueprints), a written report, and possibly an iconic model. With the recent development in rapid prototyping techniques, a prototype can certainly be a good substitute for an iconic model. This approach has the advantage of revealing problems that may be encountered during manufacturing.

1.10 PRODUCT DESIGN: THE CONCEPT OF DESIGN FOR MANUFACTURING

The conventional procedure for product design, as illustrated in Figure 1.10, used to start with an analysis of the desired function, which usually dictated the form as well as the materials of the product to be made. The design (blueprint) was then sent to the manufacturing department, where the kind and sequence of production operations were determined mainly by the form and materials of the product. In fact, the old design procedure had several disadvantages and shortcomings:

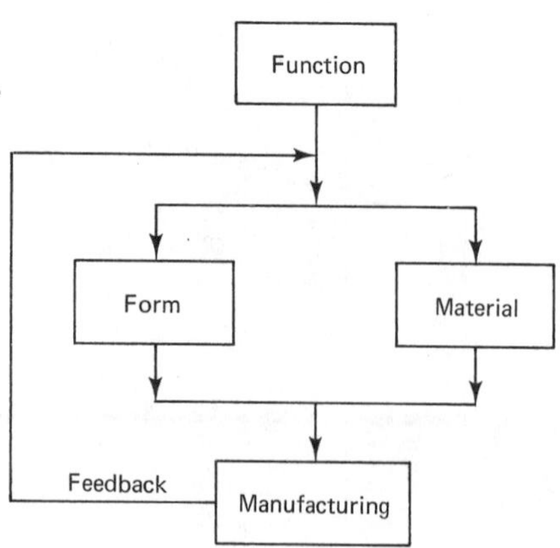

FIGURE 1.10
The old procedure for product design

1. In some cases, nice-looking designs were impossible to make; in many other cases, the designs had to be modified so that they could be manufactured.
2. Preparing the design without considering the manufacturing process to be carried out and/or the machine tools available would sometimes result in a need for special-purpose, expensive machine tools. The final outcome was an increase in the production cost.
3. When the required production volume was large, parts had to be specially designed to facilitate operations involved in mass production (such as assembly).
4. A group of different products produced by the same manufacturing process has common geometric characteristics and features that are dictated by the manufacturing process employed (forgings, for example, have certain characteristic design features that are different from those of castings, extrusions, or stampings). Ignoring the method of manufacturing during the design phase would undermine these characteristic design features and thus result in impractical or faulty design.

Because of these reasons and also because of the trend of integrating the activities in a manufacturing corporation, the modern design procedure takes into consideration the method of manufacturing during the design phase. As can be seen in Figure 1.11, design, material, and manufacturing are three interactive, interrelated elements that form the manufacturing system, whose prime inputs are conceptual products (and/or functions) and whose outputs are manufactured products. In fact, the barriers and borders between the design and manufacturing departments are fading

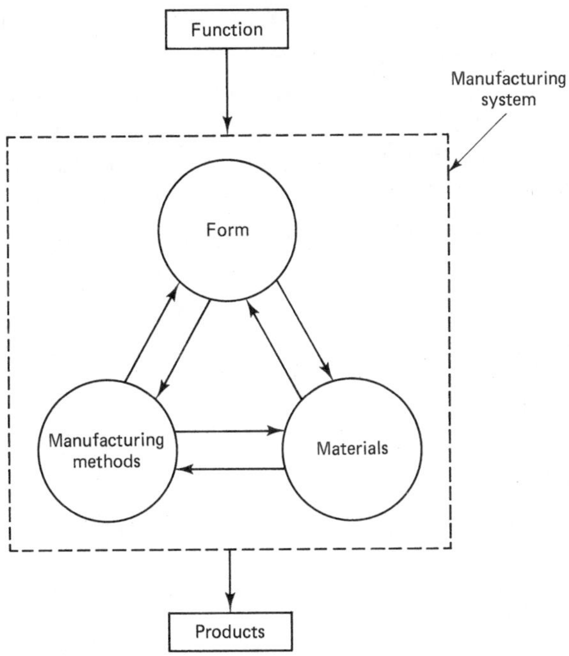

FIGURE 1.11 The new concept of a manufacturing system for achieving rational product designs

out and will eventually disappear. The tasks of the designer and those of the manufacturing engineer are going to be combined and done by the same person. It is, therefore, the mission of this text to emphasize concepts like *design for manufacturing* and to promote the systems approach for product design.

Review Questions

1. What is the definition of *manufacturing?*
2. Is there any relationship between the status of manufacturing in a nation and the standard of living of the people in that nation? Explain why.
3. Explain the different approaches for obtaining a desired shape and give examples of some manufacturing processes that belong to each group.
4. List the different types of production and explain the main characteristics of each. Also mention some suitable applications for each type.
5. Explain the meaning of the term *tolerance.*
6. How do we scientifically describe the tightness or looseness of two mating parts?
7. What concepts did Eli Whitney establish to ensure trouble-free running of the mass production of multicomponent products?
8. What is meant by the *production turn?* What role does marketing play in this cycle?
9. Using the concept of production turn, how can we maximize the profits of a company by two different methods?
10. Explain the stages involved in the life cycle of a product.
11. What is the significance of the product life cycle during the phase of planning for the production of new products?
12. What is the *S curve?* Explain an American success story in employing it.
13. Give some examples of transfer from one technology development curve to another.
14. What are the stages involved in the design process? Explain each briefly.
15. What is meant by *trade-offs?* How can these be achieved during the decision-making stage?
16. Explain the old approach for product design. What are its disadvantages?
17. Explain the concept of *design for manufacturing.* Why is it needed in modern industries?

Chapter 2

Concurrent Engineering

INTRODUCTION

Concurrent engineering is a manufacturing philosophy that involves managing the product development process with the aim of getting new products with the highest quality at the best competitive price in the least time to the market. It has proven to be a key factor for the survival and, more importantly, for the prosperity of companies that are clever enough to adopt its methodology and tools (Motorola and Hewlett-Packard are good examples). In fact, many companies can, in good faith, argue that they have been using this methodology in one form or another for some time—consider the efforts of corporations like Xerox, Hewlett-Packard, and Ford in the late 1970s to review and revise their product design practices versus those of their foreign competitors. An important milestone in the history of concurrent engineering is considered to be the report issued in December 1988 by the Defense Advanced Research Projects Agency (DARPA) as a result of a study to improve concurrency in the product design process, a study that lasted more than five years. Many professionals in this field rightfully believe that the DARPA report is the true foundation for the concept of concurrent engineering. Many terms were and still are (though to a lesser degree) used to describe this methodology. Examples include *team design, simultaneous engineering,* and *integrated product development.* The term *concurrent engineering* was first coined by the Institute for Defense Analysis (IDA), which also provided the following definition:

> Concurrent engineering is a systematic approach to integrated, concurrent design of products and their related processes, including manufacture and support. This approach is intended to cause the developers,

> from the outset, to consider all elements of the product life cycle from concept through disposal, including quality, cost, schedule, and user requirements.

A good way to understand this new concept and what it means is to compare the product development process in the traditional engineering approach, which is usually referred to as *serial* or *sequential engineering,* with the one in the concurrent engineering environment. In serial engineering, a team of qualified professional engineers designs a product without much interaction with or input from other departments within the corporation such as manufacturing, sales, or customer service. A model or prototype is then fabricated in the prototype workshop based on the documented design produced by the team. Note that the environment in the model shop is ideal and is different from the real one on the shop floor during production. Weeks or even months after releasing the design, the testing department receives the model and carries out acceptance tests to make sure that the model conforms to the documented design and also meets the criteria established and agreed upon for the functioning and performance of the product. As you may expect, alterations and modification and/or revisions of the design are needed in most cases as a result of the absence of inputs from other departments during the design process. As a consequence, revisions call for new designs that, in turn, require the fabrication and testing of new prototypes.

This obviously time-consuming cycle may have to be repeated a few times in order to achieve the desired goals. Such a cycle prolongs the new product development process to various degrees. Depending upon the complexity of the product and the number of iterations, the delay can be excessive, thus causing damage to the marketing strategy and sales of the new product. It is, therefore, clear that the absence of communication between the different departments starting early in the initial phase of the design process and continuing throughout that process would result in a larger number of design iterations and delays in releasing the product to the market. On the contrary, in a concurrent engineering environment, all relevant departments, such as design, manufacturing, R & D, and marketing, become involved and participate in the design process from its very beginning. This interaction reduces to a large extent or even eliminates design iterations, thus compressing the development cycle with the final outcome of reducing the time-to-market for a new product. The products can, therefore, be turned over at a much faster pace. Bearing in mind that the larger portion of profit occurs in the early part of the cycle for successful products, it would consequently be possible to allow new products to be retired at nearly their optimum profitability. Also, because customers are consulted (through the marketing and customer service departments) early in the product development process, the new product would most probably penetrate the market easily because it would correctly meet customers' expectations in terms of both function and quality. A good example is the case of the Boeing 777 jetliner. During the initial development stage, Boeing called representatives of its customers, including BOAC (the British-government-owned airlines), although Britain is a part of the consortium building the air bus. By doing so, Boeing got the praise and the support of its customers all over the world (including BOAC) in the form of orders of the new product under development. More importantly, some serious modifications in the design were made in the very early phase, thus saving a lot of money and effort if they had not been done. The choice of the height of the wings above ground level is a clear example. Looking at the initial conceptual design, the customers realized that the wing height was too high to the extent that it would create difficulties during fueling and would require the use of a special fueling truck. Boeing was promptly advised to lower the level of the wings so that the currently used fueling trucks could be easily and successfully employed.

2.1 REASONS FOR ADOPTING CONCURRENT ENGINEERING

Modern manufacturing industries are facing many challenges, such as global competition and fast-changing consumer demands. These and other challenges call for the adoption of the concurrent engineering methodology. Following is a list of some of the challenges that can be successfully met by concurrent engineering:

1. Increasing product complexities that prolong the product development process and make it more difficult to predict the impact of design decisions on the functionality and performance of the final product.
2. Increasing global competitive pressures that result from the emerging concept of reengineering (which enabled many Asian countries to produce extremely cost-effective products because the cost of R & D in this case is almost zero). This definitely creates the need for a cost-effective product development cycle.

3. The need for rapid response to fast-changing consumer demands. This phenomenon calls for the need to continuously listen to the "voice" of the consumer—one of the solid principles of concurrent engineering methodology.

4. The need for shorter product life cycles. This phenomenon necessitates the introduction of new products to the market at a very high pace—something that can only be achieved by compressing the product development cycle. Consider the changes that the old-fashioned mechanical typewriter has undergone. Its life cycle was 20 to 30 years, which then decreased to 10 years for an electromechanical typewriter and finally to only 18 months for a word processor. This example clearly illustrates the need to solve the problem of time-to-market pressure.

5. Large organizations with several departments working on developing numerous products at the same time. The amount of data exchanged between these departments is extremely large, and unless properly managed in a rational manner, the flow and transfer of information is not fast or easy (i.e., a piece of information needed by a certain department may be passed to another one). The final outcome would certainly be delays in the process of product development and products that will not appear in the market at the scheduled time.

6. New and innovative technologies emerging at a very high rate, thus causing the new products to be technologically obsolete within a short period of time. This phenomenon is particularly evident in the electronics industry, where the life cycle of a typical product is in months (it used to be years during the 1980s). As a consequence, new products must appear in the market at a very high pace—something that definitely necessitates a shorter product development cycle.

2.2 BENEFITS OF CONCURRENT ENGINEERING

The benefits of adopting concurrent engineering are numerous and positively affect the various activities in a corporation. Following is a summary of the important ones:

1. Because the customer is consulted during the early product development process, the product will appear on the market with a high level of quality and will meet the expectations of the customer. The product introduction region (or start-up) of the product life cycle (see Figure 1.4) will be very short. The sales volume will, therefore, attain maturity in a very short time. As a consequence, large revenues and profits would be achieved during the early phase of the product life cycle. This is very important because products become technologically obsolete very quickly as a result of fast-emerging, innovative technology.

2. Adopting concurrent engineering will result in improved design quality, which is measured by the number of design changes made during the first six months after releasing a new product to the market. These design changes are extremely expensive unless caught early during the product development process. The lower the

number of these changes, the more robust the design of the product is. In a concurrent engineering environment, these design changes would evidently be minimal.

3. Reduced product development and design times will result from listening to the voice of the customer and from transferring information between the various departments involved, including those downstream. This benefit is, in fact, a consequence of the reduction in the number of design iterations necessary to achieve optimum product design. Another factor is forsaking sequential methods of product development and replacing them with concurrent ones.
4. Reduced production cost is a consequence of the preceding two benefits—namely, the reduction in the number of design changes after releasing the product and the reduction in the time of the product development process. Reduced cost, of course, provides a manufacturing company with a real advantage in meeting global competitive pressures.
5. Elimination of delays when releasing the product to the market will guarantee a good market share for the new product. Also, it has been proven that delays in releasing the product will result in market loss of revenues.
6. As a result of the reduced design time and effort, new products will be pumped into the market more frequently, which is, indeed, the advantage that Japanese automakers have over their American counterparts. They can produce more different models, with smaller production volumes and shorter life cycles.
7. Increased reliability and customer satisfaction will result from delivering the product "right the first time" and will also enhance the credibility of the manufacturing company.

2.3 FACTORS PREVENTING THE ADOPTION OF CONCURRENT ENGINEERING

Now, if adopting concurrent engineering results in all the benefits just listed, why isn't it widely applied in all manufacturing corporations? The reason is that concurrent engineering is based on a manufacturing philosophy that requires breaking barriers between departments and establishing multidisciplinary teams. This philosophy clearly contradicts the authoritative culture that is currently dominant in the industrial establishment. The threat of loss of power and authority makes middle management and bureaucrats resistant to the idea of implementing concurrent engineering. There is also a natural resistance to anything new inherent in the minds of some people. Another factor may be the need to build an excellent communication infrastructure for facilitating the flow of information throughout the product development cycle. Apparently, a lot of money and effort must be invested to create an adequate information system—something that many companies either cannot afford or do not want to do. Yet another factor that holds up the implementation of concurrent engineering is the temptation to come up with temporary short-run solutions to the problem of decreasing revenues, without any regard to strategic planning and long-term goals. Examples include cutting the work force to increase profits (a faulty, shortsighted approach that would cause a company to lose trained employees who are needed to enable quick, on-schedule product delivery) and cutting the price without any basis (a solution that would inevitably eliminate or reduce the profits).

2.4 THE FOUR PILLARS OF CONCURRENT ENGINEERING

The implementation of concurrent engineering is based on managing forces of change and using them as resources or tools in four arenas for efficient, fast, and economical product development. These arenas—organization, communication infrastructure, requirements, and product development—are the pillars on which the methodology of concurrent engineering rests. Let us examine each of them and see how each can be managed.

Organization

This arena includes the managers, product development teams, and support teams (i.e., the organization itself and the interactions of its components). The role of management is vital and includes not only motivating people to change their work habits to match the concurrent engineering environment but also ensuring unhindered exchange of information between the different disciplines. In fact, management can be used as one of the forces of change or tools that guarantees continuous improvement of the product development process, as will be discussed later.

Communication Infrastructure

The communication infrastructure encompasses the hardware, software, and expertise that together form a system that allows the easy transfer of information relating to product development. As you may expect, when the product complexity increases, the number of disciplines involved also increases, as does the volume of information to be transferred. The system to be established must be capable of handling the type and amount of data necessary for the product development process. It must retrieve, evaluate, and present the data in an organized format that is easy to understand and to use by team members and by management. In fact, many corporations learned the hard way that communication technologies are as important as design and manufacturing technologies for the success of a new product. The first task to handle, after purchasing the hardware, is to build a comprehensive and efficient database that has queries and that can be accessed by teams and by managers who are in charge of monitoring and evaluating the product development process. Electronic mail, interactive browsing capabilities, and other modern information transfer technology are also essential in order to eliminate the need for shoveling piles of documents and papers between the different departments and teams.

Requirements

A broad (but accurate) description of the product requirements involves all product attributes that affect customer satisfaction. Consequently, customers' needs are considered when setting the specifications for the conceptual design. This consideration would, indeed, ensure that the model or prototype meets the original goals from the start. This process of creating the conceptual specifications is extremely important and must be carried out rigorously. The more product attributes and constraints are initially specified, the fewer the problems associated with the final product design are, and the fewer the number of design changes or iterations will be. Of course, the conceptual design constraints should be defined very clearly and subjected to a continuous process of updating, evaluation, and validation. As is well known, constraints like government regulations, environmental laws, and industry and national standards are changing all the time. Continually updating these would certainly improve the product development process.

Product Development

In a concurrent engineering environment, the downstream processes of manufacturing, maintenance, customer service, sales, and so on, must be considered in the early design phase. This consideration, as previously mentioned, is a necessary condition for the implementation of concurrent engineering. The second important condition is the need for continuous improvement and optimization of the product development process. As a consequence of these two conditions, there is a continuous drive to develop, evaluate, and adopt new design methodologies. Concepts and approaches like *design for manufacturing* (DFM) and *design for assembly* (DFA) have been popular in recent days and have proven to be valuable tools in adopting and implementing concurrent engineering. The reason is that their philosophy is based on using manufacturing (or assembly) as a design constraint, thus taking downstream processes like manufacturing and assembly into full consideration during the early design phase of the product. In other words, the success of these methodologies is linked to their philosophies being compatible to (or matching) that of concurrent engineering.

Another part of the product development arena is what is sometimes referred to as the *component libraries.* The design (and manufacturing) attributes of the different components, whether standard ones that were purchased or parts that were previously designed and manufactured, are kept in a database. The availability of such a database to team members will speed up the design process by providing them with many alternatives to choose from and, more importantly, by freeing them from reinventing the wheel. Using previously tested components, maybe with very slight modification in the design, can save a lot of time and effort. Keeping a computerized database has the advantages of easy retrieval of designs and simultaneous availability to all team members. This topic will be covered later in the book in detail when we discuss group technology and computer-aided process planning.

Also part of the product development arena is the design process itself. We have already covered its stages and methodology in Chapter 1. Here, we want to emphasize again that good design has always been based on customer needs, which must be, in turn, determined by listening to customer concerns. In fact, it was for this reason that the concept of *quality function deployment* (QFD) was developed in Japan's Kyoto Shipyard in the 1980s. By including QFD in the design process, teams do not lose touch with the customer, and, consequently, the designed products will meet customers' needs and expectations. Although QFD is beyond the scope of this text, a brief discussion will enlighten those engineers who must communicate with members in charge of QFD in a multidisciplinary team. QFD seeks to identify and evaluate the meaning of the word *quality* from the customer's point of view. The approach involves constructing a matrix that is quite similar to the design decision matrix covered in Chapter 1 (see Figure 1.9). This matrix is called the *house of quality.* The attributes, functions, and characteristics that the customer wants can be clearly identified and used as input constraints or requirements for the process of designing the product as previously mentioned. To repeat, remember that one of the goals of the design team is to have a decreasing number of design changes with increasing order of design stage. In the final design stage, if the appropriate methodology is adopted, the number of changes should be zero.

2.5 FORCES OF CHANGE

Implementing concurrent engineering is based on managing some forces of change and using them as resources or tools to create the concurrent engineering environment. Following are some of these forces of change.

Technology

Technology has a very important role to play in each of the four arenas of concurrent engineering. It speeds up and optimizes the product development process, minimizes the number of design iterations, and facilitates communication and information transfer between the different teams and departments. Managers should, therefore, take full advantage of the most up-to-date technology available and avoid technology that is under development or obsolete. Unfortunately, the problem of acquiring up-to-date technology is far more complicated than it seems because of the extremely fast pace with which technology is advancing and the vast amount of different options available. Here are some tips that address the technology problem:

1. Keep engineers abreast of the latest technological developments by providing them with technical journals and periodicals, sending them to international engineering conferences and exhibitions, and ensuring a continuous learning process through workshops and short courses offered on site.
2. Ensure that the latest scientific findings are promptly employed in developing a company's technology and, therefore, result in high-quality products. Companies should focus their efforts on applied research for developing products and processes and integrate their R & D with design and development activities. (In fact, this is one of the reasons behind the success of several countries in the Far East.)

3. Try to make use of the results of government-funded research and thus save time and money spent in obtaining similar findings.
4. Definitely overcome the "not-invented-here" syndrome. Many industry people make the mistake of completely ignoring any technology that was not invented in their company. This syndrome leads to isolationism and, eventually, falling behind. It is very difficult for a company to fully develop technology starting from scratch. Acquiring technology by purchasing it, by establishing partnerships between companies, and by encouraging technology exchanges is worth exploring.

It is important here to cast light on one of industry's most difficult problems in the United States—the bad effects of having advanced technology geared toward military applications. Although there is a wealth of technological information as a result of active R & D in military industries, it is classified and, therefore, not accessible for civilian industries and commercial applications. A further obstacle is the difference in the product requirements in both cases. Although military criteria specify quality regardless of cost, civilian requirements call for both quality and cost. The picture is clear when you compare the performance and the cost of a nuclear bomber with those of a commercial jetliner.

Management

Management in a concurrent engineering environment takes its role from management in a traditional serial manufacturing corporation and goes far beyond it. A manager's role involves not only setting schedules and work expectations of engineers and assigning responsibilities but also managing changes and building an organizational structure that is flexible and can respond quickly to surprises and sudden changes in demands and requirements. You may have already concluded that managers must have a general but solid understanding of current and relevant technical issues in order to communicate effectively with multidisciplinary teams. In fact, one of the most important tasks of middle management in a concurrent engineering environment is the creation of those multidisciplinary teams in order to carry out the product development process. In summary, the traditional role of management that is based on vertical chain of command, authoritative decision making, and the "carrot-stick" model of running corporations is diminishing continuously, especially in a company that adopts the concurrent engineering philosophy. More emphasis is being placed on creating product development teams and facilitating information transfer and communication between them.

Let us look more thoroughly at the process of establishing multidisciplinary teams. As you may expect, complex tasks are handled by breaking them into less complex ones that are, in turn, dealt with simultaneously but separately with different teams. Good managers should optimize the size of each team. A team that is too large or too small creates communication problems, is less efficient, and is more expensive. Attention must also be given to the talents and the quality of team members in terms of choosing the right person for the right job. When establishing the teams, the management focus must be to concurrently execute tasks that are normally carried out sequentially and to integrate those activities that are concurrent. Consequently, an appropriate project-modeling tool must be used in order to identify and locate the patterns of information flow and interaction. There

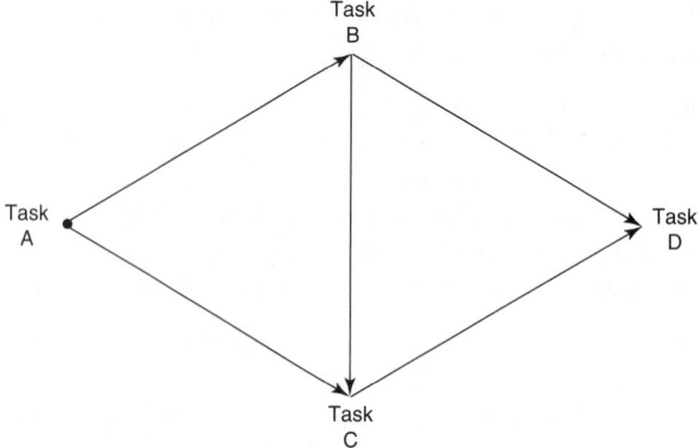

FIGURE 2.1
The PERT chart

are basically three approaches or tools—namely, the PERT chart, the GANTT chart, and the design structure matrix (DSM).

The PERT chart, which is illustrated in Figure 2.1, is basically used to determine project duration and critical path. On the other hand, the GANTT chart displays the relative positioning of tasks on a time scale, as shown in Figure 2.2. In fact, some researchers believe that the DSM method is far better in displaying the connectivity of interacting tasks and improving the product development process. It also clearly illustrates where the integration of tasks should take place.

The DSM method and its modified version have been extensively used by Smith and Eppinger of the Sloan School of Management at the Massachusetts Institute of Technology (MIT). The basic method involves representing the relationship among project tasks in a matrix form and allows for different tasks to be coupled. As can be seen in Figure 2.3, each individual task is represented by a row and by a column of a square matrix; the need for information flow between two tasks is indicated by a check mark (×). Going horizontally across a task's row, the columns under which there are check marks are those from which information must be received in order to complete the given task. On the other hand, going vertically down a task's column, the check

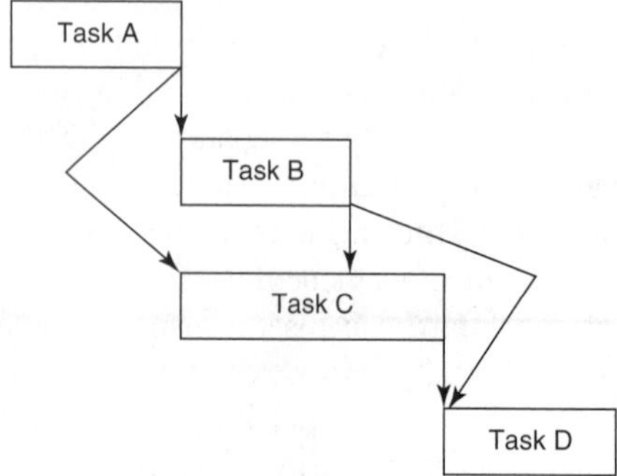

FIGURE 2.2
The GANTT chart

FIGURE 2.3 Initial phase of the design structure matrix

Tasks	A	B	C	D	E	F	G	H	I	J	K	L	M	N	O	P	Q	R	S	T	U	V
A	A	X	X	X		X		X	X	X									X		X	
B	X	B	X					X		X	X	X										X
C	X	X	C																			
D	X	X		D																		X
E				X	E	X																
F	X			X	X	F	X		X													
G						X	G															
H	X	X						H		X	X				X	X	X					
I	X								I													X
J	X	X						X		J												
K		X						X		X	K	X	X			X	X	X				X
L		X										L	X		X							
M											X	X	M							X		
N								X			X			N			X					
O								X				X			O							
P											X					P						
Q											X			X			Q	X		X		
R																	X	R				
S	X	X	X			X	X					X				X			S	X	X	X
T										X						X		X	X	T	X	
U	X											X							X	X	U	X
V	X	X		X		X			X		X					X			X		X	V

marks indicate the rows (tasks) that require output from the given column. The diagonal elements are hatched because a task cannot be coupled with itself. Now, structuring the teams, usually referred to as *product development teams* (PDTs), can be accomplished by identifying highly coupled sets of tasks. First, the rows and columns must be rearranged so as to yield "batches" of check marks where a few tasks are coupled together and where the information of PDTs is most appropriate, as shown in Figure 2.4. The process of swapping the rows and columns of the matrix requires experience and skill because it is based on trial and error. Nevertheless, the process is also amenable to computer manipulation and analysis. As you can see, however, the DSM model does not take into consideration the degree of dependence or coupling between each two tasks. Recently, Smith and Eppinger replaced the check marks by numbers indicating the strength of dependence. The eigenvalue of such a matrix would reveal the highly coupled sets of tasks.

After the teams are established, the next question is how to manage the product development project and ensure that it is on target to meet the previously agreed-upon milestones and deadlines. Again, PERT and GANTT charts can be employed to

FIGURE 2.4
Final phase of the design structure matrix

Tasks	A	F	G	D	E	I	B	C	J	K	P	H	N	O	Q	L	M	R	S	T	U	V
A	A	X		X		X	X	X	X			X							X		X	
F	X	F	X	X	X	X																
G		X	G																			
D	X			D			X															X
E		X		X	E																	
I	X					I																X
B	X						B	X	X	X		X										X
C	X						X	C														
J	X						X		J			X										
K							X		X	K	X	X			X	X	X	X				X
P										X	P											
H	X						X		X	X	X	H		X	X							
N										X		X	N		X							
O												X		O		X						
Q										X			X		Q			X		X		
L					X									X		L	X					
M										X						X	M			X		
R															X			R				
S	X	X	X			X	X				X					X			S	X	X	X
T									X		X							X	X	T	X	
U	X															X			X	X	U	X
V	X	X		X		X	X			X	X								X		X	V

achieve these goals. There are, however, some other methods for project updating that visually illustrate the project status in one integrated chart, thus ensuring that different PDTs meet their stated goals concurrently. The radar (spider) chart is a popular one. As can be seen in Figure 2.5, each task or area of activity is represented by a radial line and for a one-year period. Thus, one look at the chart is enough to see whether or not the separate goals in the different areas are met. In the ideal case, when all tasks are performed exactly according to the planned schedule, this chart will end up having concentric circles at the various time periods, as indicated by the dashed lines in Figure 2.5. On the other hand, the bug chart is a plot of project expenditures versus product goals or milestones, which are indicated on the time axis as shown in Figure 2.6. The scales of both axes are adjusted so that a project that is on track is represented by a straight line making 45° with both axes. Although this chart has the advantages of indicating project-cost updating and individual milestones, it is sometimes misleading (a delay in purchasing supplies, for example, might seem or be interpreted as a positive indication). Further details are beyond the scope of this text, and interested readers are advised and encouraged to seek specialized books on the subject.

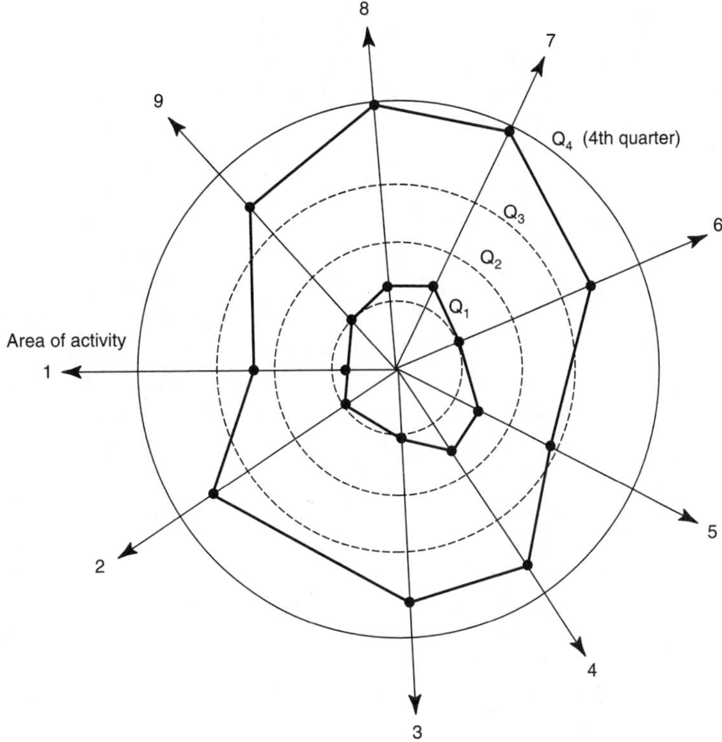

FIGURE 2.5
The radar (spider) chart

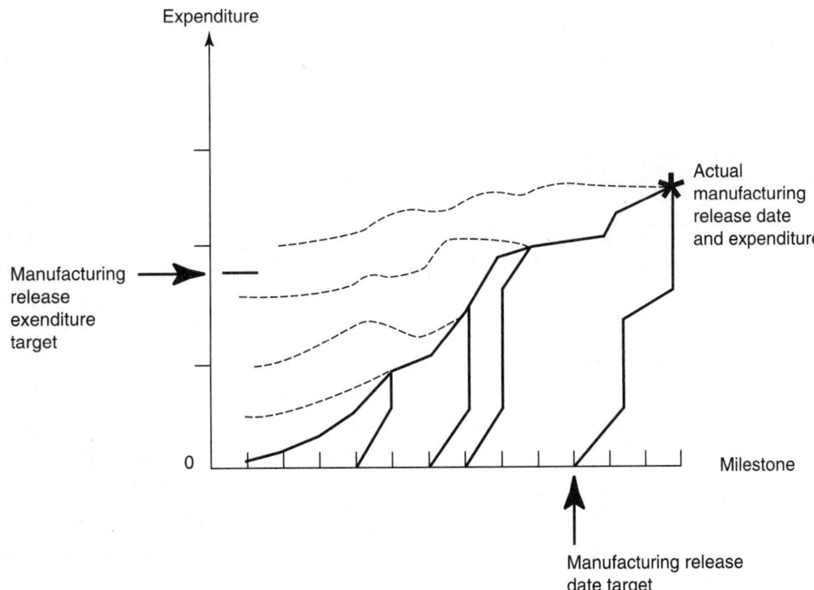

FIGURE 2.6
The bug chart

Tools

There is an extremely large number of tools for handling various tasks in the different arenas. The selection of the right tool for the right job is, therefore, not easy. In addition, most of these tools are undergoing a rapid, continuous, never-ending evolution. Tools that were new in the 1980s are now technically obsolete, which indicates the need to continually upgrade and replace a company's acquired tools. For example, structured analysis software tools manage information systems and present complex systems in a clear, easy-to-comprehend way. Instead of the old-fashioned written flowcharting procedures, this software provides graphical representation of any complex operation by a network of elements (each stands for a particular function) and by arrows indicating data flow and interaction between those elements. It also enables the elimination of redundant loops, thus making an operation more efficient.

Another example is tools used for design automation. They find increasing application in manufacturing corporations, and it is anticipated that by the year 2000 about 80 percent of all designs will be electronically done using these tools. Also, integration of these islands of automation (i.e., engineering departments in a manufacturing firm) is the trend in the 1990s, where local-area networks (LANs) are extensively used to transfer information from one department to another in a standardized format.

The adoption of new tools creates the problem of changing the responsibilities and nature of the jobs of employees, who will need retraining and, sometimes, have to be swapped. Also, based on the preceding discussion, the level of automation used must be appropriate for the company, and automated PDTs must be integrated into a system.

It is always important to remember that the forces of change discussed herein are just examples and that there can be other forces of change depending upon the nature of the manufacturing corporation and its production. Nevertheless, in all cases and regardless of the tools used, the change from serial manufacturing to concurrent engineering must be well planned and managed so as to take place gradually and smoothly and must always be monitored by management. In fact, abrupt changes and employee dissatisfaction are two factors that can impede the implementation of concurrent engineering.

2.6 A SUCCESS STORY: NIPPONDENSO

Now that you understand the arenas of concurrent engineering and the forces of change, it is time to look at a case study indicating how concurrent engineering was successfully implemented and resulted in solving tough problems that were facing one of the world's largest manufacturers of automotive parts. The original report was given in a paper entitled "Nippondenso Co. Ltd: A Case Study of Strategic Product Design," authored by Daniel E. Whitney and presented at the Collaborative Engineering Conference held at MIT in October 1993. This paper contained a wealth of information and was based on seven personal visits by the author to Nippondenso Co. Ltd. during the period 1974 to 1991, as well as on interviews with the company's personnel and papers published by the engineering staff. The information has been rearranged here, however, so as to draw parallelism with the previously mentioned concurrent engineering model and its four arenas.

Nippondenso Co. Ltd. is one of the world's largest manufacturers of automotive components, including air conditioners, heaters, relays, alternators, radiators, plus meters, diesel components, filters, controls, brake systems, and entertainment equipment. The company has 20 plants in 15 foreign countries in addition to 10 plants in Japan. In 1991, almost 43,000 people were employed by the company worldwide. Nippondenso is the first-tier supplier to Toyota and other Japanese and foreign car companies, and its sales amounted to about $10 billion dollars in 1989. Now that you have a clear idea about the size of this company and the diversity of its products, let us see how they created a concurrent engineering environment. Following is the company's approach in each of the previously mentioned arenas.

Organization

Nippondenso's philosophy is based on developing the product and the process for making it simultaneously. Consequently, multidisciplinary teams are formed through representation from various departments like production engineering, machines and tools, product design, and so on. Teams are small at the beginning but become larger as the project proceeds from the concept phase to the detailed-design phase. Top management promptly steps in when a crisis occurs and when a crucial decision needs to be made. Of course, a parallel-task approach is employed by overlapping some of the design steps.

Requirements

In addition to product performance specifications and production cost targets, there are other severe constraints dictated by the nature of the business of Nippondenso as a supplier to large auto manufacturers (i.e., the need to meet ordering patterns). The requirements of customers (like Toyota) include delivering extremely large amounts of products on a just-in-time (JIT) basis, with high variety and an unpredictable model mix that is always changing. A further constraint is to achieve all these goals with little or no changeover time. As you will see later, defining customer requirements helped Nippondenso to address the problems in a rational, thoughtful manner.

Communication Infrastructure

Nippondenso built an excellent system for information exchange. It is used to integrate the different machine tools throughout the plant through local- or wide-area networks that are, in turn, linked with the engineering departments dealing with computer control, scheduling, quality monitoring, and the like. Any change in data by a team member is promptly made available to all members of other teams, thus breaking the barriers between departments and between teams.

Product Development

The two most important elements upon which Nippondenso's approach in the product development arena is based include developing the product and its manufacturing processes simultaneously and developing new product design methodologies. In fact, this approach is credited for enabling Nippondenso to meet customer requirements.

In order to meet the challenge of high production volume and high variety, the first step for Nippondenso was standardization after negotiating with customers and listening to their concerns. The next step was to design the products intelligently so as to achieve the desired flexibility during assembly, rather than employing complex and expensive production methods. In other words, their philosophy was based on using assembly rather than manufacturing to make different models. High variety was achieved by producing several versions of each component in the product and then assembling the appropriate component's versions into any desired model. Thus, an extremely large number of combinations of component versions resulted in a large number of possible models. Moreover, this approach also ensured quick changeover from one model to another.

At this point, the basic concept of concurrent engineering has been thoroughly demonstrated. Interested readers are encouraged to consult more specialized books on the subject (see the titles provided in the references at the back of the book).

Review Questions

1. Define the term *concurrent engineering* and elaborate on its meaning.
2. In what way does a concurrent engineering environment differ from that of serial manufacturing?
3. How did concurrent engineering come into being?
4. What are the reasons for adopting concurrent engineering?
5. Discuss three of these reasons in detail.
6. List the benefits of adopting concurrent engineering and discuss three of them in detail.
7. If concurrent engineering is so beneficial, why don't all manufacturing companies adopt it?
8. What are the four pillars on which concurrent engineering rests?
9. What is the difference between the role of management in a concurrent engineering environment and that role in conventional serial manufacturing?
10. How does the product development process differ in a concurrent engineering environment from that in conventional serial manufacturing?
11. Explain why new concepts like DFM, DFA, and QFD are important and very useful when implementing concurrent engineering.

Chapter 3

Casting and Foundry Work

INTRODUCTION

Definition. The word *casting* is used both for the process and for the product. The process of casting is the manufacture of metallic objects (castings) by melting the metal, pouring it into a mold cavity, and allowing the molten metal to solidify as a casting whose shape is a reproduction of the mold cavity. This process is carried out in a *foundry,* where either ferrous (i.e., iron-base) or nonferrous metals are cast.

Casting processes have found widespread application, and the foundry industry is considered to be the sixth largest in the United States because it produces hundreds of intricately shaped parts of various sizes like plumbing fixtures, furnace parts, cylinder blocks of automobile and airplane engines, pistons, piston rings, machine tool beds and frames, wheels, and crankshafts. In fact, the foundry industry includes a variety of casting processes that can be classified in one of the following three ways:

1. By the mold material and/or procedure of mold production
2. By the method of filling the mold
3. By the metal of the casting itself

Historical Background. At the dawn of the metal age, human knowledge was not advanced enough to generate the high temperatures necessary for smelting metals. Therefore, because casting was not possible, metals were used as found or heated to a soft state and worked into shapes. The products of that era are exemplified by the copper pendant from Shanidar Cave (northeast of Iraq), which dates back to 9500 B.C. and which was shaped by hammering a piece of native metal and finishing with abrasives. Later, copper-smelting techniques were developed, and copper castings were produced in Mesopotamia as early as 3000 B.C. The art of casting was then refined by the ancient Egyptians, who innovated the "lost-wax" molding process. During the Bronze Age, foundry work flourished in China, where high-quality castings with intricate shapes could be produced. The Chinese developed certain bronze alloys and mastered the lost-wax process during the Shang dynasty. Later, that art found its way to Japan with the introduction of Buddhism in the sixth century. There were also some significant achievements in the West, where the Colossus of Rhodes, a statue of the sun god Helios weighing 360 tons, was considered to be one of the seven wonders of the world. That bronze statue was cast in sections, which were assembled later, and stood 105 feet high at the entrance of the harbor of Rhodes.

Although iron was known in Egypt as early as 4000 B.C., the development of cast iron was impossible because the high melting temperature needed was not achievable then and pottery vessels capable of containing molten iron were not available. The age of cast iron finally arrived in 1340 when a flow oven (a crude version of the blast furnace) was erected at Marche-Les-Dames in Belgium. It was capable of continuous volume production of molten iron. Ferrous foundry practice developed further with the invention of the cupola furnace by John Wilkenson in England. This was followed by the production of black-heart malleable iron in 1826 by Seth Boyden and the development of metallography by Henry Sorby of England. The relationship between the properties and the microstructure of alloys became understood, and complete control of the casting process became feasible based on this knowledge. Nevertheless, forming processes developed more rapidly than foundry practice because wrought alloys could better meet a wider range of applications. Nodular cast iron, which possesses both the castability of cast iron and the impact strength of steel, was introduced in 1948, thus paving the way for castings to compete more favorably with wrought alloys.

3.1 CLASSIFICATIONS OF CASTING BY MOLD MATERIAL

Molds can be either *permanent* or *nonpermanent.* Permanent molds are made of steel, cast iron, and even graphite. They allow large numbers of castings to be produced successively without changing the mold. A nonpermanent mold is used for one pouring only. It is usually made of a silica sand mixture but sometimes of other refractory materials like chromite and magnesite.

Green Sand Molds

Molding materials. Natural deposits taken from water or riverbeds are used as molding materials for low-melting-point alloys. Thus, the material is called *green* sand, meaning unbaked or used as found. These deposits have the advantages of availability and low cost, and they provide smooth as-cast surfaces, especially for light, thin jobs. However, they contain 15 to 25 percent clay, which, in turn, includes some organic impurities that markedly reduce the fusion temperatures of the natural sand mixture, lower the initial binding strength, and require a high moisture content (6 to 8 percent). Therefore, synthetic molding sand has been developed by mixing a cleaned pure silica sand base, in which grain structure and grain-size distribution are controlled, with up to 18 percent combined fireclay and bentonite and only about 3 percent moisture. Because the amount of clay used as a binding material is minimal, synthetic molding sand has higher refractoriness, higher green (unbaked) strength, better permeability, and lower moisture content. The latter advantage results in the evolution of less steam during the casting process. Thus, control of the properties of the sand mixture is an important condition for obtaining good castings. For this reason, a sand laboratory is usually attached to the foundry to determine the properties of molding sands prior to casting. Following are some important properties of a green sand mixture:

1. **Permeability.** Permeability is the most important property of the molding sand and can be defined as the ability of the molding sand to allow gases to pass through. This property depends not only on the shape and size of the particles of the sand base but also on the amount of the clay binding material present in the mixture and on the moisture content. The permeability of molds is usually low when casting gray cast iron and high when casting steel.

2. **Green compression strength of a sand mold.** Green strength is mainly due to the clay (or bentonite) and the moisture content, which both bind the sand particles together. Molds must be strong enough not to collapse during handling and transfer and must also be capable of withstanding pressure and erosion forces during pouring of the molten metal.

3. **Moisture content.** Moisture content is expressed as a percentage and is important because it affects other properties, such as the permeability and green strength. Excessive moisture content can result in entrapped steam bubbles in the casting.

4. **Flowability.** Flowability is the ability of sand to flow easily and fill the recesses and the fine details in the pattern.

5. **Refractoriness.** Refractoriness is the resistance of the molding sand to elevated temperatures; that is, the sand particles must not melt, soften, or sinter when they come in contact with the molten metal during the casting process. Molding sands with poor refractoriness may burn when the molten metal is poured into the mold. Usually, sand molds should be able to withstand up to 3000°F (1650°C).

Sand molding tools. Sand molds are made in *flasks,* which are bottomless containers. The function of a flask is to hold and reinforce the sand mold to allow handling and manipulation. A flask can be made of wood, sheet steel, or aluminum and consists of two parts: an upper half called the *cope* and a lower half called the *drag.* The standard flask is rectangular, although special shapes are also in use. For proper alignment of the two halves of the mold cavity when putting the cope onto the drag prior to casting, flasks are usually fitted with guide pins. When the required casting is high, a middle part, called the *cheek,* is added between the drag and the cope. Also, when a large product is to be cast, a pit in the ground is substituted for the drag; the process is then referred to as *pit molding.*

Other sand molding tools can be divided into two main groups:

1. Tools (such as molders, sand shovels, bench rammers, and the like) used for filling the flask and ramming the sand
2. Tools (such as draw screws, draw spikes, trowels, slicks, spoons, and lifters) used for releasing and withdrawing the pattern from the mold and for making required repairs on or putting finishing touches to the mold surfaces

Patterns for sand molding. The mold cavity is the impression of a *pattern,* which is an approximate replica of the exterior of the desired casting. Permanent patterns (which are usually used with sand molding) can be made of softwood like pine, hardwood like mahogany, plastics, or metals like aluminum, cast iron, or steel. They are made in special shops called *pattern shops.* Wood patterns must be made of dried or seasoned wood containing less than 10 percent moisture to avoid warping and distortion of the pattern if the wood dries out. They should not absorb any moisture from the green molding sand. Thus, the surfaces of these patterns are painted and coated with a waterproof varnish. A single-piece wood pattern can be used for making 20 to 30 molds, a plastic pattern can be used for 20,000 molds, and a metal pattern can be used for up to 100,000 molds, depending upon the metal of the pattern. In fact, several types of permanent patterns are used in foundries. They include the following:

1. **Single or loose pattern.** This pattern is actually a single copy of the desired casting. Loose patterns are usually used when only a few castings are required or when prototype castings are produced.
2. **Gated patterns.** These are patterns with gates in a runner system. They are used to eliminate the hand-cutting of gates.
3. **Match-plate patterns.** Such patterns are used for large-quantity production of smaller castings, where machine molding is usually employed. The two halves of the pattern, with the line of separation conforming to the parting line, are permanently mounted on opposite sides of a wood or metal plate. This type of pattern always incorporates the gating system as a part of the pattern.
4. **Cope-and-drag pattern plates.** The function of this type of pattern is similar to that of the match-plate patterns. Such a pattern consists of the cope and drag parts of the

pattern mounted on separate plates. It is particularly advantageous for preparing molds for large and medium castings, where the cope and drag parts of the mold are prepared on different molding machines. Therefore, accurate alignment of the two halves of the mold is necessary and is achieved through the use of guide and locating pins and bushings in the flasks.

In order for a pattern to be successfully employed in producing a casting having the desired dimensions, it must not be an exact replica of the part to be cast. A number of allowances must be made on the dimensions of the pattern:

1. **Pattern drafts.** This is a taper of about 1 percent that is added to all surfaces perpendicular to the parting line in order to facilitate removal of the pattern from the mold without ruining the surfaces of the cavity. Higher values of pattern draft are employed in the case of pockets or deep cavities.

2. **Shrinkage allowance.** Because molten metals shrink during solidification and contract with further cooling to room temperature, linear dimensions of patterns must be made larger to compensate for that shrinkage and contraction. The value of the shrinkage allowance depends upon the metal to be cast and, to some extent, on the nature of the casting. The shrinkage allowance is usually taken as 1 percent for cast iron, 2 percent for steel, 1.5 percent for aluminum, 1.5 percent for magnesium, 1.6 percent for brass, and 2 percent for bronze. In order to eliminate the need for recalculating all the dimensions of a casting, pattern makers use a shrink rule. It is longer than the standard 1-foot rule; its length differs for the different metals of the casting.

3. **Machine finish allowance.** The dimensions on a casting are oversized to compensate for the layer of metal that is removed through subsequent machining to obtain better surface finish.

4. **Distortion allowance.** Sometimes, intricately shaped or slender castings distort during solidification, even though reproduced from a defect-free pattern. In such cases, it is necessary to distort the pattern intentionally to obtain a casting with the desired shape and dimensions.

Cores and core making. *Cores* are the parts of the molds that form desired internal cavities, recesses, or projections in castings. A core is usually made of the best quality of sand to have the shape of the desired cavity and is placed into position in the mold cavity. Figure 3.1 shows the pattern, mold, and core used for producing a short pipe with two flanges. As you can see, projections, called *core prints,* are added to both sides of the pattern to create impressions that allow the core to be supported and held at both ends. When the molten metal is poured, it flows around the core to fill the rest of the mold cavity. Cores are subjected to extremely severe conditions, and they must, therefore, possess very high resistance to erosion, exceptionally high strength, good permeability, good refractoriness, and adequate collapsibility (i.e., the rapid loss of strength after the core comes in contact with the molten metal). Because a core is surrounded by molten metal from all sides (except the far ends) during casting, gases have only a small area through which to escape. Therefore, good permeability is sometimes

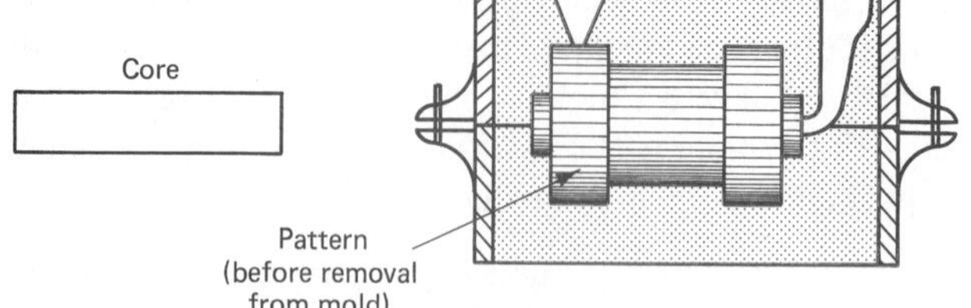

FIGURE 3.1
The pattern, mold, and core used for producing a short pipe

assisted by providing special vent holes to allow gases to escape easily. Another required characteristic of a core is the ability to shrink in volume under pressure without cracking or failure. The importance of this characteristic is obvious when you consider a casting that shrinks onto the core during solidification. If the core is made hard enough to resist the shrinkage of the casting, the latter would crack as a result of being hindered from shrinking. Figure 3.2 is a photograph of a sand core for an automotive cam tunnel.

Core sand is a very pure, fine-grained silica sand that is mixed with different binders, depending upon the casting metal with which it is going to be used. The binder used with various castings includes fireclay, bentonite, and sodium silicate (inorganic binders), as well as oils (cottonseed or linseed oil), molasses, dexstrin, and polymeric resins (organic binders).

Cores are usually made separately in *core boxes,* which involve cutting or machining cavities into blocks of wood, metal, or plastic. The surfaces of each cavity must be very smooth, with ample taper or draft, to allow easy release of the green (unbaked) core. Sometimes, a release agent is applied to the surfaces of the cavity. Core sand is rammed into the cavity, and the excess is then struck off evenly with the top of the core box. Next, the green core is carefully rolled onto a metal plate and is baked in an oven. Intricate cores are made of separate pieces that are pasted together after baking. Sometimes, cores are reinforced with annealed low-carbon steel wires or even

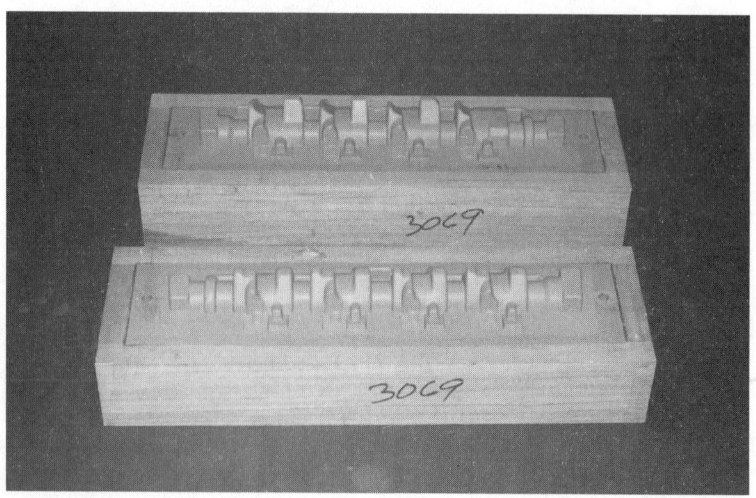

FIGURE 3.2
Core for an automotive cam tunnel

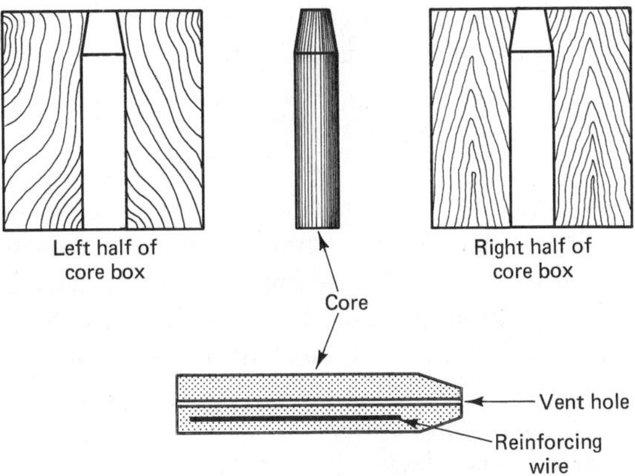

FIGURE 3.3
A simple core and its corresponding core box

cast-iron grids (in the case of large cores) to ensure coherence and stability. Figure 3.3 illustrates a simple core and its corresponding core box.

Large, round cores can be made by means of sweeps or templates, and drawing sweeps are employed to produce large cores that are not bodies of revolution. Various machines may also be employed in the core-making process. These include die extruders, jolt-squeeze machines, sandslingers, and pneumatic core blowers. Large cores are handled in the foundry and placed into the mold by means of a crane.

Gating systems. Molds are filled with molten metal by means of channels, called *gates,* cut in the sand of the mold. Figure 3.4 illustrates a typical *gating system,* which includes a pouring basin, a down sprue, a sprue base(well), a runner, and in-gates. The design of the gating system is sometimes critical and should, therefore, be based on the theories of fluid mechanics, as well as the recommended industrial practice. In fact, a gating system must be designed so that the following are ensured:

1. A continuous, uniform flow of molten metal into the mold cavity must be provided without any turbulence.

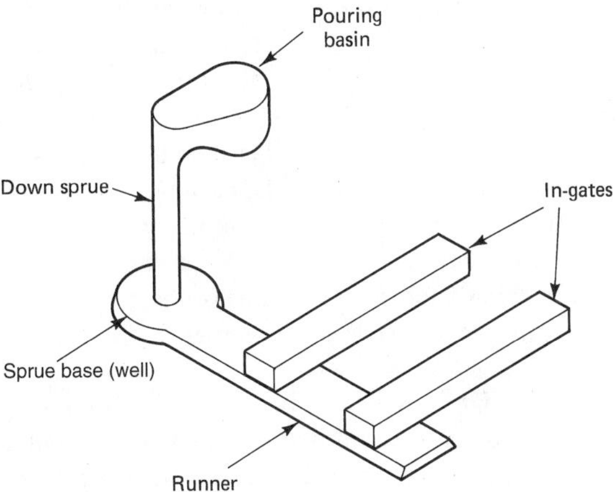

FIGURE 3.4
A typical gating system

2. A reservoir of molten metal that feeds the casting to compensate for the shrinkage during solidification must be maintained.
3. The molten metal stream must be prevented from separating from the wall of the sprue.

Let us now break down the gating system into its components and discuss the design of each of them. The *pouring basin* is designed to reduce turbulence. The molten metal from the ladle must be poured into the basin at the side that does not have the tapered sprue hole. The hole should have a projection with a generous radius around it, as shown in Figure 3.4, in order to eliminate turbulence as the molten metal enters the sprue. Next, the *down sprue* should be made tapered (its cross-sectional area should decrease when going downward) to prevent the stream of molten metal from separating from its walls, which may occur because the stream gains velocity as it travels downward and, therefore, contracts (remember the continuity equation in fluid mechanics, $A_1V_1 = A_2V_2$). The important and critical element of the gating system is the *in-gate,* whose dimensions affect those of all other elements. Sometimes, the cross-sectional area of the in-gate is reduced in the zone adjacent to the *sprue base* to create a "choke area" that is used mainly to control the flow of molten metal and, consequently, the pouring time. In other words, it serves to ensure that the rate of molten-metal flow into the mold cavity is not higher than that delivered by the ladle and, therefore, keeps the gating system full of metal throughout the casting operation. On the other hand, gas contamination, slag inclusions, and the like should be eliminated by maintaining laminar flow. Accordingly, the Reynolds number (R_n) should be checked throughout the gating system (remember that the flow is laminar when $R_n < 2000$). Use must also be made of Bernoulli's equation to calculate the velocity of flow at any cross section of the gating system.

In some cases, when casting heavy sections or high-shrinkage alloys, extra reservoirs of molten metal are needed to compensate continually for the shrinkage of the casting during solidification. These molten-metal reservoirs are called *risers* and are attached to the casting at appropriate locations to control the solidification process. The locations of the feeding system and the risers should be determined based on the phenomenon that sections most distant from those molten-metal reservoirs solidify first. Risers are molded into the cope half of the mold to ensure gravity feeding of the molten metal and are usually open to the top surface of the mold. In that case, they are referred to as *open risers.* When they are not open to the top of the mold, they are then called *blind risers.* Risers can also be classified as *top risers* and *side risers,* depending upon their location with respect to the casting.

Another way to achieve directional solidification is the use of *chills;* these involve inserts of steel, cast iron, or copper that act as a "heat sink" to increase the solidification rate of the metal at appropriate regions of the casting. Depending upon the shape of the casting, chills can be external or internal.

Molding processes. Green sand can be molded by employing a variety of processes, including some that are carried out both by hand and with molding machines. Following is a brief survey of the different green sand molding methods:

1. **Flask molding.** Flask molding is the most widely used process in both hand- and machine-molding practices. Figure 3.5 illustrates the procedure for simple hand-molding using a single (loose) pattern. First, the lower half of the pattern is placed on a molding board and surrounded by the drag. The drag is then filled with sand (using a shovel) and rammed very firmly. Ventilation holes are made using a steel wire, but these should not reach the pattern. The drag is turned upside down to bring the parting plane up so that it can be dusted. Next, the other half of the pattern is placed in position to match the lower half, and the cope is located around it, with the eyes of the cope fitted to the pins of the drag. Sand is shoveled into the cope and rammed firmly, after using a sprue pin to provide for the feeding passage. Ventilation holes are made in the cope part of the mold in the same way they were made in the other half. The pouring basin is cut around the head of the sprue pin using a trowel, and the sprue pin is pulled out of the cope. The cope is then carefully lifted off the drag and turned so that the parting plane is upward. The two halves of the pattern are removed from both the cope and the drag. The runner and/or gate are cut from the mold cavity to the sprue in the drag part of the mold. Then, any damages are repaired by slightly wetting the location and using a slick. The cope is then carefully placed on the drag to assemble the two halves of the

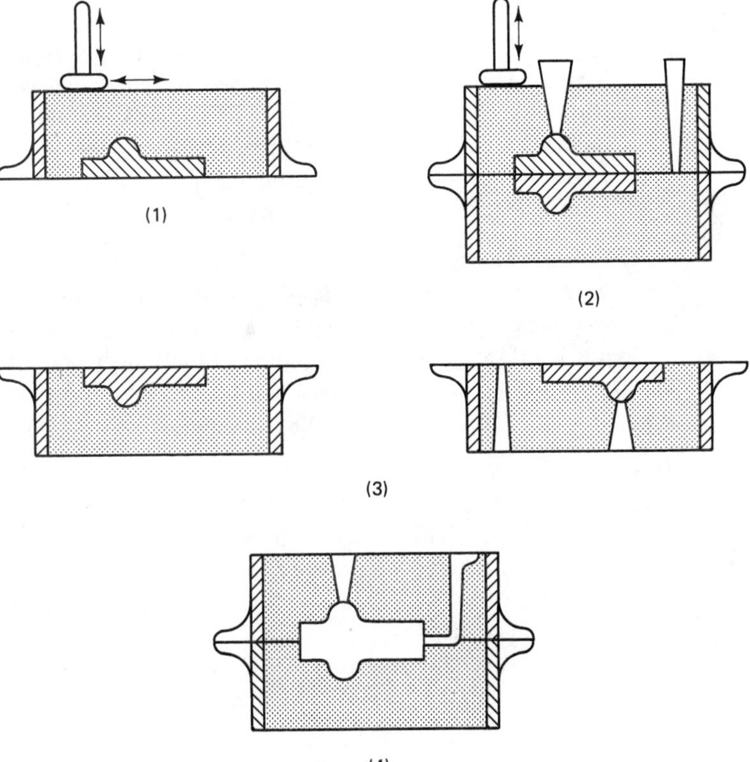

FIGURE 3.5 The procedure of flask molding using a single (loose) pattern

mold. Finally, the cope and the drag are fastened together by means of shackles or bolts to prevent the pressure created by the molten metal (after pouring) from separating them. Enough weight can be placed on the cope as an alternative to using shackles or bolts. In fact, the pressure of the molten metal after casting can be given by the following equation:

$$p = w \times h \tag{3.1}$$

where: p is the pressure

w is the specific weight of the molten metal

h is the height of the cope

The force that is trying to separate the two halves of the mold can, therefore, be given by the following equation:

$$F = p \times A \tag{3.2}$$

where: F is the force

A is the cross-sectional area of the casting (including the runner, gates, etc.) at the parting line

2. **Stack molding.** Stack molding is best suited for producing a large number of small, light castings while using a limited amount of floor space in the foundry. As can be seen in Figure 3.6a and b, there are two types of stack molding: *upright* and *stepped.* In upright stack molding, 10 to 12 flask sections are stacked up. They all have a common sprue that is employed in feeding all cavities. The drag cavity is always molded in the upper surface of the flask section, whereas the cope cavity is molded in the lower surface. In stepped stack molding, each section has its own sprue and is, therefore, offset from the one under it to provide for the pouring basin. In this case, each mold is cast separately.

3. **Sweep molding.** Sweep molding is used to form the surfaces of the mold cavity when a large-size casting must be produced without the time and expenses involved in making a pattern. A sweep that can be rotated around an axis is used for producing a surface of revolution, contrary to a drawing sweep, which is pushed axially while being guided by a frame to produce a surface having a constant section along its length (see discussion of the extrusion process in Chapter 5).

4. **Pit molding.** Pit molding is usually employed for producing a single piece of a large casting when it would be difficult to handle patterns of that size in flasks. Molding is done in specially prepared pits in the ground of the foundry. The bottom of the pit is often covered with a layer of coke that is 2 to 3 inches (50 to 75 mm) thick. Then, a layer of sand is rammed onto the coke to act as a "bed" for the mold. Vent pipes connect the coke layer to the ground surface. Molding is carried out as usual, and molds are almost always dried before pouring the molten metal. This drying is achieved by means of a portable mold drier. A cope that is also dried is then placed on the pit, and a suitable weight or a group of weights are located on the cope to prevent it from floating when the molten metal is poured.

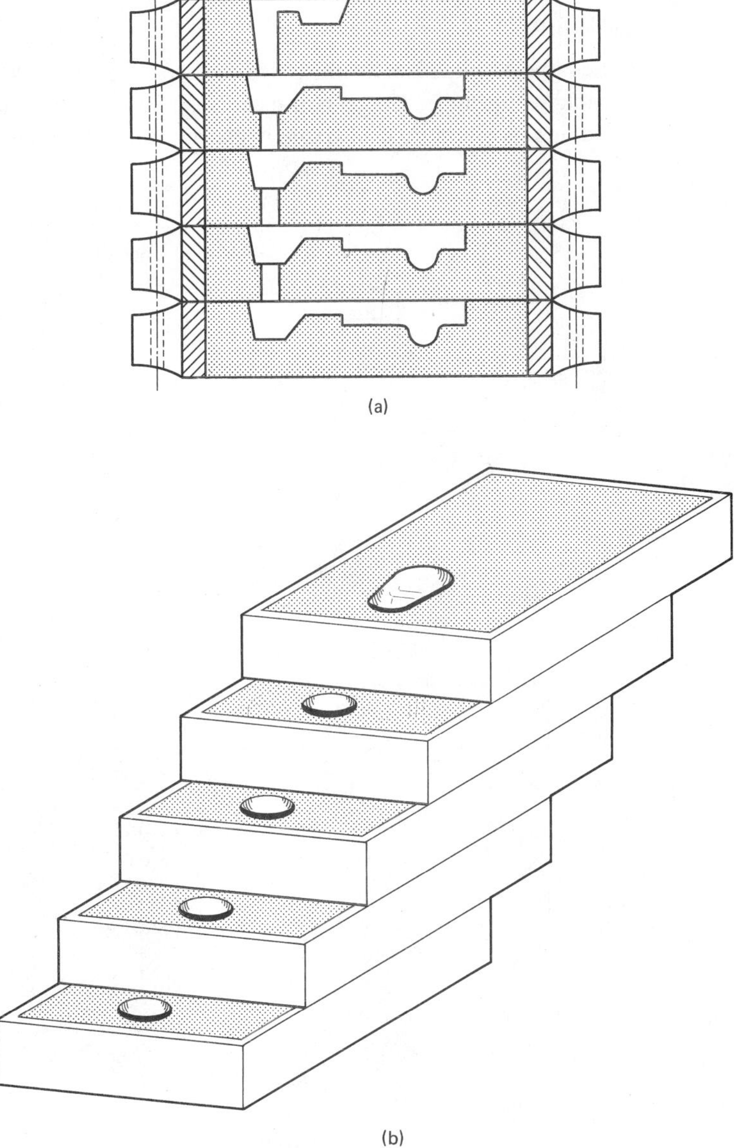

FIGURE 3.6
The two types of stack molding: (a) upright; (b) stepped

Molding machines. The employment of molding machines results in an increase in the production rate, a marked increase in productivity, and a higher and more consistent quality of molds. The function of these machines is to pack the sand onto the pattern and draw the pattern out from the mold. There are several types of molding machines, each with a different way of packing the sand to form the mold. The main types include squeezers, jolt machines, and sandslingers. There are also some machines, such as jolt-squeeze machines, that employ a combination of the working principles of two of the main types. Following is a brief discussion of the three main types of molding machines (see Figure 3.7):

1. **Squeezers.** Figure 3.7a illustrates the working principle of the squeezer type of molding machine. The pattern plate is clamped on the machine table, and a flask is put into position. A sand frame is placed on the flask, and both are then filled with

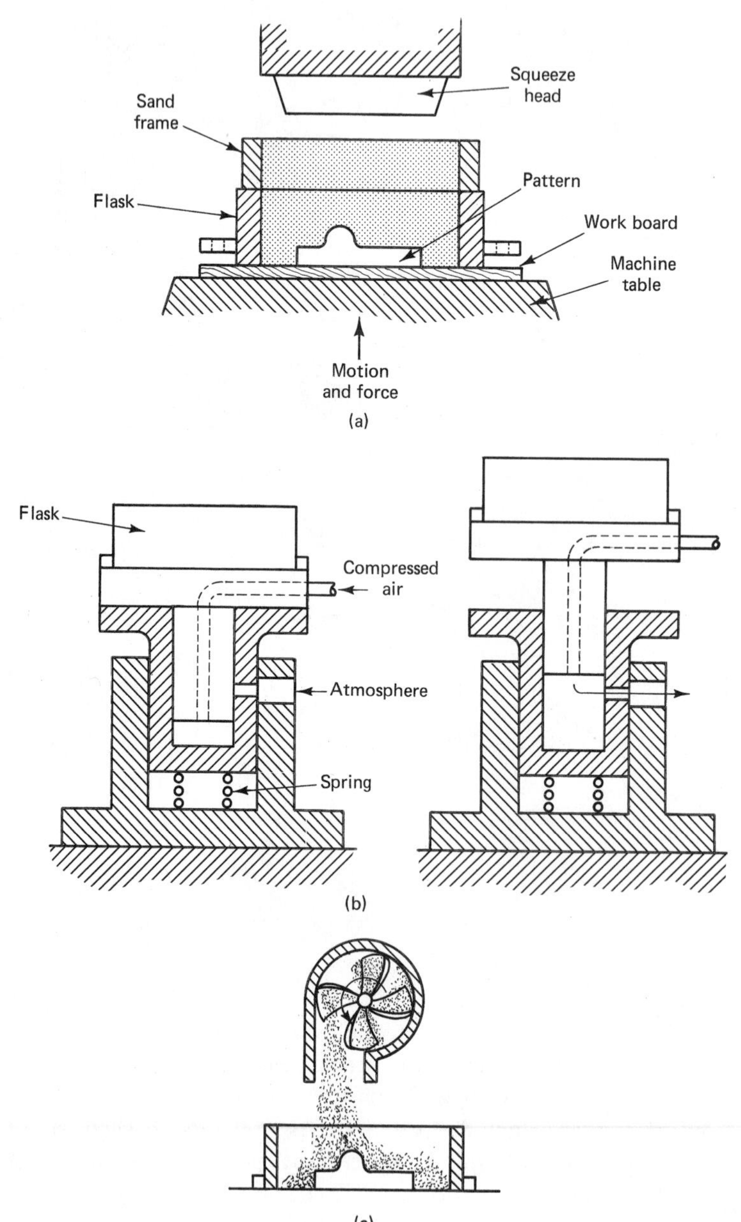

FIGURE 3.7 Molding machines: (a) squeezer; (b) jolt machine; (c) sandslinger

sand from a hopper. Next, the machine table travels upward to squeeze the sand between the pattern plate and a stationary head. The squeeze head enters into the sand frame and compacts the sand so that it is level with the edge of the flask.

2. **Jolt machines.** Figure 3.7b illustrates the working principle of the jolt type of molding machine. As can be seen, compressed air is admitted through the hose to a pressure cylinder to lift the plunger (and the flask, which is full of sand) up to a certain height, where the side hole is uncovered to exhaust the compressed air. The plunger then falls down and strikes the stationary guiding cylinder. The shock wave resulting from each of the successive impacts contributes to packing the molding sand in the flask.

3. **Sandslingers.** Figure 3.7c shows a sandslinger. This type of machine is employed in molding sand in flasks of any size, whether for individual or mass production of molds. Sandslingers are characterized by their high output, which amounts to 2500 cubic feet (more than 60 cubic meters) per hour. As can be seen, molding sand is fed into a housing containing an impeller that rotates rapidly around a horizontal axis. Sand particles are picked up by the rotating blades and thrown at a high speed through an opening onto the pattern, which is located in the flask.

No matter what type of molding machine is used, special machines are employed to draw the pattern out of the mold. Basically, these machines achieve that goal by turning the flask (together with the pattern) upside down and then lifting the pattern out of the mold. Examples of these machines include roll-over molding machines and rock-over pattern-draw machines.

Sand conditioning. The molding sand, whether new or used, must be conditioned before being used. When used sand is to be recycled, lumps should be crushed and then metal granules or small parts removed (a magnetic field is employed in a ferrous foundry). Next, sand (new or recycled) and all other molding constituents must be screened in shakers, rotary screens, or vibrating screens. Molding materials are then thoroughly mixed in order to obtain a completely homogeneous green sand mixture. The more uniform the distribution, the better the molding properties (like permeability and green strength) of the sand mixture will be.

Mixing is carried out in either continuous-screw mixers or vertical-wheel mullers. The mixers mix the molding materials by means of two large screws or worm gears; the mullers are usually used for batch-type mixing. A typical muller is illustrated in Figure 3.8. It consists primarily of a pan in which two wheels rotate about their own horizontal axis as well as about a stationary vertical shaft. Centrifugal mullers are also in use, especially for high production rates.

Dry Sand Molds

As previously mentioned, green sand molds contain up to 8 percent water, depending upon the kind and percentage of the binding material. Therefore, this type of mold can be used only for small castings with thin walls; large castings with thick walls would heat the mold, resulting in vaporization of water, which would, in turn, lead to bubbles

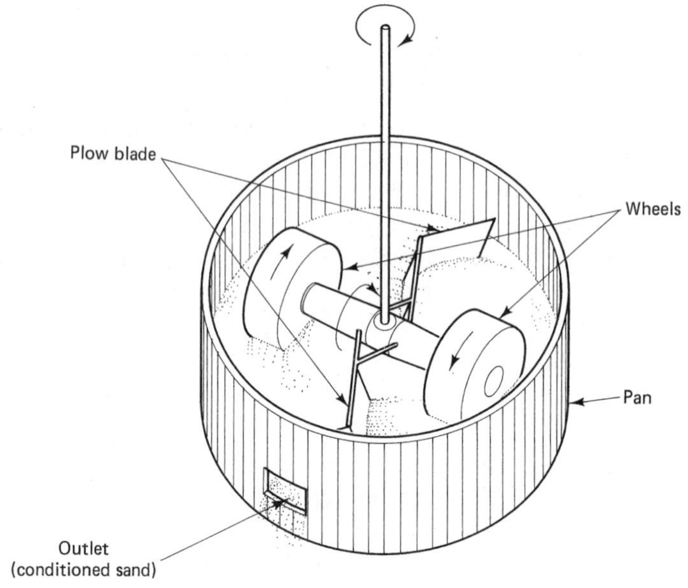

FIGURE 3.8
A muller for sand conditioning

in the castings. For this reason, molds for large castings should be dried after they are made in the same way as green sand molds. The drying operation is carried out in ovens at temperatures ranging from 300°F to 650°F (150°C to 350°C) for 8 up to 48 hours, depending upon the kind and amount of binder used.

Core-Sand Molds

When the mold is too big to fit in an oven, molds are made by assembling several pieces of sand cores. Consequently, patterns are not required, and core boxes are employed instead to make the different sand cores necessary for constructing the mold. Because core-sand mixtures (which have superior molding properties) are used, very good quality and dimensional accuracy of the castings are obtained.

Cement-Bonded Sand Molds

A mixture of silica sand containing 8 to 12 percent cement and 4 to 6 percent water is used. When making the mold, the cement-bonded sand mixture must be allowed to harden first before the pattern is withdrawn. The obtained mold is then allowed to cure for about 3 to 5 days. Large castings with intricate shapes, accurate dimensions, and smooth surfaces are usually produced in this way, the only shortcoming being the long time required for the molding process.

Carbon Dioxide Process for Molding

Silica sand is mixed with a binder involving a solution of sodium silicate (water glass) amounting to 6 percent. After the mold is rammed, carbon dioxide is blown through the sand mixture. As a result, the gel of silica binds the sand grains together, and no drying is needed. Because the molds are allowed to harden while the pattern is in position, high dimensional accuracy of molds is obtained.

Plaster Molds

A plaster mold is appropriate for casting silver, gold, magnesium, copper, and aluminum alloys. The molding material is a mixture of fine silica sand, asbestos, and plaster of paris as a binder. Water is added to the mixture until a creamy slurry is obtained, which is then employed in molding. The drying process should be very slow to avoid cracking of the mold.

Loam Molds

The loam mold is used for very large jobs. The basic shape of the desired mold is constructed with bricks and mortar (just like a brick house). A loam mixture is then used as a molding material to obtain the desired fine details of mold. Templates, sweeps, and the like are employed in the molding process. The loam mixture used in molding consists of 50 percent or more of loam, with the rest being mainly silica sand. Loam molds must be thoroughly dried before pouring the molten metal.

Shell Molds

In shell molding, a thin mold is made around a heated-metal pattern plate. The molding material is a mixture of dry, fine silica sand (with a very low clay content) and 3 to 8 percent of a thermosetting resin like phenolformaldehyde or ureaformaldehyde. Conventional dry-mixing techniques are used for obtaining the molding mixture. Specially prepared resin-coated sands are also used.

When the molding mixture drops onto the pattern plate, which is heated to a temperature of 350°F to 700°F (180°C to 375°C), a shell about 1/4 inch (6 mm) thick is formed. In order to cure the shell completely, it must be heated at 450°F to 650°F (230°C to 350°C) for about 1 to 3 minutes. The shell is then released from the pattern plate by ejector pins. To prevent sticking of the baked shell, sometimes called the *biscuit,* to the pattern plate, a silicone release agent is applied to the plate before the molding mixture drops onto it. Figure 3.9 is a photograph of a pattern of a crankshaft used in shell molding.

Shell molding is suitable for mass production of thin-walled, gray cast-iron (and aluminum-alloy) castings having a maximum weight between 35 and 45 pounds (15 and 20 kg). However, castings weighing up to 1000 pounds (450 kg) can be made by employing shell molding on an individual basis. The advantages of shell molding include good surface finish, few restrictions on casting design, and the fact that this process renders itself suitable for automation.

Ceramic Molds

In the ceramic molding process, the molding material is actually a slurry consisting of refractory grains, ceramic binder, water, alcohol, and an agent to adjust the pH value (see discussion of slurry casting in Chapter 7). The slurry is poured around the

FIGURE 3.9 A pattern of a crankshaft used in shell molding

permanent (reusable) pattern and is allowed to harden when the pattern is withdrawn. Next, the mold is left to dry for some time and then is fired to gain strength. In fact, ceramic molds are usually preheated before pouring the molten metal. For this reason, they are suitable for casting high-pouring-temperature alloys. Excellent surface finish and very close tolerances of the castings are among the advantages of this molding process and lead to the elimination of the machining operations that are usually performed on castings. Therefore, ceramic molds are certainly advantageous when casting precious or difficult-to-machine metals as well as for making castings with great shape intricacy.

Precision Molds (Investment Casting)

Precision molding is used when castings with intricate shapes, good dimensional accuracy, and very smooth surfaces are required. The process is especially advantageous for high-melting-point alloys as well as for difficult-to-machine metals. It is also most suitable for producing small castings having intricate shapes, such as the group of investment castings shown in Figure 3.10. A nonpermanent pattern that is usually made of wax must be prepared for each casting. Therefore, the process is sometimes referred to as the *lost-wax process*. Generally, the precision molding process involves the following steps (see Figure 3.11):

1. A heat-disposable pattern, together with its gating system, is prepared by injecting wax or plastic into a die cavity.

2. A pattern assembly that is composed of a number of identical patterns is made. Patterns are attached to a runner bar made of wax or plastic in much the same manner as leaves are attached to branches. A ceramic pouring cup is also attached to the top of the pattern assembly, which is sometimes referred to as the *tree* or *cluster* (see Figure 3.11a).

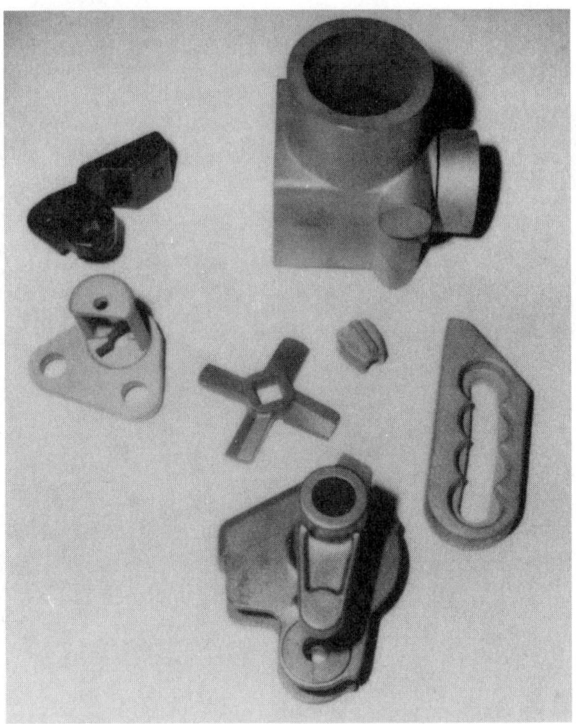

FIGURE 3.10
A group of investment castings *(Courtesy of Fansteel ESCAST, Addison, Illinois)*

3. The tree is then invested by separately dipping it into a ceramic slurry that is composed of silica flour suspended in a solution of ethyl silicate and sprinkling it with very fine silica sand. A self-supporting ceramic shell mold about 1/4 inch (6 mm) thick is formed all around the wax assembly (see Figure 3.11b). Alternatively, a thin ceramic precoating is obtained, and then the cluster is placed in a flask and a thick slurry is poured around it as a backup material.
4. The pattern assembly is then baked in an oven or a steam autoclave to melt out the wax (or plastic). Therefore, the dimensions of the mold cavity precisely match those of the desired product.
5. The resulting shell mold is fired at a temperature ranging from 1600°F to 1800°F (900°C to 1000°C) to eliminate all traces of wax and to gain reasonable strength.
6. The molten metal is poured into the mold while the mold is still hot, and a cluster of castings is obtained (see Figure 3.11c).

Today, the lost-wax process is used in manufacturing large objects like cylinder heads and camshafts. The modern process, which is known as the *lost-foam method,* involves employing a styrofoam replica of the finished product, which is then coated with a refractory material and located in a box, where sand is molded around it by vibratory compaction. When the molten metal is finally poured into the mold, the styrofoam vaporizes, allowing the molten metal to replace it.

FIGURE 3.11
Steps involved in investment casting: (a) a cluster of wax patterns; (b) a cluster of ceramic shells; (c) a cluster of castings *(Courtesy of Fansteel ESCAST, Addison, Illinois)*

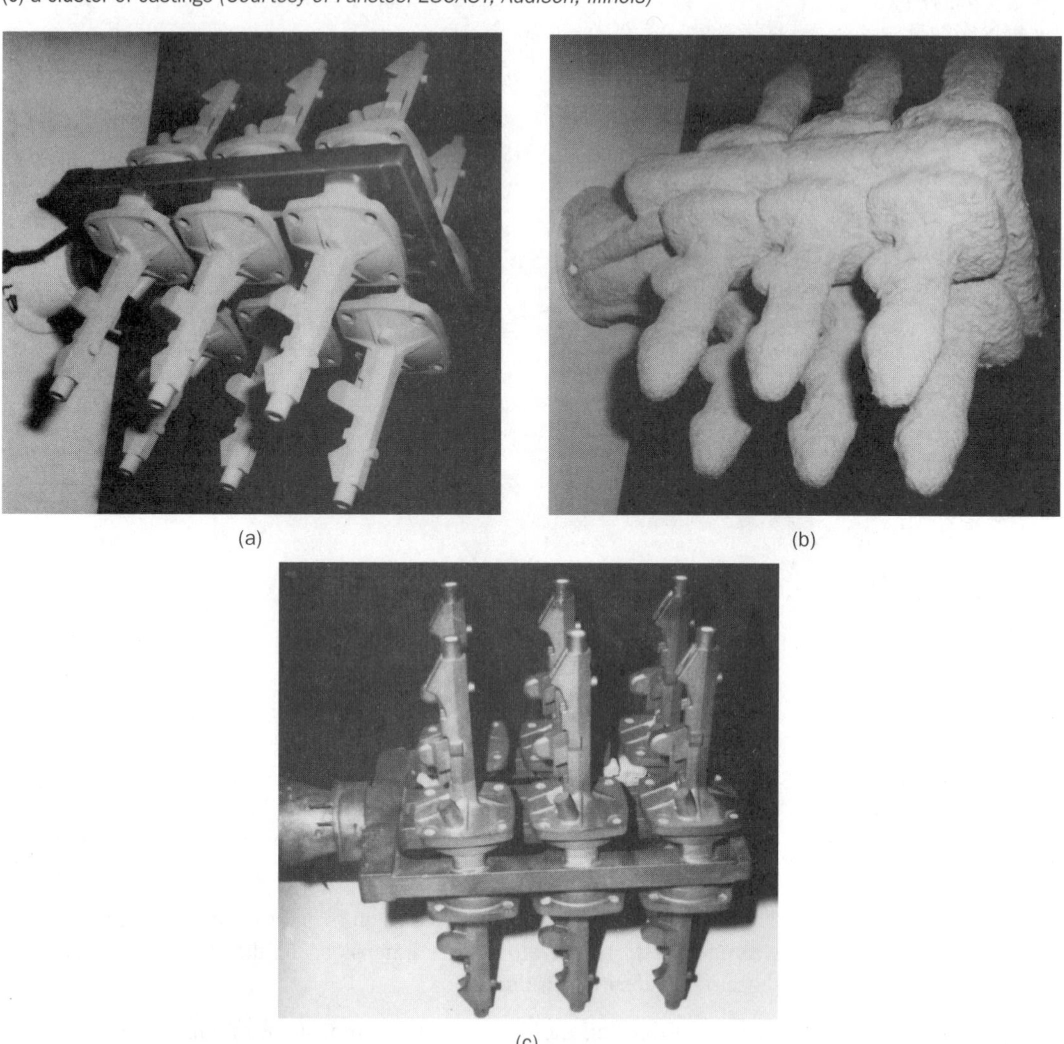

(a) (b) (c)

Graphite Molds

Graphite is used in making molds to receive alloys (such as titanium) that can be poured only into inert molds. The casting process must be performed in a vacuum to eliminate any possibility of contaminating the metal. Graphite molds can be made either by machining a block of graphite to create the desired mold cavity or by compacting a graphite-base aggregate around the pattern and then sintering the obtained mold at a temperature of 1800°F to 2000°F (1000°C to 1120°C) in a reducing atmosphere (see Chapter 7). In fact, graphite mold liners have found widespread industrial application in the centrifugal casting of brass and bronze.

Permanent Molds

A permanent mold can be used repeatedly for producing castings of the same form and dimensions. Permanent molds are usually made of steel or gray cast iron. Figure 3.12a and b shows a permanent mold made of alloy steel for molding a cylinder block. Each mold is generally made of two or more pieces that are assembled together by fitting and clamping. Although the different parts of the mold can be cast to their rough contours, subsequent machining and finishing operations are necessary to eliminate the possibility of the casting's sticking to the mold. Simple cores made of metal are frequently used. When complex cores are required, they are usually made of sand or plaster, and the mold is said to be *semipermanent.*

Different metals and alloys can successfully be cast in permanent molds. They include aluminum alloys, magnesium alloys, zinc alloys, lead, copper alloys, and cast

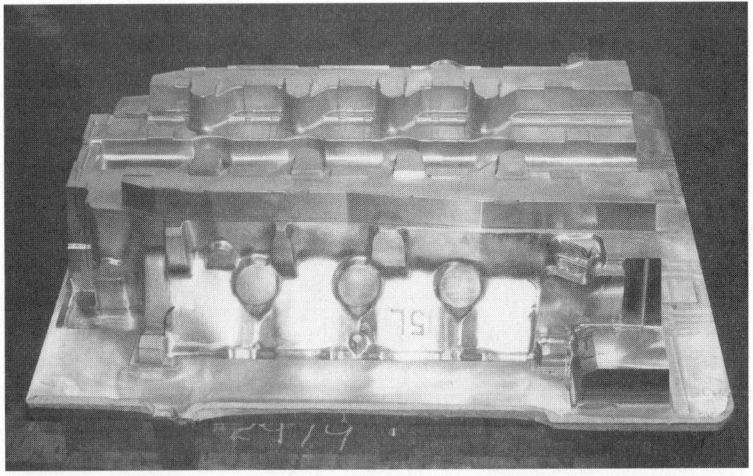

(a)

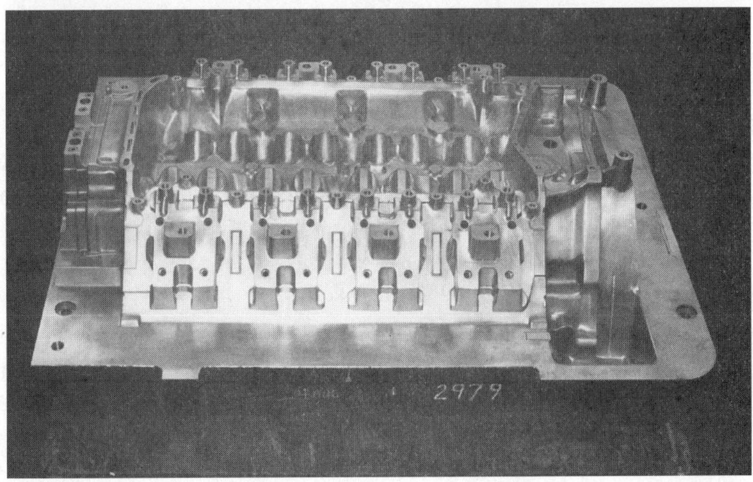

(b)

FIGURE 3.12 A permanent mold made of alloy steel for casting a cylinder block: (a) drag; (b) cope

irons. It is obvious that the mold should be preheated to an appropriate temperature prior to casting. In fact, the operating temperature of the mold, which depends upon the metal to be cast, is a very important factor in successful permanent-mold casting.

Based on the preceding discussion, we can expect the mold life to be dependent upon a number of interrelated factors, including the mold material, the metal to be cast, and the operating temperature of the mold. Nevertheless, it can be stated that the life of a permanent mold is about 100,000 pourings or more when casting zinc, magnesium, or aluminum alloys and not more than 20,000 pourings for copper alloys and cast irons. However, mold life can be extended by spraying the surface of the mold cavity with colloidal refractories suspended in liquids.

The advantages of permanent-mold casting include substantial increases in productivity (a mold does not have to be made for each casting), close tolerances, superior surface finish, and improved mechanical properties of the castings. A further advantage is the noticeable reduction in the percentage of rejects when compared with the conventional sand-casting processes. Nevertheless, the process is economically feasible for mass production only. There is also a limitation on the size of parts produced by permanent-mold casting. A further limitation is that not all alloys are suited to this process.

3.2 CLASSIFICATIONS OF CASTING BY METHOD OF FILLING THE MOLD

For all types of molds that we have discussed, the molten metal is almost always fed into the mold only by the action of gravity. Therefore, the casting process is referred to as *gravity casting*. There are, however, other special ways of pouring or feeding the molten metal into the desired cavities. These casting methods are generally aimed at forcing the molten metal to flow and fill the fine details of the mold cavity while eliminating the internal defects experienced in conventional gravity casting processes. Following is a survey of the commonly used special casting processes.

Die Casting

Die casting involves forcing the molten metal into the cavity of a steel mold, called a *die,* under very high pressure (1000 to 30,000 pounds per square inch, or about 70 to 2000 times the atmospheric pressure). In fact, this characteristic is the major difference between die casting and permanent-mold casting, where the molten metal is fed into the mold either by gravity or at low pressures. Die casting may be classified according to the type of machine used. The two principal types are *hot-chamber* machines and *cold-chamber* machines.

Hot-chamber machines. The main components of the hot-chamber die casting machine include a steel pot filled with the molten metal to be cast and a pumping system that consists of a pressure cylinder, a plunger, a gooseneck passage, and a nozzle. With the plunger in the up position, as shown in Figure 3.13a, the molten metal flows by gravity through the intake ports into the submerged hot chamber. When the plunger is

FIGURE 3.13
The hot-chamber die casting method: (a) filling the chamber; (b) metal forced into the die cavity

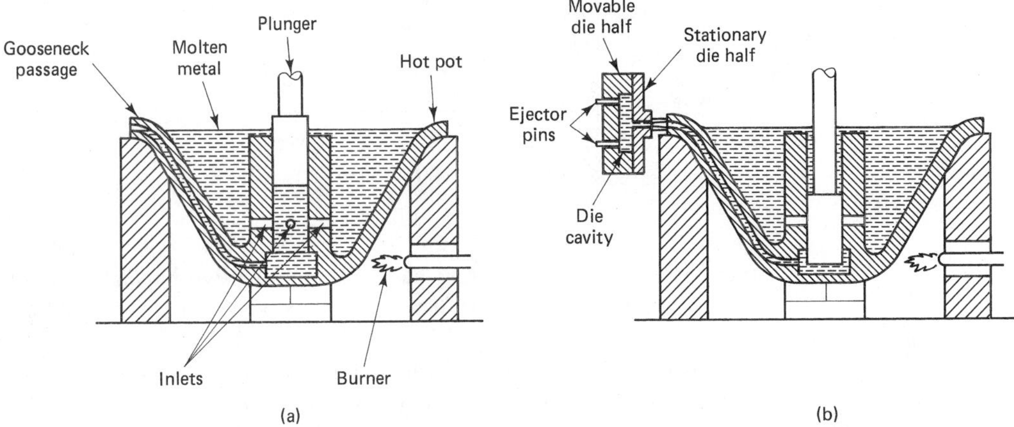

pushed downward by the power cylinder (not shown in the figure), it shuts off the intake port. Then, with further downward movement, the molten metal is forced through the gooseneck passage and the nozzle into the die cavity, as shown in Figure 3.13b. Pressures ranging from 700 to 2000 pounds per square inch (50 to 150 atmospheres) are quite common to guarantee complete filling of the die cavity. After the cavity is full of molten metal, the pressure is maintained for a preset dwell time to allow the casting to solidify completely. Next, the two halves of the die are pushed apart, and the casting is knocked out by means of ejector pins. The die cavity is then cleaned and lubricated before the cycle is repeated.

The advantages of hot-chamber die casting are numerous. They include high production rates (especially when multicavity dies are used), improved productivity, superior surface finish, very close tolerances, and the ability to produce intricate shapes with thin walls. Nevertheless, the process has some limitations. For instance, only low-melting-point alloys (such as zinc, tin, lead, and the like) can be cast because the components of the pumping system are in direct contact with the molten metal throughout the process. Also, die casting is usually only suitable for producing small castings that weigh less than 10 pounds (4.5 kg).

Cold-chamber machines. In the cold-chamber die casting machine, the molten-metal reservoir is separate from the casting machine, and just enough for one shot of molten metal is ladled every stroke. Consequently, the relatively short exposure of the shot chamber and the plunger to the molten metal allows die casting of aluminum, magnesium, brass, and other alloys having relatively high melting points. In the sequence of operations in cold-chamber die casting, the molten metal is first ladled through the pouring hole of the shot chamber while the two halves of the die are closed and locked together, as shown in Figure 3.14. Next, the plunger moves forward to close off the pouring hole and then forces the molten metal into the die cavity. Pressures in the shot chamber may go over 30,000 pounds per square inch (2000

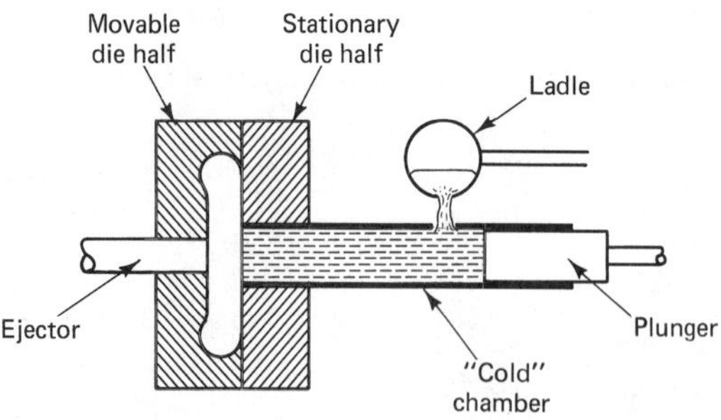

FIGURE 3.14
The cold-chamber die casting method

atmospheres). After the casting has solidified, the two halves of the die are opened, and the casting, together with the gate and the slug of excess metal, are ejected from the die.

It is not difficult to see that large parts weighing 50 pounds (23 kg) can be produced by cold-chamber die casting. The process is very successful when casting aluminum alloys, copper alloys, and high-temperature aluminum-zinc alloys. However, this process has a longer cycle time when compared with hot-chamber die casting. A further disadvantage is the need for an auxiliary system for pouring the molten metal. It is mainly for this reason that vertical cold-chamber machines were developed. As can be seen in Figure 3.15, such a machine has a transfer tube that is submerged into molten metal. It is fed into the shot chamber by connecting the die cavity to a vacuum tank by means of a special valve. The molten metal is forced into the die cavity when the plunger moves upward.

Centrifugal Casting

Centrifugal casting refers to a group of processes in which the forces used to distribute the molten metal in the mold cavity (or cavities) are caused by centrifugal acceleration. Centrifugal casting processes can be classified as *true centrifugal* casting,

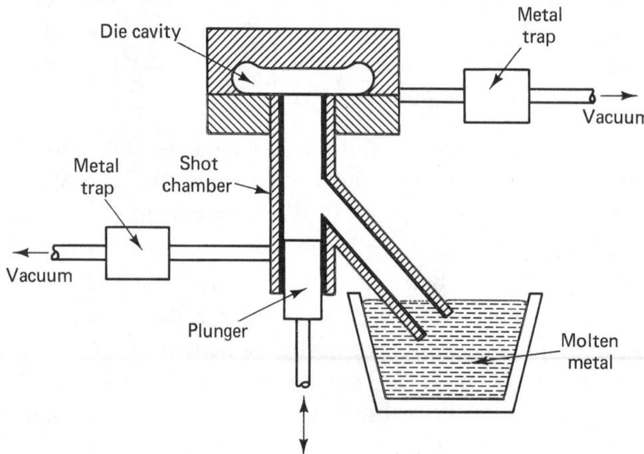

FIGURE 3.15
A vertical cold-chamber die casting machine

semicentrifugal casting, and the *centrifuging* method. Each of these processes is briefly discussed next.

True centrifugal casting. True centrifugal casting involves rotating a cylindrical mold around its own axis, with the revolutions per minute high enough to create an effective centrifugal force, and then pouring molten metal into the mold cavity. The molten metal is pushed to the walls of the mold by centrifugal acceleration (usually 70 to 80 times that of gravity), where it solidifies in the form of a hollow cylinder. The outer shape of the casting is given by the mold contour, while the diameter of the inner cylindrical surface is controlled by the amount of molten metal poured into the mold cavity. The machines used to spin the mold may have either horizontal or vertical axes of rotation. Short tubes are usually cast in vertical-axis machines, whereas longer pipes, like water supply and sewer pipes, are cast using horizontal-axis machines. The basic features of a true centrifugal casting machine with a horizontal axis are shown in Figure 3.16.

Centrifugal castings are characterized by their high density, refined fine-grained structure, and superior mechanical properties, accompanied by a low percentage of rejects and, therefore, a high production output. A further advantage of the centrifugal casting process is the high efficiency of metal utilization due to the elimination of sprues and risers and the small machining allowance used.

Semicentrifugal casting. Semicentrifugal casting is quite similar to the preceding type, the difference being that the mold cavity is completely filled with the molten metal. But because centrifugal acceleration is dependent upon the radius, the central core of the casting is subjected to low pressure and is, therefore, the region where entrapped air and inclusions are present. For this reason, the semicentrifugal casting process is recommended for producing castings that are to be subjected to subsequent machining to remove their central cores. Examples include cast track wheels for tanks,

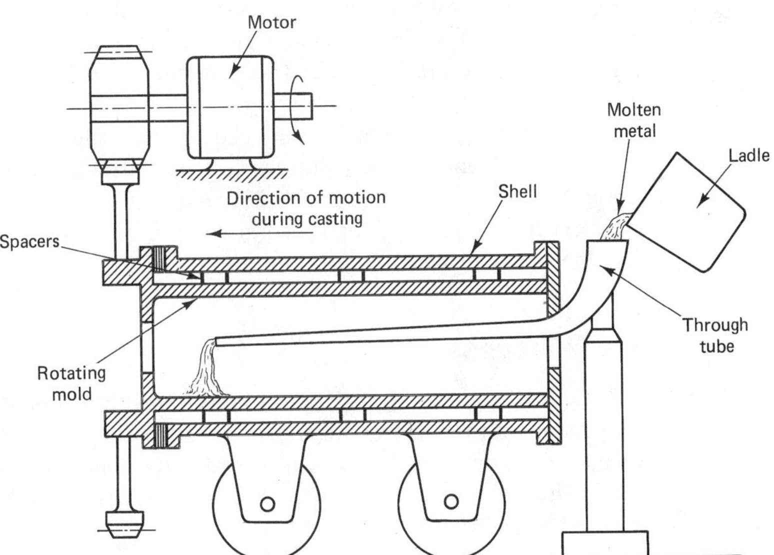

FIGURE 3.16
A true centrifugal casting machine

tractors, and the like. A sand core is sometimes used to form the central cavity of the casting in order to eliminate the need for subsequent machining operations.

Centrifuging. In the centrifuging method, a number of mold cavities are arranged on the circumference of a circle and are connected to a central down sprue through radial gates. Next, molten metal is poured, and the mold is rotated around the central axis of the sprue. In other words, each casting is rotated around an axis off (shifted from) its own center axis. Therefore, mold cavities are filled under high pressure, so the process is usually used for producing castings with intricate shapes; the increased pressure on the casting during solidification allows the fine details of the mold to be obtained.

Continuous Casting

The continuous casting process is gaining widespread industrial use, especially for high-quality alloy steel. In fact, the process itself passed through a few evolutionary stages. Although it was originally developed for producing cast-iron sheets, an up-to-date version is now being used for casting semifinished products that are to be processed subsequently by piercing, forging, extrusion, and the like.

The continuous casting process basically involves controlling the flow of a stream of molten metal that comes out from a water-cooled orifice in order to solidify and form a continuous strip (or rod). The new version of this process is usually referred to as *rotary continuous casting* because the water-cooled mold (orifice) is always oscillating and rotating at about 120 revolutions per minute during casting. Figure 3.17 illustrates the principles of rotary continuous casting. The steel is melted, refined, and degassed and its chemical composition controlled before it is transferred and poured into the caster (tundish). The molten metal then enters the rotating mold tangent to the edge through the bent tube. The centrifugal force then forces the steel against the mold wall, while lighter inclusions and impurities remain in the center of the vortex, where they are removed by the operator. Solidification of the metal flowing out of the mold continues at a precalculated rate. The resulting bar is then cut by a circular saw that is traveling downward at the same speed as the bar. The bar is tilted and loaded onto a conveyor to transfer it to the cooling bed and the rolling mill.

The continuous casting process has the advantages of very high metal yield (about 98 percent, compared with 87 percent in conventional ingot-mold practice), excellent quality of cast, controlled grain size, and the possibility of casting special cross-sectional shapes.

The V-Process

The vacuum casting process (V-process for short) involves covering the two halves of the pattern with two plastic films that are 0.005 inch (0.125 mm) thick by employing vacuum forming (see chapter 8). The pattern is then removed, and the two formed-plastic sheets are tightened together to form a mold cavity that is surrounded by a flask filled with sand (there is no need for a binder). This mold cavity is kept in a vacuum as the molten metal is poured to assist and ensure easy flow.

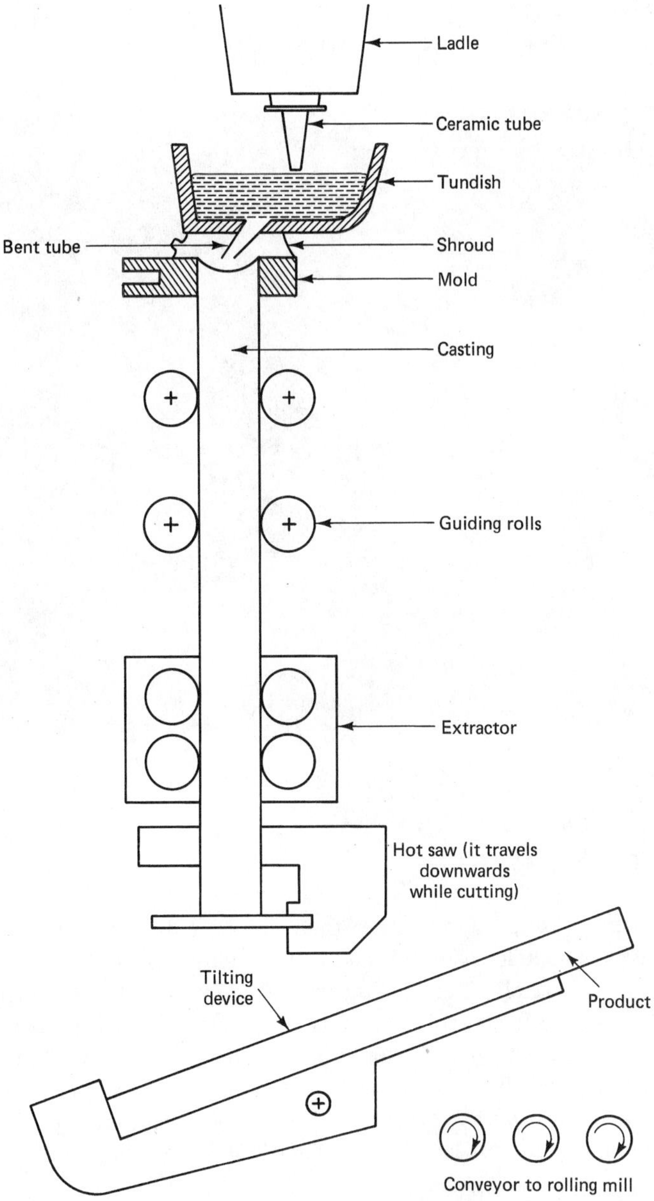

FIGURE 3.17
The principles of rotary continuous casting

The V-process, developed in Japan in the early 1970s, offers many advantages, such as the elimination of the need for special molding sands with binders and the elimination of the problems associated with green sand molding (like gas bubbles caused by excess humidity). Also, the size of risers, vents, and sprues can be reduced markedly, thus resulting in an increase in the efficiency of material utilization. Figure 3.18 shows a plastic mold being prepared for the V-process.

FIGURE 3.18
A plastic mold being prepared for the V-process *(Courtesy of Spectrum Casting, Inc., Flint, Michigan)*

3.3 CLASSIFICATIONS OF CASTING BY METAL TO BE CAST

When classified by metal, castings can be either *ferrous* or *nonferrous.* The ferrous castings include cast steels and the family of cast irons, whereas the nonferrous castings include all other metals, such as aluminum, copper, magnesium, titanium, and their alloys. Each of these metals and alloys is melted in a particular type of foundry furnace that may not be appropriate for melting other metals and alloys. Also, molding methods and materials, as well as fluxes, degassers, and additives, depend upon the metal to be cast. Therefore, this classification method is popular in foundry work. Following is a brief discussion of each of these cast alloys.

Ferrous Metals

Cast steels. Steels are smelted in open-hearth furnaces, convertors, electric-arc furnaces, and electric-induction furnaces. Cast steels can be either plain-carbon, low-alloy, or high-alloy steel. However, plain-carbon cast steel is the most commonly produced type. When compared with cast iron, steel certainly has poorer casting properties—namely, higher melting point, higher shrinkage, and poorer fluidity. Steels are also more susceptible to hot and cold cracks after the casting process. Therefore, cast steels are almost always subjected to heat treatment to relieve the internal stresses and improve the mechanical properties.

In order to control the oxygen content of molten steels, aluminum, silicon, or manganese is used as a deoxidizer. Aluminum is the most commonly used of these elements because of its availability, low cost, and effectiveness.

There is an important difference between cast-steel and wrought products. This involves the presence of a "skin," or thin layer, just below the surface of a casting, where scales, oxides, and impurities are concentrated. Also, this layer may be chemically or structurally different from the base metal. Therefore, it has to be removed by machining in a single deep cut, which is achieved through reducing the cutting speed to half of the conventionally recommended value.

Gray cast iron. Gray cast iron is characterized by the presence of free graphite flakes when its microstructure is examined under the microscope. This kind of microstructure is, in fact, responsible for the superior properties possessed by gray cast iron. For instance, this dispersion of graphite flakes acts as a lubricant during machining of gray cast iron, thus eliminating the need for machining lubricants and coolants. When compared with any other ferrous cast alloy, gray cast iron certainly possesses superior machinability. The presence of those graphite flakes is also the reason for its ability to absorb vibrations. The compressive strength of this iron is normally four times its tensile strength. Thus, gray cast iron has found widespread application in machine tool beds (bases) and the like. On the other hand, gray cast iron has some disadvantages and limitations, such as its low tensile strength, brittleness, and poor weldability. Nevertheless, gray cast iron has the lowest casting temperature, least shrinkage, and the best castability of all cast ferrous alloys.

The cupola is the most widely used foundry furnace for producing and melting gray cast iron. The chemical composition, microstructure, and, therefore, the properties of the obtained castings are determined by the constituents of the charge of the cupola furnace. Thus, the composition and properties of gray cast iron are controlled by changing the percentages of the charge constituents and also by adding *inoculants* and alloying elements. Commonly used inoculants include calcium silicide, ferrosilicon, and ferromanganese. An inoculant is added to the molten metal (either in the cupola spout or ladle) and usually amounts to between 0.1 and 0.5 percent of the molten iron by weight. It acts as a deoxidizer and also hinders the growth of precipitated graphite flakes. It is important for a product designer to remember that the properties of a gray cast-iron product are also dependent upon the dimensions (the thicknesses of the walls) of that product because the cooling rate is adversely affected by the cross section of the casting. Actually, the cooling rate is high for small castings with thin walls, sometimes yielding white cast iron. For this reason, gray cast iron must be specified by the strength of critical cross sections.

White cast iron. When the molten cast-iron alloy is rapidly chilled after being poured into the mold cavity, dissolved carbon does not have enough time to precipitate in the form of flakes. Instead, it remains chemically combined with iron in the form of cementite. This material is primarily responsible for the whitish crystalline appearance of a fractured surface of white cast iron. Cementite is also responsible for the high hardness, extreme brittleness, and excellent wear resistance of this kind of cast iron. Industrial applications of white cast iron involve components subjected to abrasion. Sometimes, gray cast iron can be chilled to produce a surface layer of white cast iron in order to combine the advantageous properties of the two types of cast iron. In this case, the product metal is usually referred to as *chilled* cast iron.

Ductile cast iron. Ductile cast iron is also called *nodular* cast iron and *spheroidal-graphite* cast iron. It is obtained by adding trace amounts of magnesium to a very pure molten alloy of gray cast iron that has been subjected to desulfurization. Sometimes, a small quantity of cerium is also added to prevent the harmful effects of impurities like aluminum, titanium, and lead. The presence of magnesium and cerium causes the graphite to precipitate during solidification of the molten alloy in the form of small spheroids, rather than flakes as in the case of gray cast iron. This microstructural change results in a marked increase in ductility, strength, toughness, and stiffness of ductile iron, as compared with gray cast iron, because the stress concentration effect of a flake is far higher than that of a spheroid (remember what you learned in fracture mechanics). The disadvantages of ductile iron, as compared with gray cast iron, include lower damping capacity and thermal conductivity. Ductile iron is used for making machine parts like axles, brackets, levers, crankshafts, housings, die pads, and die shoes.

Compacted-graphite cast iron. Compacted-graphite (CG) cast iron falls between gray and ductile cast irons, both in its microstructure and mechanical properties. The free graphite in this type of iron takes the form of short, blunt, and interconnected flakes. The mechanical properties of CG cast iron are superior to those of gray cast iron but are inferior to those of ductile cast iron. The thermal conductivity and damping capacity of CG cast iron approach those of gray cast iron. Compacted-graphite cast iron has some application in the manufacture of diesel engines.

Malleable cast iron. Malleable cast iron is obtained by two-stage heat treatment of white cast iron having an appropriate chemical composition. The hard white cast iron becomes malleable after the heat treatment due to microstructural changes. The combined carbon separates as free graphite, which takes the form of nodules. Because the raw material for producing malleable iron is actually white cast iron, there are always limitations on casting design. Large cross sections and thick walls are not permitted because it is difficult to produce a white cast-iron part with these geometric characteristics.

The two basic types of malleable cast iron are the *pearlitic* and the *ferritic* (blackheart). Although the starting alloy for both types is the same (white cast iron), the heat treatment cycle and the atmosphere of the heat-treating furnace are different in each case. Furnaces with oxidizing atmospheres are employed for producing pearlitic malleable cast iron, whereas furnaces with neutral atmospheres are used for producing ferritic malleable cast iron. When comparing the properties of these two types, the ferritic grades normally have higher ductility and better machinability but lower strength and hardness. Pearlitic grades can, however, be subjected to further surface hardening when the depth of the hardened layer is controlled.

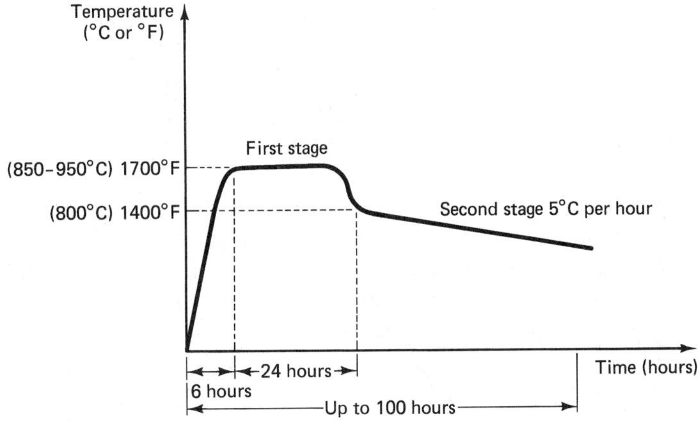

FIGURE 3.19 The heat treatment sequence for producing malleable cast iron

Figure 3.19 shows the heat treatment sequence for producing malleable cast iron. Referred to as the *malleabilizing cycle,* it includes two stages, as shown in Figure 3.19. In the first stage, the casting is slowly heated to a temperature of about 1700°F (950°C) and is kept at that temperature for about 24 hours. In the second stage, the temperature is decreased very slowly at a rate of 5°F to 9°F (3°C to 5°C) per hour from a temperature of 1400°F (800°C) to a temperature of 1200°F (650°C), where the process ends and the casting is taken out of the furnace. The whole malleabilizing cycle normally takes about 100 hours.

Malleable cast iron is usually selected when the engineering application requires good machinability and ductility. Excellent castability and high toughness are other properties that make malleable cast iron attractive as an engineering material. Typical applications of malleable cast iron include flanges, pipe fittings, and valve parts for pressure service at elevated temperatures, steering-gear housings, mounting brackets, and compressor crankshafts and hubs.

Alloyed cast irons. Alloying elements like chromium, nickel, and molybdenum are added to cast irons to manipulate the microstructure of the alloy. The goal is to improve the mechanical properties of the casting and also to impart some special properties to it, like resistance to wear, corrosion, and heat. A typical example of alloyed irons is the white cast iron containing nickel and chromium that is used for corrosion-resistant (and abrasion-resistant) applications like water pump housings and grinding balls (in a ball mill).

Nonferrous Metals

Cast aluminum and its alloys. Aluminum continues to gain wide industrial application, especially in the automotive and electronics industries, because of its distinguished strength-to-weight ratio and its high electrical conductivity. Alloying elements can be added to aluminum to improve its mechanical properties and metallurgical characteristics. Silicon, magnesium, zinc, tin, and copper are the elements most commonly alloyed with aluminum. In fact, most metallic elements can be alloyed with aluminum, but commercial and industrial applications are limited to those just mentioned.

A real advantage of aluminum is that it can be cast by almost all casting processes. Nevertheless, the common methods for casting aluminum include die casting, gravity casting in sand and permanent molds, and investment casting (the lost-foam process).

The presence of hydrogen when melting aluminum always results in unsound castings. Typical sources of hydrogen are the furnace atmosphere and the charge metal. When the furnace has a reducing atmosphere because of incomplete combustion of the fuel, carbon monoxide and hydrogen are generated and absorbed by the molten metal. The presence of contaminants like moisture, oil, or grease, which are not chemically stable at elevated temperatures, can also liberate hydrogen. Unfortunately, hydrogen is highly soluble in molten aluminum but has limited solubility in solidified aluminum. Therefore, any hydrogen that is absorbed by the molten metal is liberated or expelled during solidification, causing porosity. Hydrogen may also react with (and reduce) metallic oxides to form water vapor, which again causes porosity. Thus, hydrogen must be completely removed from molten aluminum before casting. This is achieved by using appropriate *degassers*. Chlorine and nitrogen are considered to be the traditional degassers for aluminum. Either of these is blown through the molten aluminum to eliminate any hydrogen. However, because chlorine is toxic and nitrogen is not that efficient, organic chloride fluxing compounds (chlorinated hydrocarbons) are added to generate chlorine within the melt. They are commercially available in different forms, such as blocks, powders, and tablets; the most commonly used fluxing degasser is perhaps hexachlorethane. Another source of problems when casting aluminum is iron, which dissolves readily in molten aluminum. Therefore, care must be taken to spray (or cover) iron ladles and all iron surfaces that come into direct contact with the molten aluminum with a ceramic coating. This extends the service life of the iron tools used and also results in sound castings.

The most important cast-aluminum alloys are those containing silicon, which serves to improve the castability, reduce the thermal expansion, and increase the wear resistance of aluminum. Small additions of magnesium make these alloys heat treatable, thus allowing the final properties of the castings to be controlled. Aluminum-silicon alloys (with 5 to 13 percent silicon) are used in making automobile parts (e.g., pistons) and aerospace components.

Aluminum-copper alloys are characterized by their very high tensile-strength-to-weight ratio. They are, therefore, mainly used for the manufacture of premium-quality aerospace parts. Nevertheless, these alloys have poorer castability than the aluminum-silicon alloys. Also, amounts of the copper constituent in excess of 12 percent make the alloy brittle. Copper additions of up to 5 percent are usually used and result in improved high-temperature properties and machinability.

Additions of magnesium to aluminum result in improved corrosion resistance and machinability, higher strength, and attractive appearance of the casting when anodized. However, aluminum-magnesium alloys are generally difficult to cast. Zinc is also used as an alloying element, and the aluminum-zinc alloys have good machinability and moderately high strength. But these alloys are generally prone to hot cracking and have poorer castability and high shrinkage. Therefore, zinc is usually alloyed with aluminum in combination with other alloying elements and is employed in such cases for promoting very high strength. Aluminum-tin alloys are also in use. They possess high load-carrying capacity and fatigue strength and are, therefore, used for making bearings and bushings.

Cast copper alloys. The melting temperatures of cast copper alloys are far higher than those of aluminum, zinc, or magnesium alloys. Cast copper alloys can be grouped according to their composition as follows:

1. Pure copper and high-copper alloys
2. Brasses (alloys including zinc as the principal alloying element)
3. Bronzes (alloys including tin as the principal alloying element)
4. Nickel silvers, including copper-nickel alloys and copper-nickel-zinc alloys

Cast copper alloys are melted in crucible furnaces, open-flame furnaces, induction furnaces, or indirect-arc furnaces. The selection of a furnace depends upon the type of alloy to be melted, as well as the purity and quantity required. In melting pure copper, high-copper alloys, bronzes, or nickel silver, precautions must be taken to prevent contamination of the molten metal with hydrogen. It is recommended that the atmosphere of the furnace be slightly oxidizing and also that a covering flux be used. Prior to casting, however, the molten metal should be *deoxidized* by adding phosphorus in the form of a phosphorous copper flux. On the other hand, brass is usually not susceptible to hydrogen porosity. The problem associated with melting brass is the vaporization and oxidation of the zinc. As a remedy, the atmosphere of the furnace should be slightly reducing. Also, a covering flux should be used to prevent vaporization of the zinc; a deoxidizing flux (like phosphorous copper) is then added immediately prior to pouring. The applications of cast-copper alloys include pipe fitting, ornaments, propeller hubs and blades, steam valves, and bearings.

Zinc alloys. The family of zinc alloys is characterized by low melting temperatures. Zinc alloys also possess good fluidity. Therefore, they can be produced in thin sections by submerged-hot-chamber die casting. Alloying elements employed include aluminum, copper, and magnesium.

Magnesium alloys. The main characteristic of magnesium is its low density, which is lower than that of any other commercial metal. The potential uses of magnesium are many because it is readily available as a component of seawater and most of its disadvantages and limitations can be eliminated by alloying. Magnesium alloys usually are cast in permanent molds or are produced by hot-chamber die casting.

3.4 FOUNDRY FURNACES

Various furnaces are employed for smelting different ferrous and nonferrous metals in foundry work. The type of foundry furnace to be used is determined by the kind of metal to be melted, the hourly output of molten metal required, and the purity desired. Following is a brief review of each of the commonly used foundry furnaces.

Cupola Furnaces

Structure. The cupola is the most widely used furnace for producing molten gray cast iron. A sketch of a cupola furnace is given in Figure 3.20. As can be seen, the cupola is a shaft-type furnace whose height is three to five times its diameter. It is constructed of a steel plate that is about 3/8 inch (10 mm) thick and that is internally lined with refractory fireclay bricks. The whole structure is erected on legs, or columns. Toward the top of the furnace is an opening through which the charge is fed. Air, which is needed for the combustion, is blown through the tuyeres located about 36 inches (900 mm) above the bottom of the furnace. Slightly above the bottom and in the front are a tap hole and spout to allow molten cast iron to be collected. There is also a slag hole located at the rear and above the level of the tap hole

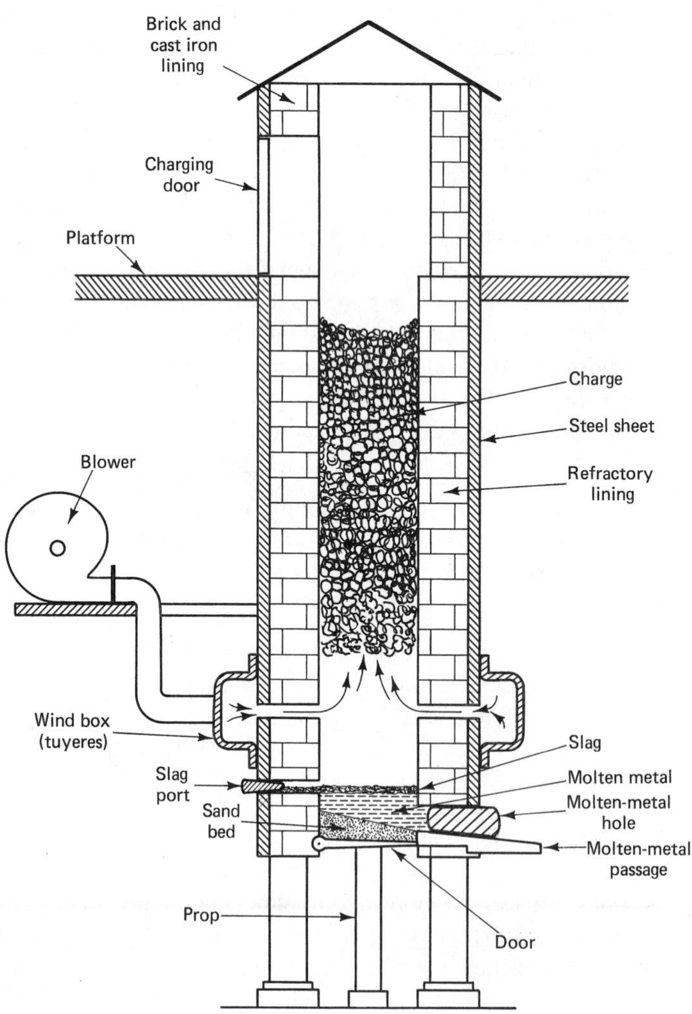

FIGURE 3.20
A cupola furnace

(because slag floats on the surface of molten iron). The bottom of the cupola is closed with drop doors to dump residual coke or metal and also to allow for maintenance and repair of the furnace lining.

Operation. A bed of molding sand is first rammed on the bottom to a thickness of about 6 inches (150 mm) or more. A bed of coke about 40 inches (1.0 m) thick is next placed on the sand. The coke is then ignited, and air is blown at a lower-than-normal rate. Next, the charge is fed into the cupola through the charging door. Many factors, such as the charge composition, affect the final structure of the gray cast iron obtained. Nevertheless, it can generally be stated that the charge is composed of 25 percent pig iron, 50 percent gray cast-iron scrap, 10 percent steel scrap, 12 percent coke as fuel, and 3 percent limestone as flux. These constituents form alternate layers of coke, limestone, and metal. Sometimes, ferromanganese briquettes and inoculants are added to the charge to control and improve the structure of the cast iron produced.

Direct Fuel-Fired Furnaces (Reverberatory Furnaces)

The direct fuel-fired furnace, or reverberatory furnace, is used for the batch-type melting of bronze, brass, or malleable iron. The burners of the furnace are fired with pulverized coal or another liquid petroleum product. Figure 3.21 shows that the roof of the reverberatory furnace reflects the flame onto the metal placed on the hearth, thus heating the metal and melting it. The gaseous products of combustion leave the furnace through the flue duct. The internal surface of the furnace is lined with fire bricks, and there are charging and tap holes. When iron is melted, the fuel-air ratio is adjusted to produce a completely white iron without free graphite flakes because they lower the properties of the resulting malleable iron.

Crucible (Pot) Furnaces

Nonferrous metals like bronzes, brasses, aluminum, and zinc alloys are usually melted in a crucible, or pot, furnace. Crucible furnaces are fired by liquid, gaseous, or pulverized solid fuel. Figure 3.22 shows that the products of combustion in a crucible furnace

FIGURE 3.21
A reverberatory furnace

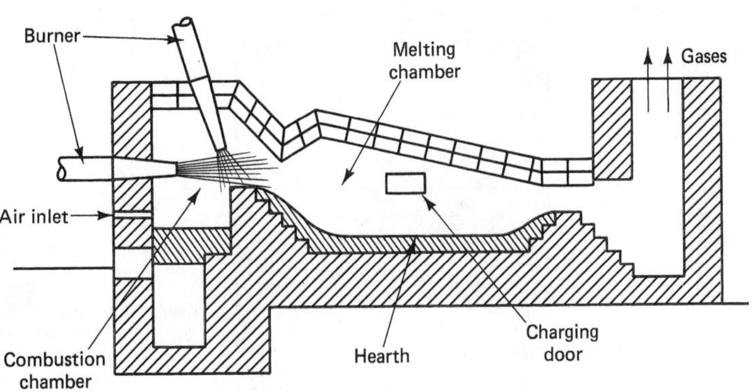

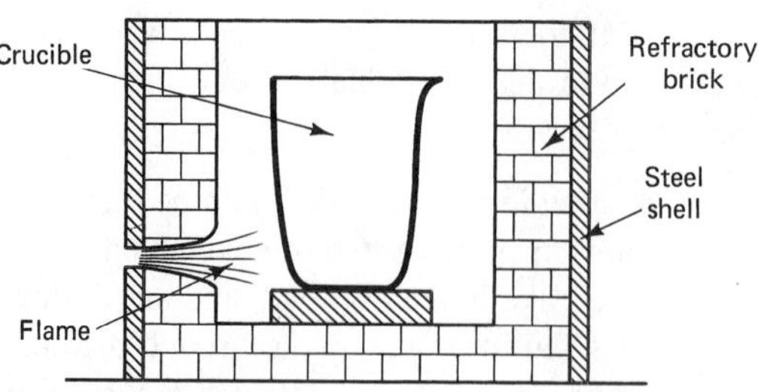

FIGURE 3.22
A crucible furnace

do not come in direct contact with the molten metal, thus enabling the production of quality castings. Crucible furnaces can be stationary or tilting. When the stationary type is employed, crucibles are lifted out by tongs and are then carried in shanks. On the other hand, crucibles with long pouring lips are always used with the tilting type.

Crucibles are made of either refractory material or alloy steels (containing 25 percent chromium). Refractory crucibles can be of the clay-graphite ceramic-bonded type or the silicon-carbide carbon-bonded type. The first type is cheaper, while the second one is more popular in industry. Ceramic crucibles are used when melting aluminum, bronze, or gray cast iron, whereas brasses are melted in alloy steel crucibles. Different alloys must not be melted in the same crucible to avoid contamination of the molten metal.

Electric Furnaces

An electric furnace is usually used when there is a need to prevent the loss of any constituent element from the alloy and when high purity and consistency of casting quality are required. An electric furnace is also employed when melting high-temperature alloys. In all types of electric furnaces, whether they are electric-arc, resistance, or induction furnaces, the electric energy is converted into heat.

Electric-arc furnace. The electric-arc furnace is the most commonly used type of electric furnace. Figure 3.23 is a sketch of an electric-arc furnace. The heat generated by an electric arc is transferred by direct radiation or by reflected radiation off the in-

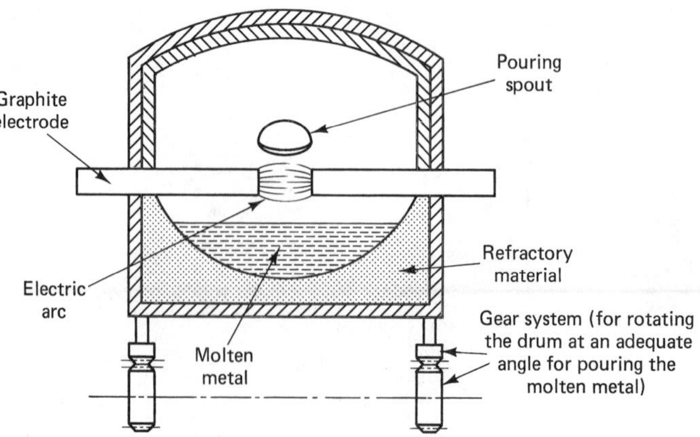

FIGURE 3.23
An electric-arc furnace

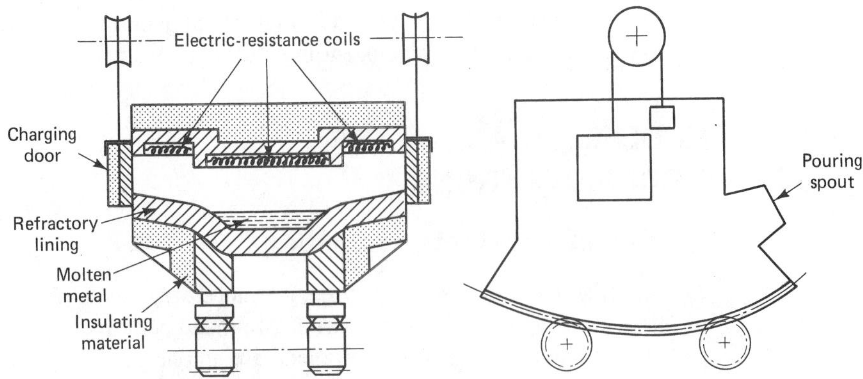

FIGURE 3.24 An electric-resistance furnace

ternal lining of the furnace. The electric arc is generated about midway between two graphite electrodes. In order to control the gap between the two electrodes and, accordingly, control the intensity of heat, one electrode is made stationary and the other one movable. Electric-arc furnaces are used mainly for melting steels and, to a lesser extent, gray cast iron and some nonferrous metals.

Resistance furnace. The resistance furnace is employed mainly for melting aluminum and its alloys. Figure 3.24 indicates the basic features of a typical resistance furnace. The solid metal is placed on each of the two inclined hearths and is subjected to heat radiation from the electric-resistance coils located above it. When the metal melts, it flows down into a reservoir. The molten metal can be poured out through the spout by tilting the whole furnace.

Induction furnace. The induction furnace has many advantages, including evenly distributed temperatures within the molten metal, flexibility, and the possibility of controlling the atmosphere of the furnace. In addition, the motor effect of the electromagnetic forces helps to stir the molten metal, thus producing more homogeneous composition. Induction furnaces are used to melt steel and aluminum alloys. Figure 3.25 shows the construction of a typical induction furnace. It basically involves an electric-induction coil that is built into the walls of the furnace. An alternating current in the coil induces current in any metallic object that obstructs the electromagnetic

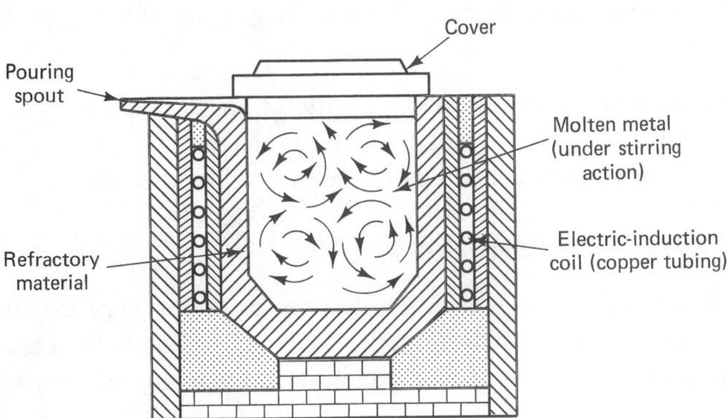

FIGURE 3.25 An electric-induction furnace

flux. Furnaces of both high- and low-frequency current are successfully used in industry to induce alternating current in solid metal to melt it.

3.5 CASTING DEFECTS AND DESIGN CONSIDERATIONS

Common Defects in Castings

In order to obtain a sound casting, it is necessary to control adequately the various factors affecting the casting process. Casting and pattern designs, molding procedure, and melting and pouring of molten metal are among the factors affecting the soundness of a casting. Following is a survey of the commonly experienced defects in castings.

Hot tears. Hot tears can appear on the surface or through cracks that initiate during cooling of the casting. They usually are in locations where the metal is restrained from shrinking freely, such as a thin wall connecting two heavy sections.

Cold shut. A cold shut is actually a surface of separation within the casting. It is believed to be caused by two "relatively cold" streams of molten metal meeting each other at that surface.

Sand wash. A sand wash can be described as rough, irregular surfaces (hills and valleys) of the casting that result from erosion of the sand mold. This erosion is, in turn, caused by the metal flow.

Sand blow. A sand blow is actually a surface cavity that takes the form of a very smooth depression. It can be caused by insufficient venting, lack of permeability, or a high percentage of humidity in the molding sand.

Scab. A scab is a rough "swollen" location in the casting that has some sand embedded in it. Such a defect is usually encountered when the molding sand is too fine or too heavily rammed.

Shrinkage porosity (or cavity). A shrinkage porosity is a microscopic or macroscopic hole formed by the shrinkage of spots of molten metal that are encapsulated by solidified metal. It is usually caused by poor design of the casting.

Hard spots. Hard spots are hard, difficult-to-machine areas that can occur at different locations.

Deviation of the chemical composition from the desired one. Deviation may be due to the loss of a constituent element (or elements) during the melting operation. It may also be caused by contamination of the molten metal.

Design Considerations

A product designer who selects casting as the primary manufacturing process should make a design not only to serve the function (by being capable of withstanding the loads and the environmental conditions to which it is going to be subjected during its service life) but also to facilitate or favor the casting process. Following are some design considerations and guidelines.

Promote directional solidification. When designing the mold, be sure that the risers are properly dimensioned and located to promote directional solidification of the casting toward the risers. In other words, the presence of large sections or heat masses in locations distant from the risers should be avoided, and good rising practice as previously discussed should be followed. Use can also be made of chills to promote directional solidification. Failure to do so may result in shrinkage cavities (porosity) or cracks in those large sections distant from the risers. It is also very important to remember that a riser will not feed a heavy section through a lighter section.

Ensure easy pattern drawing. Make sure that the pattern can easily be withdrawn from the nonpermanent mold (this does not apply to investment casting). This can be achieved through rational selection of the parting line as well as by providing appropriate pattern draft wherever needed. In addition, undercuts or protruding bosses (especially if their axes do not fall within the parting plane) and the like should be avoided. Nevertheless, remember that undercuts can be obtained, if necessary, by using cores.

Avoid the shortcomings of columnar solidification. Dendrites often start to form on the cold surface of a mold and then grow to form a columnar casting structure. This almost always results in planes of weakness at sharp corners, as illustrated in Figure 3.26a. Therefore, rounding the edges is a must for eliminating the development of planes of weakness, as shown in Figure 3.26b. Rounded edges are also essential for smooth laminar flow of the molten metal.

Avoid hot spots. Certain shapes, because of their effect on the rate of heat dissipation during solidification, tend to promote the formation of shrinkage cavities. This is always the case at any particular location where the rate of solidification is slower than that at the surrounding regions of the casting. The rate of solidification (and the rate of heat dissipation to start with) is slower at locations having a low ratio of surface area to volume. Such locations are usually referred to as *hot spots* in foundry work. Unless precautions are taken during the design phase, hot spots and, consequently, shrinkage cavities are likely to occur at the L, T, V, Y, and + junctions, as illustrated in Figure 3.27a. Shrinkage cavities can be avoided by modifying the design, as shown in

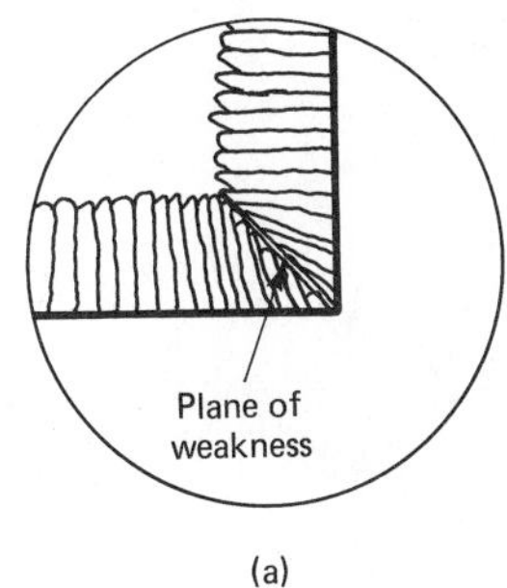

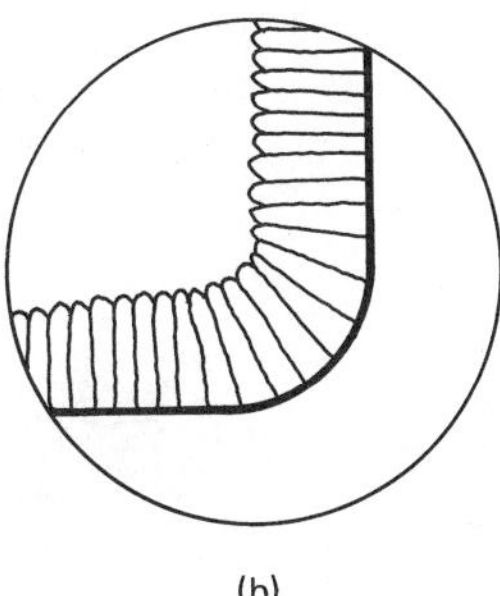

FIGURE 3.26 Columnar solidification and planes of weakness: (a) poor design (sharp corner); (b) rounded edges to eliminate planes of weakness

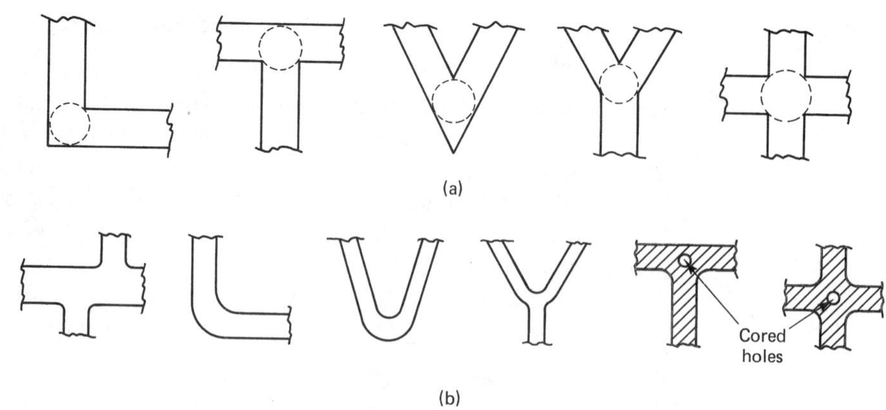

FIGURE 3.27 Hot spots: (a) poor design, yielding hot spots; (b) better design, eliminating hot spots

Figure 3.27b. Also, it is always advisable to avoid abrupt changes in sections and to use taper (i.e., make the change gradual), together with generous radii, to join thin to heavy sections, as shown in Figure 3.28.

Avoid the causes of hot tears. Hot tears are casting defects caused by tensile stresses as a result of restraining a part of the casting. Figure 3.29a and b shows locations where hot tears can occur and a recommended design that would eliminate their formation.

Distribute the masses of a section to save material. Cast metals are generally weaker in tension in comparison with their compressive strengths. Nonetheless, the casting process offers the designer the flexibility of distributing the masses of a section with a freedom not readily available when other manufacturing processes are employed. Therefore, when preparing a design of a casting, try to distribute masses in such a manner as to lower the magnitude of tensile stresses in highly loaded areas of the cross section and to reduce material in lightly loaded areas. As can be seen in Figure 3.30, a T section or an I beam is more advantageous than just a round or square one when designing a beam that is to be subjected to bending.

Avoid thicknesses lower than the recommended minimum section thickness. The minimum thickness to which a section of a casting can be designed depends upon such factors as the material, the size, and the shape of the casting as well as the specific

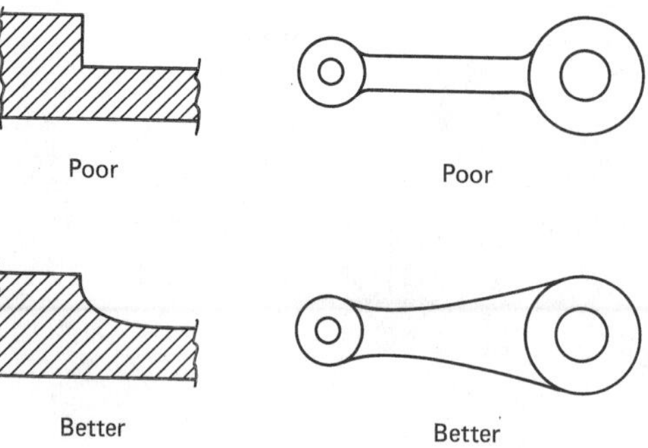

FIGURE 3.28 Avoiding abrupt changes in sections

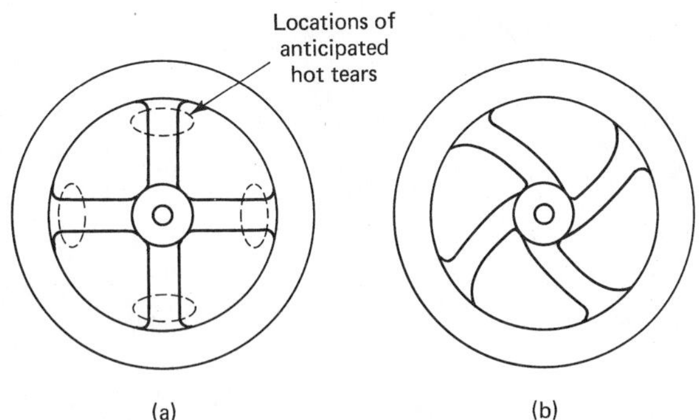

FIGURE 3.29 Hot tears: (a) a casting design that promotes hot tears; (b) recommended design to eliminate hot tears

casting process employed (i.e., sand casting, die casting, etc.). In other words, strength and rigidity calculations may prove a thin section to be sufficient, but casting considerations may require adopting a higher value for the thickness so that the cast sections will fill out completely. This is a consequence of the fact that a molten metal cools very rapidly as it enters the mold and may become too cold to fill a thin section far from the gate. A minimum thickness of 0.25 inch (6 mm) is suggested for design use when conventional steel casting techniques are employed, but wall thicknesses of 0.060 inch (1.5 mm) are quite common for investment castings. Figure 3.31 indicates the relationship between the minimum thickness of a section and its largest dimension. It should be pointed out that for a given thickness, steel flows best in a narrow rather than in a wide web. For cast-iron and nonferrous castings, the recommended values for minimum thicknesses are much lower than those for steel castings having the same shape and dimensions.

Strive to make small projections in a large casting separate. As can be seen in Figure 3.32a, a small projection may be subjected to more accidental knocks than a large

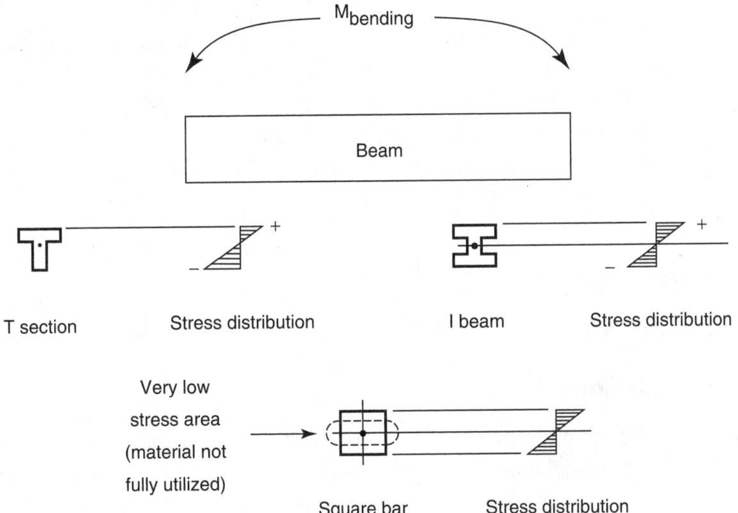

FIGURE 3.30 Distribution of masses to reduce weight

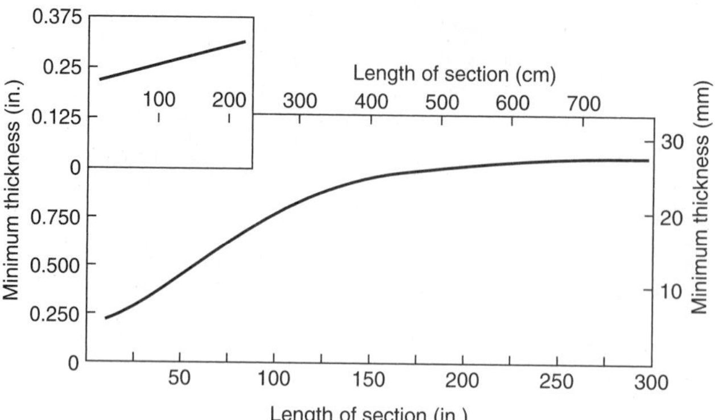

FIGURE 3.31 Minimum thickness of cast steel sections as a function of their largest dimension *(Adapted from* Steel Castings Handbook, *5th ed. Rocky River, Ohio: Steel Founders Society of America, 1980)*

casting, and if it gets broken, the whole casting will be scrapped. It is, therefore, highly recommended to make the small projection separate and attach it to the large casting by an appropriate mechanical joining method, as shown in Figure 3.32b.

Strive to restrict machined surfaces. Whereas some castings are used in their entirely as-cast condition, some others may require one or more machining operations. It is the task of the designer to ensure that machining is performed only on areas where it is absolutely necessary. An example of cases where the machining needed involves bearing surfaces is shown in Figure 3.33.

Use reinforcement ribs to improve the rigidity of thin, large webs. A common use of brackets or reinforcement ribs is to provide rigidity to thin, large webs (or the like) as an alternative to increasing the thickness of the webs. The ribs should be as thin as possible (i.e., minimum permissible thickness) and should also be staggered, as shown in Figure 3.34. Always remember that parabolic ribs are better than straight ribs in terms of economy and uniformity of stress.

Consider the use of cast-weld construction to eliminate costly cored design. The design of some products necessitates the use of complicated steel-wire-reinforced cores that are difficult to reach and remove after casting, thus leaving the surfaces unclean. An example, a steam ring, is shown in Figure 3.35a. The alternative design would be to employ a simple cut plate that is welded into the casting to produce the cast-weld construction shown in Figure 3.35b.

3.6 CLEANING, TESTING, AND INSPECTION OF CASTINGS

Cleaning

The cleaning process involves the removal of the molding sand adhering to a casting. It also includes the elimination of gates, runners, and risers. Generally, surface cleaning can be carried out in rotary separators or by employing sand-blasting and/or metal-

FIGURE 3.32
Large casting with a small projection: (a) as an integral part; (b) two separate parts

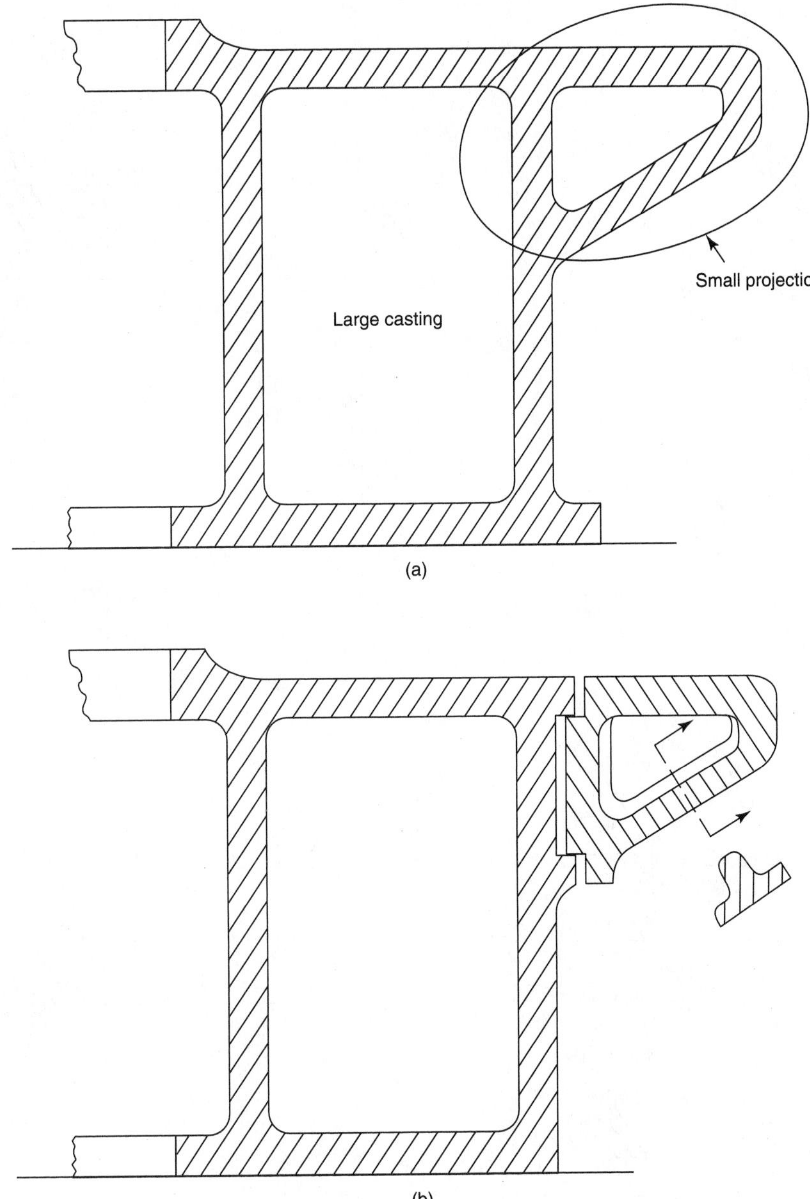

lic shot-blasting machines. The latter two machines use sand particles or shots traveling at high velocities onto the surface of the casting to loosen and remove the adhering sand. As you may expect, these machines are particularly suitable when cleaning medium and heavy castings. On the other hand, rotary separators are advantageous for cleaning light castings. A separator is actually a long, large-diameter drum that rotates around its horizontal axis into which the castings are loaded together with jack

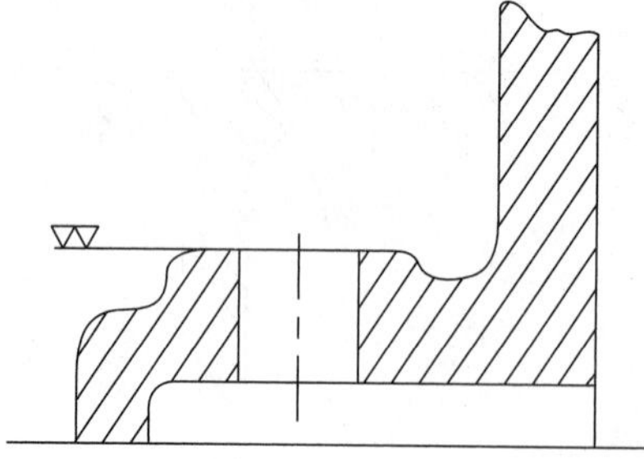
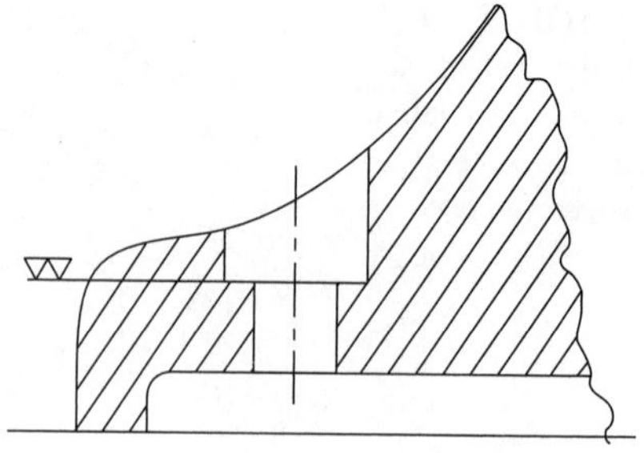

FIGURE 3.33
Restriction of surfaces to be machined

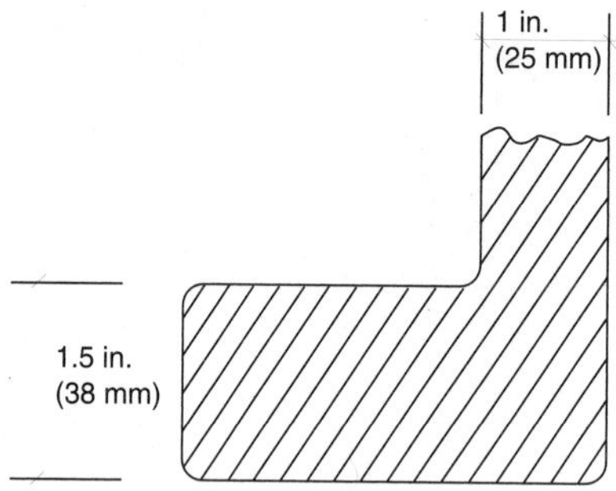

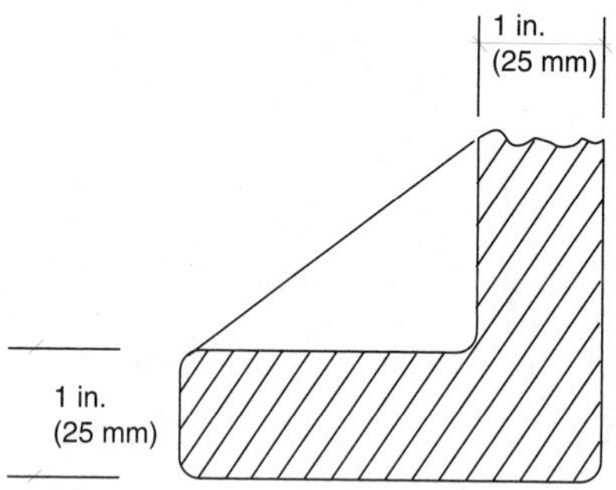

FIGURE 3.34
Use of reinforcement ribs

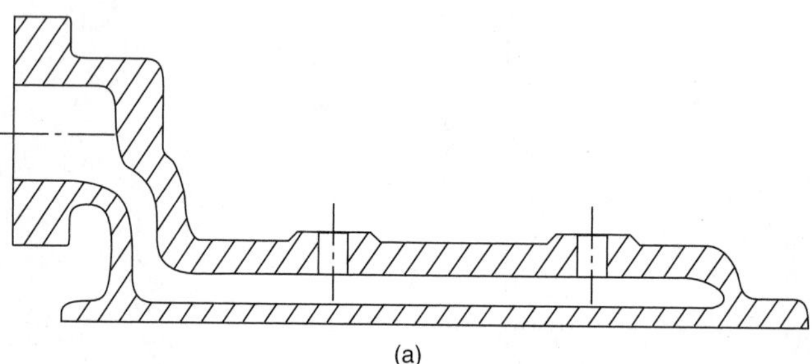

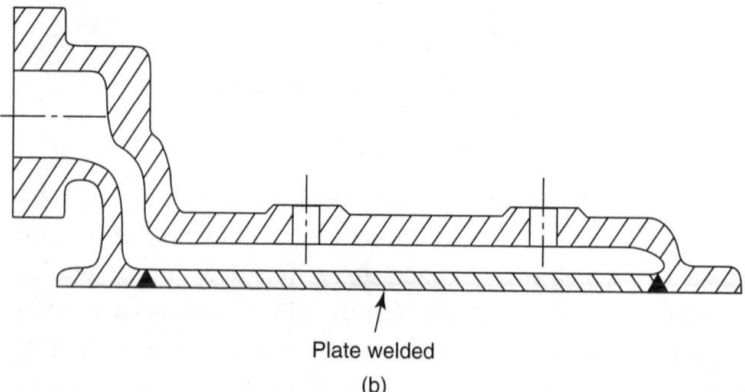

FIGURE 3.35
The design of a steam ring: (a) cast construction; (b) cast-weld construction

stars made of white cast iron. A further advantage of rotary separators is that they automatically break off gate-and-runner systems and, often, risers.

Testing and Inspection

Like any other manufactured parts, castings must be subjected to thorough quality control in order to separate defective products and to reduce the percentage of rejects through identifying the defects and tracing their sources. Following are some of the commonly used tests and inspection methods.

Testing of the mechanical properties of the casting. Standard tension and hardness tests are carried out to determine the mechanical properties of the metal of the casting in order to make sure that they conform to the specifications.

Inspection of the dimensions. Dimensions must fall within the specified limits. Therefore, measuring tools and different kinds of gages (e.g., snap, progressive, plug, template) are used to check that the dimensions conform to the blueprint.

Visual examination. Visual inspection is used to reveal only very clear defects. However, it is still commonly used in foundries.

Hydraulic leak testing. The hydraulic leak test is used to detect microscopic shrinkage porosity. Various penetrants and testing methods are now available. Details are given in the American Society for Testing and Materials (ASTM) standards, designation E165.

Nondestructive testing. There are several nondestructive testing methods that detect microscopic and hair cracks. They involve ultrasonic testing, magnetic particle inspection, eddy current testing, and radiography.

Testing for metal composition. Several methods are employed to determine the chemical composition accurately and to assure product quality. The classical method used to be "wet analysis" (i.e., employing acids and reagents in accurate chemical analysis). However, because this method is time-consuming, it is being replaced by methods like emission spectroscropy, X-ray fluorescence, and atomic absorption spectroscopy.

3.7 CASTABILITY (FLUIDITY)

The ability of the molten metal to flow easily without premature solidification is a major factor in determining the proper filling of the mold cavity. This important property is referred to as *castability* or, more commonly, *fluidity*. The higher the fluidity of a molten metal, the easier it is for that molten metal to fill thin grooves in the mold and exactly reproduce the shape of the mold cavity, thereby successfully producing castings with thinner sections. Poor fluidity leads to casting defects such as incomplete filling or misruns, especially in the thinner sections of a casting. Because fluidity is dependent mainly upon the viscosity of the molten metal, it is clear that higher temperatures improve the fluidity of molten metal and alloys, whereas the presence of impurities and nonmetallic inclusions adversely affects it.

FIGURE 3.36
Details of the test for measuring fluidity

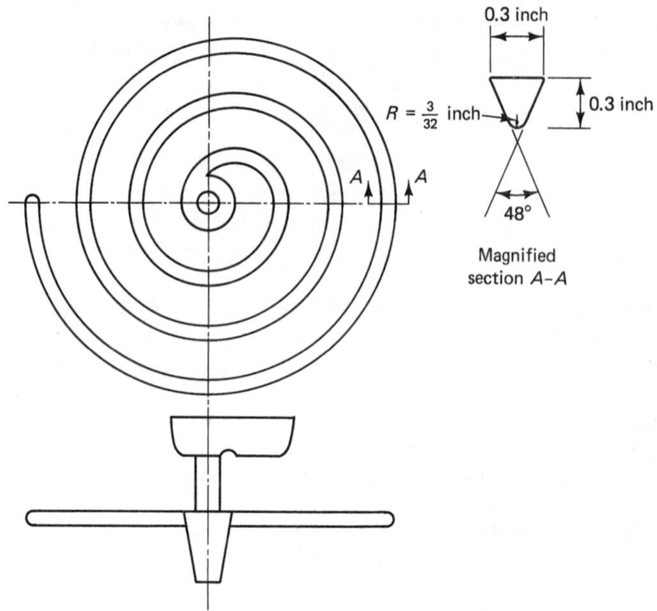

Several attempts have been made to quantify and measure the fluidity of metals. A commonly used standard test involves pouring the molten metal into a basin so that it flows along a spiral channel of a particular cross section, as shown in Figure 3.36. Both the basin and the channel are molded in sand, and the fluidity value is indicated by the distance traveled by the molten metal before it solidifies in the spiral channel.

Review Questions

1. What is meant by the word *casting?*
2. What are the constituents of green molding sand?
3. List some of the important properties that green sand must possess.
4. What is a *flask?* What is its function? List the parts that form a flask.
5. Explain the meaning of the word *pattern.*
6. List some of the materials used in making patterns.
7. List the different types of permanent patterns used in foundries.
8. What are the different pattern allowances? Discuss the function of each.
9. What are *cores?* How are they made?
10. What is meant by a *gating system?* What functions does it serve?
11. What are the components of a gating system?
12. What are *risers?* What function do they serve?
13. List the various green sand properties and discuss each briefly.
14. Why should weights be located on the cope in pit molding?

15. List the various molding machines and discuss the operation of each briefly.
16. Explain sand conditioning and how it is done.
17. What advantages does dry sand molding have over green sand molding?
18. When are cement-bonded sand molds recommended?
19. What is the main advantage of the carbon dioxide process for molding?
20. What metals can be cast in plaster molds?
21. When are loam molds used?
22. Describe shell molding. What are its advantages?
23. When are ceramic molds recommended?
24. Explain investment casting and why it is sometimes called the *lost-wax process.*
25. Name a metal that should be cast in a graphite mold.
26. What are the advantages of employing permanent molds? Why?
27. Can molten metals be cast directly into cavities of cold permanent molds? Why?
28. What is the main difference between the hot-chamber and the cold-chamber methods of die casting?
29. List some metals that you think can be cast by the hot-chamber method. Justify your answer.
30. List some metals that you think can be cast by the cold-chamber method. Justify your answer.
31. What are the types of centrifugal casting?
32. Differentiate between the different types of centrifugal casting and discuss the advantages and shortcomings of each type.
33. What are the products that can be manufactured by continuous casting?
34. What does the continuous casting process involve?
35. Discuss some advantages of the continuous casting process.
36. What does the V-process involve?
37. List some of the merits and advantages of the V-process.
38. Discuss some of the problems encountered in casting steels.
39. What precautions should be taken to eliminate the problems in casting steels?
40. What is gray cast iron?
41. Discuss some of the properties that make gray cast iron attractive for some engineering applications.
42. Why are inoculants added to gray cast iron?
43. Differentiate between gray cast iron and white cast iron.
44. What is meant by compacted-graphite cast iron.
45. What is ductile cast iron? How can it be obtained?
46. What is malleable cast iron? How can it be obtained? What are the limitations on producing it?
47. List some alloying elements that are added to cast iron. List some applications for alloyed cast iron.
48. What are the problems caused by hydrogen when melting and casting aluminum and how can these problems be eliminated?
49. What are the sources of hydrogen when melting aluminum?
50. List some cast aluminum alloys and discuss their applications.
51. How are cast copper alloys classified?
52. What is meant by a *deoxidizer?* Give an example.
53. List some of the characteristics and applications of cast zinc alloys.
54. List some of the characteristics and applications of cast magnesium alloys.
55. For what purpose is the cupola furnace used?
56. Describe briefly the operation and charge of the cupola furnace.

57. For what purpose is the reverberatory furnace used?

58. List some of the metals that can be melted in crucible furnaces.

59. What are the main differences in construction between the stationary and the tilting crucible furnaces?

60. List the different types of electric furnaces and mention the principles of operation in each case.

61. List the main advantages and applications of electric furnaces.

62. List some of the common defects of castings and discuss the possible causes of each defect.

63. List and discuss the main design considerations for castings.

64. List and discuss the various testing and inspection methods used for the quality control of castings.

Design Example

PROBLEM

Your company has received an order to manufacture wrenches for loosening and tightening nuts and bolts of large machines. The plant of the company involves a foundry and a machining workshop with a few basic machine tools. Here are the details of the order:

Lot size:	500 wrenches
Nut size:	2 inches (50 mm)
Required torque:	about 20 lb ft (27.12 N·m)

You are required to provide a design and a production plan (see the explanation of the word *design* in the design projects section that appears later).

Solution

Before we start solving this design problem, we should make some assumptions. For instance, consider the force that can be generated by the ordinary human hand. It will allow us to determine the length of the wrench using the following equation:

$$T = F \times \ell$$

where: T is the torque
F is the force
ℓ is the length

As can be seen from the equation, a low value of F would make the length large and thus make the handling of the wrench impractical because of the weight. On the other hand, a high value of F is not practical and may not be generated by an ordinary person. Let us take $F = 15$ pounds. Therefore,

$$\ell = \frac{20}{15} = 1.33 \text{ feet}$$

Apparently, the force acts at the middle of the fist, and we have to add a couple of inches for proper holding:

length of wrench $\cong$ 18 inches

Let us now design the section where the maximum bending moment occurs. You can assume some dimensions and determine the stress, which will serve as a guide in selecting material. Take the section as shown in Figure 3.37a. The moment of inertia of the section is

$$\begin{aligned} I &= \frac{1}{12}(0.25)(0.75)^3 + 2\left[\frac{1}{12}(0.375)(0.25)^3 + 0.375 \times 0.25 \times 0.5\right] \\ &= 0.008789 + 0.00098 + 0.046875 \\ &= 0.056655 \text{ in.}^4 \end{aligned}$$

Note that the minimum thickness for steel casting was adhered to. Now, determine the stress:

$$\text{max. stress} = \frac{20 \times 12 \times 1.25}{2 \times 0.056655} = 2648 \text{ lb/in.}^2$$

That value is very low, and we should try to reduce the section and save material. It is always a good idea to make use of spreadsheets to change the dimensions and get the stresses acting in each case. Now, take the section as shown in Figure 3.37b:

$$\begin{aligned} I &= \frac{1}{12}(0.25)(0.5)^3 + 2\left[\frac{1}{12}(0.375)(0.25)^3 + 0.375 \times 0.25 \times 0.375^2\right] \\ &= 0.002604 + 0.00098 + 0.026367 \\ &= 0.038771188 \text{ in.}^4 \end{aligned}$$

$$\text{max. stress} = \frac{20 \times 12}{0.038771188} \times \frac{1.25}{2} = 3868 \text{ lb/in.}^2$$

As can be seen, we took the minimum thickness to be 0.25 inch, which is the recommended value for conventional castings of steels.

The material should be low-carbon steel having 0.25 percent carbon in order to possess enough ductility. Also, the steel should be thoroughly killed. A recommended material is ASTM A27-77, grade U60-30, which has a yield strength of 30 ksi. When taking a factor of safety of 4, the allowable stress would be 7500 lb/in.^{2}, which is higher than the obtained value of the working stress.

Now, in order to calculate the thickness of the wrench, let us calculate the bearing stress on the nut. A reasonable estimate of the force on the surface of the nut is

$$\frac{20 \times 12}{0.75} = 320 \text{ pounds}$$

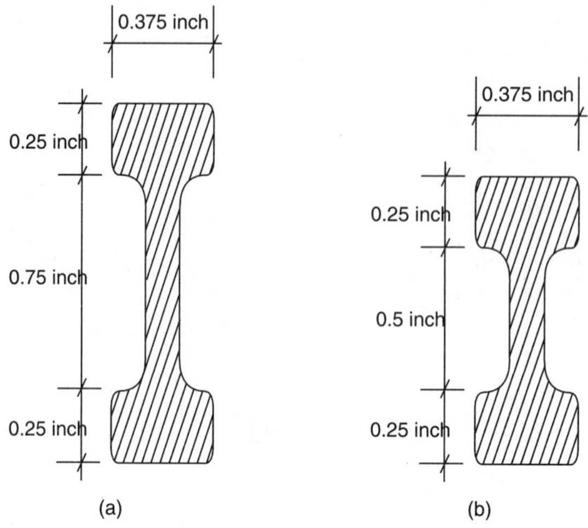

FIGURE 3.37
Cross section of the wrench: (a) first attempt; (b) second attempt

This is based on the assumption that the torque is replaced by two opposite forces having a displacement of 0.75 inch between the lines of action. Thus,

$$\text{bearing stress} = \frac{320}{0.75 \times t} = 7500$$

$$t = 0.056 \text{ inch}$$

Take it as 0.5 inch to facilitate casting the part.

Because all dimensions are known, a detailed design can be prepared, as shown in Figure 3.38. Notice the surface finish marks indicating the surfaces to be machined.

FIGURE 3.38
A wrench manufactured by casting

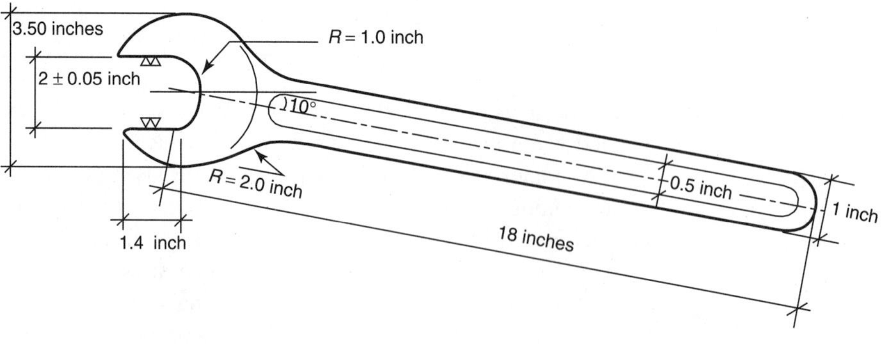

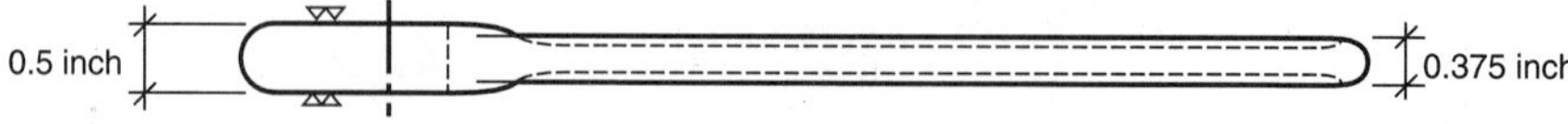

As previously mentioned, conventional sand casting is to be employed, using a cope-and-drag pattern plate to cast two wrenches per flask. We can use a single down sprue to feed the narrow end of the wrench and a riser at the other end. The parting line will pass through the web of the *I* section.

Design Projects

Whenever the word *design* is mentioned hereafter, you should provide, *at least,* the following:

- Two neatly dimensioned graphical projections of the product (i.e., a blueprint ready to be released to the workshop for actual production), including fits (if applicable), tolerances, surface finish marks, and so on
- Material selection with rational justification
- Selection of the specific manufacturing processes required, as well as their sequence in detail
- Simple but necessary calculations to check the stresses at the critical sections

1. Design a bracket for a screw C-clamp that has the following characteristics:

Maximum clamping force:	22 pounds (100 N)
Clamping gap:	3 inches (7.5 cm)
Distance between centerline of screw and inner surface of bracket:	2 inches (5 cm)
Root diameter of screw:	0.25 inch (6 mm)

Assume that manufacturing is by casting and that production volume is 4000 pieces.

2. Design a flat pulley. Its outer diameter is 36 inches (90 cm), and it is to be mounted on a shaft that is 2½ inches (6.25 cm) in diameter. Its width is 10 inches (25 cm), and it has to transmit a torque of 3000 lb ft (4000 N·m). Assume that 500 pieces are required. Will the design change if only 3 pieces are required?

3. A connecting lever has two short bosses, each at one of its ends and each with a vertical hole that is 3/4 inch (19 mm) in diameter. The lever is straight, and the horizontal distance between the centers of the holes is 8 inches (200 mm). The lever during functioning is subjected to a bending moment of 50 lb ft (67.8 N·m) that acts in the plane formed by the two vertical axes. Provide a detailed design for this lever if it is to be produced by casting and

a. When only 100 pieces are required

b. When 10,000 pieces are required

4. Design a micrometer frame for each of the following cases:
   a. The gap of the micrometer is 1.0 inch (25 mm), and the distance from the axis of the barrel to the inner side of the frame is 1.5 inches (37.5 mm). The maximum load on the anvil is 22 lb (100 N).
   b. The gap of the micrometer is 6 inches (150 mm), and the distance from the axis of the barrel to the inner side of the frame is 4.0 inches (100 mm). The maximum load on the anvil is 22 lb (100 N).

   Assume that production volume is 4000 pieces and that one of the various casting processes is used.

   TIP: Base your design on rigidity. The maximum deflection must not exceed 0.1 of the smallest reading of the micrometer.

5. A pulley transmits a torque of 600 lb ft (813.6 N·m) to a shaft that is 1¼ inches (31 mm) in diameter. The outer diameter of the pulley is 10 inches (250 mm), and it is to be driven by a flat belt that is 2 inches (50 mm) in width. Design this pulley if it is to be manufactured by casting and 500 pieces are required.

6. Design a hydraulic jack capable of lifting 1 ton and having a stroke of 6 inches (150 mm). The jack is operated by a manual displacement (plunger) pump that pumps oil from a reservoir into the high-pressure cylinder through two spring-actuated nonreturn valves to push the ram upward. The reservoir and the high-pressure cylinder are also connected by a conduit, but the flow of oil is obstructed by a screw that, when unscrewed, relieves the pressure of the cylinder by allowing high-pressure oil to flow back into the reservoir and the ram then to be pushed downward. Provide a workshop drawing for each component, as well as an assembly drawing for the jack. Steel balls and springs are to be purchased. Assume production volume is 5000 pieces.

7. Design a table for the machine shop. That table should be 4 feet (1.2 m) in height, with a surface area of 3 by 3 feet (0.90 by 0.9 m), and should be able to carry a load of half a ton. Assume production volume is 2000 pieces.

8. Design a little wrench for loosening and tightening nuts and bolts of a bicycle. The nut size is 5/8 inch (15 mm), and the required torque is about 1.0 lb ft (1.356 N·m). Assume production volume is 10,000 pieces.

9. A straight-toothed spur-gear wheel transmits 1200 lb ft (1627 N·m) of torque to a steel shaft that is 2 inches (50 mm) in diameter. The pitch diameter of the gear is 8 inches (200 mm), its width is 3 inches (75 mm), and the base diameter is 7.5 inches (187.5 mm). Design this gear's blank. Assume production volume is 4000 pieces.

10. Design a frame for an open-arch (C-type) screw press that can deliver a load of up to 2 tons. The open gap is 2 feet (600 mm), and the bed on which workpieces are placed is 12 by 12 inches (300 by 300 mm). Assume that the base diameter of the screw thread is 1½ inches (37.5 mm).

11. Design a hydraulic cylinder for earth-moving equipment. It can generate a maximum force of 2 tons and has a stroke of 4 feet (1200 mm). Although the maximum force is generated only when the plunger rod is moving out, the cylinder is double acting and generates a force of 1 ton during its return stroke. Expected production volume is 2000 pieces, and the pistons, oil rings, and so on, are going to be purchased from vendors.

12. Design a safety valve to be mounted on a high-pressure steam boiler. The pipe on which it will be mounted has a bore diameter of 2 inches (50 mm). The pressure inside the boiler should not exceed 50 folds of the atmospheric pressure. Expected production volume is 5000 pieces, and the stems, springs, bolts, and gaskets are going to be purchased from vendors.

Chapter 4

Joining of Metals

INTRODUCTION

When two parts of metal are to be attached together, the resulting joint can be made dismountable (using screws and the like), or it can be made permanent by employing riveting, welding, or brazing processes. The design of dismountable joints falls beyond the scope of this text and is covered in machine design. It is, therefore, the aim of this chapter to discuss the design and production of permanent joints when various technologies and methods are applied. Because the same equipment used in welding is also sometimes employed in the cutting of plates, thermal cutting processes will also be discussed in this chapter.

4.1 RIVETING

The process of *riveting* involves inserting a ductile metal pin through holes in two or more sheet metals and then forming over (heading) the ends of the metal pin so as to secure the sheet metals firmly together. This process can be performed either cold or hot, and each rivet is usually provided with one preformed head. Figure 4.1a and b indicates the sequence of operations in riveting, while Figure 4.2 illustrates different shapes of preformed rivet heads.

4.2 WELDING

Welding is the joining of two or more pieces of metal by creating atom-to-atom bonds between the adjacent surfaces through the application of heat, pressure, or both. In order for a welding technique to be industrially applicable, it must be reasonable in cost, yield reproducible or consistent weld quality, and, more importantly, produce

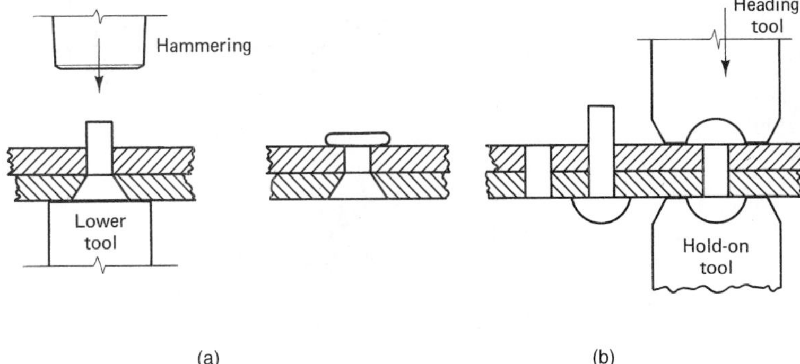

FIGURE 4.1 Sequence of operations in riveting: (a) flat-head rivet; (b) regular rivet

joints with properties comparable to those of the base material. Various welding techniques have been developed that are aimed at achieving these three goals. However, no matter what welding method is used, the interface between the original two parts must disappear if a strong joint is to be obtained. Before we discuss the different methods employed to make those surfaces disappear, let us discuss joint design and preparation.

Joint Design and Preparation

A weld joint must be designed to withstand the forces to which it is going to be subjected during its service life. Therefore, the joint design is determined by the type and magnitude of the loading that is expected to act on the weldment. In other words, selection of the type of joint has to be made primarily on the basis of load requirement. As Figure 4.3a through e shows, there are five types of weld joints: butt, lap, corner, T, and edge. Following is a discussion of each of these different types of joints.

Butt joint. The butt joint involves welding the edges or end faces of the two original parts, as shown in Figure 4.3a. Therefore, the two parts must be aligned in the same plane. Usually, when the thickness of the parts falls between 1/8 and 3/8 inch (about 3 and 9 mm), the two parts are welded without any edge preparation. This type of weld is referred to as a *square* weld and can be either single or double, depending upon the thickness of the metal, as shown in Figure 4.4a. As can be seen in Figure 4.4b through e, the edges of thicker parts should be prepared with single or double bevels or V-, J-, or U-grooves to allow adequate access to the root of the joint. Usually, it is recommended to adopt the single or double U-groove when the thickness of the parts is more than 0.8 inch (20 mm).

Lap joint. We can see in Figure 4.3b that the lap joint is produced by fillet welding overlapping members; the amount of overlap is normally taken to be about three to

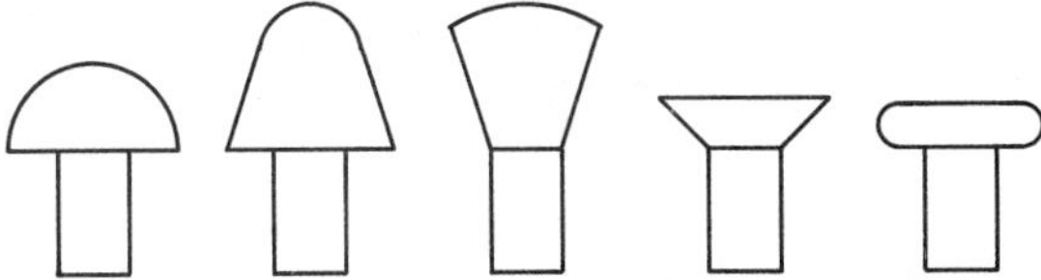

FIGURE 4.2 Different shapes of preformed rivet heads

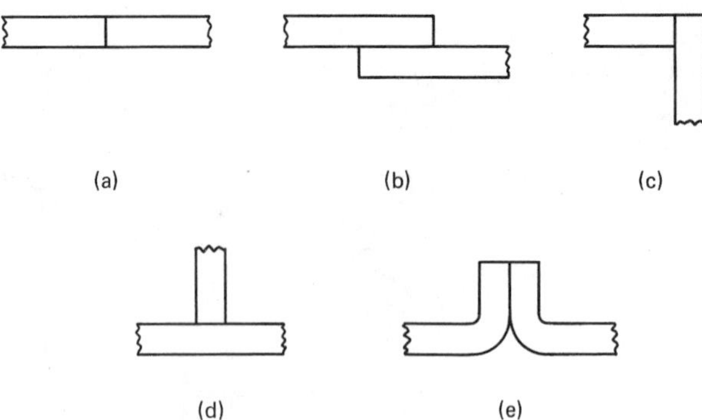

FIGURE 4.3 Types of weld joints: (a) butt joint; (b) lap joint; (c) corner joint; (d) T-joint; (e) edge joint

five times the thickness of the member. The fillet weld can be continuous and may also be of the plug or slot type, as shown in Figure 4.5.

Corner joint. Figure 4.3c illustrates the corner joint, which can be welded with or without edge preparation (see Figure 4.4 for the various possible edge preparations).

T-joint. The T-joint is shown in Figure 4.3d. T-joints that will be subjected to light static loads may not require edge preparation. On the other hand, edge preparations (again see Figure 4.4) are often necessary for greater metal thicknesses or when the joint is to be subjected to relatively high, alternating, or impulsive loading.

Edge joint. The edge joint is usually used when welding thin sheets of metal with a thickness of up to 1/8 inch (3 mm). Notice in Figure 4.3e that the edges of the members must be bent before the welding process is carried out.

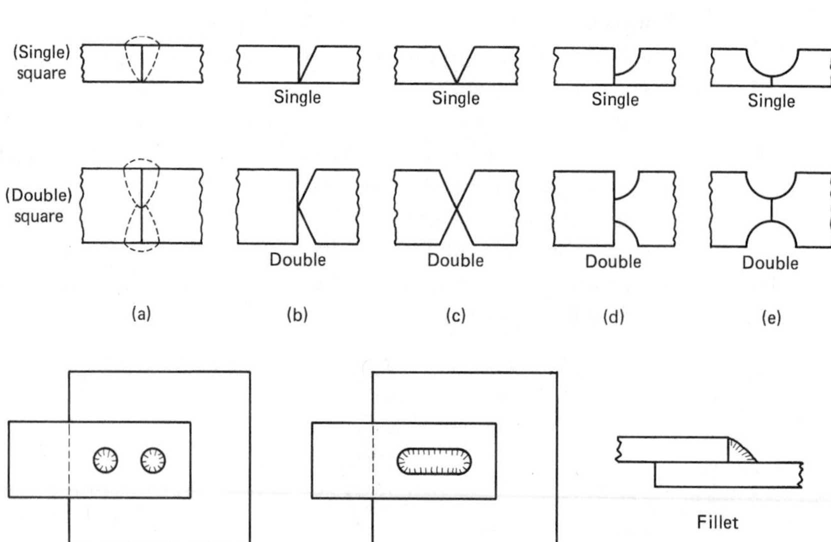

FIGURE 4.4 Different edge preparation for butt welding: (a) square; (b) bevel; (c) V-groove; (d) J-groove; (e) U-groove

FIGURE 4.5 Basic types of fusion lap welds

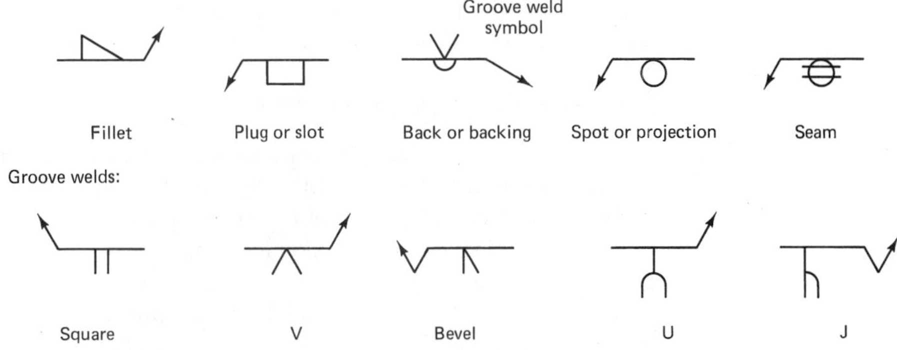

FIGURE 4.6
Weld symbols

Weld Symbols and Identification

Figure 4.6 shows the different weld symbols, whereas Figure 4.7 shows the standard identification of welds employed in design drawings.

Classification of the Welding Processes

Different methods can be used for classifying industrial welding processes. Each method is employed to form groups of welding processes, with each group having something in common. For instance, welding processes can be classified according to the source of energy required to accomplish welding. In such a case, it is obvious that there are four main groups: mechanical, electrical, chemical, or optical. Welding processes can also be classified by the degree of automation adopted, which yields three groups: manual, semiautomatic, and automatic. The most commonly used method of classification is according to the state of the metal at the locations being

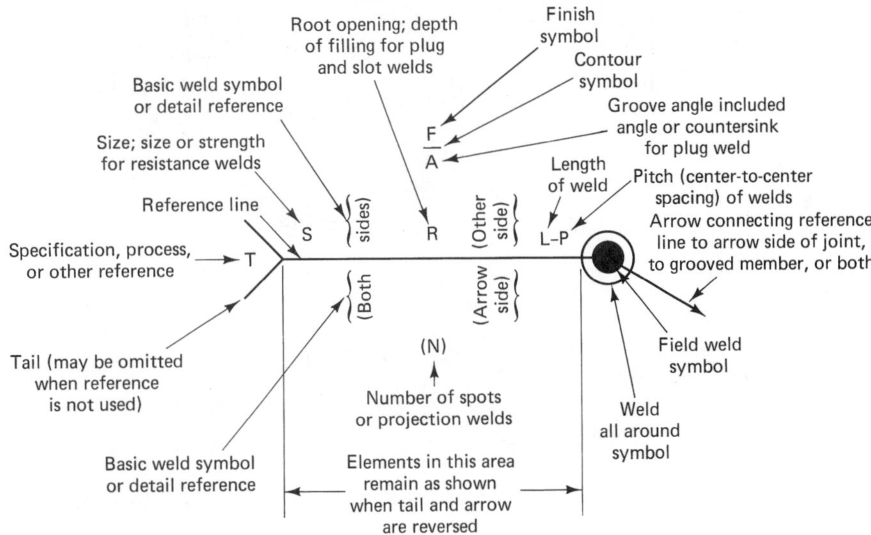

FIGURE 4.7
Standard identification of welds

welded, thus splitting the welding processes into two main categories: pressure welding and fusion welding. We now discuss each of these two categories in detail.

Pressure Welding Processes

Pressure welding involves processes in which the application of external pressure is indispensable to the production of weld joints formed either at temperatures below the melting point (solid-state welding) or at temperatures above the melting point (fusion welding). In both cases, it is important to have very close contact between the atoms of the parts that are to be joined. The atoms must be moved together to a distance that is equal to or less than the equilibrium interatomic-separation distance. Unfortunately, there are two obstacles that must be overcome so that successful pressure welding can be carried out and a sound weldment can be obtained. First, surfaces are not flat when viewed on a microscopic scale. Consequently, intimate contact can be achieved only where peaks meet peaks, as can be seen in Figure 4.8, and the number of bonds would not be enough to produce a strong welded joint. Second, the surfaces of metals are usually covered with oxide films that inhibit direct contact between the two metal parts to be welded. Therefore, those oxide and nonmetallic films must be removed (cleaned with a wire brush) before welding in order to ensure a strong welded joint. Pressure welding processes are applied primarily to metals possessing high ductility or those whose ductility increases with increasing temperatures; thus, the peaks that keep the surfaces of the two metallic members apart are leveled out under the action of mechanical stresses or the combined effect of high temperatures and mechanical stresses. In fact, a wide variety of pressure welding processes are used in industry. The commonly used ones are listed in Figure 4.9.

Cold-pressure welding. Cold-pressure welding is a kind of solid-state welding used for joining sheets, wires, and small electric components. As previously discussed, the surfaces to be welded must be cleaned with a wire brush to remove the oxide film and must be carefully degreased before welding. As Figure 4.10 shows, a special tool is used to produce localized plastic deformation, which results in coalescence between the two parts. This process, which can replace riveting, is usually followed by

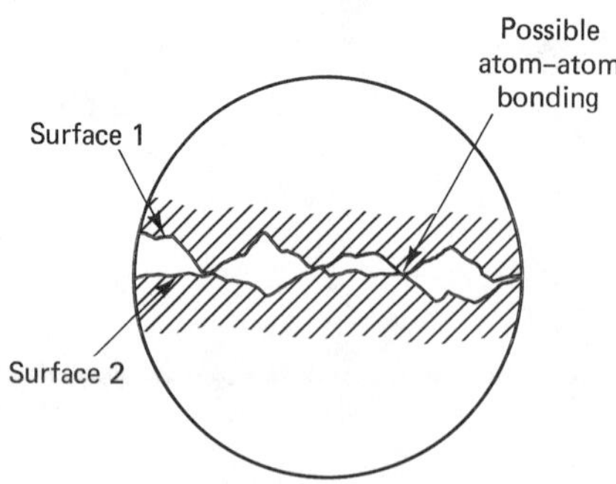

FIGURE 4.8
A microscopic view of two mating surfaces

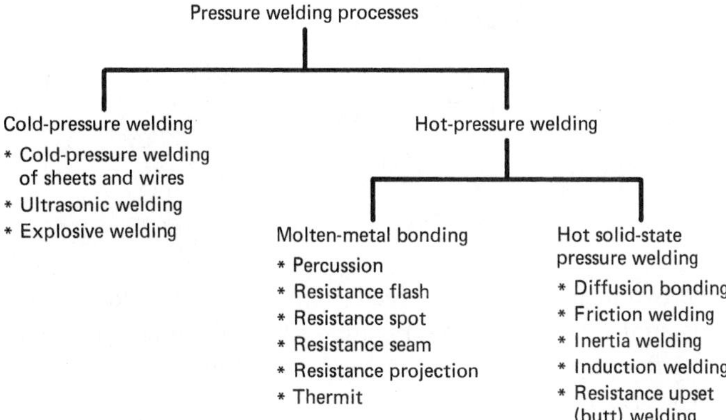

FIGURE 4.9 Classification of the commonly used pressure welding processes

annealing of the welded joint. Figure 4.10 also shows that recrystallization takes place during the annealing operation. This is added to diffusion, which finally results in complete disappearance of the interface between the two parts.

Cold-pressure welding of wires is performed by means of a special-purpose machine. Figure 4.11 illustrates the steps involved in this process. As can be seen, the wires' ends are clamped and pressed repeatedly against each other in order to ensure adequate plastic deformation. The excess upset metal is then trimmed by the sharp edges of the gripping jaws. This technique is used when welding wires of nonferrous metals such as aluminum, copper, or aluminum-copper alloys.

Explosive welding. Explosive welding is another technique that produces solid-state joints and, therefore, eliminates the problems associated with fusion welding methods, like the heat-affected zone and the microstructural changes. The process is based on using high explosives to generate extremely high pressures that are, in turn, used to combine flat plates or cylindrical shapes metallurgically. Joints of dissimilar metals and/or those that are extremely difficult to combine using conventional methods can easily be produced by explosive welding.

During explosive welding, a jet of soft (or fluidlike) metal is formed (on a microscopic scale) and breaks the oxide film barrier to bring the two metal parts into intimate contact. That metal jet is also responsible for the typical wavy interface between

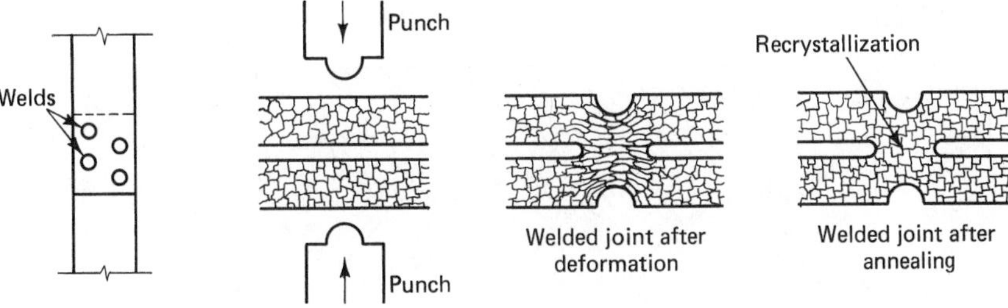

FIGURE 4.10 Cold-pressure welding of sheets

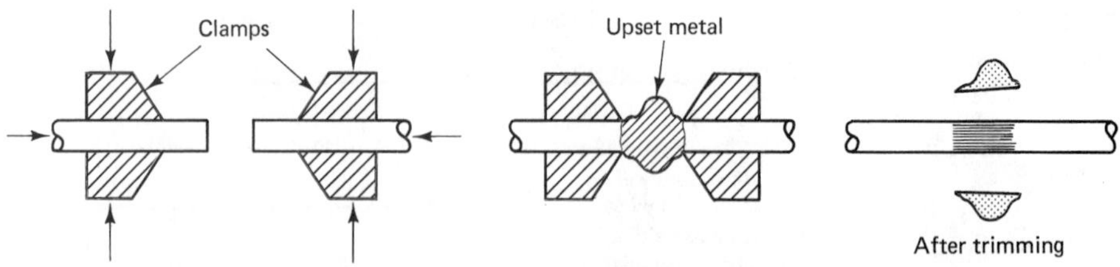

FIGURE 4.11 Cold-pressure welding of wires

the two metal parts, thus creating mechanical interlocking between them and, finally, resulting in a strong bond. Figure 4.12 illustrates an arrangement for explosive welding two flat plates, and Figure 4.13 is a magnified sketch of the wavy interface between explosively welded parts.

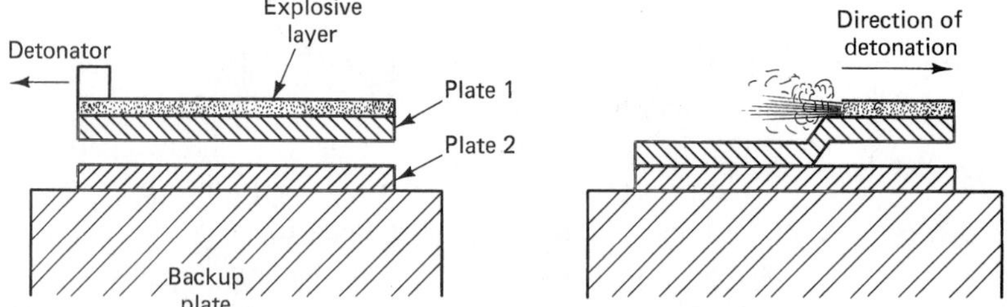

FIGURE 4.12 An arrangement for explosive welding two flat plates

Explosive welding and explosive cladding are popular in the manufacture of heat exchangers and chemical-processing equipment. Armored and reinforced composites with a metal matrix are also produced by explosive welding. Nevertheless, a clear limitation is that the process cannot be used successfully for welding hard, brittle metals. Research is being carried out in this area, and new applications are continuously introduced.

Ultrasonic welding. The ultrasonic welding method of solid-state welding is commonly used for joining thin sheets or wires of similar or dissimilar metals in order to obtain lap-type joints. Mechanical vibratory energy with ultrasonic frequencies is applied along the interfacial plane of the joint, while a nominal static stress is applied, normal to that interface, to clamp the two components together. Oscillating shear

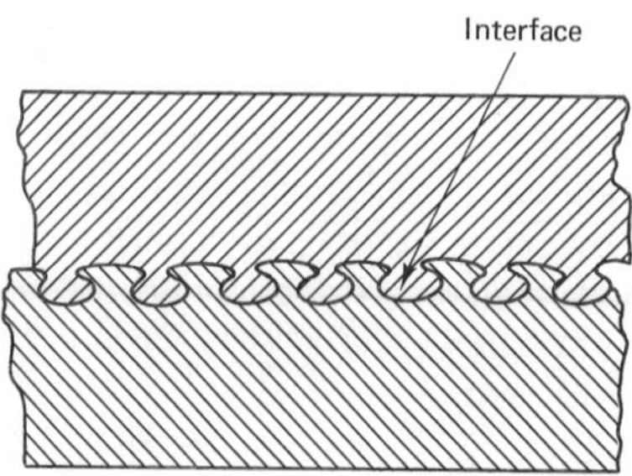

FIGURE 4.13 A sketch of the wavy interface between explosively welded parts

stresses are, therefore, initiated at the interface and disperse surface films, allowing intimate contact between the two metals, and, consequently, producing a strong joint. Ultrasonic welding does not involve the application of high pressures or temperatures and is accomplished within a short time. Therefore, this process is especially suitable for automation and has found widespread application in the electrical and microelectronics industries in the welding of thin metal foils for packaging and splicing and in the joining of dissimilar materials in the fabrication of nuclear reactor components. It must be noted, however, that the process is restricted to joining thin sheets or fine wires. Nevertheless, this restriction applies only to thinner pieces, and the process is often used in welding thin foils to thicker sheets.

Different types of ultrasonic welding machines are available, each constructed to produce a certain type of weld, such as spot, line, continuous seam, or ring. A sketch of a spot-type welding machine that is commonly used in welding microcircuit elements is illustrated in Figure 4.14. As we can see, the machine consists basically of a frequency convertor that transforms the standard 60-Hz (or 50-Hz in Europe) electric current into a high-frequency current (with a fixed frequency in the range of 15 to 75 kHz), a transducer that converts the electrical power into elastic mechanical vibrations, and a horn that magnifies the amplitude of these vibrations and delivers them to the weld zone. Other associated elements include the anvil, a force-application and clamping device, a sonotrode (as compared with the electrode in resistance welding), and appropriate controls to set up optimal values for the process variables, such as vibratory power and weld time.

Friction welding. In friction welding, a type of hot solid-state welding, the parts to be welded are tightly clamped, one in a stationary chuck and the other in a rotatable chuck that is mounted on a spindle. External power is employed to drive the spindle at a constant speed, with the two parts in contact under slight pressure. Kinetic energy is converted to frictional heat at the interface. When the mating edges of the workpieces attain a suitable temperature (in the forging range) that permits easy plastic flow, the spindle rotation is halted, and high axial pressure is applied to plastically deform the metal, obtain intimate contact, and produce a strong, solid weld. This is clearly shown in Figure 4.15, which indicates the stages involved in friction welding.

Several advantages have been claimed for the friction welding process. These include simplicity, high efficiency of energy utilization, and the ability to join similar as

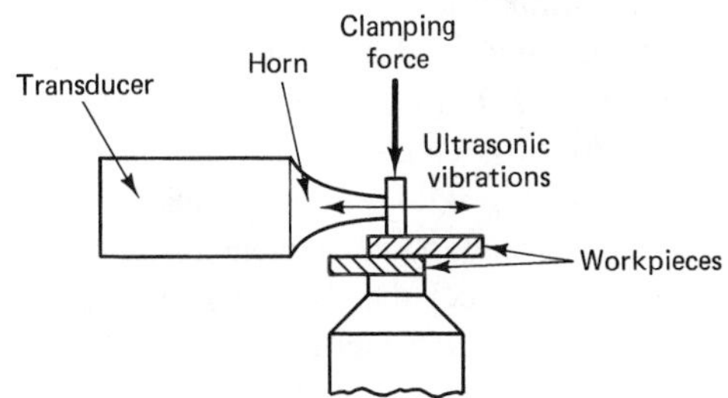

FIGURE 4.14 A sketch of an ultrasonic spot-type welding machine

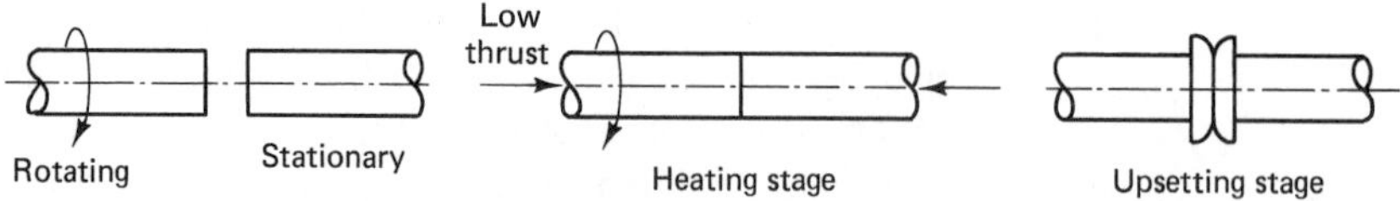

FIGURE 4.15 Stages involved in friction welding

well as dissimilar metal combinations that cannot be joined by conventional welding methods (e.g., aluminum to steel or aluminum to copper). Also, since contaminants and oxide films are carried away from the weld area where grain refinement takes place, a sound bond is obtained and usually has the same strength as the base metal. Nevertheless, a major limitation of the process is that at least one of the two parts to be joined must be a body of revolution around the axis of rotation (like a round bar or tube). A further limitation is that only forgeable metals that do not suffer from hot shortness can successfully be friction welded. Also, care must be taken during welding to ensure squareness of the edges of workpieces as well as concentricity of round bars or tubes.

Inertia welding. Inertia welding is a version of friction welding that is recommended for larger workpieces or where high-strength alloys (i.e., superalloys) are to be joined together. Inertia welding, as the name suggests, efficiently utilizes the kinetic energy stored in a rotating flywheel as a source for heating and for much of the forging of the weld. As is the case with friction welding, the two workpieces to be inertia welded are clamped tightly in stationary and rotatable chucks, the difference being that the rotatable chuck is rigidly coupled to a flywheel in the case of inertia welding. The process involves rotating the flywheel at a predetermined angular velocity and then converting the kinetic energy of the freely rotating flywheel to frictional heat at the weld interface by applying an axial load to join the abutting ends under controlled pressure. The process requires shorter welding time than that taken in conventional friction welding, especially for larger workpieces. Examples of inertia-welded components include hydraulic piston rods for agricultural machinery, carbon steel shafts welded to superalloy turbocharger wheels, and bar stock welded to small forgings.

Induction welding. As the name suggests, induction welding is based on the phenomenon of induction. We know from physics (electricity and magnetism) that when an electric current flows in an inductor coil, another electric current is induced in any conductor that intersects with the magnetic flux. In induction welding, the source of heat is the resistance, at the abutting workpieces' interface, to the flow of current induced in the workpieces through an external induction coil. Figure 4.16 illustrates the principles of induction welding. For efficient conversion of electrical energy into heat energy, high-frequency current is employed, and the process is usually referred to as *high-frequency induction welding* (HFIW). Frequencies in the range of 300 to 450 kHz are commonly used in industry, although frequencies as low as 10 kHz are also in use. It is always important to remember the "skin effect" when designing an induction-welded joint. This effect refers to the fact that the electric current flows superficially (i.e., near the surface). In fact, the depth of the layer through which the current flows is dependent mainly upon the frequency and the electromagnetic properties of the

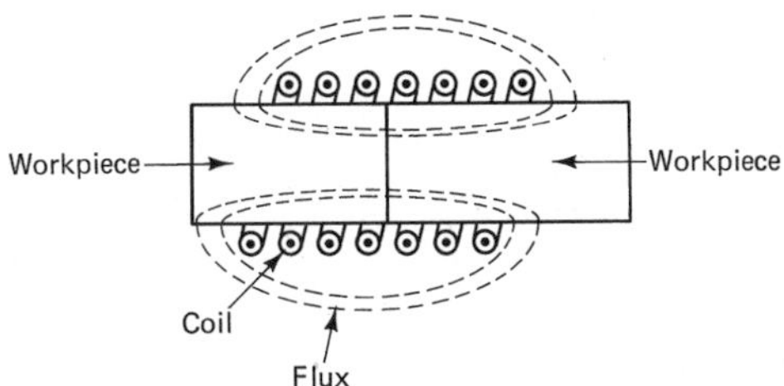

FIGURE 4.16
Principles of induction welding

workpiece metal. Industrial applications of induction welding include butt welding of pipes and continuous-seam welding for the manufacture of seamed pipes.

Thermit welding. Thermit welding makes use of an exothermic chemical reaction to supply heat energy. That reaction involves the burning of thermit, which is a mixture of fine aluminum powder and iron oxide in the form of rolling-mill scale, mixed at a ratio of about 1 to 3 by weight. Although a temperature of 5400°F (3000°C) may be attained as a result of the reaction, localized heating of the thermit mixture up to at least 2400°F (1300°C) is essential in order to start the reaction, which can be given by the following chemical formula:

$$8Al + 3Fe_3O_4 \rightarrow 9Fe + 4Al_2O_3 + \text{heat} \tag{4.1}$$

As we can see from the formula, the outcome is very pure molten iron and slag. In fact, other oxides are also used to produce pure molten metals; these include chromium, manganese, or vanadium, depending upon the parent metals to be welded.

Usually, the thermit welding process requires the application of pressure in order to achieve proper coalescence between the parts to be joined. However, fusion thermit welding is also used; it does not require the application of force. In this case, the resulting molten metal is a metallurgical joining agent and not just a means for heating the weld area.

Thermit welding is used in joining railroad rails, pipes, and thick steel sections, as well as in repairing heavy castings. The procedure involves fitting a split-type refractory mold around the abutting surfaces to be welded, igniting the thermit mixture using a primer (ignition powder) in a special crucible, and, finally, pouring the molten metal (obtained as a result of the reaction) into the mold. Because the temperature of the molten metal is about twice the melting point of steel, the heat input is enough to fuse the abutting surfaces, which are usually pressed together to give a sound weld.

Diffusion bonding. Diffusion bonding is a solid-state welding method in which the surfaces to be welded are cleaned and then maintained at elevated temperatures under appropriate pressure for a long period of time. No fusion occurs, deformation is limited, and bonding takes place principally due to diffusion. As we know from metallurgy, the process parameters are pressure, temperature, and time, and they should be adjusted to achieve the desired results.

Butt welding. Butt welding belongs to the resistance welding group, which also consists of the spot, seam, projection, percussion, and flash welding processes. All of these

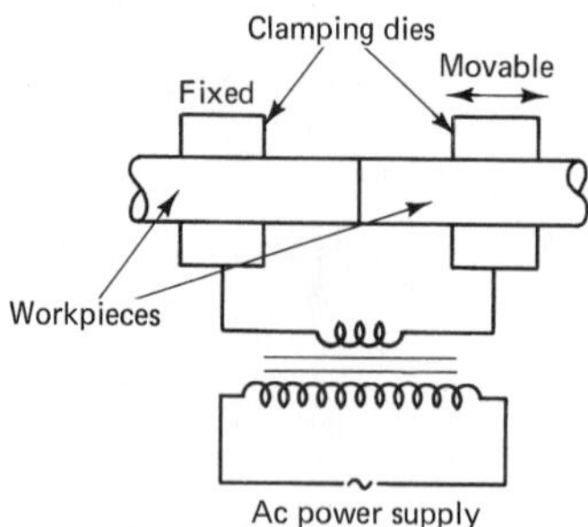

FIGURE 4.17
Upset-butt welding

operate on the same principle, which involves heating the workpieces as a result of being a part of a high-amperage electric circuit and then applying external pressure to accomplish strong bonding. Consequently, all the resistance welding processes belong to the larger, more general group of pressure welding; without the application of external pressure, the weld joint cannot be produced.

In butt welding, sometimes called *upset-butt* welding or just *upset* welding, the parts are clamped and brought in solid contact, and low-voltage (1 to 3 V) alternating current is switched on through the contact area, as illustrated in Figure 4.17. As a result of the heat generated, the metal in the weld zone assumes a plastic state (above the solidus) and is gradually squeezed and expelled from the contact area. When enough upset metal becomes evident, the current is switched off and the welded parts are released. Figure 4.18 indicates a typical upset welding cycle. Note that upset welding would not be successful for larger sections because these cannot be uniformly heated and require extremely high-amperage current. Therefore, the process is limited to welding wires and rods up to 3/8 inch (about 10 mm) in diameter. Also, a sound joint can be ensured only when the two surfaces being welded together have the same cross-sectional area as well as negligible or no eccentricity.

Flash welding. Flash welding is somewhat similar to upset welding. The equipment for flash welding includes a low-voltage transformer (5 to 10 V), a current timing

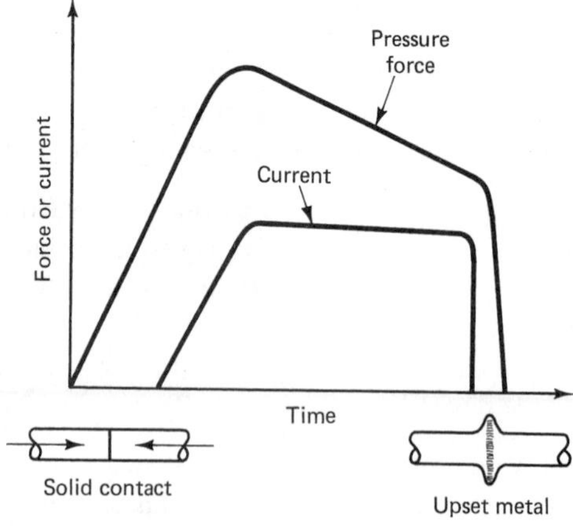

FIGURE 4.18
A typical upset welding cycle

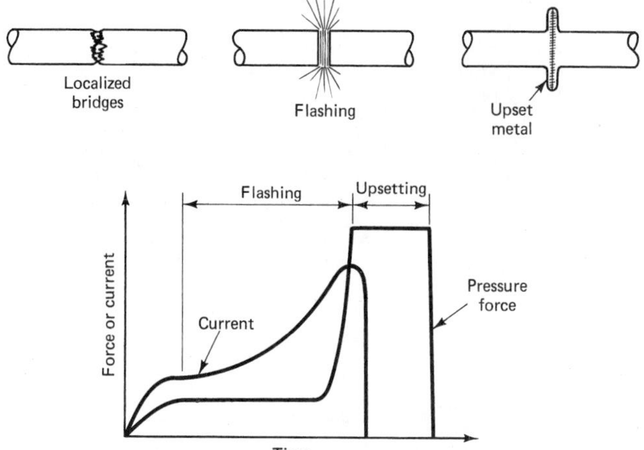

FIGURE 4.19
Stages in a flash welding cycle

device, and a mechanism to compress the two workpieces against each other. Figure 4.19 illustrates the different stages involved in a flash welding cycle. We can see that the pressure applied at the beginning is low. Therefore, there are a limited number of contact points that act as localized bridges for the electric current. Consequently, metal is heated at those points when the current is switched on, and the temperature increases with the increasing current until it exceeds the melting point of the metal. At this stage, the molten metal is expelled from the weld zone, causing "flashing." New bridges are formed and move quickly across the whole interface, resulting in uniform heating all over. When the whole contact area is heated above the liquidus line, electric current is switched off, and the pressure is suddenly increased to squeeze out the molten metal, upset the abutted parts, and weld them together.

Flash welding is used for joining large sections, rails, chain links, tools, thin-walled tubes, and the like. It can also be employed for welding dissimilar metals. The advantages claimed for the process include its higher productivity and its ability to produce high-quality welds. The only disadvantage is the loss of some metal in flashing.

Percussion welding. In percussion welding, a method of resistance welding, a high-intensity electric current is discharged between the parts before they are brought in solid contact. This results in an electric arc in the gap between the two surfaces. That electric arc lasts only for about 0.001 second and is enough to melt the surfaces to a depth of a few thousandths of an inch. The two parts are then impacted against each other at a high speed to obtain a sound joint. The major limitation of this process is the cross-sectional area of the welded joint. It should not exceed 0.5 square inch (300 mm^2) in order to keep the intensity of current required at a practical level. In industry, percussion welding is limited to joining dissimilar metals that cannot be welded otherwise.

Spot welding. Figure 4.20a illustrates the principles of operation of spot welding, a resistance welding process. Electric current is switched on between the welding electrodes

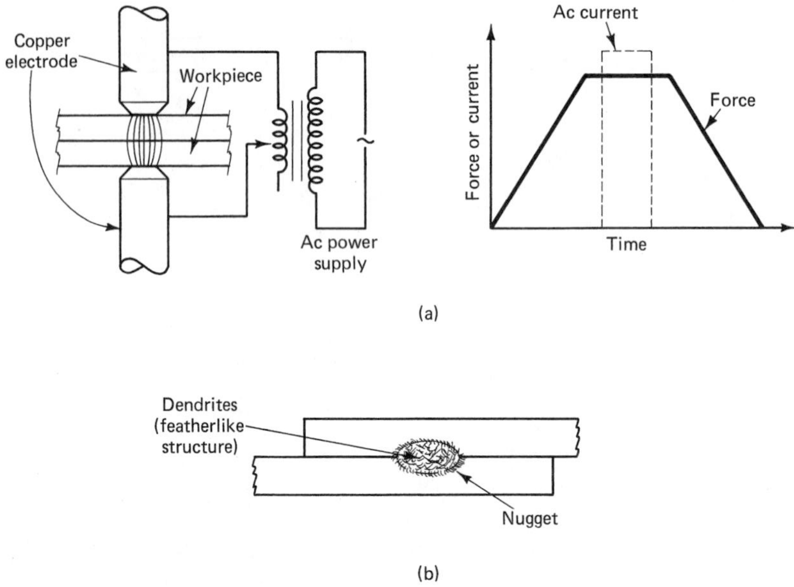

FIGURE 4.20 Resistance spot welding: (a) principles of operation; (b) a cross section through a spot weld

to flow through the lapped sheets (workpieces) that are held together under pressure. As can be seen in Figure 4.20b, the metal fuses in the central area of the interface between the two sheets and then solidifies in the form of a nugget, thus providing the weld joint. Heat is also generated at the contact areas between the electrodes and the workpieces. Therefore, some precautions must be taken to prevent excessive temperatures and fusing of the metal at those spots. The electrodes used must possess good electrical and thermal conductivities. They are usually hollow and are water-cooled. In addition, areas of workpieces in contact with the electrodes must be cleaned carefully.

Spot welding is the most widely used resistance welding process in industry. Carbon steel sheets having a thickness up to 0.125 inch (4 mm) can be successfully spot welded. Spot-welding machines have ratings up to more than 600 kVA and use a voltage of 1 to 12 V obtained from a step-down transformer. Multispot machines are used, and the process can be fully automated. Therefore, spot welding has found widespread application in the automobile, aircraft, and electronics industries, as well as in sheet metal work.

Seam welding. Seam welding and projection welding are modifications of spot welding. In seam welding, the lapped sheets are passed between rotating circular electrodes through which the high-amperage current flows, as shown in Figure 4.21. Electrodes vary in diameter from less than 2 up to 14 inches (40 to 350 mm), depending upon the curvature of the workpieces being welded. Welding current as high as 5000 A may be employed, and the pressing force acting upon the electrodes can go up to 6 kN (more than half a ton). A welding speed of about 12 feet per minute (4 m/min.) is quite common. Seam welding is employed in the production of pressure-tight joints used in containers, tubes, mufflers, and the like. Advantages of this process include low cost, high

FIGURE 4.21
Principles of seam welding

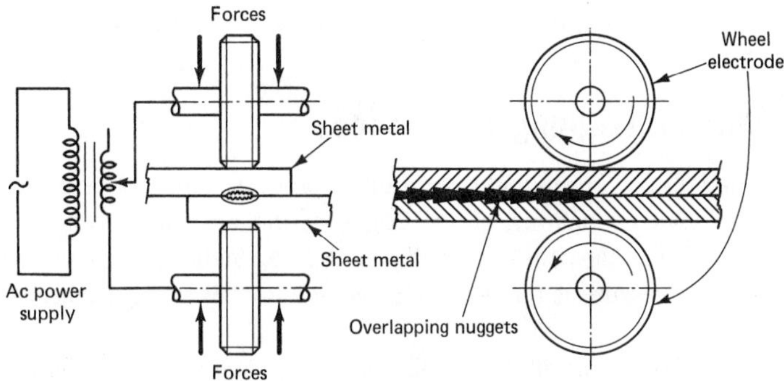

production rates, and suitability for automation. Nevertheless, the thickness of the sheets to be seam welded is limited to 0.125 inch (4 mm) in the case of carbon steels and much less for more conductive alloys due to the extremely high amperage required (0.125-inch-thick steel sheets require 19,000 A, whereas aluminum sheets having the same thickness would require 76,000 A).

Projection welding. In projection welding, one of the workpieces is purposely provided with small projections so that current flow and heating are localized at those spots. The projections are usually produced by die pressing, and the process calls for the use of a special upper electrode. Figure 4.22 illustrates an arrangement of two parts to be projection welded, as well as the resulting weld nugget. As you may expect, the projections collapse under the externally applied force after sufficient heating, thus yielding a well-defined, fused weld nugget. When the current is switched off, the weld cools down and solidification takes place under the applied force. The electrode force is then released, and the welded workpiece is removed. As is the case with spot welding, the entire projection welding process takes only a fraction of a second.

Projection welding has some advantages over conventional spot welding. For instance, sheets that are too thick to be joined by spot welding can be welded using this process. Also, the presence of grease, dirt, or oxide films on the surface of the workpieces has less effect on the weld quality than in the case of spot welding. A further

FIGURE 4.22
An arrangement for projection welding two parts

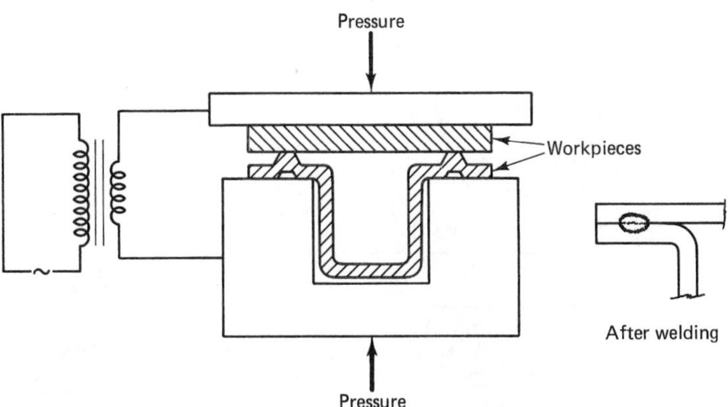

advantage of projection welding is the accuracy of locating welds inherent in that process.

Fusion Welding Processes

Fusion welding includes a group of processes that all produce welded joints as a result of localized heating of the edges of the base metal above its melting temperature, wherein coalescence is produced. A filler metal may or may not be added, and no external pressure is required. The welded joint is obtained after solidification of the fused weld pool.

In order to join two different metals together by fusion welding, they must possess some degree of mutual solubility in the solid state. In fact, metals that are completely soluble in the solid state exhibit the highest degree of weldability. Metals with limited solid solubility have lower weldability, and metals that are mutually insoluble in the solid state are completely unweldable by any of the fusion welding methods. In that case, an appropriate pressure welding technique should be employed. An alternative solution is to employ an intermediate metal that is soluble in both base metals.

Metallurgy of fusion welding. Before surveying the different fusion welding processes, let us discuss the metallurgy of fusion welding. Important microstructural changes take place in and around the weld zone during and after the welding operation. Such changes in the microstructure determine the mechanical properties of the welded joint. Therefore, a study of the metallurgy of fusion welding is essential for good design of welded joints, as well as for the optimization of the process parameters.

During fusion welding, three zones can be identified, as shown in Figure 4.23, which indicates a single V-weld in steel after solidification and the corresponding temperature distribution during welding. In the first zone, called the *fusion zone,* the base metal and deposited metal (if a filler rod is used) are brought to the molten state during welding. Therefore, when this zone solidifies after welding, it generally has a columnar dendritic structure with haphazardly oriented grains. In other words, the microstructure of this zone is quite similar to that of the cast metal. Nevertheless, if the molten metal is overheated during welding, this results in an acicular structure that is brittle, has low strength, and is referred to as the *Widmanstatten structure.* Also, the

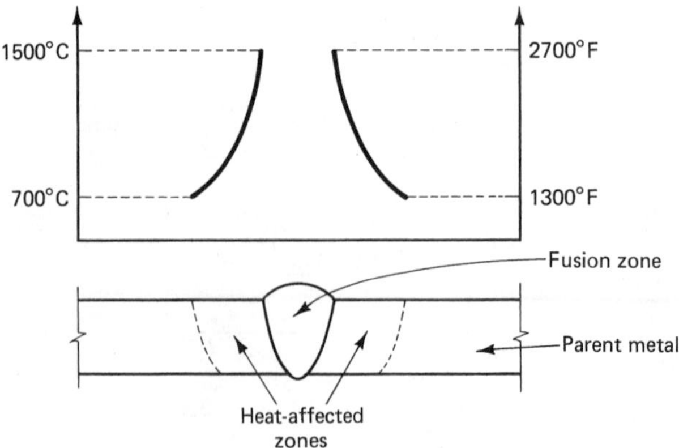

FIGURE 4.23
The three zones in a fusion-welded joint and the temperature distribution during welding

chemical composition of the fusion zone may change, depending upon the kind and amount of filler metal added.

The second zone, which is referred to as the *heat-affected zone* (HAZ), is that portion of the base metal that has not been melted. Therefore, its chemical composition before and after welding remains unchanged. Nevertheless, its microstructure is always altered because of the rapid heating during welding and subsequent cooling. In fact, the HAZ is subjected to a normalizing operation during welding and may consequently undergo phase transformations and precipitation reactions, depending upon the nature (chemical composition and microstructure) of the base metal. The size of the HAZ is dependent upon the welding method employed and the nature of the base metal. This can be exemplified by the fact that the HAZ is 0.1 inch (2.5 mm) when automatic submerged arc welding is used, ranges from 0.2 to 0.4 inch (5 to 10 mm) for shielded-metal arc welding, and may reach 1 inch (25 mm) in conventional gas welding. This evidently affects the microstructure of the weld, which is generally fine-grained. The effect of these structural changes on the mechanical properties of the weld differs for different base metals. For instance, the structural changes have negligible effect on the mechanical properties of low-carbon steel, regardless of the welding method used. On the contrary, when welding high-carbon alloy steel, hardened structures like maternsite are formed in the HAZ of the weld that result in a sharp reduction in the ductility of the welded joint and/or crack formation. (Remember the effect of alloying elements on the critical cooling rate in the TTT diagram that you studied in metallurgy.)

The third zone involves the unaffected *parent metal* adjacent to the HAZ that is subjected to a temperature below AC_3 (a critical temperature) during welding. In this zone, no structural changes take place unless the base metal has been subjected to plastic deformation prior to welding, in which case recrystallization and grain growth would become evident.

Arc welding. Arc welding is based on the thermal effect of an electric arc that is acting as a powerful heat source to produce localized melting of the base metal. The electric arc is, in fact, a sustained electrical discharge (of electrons and ions in opposite directions) through an ionized, gaseous path between two electrodes (i.e., the anode and the cathode). In order to ionize the air in the gap between the electrodes so that the electric arc can consequently be started, a certain voltage is required. (The voltage required depends upon the distance between the electrodes.) The ionization process results in the generation of electrons and positively charged ions. Next, the electrons impact on the anode, and the positively charged ions impact on the cathode. The collisions of these particles, which are accelerated by the arc voltage, transform the kinetic energy of the particles into thermal and luminous energy, and the temperature at the center of the arc can reach as high as 11,000°F (6000°C). Actually, only a comparatively low potential difference between the electrodes is required to start the arc. For instance, about 45 V is usually sufficient for direct current (dc) welding equipment, and up to 60 V for an alternating current (ac) welder. Also, the voltage drops after the arc is started, and a stable arc can then be maintained with a voltage in the range of 15 to 30 V. Generally, arc welding involves using a metal electrode rod and attaching the other electrode to the workpiece. The electrode rod either melts during the process (consumable electrode) and provides the necessary filler metal for the weld, or the electrode does not melt and the filler metal is separately provided.

As just mentioned, either alternating current or direct current can be used in arc welding, although each has its distinct advantages. While arc stability is much better with alternating current than with direct current, the ac welding equipment is far less expensive, more compact in size, and simpler to operate. A further advantage of ac arc welding is the high efficiency of the transformer used, which goes up to 85 percent, whereas the efficiency of dc welding systems usually varies between only 30 and 60 percent.

In dc arc welding, the degree to which the work is heated can be regulated by using either straight or reversed polarity. As can be seen in Figure 4.24a and b, the cathode is the electrode rod and the anode is the workpiece in straight polarity (DCSP), whereas it is the other way around in reversed polarity (DCRP). When using DCSP, more heat is concentrated at the cathode (the electrode rod) than at the anode (the workpiece). Therefore, melting and deposition rates (of consumable electrodes) are high, but penetration in the workpiece is shallow and narrow. Consequently, DCSP is recommended when welding sheet metal, especially at higher welding speeds. With DCRP, heat is concentrated at the cathode (the workpiece) and results in deeper penetration for a given welding condition. It is, therefore, preferred for groove welds and similar applications.

During the welding operation, heat is generated in the transformer as well as in other elements of the welding circuit, resulting in a temperature rise that may cause damage to those elements. There is, therefore, a time limitation when using the welding equipment at a given amperage. That time limitation is usually referred to as the *rated duty cycle.* Consider the following numerical example. A power supply for arc welding rated at a 150-A 40-percent duty cycle means that it can be used only 40 percent of the time when welding at 150 A. The idle or unused time is required to allow the equipment to cool down. The percentage of duty cycles at currents other than the rated current can be calculated using the following equation:

$$\%\text{ duty cycle} = \left(\frac{\text{rated current}}{\text{load current}}\right)^2 \times \text{rated duty cycle} \tag{4.2}$$

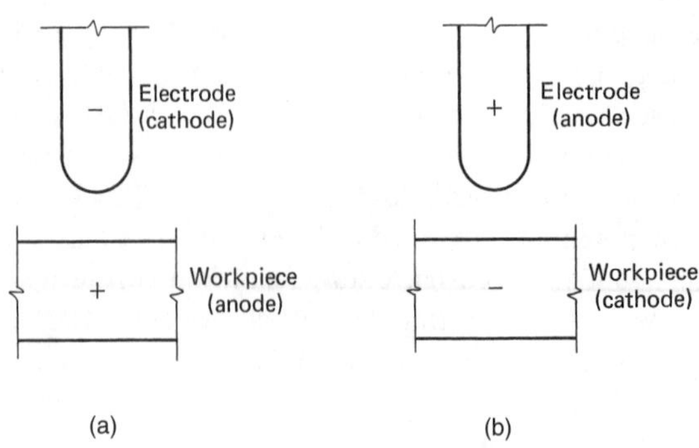

FIGURE 4.24
Straight and reversed polarities in dc arc welding: (a) dc straight polarity (DCSP); (b) dc reversed polarity (DCRP)

Therefore, for this power supply, the percentage of the duty cycle at 100 A is as follows:

$$\% \text{ duty cycle at } 100 \text{ A} = \left(\frac{150}{100}\right)^2 \times 40\% = 90\%$$

There are various types of arc welding. They include the following methods:

1. Shielded-metal arc welding (SMAW)
2. Carbon arc welding (CAW)
3. Flux-cored arc welding (FCAW)
4. Stud arc welding (SW)
5. Submerged arc welding (SAW)
6. Gas-metal arc welding (GMAW, usually called MIG)
7. Gas-tungsten arc welding (GTAW, usually called TIG)
8. Plasma arc welding (PAW)

In addition, there is another welding process, electroslag welding (EW), which is not based on the phenomenon of the electric arc but, nevertheless, employs equipment similar to that used in gas-metal arc, flux-cored arc, or submerged arc welding.

1. **Shielded-metal arc welding.** Shielded-metal arc welding (SMAW) is a manual arc welding process that is sometimes referred to as *stick welding*. The source of heat for welding is an electric arc maintained between a flux-covered, consumable metal electrode and the workpiece. As can be seen in Figure 4.25, which indicates the operating principles of this process, shielding of the electrode tip, weld puddle, and weld area in the base metal is ensured through the decomposition of the flux covering. A blanket of molten slag also provides shielding for the molten-metal pool. The filler metal is provided mainly by the metal core of the electrode rod.

   Shielded-metal arc welding can be used for joining thin and thick sheets of plain-carbon steels, low-alloy steels, and even some alloy steels and cast iron, provided that the electrode is properly selected and also that preheating and postheating treatments are performed. It is actually the most commonly used welding process and has found widespread application in steel construction and shipbuilding. Nevertheless, it is uneconomical and/or impossible to employ shielded-metal arc welding to join some alloys, such as aluminum alloys, copper, nickel, copper-nickel alloys, and low-melting-point alloys such as zinc, tin, and magnesium alloys.

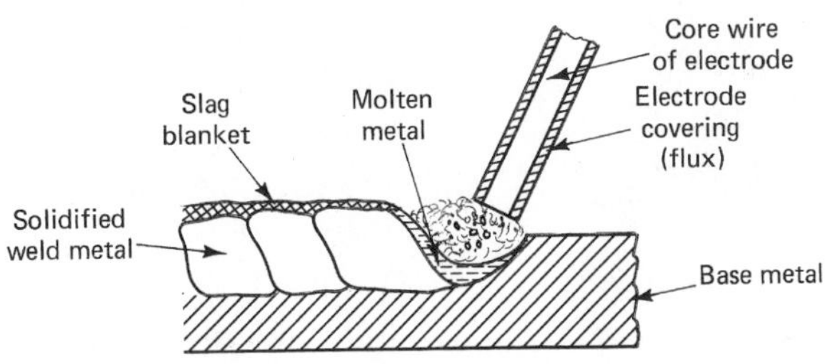

FIGURE 4.25 Operating principles of shielded-metal arc welding

Another clear shortcoming of the process is that welding must be stopped each time an electrode stick is consumed to allow mounting a new one. This results in idle time and, consequently, a drop in productivity.

The core wires of electrodes used for shielded-metal arc welding have many different compositions. The selection of a particular electrode material depends upon the application for which it is going to be used and the kind of base metal to be welded. Consumable electrodes are usually coated with flux but can also be uncoated. The metal wire can have a diameter of up to 15/32 inch (12 mm) and a length of about 18 inches (450 mm). Although various metals are used as wire materials, by far the most commonly used electrode materials involve low-carbon steel (for welding carbon steels) and low-alloy steel (for welding alloy steels). Electrodes can be bare, lightly coated, or heavily coated. The electrode covering, or coating, results in better-quality welds as it improves arc stability, produces gas shielding to prevent oxidation and nitrogen contamination, and also provides slag, which, in turn, retards the cooling rate of the weld's fusion zone. Therefore, electrode coatings are composed of substances that serve these purposes. Table 4.1 indicates the composition of typical electrode coatings, together with the function of each constituent.

2. **Carbon arc welding.** In carbon arc welding (CAW), nonconsumable electrodes made of carbon or graphite are used. Only a dc power supply can be employed, and the electric arc is established either between a single carbon electrode and the workpiece (Bernardos method) or between two carbon electrodes (independent arc method). In both cases, no shielding is provided. A filler metal may be used, especially when welding sheets with thicknesses more than 1/8 inch (3 mm). The carbon electrodes have diameters ranging from 3/8 to 1 inch (10 to 25 mm) and are used with currents that range from 200 to 600 A.

TABLE 4.1
The constituents of typical electrode coatings and their functions

Main Function	Constituent	Percentage
Gas generating	Starch Cellulose Calcium carbonate	25–40%
Slag forming	Kaolin Titanium dioxide Feldspar Asbestos	20–40%
Binding	Sodium silicate Potassium silicate	20–30%
Deoxidizing	Ferrosilicon Aluminum	5–10%
Arc stabilizing	Potassium titanate Titanium oxide	5–10%
Increasing deposition rate	Iron powder	0–40%
Improving weld strength	Different alloying elements	5–10%

Carbon arc welding is not commonly used in industry. Its application is limited to the joining of thin sheets of nonferrous metals and to brazing.

3. **Flux-cored arc welding.** Flux-cored arc welding (FCAW) is an arc welding process in which the consumable electrode takes the form of a tubular, flux-filled wire, that is continuously fed from a spool. Shielding is usually provided by the gases evolving during the combustion and decomposition of the flux contained within the tubular wire. The process is, therefore, sometimes called *inner-shielded,* or *self-shielded,* arc welding. Additional shielding may be acquired through the use of an auxiliary shielding gas, such as carbon dioxide, argon, or both. In the latter case, the process is a combination of the conventional flux-cored arc welding and the gas-metal arc welding methods and is referred to as *dual-shielded* arc welding.

   Flux-cored arc welding is generally applied on a semiautomatic basis, but it can also be fully automated. In that case, the process is normally used to weld medium-to-thick steel plates and stainless steel sheets. Figure 4.26 illustrates the operating principles of flux-cored arc welding.

4. **Stud arc welding.** Stud arc welding (SW) is a special-purpose arc welding process by which studs are welded to flat surfaces. This facilitates fastening and handling of the components to which studs are joined and meanwhile eliminates the drilling and tapping operations that would have been required to achieve the same goal. Only dc power supplies are employed, and the process also calls for the use of a special welding gun that holds the stud during welding. Figure 4.27 shows the stages involved in stud arc welding, a process that is entirely controlled by the timer of the gun. As can be seen in the figure, shielding is accomplished through the use of a ceramic ferrule that surrounds the end of the stud during the process. Stud arc welding requires a low degree of welding skill, and the whole welding cycle usually takes less than a second.

5. **Submerged arc welding.** Submerged arc welding (SAW) is a fairly new automatic arc welding method in which the arc and the weld area are shielded by a blanket of a fusible granular flux. A bare electrode is used and is continuously fed by a special mechanism during welding. Figure 4.28 shows the operating principles of submerged arc welding. As can be seen from the figure, the process is used to join flat plates in the horizontal position only. This limitation is imposed by the nature of the flux and the way it is fed.

   As is the case with previously discussed arc welding processes, gases evolve as a result of combustion and decomposition of the flux, due to the high temperature

FIGURE 4.26
Operating principles of flux-cored arc welding

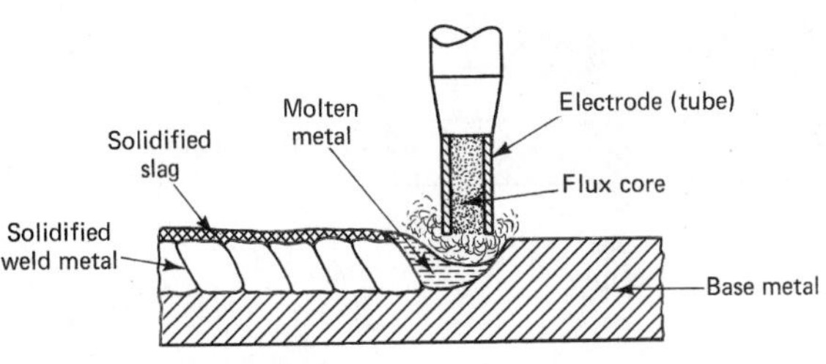

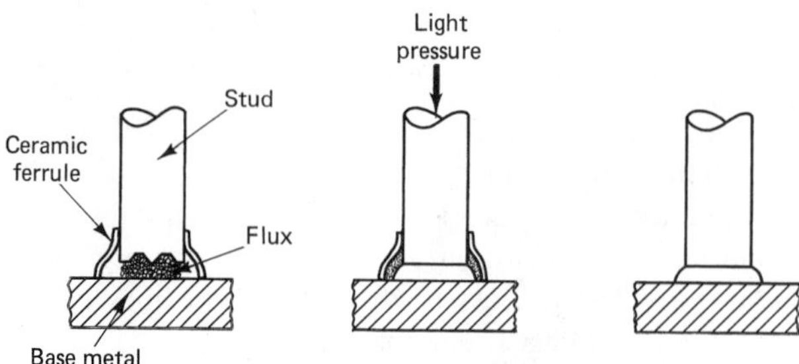

FIGURE 4.27 Stages involved in stud arc welding

of the arc, and form a pocket, or gas bubble, around the arc. As Figure 4.29 shows, this gas bubble is sealed from the arc by a layer of molten flux. This isolates the arc from the surrounding atmosphere and, therefore, ensures proper shielding.

The melting temperature of the flux must be lower than that of the base metal. As a result, the flux always solidifies after the metal, thus forming an insulating layer over the solidifying molten metal pool. This retards the solidification of the fused metal and, therefore, allows the slag and nonmetallic inclusions to float off the molten pool. The final outcome is always a weld that is free of nonmetallic inclusions and entrapped gases and has a homogeneous chemical composition. The flux should also be selected to ensure proper deoxidizing of the fused metal and

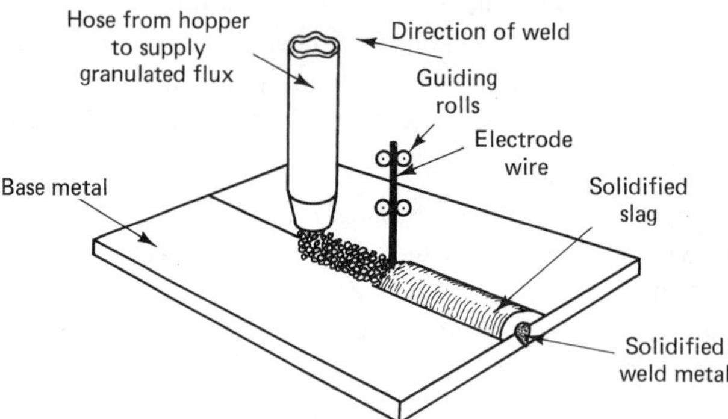

FIGURE 4.28 Operating principles of submerged arc welding

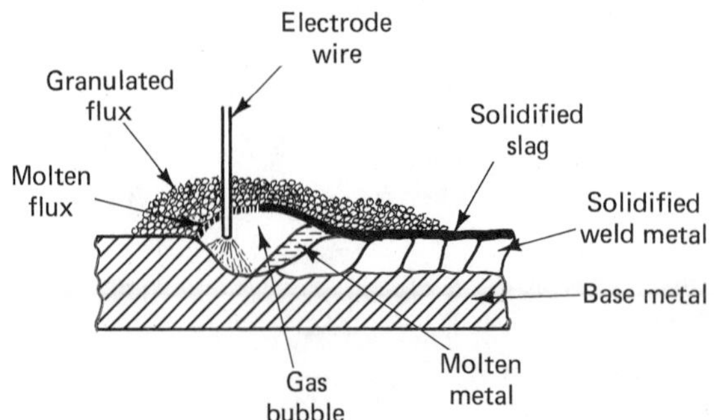

FIGURE 4.29 The mechanics of shielding in submerged arc welding

should contain additives that make up for the elements burned and lost during the welding process.

Electric currents commonly used with submerged arc welding range between 3000 and 4000 A. Consequently, the arc obtained is extremely powerful and is capable of producing a large molten-metal pool as well as achieving deeper penetration. Other advantages of this process include its high welding rate, which is five to ten times that produced by shielded-metal arc welding, and the high quality of the welds obtained.

6. **Gas-metal arc welding.** The GMAW process is commonly called *metal–inert-gas* (MIG) welding. It employs an electric arc between a solid, continuous, consumable electrode and the workpiece. As can be seen in Figure 4.30, shielding is obtained by pumping a stream of chemically inert gas, such as argon or helium, around the arc to prevent the surrounding atmosphere from contaminating the molten metal. (The electrode is bare, and no flux is added.) Dry carbon dioxide can sometimes be employed as a shielding gas, yielding fairly good results.

   Gas-metal arc welding is generally a semiautomatic process. However, it can also be applied automatically by machine. In fact, welding robots and numerically controlled MIG welding machines have gained widespread industrial application. The gas-metal arc welding process can be used to weld thin sheets as well as relatively thick plates in all positions, and the process is particularly popular when welding nonferrous metals such as aluminum, magnesium, and titanium alloys. The process is also used for welding stainless steel and critical steel parts.

   The penetration for gas-metal arc welding is controlled by adopting DCRP and adjusting the current density. The higher the current density is, the greater the penetration is. The kind of shielding gas used also has some effect on the penetration. For instance, helium gives the maximum penetration; carbon dioxide, the least; argon, intermediate penetration. Thus, it is clear that higher current densities and the

FIGURE 4.30
Operating principles of gas-metal arc welding

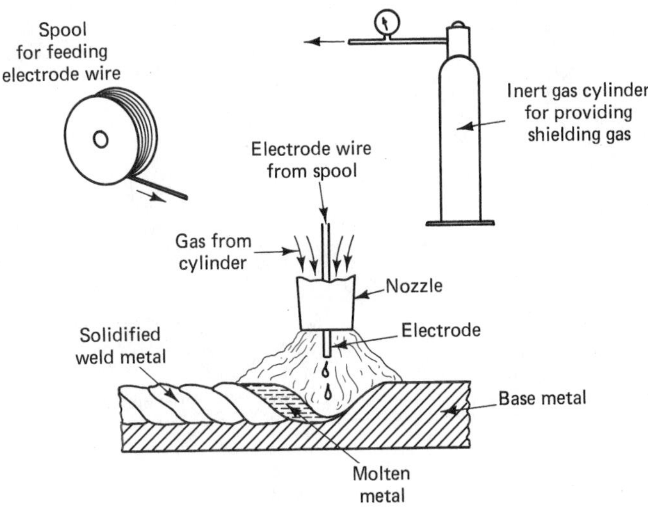

appropriate shielding gas can be employed in welding thick plates, provided that the edges of these plates are properly prepared.

The electrode wires used for MIG welding must possess close dimensional tolerances and a consistent chemical composition appropriate for the desired application. The wire diameter varies between 0.02 and 0.125 inch (0.5 and 3 mm). Usually, MIG wire electrodes are coated with a very thin layer of copper to protect them during storage. The electrode wire is available in the form of a spool weighing from 2½ to 750 pounds (1 to 350 kg). As you may expect, the selection of the composition of the electrode wire for a given material depends upon other factors, such as the kind of shielding gas used, the conditions of the metal being welded (i.e., whether there is an oxide film, grease, or contaminants), and, finally, the required properties of the weldment.

7. **Gas-tungsten arc welding.** Gas-tungsten arc welding (GTAW), which is usually called *tungsten–inert-gas* (TIG) welding, is an arc welding process that employs the heat generated by an electric arc between a nonconsumable tungsten electrode and the workpiece. Figure 4.31 illustrates the operating principles of this process. As can be seen, a filler rod may (or may not) be fed to the arc zone. The electrode, arc, weld puddle, and adjacent areas of the base metal are shielded by a stream of either argon or helium to prevent any contamination from the atmosphere. TIG welding is normally applied manually and requires a relatively high degree of welder skill. It can also be fully automated, in which case the equipment used drives the welding torch at a preprogrammed path and speed, adjusts the arc voltage, and starts and stops it.

   Gas-tungsten arc welding is capable of welding nonferrous and exotic metals in all positions. The list of metals that can be readily welded by this process is long and includes alloy steels, stainless steels, heat-resisting alloys, refractory metals, aluminum alloys, magnesium alloys, titanium alloys, copper and nickel alloys, and steel coated with low-melting-point alloys. The process is recommended for

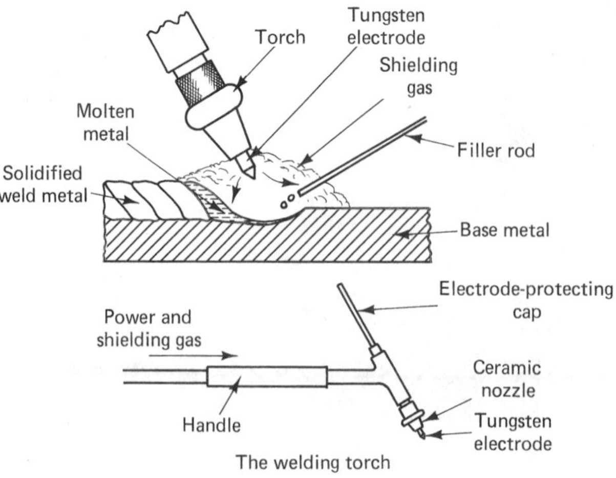

FIGURE 4.31
Operating principles of gas-tungsten arc welding

welding very thin sheets, as thin as 0.005 inch (about 0.125 mm), for the root and hot pass on tubing and pipes, and wherever smooth, clean welds are required (e.g., in food-processing equipment). Ultrahigh-quality welds can be obtained in the nuclear, rocketry, and submarine industries by employing a modified version of TIG welding that involves placing carefully selected and prepared inserts in the gap between the sections to be joined and then completely fusing the inserts together with the edges of base metal using a TIG torch.

All three types of current supplies (i.e., ac, DCSP, and DCRP) can be used with gas-tungsten arc welding, depending upon the metal to be welded. Thin sheets of aluminum or magnesium alloys are best welded by using DCRP, which prevents burn-throughs, as previously explained. Nevertheless, it is recommended that an ac power supply be used when welding normal sheets of aluminum and magnesium. DCSP is best suited for welding high-melting-point alloys such as alloy steels, stainless steels, heat-resisting alloys, copper alloys, nickel alloys, and titanium. In addition to these considerations, DCRP is also helpful in removing surface oxide films due to its cleaning action (the impacting of ions onto the surface like a grit blasting).

8. **Plasma arc welding.** Figure 4.32 is a sketch of the torch employed in plasma arc welding (PAW). The electric arc can take either of two forms: a transferred arc that is a constricted arc between a tungsten electrode and the workpiece or a nontransferred arc between the electrode and the constricting nozzle. The gas flowing around the arc heats up to extremely high temperatures like 60,000°F (33,000°C) and becomes, therefore, ionized and electrically conductive; it is then referred to as *plasma.* The main shielding is obtained from the hot ionized gas emerging from the nozzle. Additional inert-gas shielding can be used when high-quality welds are required. In fact, plasma arc welding can be employed to join almost all metals in all positions, although it is usually applied to thinner metals. Generally, the process is applied manually and requires some degree of welder skill; however, the process is sometimes automated in order to increase productivity.

Electroslag welding. Electroslag welding (ESW), which was developed by the Russians, is not an arc welding process but requires the use of equipment similar to that

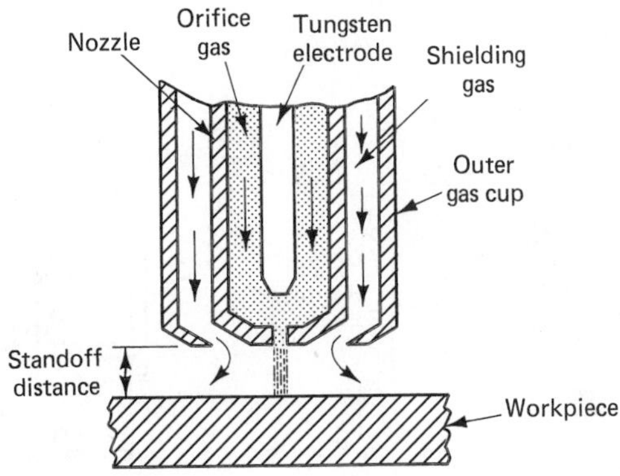

FIGURE 4.32 The torch employed in plasma arc welding

used in arc welding. Although an electric arc is used to start the process, heat is continuously generated as a result of the current flow between the electrode (or electrodes) and the base metal through a pool of molten slag (flux). As we will see later, the molten-slag pool also serves as a protective cover for the fused-metal pool.

The electroslag welding process is shown in Figure 4.33. As can be seen in the figure, the parts to be joined are set in the vertical position, with a gap of 1/2 to 1½ inches (12 to 37 mm) between their edges. (The gap is dependent upon the thickness of the parts.) The welding electrode (or electrodes) and the flux are fed automatically into the gap, and an arc is established between the electrodes and the steel backing plate to provide the initial molten-metal and slag pools. Next, the electrical resistivity of the molten slag continuously produces the heat necessary to fuse the flux and the filler and the base metals. Water-cooled copper plates travel upward along the joint, thus serving as dams and cooling the fused metal in the cavity to form the weld.

Electroslag welding is very advantageous in joining very thick parts together in a single pass without any need for beveling the edges of those parts. Therefore, the process is widely used in industries that fabricate beds and frames for heavy machinery, drums, boilers, and the like.

Gas welding. Gas welding refers to a group of oxyfuel gas processes in which the edges of the parts to be welded are fused together by heating them with a flame obtained from the combustion of a gas (such as acetylene) in a stream of oxygen. A filler metal is often introduced into the flame to melt and, together with the base metal, form the weld puddle. Gas welding is usually applied manually and requires good welding skill. Common industrial applications involve welding thin-to-medium sheets and sections of steels and nonferrous metals in all positions. Gas welding is also widely used in repair work and in restoring cracked or broken components.

The fuel gases used for producing the flame during the different gas welding processes include acetylene, hydrogen, natural gas (94 percent methane), petroleum gas, and vaporized gasoline and kerosene. However, acetylene is the most commonly used gas for gas welding because it can provide a flame temperature of about 5700°F (3150°C). Unfortunately, acetylene is ignited at a temperature as low as 790°F (420°C) and becomes explosive in nature at pressures exceeding 1.75 atmospheres. Therefore, it is stored in metal cylinders, in which it is dissolved in acetone under a pressure of

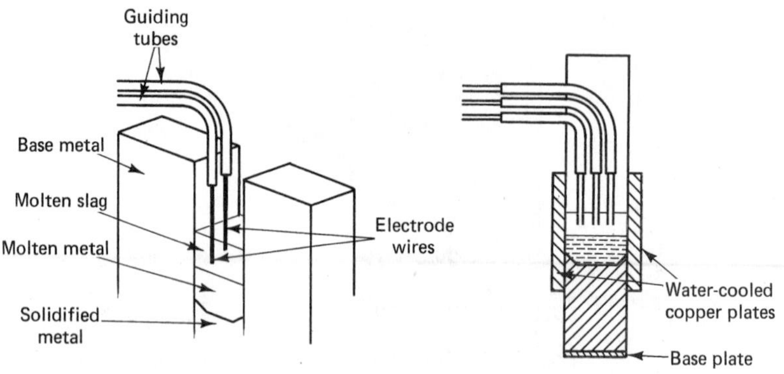

FIGURE 4.33
The electroslag welding process

about 19 atmospheres. For more safety, acetylene cylinders are also filled with a porous filler (such as charcoal) in order to form a system of capillary vessels that are then saturated with the solution of acetylene in acetone.

The oxygen required for the gas welding process is stored in steel cylinders in the liquid state under a pressure of about 150 atmospheres. It is usually prepared in special plants by liquefying air and then separating the oxygen from the nitrogen.

The equipment required in gas welding, as shown in Figure 4.34, includes oxygen and acetylene cylinders, regulators, and the welding torch. The regulators serve to reduce the pressure of the gas in the cylinder to the desired working value and keep it that way throughout the welding process. Thus, the proportion of the two gases is controlled, which determines the characteristics of the flame. Next, the welding torch serves to mix the oxygen and the acetylene together and discharges the mixture out at the tip, where combustion takes place.

Depending upon the ratio of oxygen to acetylene, three types of flames can be obtained: neutral, reducing, and oxidizing. Figure 4.35 is a sketch of a typical oxyacetylene welding flame. As can be seen, the welding flame consists of three zones: the inner luminous cone at the tip of the torch, the reducing zone, and the oxidizing zone.

The first zone, the *luminous cone,* consists of partially decomposed acetylene as a result of the following reaction:

$$C_2H_2 \rightarrow 2C + H_2$$

The carbon particles obtained are incandescent and are responsible for the white luminescence of that brightest part of the flame. Those carbon particles are partly oxidized in the second zone, the *reducing zone,* yielding carbon monoxide and a large amount of heat that brings the temperature up to about 5400°F (3000°C). Gases like hydrogen and carbon monoxide are capable of reducing oxides. Next, complete combustion of those gases yields carbon dioxide and water vapor that together with the excess oxygen (if any) result in the third zone, the *oxidizing zone.* Those gases, however, form a shield that prevents the atmosphere from coming in contact with the molten-metal pool.

As can be expected, the extent (as well as the appearance) of each of the zones depends upon the type of flame (i.e., the oxygen-to-acetylene ratio). When the ratio is about 1, the flame is *neutral* and distinctively has the three zones just outlined. If the oxygen-to-acetylene ratio is less than 1, a *reducing,* or *carbonizing,* flame is obtained. In this case, the luminous cone is longer than that obtained with the neutral

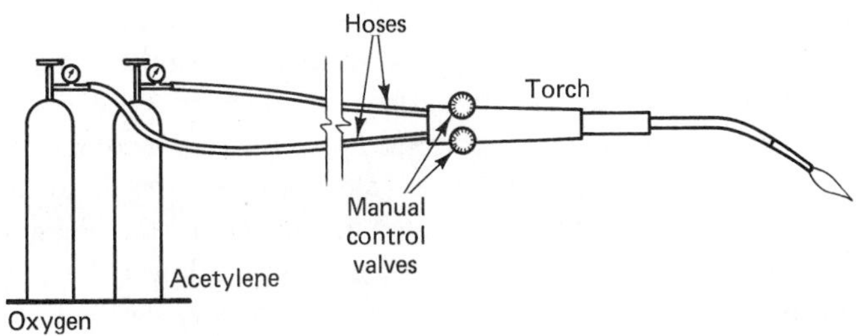

FIGURE 4.34
The equipment required in gas welding

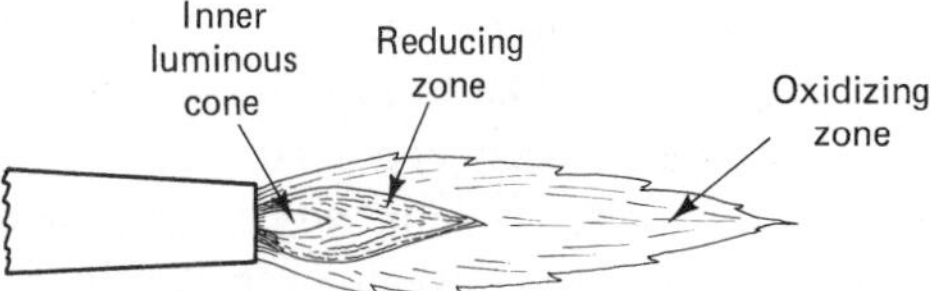

FIGURE 4.35
A sketch of a typical oxyacetylene welding flame

flame, and the outline of the flame is not sharp. This type of flame is employed in welding cast iron and in hard-surfacing with high-speed steel and cemented carbides. The third type of flame, the *oxidizing* flame, is obtained when the oxygen-to-acetylene ratio is higher than 1. In this case, the luminous cone is shorter than that obtained with the neutral flame, and the flame becomes light blue in color. The oxidizing flame is employed in welding brass, bronze, and other metals that have great affinity to hydrogen.

Another method that utilizes the heat generated as a result of the combustion of a fuel gas is known as pressure-gas welding. As the name suggests, this is actually a pressure welding process in which the abutting edges to be welded are heated with an oxyacetylene flame to attain a plastic state; then, coalescence is achieved by applying the appropriately high pressure. In order to ensure uniform heating of the sections, a multiple-flame torch that surrounds the sections is used. The shape of that torch is dependent upon the outer contour of the sections to be welded, and the torch is usually made to oscillate along its axis. Upsetting is accomplished by a special pressure mechanism. This method is sometimes used for joining pipeline mains, rails, and the like.

Electron-beam welding. Electron-beam welding (EBW) was developed by Dr. Jacques Stohr (CEA–France, the atomic energy commission) in 1957 to solve a problem in the manufacturing of fuel elements for atomic power generators. The process is based upon the conversion of the kinetic energy of a high-velocity, intense beam of electrons into thermal energy as the accelerated electrons impact on the joint to be welded. The generated heat then fuses the interfacing surfaces and produces the desired coalescence.

Figure 4.36 shows the basic elements and working principles of an electron-beam welding system. The system consists of an electron-beam gun (simply an electron emitter such as a hot filament) that is electrically placed at a negative potential with respect to an anode and that together with the workpiece is earth-grounded. A focus coil (i.e., an electromagnetic lens) is located slightly below the anode in order to bring the electron beam into focus upon the work. This is achieved by adjusting the current of the focus coil. Additional electromagnetic coils are provided to deflect the beam from its neutral axis as required. Because the electrons impacting the work travel at an ultra-high velocity, the process should be carried out in a vacuum in order to eliminate any resistance to the traveling electrons. Pressures on the order of 10^{-4} torr (1 atmosphere = 760 torr) are commonly employed, although pressures up to almost atmospheric can be used. Nevertheless, it must be noted that the higher the pressure is, the wider and more dispersed the electron beam becomes, and the lower the energy density is. (Energy density is the number of kilowatts per unit area of the spot being welded.)

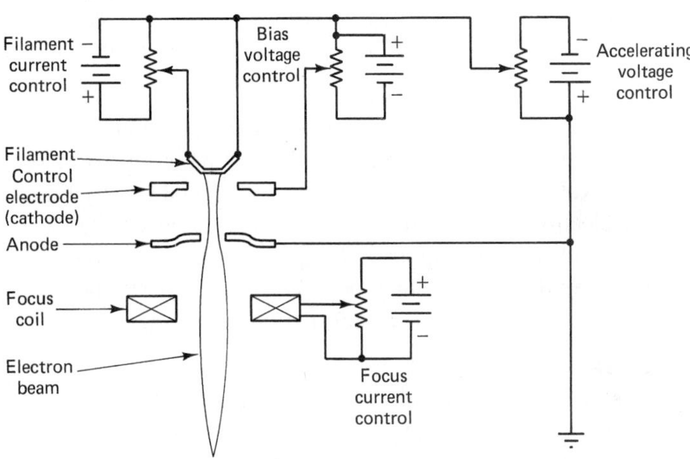

FIGURE 4.36 The basic elements and working principles of an electron-beam welding system

Electron-beam welding machines can be divided into two groups: low-voltage and high-voltage machines. Low-voltage machines are those operating at accelerating voltages up to 60 kV, whereas high-voltage machines operate at voltages up to 200 kV. Although each of these two types has its own merits, the main consideration should be the beam-power density, which is, in turn, dependent upon the beam power and the (focused) spot size. In the early days of electron-beam welding, machines were usually built to have a rating of 7.5 kW and less. Today, a continuous-duty rating of 60 kW is quite common, and the trend is toward still higher ratings.

There are several advantages to the electron-beam welding process. They include the following five:

1. Because of the high intensity of the electron beam used, the welds obtained are much narrower, and the penetration in a single pass is much greater than that obtained by conventional fusion welding processes.
2. The high intensity of the electron beam can also develop and maintain a borehole in the workpiece, thus yielding a parallel-sided weld with a very narrow heat-affected zone. As a consequence, the welds produced by this method have almost no distortion, have minimum shrinkage, and are stronger than welds produced by conventional fusion welding processes.
3. Because parallel-sided welds are obtained by this process, there is no need for edge preparation of the workpieces (such as V- or J-grooves). Square butt-type joints are commonly produced by electron-beam welding.
4. High welding speeds can be obtained with this process. Speeds up to 200 inches per minute (0.09 m/s) are common, resulting in higher productivity.
5. Because the process is usually performed in a vacuum chamber at pressures on the order of 10^{-4} torr, the resulting weld is excellent, is metallurgically clean, and has an extremely low level of atmospheric contamination. Therefore, electron-beam

welding is especially attractive for joining refractory metals whose properties are detrimentally affected by even low levels of contamination.

Because of the ultrahigh quality of the joints produced by electron-beam welding, the process has found widespread use in the atomic power, jet engine, aircraft, and aerospace industries. Nevertheless, the time required to vacuum the chamber before each welding operation results in reduced productivity, and, therefore, the high cost of the electron-beam welding equipment is not easily justified. This apparently kept the process from being applied in other industries until it was automated. Today, electron-beam welding is becoming popular for joining automotive parts such as gear clusters, valves, clutch plates, and transmission components.

Laser-beam welding. The term *laser* stands for light amplified by stimulated emission of radiation. It is, therefore, easy to see that a laser beam is actually a controlled, intense, highly collimated, and coherent beam of light. In fact, a laser beam proved to be a unique source of high-intensity energy that can be used in fusing metals to produce welded joints having very high strength.

Figure 4.37 shows the working principles of laser-beam welding (LBW). In this laser system, energy is pumped into a laser medium to cause it to fluoresce. This fluorescence, which has a single wavelength or color, is trapped in the laser medium (laser tube) between two mirrors. Consequently, it is reflected back and forth in an optical resonator path, resulting in more fluorescence, which amplifies the intensity of the light beam. The amplified light (i.e., the laser beam) finds its way out through the partly transparent mirror, which is called the *output mirror.* The laser medium can be a solid, such as a crystal made of yttrium aluminum garnet (YAG). It can also be a gas, such as carbon dioxide, helium, or neon. In the latter case, the pumping energy input is usually introduced directly by electric current flow.

Let us now consider the mechanics of laser-beam welding. The energy intensity of a laser beam is not high enough to fuse a metal such as steel or copper. Therefore, it must be focused by a highly transparent lens at a very tiny spot, 0.01 inch (0.25 mm) in diameter, in order to increase the intensity of energy up to a level of 10 million W per square inch (15,500 W/mm^2) at the focal point. The impacting laser energy is

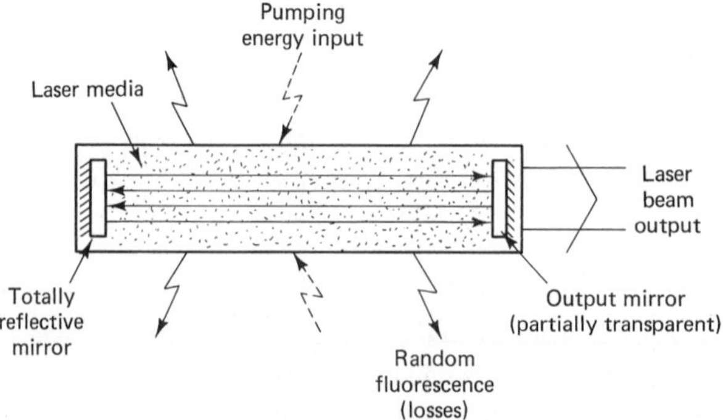

FIGURE 4.37
The working principles of a laser-beam welding system

converted into heat as it strikes the surface of a metal, causing instantaneous fusion of the metal at the focal point. Next, a cylindrical cavity, known as a *keyhole,* that is full of vaporized, ionized metallic gas is formed and is surrounded by a narrow region of molten metal. As the beam moves relative to the workpiece, the molten metal fills behind the keyhole and subsequently cools and solidifies to form the weld. It is worth mentioning that a stream of a cooling (and shielding) gas should surround the laser beam to protect the focusing lens from vaporized metal. Usually, argon is used for this purpose because of its low cost, although helium is actually the best cooling gas.

In spite of the high initial capital cost required, laser-beam welding has gained widespread industrial application because of several advantages that the process possesses. Among these advantages are the following six:

1. Based on the preceding discussion of the mechanics of laser-beam welding, we would always expect to have a very narrow heat-affected zone with this welding method. Consequently, the chemical, physical, and mechanical properties of the base metal are not altered, thus eliminating the need for any postwelding heat treatment.
2. The ultrahigh intensity of energy of the laser beam at the focal point allows metals having high melting points (refractory metals) to be welded.
3. The process can be successfully used to weld both nonconductive as well as magnetic materials that are almost impossible to join even with electron-beam welding.
4. The laser beam can be focused into a chamber through highly transparent windows, thus rendering laser-beam welding suitable for joining radioactive materials and for welding under sterilized conditions.
5. The process can be used for welding some materials that have always been considered unweldable.
6. The process can be easily automated. Numerically controlled laser-beam welding systems are quite common and are capable of welding along a complex contour.

Since the Apollo project, laser-beam welding has become popular in the aerospace industry. Today, the process is mainly employed for joining exotic metals such as titanium, tantalum, zirconium, columbium, and tungsten. The process is especially advantageous for making miniature joints as in tiny pacemaker cans, integrated-circuit packs, camera parts, and batteries for digital watches. Nevertheless, laser-beam welding is not recommended for joining brass, zinc, silver, gold, or galvanized steel.

Welding Defects

In fusion welding processes, considerable thermal stresses develop during heating and subsequent cooling of the workpiece, especially with those processes that result in large heat-affected zones. Also, metallurgical changes and structural transformations take place in the weld puddle as well as in the heat-affected zone, and these may be accompanied by changes in the volume. Therefore, if no precautions are taken, defects that are damaging to the function of the weldment may be generated. It is the combined duty of the manufacturing engineer, the welder, and the inspector to make sure that all weldments are free from all kinds of defects. Following is a brief survey of the common kinds of welding defects.

Distortion. Distortion, warping, and buckling of the welded parts are welding defects involving deformation (which can be plastic) of the structures as a result of residual stresses. They come as a result of restraining the free movement of some parts or members of the welded structure. They can also result from nonuniform expansion and shrinkage of the metal in the weld area as a consequence of uneven heating and cooling. Although it is possible to predict the magnitude of the residual stresses in some simple cases (e.g., butt welding of two plates), an analysis to predict the magnitude of these stresses and to eliminate distortion in the common case of a welded three-dimensional structure is extremely complicated. Nevertheless, here are some recommendations and guidelines to follow to eliminate distortion:

1. Preheat the workpieces to a temperature dependent on the properties of the base metal in order to reduce the temperature gradient.
2. Clamp the various elements (to be welded) in a specially designed rigid welding fixture. Although no distortion occurs with this method, there are always inherent internal stresses. The internal stresses can be eliminated by subsequent stress-relieving heat treatment.
3. Sometimes, it is adequate just to tack-weld the elements securely in the right position (relative to each other) before actual-strength welds are applied. It is also advisable to start by welding the section least subject to distortion first in order to form a rigid skeleton that contributes to the balance of assembly.
4. Create a rational design of weldments (e.g., apply braces to sections most likely to distort).

Porosity. Porosity can take the form of elongated blowholes in the weld puddle, which is known as *wormhole porosity,* or of scattered tiny spherical holes. In both cases, porosity is due mainly to either the evolution of gases during welding or the release of gases during solidification as a result of their decreasing solubility in the solidifying metal. Excess sulfur or sulfide inclusions in steels are major contributors to porosity because they generate gases that are often entrapped in the molten metal. Other causes of porosity include the presence of hydrogen (remember the problem caused by hydrogen in casting), contamination of the joint, and contaminants in the flux. Porosity can be eliminated by maintaining clean workpiece surfaces, by properly conditioning the electrodes, by reducing welding speed, by eliminating any moisture on workpieces, and, most importantly, by avoiding the use of a base metal containing sulfur or electrodes with traces of hydrogen.

Cracks. Welding cracks can be divided into two main groups: fusion zone cracks and heat-affected zone cracks. The first group includes longitudinal and transverse cracks as well as cracks appearing at the root of the weld bead. This type of cracking is sometimes called *hot cracking* because it occurs at elevated temperatures just after the molten metal starts to solidify. It is especially prevalent in ferrous alloys with high percentages of sulfur and phosphorus and in alloys having large solidification ranges.

The second type of cracking, heat-affected zone cracks, is also called *cold cracking*. This defect is actually due to aggravation by excessive brittleness of the heat-affected zone that can be caused by hydrogen embrittlement or by martensite formation as a result of rapid cooling, especially in high-carbon and alloy-steel welded joints. (Remember the effect of alloying elements on the TTT curve; they shift it to the right, thus decreasing the critical cooling rate.) Cold cracks can be eliminated by using a minimum potential source of hydrogen and by controlling the cooling rate of the welded joint to keep it at a minimum (e.g., keep joints in a furnace after welding or embed them in sand).

The use of multiple passes in welding can sometimes eliminate the need for prewelding or postwelding heat treatment. Each pass would provide a sort of preheating for the pass to follow. This technique is often effective in the prevention of weld cracks.

Slag inclusions. Slag entrapment in the weld zone can occur in single-pass as well as in multipass welds. In single-pass arc welding, slag inclusions are caused by improper manipulation of the electrode and/or factors such as too high a viscosity of the molten metal or too rapid solidification. Some slag pushed ahead of the arc is drawn down by turbulence into the molten-metal pool, where it becomes entrapped in the solidifying weld metal. In multipass welds, slag inclusions are caused by improper removal of the slag blanket after each pass.

Lack of fusion. Lack of fusion, shown in Figure 4.38, can result from a number of causes. These include inadequate energy input, which leads to insufficient temperature rise; improper electrode manipulation; and failure to remove oxide films and clean the weld area prior to welding.

Lack of penetration. Lack of penetration, shown in Figure 4.39, is due to a low energy input, the wrong polarity, or a high welding speed.

Undercutting. Undercutting, shown in Figure 4.40, is a result of a high energy input (excessive current in arc welding), which, in turn, causes the formation of a recess. As we know, such sharp changes in the weld contour act as stress raisers and often cause premature failure.

FIGURE 4.38
Lack of fusion

FIGURE 4.39
Lack of penetration

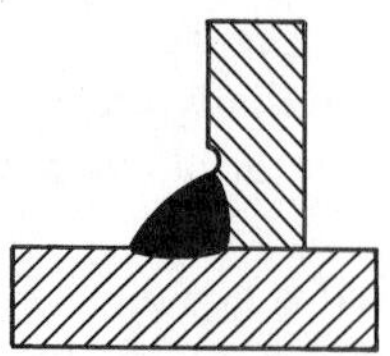

FIGURE 4.40
Undercutting

FIGURE 4.41
Underfilling

Underfilling. Underfilling, shown in Figure 4.41, involves a depression in the weld face below the surface of the adjoining base metal. More filler metal has to be added in order to prevent this defect.

Testing and Inspection of Welds

Welds must be evaluated by being subjected to testing according to codes and specifications that are different for different countries. The various types of tests can be divided into two groups: destructive and nondestructive. Destructive testing always results in destroying the specimen (the welded joint) and rendering it unsuitable for its design function. Destructive tests can be mechanical, metallurgical, or chemical. We next review various destructive and nondestructive testing methods.

Visual inspection. Visual inspection involves examination of the weld by the naked eye and checking its dimensions by employing special gages. Defects such as cracks, porosity, undercuts, underfills, or overlaps can be revealed by this technique.

Mechanical tests. Mechanical tests are generally similar to the conventional mechanical tests, the difference being the shape and size of the test specimen. Tensile, bending, impact, and hardness tests are carried out. Such tests are conducted either on the whole welded joint or on the deposited metal only.

Metallurgical tests. Metallurgical tests involve metallurgical microstructure and macrostructure examination of specimens. Macrostructure examination reveals the depth of penetration, the extent of the heat-affected zone, and the weld bead shape, as well as hidden cracks, porosity, and slag inclusions. Microstructure examination can show the presence of nitrides, martensite, or other structures that cause metallurgically oriented welding problems.

Chemical tests. Chemical tests are carried out to ensure that the composition of the filler metal is identical to that specified by the manufacturing engineer. Some are crude tests, such as spark analysis or reagent analysis; however, if accurate data are required, chemical analysis or spectrographic testing must be carried out.

Radiographic inspection. Radiographic inspection is usually performed by employing industrial X rays. This technique can reveal hidden porosity, cracks, and slag inclusions. It is a nondestructive test that does not destroy the welded joint. High-penetration X rays are sometimes also employed for inspecting weldments having thicknesses up to 1½ inches (37 mm).

Pressure test. Hydraulic (or air) pressure is applied to welded conduits that are going to be subjected to pressure during their service lives to check their tightness and durability.

Ultrasonic testing. Ultrasonic waves with frequencies over 20 kHz are employed to detect various kinds of flaws in the weld, such as the presence of nonmetallic inclusions, porosity, and voids. This method is reliable even for testing very thick parts.

Magnetic testing. As we know from physics, the lines of magnetic flux are distorted in such a way as to be concentrated at the sides of a flaw or a discontinuity, as seen in Figure 4.42a and b. This test, therefore, involves magnetizing the part and then using fine iron-powder particles that were uniformly dispersed on the surface of the part to reveal the concentration of the flux lines at the location of the flaw. This method is successful in detecting superficial hair cracks and pores in ferrous metal.

Ammonia penetrant test. The ammonia penetrant test is used to detect any leakage from welded vessels. It involves filling the vessel with a mixture of compressed air and ammonia and then wrapping it with paper that has been impregnated in a 5 percent solution of mercuric nitrate. Any formation of black spots is an indication of leakage.

Fluorescent penetrant test. The part is immersed for about half an hour in oil (or an oil mixture) and then dipped in magnesia powder. The powder adheres at any crack location.

Design Considerations

As soon as the decision is made to fabricate a product by welding, the next step is to decide which welding process to use. This decision should be followed by selection of the types of joints, by determination of the locations and distribution of the welds, and, finally, by making the design of each joint. Following is a brief discussion of the factors to be considered in each design stage.

Selection of the joint type. We have previously discussed the various joint designs and realized that the type of joint depends upon the thickness of the parts to be welded. In fact, there are other factors that should also affect the process of selecting a particular type of joint. For instance, the magnitude of the load to which the joint is going

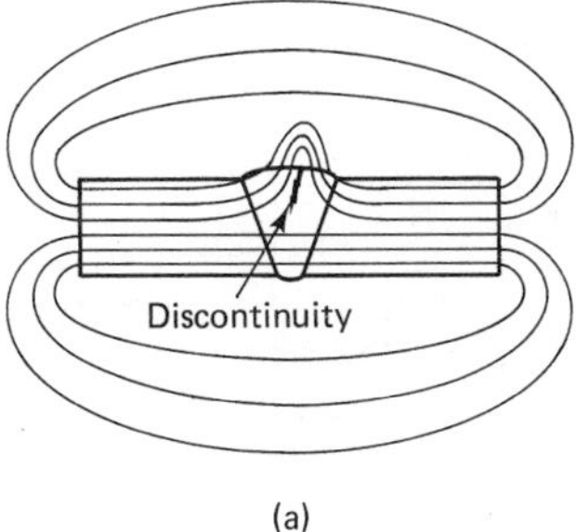

(a)

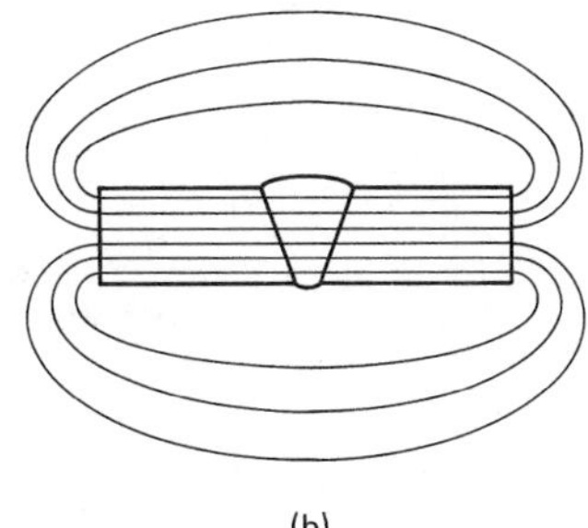

(b)

FIGURE 4.42 Magnetic testing of welds: (a) defective weld; (b) sound weld

to be subjected during its service life is one other important factor. The manner in which the load is applied (i.e., impact, steady, or fluctuating) is another factor. Whereas the square butt, simple-V, double-V, and simple-U butt joints are suitable only for usual loading conditions, the double-U butt joint is recommended for all loading conditions. On the other hand, the square-T joint is appropriate for carrying longitudinal shear under steady-state conditions. When severe longitudinal or transverse loads are anticipated, other types of joints (e.g., the single-bevel-T, the double-bevel-T, and the double-J) have to be considered. In all cases, it is obvious that cost is the decisive factor whenever there is a choice between two types of joints that would function equally well.

Location and distribution of welds. It has been found that the direction of the linear dimension of the weld with respect to the direction of the applied load has an effect on the strength of the weld. In fact, it has been theoretically and experimentally proven that a lap weld whose linear direction is normal to the direction of the applied load, as is shown in Figure 4.43a, is 30 percent stronger than a lap weld whose linear direction is parallel to the direction of the applied load, as shown in Figure 4.43b. In the first case, the maximum force F that the joint can carry without any signs of failure can be approximated by the following equation:

$$F = 0.707 \times \ell \times W \times \sigma_{\text{allowable}} \tag{4.3}$$

where: ℓ is the weld leg

W is the length of the weld

$\sigma_{\text{allowable}}$ is the allowable tensile stress of the filler material (e.g., electrode)

In the second case (Figure 4.43b), the strength of the joint is based on the fact that the throat plane of the weld is subjected to pure shear stress and is given by the following equation:

$$F = 0.707 \times \ell \times W \times \tau_{\text{allowable}} \tag{4.4}$$

where: ℓ is the weld leg

W is the length of the weld

$\tau_{\text{allowable}}$ is the allowable shear stress of the electrode

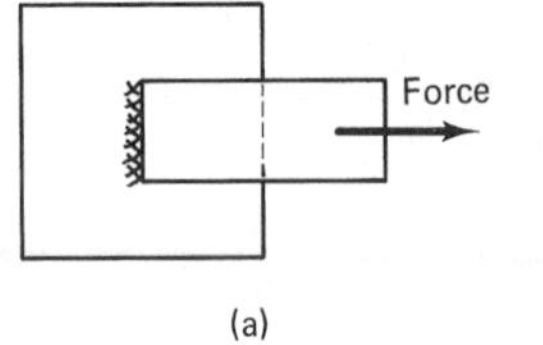

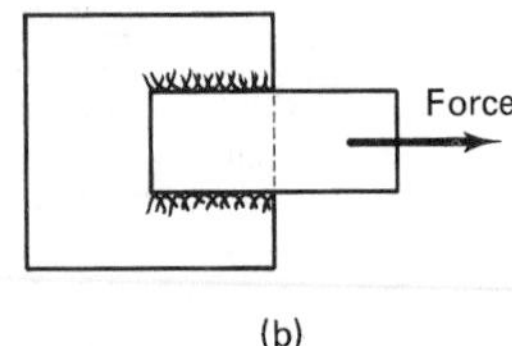

FIGURE 4.43 Location and distribution of welds: (a) weld linear direction normal to the applied load; (b) weld linear direction parallel to the applied load

From the theory of plasticity, assuming you adopt the same safety factor in both cases, it is easy to prove that

$$\tau_{\text{allowable}} = \frac{1}{\sqrt{3}} \sigma_{\text{allowable}} = 0.565\ \sigma_{\text{allowable}} \tag{4.5}$$

On the other hand, the strength of a butt-welded joint can be given by the following equation:

$$F = \ell \times W \times \sigma_{\text{allowable}} \tag{4.6}$$

where ℓ, W, and $\sigma_{\text{allowable}}$ are as previously mentioned. A product designer should, therefore, make use of this characteristic when planning the location and distribution of welds.

Another important point to consider is the prevention of any tendency of the welded elements to rotate when subjected to mechanical loads. A complete force analysis must be carried out in order to determine the proper length of each weld. Let us now consider a practical example to see the cause and the remedy for this tendency to rotate. Figure 4.44 shows an L angle welded to a plate. Any load applied through the angle will pass through its center of gravity. Therefore, the resisting forces that act through the welds will not be equal; the force closer to the center of gravity of the angle will always be larger. Consequently, if any tendency to rotate is to be prevented, the weld that is closer to the center of gravity must be longer than the other one. Using simple statics, it can easily be seen that

$$\frac{W_1}{W_2} = \frac{d_2}{d_1} \tag{4.7}$$

It is also recommended that very long welds be avoided. It has been found that two small welds, for example, are much more effective than a single long weld.

Joint design. In addition to the procedures and rules adopted in common design practice, there are some guidelines that apply to joint design:

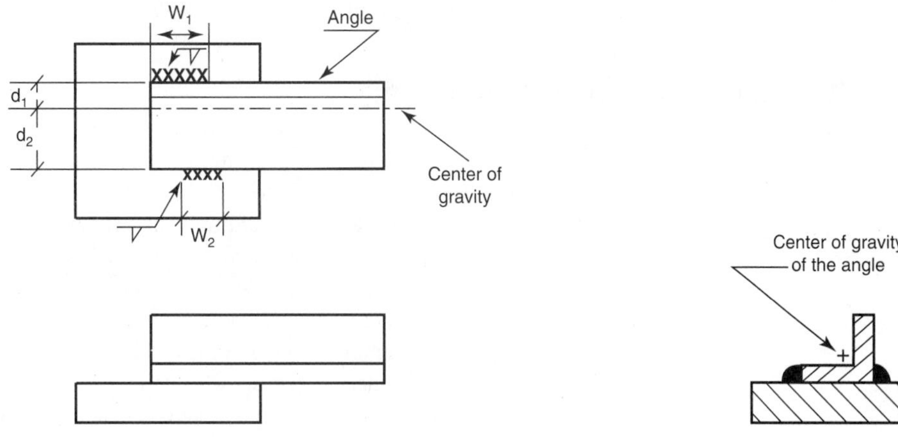

FIGURE 4.44 Preventing the tendency of the welded element to rotate by appropriate distribution of welds

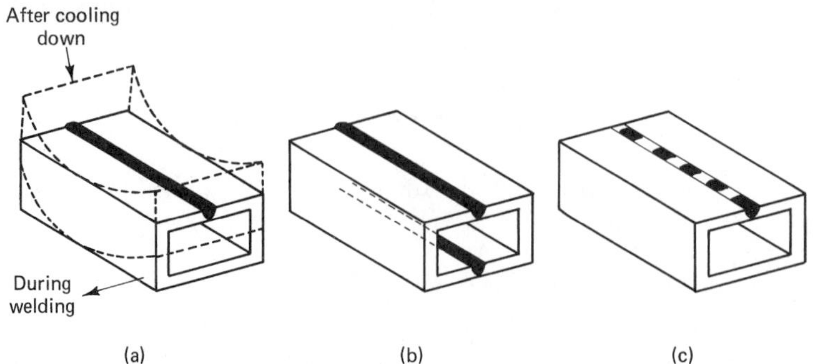

FIGURE 4.45
Designs that promote or eliminate distortion in welding:
(a) distortion caused by unbalanced weld;
(b) and (c) methods for reducing distortion

1. Try to ensure accessibility to the locations where welds are to be applied.
2. Try to avoid overhead welding.
3. Consider the heating effect on the base metal during the welding operation. Balance the welds to minimize distortion. Use short, intermittent welds. Figure 4.45a shows distortion caused by an unbalanced weld, whereas Figure 4.45b and c shows methods for reducing that distortion.
4. Avoid crevices around welds in tanks as well as grooves (and the like) that would allow dirt to accumulate. Failure to do so may result in corrosion in the welded joint.
5. Do not employ welding to join steels with high hardenability.
6. Do not weld low-carbon steels to alloy steels by the conventional fusion welding methods because they have different critical cooling rates and hence cannot be successfully welded.
7. When employing automatic welding (e.g., submerged arc), the conventional joint design of manual welding should be changed. Wider Vs (for butt joints) are used, and single-pass welds replace multipass welds.

4.3 SURFACING AND HARD-FACING

Surfacing involves the application of a thin deposit on the surface of a metallic workpiece by employing a welding method such as oxyacetylene-gas welding, shielded-metal arc welding, or automatic arc welding. The process is carried out to increase the strength, the hardness, and the resistance to corrosion, abrasion, or wear. For the last reason, the process is commonly known as *hard-facing*.

Good hard-facing practice should be aimed at achieving a strong bond between the deposit and the base metal and also at preventing the formation of cracks and other defects in the deposited layer. Therefore, the deposited layer should not generally exceed 3/32 inch (2 mm) and will rarely exceed 1/4 inch (6 mm). Also, the base metal should be heated to a temperature of 500°F to 950°F (350°C to 500°C) to ensure a good metallurgical bond and to allow the deposited layer to cool down slowly.

Hard-facing permits the use of very hard wear- and corrosion-resisting compounds. The materials used in this process are complex. They involve hard compounds, like carbides and borides, that serve as the wear-resisting elements, and a tough matrix composed of air-hardening steel or iron-base alloys. Such deposited materials increase the service life of a part three- or fourfold. The process is also employed in restoring worn parts.

The process of hard-facing has found widespread application in the heavy construction equipment industry, in mining, in agricultural machinery, and in the petroleum industry. The list of parts that are usually hard-faced is long and includes, for example, the vulnerable surfaces of chemical-process vessels, pump liners, valve seats, drive sprockets, ripper teeth, shovel teeth, chutes, and the edges of coal recovery augers.

4.4 THERMAL CUTTING OF METALS

In this section, we discuss the *thermal cutting* of metals, specifically oxyfuel flame cutting and the different arc cutting processes. Although all these processes do not, by any means, fall under the topic of *joining* (the action involved is opposite to that of joining), they employ the same equipment as the corresponding welding process in each case. The thermal cutting processes are not alternatives to sawing but rather are used for cutting thick plates, 1 to 10 inches (25 to 250 mm) thick, as well as for difficult-to-machine materials. Thermal cutting may be manual, using a hand-operated cutting torch (or electrode), or the cutting element can be machine driven by a numerically controlled system or by special machines called *radiographs.*

Oxyfuel Cutting

Oxyfuel cutting (OFC) is similar to oxyfuel welding except that an oxidizing flame must always be used. The process is extensively used with ferrous metal having thicknesses up to 10 inches (250 mm). During the process, red-hot iron, directly subjected to the flame, is oxidized by the extra oxygen in the flame; it then burns up, leaving just ashes or slag. Also, the stream of burning gases washes away any molten metal in the region being cut. Generally, there is a relationship between the speed of travel of the torch or electrode and the smoothness of the cut edge: The higher the speed of travel, the coarser the cut edge.

Although acetylene is commonly used as a fuel in this process, other gases are also employed, including butane, methane, propane, natural gas, and a newly developed gas with the commercial name Mapp. Hydrogen is sometimes used as a fuel, especially underwater to provide a powerful preheating flame. In this case, compressed air is used to keep water away from the flame.

The oxyfuel cutting process can be successfully employed only when the ignition temperature of the metal being cut is lower than its melting point. Another condition for the successful application of the process involves ensuring that the melting points of the formed oxides are lower than that of the base metal itself. Therefore, oxyfuel cutting is not recommended for cast iron because its ignition temperature is higher than its melting point. The process is also not appropriate for cutting stainless steel, high-alloy chromium and chrome-nickel alloys, and nonferrous alloys because the melting points of the oxides of these metals are higher than the melting points of the metals themselves.

Arc Cutting

There are several processes based upon utilization of the heat generated by an electric arc. These arc cutting processes are generally employed for cutting nonferrous metals, medium-carbon steel, and stainless steel.

Conventional arc cutting. Conventional arc cutting is similar to shielded-metal arc welding. It should always be remembered, however, that the electrode enters the gap of the cut, so the coating must serve as an insulator to keep the electric arc from shorting out. Consequently, electrodes with coatings containing iron powder are not recommended for use with this process.

Air arc cutting. Air arc cutting involves preheating the metal to be cut by an electric arc and blowing out the resulting molten metal by a stream of compressed air. The arc-air torch is actually a steel tube through which compressed air is blown.

Oxygen arc cutting. Oxygen arc cutting is similar to air arc cutting except that oxygen is blown instead of air. The process is capable of cutting cast irons and stainless steels with thicknesses up to 2 inches (50 mm).

Carbon arc cutting. In carbon arc cutting, a carbon or graphite electrode is used. The process has the disadvantage of consuming that electrode quickly, especially if continuous cutting is carried out.

Tungsten arc cutting. The electrode used in tungsten arc cutting is made of tungsten and has, therefore, a service life that is far longer than that of the carbon or graphite electrodes. Tungsten arc cutting is commonly employed for stainless steel, copper, magnesium, and aluminum.

Air-carbon arc cutting. Air-carbon arc cutting is quite similar to carbon arc cutting, the difference being the use of a stream of compressed air to blow the molten metal (that has been fused by the arc) out of the kerf (groove). The process cuts almost all metals because its mechanics involve oxidation of the metal. Its applications involve removal of welds, removal of defective welds, and dismantling of steel structures.

Plasma arc cutting. A plasma arc is employed to cut metals in plasma arc cutting (PAC). The temperature of the plasma jet is extremely high (ten times higher than that obtained with oxyfuel), thus enabling high-speed cutting rates to be achieved. Also, as a consequence, the heat-affected zone formed along the edge of the kerf is usually less than 0.05 inch (1.3 mm). Plasma arc cutting can be used for cutting stainless steel as well as hard-to-cut alloys. A modification of the process involves using a special nozzle to generate a whirlpool of water on the workpiece, thus increasing the limit on the

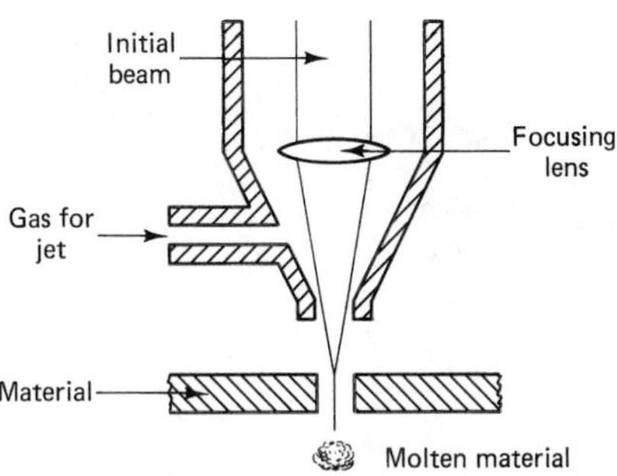

FIGURE 4.46
Laser-beam cutting of sheets and plates

thickness of the workpiece up to 3 inches (75 mm) and meanwhile improving the quality of the cut. The only limitation on plasma arc cutting is that the workpiece must be electrically conductive.

Laser-beam cutting. The basic principles of laser-beam cutting are similar to those of laser-beam welding. Nevertheless, laser cutting is achieved by the pressure from a jet of gas that is coaxial with the laser beam, as shown in Figure 4.46. The function of the gas jet is to blow away the molten metal that has been fused by the laser beam. Laser beams can be employed in cutting almost any material, including nonconductive polymers and ceramics. Also, the process is usually automated by using computerized numerical control systems to control the movements of the machine table under the laser beam so that workpieces can be cut to any desired contour. Other advantages of the laser-beam cutting process include the straight-edged kerfs obtained, the very narrow heat-affected zone that results, and the elimination of the part distortion experienced with other conventional thermal cutting processes.

4.5 BRAZING AND SOLDERING

Brazing and *soldering* are processes employed for joining solid metal components by heating them to the proper temperature and then introducing between them a molten filler alloy (brazing metal or solder). The filler alloy must always have a melting point lower than that of the base metal of the components. The filler alloy must also possess high fluidity and wettability (i.e., be able to spread and adhere to the surface of the base metal). As you may expect, the mechanics of brazing or soldering are different from those of welding. A strong brazed joint is obtained only if the brazing metal can diffuse into the base metal and form a solid solution with it. Figure 4.47a and b is a sketch of the microstructure of two brazed joints and is aimed at clarifying the mechanics of brazing and soldering.

Brazing and soldering can be employed to join carbon and alloy steels, nonferrous alloys, and dissimilar metals. The parts to be joined together must be carefully cleaned, degreased, and clamped. Appropriate flux is applied to remove any remaining oxide and to prevent any further oxidation of the metals. It is only under

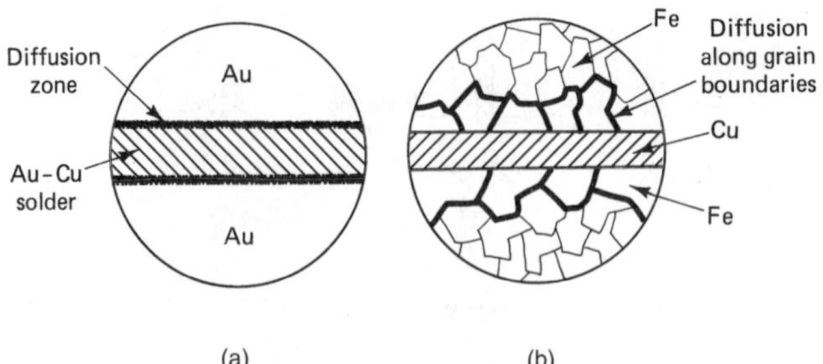

FIGURE 4.47 A sketch of the microstructure of two brazed joints: (a) gold base metal; (b) low-carbon steel base metal

such conditions that the filler metal can form a strong metallic bond with the base metal.

The main difference between soldering and brazing is the melting point of the filler metal in each case. Soft solders used in soldering have melting points below 930°F (500°C) and produce joints with relatively low mechanical strength, whereas hard solders (brazing metals) have higher melting points, up to 1650°F (900°C), and produce joints with high mechanical strength.

Soft solders are low-melting-point eutectic alloys. They are basically tin, lead, cadmium, bismuth, or zinc alloys. On the other hand, brazing filler metals are alloys consisting mainly of copper, silver, aluminum, magnesium, or nickel. Table 4.2 gives the recommended filler alloys for different base metals. The chemical composition and the field of application for the commonly used soft and hard solders are given in Table 4.3.

TABLE 4.2
Recommended filler alloys for different base metals

	Filler Alloys				
Base Metal	**Tin-Lead**	**Silver Solder**	**40% Zinc Brass**	**Nickel Silver**	**Copper**
Steel	*	*	×	*	×
Stainless steel	*	×	*	×	
Nickel alloys	*	*	×	×	
Copper (pure)	*	×	×		
Brass or bronze	*	×			
Silver (pure)	*	×		Not allowed	
Aluminum	Either Sn-Zn 10				
Al-Mg 3, Al-Si 5	or Al-Si 12				

* = may be used
× = best used

TABLE 4.3
Most commonly used soft and hard brazing filler metals

Type	Approximate Chemical Composition	Brazing or Soldering Temperature	Application
Tin solder	Sn-Pb 70	360–490°F (183–255°C)	General-purpose solder
	Sn-Pb 50	360–420°F (183–216°C)	Electrical application
	Sn-Zn 10	509–750°F (265–400°C)	Soft soldering of aluminum
Silver solder	Ag 25-50, Cu 20-40, Sn 0-35, Cd 0-20, Zn 0-20	1175–1550°F (635–845°C)	Brazing of copper alloys and silver in electronics
Brass solder	Cu-Zn 40	1670–1750°F (910–955°C)	General-purpose hard solder
Nickel silver	Cu balance, Zn 20-30, Ni 10-20	1720–1800°F (938–982°C)	Nickel alloys and steel brazing
Copper	99.9% copper	2000–2100°F (1093–1150°C)	Brazing of steel
Silumin	Al-Si 12	1080–1120°F (582–605°C)	Brazing all aluminum alloys except silumin

Fluxes

Fluxes are employed in soft soldering as well as in brazing in order to protect the cleaned surfaces of the base metal against oxidation during those processes. In addition, fluxes enable proper wetting of the surfaces of the base metals by the molten filler solders.

There are two kinds of fluxes for soft soldering operations: organic and inorganic. The inorganic fluxes are mostly aqueous solutions of zinc and/or ammonium chlorides. They must, however, be completely removed after the soldering operation because of their corrosive effect. It is, therefore, completely forbidden to use inorganic fluxes in soldering electronic components. On the other hand, organic fluxes do not have such corrosive effects and are, therefore, widely used for fine soldering in electronic circuits. The commonly used organic fluxes involve colophony, a kind of resin with a melting point between 350°F and 390°F (180°C and 200°C), as well as some fats.

The fluxes employed in brazing include combinations of borox, boric acid, borates, fluorides, and fluoborates together with a wetting agent. The flux can be in the form of a liquid, slurry, powder, or paste, depending upon the brazing method used.

Soldering Techniques

The manual soldering method involves using a hand-type soldering iron that is made of copper and has to be tinned each time before use. The iron is first heated to a temperature of about 570°F (300°C), and its tip is then dipped into the flux and tinned with the solder. Next, the iron is used for heating the prepared surfaces of the base metal and for melting and distributing the soft solder. When the solder solidifies, it forms the required solder seam.

Several other methods are also used for soldering. These include dip soldering and induction soldering as well as the use of guns (blowtorches). Nevertheless, electric soldering irons are still quite common.

Brazing Techniques

The selection of a preferred brazing method has to be based on the size and shape of the joined components, the base metal of the joint, the brazing filler metal to be used, and the production rate. When two brazing techniques are found to be equally suitable, cost is the deciding factor. The following brazing methods are commonly used in industry.

Torch brazing. Torch brazing is still the most commonly used method. It is very similar to oxyfuel flame welding in that the source of heat is a flame obtained from the combustion of a mixture of a fuel gas (e.g., acetylene) and oxygen. The process is very popular for repair work on cast iron and is usually applied manually, although it can be used on a semiautomatic basis. In this process, however, a reducing flame should be used to heat the joint area to the appropriate brazing temperature. A flux is then applied, and as soon as it melts, the filler metal (brazing alloy) is hand-fed to the joint area. When the filler metal melts, it flows into the clearance between the base components by capillary attraction. The filler metal should always be melted by the heat gained by the joint and not by directly applying the flame.

Furnace brazing. Furnace brazing is performed in either a batch or a continuous conveyor-type furnace and is, therefore, best suited for mass production. The atmosphere of the furnace is controlled to prevent oxidation and to suit the metals involved in the process. That atmosphere can be dry hydrogen, dissociated ammonia, nitrogen, argon, or any other inert gas. Vacuum furnaces are also employed, especially with brazing materials containing titanium or aluminum. Nevertheless, a suitable flux is often employed. The filler metal must be placed in the joint before the parts go inside the furnace. The filler metal can, in this case, take the form of a ring, washer, wire, powder, or paste.

Induction brazing. In induction brazing, the components to be brazed are heated by placing them in an alternating magnetic field that, in turn, induces an alternating current in the components that rapidly reverses its direction. Special coils made of copper, referred to as *inductors,* are employed for generating the magnetic field. The filler metal is often placed in the joint area before brazing but can also be hand-fed by the operator. This technique has a clear advantage, which is the possibility of obtaining a very closely controlled heating area.

Dip brazing. Dip brazing involves dipping the joint to be brazed in a molten filler metal. The latter is maintained in a special externally heated crucible and is covered with a layer of flux to protect it from oxidation. Because the filler metal coats the entire workpiece, this process is used only for small parts.

Salt-bath brazing. The source of heating in salt-bath brazing is a molten bath of fluoride and chloride salts. The filler metal is placed in the joint area before brazing and is also sometimes cladded. Next, the whole assembly is preheated to an appropriate temperature and then dipped for 1 to 6 minutes in the salt bath. Finally, the hot brazed joint is rinsed thoroughly in hot and cold water to remove any remaining flux or salt. Generally, this process is employed for brazing aluminum and its alloys. There is, however, a problem associated with the process, and that is the pollution caused by the effluent resulting from the rinsing operation.

Resistance brazing. Low-voltage, high-amperage current is used as the source of energy in resistance brazing, as is the case with spot welding. In fact, a spot welder can be employed to carry out this process, provided that the pressure is carefully adjusted so as to be just enough to secure the position of the contact where heat develops. The workpiece is held between the two electrodes, with the filler metal preloaded at the joint area. This process is normally used for brazing of electrical contacts and in the manufacture of copper transformer leads.

Design of Brazed Joints

For the proper design of brazed joints, two main factors have to be taken into consideration. The first factor involves the mechanics of the process in that the brazing filler metal flows through the joint by capillary attraction. The second factor is that the strength of the filler metal is poorer than that of the base metals. The product designer should aim for the following:

1. Ensuring that the filler metal is placed on one side of the joint and allocating a space for locating the filler metal before (or during) the process.
2. Adjusting the joint clearance in order to ensure optimum conditions during brazing. That clearance is dependent upon the filler metal used and normally takes a value less than 0.005 inch (0.125 mm), except for silumin, in which case it can go up to 0.025 inch (0.625 mm).
3. Ensuring that the distance to be traveled by the filler metal is shorter than the limit distant, as dictated by the physics of capillarity.
4. Providing enough filler metal.
5. Increasing the area of the joint because the filler metal is weaker than the base metal.

There are three types of joint-area geometries: butt, scarf, and lap. The butt joint is the weakest, and the lap is the strongest. Nevertheless, when designing lap joints, make sure that the joint overlap is more than $3t$, where t is the thickness of the thinner parent metal. Examples of some good and poor practices in the design of brazed joints are shown in Figure 4.48 as guidelines for beginners in product design. Also, always remember that brazed joints are designed to carry shear stress and not tension.

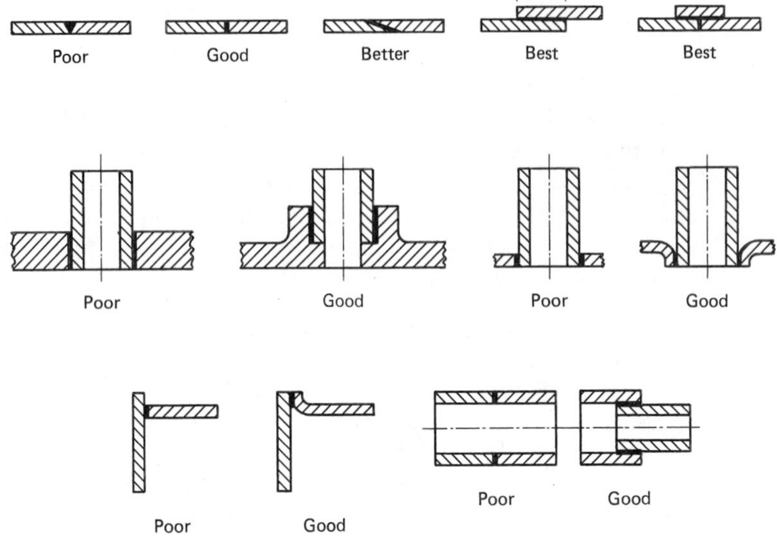

FIGURE 4.48
Good and poor practices in the design of brazed joints

4.6 STICKING OF METALS

Sticking, or *adhesive bonding,* of metals is becoming very popular in the automotive, aircraft, and packaging industries because of the advantages that this technique can offer. Thanks to the recent development in the chemistry of polymers, adhesives are now cheap, can be applied easily and quickly, and can produce reasonably strong joints. Adhesive bonding can also be employed in producing joints of dissimilar metals or combinations of metals and nonmetals like ceramics or polymers. This certainly provides greater flexibility when designing products and eliminates the need for complicated, expensive joining processes.

As we know, it is possible to stick entirely smooth metal surfaces together. It is obvious, therefore, that the sticking action is caused by adhesive forces between the sticking agent and the workpiece and not by the flowing and solidification of the sticking agent into the pores of the workpiece as occurs, for instance, with wood. In other words, adhesion represents attractive intermolecular forces under whose influence the particles of a surface adhere to those of another one. There are also many opinions supporting the theory that mechanical interlocking plays a role in bonding.

Adhesives

Structural adhesives are normally systems including one or more polymeric materials. In their unhardened state (i.e., before they are applied and cured), these adhesives can take the form of viscous liquids or solids with softening temperatures of about 212°F (100°C). The unhardened adhesive agents are often soluble in ketones, esters, and higher alcohols, as well as in aromatic and chlorine hydrocarbons. The hardened adhesives, however, resist nearly all solvents. Adhesives that find industrial application in bonding two nonmetallic workpieces include cyanacrylates, acrylics, and polyurethanes. Following is a brief description of the adhesives that are commonly used in industry.

Epoxies. Epoxies are thermosetting polymers (see Chapter 8) that require the addition of a hardener or the application of heat so that they can be cured. Epoxies are considered to be the best sticking agents because of their versatility, their resistance to solvents, and their ability to develop strong and reliable joints.

Phenolics. Phenolics are characterized by their low cost and heat resistance of up to about 930°F (500°C). They can be cured by a hardener or by heat or can be used in solvents that evaporate and thus allow setting to occur. Like epoxies, phenolics are thermosetting polymers with good strength, but they generally suffer from brittleness.

Polyamide. The polyamide group of polymers is characterized by its oil and water resistance. Polyamides are usually applied in the form of hot melts but can also be used by evaporation of solvents in which they have been dissolved. Polyamides are normally used as can-seam sealants and the like. They are also used as hot-melt for shoes.

Silicones. Silicones can perform well at elevated temperatures; however, cost and strength are the major limitations. Therefore, silicones are usually used as high-temperature sealants.

Joint Preparation

The surfaces to be bonded must be clean and degreased because most adhesives do not amalgamate with fats, oils, or wax. Joint preparation involves grinding with sandpaper, grinding and filling, sand blasting, and pickling and degreasing with trichlorethylene. Oxide films, electroplating coats, and varnish films need not be removed (as long as they are fixed to the surface). Roughening of the surface is advantageous, provided that it is not overdone.

Joint Design

There are basically three types of adhesive-bonded joints. They are shown in Figure 4.49 and include tension, shear, and peel joints. Most of the adhesives are weaker in peel and tension than in shear. Therefore, when selecting an adhesive, you should always keep in mind the types of stresses to which the joint is going to be subjected. It is also recommended that you avoid using tension and peel joints and change the design to replace these by shear joints whenever possible.

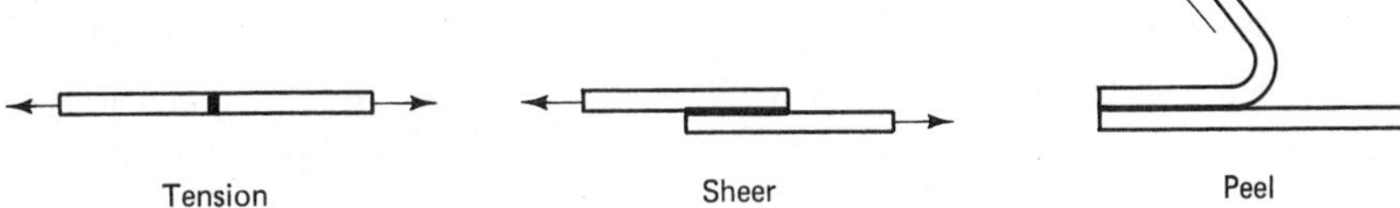

FIGURE 4.49 The three types of adhesive-bonded joints

Review Questions

1. What does the riveting process involve?
2. What are rivets usually made of?
3. List some applications of riveting.
4. Spot welding could not completely replace riveting. List some applications of riveting that cannot be done by spot welding.
5. How would you define *welding?*
6. List five types of welded joint designs and discuss suitable applications for each type.
7. What are the types of different methods for classifying the welding processes?
8. How would you break all the manufacturing methods into groups according to each of these classifying methods?
9. Explain briefly the mechanics of solid-state welding.
10. What are the two main obstacles that must be overcome so that successful pressure welding can be achieved?
11. What is cold-pressure welding? Give two examples, using sketches.
12. Discuss briefly the mechanics of explosive welding and draw a sketch to show the interface between the welded parts.
13. List some industrial applications for explosive welding.
14. Discuss briefly the mechanics of ultrasonic welding.
15. What are the typical applications of ultrasonic welding?
16. What are the different types of ultrasonic welding machines? List the main components common to all these machines.
17. What is the basic idea on which friction welding is based?
18. Explain briefly the stages involved in a friction welding operation.
19. List the various advantages claimed for friction welding.
20. List the limitations of friction welding.
21. What is the difference between friction welding and inertia welding?
22. What advantages does inertia welding have over friction welding?
23. Give examples of some parts that are fabricated by inertia welding.
24. Explain briefly the mechanics of induction welding.
25. List some of the common industrial applications of the induction welding process.
26. What is the source of energy in thermit welding? Explain.
27. How is thermit welding performed?
28. What are the common applications of thermit welding?
29. How does bonding take place in diffusion bonding?
30. List the different processes that belong to resistance welding.
31. Explain briefly the stages involved in a resistance-butt welding process.
32. Using a sketch, explain the pressure-time and current-time relationships in resistance-butt welding.
33. List some of the applications of resistance-butt welding.
34. Clarify the difference between flash welding and butt welding.
35. Draw a graph illustrating current versus time and pressure versus time in flash welding.

36. When is flash welding recommended over butt welding?
37. What is the major disadvantage of flash welding?
38. Explain the basic idea of percussion welding.
39. Explain briefly the mechanics of spot welding.
40. Draw a sketch of a section through a spot-welded joint.
41. Draw a sketch to show a typical cycle for a spot-welding machine.
42. How do you compare seam welding with spot welding?
43. List some of the advantages of seam welding.
44. What are the industrial applications of seam welding?
45. What is the basic idea of projection welding?
46. What are the advantages of projection welding?
47. What is the condition for two metals to be joined together by fusion welding?
48. How many zones can be identified in a joint produced by a conventional fusion welding process? Discuss briefly the microstructure in each of these zones.
49. For what do the letters HAZ stand?
50. Explain briefly the phenomenon of the electric arc and how it can be employed in welding.
51. What are the advantages of alternating current over direct current in arc welding?
52. What is the difference between DCSP and DCRP? When would you recommend using each of them?
53. What is meant by the *rated duty cycle?*
54. What shields the molten metal during shielded-metal arc welding?
55. What is the main shortcoming of shielded-metal arc welding?
56. List some of the functions of electrode coatings.
57. Explain briefly the Bernardos welding method.
58. What is the main feature of the electrodes in flux-cored arc welding?
59. What provides the shielding in flux-cored arc welding?
60. What is stud arc welding? How is it performed? List the main applications of this process.
61. How is shielding achieved in submerged arc welding?
62. Why must the plates to be joined by submerged arc welding be horizontal only?
63. Why does submerged arc welding always yield very high quality welds?
64. List some of the advantages of submerged arc welding.
65. What provides shielding in MIG welding?
66. Why does the MIG welding process render itself suitable for automation?
67. How can the penetration for gas-metal arc welding be controlled?
68. What is the main difference between the MIG and the TIG welding processes?
69. List some of the applications of TIG welding.
70. In TIG welding, when would you use an ac power supply and when would you use DCSP and DCRP?
71. Explain briefly the mechanics and the basic idea of plasma arc welding.
72. When is plasma arc welding most recommended?
73. Do you consider electroslag welding to be a true arc welding process?
74. How does welding take place in the electroslag welding process?
75. When is electroslag welding usually recommended?
76. What is the source of energy in oxyacetylene flame welding?

77. How is acetylene stored for use in welding operations?

78. What does the equipment required in gas welding include? Explain the function of each component.

79. What are the types of flames that can be obtained in gas welding? How is each one obtained?

80. What are the zones of a neutral flame? Discuss the effect of the oxygen-to-acetylene ratio on the nature of the flame obtained.

81. Explain briefly the operating principles of electron-beam welding.

82. What are the major limitations of the electron-beam welding process?

83. List some of the advantages of electron-beam welding.

84. What are the major applications of electron-beam welding?

85. For what do the letters in the word *laser* stand?

86. Using a sketch, explain how a laser beam capable of carrying out welding can be generated.

87. Explain briefly the mechanics of laser-beam welding.

88. What are the main advantages of laser-beam welding?

89. List some of the applications of laser-beam welding.

90. Using sketches, illustrate the commonly experienced welding defects. How can each be avoided?

91. What are the main tests for the inspection of welds? Discuss each briefly.

92. What are the factors affecting the selection of the joint type?

93. On what basis are the location and distribution of welds planned?

94. What rules would you consider when designing a welded joint?

95. What is meant by *hard-facing?*

96. What are the main applications of hard-facing?

97. List the main types of thermal cutting processes. Discuss briefly the advantages and limitations of each.

98. How do the mechanics of brazing differ from those of welding?

99. What is the main difference between brazing and soft soldering?

100. List some of the alloys used as brazing fillers and mention the base metals that can be brazed with each one.

101. List some of the commonly used soft solders.

102. What is the main function of brazing fluxes?

103. List some of the fluxes used in brazing.

104. List some of the fluxes used in soft soldering. Discuss the limitations and applications of each.

105. In soft soldering, how should the solder be fused?

106. List the different brazing techniques used in industry. Discuss the advantages and limitations of each.

107. As a product designer, what factors should you take into consideration when designing a brazed joint?

108. In what case can sticking of metals not be replaced by other welding and brazing techniques?

109. List some of the commonly used adhesives. Discuss the characteristics and common applications of each.

110. What are the types of adhesive-bonded joints? Which one is usually the strongest?

Problems

1. Two steel slabs, each 1/4 inch (6.35 mm) thick, are to be joined by two fillet welds (i.e., at both edges). If the width of each slab is 2.5 inches (62.5 mm) and the joint is to withstand a load of 35,000 pounds (156,000 N), determine the allowable tensile strength of the electrode type to be used in welding.
2. Two steel plates, each 1/4 inch (6.35 mm) thick, are to be joined by two fillet welds. If the joint is to withstand a load of 50,000 pounds (222,500 N) and an E7014 electrode (allowable tensile strength = 21,000 lb/in.^{2} i.e., 145,000 KN/m^2) is used, determine the length of weld at each edge. If the plate width is 10 inches (254 mm), how would you distribute the weld? Draw a sketch.
3. Two steel plates, each 5/16 inch thick (7.9 mm), are to be fillet-welded to a third one that is sandwiched between them. The width of each of the first two plates is 4 inches (100 mm), whereas the width of the third one is 6 inches (150 mm). The two plates overlap the third one by 6 inches (150 mm), and an E7014 electrode (allowable tensile strength = 21,000 lb/in.^{2}, 145,000 KN/m^2) is to be used. If the joint is to withstand a load of 190,000 pounds (846 KN), use a sketch to illustrate a design for this joint and provide all calculations.
4. An equal-leg-angle steel section 3 by 3 by 1/4 inch (75 by 75 by 6 mm) is to be welded to a plate using an E7014 electrode (allowable tensile strength = 21,000 lb/in.^{2} i.e., 145,000 KN/m^2). If the joint is to withstand a load of 10,000 pounds (44.5 KN) coinciding with the axis of the angle, design the joint and make a sketch indicating the distribution of the weld to eliminate any tendency of the angle to rotate.
5. Two mild steel pipes, each having a 3/4-inch (19 mm) outer diameter, are to be joined together by brazing. Assuming that the joint is to withstand an axial load of 6 tons, give a detailed design of this joint. (Take allowable shear stress of copper to be 6000 lb/in.^{2} i.e., 41,430 KN/m^2.)
6. Two mild steel sheets, each 3/32 inch (2.4 mm) thick, are to be brazed together using copper as a filler material. Calculate the strength of the joint when it is manufactured according to each of the designs given in Figure 4.48. Compare the results and recommend the design that gives maximum strength.
7. A power supply for arc welding is rated at a 150-A 30-percent duty cycle. What will be the percentage of actual time utilized in welding to the total time the power supply is on if the current employed in welding is only 125 A?

Design Example

PROBLEM

You are required to design a flat-belt pulley so that it can be fabricated by welding. The pulley is to be mounted on a shaft that is 1¼ inches (31 mm) in diameter, and the outside diameter of the rim is 10 inches (250 mm). The rim of the pulley is to provide a surface to transmit a torque of 600 lb ft (816 N·m) from a 2-inch-wide (50-mm) flat belt to the shaft. The number of pulleys required is only 5.

Solution

It is advisable to start by gathering information about guidelines for the constructional features of flat-belt pulleys (e.g., width of rim for a certain belt width and thickness of rim). Information about the safe speeds of various sizes of pulleys should also be collected.

Key. The best strategy is to design the key so that it will be the weak link in the pulley-key-shaft assembly because it is easy to replace. A suitable key material is AISI 1020 CD steel, which is commercially available as a key stock material. It has the following mechanical properties:

Ultimate Tensile Strength (UTS) = 78,000 lb/in.^{2}

yield stress = 66,000 lb/in.^{2}

yield stress in shear = 38,000 lb/in.^{2}

Consider Figure 4.50. The force acting on the key is given by

$$P = \frac{T}{r} = \frac{600 \text{ lb ft} \times 12 \text{ in/ft}}{0.625 \text{ inch}} = 11{,}520 \text{ pounds}$$

Take the key cross section to be 1/4 by 1/4 inch (6 by 6 mm), and its length ℓ in inches:

$$\text{shear stress in the key} = \frac{11{,}520 \text{ pounds}}{\frac{1}{4} \times \ell} \leq \tau_{\text{allowable}}$$

Take a safety factor of 2 for the key:

$$\tau_{\text{allowable}} = \frac{38{,}000}{2} = 19{,}000 \text{ lb/in.}^2$$

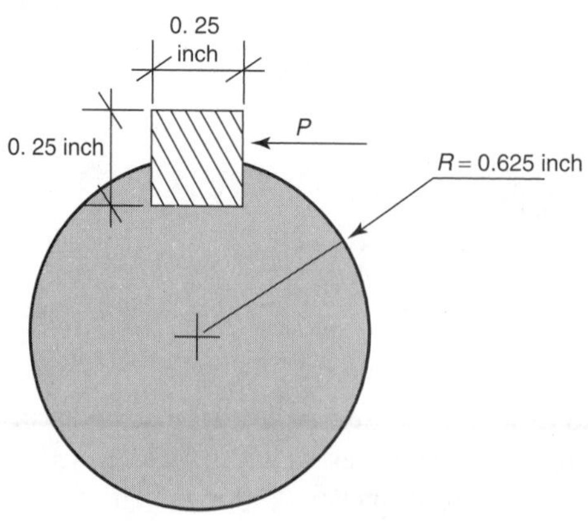

FIGURE 4.50 Forces acting on the key

Therefore,

$$\frac{11{,}520}{\frac{1}{4} \times \ell} = 19{,}000$$

and

$$\ell = 2.4 \text{ inches (60 mm)}$$

But,

$$\text{bearing stress} = \frac{11{,}520}{\frac{1}{4} \times \frac{1}{2} \times \ell} \leq \text{allowable compressive stress}$$

$$\leq \frac{66{,}000}{2}$$

$$= 33{,}000 \text{ lb/in.}^2 \text{ (safety factor of 2)}$$

Therefore,

$$\ell = 2.8 \text{ inches (70 mm)}$$

We take this value to ensure safety against both shearing and compressing loads. We should, however, round it, so the length of the key is to be 3 inches (75 mm).

Hub. Use a round seamless tube having a 2.25-inch outer diameter and 9/16-inch wall. A suitable material is AISI 1020 CD steel. Again, the length of the hub must not be less than 2.8 inches to keep the bearing stress below the allowable value. Take it as 2.875 inches.

Rim. Use a round seamless tube having a 10-inch outer diameter and 1/4-inch wall. Again, a suitable material is AISI 1020 CD steel because of its availability and ability to withstand the rubbing effect of the moving belt.

Spokes. The positioning and welding of four or five spokes would create a serious problem and necessitates the use of a complicated welding fixture. Therefore, the spokes are to be replaced in the design by a web. Use a 5/16-inch flat plate, machined to have an outer diameter of 9.43 inches and an inner diameter 2.31 inches. An appropriate material is AISI 1020 HR steel. Because weight can be a factor, it is good practice to provide six equally spaced holes in the web by machining. These can also serve as an aid in the handling and positioning of the web during welding.

Welding. Use conventional arc welding; an E7014 electrode (allowable tensile stress = 21,000 lb/in.^{2}) can be used. A fillet weld with a leg of 1/4 inch is adopted. The force is given by

$$F = \frac{\text{torque}}{\text{radius}} = \frac{7200}{1.125 \text{ inches}} = 6400 \text{ pounds}$$

FIGURE 4.51
Detailed workshop drawing of the pulley

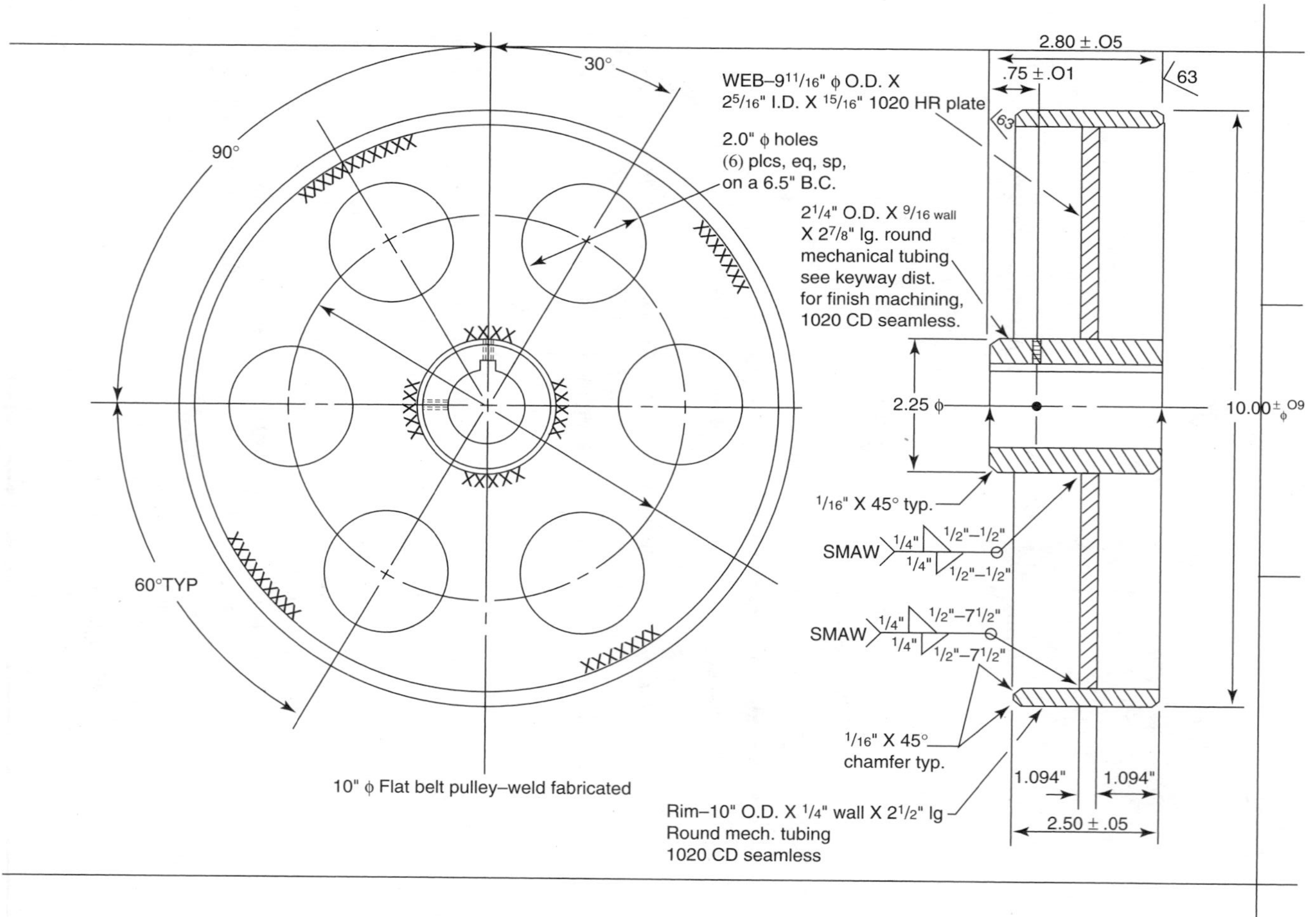

The required length of the weld is

$$\frac{6400}{\frac{1}{4} \times 0.707 \times 0.57 \times 21{,}000} = 3.03 \text{ inches (75 mm)}$$

Use 2.00 inches (50 mm) of weld on each side of the hub.

The circumference of the hub equals π times 2.25, or 7.06 inches. Space four welds, each 0.5 inch (12.5 mm) in length, equally, 90° apart around the circumference of the hub. Welds on both sides of the web should be staggered. Adopt the same welds at the rim. They should be safe because the shearing force is much lower (the radius is larger than that of the hub).

Once all the dimensions and details are known, we are in a position to construct the pulley as shown in the workshop drawing in Figure 4.51.

Design Projects

1. Design a table for the machine shop. The table should be 4 feet (1200 mm) in height, with a surface area of 3 by 3 feet (900 by 900 mm), and should be able to carry a load of half a ton. Because only two tables are required, the design should involve the use of steel angles and a plate that are to be joined together by welding.

2. Design a tank for compressed air. It has a capacity of 100 cubic feet (2.837 m^3) and can withstand an internal pressure of 40 atmospheres (ata). The number of tanks required is 50, and the tanks are going to be placed in a humid environment.

3. Design a compressed-air reservoir (tank) that is to be subjected to an extremely corrosive environment. The capacity of the tank is 30 cubic feet (0.85 m^3), and the maximum gage pressure is 70 ata, but the pressure is pulsating from zero to the maximum value about once every 5 minutes. The number of tanks required is 100.

4. A straight-toothed spur-gear wheel transmits a torque of 1200 lb ft (1632 N·m) to a 2-inch-diameter (50 mm) steel shaft (AISI 1045 CD steel). The pitch diameter of the gear is 8 inches (200 mm), its width is 3 inches (75 mm), and the base diameter is 7.5 inches (187.5 mm). Make a detailed design for the gear's blank (i.e., before the teeth are cut).

5. A mobile winch (little crane) can be moved on casters. It has a capacity of lifting 1 ton for 3 feet (0.9 m) about ground. The lifting arm can be extended, and the winch can then lift 1/2 ton for up to 6 feet (1.8 m). Knowing that the production volume is 4000 units and that casters and hydraulic pressure cylinders are to be purchased from vendors, provide a detailed design and include full specifications of the parts to be purchased.

6. The lifting arm for a crane is 60 feet (about 20 m), and its lifting capacity is 1 ton. It is to be used in construction work and to be subjected to humidity, dirt, and so on. Provide a detailed design for this arm using steel angles that are to be welded together.

7. Design a frame for a hydraulic press for fabrication by welding. The height of the cross arm is 12 feet (about 4 m). The cross arm is mounted (by welding) on two vertical columns that are, in turn, welded to the base. The press can produce a maximum load of 200 tons by means of a hydraulic cylinder attached to the cross arm (below it), and the stroke is 12 inches (300 mm).

   TIP: The energy absorbed when the frame deforms should not exceed 2 percent of the total energy output of the press.

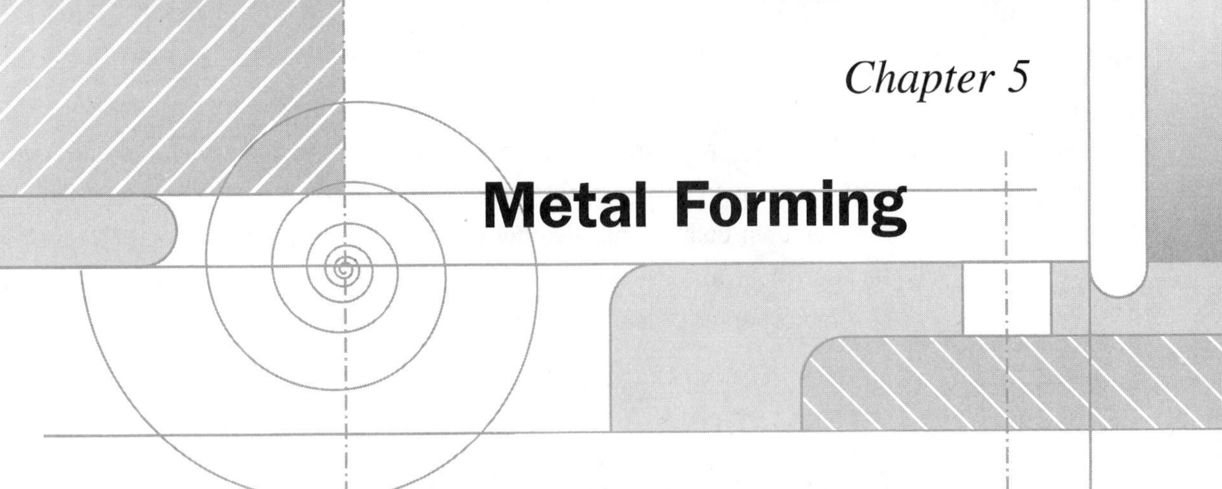

Chapter 5

Metal Forming

INTRODUCTION

Metal forming processes have gained significant attention since World War II as a result of the rapid increase in the cost of raw materials. Whereas machining processes involve the removal of portions of the stock material (in the form of chips) in order to achieve the required final shape, metal forming processes are based upon the plastic deformation and flow of the billet material in its solid state so as to take the desired shape. Consequently, metal forming processes render themselves more efficient with respect to raw material utilization than machining processes, which always result in an appreciable material waste.

In fact, although metal forming techniques were employed in manufacturing only semifinished products (like sheets, slabs, and rods) in the past, finished products that require no further machining can be produced today by these techniques. This was brought about by the recent developments in working methods, as well as by the construction features of the forming machines employed. Among the advantages of these up-to-date forming techniques are high productivity and very low material waste. Therefore, more designers tend to modify the construction of the products manufactured by other processes to use forming. Also, bearing in mind that metal forming methods are still being used for producing semifinished products, it is evident that the vast majority of all metal products are subjected to forming, at least at one stage during their production. This latter fact clearly manifests the importance of the metal forming methods.

Generally, metal forming involves both billet and sheet metal forming. However, it has been a well-accepted convention to divide those processes into two main groups: bulk (or massive) forming and sheet metal working. In this chapter, only bulk forming processes (e.g., forging, cold forming, and rolling) are covered; Chapter 6 deals with the working of sheet metal.

5.1 PLASTIC DEFORMATION

Factors Affecting Plastic Deformation

During any forming process, the material plastically flows while the total volume of the workpiece remains substantially constant. However, there are some marked changes that take place on a microscopic scale within the grains and the crystal lattice of the metal, resulting in a corresponding change in the properties of the material. This latter change can be explained in view of the dislocation theory, which states that the plastic deformation and flow of metal are caused by movement and transfer of dislocations (defects in the crystal lattice) through the material with the final outcome of either piling up or annihilating them. Following are some factors that affect plastic deformation by influencing the course of dislocations.

Impurities and alloying additives. It is well known that pure metals possess higher plasticity than their alloys. The reason is that the presence of structural components and chemical compounds impedes the transfer and migration of dislocations, resulting in lower plasticity.

Temperature at which deformation takes place. As a rule, the plasticity of a metal increases with temperature, whereas its resistance to deformation decreases. The higher the temperature, the higher the plasticity and the lower the yield point. Moreover, no work-hardening occurs at temperatures above the recrystallization temperature. This should be expected because recrystallization denotes the formation and growth of new grains of metal from the fragments of the deformed grains, together with restoring any distortion in the crystal lattice. Consequently, strength values drop to the level of a nonwork-hardened state, whereas plasticity approaches that of the metal before deformation. In fact, a forming process is termed *hot* if the temperature at which deformation takes place is higher than the recrystallization temperature. Lead that is formed at room temperature in summer actually undergoes *hot forming* because the recrystallization temperature for lead is 39.2°F (4°C). When deformation occurs at a temperature below the recrystallization temperature of the metal, the process is termed *cold forming.* Cold forming processes are always accompanied by work-hardening due to the piling up of dislocations. As a result, strength and hardness increase while both ductility and notch toughness decrease. These changes can be removed by heat treatment (annealing). On the other hand, when hot forming a metal, the initial dendritic structure (the primary structure after casting) disintegrates and deforms, and its crystals elongate in the direction of the metal flow. The insoluble impurities like nonmetallic inclusions (around the original grain boundaries) are drawn and squeezed between the elongated grains. This texture of flow lines is usually referred to as the *fibrous macrostructure*. This fibrous macrostructure is permanent and cannot be removed by heat treatment or further working. As a result, there is always anisotropy of mechanical properties; strength and toughness are better in the longitudinal direction of fibers. Also, during hot forming, any voids or cracks around grain boundaries are closed, and the metal welds together, which, in turn, results in improvements in the mechanical properties of the metal.

Rate of deformation. It can generally be stated that the *rate of deformation* (strain rate) in metal working adversely affects the plasticity of the metal (i.e., an increase in the deformation rate is accompanied by a decrease in plasticity). Because it takes the process of recrystallization some time to be completed, that process will not have enough time for completion when deformation occurs at high strain rates. Therefore, greater resistance of the metal to deformation should be expected. This does not mean that the metal becomes brittle.

State of stress. A *state of stress* at a point can be simply described by the magnitudes and directions of the principal stresses (a stress is a force per unit area) acting on planes that include the point in question. The state of stress is, in fact, a precise and scientific expression for the magnitudes and the directions of the external forces acting on the metal. All possible states of stress can be reduced to only nine main systems, as shown in Figure 5.1. These nine cases can, in turn, be divided into three groups. The

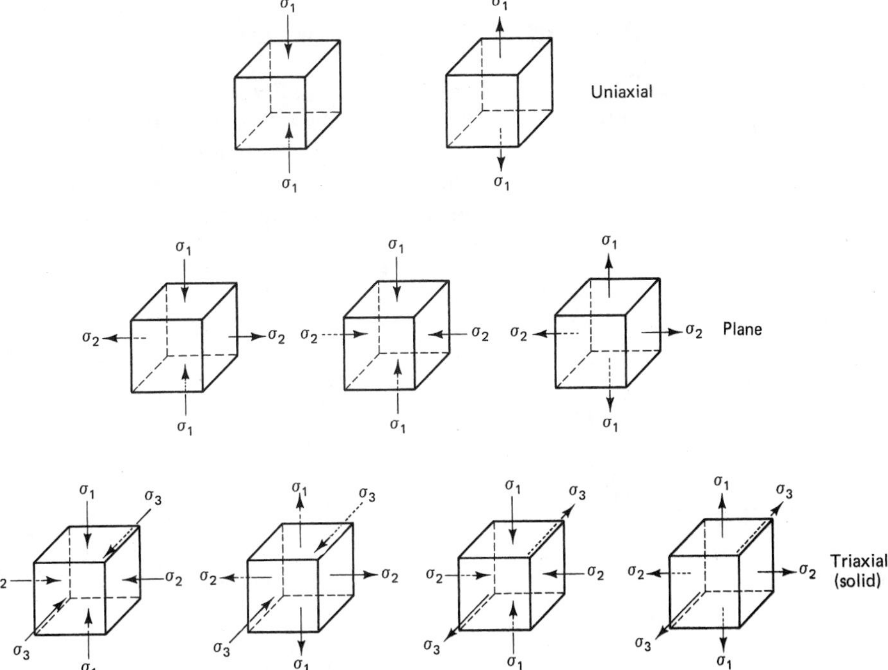

FIGURE 5.1
The nine main systems of the state of stress

first group includes two systems that are characterized by the absence of stress (forces) along two directions, and the stress system is therefore called *uniaxial.* This is the case when stretching sheet metal having a length that considerably exceeds its width. In each of the three systems included in the second group, it is clear that a stress along only one of the directions is absent. Because the other two directions (stresses) form a plane, each of these systems is referred to as a *plane-stress state.* It may approximately be represented by stretching of a thin sheet in two or more directions. The remaining group indicates the state of stress of a body, where there are stresses acting along all three directions in space, yielding the term *triaxial.* In fact, most of the bulk forming operations (forging, rolling, and wire drawing) cause states of stress that belong to this latter group.

Load and Energy Requirement

The force required for deforming a given metal (at any unchanged desired temperature and at usual strain-rate levels) is dependent upon the *degree of deformation,* which is the absolute value of the natural logarithm of the ratio of the final length of the billet to its original length. On the other hand, the energy consumed throughout the forming process is equivalent to the area under the load-deformation curve for that forming process. Therefore, that energy can be calculated if the relationship between the load and the deformation is known. Figure 5.2 shows the degree of deformation and the energy consumed in an upsetting operation. It must be noted that in both hot and cold working, there is an upper limit for the degree of deformation (especially in cold working) above which cracks and discontinuities in the workpiece initiate.

Preheating the Metal for Hot Forming

Before being subjected to hot forming processes, ingots (or billets) should be uniformly heated throughout their cross sections, without overheating or burning the metal at the surface. This is particularly important when forming steels. Attention must also be given to the problems of decarburization and formation of scale in order to bring them to a minimum. The thermal gradient is another important factor that affects

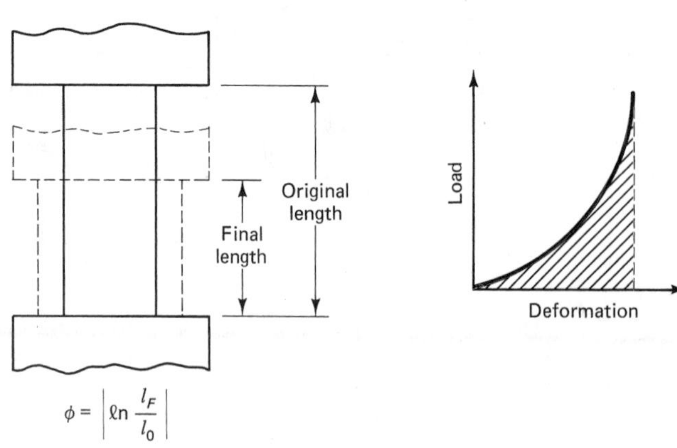

FIGURE 5.2
The degree of deformation and the energy consumed in upsetting

the soundness of the deformed part. If the temperature gradient is high, thermal stresses may initiate and can cause internal cracks. This usually happens when a portion of the metal is above the critical temperature of metal (AC_1 or AC_3) while the rest of the billet is not. The larger the cross section of the billet and the lower its coefficient of thermal conductivity, the steeper the temperature gradient will be, and the more liable to internal cracking during heating the billet becomes. In the latter case, the rate of heating should be kept fairly low (about 2 hours per inch of section of the billet) in order not to allow a great difference to occur between the temperatures at the surface and the core of the billet. The metal must then be "soaked" at the maximum temperature for a period of time long enough to ensure uniformity of temperature.

The maximum temperature to which the billet is heated before forming differs for different metals. There is usually an optimum range of temperatures within which satisfactory forming is obtained because of increased plasticity and decreased resistance to deformation. Nevertheless, any further increase in temperature above that range may, on the contrary, result in a defective product. Burned metal and coarse grain structure are some of the defects encountered when a metal is excessively heated.

The ingots may be heated in soaking pits, forge hearths, chamber furnaces, or car-bottom furnaces, which are all heated with gas. Rotary hearth furnaces represent another type of heating furnace. In mass production or automated lines, small objects (billets) are heated using electric current and the phenomenon of induction. This induction-heating method is quick and keeps the surfaces of the billets clean, and temperatures can be accurately controlled. Moreover, physical equipment requires limited floor space and can be fully automated.

Friction and Lubrication in Working of Metals

Friction plays an important role in all metal forming processes and is generally considered to be undesirable because it has various harmful effects on the forming processes, on the properties of products, and on the tool life. During the deformation of a metal, friction occurs at the contact surface between the flowing metal and the tool profile. Consequently, the flow of the metal is not homogeneous, which leads to the initiation of residual stresses, with the final outcome being an unsound product with inferior surface quality. Also, friction increases the pressure acting on the forming tool (as well as the power and energy consumed) and thus results in greater wear of the tools.

Friction in metal forming is drastically different from the conventional Columb's friction because extremely high pressure between the mating bodies (tool and workpiece) is involved. Recent theories on friction in metal forming indicate that it is actually the resistance to shear of a layer, where intensive shear stress is generated as a result of relative displacement between two bodies. When these bodies have direct metal-to-metal contact, slipping and shear flow occur in a layer adjacent to the contact interface. But, if a surface of contact is coated with a material having low shear resistance (a lubricant that can be solid or liquid), slipping takes place through that layer of lubricant and, therefore, has low resistance. This discussion indicates clearly that the magnitude of the friction force is determined by the mechanical properties (yield point in shear) of the layer where actual slipping occurs. Hence, it is evident that a metal having a low yield point in shear, such as lead, can be used as a lubricant when forming metals having relatively high yield strength in compression. Figure 5.3 shows the shear layer in three different cases: solid lubrication, dry sticking friction, and hydrodynamic (liquid) lubrication.

In order to reduce friction and thus eliminate its harmful effects, lubricants are applied to the tool-workpiece interface in metal forming processes. The gains include lower load and energy requirement, prevention of *galling* or sticking of the workpiece metal onto the tool, better surface finish of products, and longer tool life. An important consideration when selecting a lubricant is its activity (i.e., its ability to adhere strongly to the surface of the metal). The activity of a lubricant can, however, be enhanced by adding material with high capability of adsorption, such as fat acids. Among other factors to be considered are thermal stability, absence of poisonous fumes, and complete burning during heat treatment of the products.

In cold forming processes, vegetable and mineral oils as well as aqueous emulsions are employed as lubricants. These have the advantage of acting as coolants, eliminating excessive heat and thus reducing the temperature of the tool. Solid polymers, waxes, and solid soaps (sodium or calcium stearates) are also widely used in cold-metal working.

For relatively high temperature applications, chlorinated organic compounds and sulfur compounds are used. Solid lubricants like molybdenum disulfide and graphite possess low-friction properties up to elevated temperatures and are, therefore, used as solid lubricants in hot forming. Graphite is sometimes dispersed in grease, especially in hot forging ferrous materials. Lately, use has been made of molten glass as a lubricant when alloy steels and special alloys are hot formed. The glass is added in the form of powder between the die and a hot billet. The advantages of molten glass include low friction, excellent surface finish, and improved tool life.

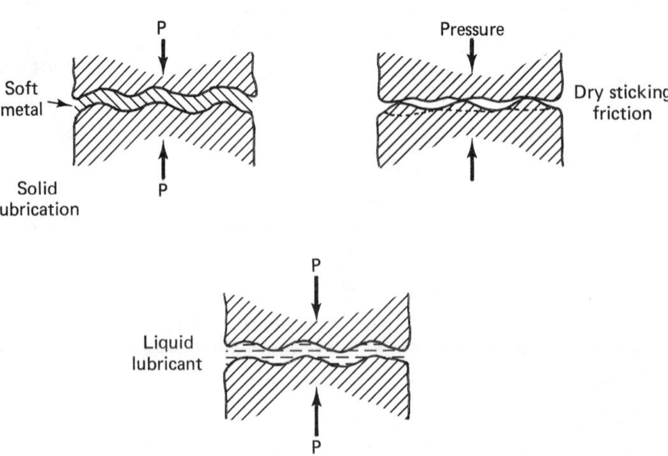

FIGURE 5.3 The shear layer in three different cases

Cold Forming Versus Hot Forming

Cold forming has its own set of advantages and disadvantages, as does hot forming, and, therefore, each renders itself appropriate for a certain field of applications. For instance, cold forming will enhance the strength of the workpiece metal, improve the quality of the surface, and provide good dimensional accuracy, but the plastic properties of the metal (elongation percentage and reduction-in-area percentage) and the impact strength drop. Therefore, the final properties of cold-formed products are obtained as required by adjusting the degree of deformation and the parameters of the postheating treatment process. Because the loads involved in cold forming are high, this technique is generally employed in the manufacture of small parts of soft, ductile metals, such as low-carbon steel. Also, large quantities must be produced to justify the high cost of tooling involved. Nevertheless, if the products are to be further processed by machining, the increased hardness caused by cold working is a real advantage because it results in better machinability. Therefore, cold-rolled plates and cold-drawn bars are more suitable for machining purposes than hot-formed ones.

On the other hand, the yield strength of a metal drops significantly at elevated temperatures, and no work-hardening occurs. Consequently, hot forming processes are used when high degrees of deformation are required and/or when forming large ingots or billets because the loads and energies needed are far lower than those required in cold forming. Moreover, hot forming refines the grain structure, thus producing softer and more ductile parts suitable for further processing by cold forming processes. However, high temperatures affect the surface quality of products, giving oxidation and scales.

Decarburization may also occur in steels, especially when hot forming high-carbon steel. The scales, oxides, and decarburized layers must be removed by one or more machining processes. This slows down the production, adds machining costs, and yields waste material, resulting in lower efficiency of material utilization. A further limitation of hot forming is reduced tool life due to the softening of tool surfaces at elevated temperatures and the rubbing action of the hot metal while flowing. This actually subjects the tools to thermal fatigue, which shortens their life.

5.2 ROLLING

Hot rolling is the most widely used metal forming process because it is employed to convert metal ingots to simple stock members called *blooms* and *slabs*. This process refines the structure of the cast ingot, improves its mechanical properties, and eliminates the hidden internal defects. The process is termed *primary rolling* and is followed by further hot rolling into plates, sheets, rods, and structural shapes. Some of these may be subjected to cold rolling to enhance their strength, obtain good surface finish, and ensure closer dimensional tolerances. Figure 5.4 illustrates the sequence of operations involved in manufacturing rolled products.

FIGURE 5.4
Sequence of operations involved in manufacturing rolled products

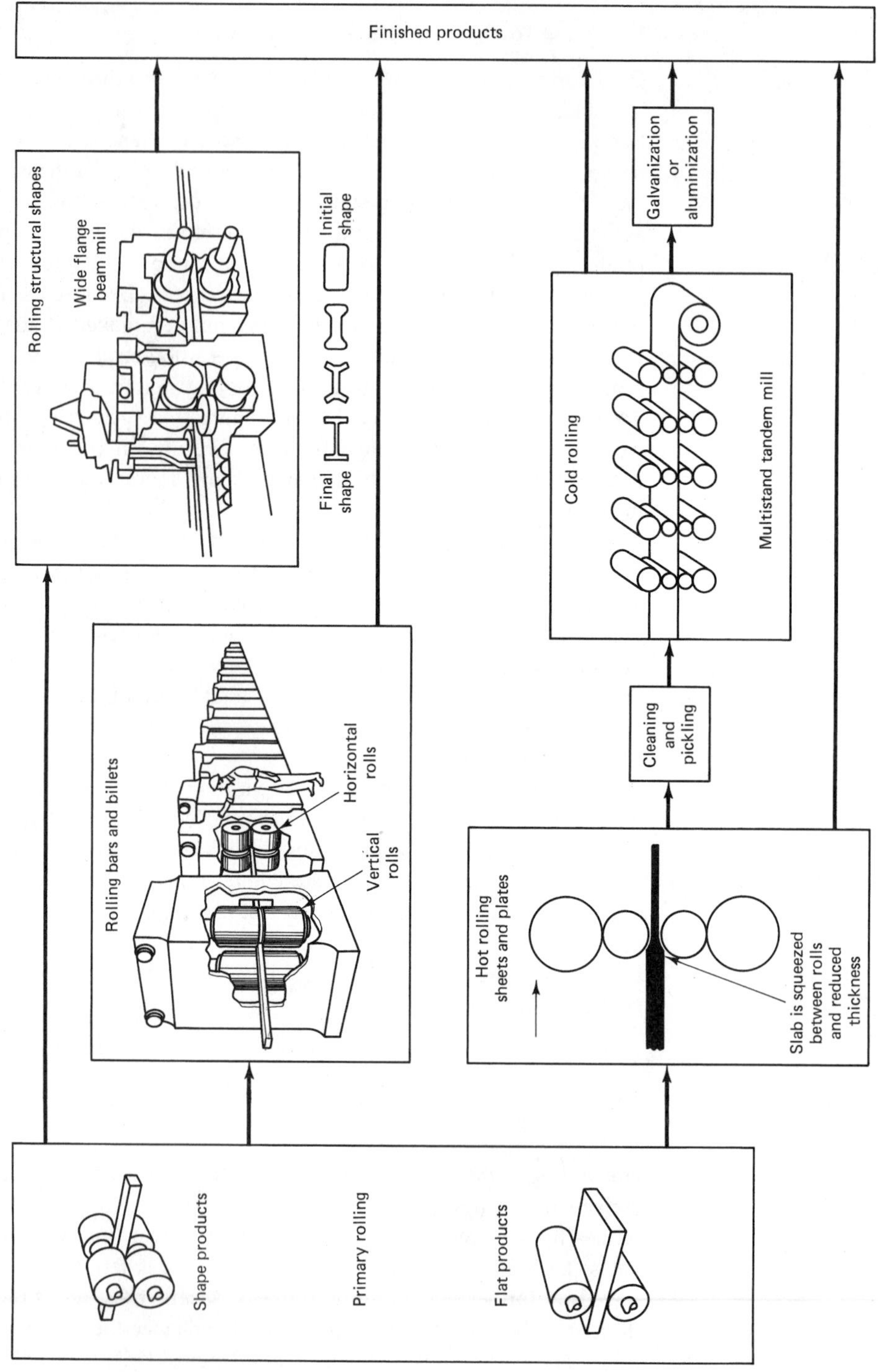

Principles of Rolling

The process of *rolling* consists of passing the metal through a gap between rolls rotating in opposite directions. That gap is smaller than the thickness of the part being worked. Therefore, the rolls compress the metal while simultaneously shifting it forward because of the friction at the roll-metal interfaces. When the workpiece completely passes through the gap between the rolls, it is considered fully worked. As a result, the thickness of the work decreases while its length and width increase. However, the increase in width is insignificant and is usually neglected. As can be seen in Figure 5.5, which shows the rolling of a plate, the decrease in thickness is called *draft,* whereas the increase in length and the increase in width are termed *absolute elongation* and *absolute spread,* respectively. Two other terms are the *relative draft* and the *coefficient of elongation,* which can be given as follows:

$$\text{relative draft } \varepsilon = \frac{\Delta h \times 100}{h_o} = \frac{h_o - h_f}{h_o} \times 100 \tag{5.1}$$

$$\text{coefficient of elongation } \eta = \frac{l_f}{l_o} \tag{5.2}$$

But because the volume of the work is constant, it follows that

$$\eta = \frac{h_o \times b_o}{h_f \times b_f} = \frac{A_o}{A_f} \tag{5.3}$$

Equation 5.3 indicates that the coefficient of elongation is adversely proportional to the ratio of the final to the original cross-sectional areas of the work.

As can be seen in Figure 5.6, the metal is deformed in the shaded area, or *deformation zone.* The metal remains unstrained before this area and does not undergo any further deformation after it. It can also be seen that the metal undergoing deformation is in contact with each of the rolls along the arc AB, which is called the *arc of contact.* It corresponds to a central angle, α, that is, in turn, called the *angle of contact,* or *angle of bite.* From the geometry of the drawing and by employing simple trigonometry, it can be shown that

$$\cos \alpha = 1 - \frac{h_o - h_f}{2R} = 1 - \frac{\Delta h}{2R} \tag{5.4}$$

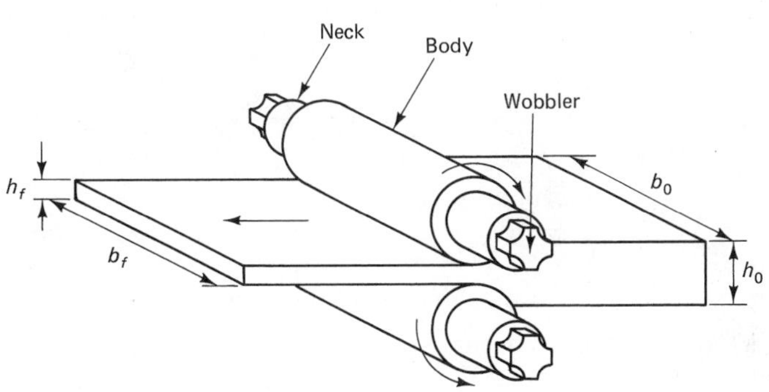

FIGURE 5.5
Simple rolling of a plate

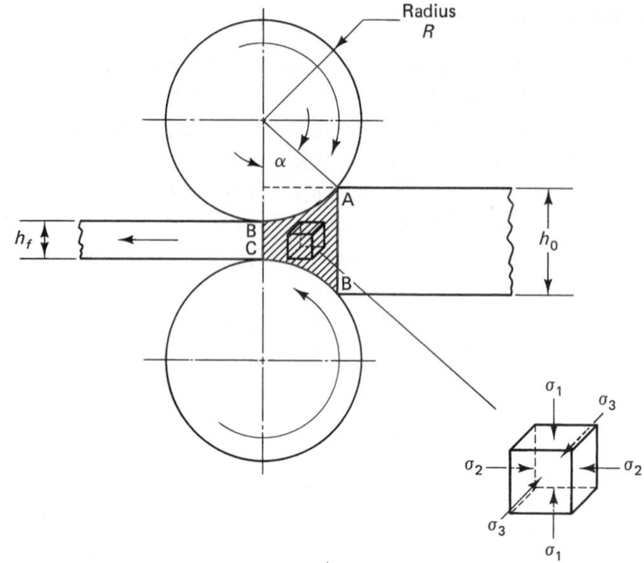

FIGURE 5.6 The deformation zone, state of stress, and angle of contact in rolling

Equation 5.4 gives the relationship between the geometrical parameters of the rolling process, the angle of contact, the draft, and the radius of the rolls. Note that in order to ensure that the metal will be shifted by friction, the angle of contact must be less than β, the angle of friction, where $\tan \beta = \mu$ (the coefficient of friction between roll surface and metal). In fact, the maximum permissible value for the angle of contact depends upon other factors, such as the material of the rolls, the work being rolled, and the rolling temperature and speed. Table 5.1 indicates the recommended maximum angle of contact for different rolling processes.

Load and Power Requirement

As can also be seen in Figure 5.6, the main stress system in the deformation zone in a rolling process is triaxial compression, with the maximum (principal) stress acting normal to the direction of rolling. The deformed metal is exerting an equal counterforce on each of the rolls to satisfy the equilibrium conditions. Therefore, this force normal to the direction of rolling is important when doing the design calculations for the rolls as well as the mill body. It is also important in determining the power consumption in

TABLE 5.1 Maximum allowable angle of contact for rolling

Rolling Process	Maximum Allowable Angle of Contact
Rolling of blooms and heavy sections	24°–30°
Hot rolling of sheets and strips	15°–20°
Cold rolling of lubricated sheets	2°–10°

a rolling process. Unfortunately, the exact determination of that rolling load and power consumption is complicated and requires knowledge of theory of plasticity as well as calculus. Nevertheless, a first approximation of the roll load can be given by the following simple equation:

$$F = \overline{Y} \times b \times \sqrt{R \times \Delta h} \tag{5.5}$$

where $\overline{Y}$ is the average (plane-strain) yield stress assuming no spread and is equal to $1.15Y$, where Y is the mean yield stress of the metal. Therefore, Equation 5.5 should take the form

$$F = 1.15Y \times b \times \sqrt{R \times \Delta h} \tag{5.6}$$

Equation 5.6 neglects the effect of friction at the roll-work interface and, therefore, gives lower estimates of the load. Based on experiments carried out on a wide range of rolling mills, this equation can be modified to account for friction by multiplying by a factor of 1.2. The modified equation is

$$F = 1.2 \times 1.15Y \times b \times \sqrt{R \times \Delta h} \tag{5.7}$$

The power consumed in the process cannot be obtained easily; however, a rough estimate in low-friction conditions is given by

$$\text{hp} = \frac{\overline{Y} \times b \times R \times \Delta h \times \omega}{550} \tag{5.8}$$

where ω is the angular velocity of rolls in radians per second, and $\overline{Y}$, b, R and Δh are all in English units.

Rolling Mills

A rolling mill includes one or more roll stands and a main drive motor, reducing gear, stand pinion, flywheel, and coupling gear between the units. The roll stand is the main part of the mill, where the rolling process is actually performed. It basically consists of housings in which antifriction bearings that are used for carrying (mounting) the rolls are fitted. Moreover, there is a screw-down mechanism to control the gap between the rolls and thus the required thickness of the product.

Depending upon the profile of the rolled product, the body of the roll may be either smooth for rolling sheets (plates or strips) or grooved for manufacturing shapes such as structural members. A roll consists of a body, two necks (one on each side), and two wobblers (see Figure 5.5). The body is the part that contacts and deforms the metal of the workpiece. The necks rotate in bearings that act as supports, while the wobblers serve to couple the roll to the drive. Rolls are usually made from high-quality steel and sometimes from high-grade cast iron to withstand the very severe service conditions to which the rolls are subjected during the rolling process, such as combined bending and torque, friction and wear, and thermal effects. Gray cast-iron rolls are employed in roughing passes when hot rolling steel. Cast- or forged-steel rolls are used in blooming, slabbing, and section mills as well as in cold-rolling mills. Forged rolls are stronger and tougher than cast rolls. Alloy-steel rolls made of chrome-nickel or chrome-molybdenum steels are used in sheet mills.

Classification of Rolling Mills

Rolling mills are classified according to the number and arrangement of the rolls in a stand. Following are the five main types of rolling mills, as shown in Figure 5.7a through e.

Two-high rolling mills. Two-high rolling mills, the simplest design, have a two-high stand with two horizontal rolls. This type of mill can be nonreversing (unidirectional), where the rolls have a constant direction of rotation, or reversing, where the rotation and direction of metal passage can be reversed.

Three-high rolling mills. Three-high rolling mills have a three-high stand with three rolls arranged in a single vertical plane. This type of mill has a constant direction of rotation, and it is not required to reverse that direction.

Four-high rolling mills. In sheet rolling, the rolls should be designed as small as possible in order to reduce the rolling force F of the metal on the rolls and the power requirement. If such small-diameter rolls are used alone, they will bend and result in nonuniform thickness distribution along the width of the sheet, as shown in Figure 5.8. For this reason, another two backup rolls are used to minimize bending and increase the rigidity of the system. The four rolls are arranged above one another in a vertical plane. Also, the backup rolls always have larger diameters than those of the working rolls.

Multihigh rolling mills (Sendzimir mills). Multihigh rolling mills are used particularly in the manufacture of very thin sheets, those with a thickness down to 0.0005 inch (0.01 mm) and a width up to 80 inches (2000 mm), into coils. In this case, the working rolls must have very small diameters (to reduce load and power consumption, as explained before), usually in the range of 3/8 inch (10 mm) up to 1.25 inches (30 mm).

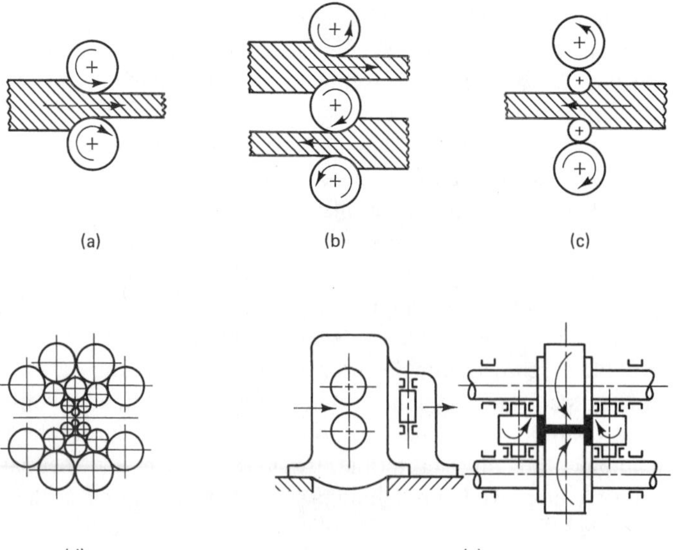

FIGURE 5.7
The five main types of rolling mills: (a) two-high rolling mill; (b) three-high rolling mill; (c) four-high rolling mill; (d) multihigh rolling mill; (e) universal rolling mill

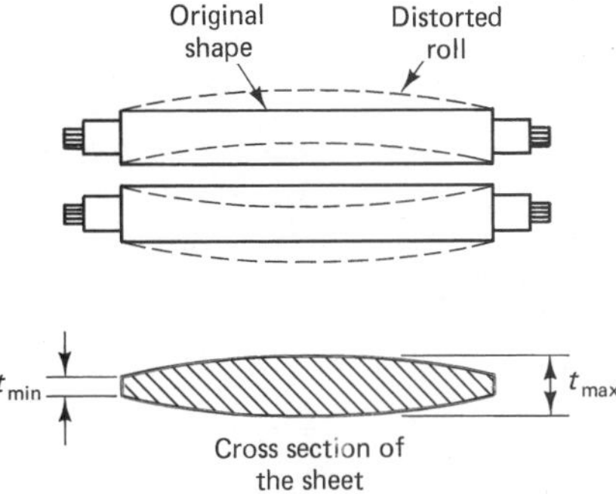

FIGURE 5.8 Rolling thin sheets with small-diameter rolls

Such small-diameter working rolls make a drive practically impossible. They are, therefore, driven by friction through an intermediate row of driving rolls that are, in turn, supported by a row of backup rolls. This arrangement involves a cluster of either 12 or 20 rolls, resulting in exceptional rigidity of the whole roll system and almost complete absence of working-roll deflections. An equivalent system that is sometimes used is the planetary rolling mill, in which a group of small-diameter working rolls rotate around a large, idle supporting roll on each side of the work.

Universal rolling mills. Universal rolling mills are used for producing blooms from ingots and for rolling wide-flange H beams (Gray's beams). In this type of mill, there are vertical rolls in addition to the horizontal ones. The vertical rolls of universal mills (for producing structural shapes) are idle and are arranged between the bearing chocks of the horizontal rolls in the vertical plane.

The Range of Rolled Products

The range of rolled products is standardized in each country in the sense that the shape, dimensions, tolerances, properties, and the like are given in a standard specifications handbook that differs from country to country. The whole range of rolled products can generally be divided into the following four groups.

Structural shapes or sections. The first group includes general-purpose sections like round and square bars; angles; channel, H, and I beams; and special sections (with intricate shapes) like rails and special shapes used in construction work and industry. Figure 5.9 shows a variety of sections that belong to this group. These products are rolled in either rail mills or section mills, where the body of each roll has grooves called *passes* that are made in the bodies of the upper and lower rolls in such a manner as to lie in the same vertical plane. They are used to impart the required shape to the work. This process is carried out gradually (i.e., the stock is partly deformed at each stand, or pass, in succession). The skill of a rolling engineer is to plan and construct the details of a system of successive passes that ensures the adequate rolling of blanks into the desired shape. This operation is called *roll pass design.* Figure 5.10a illustrates the roll passes for producing rails; Figure 5.10b, those for producing an I beam.

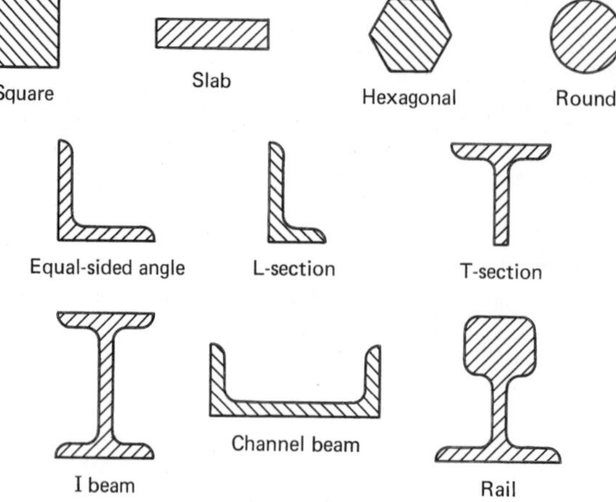

FIGURE 5.9
Some structural shapes or sections produced by rolling

Plates and sheets. Plates and sheets are produced in plate and sheet mills for the hot rolling of metal and in cold reduction mills for the production of cold-rolled coils, where multihigh rolling mills are employed, as previously mentioned. This group of products is classified according to thickness. A flat product with a width ranging from 5/32 inch (4 mm) up to 4 inches (100 mm) is called a *plate,* whereas wider and thinner flat stocks are called *sheets.*

Special-purpose rolled shapes. This group includes special shapes, one-piece rolled wheels, rings, balls, ribbed tubes, and die-rolled sections in which the cross section of the bar varies periodically along its length. These kinds of bars are used in the machine-building industry and in the construction industry for reinforcing concrete beams and columns. Figure 5.11a shows the sequence of operations in manufacturing a rolled wheel for railway cars; Figure 5.11b, the wheel during the final stage in the rolling mill.

Seamless tubes. The process of manufacturing seamless tubes involves two steps:

1. Piercing an ingot or a roughened-down round blank to form a thick-walled shell
2. Rolling the obtained shell into a hollow thin-walled tube having the desired diameter and wall thickness

In the first step, the solid blank is center-drilled at one end, heated to the appropriate temperature, and then placed in the piercing mill and forced into contact with the working rolls. There are several types of piercing mills, but the commonly used one has barrel-shaped rolls. As Figure 5.12 shows, the axes of the two rolls are skew lines, each deviating with a small angle from the direction of the blank axis. Also, the two rolls rotate in the same direction, forcing the blank to rotate and proceed against a mandrel. A hole is formed and becomes larger; finally, a rough tube is obtained. The milling stand is provided with side rollers for guiding the blank and the formed rough tube during this operation. In the second step, the hollow shell (rough tube) is usually forced over another mandrel, and the combination is longitudinally rolled at their hot

FIGURE 5.10
Roll passes: (a) for producing rails; (b) for producing an I beam

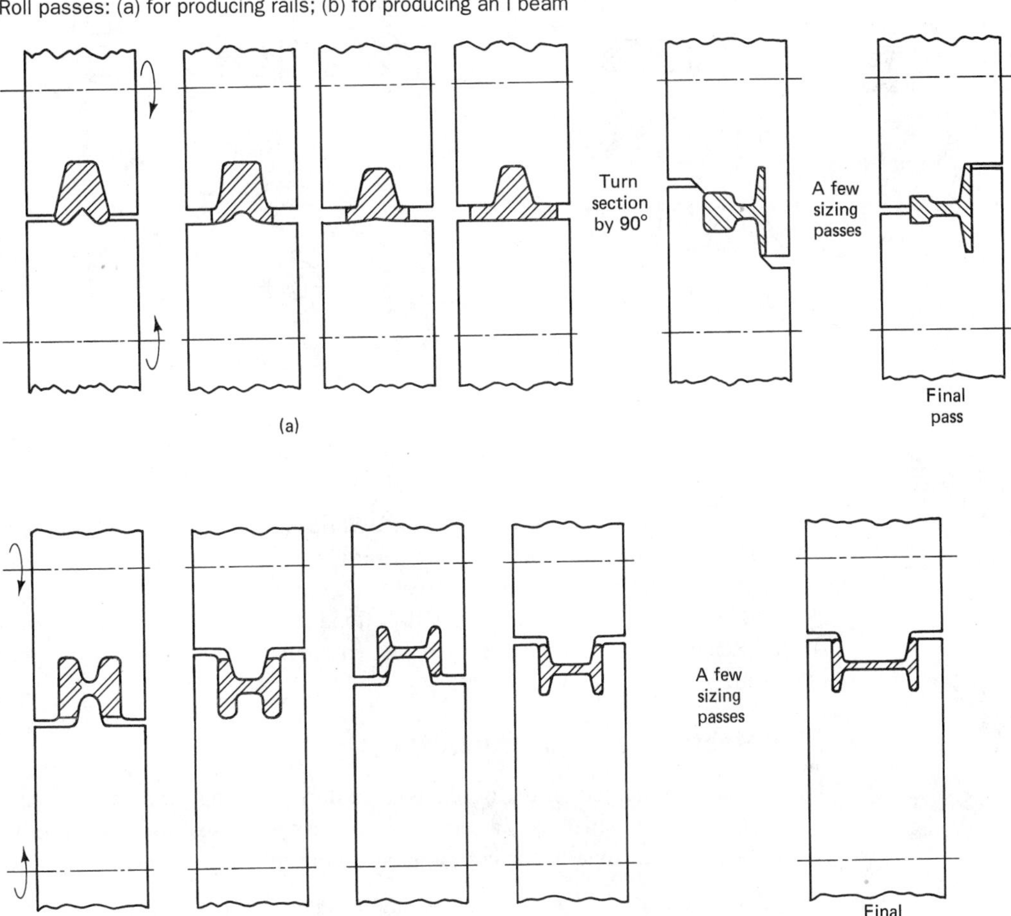

state between grooved rolls. Mills of different types are used, including continuous, automatic, and pilger mills. Finally, a sizing operation may be performed, between sizing rolls and without the use of a mandrel, at room temperature in order to improve the properties and finish of the tubes.

Lubrication in Rolling Processes

Friction plays a very important role in a rolling process and has some beneficial effects, provided that it is not excessive. In fact, it is responsible for shifting the work between the rolls and should not, therefore, be eliminated or reduced below an appropriate level. This is an important point to be taken into account when choosing a lubricant for a rolling process.

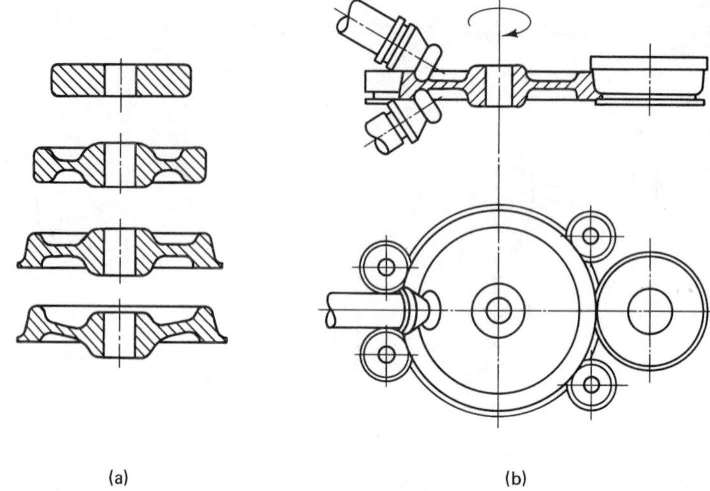

FIGURE 5.11 The production of a railway car wheel: (a) sequence of stages; (b) wheel in final stage in mill

In the cold rolling of steel, fluid lubricants of low viscosity are employed, but paraffin is suitable for nonferrous materials like aluminum or copper alloys to avoid staining during subsequent heat treatment. On the other hand, hot rolling is often carried out without lubricants but with a flood of water to generate steam and break up the scales that are formed. Sometimes, an emulsion of graphite or graphited grease is used.

Defects in Rolled Products

A variety of defects in the products arise during rolling processes. A particular defect is usually associated with a particular process and does not arise in other processes. Following are some of the common defects in rolled products.

Edge cracking. *Edge cracking* occurs in rolled ingots, slabs, or plates and is believed to be caused by either limited ductility of the work metal or uneven deformation, especially at the edges.

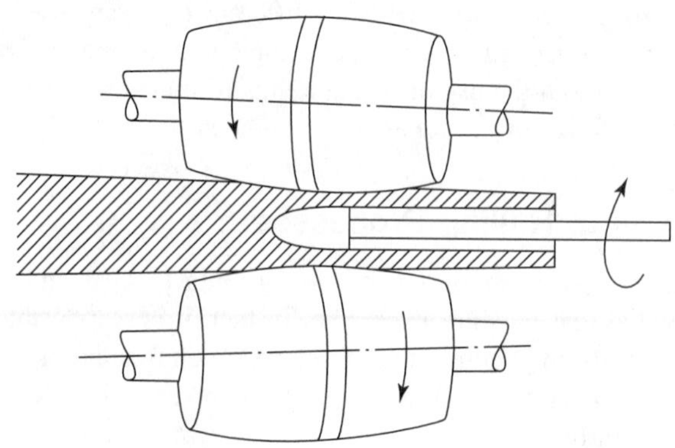

FIGURE 5.12 The production of seamless tubes by rolling

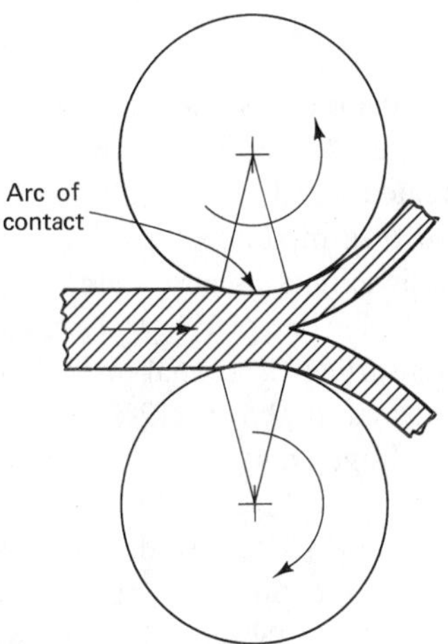

FIGURE 5.13
Alligatoring when rolling aluminum slabs

Alligatoring. Figure 5.13 shows the defect of *alligatoring,* which is less common than it used to be. It usually occurs in the rolling of slabs (particularly aluminum alloys), where the workpiece splits along a horizontal plane on exit, with the top and bottom parts following the rotation of their respective rolls. This defect always occurs when the ratio of slab thickness to the length of contact falls within the range 1.4 to 1.7.

Folds. *Folds* are defects occurring during plate rolling when the reduction per pass is too small.

Laminations. *Laminations* associated with cracking may develop when the reduction in thickness is excessive.

5.3 METAL DRAWING

Drawing is basically a forming process that involves pulling a slender semifinished product (like wire, bar stock, or tube) through a hole of a drawing die. The dimensions of that hole are smaller than the dimensions of the original material. Metals are usually drawn in their cold state, and the required shape may be achieved in a single drawing operation or through several successive drawing operations, in which case the diameters of the holes are successively decreasing. Sometimes, annealing is carried out between the drawing operations to relieve the metal from work-hardening. Accurate dimensions, good surface quality, increased strength and hardness, and the possibility of producing very small sections are some advantages of the drawing process. The drawing process has, therefore, wide industrial application and is used for manufacturing thin wires, thin-walled tubes, and components with sections that cannot be made except by machining. It is also used for sizing hot-rolled sections.

Preparing the Metal for Drawing

Before being subjected to the drawing process, metal blanks (wires, rods, or tubes) are heat treated and then cleaned of scales that result from that operation. Descaling is usually done by pickling the heat-treated metal in acid solutions. Steels are pickled in either sulfuric or hydrochloric acid or a mixture of both; copper and brass blanks are treated in sulfuric acid, whereas nickel and its alloys are cleaned in a mixture of sulfuric acid and potassium bichromate. After pickling, the metal is washed to remove any traces of acid or slag from its surface. The final operation before drawing is drying the washed blanks at a temperature above 212°F (100°C). This eliminates the moisture and a great deal of the hydrogen dissolved in the metal, thus helping to avoid pickling brittleness.

If steel is to be subjected to several successive drawing passes, its surface should then be conditioned for receiving and retaining the drawing lubricant. Conditioning is performed directly after pickling and can take the form of sulling, coppering, phosphating, or liming. In sulling, the steel rod is given a thin coat of iron hydroxide, which combines with lime and serves as a carrier for the lubricant. Phosphating involves applying a film of iron, manganese, or zinc phosphates to which lubricants stick very well. Liming neutralizes the remaining acid and forms a vehicle for the lubricant. Coppering is used for severe conditions and is achieved by immersing the steel rods (or wires) in a solution of vitriol. All conditioning operations are followed by drying at a temperature of about 650°F (300°C) in special chambers.

Wire Drawing

Drawing dies. A *die* is a common term for two parts: the die body and the die holder. Die bodies are made of cemented carbides or hardened tool steel, whereas die holders are made of good-quality tool steel that possesses high toughness. The constructional details of a die are shown in Figure 5.14. It can be seen from the figure that the die opening involves four zones: entry, working zone, die bearing, and exit. The *entry zone* allows the lubricant to reach the working zone easily and also protects the wire (or rod) against scoring by sharp edges. The *working zone* is conical in shape and has an apex

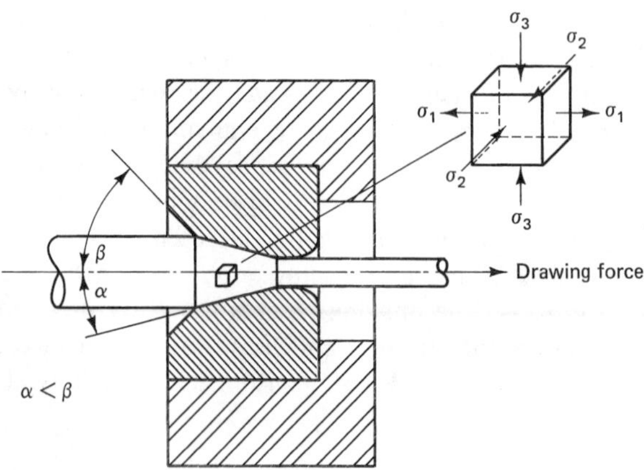

FIGURE 5.14
The constructional details of a drawing die

angle that ranges between 6° and 24°, depending upon the type of work and the metal being drawn. The *die bearing,* sometimes called the *land,* is a short cylindrical zone in which a sizing operation is performed to ensure accuracy of the shape and dimensions of the end product. The *exit zone* provides back relief to avoid scoring of the drawn wire (or rod). In a wire-drawing operation, the end of the wire is pointed by swaging and then fed freely into the die hole so that it appears behind the die. This pointed end is gripped by the jaws of a carriage that pull the wire through the die opening, where it undergoes reduction in cross-sectional area and elongation in length.

Draw benches. A wire-drawing operation usually involves the use of multidie draw benches, where the wire passes through a series of draw plates. First, the wire leaves the coil and passes through the first drawing die. Then, it is wound two or three turns around a capstan (drum) before it enters the next drawing die. A typical draw bench of this type with six draw plates is shown in Figure 5.15. In practice, a bench may include from 2 up to 22 draw plates, and the wire leaving the last die may attain a velocity of 9800 feet per minute (50 m/s). The capstan drives are designed to provide not only forward pull after each pass but also backward pull to the wire before it enters the next drawing die.

Lubrication. Lubrication reduces the required drawing force and the energy consumed during the process, increases the service life of the die, and allows a smoother wire surface to be obtained. Various kinds of soap are used as lubricants in wire-drawing processes. Examples are sodium soap or calcium stearate, which is picked up by the wire from a soap box adjacent to the die. Although they are difficult to apply and remove, polymers are also used as solid lubricants, especially in severe conditions, as in the case of drawing hard alloys or titanium. Various kinds of mineral and vegetable oils containing fatty or chlorinated additives are also used as drawing lubricants.

Mechanics of wire drawing. The state of stress during the wire-drawing process (see Figure 5.14) involves compressive forces along two of the directions and tension along the third one. An approximate but simple estimate of the drawing force can be given by the following equation:

$$F = a_f \times Y \times \ell n\left(\frac{a_o}{a_f}\right) \tag{5.9}$$

FIGURE 5.15
A typical multidie draw bench

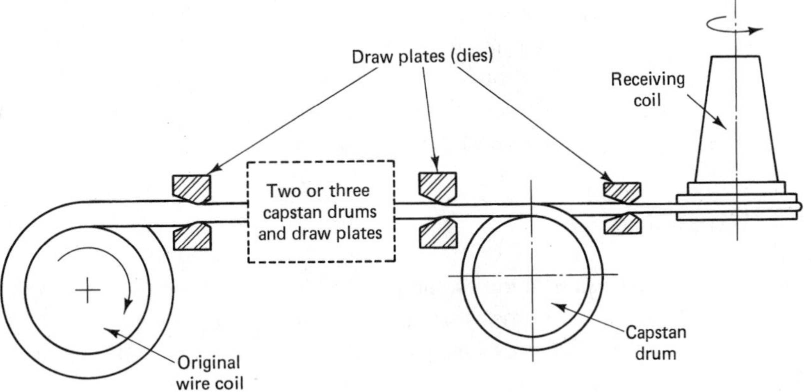

where: a_o is the original area

a_f is the final area

Y is the mean yield stress of the metal

In Equation 5.9, the ratio a_o/a_f is called the *coefficient of elongation,* or simply the *drawing ratio.* In industrial practice, it is usually about 1.25 up to 1.3. Another conjugate term that is used in drawing processes is the reduction, given by the following equation:

$$\text{reduction } r = \frac{a_o - a_f}{a_o} \times 100 \tag{5.10}$$

The theoretically obtained maximum value for the reduction is 64 percent; however, it usually does not exceed about 40 percent in industry.

Defects in wire drawing. Structural damage in the form of voids or cracks occurs in different forms in wire-drawing processes under certain conditions. Following are some of the defects encountered:

1. Internal bursts in wire, taking the form of repeating internal cup and cone fractures (cuppy wire), usually occur when drawing heavily cold-worked copper under conditions of light draft and very large die angles.
2. Similar centerline arrowhead fractures occur if the blank is a sheet and when the die angle and reduction produce severe tension on the centerline.
3. Transverse surface cracking may occur as a result of longitudinal tension stresses in the surface layers.

Tube Drawing

Diameter and thickness of pipes can be reduced by drawing. Figure 5.16 illustrates the simplest type of tube drawing. The final tube thickness is affected by two contradicting factors. The longitudinal stress tends to make the wall thinner, whereas the circumferential stress thickens it. If a large die angle is used, the thinning effect will dominate.

The technique shown in Figure 5.17 of using a fixed plug reduces the tube diameter and controls its thickness. However, a disadvantage of this type of tube drawing is the limitation imposed on the length of the tube by the length of the mandrel. When tubes having longer length are to be drawn, a floating mandrel like that shown in Figure 5.18 is then employed. Another method that has gained widespread application is using a removable mandrel like that shown in Figure 5.19.

5.4 EXTRUSION

Extrusion involves forcing a billet that is enclosed in a container through an opening whose cross-sectional area and dimensions are smaller than those of the original billet. The cross section of the extruded metal will conform to that of the die open-

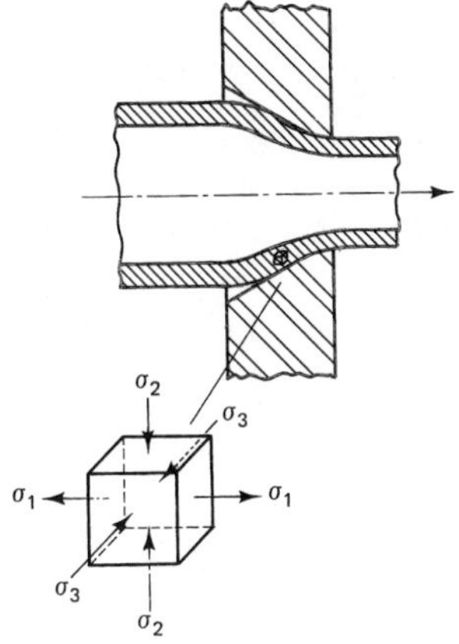

FIGURE 5.16
Simplest type of tube drawing

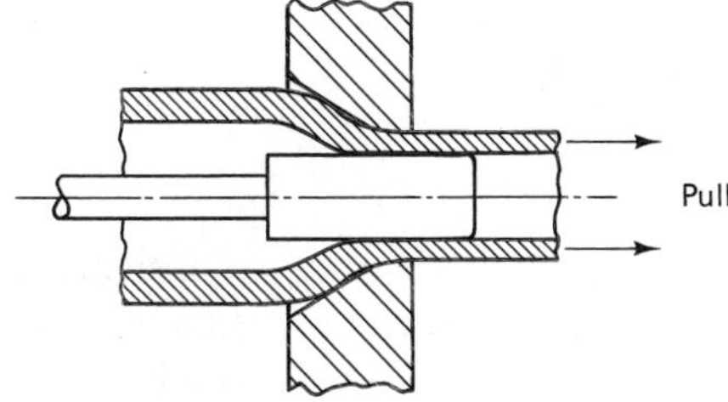

FIGURE 5.17
Tube drawing using a fixed plug

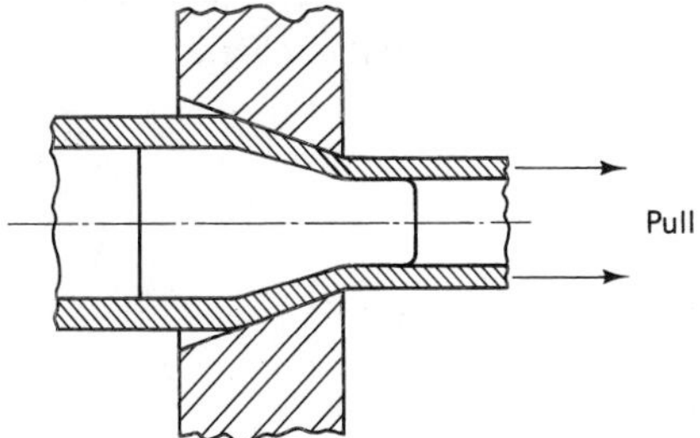

FIGURE 5.18
Tube drawing using a floating mandrel

ing. Historically, extrusion was first used toward the end of the eighteenth century for producing lead pipes. It later gained widespread industrial application for processing nonferrous metals and alloys like copper, brass, aluminum, zinc, and magnesium. Recently, with the modern developments in extrusion techniques, lubricants, and tooling, other metals, such as steels, titanium, refractory metals, uranium, and thorium, can also be extruded successfully. The stock used for extrusion is mainly a cast ingot or a rolled billet. Any surface defects in the original billets must be removed by sawing, shearing, turning, or any other appropriate machining operation before the ex-

FIGURE 5.19
Tube drawing using a removable mandrel

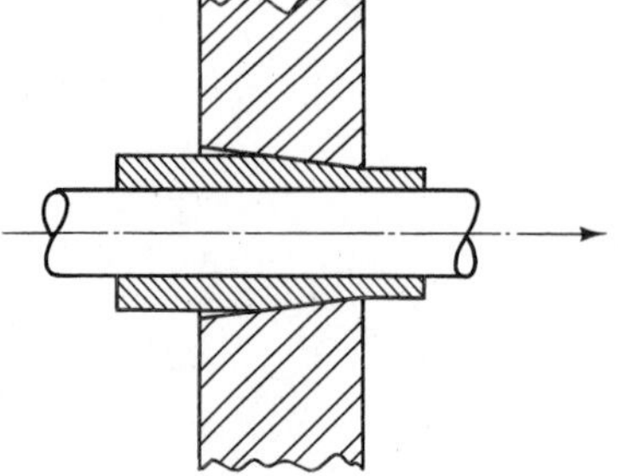

trusion process is performed. Extrusion carried out when the billets are at their cold state is known as *cold extrusion;* when they are at elevated temperatures, it is known as *hot extrusion.* In this latter case, the container, the die, and the pressing plunger must be heated to a temperature of about 650°F (350°C) prior to each extrusion cycle.

Types of Extrusion

Direct extrusion. Direct extrusion is used in the manufacture of solid and hollow slender products and for structural shapes that cannot be obtained by any other metal forming process. Figure 5.20 illustrates the working principles of this method, and Figure 5.21 shows the details of an extrusion die arrangement for producing channel sections. As can be seen, during an extrusion process a billet is pushed out of the die by a plunger and then slides along the walls of the container as the operation proceeds. At the end of the stroke, a small piece of metal (stub-end scrap) remains unextruded in the container.

FIGURE 5.20
Principles of direct extrusion for producing solid objects

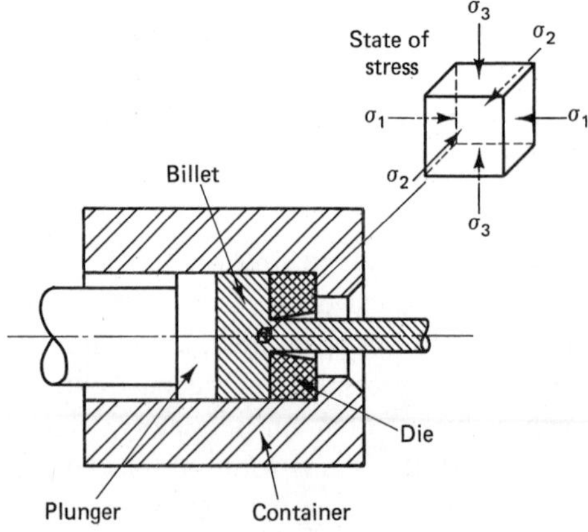

The extruded product is separated by shearing, and the stub-end is then ejected out of the container after the plunger is withdrawn. Also, the leading end of the extruded product does not undergo enough deformation. It is, therefore, poorly shaped and must be removed as well. Obviously, the efficiency of material utilization in this case is low, and the waste can amount to 10 or even 15 percent, as opposed to rolling, where the waste is only 1 to 3 percent. This makes the productivity of direct extrusion quite inferior to that of rolling.

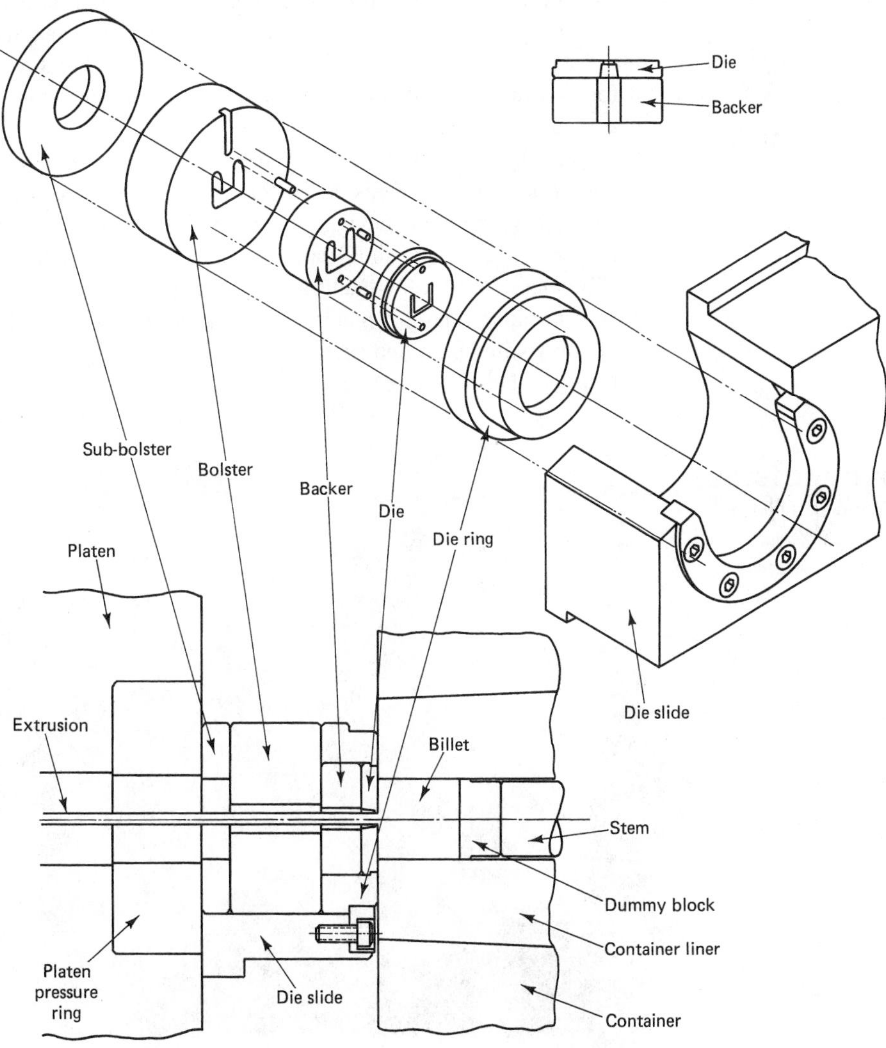

FIGURE 5.21 Typical extrusion die arrangement for producing channel sections

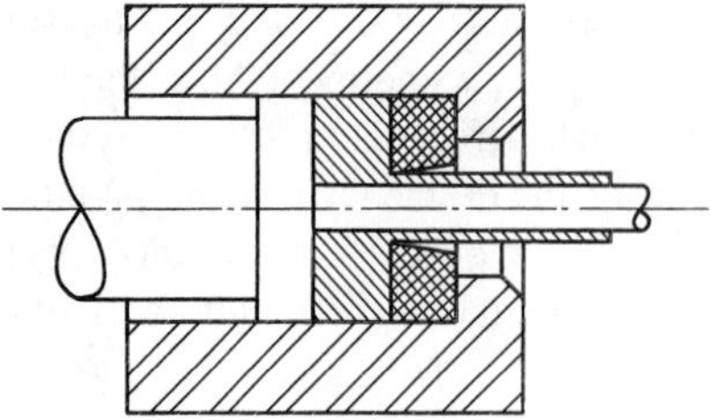

FIGURE 5.22
Direct extrusion for producing hollow objects

Figure 5.22 illustrates the technique used for producing hollow sections and tubes. As can be seen, a mandrel or a needle passes freely through a hole in the blank and the die opening. If the die opening is circular, an annular clearance between the die opening and the mandrel results. When the metal is extruded through the annular clearance, it forms a tube. A hole has to be pierced or drilled into the original blank before it is extruded.

Based on this discussion, it is clear that the conventional extrusion process has the advantages of high-dimensional accuracy and the possibility of producing complex sections from materials having poor plasticity. On the other hand, its disadvantages include low productivity, short tool life, and expensive tooling. Therefore, the process is usually employed for the manufacture of complex shapes with high-dimensional accuracy, especially when the material of the product has a low plasticity. Figures 5.23 and 5.24 show some extruded sections and parts, and Figure 5.25 shows some final products assembled from extruded sections.

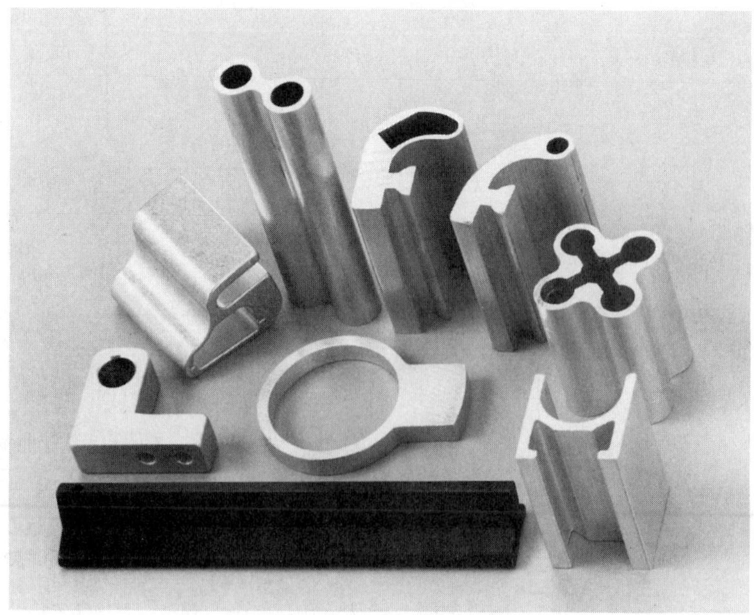

FIGURE 5.23
Some extruded sections *(Courtesy of Midwest Aluminum, Inc., Kalamazoo, Michigan)*

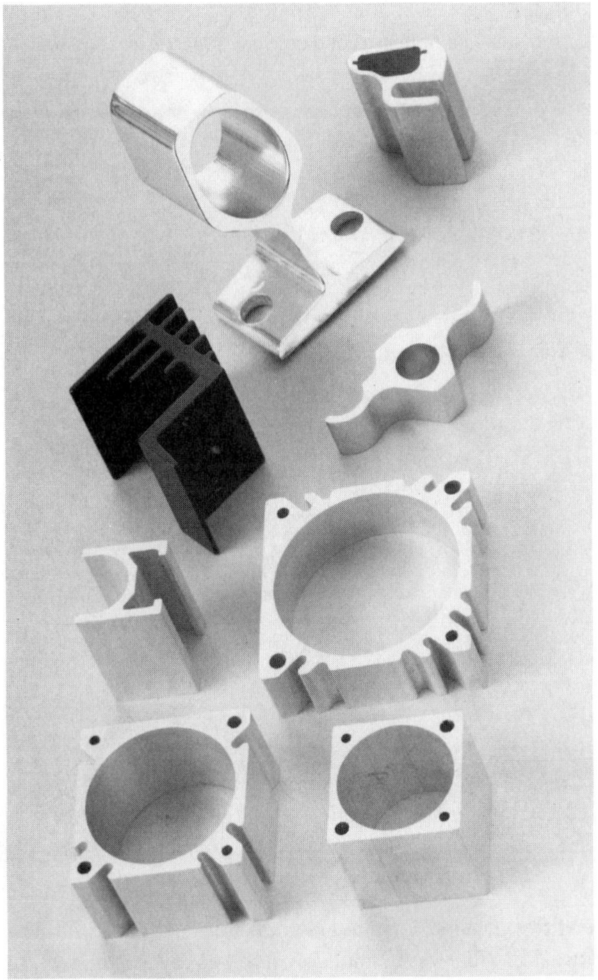

FIGURE 5.24
Some extruded parts
(Courtesy of Midwest Aluminum, Inc., Kalamazoo, Michigan)

Indirect extrusion. In indirect extrusion, the extrusion die is mounted on a hollow ram that is pushed into the container. Consequently, the die applies pressure to the billet, which undergoes plastic deformation. As shown in Figure 5.26, the metal flows out of the die opening in a direction opposite to the ram motion. There is almost no sliding motion between the billet and the container walls. This eliminates friction, and the extrusion load will be lower than that required in forward direct extrusion by about 30 percent. Also, the amount of waste scrap is reduced to only 5 percent. Nevertheless, indirect extrusion finds only limited application due to the complexity and the cost of tooling and press arrangement required.

Another indirect extrusion method, usually called *backward* or *reverse extrusion,* used in manufacturing hollow sections is shown in Figure 5.27. In this case, the metal is extruded through the gap between the ram and the container. As in indirect extrusion for solid objects, the ram and the product travel in opposite directions.

FIGURE 5.25
Some products assembled from extruded sections *(Courtesy of Midwest Aluminum, Inc., Kalamazoo, Michigan)*

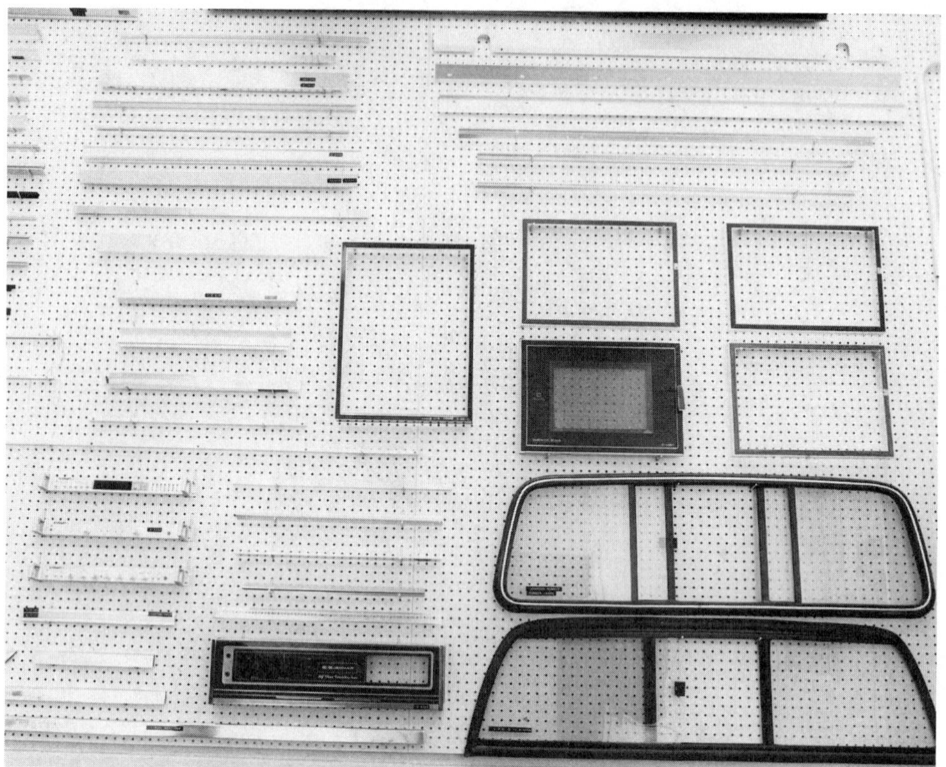

Hydrostatic extrusion. A radical development that eliminates the disadvantages of cold extrusion (like higher loads) involves hydrostatic extrusion. Figure 5.28 illustrates the basic principles of this process, where the billet is shaped to fit the die and surrounded by a high-pressure hydraulic fluid in a container. When the plunger is pressed, it increases the pressure inside the container, and the resulting high pressure forces the billet to flow through the die. Friction between the billet and the container is thus eliminated, whereas friction between the billet and the die is markedly re-

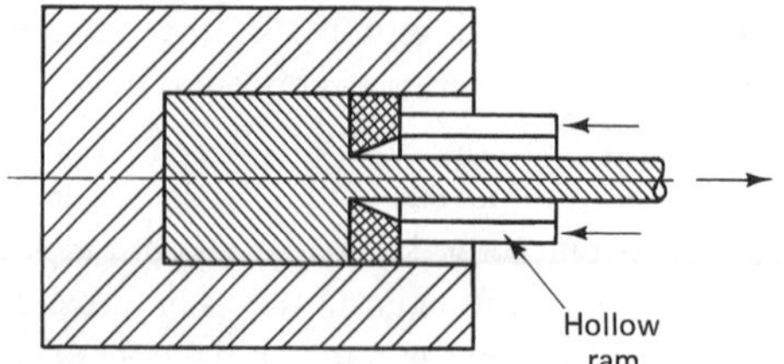

FIGURE 5.26
Indirect extrusion for producing solid objects

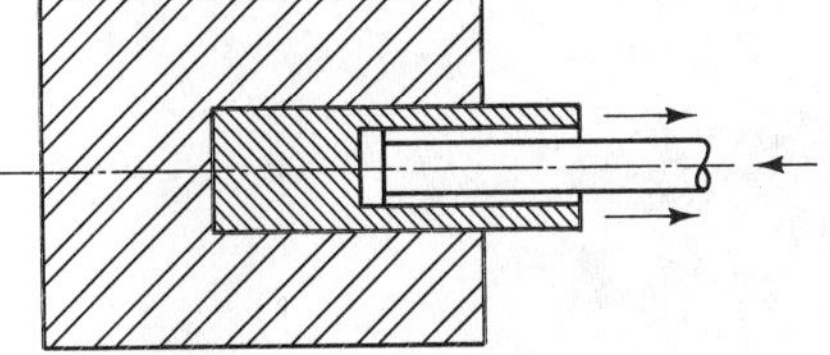

FIGURE 5.27 Indirect (backward) extrusion for producing hollow objects

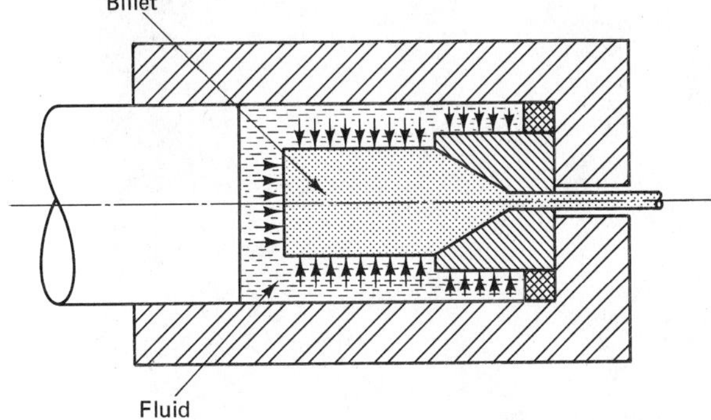

FIGURE 5.28 Principles of hydrostatic extrusion

duced. Also, the buckling effect of longer billets is eliminated because virtually the entire length of the billet is subjected to hydrostatic pressure. This makes it possible to extrude very long billets.

Impact extrusion. Impact extrusion involves striking a cold *slug* of soft metal (like aluminum) that is held in a shallow die cavity with a rapidly moving punch, thus causing the metal to flow plastically around the punch or through the die opening. The slug itself is a closely controlled volume of metal that is lubricated and located in the die cavity. The press is then activated, and the high-speed punch strikes the slug. A finished impacted product is extruded with each stroke of the press. These products are not necessarily cylindrical with a circular cross section. In fact, the range of shapes possible is very broad, including even irregular symmetrical shapes, as shown in Figures 5.29 and 5.30. There are three types of the impact extrusion processes: forward, reverse, and combination (the names referring to the direction of motion of the deforming metal relative to that of the punch).

Figure 5.31 illustrates the basic principles of *reverse* impact extrusion. It is used for manufacturing hollow parts with forged bases and extruded sidewalls. The flowing metal is guided only initially; thereafter, it goes by its own inertia. This results in the elimination of friction and, therefore, an appreciable reduction in the load and energy required. A further advantage is the possibility of producing thinner walls.

The principles of *forward* impact extrusion are illustrated in Figure 5.32. It is mainly employed in producing hollow or semihollow products with heavy flanges and multiple diameters formed on the inside and outside. Closer wall tolerances, larger slenderness ratios, better concentricities, and sound thinner sections are among the advantages of this process.

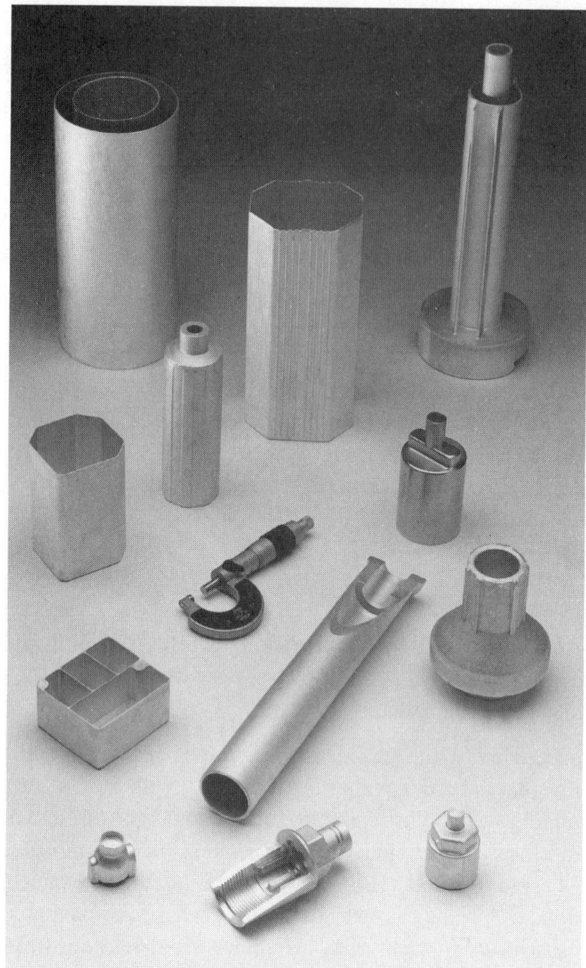

FIGURE 5.29
Some shapes produced by impact extrusion *(Courtesy of Metal Impact Corporation, Rosemont, Illinois)*

Complex shapes can be produced by a *combination* of the two preceding processes, which are performed simultaneously in the same single stroke, as shown in Figure 5.33. Like the other impact extrusion methods, this process has the advantage of cleaner product surfaces, elimination of trimming or further machining operations, and higher strength of the parts obtained.

Mechanics of Extrusion

We can clearly see (from Figure 5.20) that an element of the deforming metal being extruded is subjected to a state of stress involving triaxial compression. This all-around high pressure results in a marked improvement in the plasticity of the metal. Consequently, extrusion can be employed when working metal having poor plasticity, as opposed to rolling or wire drawing, where only ductile metals can be formed (worked).

FIGURE 5.30 Some components produced by impacting *(Courtesy of Metal Impact Corporation, Rosemont, Illinois)*

Load requirement. For the sake of simplicity, it is sometimes assumed that the processes involve ideal deformation without any friction. The extrusion pressure can then be given by the following equation:

$$p_{\text{extrusion}} = Y \times \ell n \frac{a_o}{a_f} = Y \times \ell n R \tag{5.11}$$

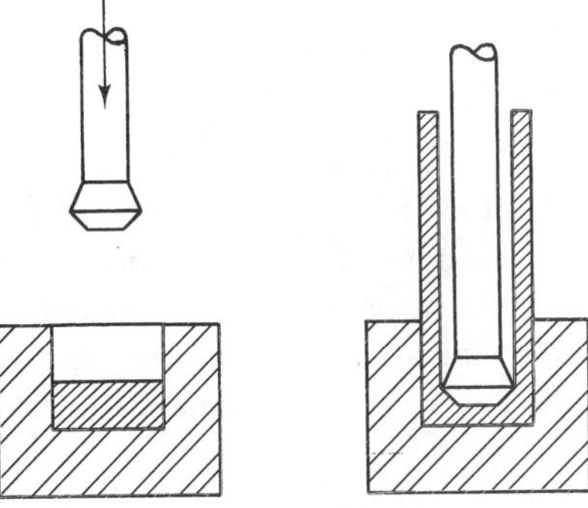

FIGURE 5.31 Principles of reverse impact extrusion

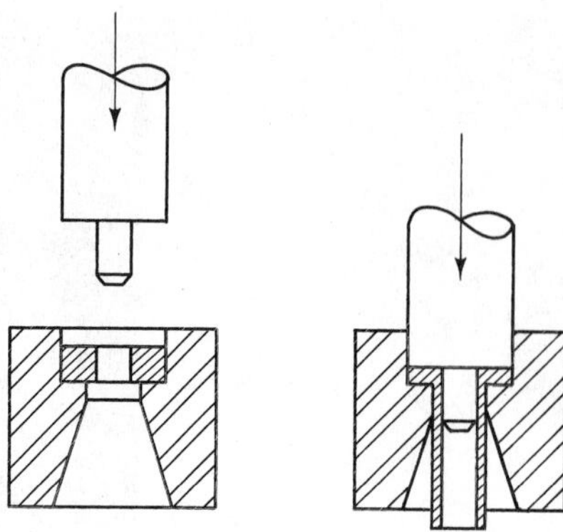

FIGURE 5.32 Principles of forward impact extrusion

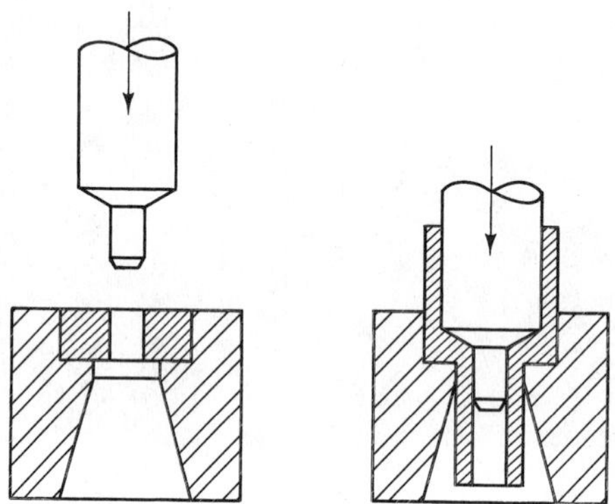

FIGURE 5.33 Combination impacting

where: a_o is the original cross-sectional area

a_f is the final cross-sectional area after extrusion

R is the extrusion ratio

Y is the mean yield stress of the metal

The extrusion load is, therefore,

$$F = p \times a_o \tag{5.12}$$

These equations are used to give only rough estimates because actual extrusion processes involve friction and the lack of homogeneous deformation of the metal, as will be seen later. Therefore, research workers developed several empirical formulas to give the extrusion pressure as a function of the extrusion ratio and the mechanical properties of the metal. A convenient formula was proposed by W. Johnson (the eminent British researcher in the area of metal forming) as follows:

$$p_{\text{extrusion}} = \overline{Y}\left(0.8 + 1.5\ell n \frac{1}{1 - r}\right) \tag{5.13}$$

In Equation 5.13, r is the reduction given by

$$\text{reduction } r = \frac{a_o - a_f}{a_o} \tag{5.14}$$

Metal flow and deformation. To study metal flow, let us consider extruding a split billet involving two identical halves, with a rectangular grid engraved on the meridional plane of each half. The separation surface is covered with lanolin or a similar appropriate material to prevent welding or sticking of the two halves during the process. After extruding the split billet, the two halves are separated, and the distortion of the grid can be investigated. Figure 5.34 shows the grid after extrusion. We can see that the units of the grid, which were originally square in shape, became parallelograms, trapezoids, and other shapes. The following can also be observed:

1. The velocity of the core is greater than that of the outer layers.
2. The outer layers are deformed to a larger degree than the core.
3. The leading end of the extruded part is almost undeformed.
4. The metal adjacent to the die does not flow easily, leading to the initiation of zones where little deformation occurs. These zones are called *dead-metal zones.*

In fact, the preceding method for studying the metal flow is usually used with models made of wax, plasticine, and lead to predict any defect that may occur during the actual process so that appropriate precautions can be taken in advance.

Lubrication in Extrusion

Friction at the billet-die and billet-container interfaces increases the load and the power requirement and reduces the service life of the tooling. For these reasons, lubricants are applied to the die and container walls.

As in wire drawing, soaps and various oils containing chlorinated additives or graphite are used as lubricants in cold extrusion of most metals, whereas lanolin is usually used for the softer ones. For hot extrusion of mild steel, graphite is an adequate lubricant. It is not, however, recommended for high-temperature extrusions, such as extruding molybdenum at 3250°F (1800°C); in this case, glass is the most successful lubricant.

Defects in Extruded Products

Defects in extruded parts usually fall into one of three main categories: surface or internal cracking, sinking (piping), and skin-inclusion defects. *Cracking* is caused by secondary tensile stresses acting within a material having low plasticity. Cracking can occur on the surface of a relatively brittle material during the extrusion process, and it

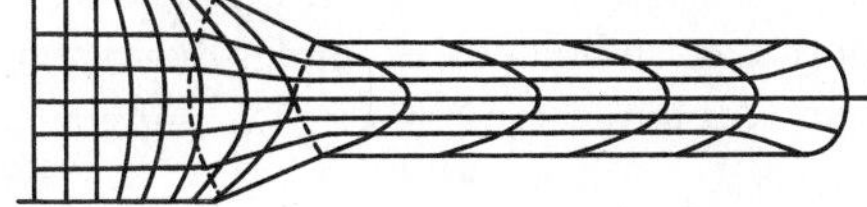

FIGURE 5.34 Distorted grid indicating metal flow in extrusion

may also occur in the form of fire-tree or central bursts when extruding materials like bismuth, magnesium, 60/40 brass, steel, and brittle aluminum alloys. *Piping* involves sinking of the material at the rear of the stub-end. This defect is usually encountered toward the end of the extrusion stroke, especially when the original billets are relatively short. *Skin-inclusion defects* may take different forms, depending upon the degree of lubrication and the hardness of the surface layer of the original stock. When extruding lubricated billets of high-copper alloys, the surface skin will slide over the container wall and then penetrate the billet, as illustrated in Figure 5.35, where the three different extrusion defects are sketched.

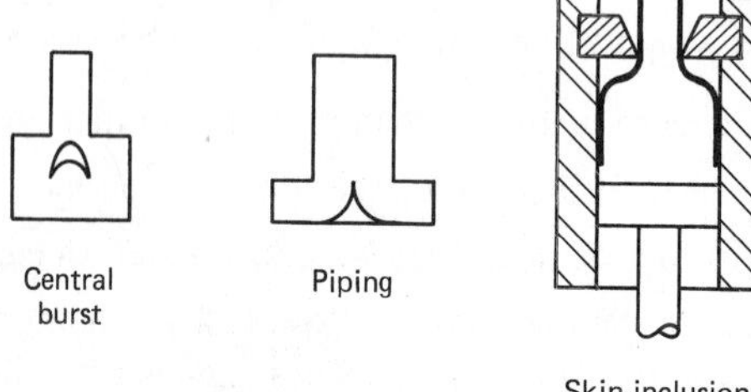

FIGURE 5.35 Three different defects occurring in an extrusion process

Design Considerations

Conventional extrusions. When making parts that have constant cross sections, the extrusion process is usually more economical and faster than machining, casting, or fabricating the shapes by welding (or riveting). Also, the designer of the extruded section is relatively free to put the metal where he or she wants. Nevertheless, there are some design guidelines that must be taken into consideration when designing an extruded section:

1. The *circle size* (i.e., the diameter of the smallest circle that will enclose the extrusion cross section) can be as large as 31 inches (775 mm) when extruding light metals.

2. Solid shapes are the easiest to extrude. Semihollow and hollow shapes are more difficult to extrude, especially if they have thin walls or include abrupt changes in wall thickness.

3. Wall thicknesses must be kept uniform. If not, all transitions must be streamlined by generous radii at the thick-thin junctions.

4. Sharp corners at the root of a die tongue should be avoided when extruding semihollow sections.

5. A complicated section should be broken into a group of simpler sections that are assembled after the separate extrusion processes. In such a case, the sections should be designed to simplify assembly; for example, they should fit, hook, or snap together. Screw slots or slots to receive other tightening material, such as plastic, may also be provided.

Figure 5.36 illustrates some recommended designs for assembling extruded aluminum sections. Figure 5.37 illustrates and summarizes some recommended designs as well as those to be avoided as general guidelines for beginning designers.

Aluminum impact extrusions. In order to accomplish good designs of aluminum impact extrusions, all factors associated with and affecting the process must be taken into account. Examples are alloy selection, tool design, lubrication, and, of course, the general consideration of mechanical design. Following are some basic guidelines and design examples:

1. Use alloys that apply in the desired case and have the lowest strength.
2. An impact extrusion should be symmetrical around the punch.
3. Threads, cuts, projections, and the like are made by subjecting the impact extrusions to further processing.
4. For reverse extrusions, the ratio of maximum length to internal diameter must not exceed 8 to avoid failure of long punches.
5. A small outer-corner radius must be provided for a reverse extrusion, but the inner-corner radius must be kept as small as possible (see Figure 5.38a).
6. The thickness of the bottom near the wall must be 15 percent greater than the thickness of the wall itself to prevent shear failure (see Figure 5.38b).
7. The inside bottom should not be completely flat. To avoid the possibility of the punch skidding on the billet, only 80 percent of it at most can be flat (see Figure 5.38c).

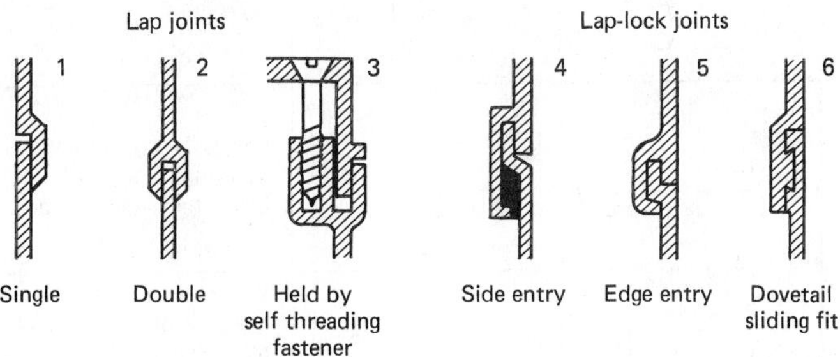

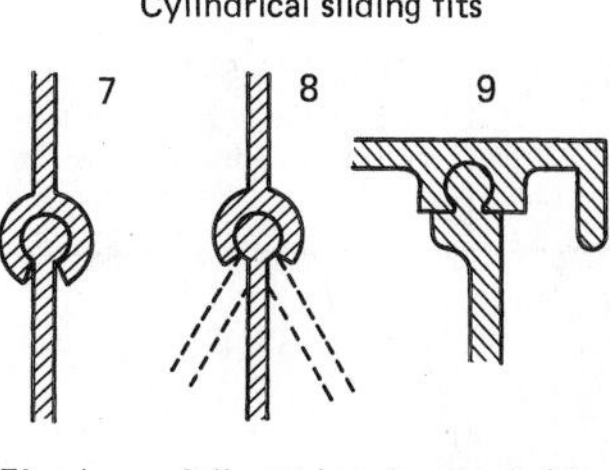

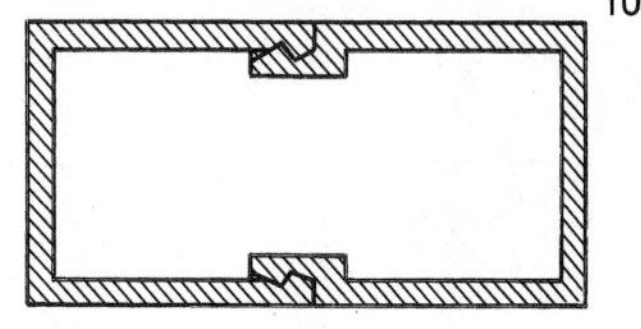

FIGURE 5.36 Some recommended designs for assembling extruded aluminum sections *(Courtesy of the Aluminum Association, Inc., Washington, D.C.)*

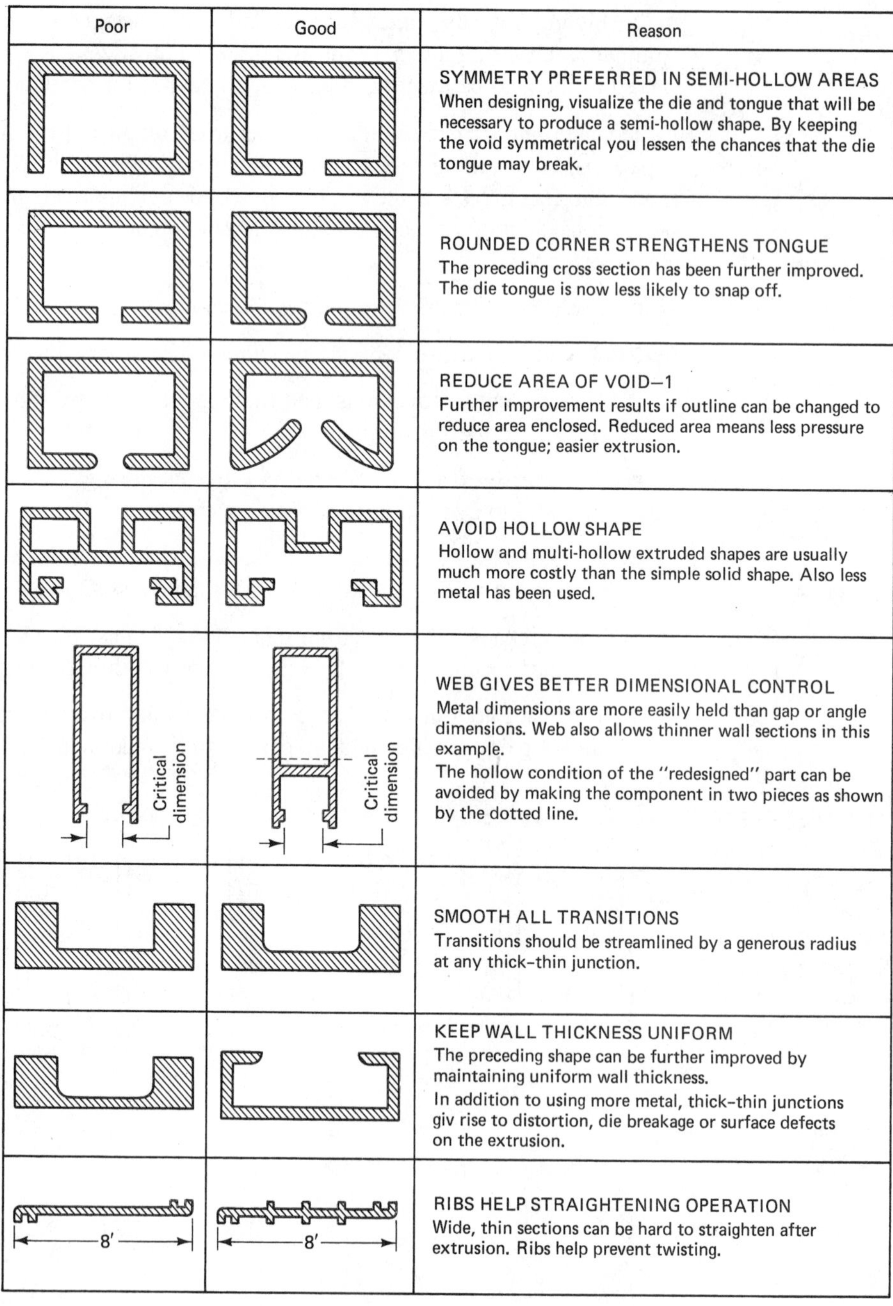

Poor	Good	Reason
		SYMMETRY PREFERRED IN SEMI-HOLLOW AREAS When designing, visualize the die and tongue that will be necessary to produce a semi-hollow shape. By keeping the void symmetrical you lessen the chances that the die tongue may break.
		ROUNDED CORNER STRENGTHENS TONGUE The preceding cross section has been further improved. The die tongue is now less likely to snap off.
		REDUCE AREA OF VOID–1 Further improvement results if outline can be changed to reduce area enclosed. Reduced area means less pressure on the tongue; easier extrusion.
		AVOID HOLLOW SHAPE Hollow and multi-hollow extruded shapes are usually much more costly than the simple solid shape. Also less metal has been used.
Critical dimension	Critical dimension	WEB GIVES BETTER DIMENSIONAL CONTROL Metal dimensions are more easily held than gap or angle dimensions. Web also allows thinner wall sections in this example. The hollow condition of the "redesigned" part can be avoided by making the component in two pieces as shown by the dotted line.
		SMOOTH ALL TRANSITIONS Transitions should be streamlined by a generous radius at any thick-thin junction.
		KEEP WALL THICKNESS UNIFORM The preceding shape can be further improved by maintaining uniform wall thickness. In addition to using more metal, thick-thin junctions giv rise to distortion, die breakage or surface defects on the extrusion.
8′	8′	RIBS HELP STRAIGHTENING OPERATION Wide, thin sections can be hard to straighten after extrusion. Ribs help prevent twisting.

FIGURE 5.37 Some design considerations for conventional extrusions *(Courtesy of the Aluminum Association, Inc., Washington, D.C.)*

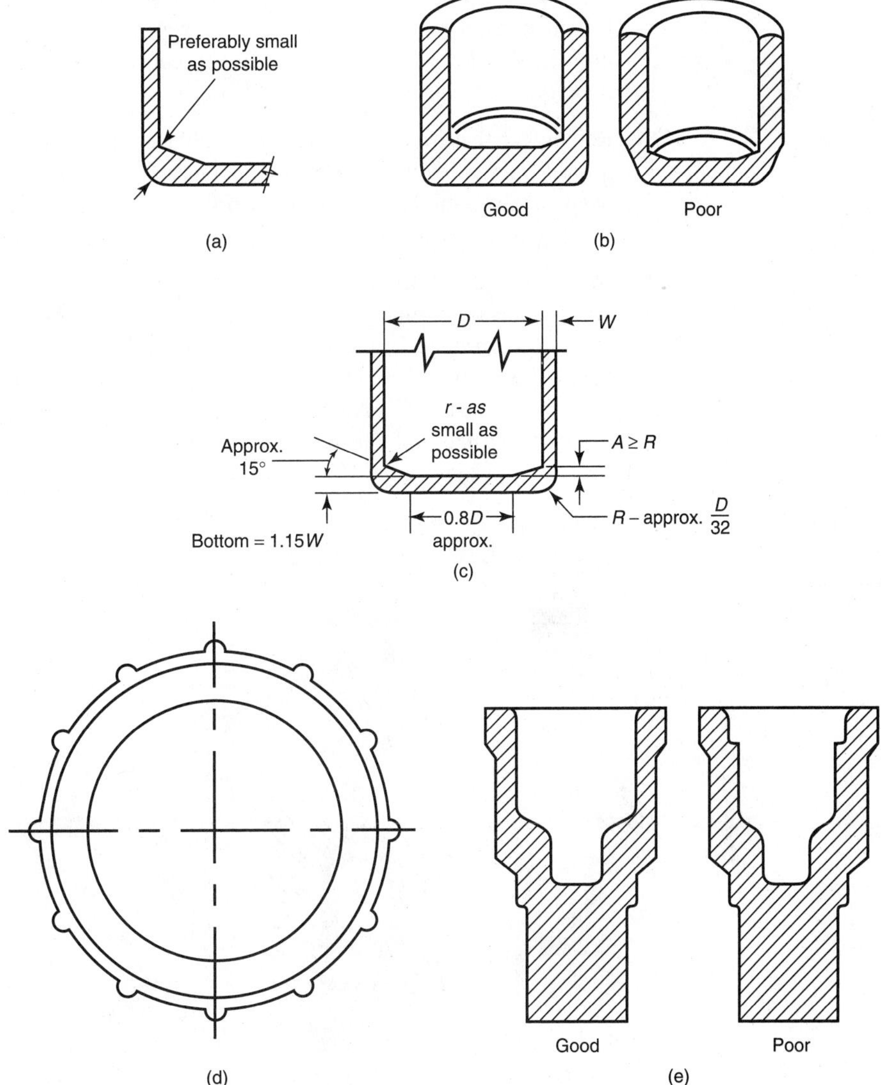

FIGURE 5.38 Some design considerations for impact extrusions: (a) corner radii for reverse extrusion; (b) thickness of the bottom near the wall; (c) inside bottom; (d) ribs; (e) multiple-diameter parts *(Courtesy of the Aluminum Association, Inc., Washington, D.C.)*

8. External and internal bosses are permitted, provided that they are coaxial with the part. However, the diameter of the internal boss should not be more than 1/4 of the internal diameter of the shell.

9. Longitudinal ribs, whether external or internal, on the full length of the impact extrusion are permitted. They should preferably be located in a symmetrical distribution. However, the height of each rib must not exceed double the thickness of the wall of the shell (see Figure 5.38d). The main function of ribs is to provide stiffness to the walls of shells. They are also sometimes used for other reasons, such as to provide locations for drilling and tapping (to assemble the part), to enhance cooling by radiation, and to provide an appropriate gripping surface.

10. An impact extrusion can have a wall with varying thickness along its length (i.e., it can be a multiple-diameter part). However, internal steps near the top of the product should be avoided because they cause excessive loading and wear of the punch (see Figure 5.38e).

11. Remember that it is sometimes impossible to obtain the desired shape directly by impacting. However, an impact extrusion can be considered as an intermediate product that can be subjected to further working or secondary operations like machining, flange upsetting, nosing and necking, or ironing (see Figure 5.39a through c).

Again, in addition to the preceding guidelines, general rules of mechanical design as well as common engineering sense are necessary for obtaining a successful design for the desired product. It would also be beneficial for the beginner to look at various designs of similar parts and to consult with experienced people before starting the design process. Given in Figure 5.40 are sketches reflecting good design practice for some impact-extruded tubular parts and shells.

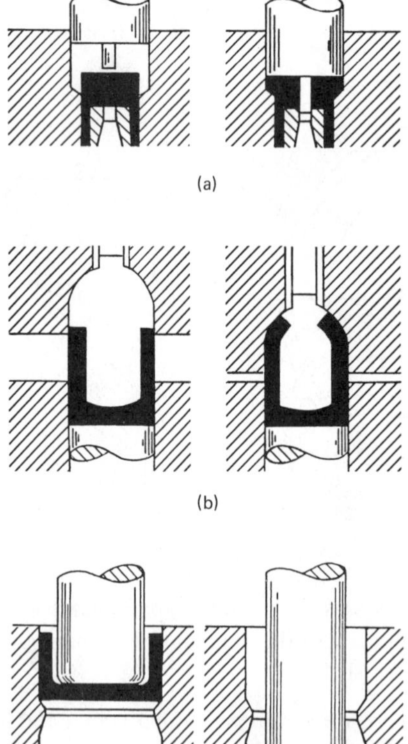

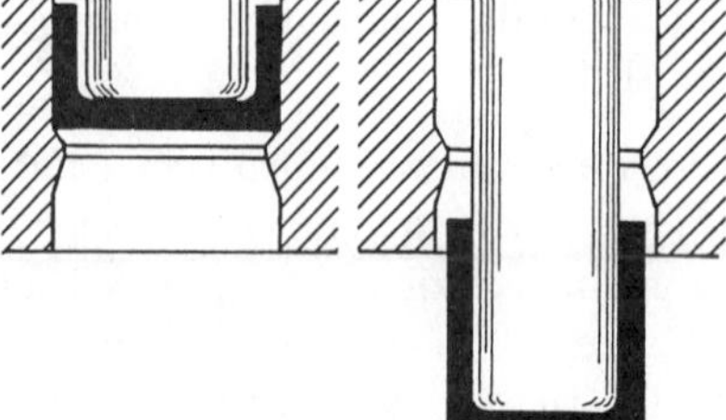

FIGURE 5.39 Some secondary operations after impact extruding: (a) flange upsetting; (b) nosing; (c) ironing *(Courtesy of the Aluminum Association, Inc., Washington, D.C.)*

FIGURE 5.40 Sketches reflecting good design practice for some impact-extruded tubular parts and shells *(Courtesy of Aluminum Association, Inc., Washington, D.C.)*

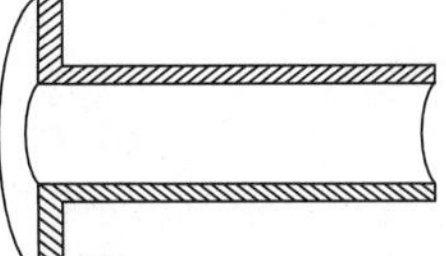

Flanged tube with open end.

Cup and tube assembly, extruded as single piece.

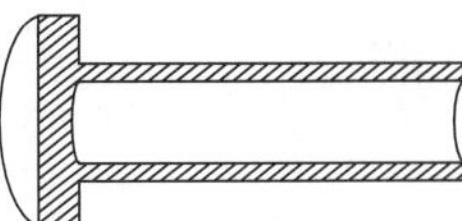

Flange end closed.

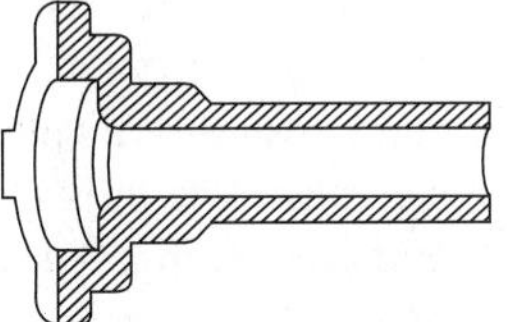

Flange with multiple step-down diameters.

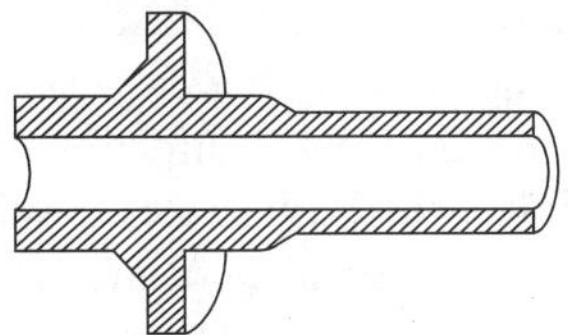

Flanged tube with multiple diameters.

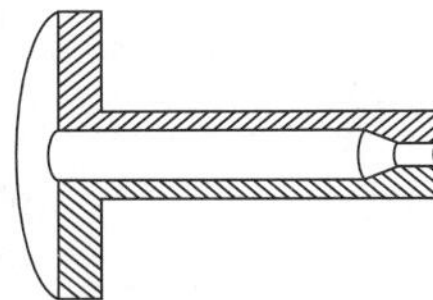

Partially closed end tube with heavy flange.

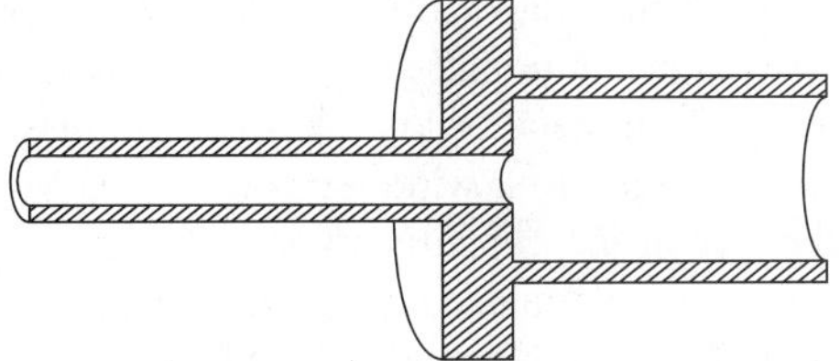

Combination impact with the flange at the midpoint. Such an impact also serves as a transition from one diameter to another. Wall thicknesses can also be varied.

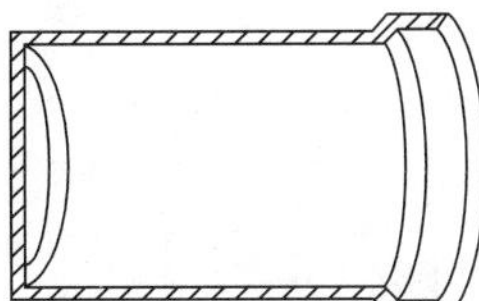

Uniform wall thickness with flanged end open.

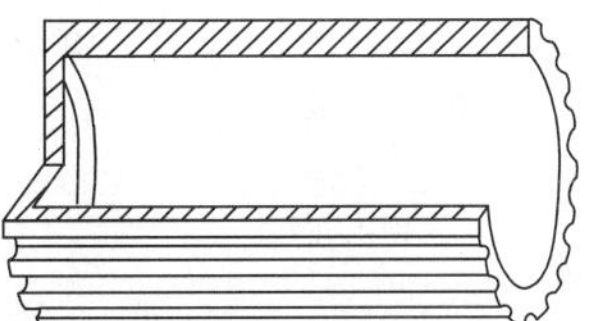

Outside longitudinal ribs can be spaced equally or in symmetrical patterns. Ribs may be extended to become cooling fins.

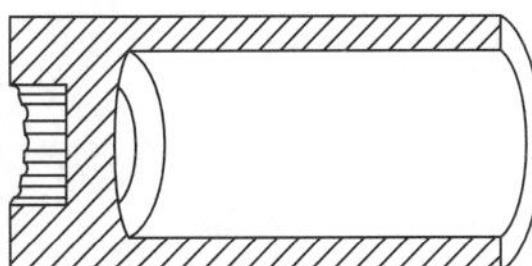

Short recessed ribs in bottom can be used for tool insertions, drive, etc.

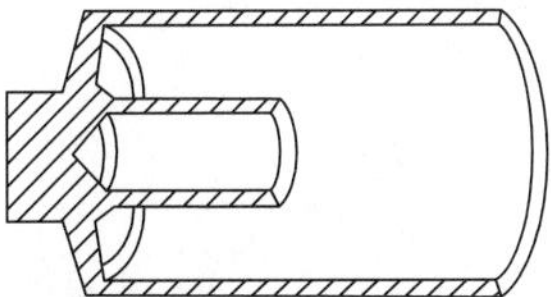

An external boss can be combined with an internal center tube.

An integral center tube can be formed so that assembly and machining are not needed.

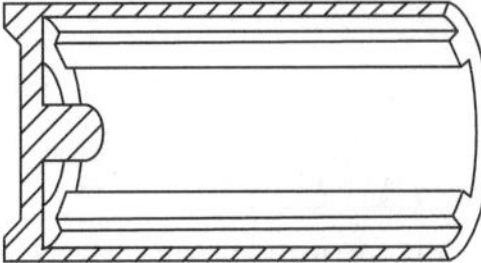

Inside bosses can be produced as integral parts of the closed end. The side wall can have longitudinal internal ribs.

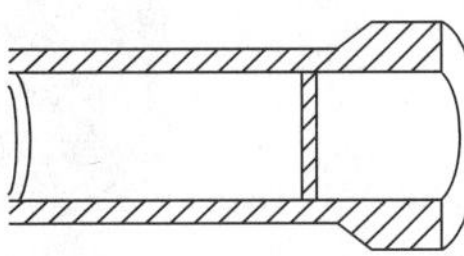

Combination impact.

5.5 FORGING

The term *forging* is used to define the plastic deformation of metals at elevated temperatures into predetermined shapes using compressive forces that are exerted through dies by means of a hammer, a press, or an upsetting machine. Like other metal forming processes, forging refines the microstructure of the metal, eliminates the hidden defects such as hair cracks and voids, and rearranges the fibrous macrostructure to conform with the metal flow. It is mainly the latter factor that gives forging its merits and advantages over casting and machining. By successful design of the dies, the metal flow during the process can be employed to promote the alignment of the fibers with the anticipated direction of maximum stress. A typical example is shown in Figure 5.41, which illustrates the fibrous macrostructure in two different crankshafts produced by machining from a bar stock and by forging. As can be seen, the direction of the fibers in the second case is more favorable because the stresses in the webs when the crankshaft is in service coincide with the direction of fibers where the strength is maximum.

A large variety of materials can be worked by forging. These include low-carbon steels, aluminum, magnesium, and copper alloys, as well as many of the alloy steels and stainless steels. Each metal or alloy has its own plastic forging temperature range. Some alloys can be forged in a wide temperature range, whereas others have narrow ranges, depending upon the constituents and the chemical composition. Usually, the forging temperatures recommended for nonferrous alloys and metals are much lower than those required for ferrous materials. Table 5.2 indicates the range of forging temperatures for the commonly used alloys.

Forged parts vary widely in size ranging from a few pounds (less than a kilogram) up to 300 tons (3 MN) and can be classified into small, medium, and heavy forgings.

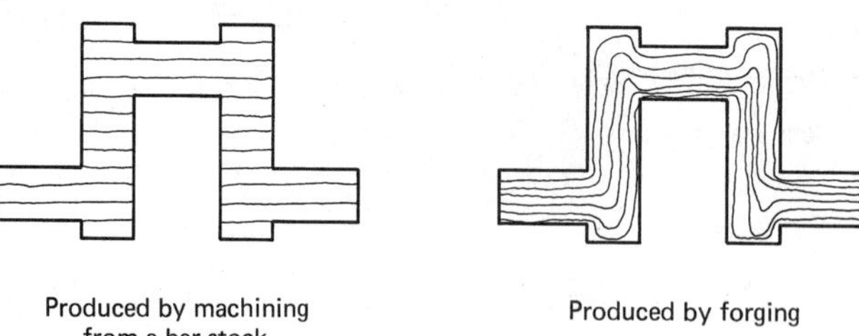

FIGURE 5.41 The fibrous macrostructure in two crankshafts produced by machining and by forging

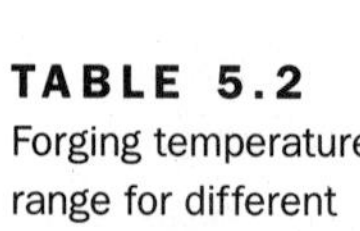

TABLE 5.2 Forging temperature range for different metals

Metal	Forging Temperature
Low-carbon steel	1450–2550°F (800–1400°C)
Aluminum	645–900°F (340–480°C)
Magnesium	645–800°F (340–430°C)
Copper	800–1900°F (430–1040°C)
Brass	1100–1700°F (590–930°C)

Small forgings are illustrated by small tools such as chisels and tools used in cutting and carving wood. Medium forgings include railway-car axles, connecting rods, small crankshafts, levers, and hooks. Among the heavier forgings are shafts of power-plant generators, turbines, and ships, as well as columns of presses and rolls for rolling mills. Small and medium forgings are forged from rolled sections (bar stocks and slabs) and blooms, whereas heavier parts are worked from ingots.

All forging processes fall under two main types: open-die forging processes, in which the metal is worked between two flat dies, and closed-die forging processes, in which the metal is formed while being confined in a closed impression of a die set.

Open-Die Forging

Open-die forging is sometimes referred to as *smith forging* and is actually a development or a modern version of a very old type of forging, blacksmithing, that was practiced by armor makers and crafts people. Blacksmithing required hand tools and was carried out by striking the heated part repeatedly by a hammer on an anvil until the desired shape was finally obtained. Nowadays, blacksmith forging is used only when low production of light forgings is required, which is mainly in repair shops. Complicated shapes having close tolerances cannot be produced economically by this process.

The modern version of blacksmithing, open-die forging, involves the substitution of a power-actuated hammer or press for the arm, hand hammer, and anvil of the smith. This process is used for producing heavy forgings weighing up to more than 300 tons, as well as for producing small batches of medium forgings with irregular shapes that cannot be produced by modern closed-die forging. The skill of the operator plays an important role in achieving the desired shape of the part by manipulating the heated metal during the period between successive working strokes. Accordingly, the shape obtained is just an approximation of the required one, and subsequent machining is always used in order to produce the part that accurately conforms to the blueprint provided by the designer.

Open-die forging operations. A smith-forging process usually consists of a group of different operations. Among the operations employed in smith forging are upsetting, drawing out, fullering, cutting off, and piercing. The force and energy required differ considerably from one operation to another, depending upon the degree of "confinement" of the metal being worked. Following is a brief description of some of these operations:

1. **Upsetting.** Upsetting involves squeezing the billet between two flat surfaces, thus reducing its height due to the increase in the cross-sectional area. As can be seen in Figure 5.42a, the state of stress is uniaxial compression. In practice, however, the billets' surfaces in contact with the die are subjected to substantial friction forces that impede the flow of the neighboring layers of metal. This finally results in a heterogeneous deformation and in barreling of the deformed billet. To obtain uniform deformation, the billet-die interfaces must be adequately lubricated.

2. **Drawing out.** In drawing out, the workpiece is successively forged along its length between two dies having limited width. This results in reducing the cross-sectional

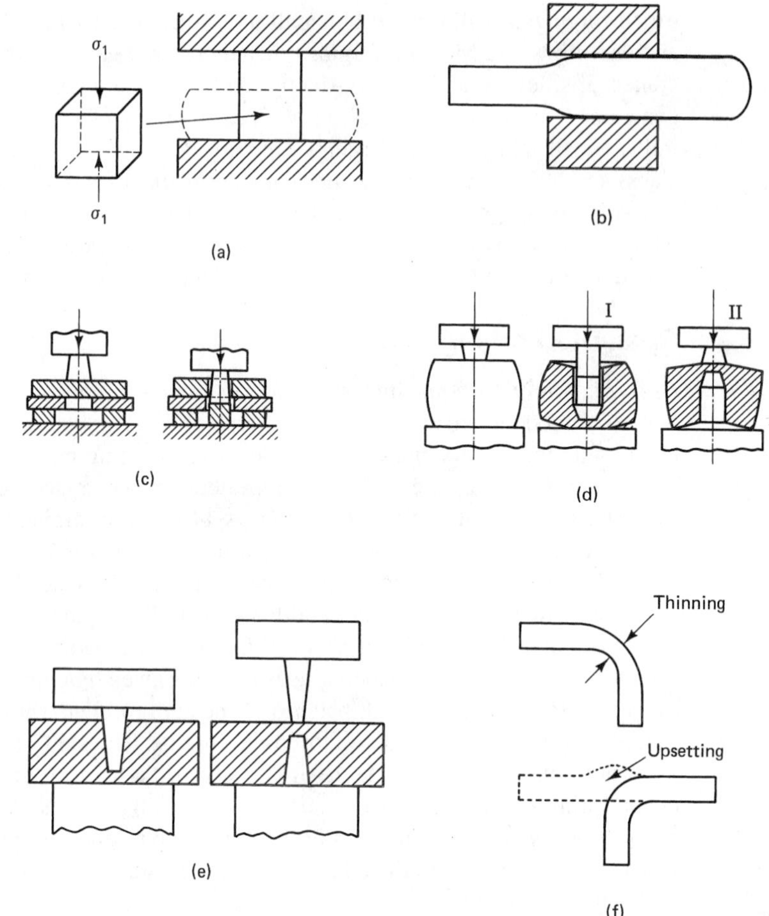

FIGURE 5.42 Various smith-forging operations: (a) upsetting; (b) drawing out; (c) piercing a short billet; (d) piercing a long billet; (e) cutting off; (f) bending

area of the workpiece while increasing its length, as shown in Figure 5.42b. This operation can be performed by starting either at the middle or at the end of the workpiece. A large reduction in the cross-sectional area can be achieved by reducing the feed of the workpiece. The bite (i.e., the length of feed before the working stroke) ranges between 40 and 75 percent of the width of the forging die.

3. **Piercing operation.** A piercing operation is performed in order to obtain blind or through holes in the billet. A through hole can be pierced directly in a short billet in a single stroke by employing a punch and a supporting ring, as shown in Figure 5.42c. On the other hand, billets with large height-to-diameter ratios are pierced while located directly on the die with the help of a piercer and possibly an extension piece as well, as shown in Figure 5.42d. In this latter case, the diameter of the piercer must not exceed 50 percent of that of the billet. For larger holes, hollow punches are employed. Also, holes can be enlarged by tapered punches.

4. **Cutting off.** Cutting off involves cutting the workpiece into separate parts using a forge cutter or a suitable chisel. This is usually done in two stages, as can be seen in Figure 5.42e.

5. **Bending.** In bending, thinning of the metal occurs on the convex side at the point of localized bending (where bending actually takes place). It is, therefore, recommended to upset the metal at this location before bending is performed, as shown in Figure 5.42f, in order to obtain a quality bend.

Examples of open-die forged parts. As mentioned before, a part may require a series of operations so that it can be given the desired shape by smith forging. Following are some examples of smith-forged industrial components, together with the steps involved in the manufacture of each part:

1. **Large motor shaft.** First, 24-inch-square (60 cm) steel ingots are rolled into square blooms, each having a 12-inch (30-cm) side. The blooms are then heated and hammered successively across the corners until the workpiece is finally rounded to a diameter of 10 inches (25 cm). These steps are illustrated in Figure 5.43.

2. **Flange coupling.** The sequence of operations is illustrated in Figure 5.44. There are two operations or stages involved, upsetting and heading. In heading, the flow of metal of most of the billet is restricted by using a ring-shaped tool. This process allows excellent grain flow to be obtained, which is particularly advantageous in carrying tangential loads.

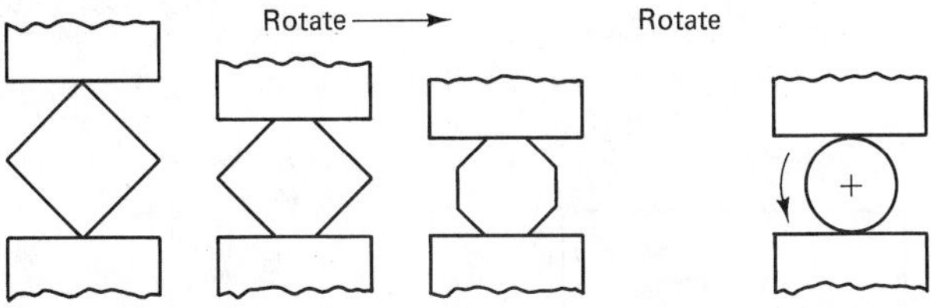

FIGURE 5.43 The production of a large motor shaft by smith forging

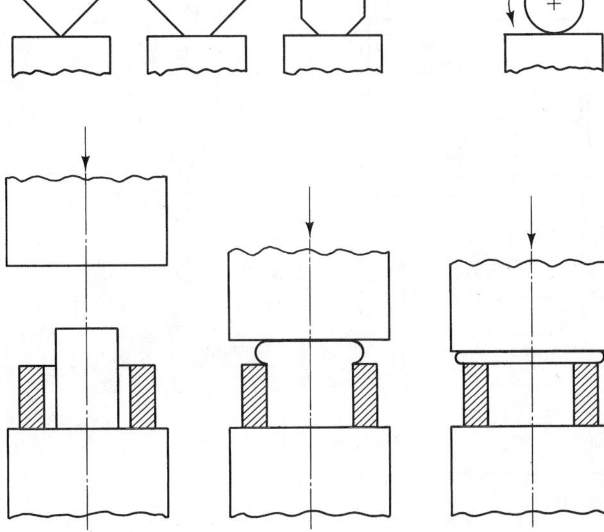

FIGURE 5.44 The production of flange coupling by smith forging

3. **Rings.** A billet is first upset and is then subjected to a piercing operation. This is followed by an expanding operation using a mandrel to reduce the thickness of the ring and increase its diameter as required. Larger rings are usually expanded on a saddle. The steps involved in the process of ring forging are illustrated in Figure 5.45.

Equipment for smith forging. Smaller billets are usually smith-forged using pneumatic-power hammers. Larger components are worked in steam-power hammers (or large pneumatic hammers), whereas very large and heavy parts are produced by employing hydraulic presses. Following is a brief description of smith-forging equipment:

1. **Steam-power hammers.** A steam-power hammer consists mainly of the moving parts (including the ram, the rod, and the piston); a lifting and propelling device, which is a double-acting high-pressure steam cylinder; the housing or frame, which can be either an arch or an open type; and the anvil. Figure 5.46 illustrates the working principles. First, the piston and the other moving parts are raised by admitting steam into the lower side of the cylinder (under the piston) through the

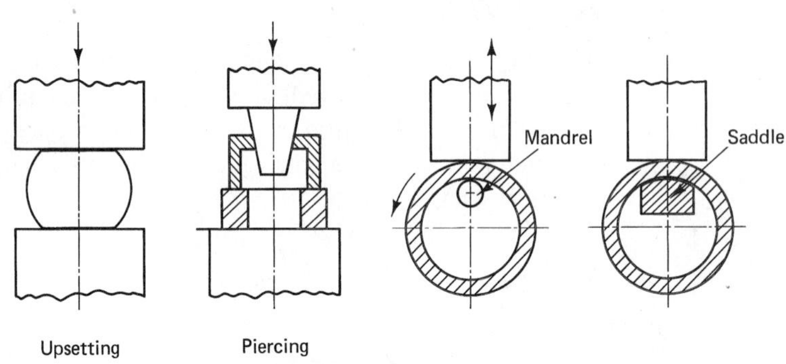

FIGURE 5.45 The production of large rings by smith forging

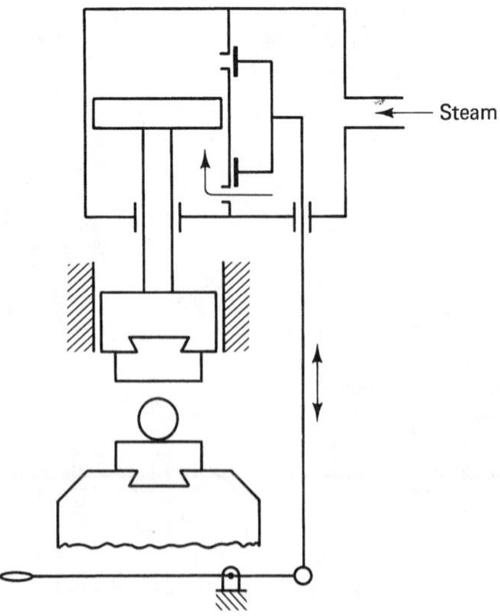

FIGURE 5.46 The working principles of a steam-power hammer

sliding valve. When a blow is required, the lever is actuated; the sliding valve is accordingly shifted to admit steam to the upper side of the cylinder (above the piston) and exhaust the steam that was in the lower side, thus pushing the moving parts downward at a high speed. In steam-power hammers, the velocity of impact can be as high as 25 feet per second (3 m/s), whereas the mass of the moving parts can be up to 11,000 slugs (5000 kg). The amount of energy delivered per blow is, therefore, extremely large and can be expressed by the equation:

$$E = \tfrac{1}{2} mV^2 \tag{5.15}$$

where: E is the energy
m is the mass of the moving parts
V is the impact velocity

Nevertheless, not all of that energy is consumed in the deformation of the workpiece. The moving parts rebound after impact, and the anvil will try to move in the opposite direction, thus consuming or actually wasting a fraction of the blow energy. The ratio between the energy absorbed in deforming the metal to that delivered by the blow is called the *efficiency* of a hammer and can be given by the following equation:

$$\eta = \frac{M}{M + m}(1 - K^2) \tag{5.16}$$

where: M is the mass of the anvil
K is a factor that depends upon the elasticity of the billet

The harder and more elastic the billet is, the higher that factor will be, and the lower the efficiency becomes. In addition, the hammer efficiency depends upon the ratio $M/(M + m)$, or actually the ratio between the masses of the anvil and the moving parts, which is taken in practice between 15 and 20. On the other hand, the value of K ranges between 0.05 and 0.25.

2. **Pneumatic-power hammers.** There are two kinds of pneumatic-power hammers. The first kind includes small hammers in which the air compressor is built in; they usually have open frames because their capacity is limited. The second kind of pneumatic hammer is generally similar to a steam-power hammer in construction and operation, the only difference being that steam is replaced by compressed air (7 to 8 times the atmospheric pressure). As is the case with steam, this necessitates separate installation for providing compressed air. Pneumatic hammers do not have some of the disadvantages of steam hammers, such as dripping of water resulting from condensation of leakage steam onto the hot billet. This may result in cracking of the part, especially when forging steel.

3. **Hydraulic presses.** Heavy forgings are worked in hydraulic presses. The press installation is composed of the press itself and the hydraulic drive. Presses capable of providing a force of 75,000 tons (750 MN) are quite common. Still, hydraulic presses that are commonly used in the forging industry have capacities ranging between 1000 tons (10 MN) and 10,000 tons (100 MN). These presses can success-

fully handle forgings weighing between 8 and 250 tons. The large-capacity presses require extremely high oil pressure in the hydraulic cylinders (200 to 300 times the atmospheric pressure). Because no pump can deliver an adequate oil discharge at that pressure level, this process is usually overcome by employing accumulators and intensifiers that magnify the oil pressure delivered by the pump by a factor of 40 or even 60.

Planning the production of a smith-forged part. Before actually smith forging a part, all the details of the process must be thoroughly planned. This involves preparation of the design details, calculation of the dimensions and the weight of the stock and of the product, choosing the forging operations as well as their sequence, choosing tools and devices that will be used, and thinking about the details of the heating and cooling cycles.

The first step in the design process is to draw the finished part and then obtain the drawing of the forging by adding a machining as well as a forging allowance all around. The machining allowance is the increase in any dimension to provide excess metal that is removed by machining. This subsequent machining is required to remove scales and the chilled, defected surface layers. The forging allowance is added mainly to simplify the shape of the as-forged part. It is always recommended to make the shape of a forging symmetrical and confined by plane and cylindrical surfaces. At this stage, a suitable tolerance is assigned to each dimension to bring the design process to an end.

The next step is to choose the appropriate equipment. Two factors affect the decision: the size of the forging and the rate of deformation (strain rate). Usually, forgings weighing 2 tons or more are forged in hydraulic presses. Also, small forging made of high-alloy steels and some nonferrous alloys must be forged on a press because they are sensitive to high strain rates that arise when using power-hammer forging. At this point, the manufacturing engineer is in a position to decide upon operations, tools, devices, and the like needed to accomplish the desired task.

Closed-Die Forging

Closed-die forging involves shaping the hot forging stock in counterpart cavities or impressions that have been machined into two mating halves of a die set. Under impact (or squeezing), the hot metal plastically flows to fill the die cavity. Because the flow of metal is restricted by the shape of the impressions, the forged part accurately conforms to the shape of the cavity, provided that complete filling of the cavity is achieved. Among the various advantages of closed-die forging are the greater consistency of product attributes than in casting, the close tolerances and good surface finish with minimum surplus material to be removed by machining, and the greater strength at lower unit weight compared with castings or fabricated parts. In fact, the cost of parts produced by machining (only) is usually two to three times the cost of closed-die forgings. Nevertheless, the high cost of forging dies (compared with patterns, for example) is the main shortcoming of this process, especially if intricate shapes are to be produced. Therefore, the process is recommended for mass or large-lot production of steel and nonferrous components weighing up to about 900 pounds (350 kg).

Generally, there are two types of closed-die forging: conventional (or flash) die forging and flashless die forging. In conventional *flash* die forging, the volume of the slug has to be slightly larger than that of the die cavity. The surplus metal forms a flash (fin) around the parting line. In *flashless* forging, no fin is formed, so the process consequently calls for accurate control of the volume of the slug. If the slug is smaller than the required final product, proper filling of the die cavity is not achieved. On the other hand, when the size of the slug is bigger than that of the desired forging, excessive load buildup will eventually result in the breaking of the tooling and/or equipment. Accordingly, flashless-forging dies are fitted with load-limiting devices to keep the generated load below a certain safe value in order to avoid breakage of the tooling.

In addition to shaping the metal in die cavities, the manufacturing cycle for a die-forged part includes some other related operations, such as cutting or cropping the rolled stock into slugs or billets, adequately heating the slugs, forging the slugs, trimming the flash (in conventional forging), heat treating the forgings, descaling, and, finally, inspecting or quality controlling. The forging specifications differ from one country to another; however, in order to ensure the product quality, one or more of the following acceptance tests must be passed:

1. Chemical composition midway between the surface and the center
2. Mechanical properties
3. Corrosion tests
4. Nondestructive tests like magnetic detection of surface or subsurface hair cracks
5. Visual tests such as macroetch and macroexamination and sulfur painting for steel

Closed-die forging processes can be carried out using drop forging hammers, mechanical crank presses, and forging machines. Factors such as product shape and tolerances, quantities required, and forged alloys play an important role in determining the best and most economical equipment to be employed in forging a desired product as each of the processes has its own advantages and limitations. Following is a brief description of the different techniques used in closed-die forging.

Drop forging. In drop forging, a type of closed-die forging, the force generated by the hammer is caused by gravitational attraction resulting from the free fall of the ram. The ram may be lifted by a single-acting steam (or air) cylinder or by friction rollers that engage a board tightly fastened to the ram. In this latter type, called a *board hammer,* once the ram reaches a predetermined desired height, a lever is actuated, the rollers retract, and the board and ram fall freely to strike the workpiece. Figure 5.47 illustrates the working principles. Whether a board hammer or single-acting steam hammer is used, accurate matching of the two halves of the die (i.e., the impressions) must be ensured. Therefore, the hammers employed in drop forging are usually of the double-housing (or arch) type and are provided with adequate ram guidance. The desired alignment of the two halves of the die is then achieved by wedging the upper half of the die onto the ram and securing the lower half onto a bolster plate that is, in turn, tightly mounted on the anvil. Also, the ratio of the weights of the anvil and the moving parts can go as high as 30 to 1 to ensure maximum efficiency and trouble-free impact.

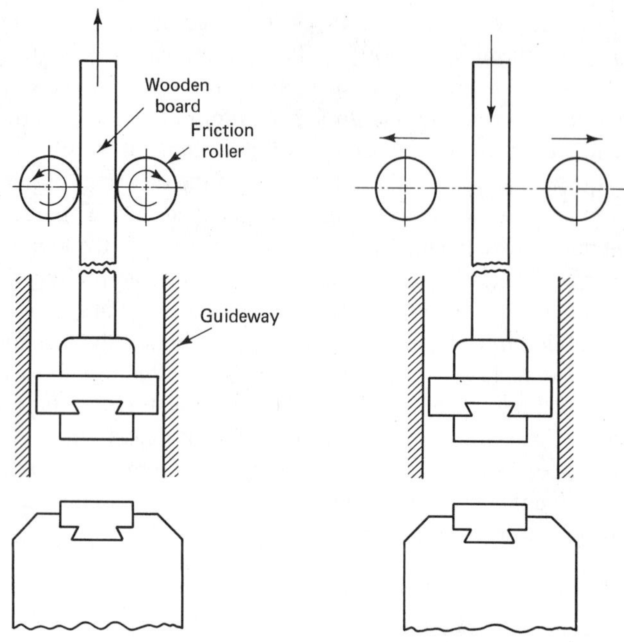

FIGURE 5.47 The working principles of a board hammer

Drop-forging dies can have one, two, or several impressions, depending upon the complexity of the required product. Simple shapes like gears, small flywheels, and straight levers are usually forged in dies with one or two impressions, whereas products with intricate shapes are successively worked in multiple-impression dies, thus making it possible to preshape a forging before it is forged into its final form. Operations like edging, drawing out, fullering, and bending are performed, each in its assigned impression. Finally, the desired shape is imparted to the metal in a finishing impression that has exactly the same shape as the desired product; its dimensions are slightly larger because shrinkage due to cooling down must be taken into account. As can be seen in Figure 5.48, a gutter for flash is provided around the finishing impressions. When properly designed, the gutter provides resistance to the flow of metal into

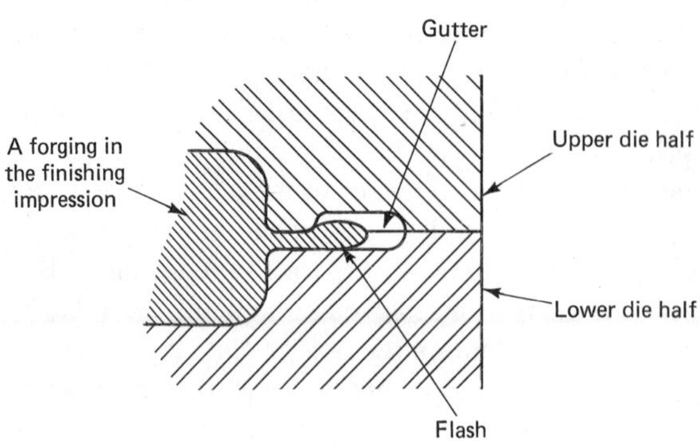

FIGURE 5.48 A gutter providing a space for excess metal

it, thus preventing further flow from the impression and forcing the metal to fill all the details, such as corners (which are the most difficult portions to fill).

The drop-forging process may involve several blows so that the desired final shape of the forged part can be obtained. Lubricants are applied to ensure easy flow of the metal within the cavity and to reduce friction and die wear. As many as four blows may be needed while the part is in the finishing impressions, and the part should be lifted slightly between successive blows to prevent overheating of the die. Finally, the gas pressure forces the part out of the die. The number of blows delivered when the part is in the different preshaping impressions is 1½ to 2 times the number of blows while the part is in the finishing impression. This sequence of drop-forging operations is shown when forging a connecting rod. As can be seen in Figure 5.49, the heated stock is first placed in the fullering impression and then hammered once or twice to obtain local spreading of the metal on the expanse of its cross section. The stock is then transferred to the edging impression, where the metal is redistributed along its length in order to properly fill the finishing die cavities (i.e., metal is "gathered" at certain predetermined points and reduced at some other ones). This is usually achieved through a series of blows, together with turnovers of the metal, as required. The next operation in this sequence is bending, which may or may not be needed, depending upon the design of the product. The stock is then worked in the semifinishing, or blocking, impression before it is finally forged into the desired shape in the finishing impression. We can see that the blocking operation contributes to reducing the tool wear in the finishing impression by giving the part its general shape.

Press forging. Press forging, which is usually referred to as *hot pressing,* is carried out using mechanical (crank-type) or hydraulic presses. These exert force at relatively slow ram travel, resulting in steadily applied pressure instead of impacting pressure.

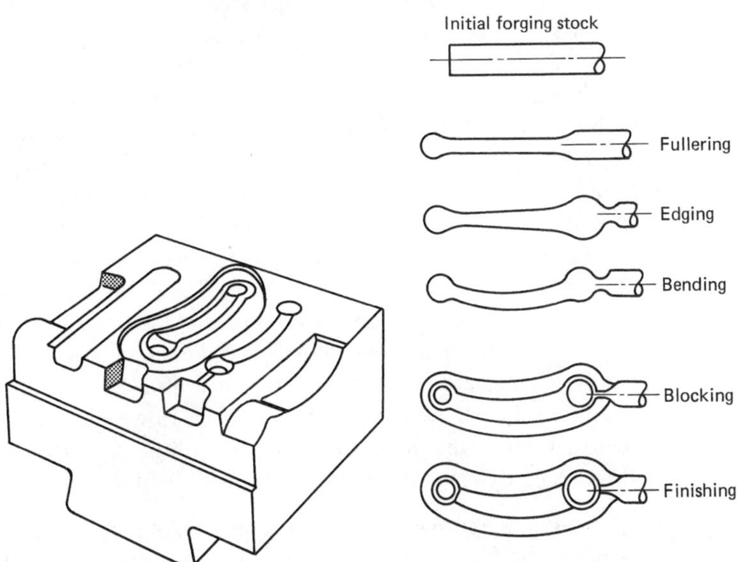

FIGURE 5.49
A multiple-impression die and the forging sequence for a connecting rod

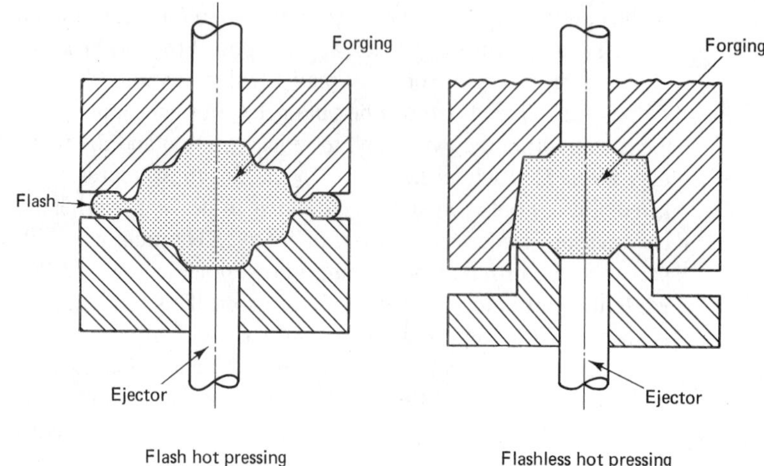

FIGURE 5.50
Flash and flashless hot pressing

The nature of metal deformation during hot pressing is, therefore, substantially different from that of drop forging. Under impact loading, the energy is transmitted into only the surface layers of the workpiece, whereas, under squeezing (steadily applied pressure), deformation penetrates deeper so that the entire volume of the workpiece simultaneously undergoes plastic deformation. Although multiple-impression dies are used, it is always the goal of a good designer to minimize the number of impressions in a die. It is also considered good industrial practice to use shaped blanks or preforms, thus enabling the part to be forged in only a single stroke.

Hot pressing involves both flash as well as flashless forging. In both cases, the forged part is pushed out of the die cavity by means of an ejector, as is illustrated in Figure 5.50. Examples of some hot-pressed parts are shown in Figure 5.51, which also shows the sequence of operations, the production rate, the estimated die life, and the approximate production cost.

A characterizing feature of hot pressing is the accurate matching of the two halves of a die due to the efficient guidance of the ram. Also, the number of working strokes per minute can be as high as 40 or even 50. There is also the possibility of automating the process through mechanization of blank feeding and of forging removal. It can, therefore, clearly be seen that hot pressing has higher productivity than drop forging and yields parts with greater accuracy in terms of tolerances within 0.010 to 0.020 inch (0.2 up to 0.5 mm), less draft, and fewer design limitations. Nevertheless, the initial capital cost is higher compared with drop forging because the cost of a crank press is always higher than that of an equivalent hammer and because the process is economical only when the equipment is efficiently utilized. The difficulty of descaling the blanks is another shortcoming of this process. However, this disadvantage can be eliminated by using hydraulic descaling (using a high-pressure water jet) or can be originally avoided by using heating furnaces with inert atmosphere.

Die forging in a horizontal forging machine. Although originally developed for heading operations, the purpose of this machine has been broadened to produce a variety of shapes. For instance, all axisymmetric parts such as rods with flanges (with through

FIGURE 5.51
Examples of hot-pressed parts

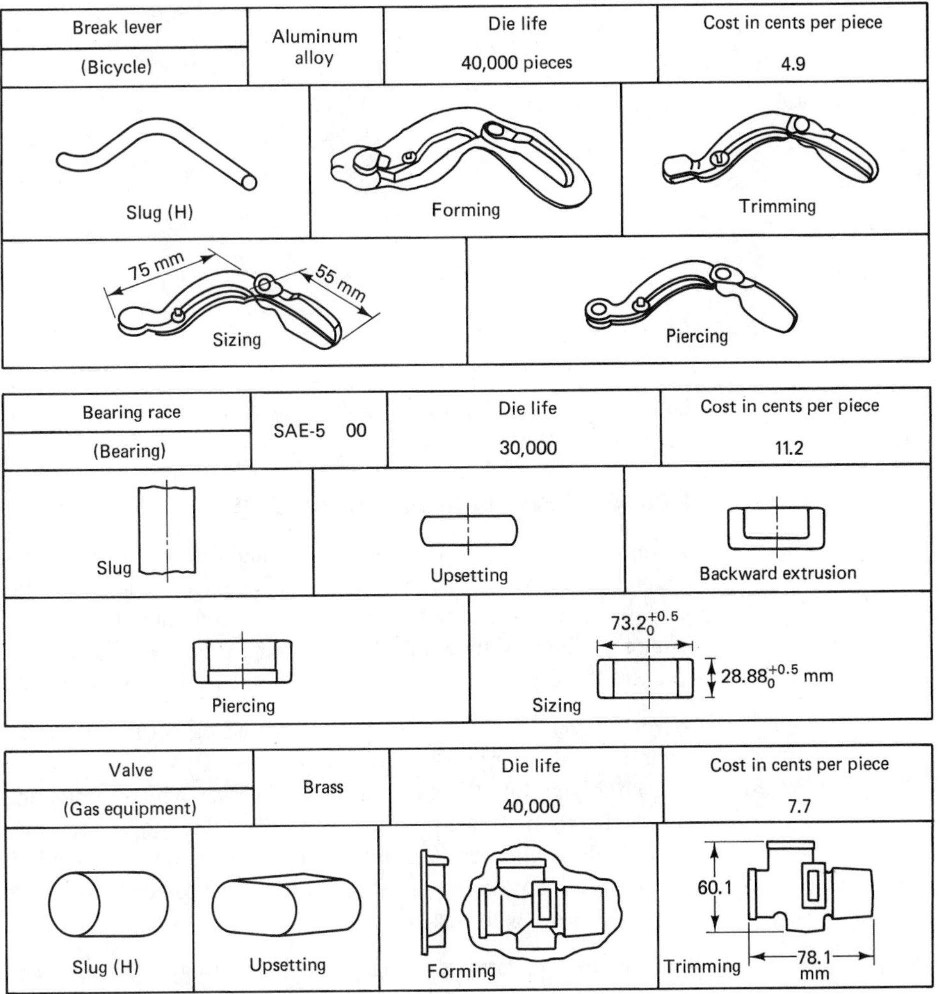

and blind holes) and/or side projections are commonly produced on horizontal forging machines. A rolled stock is cut to length, heated in a heating unit, and automatically fed to the machine. As can be seen in Figure 5.52, the hot part is then held by stationary grips (actually a split die) and upset by an upsetting ram or header. The process involves mainly upsetting and gathering where the blank is first upset; then metal flows to fill the die cavity, as opposed to drop forging, where it is spread or flattened. In the return stroke, the upsetting ram retracts, and the part is removed or transferred to the next impression of the horizontal forging machine. It is obvious that a part can be forged in one or several cavities, depending upon the complexity of its shape.

The main advantage of this process is the high production rate (up to 5000 parts per hour) due to the fact that it can be fully automated. Further advantages include the

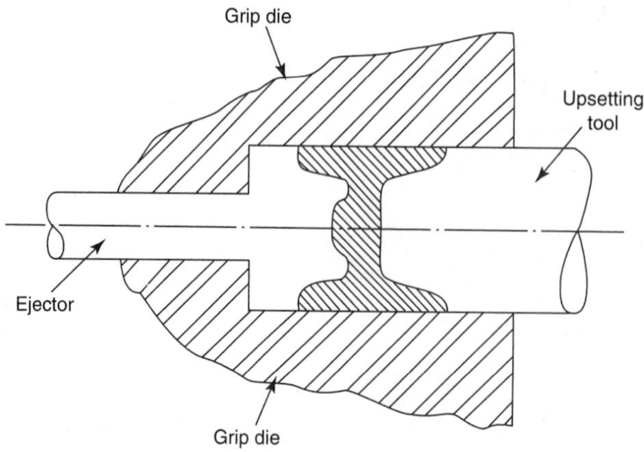

FIGURE 5.52
Die forging in a horizontal forging machine

elimination of the flash and the forging draft and the high efficiency of material utilization because the process involves little or no waste.

Recent Developments in Forging

Warm forging, high-energy-rate forging, and forming of metals in their mushy state are among the important developments in forging technology. These newly developed processes are usually carried out to obtain intricate shapes or unique structures that cannot be obtained by conventional forging processes. Following is a brief description of each of these processes, together with their advantages and disadvantages.

Warm forging. Warm forging involves forging of the metal at a temperature somewhat below the recrystallization temperature. This process combines some advantages of both the hot and the cold forming processes while eliminating their shortcomings. On one hand, increased plasticity and lower load requirements are caused by the relatively high forging temperature. On the other hand, improved mechanical properties, less scaling, and longer die life are due to the lower temperatures used as compared with those used with hot forging.

High-energy-rate forging. The conventional forging process takes some time, during which the hot metal cools down and its resistance to deformation increases. As this does not occur with high-energy-rate forging (HERF), where the whole process is performed within a few thousandths of a second, the hot metal does not have enough time to cool down and heat is not dissipated into the surroundings. Therefore, HERF is very successful when forging intricate shapes with thin sections. A special HERF machine must be used. In fact, the *Petro-Forge* machine was developed at the Mechanical Engineering Department of Birmingham University in England for this reason, and a bulky machine with the name *Dynapak* was developed in the United States. In the first case, the machine consists mainly of an internal-combustion (IC) cylinder integrated into the structure of a high-speed press. The IC cylinder is provided with a sudden release valve that allows the platen attached to the piston to be fired instantaneously when the combustion pressure reaches a preset level. The four stages of the working cycle of the Petro-Forge are shown in Figure 5.53. In the case of the Dynapak, high-pressure nitrogen in a power cylinder is used to push the platen downward. Installations to produce and keep high-pressure gas are, therefore, required in this case.

Forging of alloys in their mushy state. Forging alloys in their *mushy state* involves plastically forming alloys in the temperature range above the solidus line. Because an alloy at that temperature consists partly of a liquid phase, a remarkable decrease in the required forging load is experienced. The process also has some other merits, such as the high processing rate and the high quality of products compared with castings. Moreover, the friction at the billet-container interface has been found to be almost negligible. Nevertheless, the process is still considered to be in its experimental stage because of the instability of alloys having low solid fractions. Recently, it was reported that progress has been made toward solving this problem at the Institute of Industrial Science, Tokyo University, where the instability was overcome

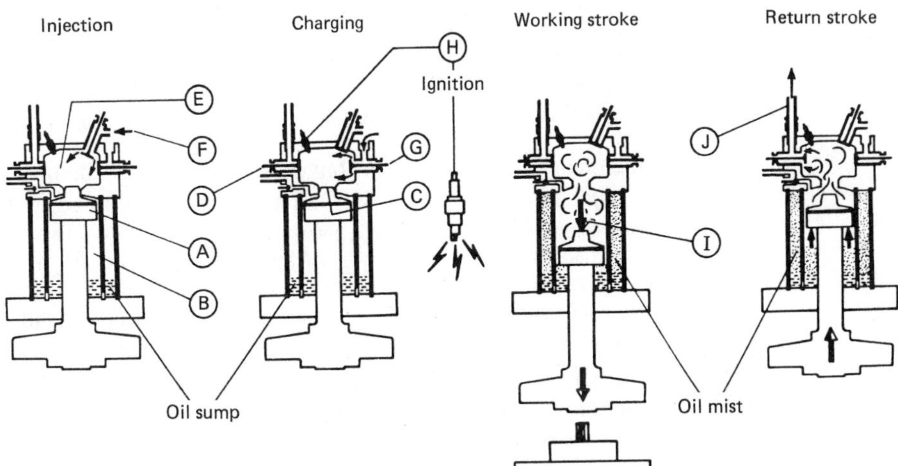

FIGURE 5.53 The working cycle of the Petro-Forge

Injection

At the beginning of the firing cycle the ram/piston assembly (A) is held at the top of its stroke by low pressure air in the back pressure chamber (B) closing the combustion chamber porting by the seal (C), this being a cylindrical projection on the top face of the piston (A). The exhaust valve (D) is open and pressure in the combustion chamber (E) is atmospheric. Upon pressing the firing button the fuel injection phase starts; the exhaust valve (D) is closed and the gaseous fuel is admitted into the combustion chamber (E) via the gas valve (F).

Charging

After closing the gas valve (F) the combustion chamber is charged by admitting compressed air through the inlet valve (G). As soon as charging is completed, the inlet valve (G) is closed and the air/gas mixture is ignited by the spark plug (H). This results in a seven to eightfold rise of the pressure in the combustion chamber (E).

Working stroke

As soon as the force due to the combustion pressure acting on the small area (I) on top of the seal (C) is sufficiently large to overcome the opposed force due to the low back pressure in the space (B) acting on the annular lower face of the piston, the piston (A) starts to move. As a result the porting between the combustion chamber (E) and the cylinder is opened and the gases are permitted to expand to act over the whole piston area. This results in a large force surge acting on the piston/ram assembly which is accelerated downwards to impinge on the workpiece.

Return stroke

During the working stroke the back pressure in space (B) is intensified and consequently acts as a return spring as soon as the forming operation is completed, thus rapidly separating the dies. The return of the ram/piston assembly to its initial position is completed by the opening of the exhaust valve (D) which permits gases to leave through the duct (J). The cycle of operation is normally completed in one second.

by dispersing a very fine alumina powder. This also yielded improved mechanical properties of forgings.

Forgeability

For the proper planning of a forging process, it is important to know the deformation behavior of the metal to be forged with regard to the resistance to deformation and any anticipated adverse effects, such as cracking. For this reason, the term *forgeability* was introduced and can be defined as the tolerance of a metal for deformation without failure. Although there is no commonly accepted standard test, quantitative assessment of the forgeability of a metal (or an alloy) can be obtained through one of the following tests.

Upsetting test. The upsetting test involves upsetting a series of cylindrical billets having the same dimensions to different degrees of deformation (reductions in height). The maximum limit of upsettability without failure or cracking (usually peripheral cracks) is taken as a measure of forgeability.

Notched-bar upsetting test. The notched-bar upsetting test is basically similar to the first test, except that longitudinal notches or serrations are made prior to upsetting. It is believed that this test provides a more reliable index of forgeability.

Hot-impact tensile test. A conventional impact-testing machine fitted with a tension-test attachment is employed. A hot bar of the metal to be studied is tested, and the impact tensile strength is taken as a measure of forgeability. This test is recommended when studying the forgeability of alloys that are sensitive to high strain rates.

Hot twist test. The hot twist test involves twisting a round, hot bar and counting the number of twists until failure. The greater the number of twists, the better the forgeability is considered to be. Using the same bar material, this test can be performed at different temperatures in order to obtain the forging temperature range in which the forgeability of a metal is maximum.

Forgeability of Some Alloys

It is obvious that the results of any of the preceding tests are affected by factors like the composition of an alloy, the presence of impurities, the grain size, and the number of phases present. These are added to the effect of temperature, which generally improves forgeability up to a certain limit, where other phases start to appear or where grain growth becomes excessive. At this point, any further increase in temperature is accompanied by a decrease in forgeability. Following is a list indicating the relative forgeability of some alloys in descending order (i.e., alloys with better forgeability are mentioned first):

1. Aluminum alloys
2. Magnesium alloys
3. Copper alloys
4. Plain-carbon steels

5. Low-alloy steels
6. Martensitic stainless steel
7. Austenitic stainless steel
8. Nickel alloys
9. Titanium alloys
10. Iron-base superalloys
11. Cobalt-base superalloys
12. Molybdenum alloys
13. Nickel-base superalloys
14. Tungsten alloys
15. Beryllium

Lubrication in Forging

In hot forging, the role of lubricants is not just limited to eliminating friction and ensuring easy flow of metal. A lubricant actually prevents the hot metal from sticking to the die and meanwhile prevents the surface layers of the hot metal from being chilled by the relatively cold die. Therefore, water spray, sawdust, or liners of relatively soft metals are sometimes employed to prevent adhesion. Mineral oil alone or mixed with graphite is also used, especially for aluminum and magnesium alloys. Graphite and/or molybdenum disulfide are widely used for plain-carbon steels, low-alloy steels, and copper alloys, whereas melting glass is used for difficult-to-forge alloys like alloy steel, nickel alloys, and titanium.

Defects in Forged Products

Various surface and body defects may be observed in forgings. The kind of defect depends upon many factors, such as the forging process, the forged metal, the tool design, and the temperature at which the process is carried out. Cracking, folds, and improper sections are generally the defects observed in forged products. Following is a brief description of each defect and its causes.

Cracking. Cracking is due to the initiation of tensile stresses during the forging process. Examples are *hot tears,* which are peripheral longitudinal cracks experienced in upsetting processes at high degrees of deformation, and *center cavities,* which occur in the primary forging of low-ductility steels. Thermal cracks may also initiate in cases when nonuniform temperature distribution prevails.

Folds. In upsetting and heading processes, folding is a common defect that is obviously caused by buckling. Folds may also be observed at the edges of parts produced by smith forging if the reduction per pass is too small.

Improper sections. Improper sections include dead-metal zones, piping, and turbulent (i.e., irregular or violent) metal flow. They are basically related to and caused by poor tool design.

Forging Die Materials

During their service life, forging dies are subjected to severe conditions such as high temperatures, excessive pressures, and abrasion. A die material must, therefore, possess adequate hardness at high temperatures as well as high toughness to be able to withstand the severe conditions. Special tool steels (hot-work steels including one or more of the following alloying additives: chromium, nickel, molybdenum, and vanadium) are employed as die materials. Die blocks are annealed, machined to make the shanks, hardened, and tempered; then, impression cavities are sunk by toolmakers.

Fundamentals of Closed-Die Forging Design

The range of forged products with respect to size, shape, and properties is very wide indeed. For this reason, it is both advisable and advantageous for the product designer to consider forging in the early stages of planning the processes for manufacturing new products. The forging design is influenced not only by its function and the properties of the material being processed but also by the kind, capabilities, and shortcomings of the production equipment available in the manufacturing facilities. Therefore, it is impossible to discuss in detail all considerations arising from the infinite combinations of the various factors. Nevertheless, some general guidelines apply in all cases and should be strictly adhered to if a sound forging is to be obtained. Following are some recommended forging design principles.

Parting line. The plane of separation between the upper and lower halves of a closed die set is called the *parting line.* The parting line can be straight, whether horizontal or inclined, or can be irregular, including more than one plane. The parting line must be designated on all forging drawings as it affects the initial cost and wear of the forging die, the grain flow that, in turn, affects the mechanical properties of the forging, and, finally, the trimming procedure and/or subsequent machining operations on the finished part. Following are some considerations for determining the shape and position of the parting line:

1. The parting line should usually pass through the maximum periphery of the forging mainly because it is always easier to spread the metal laterally than to force it to fill deep, narrow die impressions (see Figure 5.54).

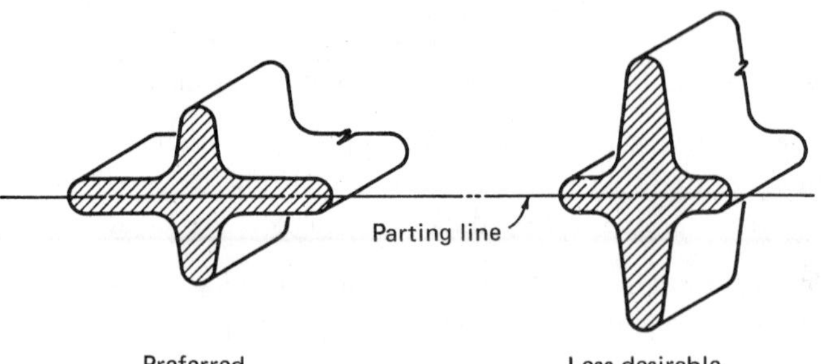

FIGURE 5.54
Recommended location of the parting line *(Courtesy of the Aluminum Association, Inc., Washington, D.C.)*

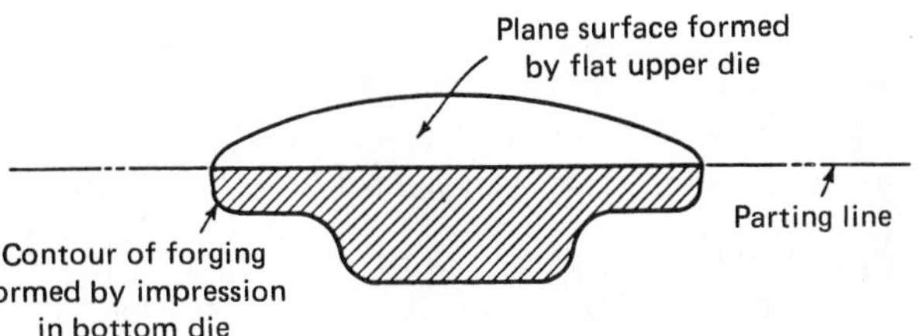

FIGURE 5.55
Flat-sided forging for simplifying the die construction *(Courtesy of the Aluminum Association, Inc., Washington, D.C.)*

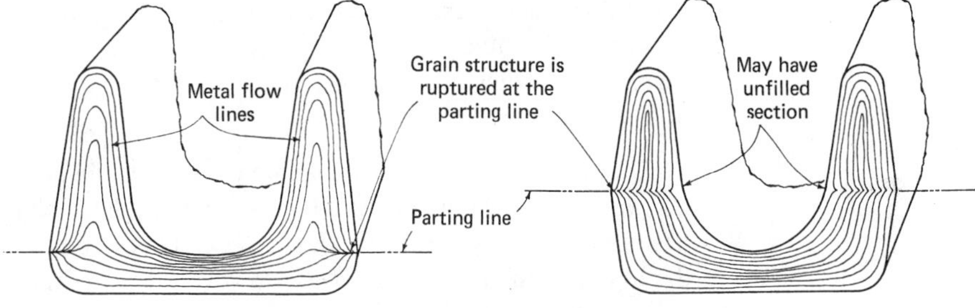

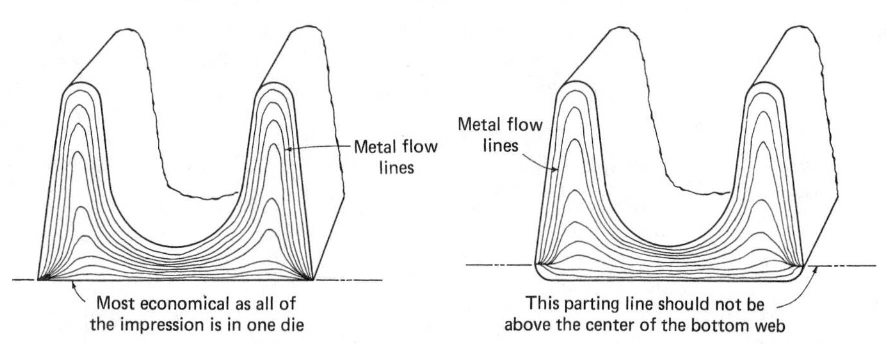

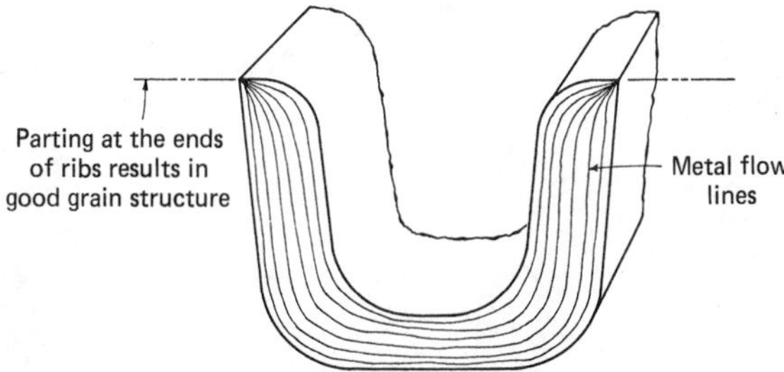

Recommended – The flow lines are smooth at stressed sections with these parting lines

FIGURE 5.56
Using the parting line to promote the alignment of the fibrous macrostructure *(Courtesy of the Aluminum Association, Inc., Washington, D.C.)*

2. It is always advantageous, whenever possible, to try to simplify the die construction if the design is to end up with flat-sided forgings (see Figure 5.55). This will markedly reduce the die cost because machining is limited to the lower die half. Also, the possibility of mismatch between die halves is eliminated.

3. If an inclined parting line must exist, it is generally recommended to limit the inclination so that it does not exceed 75°. The reason is that inclined flashes may create problems in trimming and subsequent machining.

4. A parting line should be located so that it promotes alignment of the fibrous macrostructure to fulfill the strength requirement of a forging. Because excess metal flows out of the die cavity into the gutter as the process proceeds, mislocating the parting line will probably result in irregularities, as can be seen in Figure 5.56, which indicates the fibrous macrostructures resulting from different locations of the parting line.

5. When the forging comprises a web enclosed by ribs, as illustrated in Figure 5.57, the parting line should preferably pass through the centerline of the web. It is also desirable, with respect to the alignment of fibers, to have the parting line either at the top or at the bottom surfaces. However, that desirable location usually creates manufacturing problems and is not used unless the direction of the fibrous macrostructure is critical.

6. If an irregular parting line must exist, avoid side thrust of the die, which will cause the die halves to shift away from each other sideways, resulting in matching errors. Figure 5.58 illustrates the problem of side thrust accompanying irregular parting lines, together with two suggested solutions.

Draft. *Draft* refers to the taper given to internal and external sides of a closed-die forging and is expressed as an angle from the direction of the forging stroke. Draft is required on the vast majority of forgings to avoid production difficulties, to aid in achieving desired metal flow, and to allow easy removal of the forging from the die cavity. It is obvious that

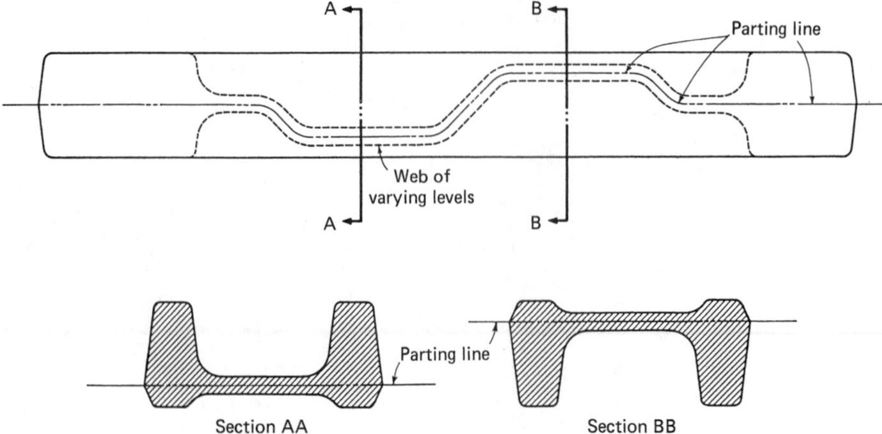

FIGURE 5.57 Location of the parting line with respect to a web *(Courtesy of the Aluminum Association, Inc., Washington, D.C.)*

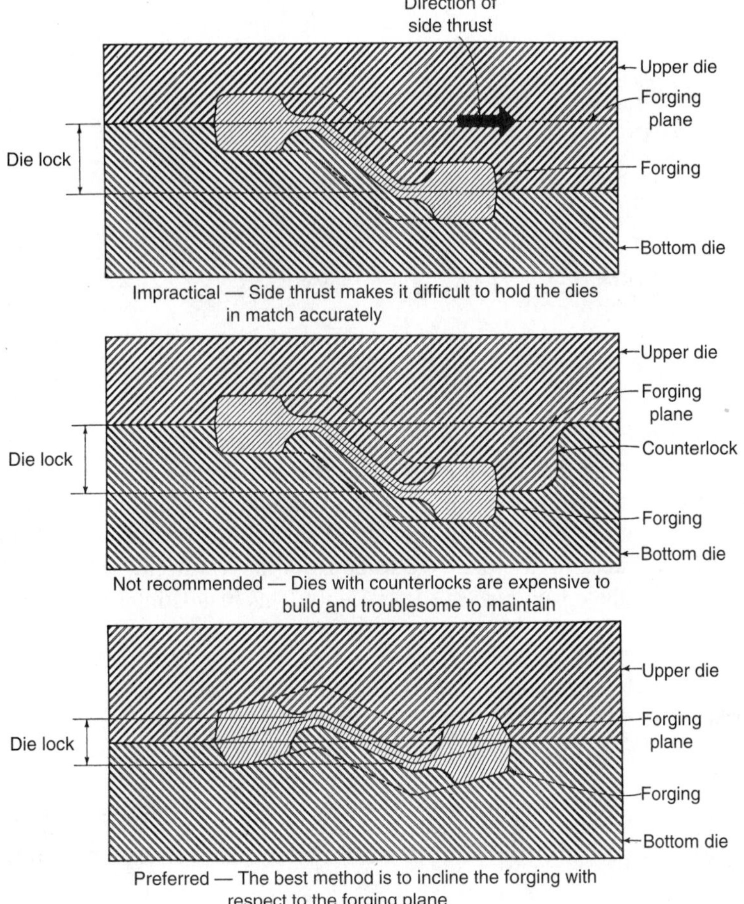

FIGURE 5.58 The problem of side thrust accompanying irregular parting lines and two suggested solutions *(Courtesy of the Aluminum Association, Inc., Washington, D.C.)*

the smaller the draft angle, the more difficult it is to remove the forging out of the die. For this reason, draft angles of less than 5° are not permitted if the part is to be produced by drop forging (remember that there is no ejector to push the part out). Standard draft angles are 7°, 5°, 3°, 1°, and 0°. A draft angle of 3° is usually used for metal having good forgeability, such as aluminum and magnesium, whereas 5° and 7° angles are used for steels, titanium, and the like. It is a recommended practice to use a constant draft all over the periphery of the forging. It is also common to apply a smaller draft angle on the outside periphery than on the inside one. This is justified in that the outer surface will shrink away from the surface of the die cavity as a result of the part's cooling down, thus facilitating the removal of the forging. Following are some useful examples and guidelines:

1. When designing the product, try to make use of the natural draft inherent in some shapes, such as curved and conical surfaces (see Figure 5.59).
2. In some cases, changing the orientation of the die cavity may result in natural draft, thus eliminating the need for any draft on the surfaces (see Figure 5.60).

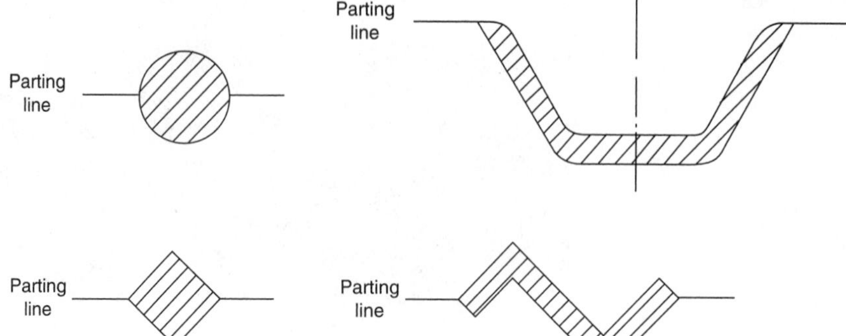

FIGURE 5.59 Examples of the natural draft inherent in some shapes

FIGURE 5.60 Examples of changing the orientation of the impression to provide natural draft

3. Sometimes, the cavity in one of the die halves (for instance, the upper) is shallower than that in the other half. This may create problems in matching the contours of the two die halves at the parting line. It is, therefore, recommended that one of the three methods illustrated in Figure 5.61a, b, or c be used. The first method involves keeping the draft the same as in the lower cavity but increasing the dimension of the upper surface of the cavity. This results in an increase in weight, and this solu-

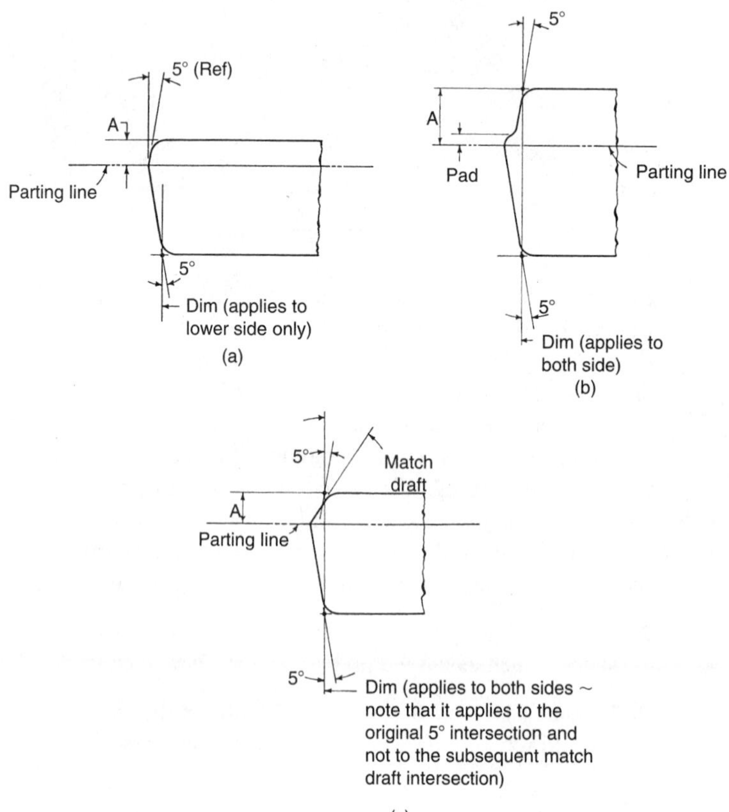

FIGURE 5.61 Methods for matching the contours of two die impressions having different depths: (a) increasing the dimension of the upper surface; (b) using a pad; (c) employing a matching draft *(Courtesy of the Aluminum Association, Inc., Washington, D.C.)*

tion is limited to smaller cavities. The second method is based on keeping the draft constant in both halves by introducing a "pad" whose height varies between 0.06 inch (1.5 mm) and 0.5 inch (12.5 mm), depending upon the size of the forging. The third method, which is more common, is to provide greater draft on the shallower die cavity; this is usually referred to as *matching draft.*

Ribs. A *rib* is a thin part of the forging that is normal to (or slightly inclined to) the forging plane. It is obvious that optimized lighter weight of a forging calls for reducing the thickness of long ribs. However, note that the narrower and longer the rib is, the higher the forging pressure is and the more difficult it is to obtain a sound rib. It is actually a common practice to keep the height-to-thickness ratio of a rib below 6, preferably at 4. The choice of a value for this ratio depends upon many factors, such as the kind of metal being processed and the forging geometry (i.e., the location of the rib, the location of the parting line, and the fillet radii). Figure 5.62 indicates the desirable rib design as well as limitations imposed on possible alternatives.

Webs. A *web* is a thin part of the forging that is passing through or parallel to the forging plane (see Figure 5.63). Although it is always desirable to keep the thickness of a web at the minimum, there are practical limits for this. The minimum thickness of webs depends on the kind of material being worked (actually on its forging temperature range), the size of forging (expressed as the net area of metal at the parting line), and on the average width. Table 5.3 indicates recommended web thickness values applicable to precision and conventional aluminum forgings. For blocking cavities, the values given in Table 5.3 must be increased by 50 percent. Also, for steels and other metals having poorer forgeability than aluminum, it is advisable to increase the values for web thickness. Thin webs may cause unfilled sections, may warp in heat treatment, and may require additional straightening operations; they even cool faster than the rest of the forging after the forging process, resulting in shrinkage, possible tears, and distortion.

FIGURE 5.62
Recommended rib design *(Courtesy of the Aluminum Association, Inc., Washington, D.C.)*

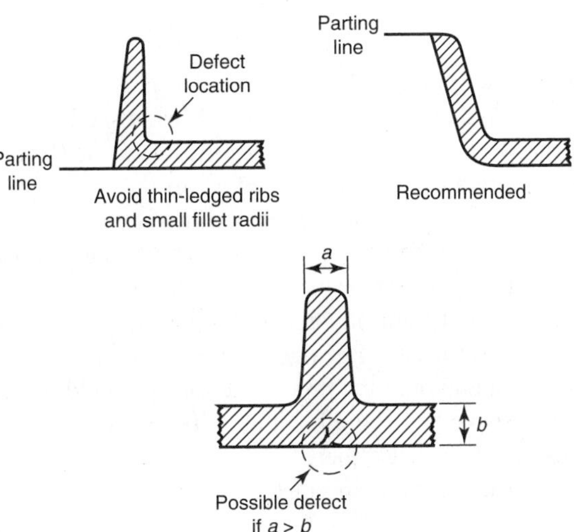

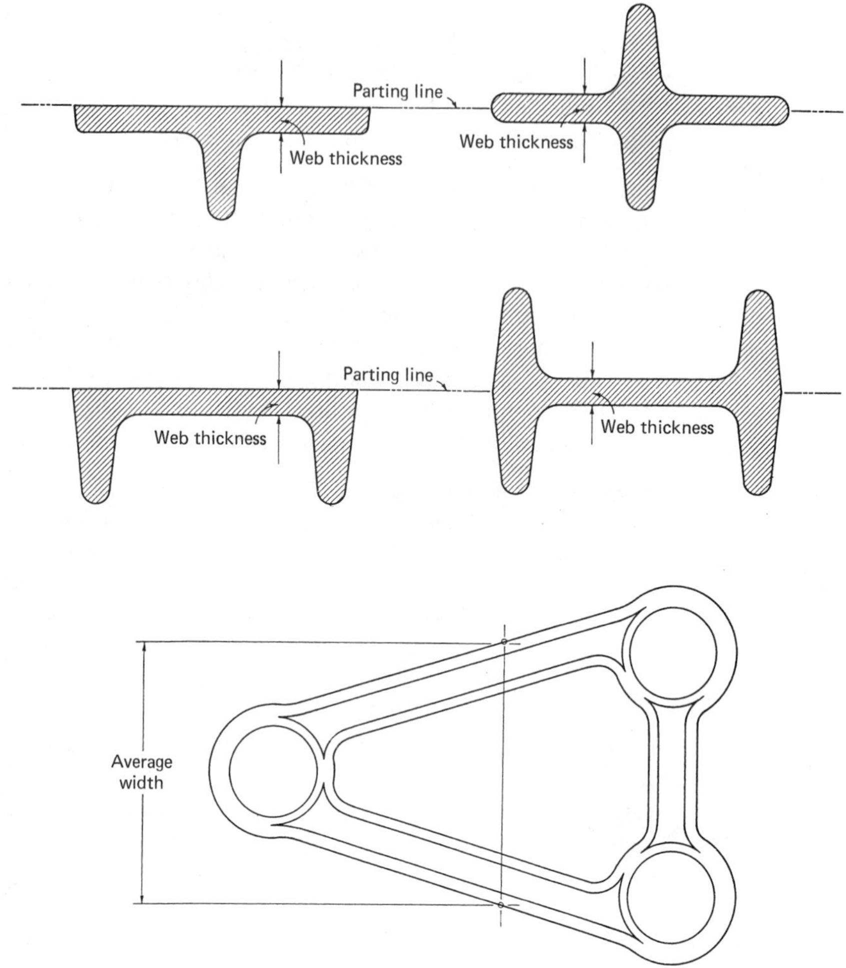

FIGURE 5.63
The shape of a web in forging *(Courtesy of the Aluminum Association, Inc., Washington, D.C.)*

Corner radii. There are two main factors that must be taken into consideration when selecting a small value for a corner radius. First, a small corner radius requires a sharp fillet in the die steel, which acts as a stress raiser; second, the smaller the corner radius, the higher the forging pressure required to fill the die cavity. In addition, some other factors affect the choice of the corner radius, such as the distance from the corner to the parting line and the forgeability of the metal being worked. The larger the distance from the parting line, the larger the corner radius should be. Also, whereas a corner radius of 0.0625 inch (1.5 mm) is generally considered adequate for aluminum forging, a corner radius of at least 0.125 inch (3 mm) is used for titanium forgings of similar shape and size. In addition, the product designer should try to keep the corner radii as consistent as possible and avoid blending different values for a given shape in order to reduce the die cost (because there will be no need for many tool changes during die sinking). Corner radii at the end of high, thin ribs are critical. A rule of thumb states

TABLE 5.3
Recommended size of minimum web thickness

Up to Average Width in. (m)	Up to Cross-Sectional Area in.^{2} (m^2)	Web Thickness in. (mm)
3 (0.075)	10 (0.00625)	0.09 (2.25)
4 (0.1)	30 (0.01875)	0.12 (3)
6 (0.15)	60 (0.0375)	0.16 (4)
8 (0.2)	100 (0.0625)	0.19 (4.75)
11 (0.275)	200 (0.125)	0.25 (6.25)
14 (0.35)	350 (0.21875)	0.31 (7.75)
18 (0.45)	550 (0.34375)	0.37 (9.25)
22 (0.55)	850 (0.53125)	0.44 (11)
26 (0.65)	1200 (0.75)	0.50 (12.5)
34 (0.85)	2000 (1.25)	0.62 (15.5)
41 (1.025)	3000 (1.875)	0.75 (18.75)
47 (1.1175)	4000 (2.50)	1.25 (31.25)
52 (1.3)	5000 (3.125)	2.00 (50)

that it is always desirable to have the rib thickness equal to twice the value of the corner radius. A thicker rib may have a flat edge with two corner radii, each equal to the recommended value. Figure 5.64 illustrates these recommendations regarding corner radii for ribs.

Fillet radii. It is of supreme importance that the product designer allow generous radii for the fillets because abrupt diversion of the direction of metal flow can result in numerous defects in the product. Figure 5.65 indicates the step-by-step initiation of forging defects and shows that small fillets result in separation of the metal from the die and initiation of voids. Although these can be filled at a later stage, laps and cold shuts will replace these voids. When the shape of the part to be forged is intricate (i.e., involving thin ribs and long, thin webs), the metal may preferentially flow into the gutter rather than into the die cavity. This results in a shear in the fibrous macrostructure and is referred to as *flow-through.* This latter defect can be avoided by using larger-than-normal fillets.

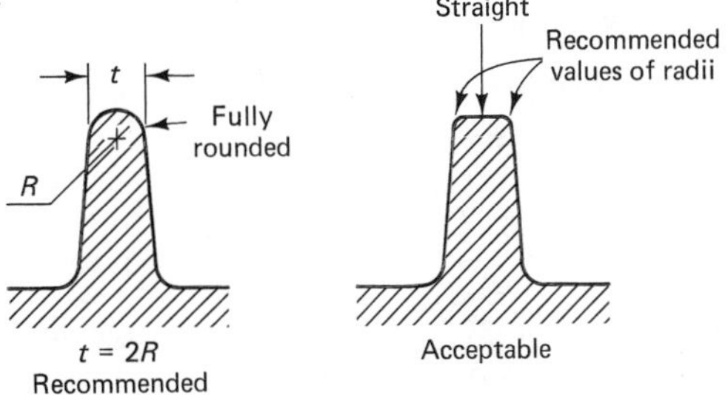

FIGURE 5.64
Recommendations regarding corner radii for ribs *(Courtesy of the Aluminum Association, Inc., Washington, D.C.)*

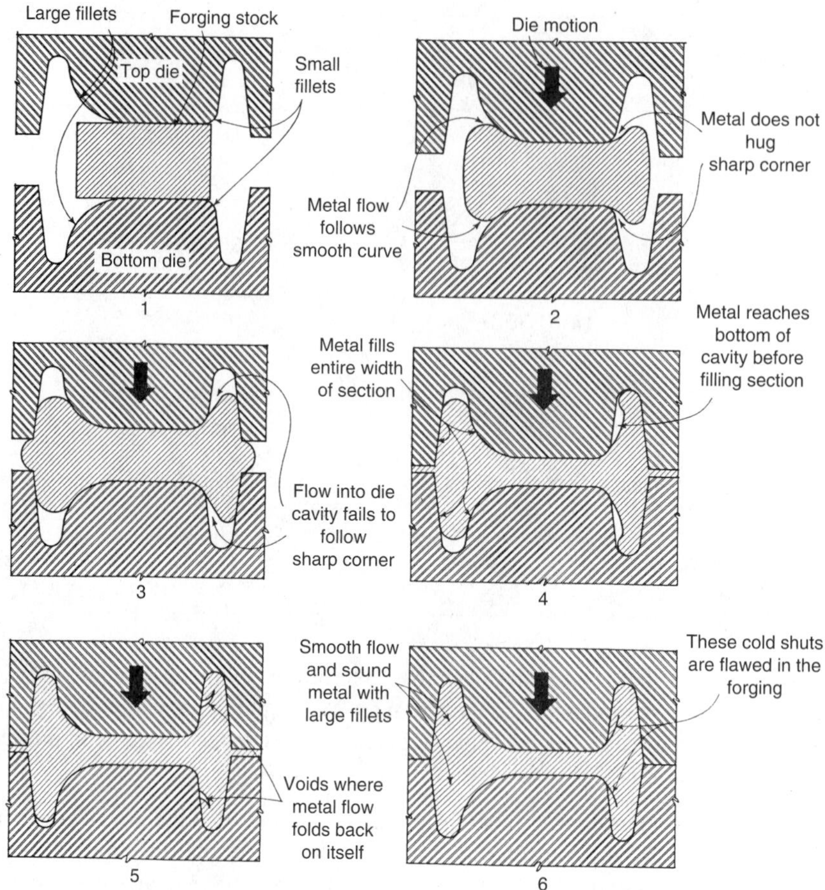

FIGURE 5.65 Defects caused by employing smaller fillet radii *(Courtesy of the Aluminum Association Inc., Washington, D.C.)*

Punchout holes. *Punchout holes* are through holes in a thin web that are produced during, but not after, the forging process. Punchouts reduce the net projected area of the forging, thus reducing the forging load required. If properly located and designed, they can be of great assistance in producing forgings with thin webs. In addition to the manufacturing advantages of punchouts, they serve functional design purposes, such as reducing the mass of a forging and/or providing clearance. Following are some guidelines regarding the design of punchouts:

1. Try to locate a punchout around the central area of a thin web, where the frictional force that impedes the metal flow is maximum.
2. Whenever possible, use a gutter around the interior periphery of a punchout. This provides a successful means for the surplus metal to escape.
3. A single large punchout is generally more advantageous than many smaller ones that have the same area. Accordingly, try to reduce the number of punchouts unless more are dictated by functional requirements.

4. Although punchouts generally aid in eliminating the problems associated with the heat treatment of forgings, it may prove beneficial to take the limitations imposed by heat treatment processes into account when designing the contour of a punchout (i.e., try to avoid irregular contours with sharp corners).

Pockets and recesses. Pockets and recesses are used to save material, promote the desirable alignment of the fibrous macrostructure, and improve the mechanical properties by reducing the thickness, thus achieving a higher degree of deformation. Following are some guidelines:

1. Recesses should never be perpendicular to the direction of metal flow.
2. Recesses are formed by punches or plugs in the dies. Therefore, the recess depth is restricted to the value of its diameter (or to the value of minimum transverse dimension for noncircular recesses).
3. Simple contours for the recesses, together with generous fillets, should be tried.

5.6 COLD FORMING PROCESSES

Cold forming processes are employed mainly to obtain improved mechanical properties, better surface finish, and closer tolerances. Several cold forming techniques have found wide industrial application. Among these are sizing, swaging, coining, and cold heading. Following is a brief description of each of them.

Sizing

Sizing (see Figure 5.66a) is a process in which the metal is squeezed in the forming direction but flows unrestricted in all transverse directions. This process is used primarily for straightening forged parts, improving the surface quality, and obtaining accurate dimensions. A sizing operation can ensure accuracy of dimensions within 0.004 up to 0.010 inch (0.1 up to 0.25 mm). Meanwhile, the pressure generated on the tools can go up to 180,000 pounds per square inch (1300 MN/m^2).

Swaging

Swaging (see Figure 5.66b) involves imparting the required shape and accurate dimensions to the entire forging (or most of it). Usually, swaging is carried out in a die where a flash is formed and subsequently removed by abrasive wheels or a trimming operation. Note that the flow of metal in the swaging process is more restricted than in sizing. Accordingly, higher forming pressures are experienced and can go up to 250,000 pounds per square inch (1800 MN/m^2).

Coining

Coining (see Figure 5.66c) is a process in which the part subjected to coining is completely confined within the die cavity (by the die and the punch). The volume of the original forging must be very close to that of the finished part. Any tangible increase in that volume may result in excessive pressures and the breakage of tools. Still, com-

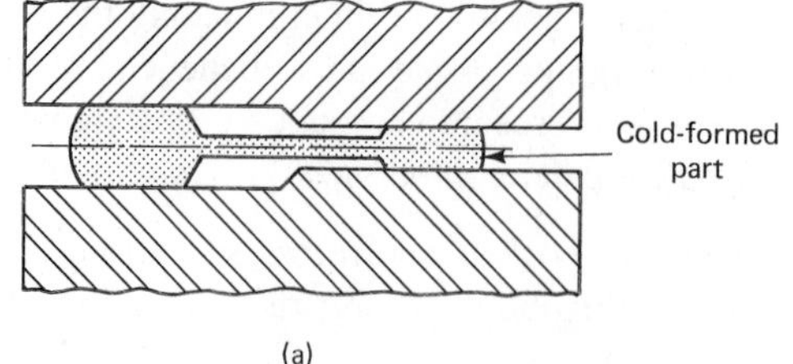

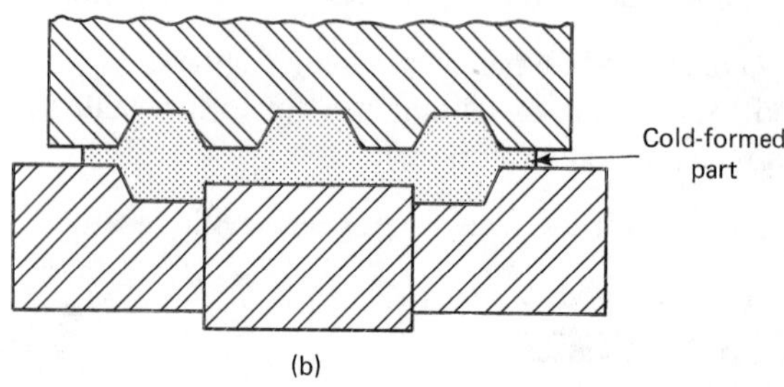

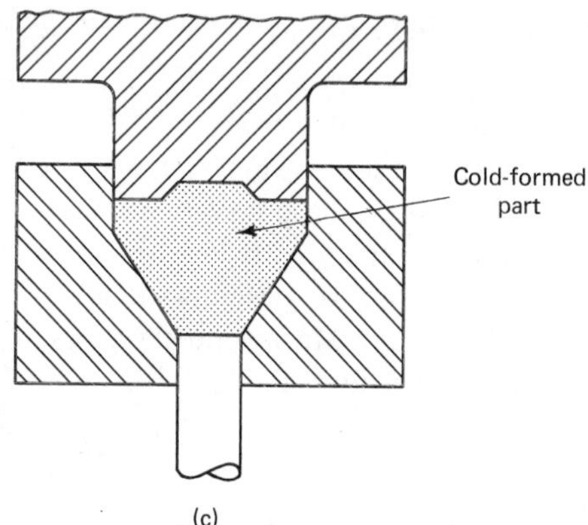

FIGURE 5.66 Cold forming processes: (a) sizing; (b) swaging; (c) coining

mon pressures (even when no problems are encountered) are in the order of 320,000 pounds per square inch (2200 MN/m^2). For this reason, coining processes (also sizing and swaging) are carried out on special presses called *knuckle presses.* The main mechanism of a knuckle press is shown in Figure 5.67. It is characterized by the ability to deliver a large force with a small stroke of the ram.

Cold Heading

Cold heading is used to manufacture bolts, rivets, nuts, nails, and similar parts with heads and collars. A group of typical products are illustrated in Figure 5.68. The main production equipment involves a multistage automatic cold header that operates on the

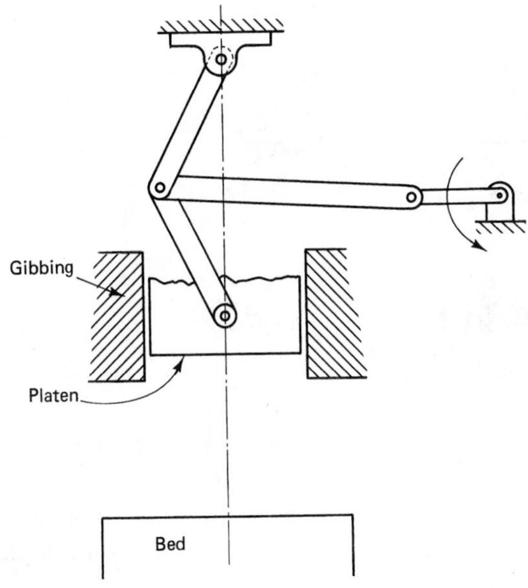

FIGURE 5.67 The working principles of a knuckle press for cold forming processes

same principle as a horizontal forging machine. Full automation and high productivity are among the advantages of this process. Products having accurate dimensions can be produced at a rate of 30 to 300 pieces per minute. Starting from coiled wires or rods made of plain-carbon steel and nonferrous metals with diameters ranging from 0.025 to 1.6 inches (0.6 to 40 mm), blanks are processed at different stations. Feeding, transfer, and ejection of the products are also automated. Figure 5.69 illustrates the different stages involved in a simple cold heading operation.

Lubrication in Cold Forming

Lubricants employed in cold forming are similar to those used in heavy wire-drawing processes. Phosphating followed by soap dipping is successful with steels, whereas only soap is considered adequate for nonferrous metals.

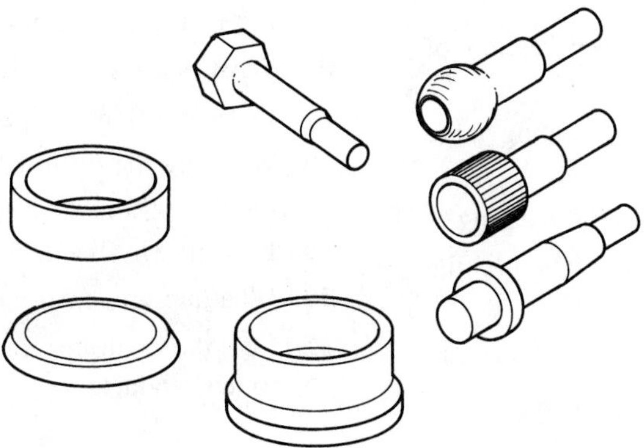

FIGURE 5.68 Some products manufactured using an automatic cold header

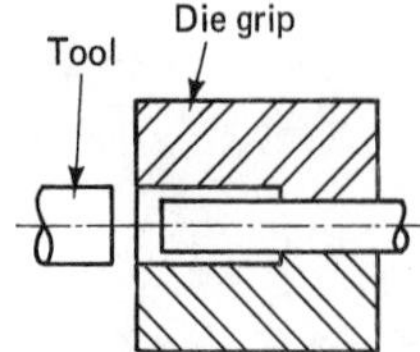

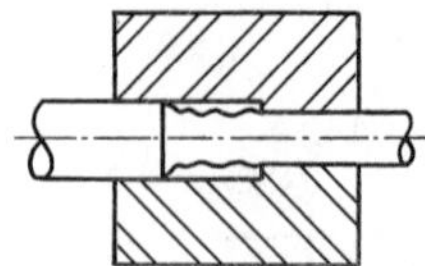
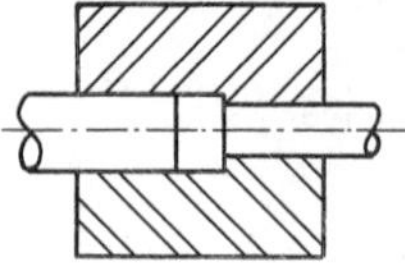

FIGURE 5.69 Different stages of a simple cold heading operation

Review Questions

1. Why have metal forming processes gained widespread industrial application since World War II?
2. What are the two main groups of metal forming processes?
3. List the different factors affecting the deformation process. Tell how each influences deformation.
4. Why are cold forming processes always accompanied by work-hardening, whereas hot forming processes are not?
5. What is meant by the *fibrous macrostructure?*
6. Are the mechanical properties of a rolled sheet isotropic? Why?
7. What is meant by the *state of stress?* List the three general types.
8. List some advantages of hot forming. What are some disadvantages?
9. List some advantages of cold forming. What are some disadvantages?
10. What may happen when a large section of steel is heated at a rapid rate? Why?
11. What should be avoided when heating large steel sections prior to hot forming?
12. Where does friction occur in metal forming?
13. What are the harmful effects of friction on the forming process?
14. Is friction always harmful in all metal forming processes?
15. Can lead be used as a lubricant when forming copper? Why?
16. When forming lead at room temperature, do you consider it cold forming? Why?
17. Why are lubricants used in metal forming processes? List some useful effects.
18. List some lubricants used in cold forming processes.
19. List some lubricants used in hot forming processes.
20. Which do you recommend for further processing by machining, a cold-worked part or a hot-worked part?
21. Is hot rolling the most widely used metal forming process? Why?
22. List some of the useful effects of hot rolling.
23. Define *rolling.*
24. What is the *angle of contact?*
25. For heavier sections, would you recommend larger angles of contact in rolling? Why?
26. What is the state of stress in rolling?
27. List the different types of rolling mills.
28. What are the different parts of a roll? What is the function of each?
29. Explain why Sendzimir mills are used.
30. What are universal mills used for?
31. List three groups included in the range of rolled products.

32. Explain, using sketches, how seamless tubes are manufactured.
33. What is *alligatoring?* What causes it?
34. Define *wire drawing.*
35. Which mechanical property should the metal possess if it is to be used in a drawing process? Why?
36. What is the state of stress in drawing?
37. List some advantages of the drawing process.
38. How is a metal prepared for a drawing process?
39. What are the different zones in a drawing die?
40. Mention the range of the apex angles (of conical shapes) used in drawing dies.
41. What material do you recommend to be used in making drawing dies?
42. Describe a draw bench.
43. What kinds of lubricants are used in drawing processes?
44. What is the *drawing ratio?*
45. Give an expression indicating the reduction achieved in a wire-drawing process.
46. Why do internal bursts occur in wire-drawing processes?
47. What are arrowhead fractures and why do they occur?
48. What is the state of stress in tube drawing?
49. Using sketches, illustrate the different techniques used in tube drawing.
50. Define *extrusion.*
51. Why can extrusion be used with metals having relatively poor plasticity?
52. List some advantages of the extrusion process.
53. What are the shortcomings and limitations of the extrusion process?
54. Using sketches, differentiate between the direct and indirect extrusion techniques.
55. Although indirect extrusion almost eliminates friction, it is not commonly used in industry. Why?
56. List the advantages of hydrostatic extrusion.
57. Compare extrusion with rolling with respect to efficiency of material utilization.
58. When is conventional direct extrusion recommended as a production process?
59. Describe impact extrusion.
60. Why is the leading end of an extruded section always sheared off?
61. What are *dead-metal zones?*
62. If hardness measurements are taken across the section (say, circular) of an extruded part, what locations will have higher hardness values? Can you plot hardness versus distance from the center?
63. What lubricants can be used in cold extrusion?
64. What material do you recommend as a lubricant when hot extruding stainless steel?
65. What defect may occur when extruding magnesium at low extrusion ratios?
66. What is *piping* and why does it occur?
67. In extrusion dies, what is meant by the *circle size?*
68. List some considerations that must be taken into account when designing a section for extrusion.
69. Why should a designer try to avoid sharp corners at the root of a die tongue? Explain using neat sketches.
70. As a product designer, you are given a very intricate section for production by extrusion. Is there any way around this problem without being forced to use a die with a very intricate construction? How?
71. List some considerations for the design of impact extrusions.
72. How can you avoid shear failure at the bottom of the wall of an impact extrusion?
73. Does forging involve just imparting a certain shape to a billet?

74. Is it just a matter of economy to produce a crankshaft by forging rather than by machining from a solid stock? Why?

75. Can a metal such as aluminum be forged at any temperature? Why?

76. List the main types of forging processes.

77. Which process is suited for the production of small batches of large parts?

78. Give examples of parts produced by each type of forging process. Support your answer with evidence.

79. What is the modern version of blacksmithing? What are the different operations involved in that process?

80. When do you recommend using a power-actuated hammer as a forging machine? Mention the type of forging process.

81. For which type of forging is a drop hammer employed?

82. For which type of forging is a crank press employed?

83. Using sketches, illustrate the different stages in manufacturing a ring by forging.

84. List the advantages that forging has over casting when producing large numbers of small parts having relatively complex shapes.

85. In the comparison of Question 84, what are the shortcomings of forging? Why don't they affect your decision in that particular case?

86. List some of the specified acceptance tests to be performed on forgings.

87. What is a board hammer used for?

88. Is it true that a closed type of forging die can have only one impression? Explain why.

89. What does *hot pressing* mean?

90. What is the advantage of HERF?

91. What are the advantages of warm forging?

92. What is meant by a *mushy state?*

93. Define *forgeability.* How can it be quantitatively assessed?

94. What is the most forgeable metal?

95. What is the main role of lubricants in hot forging?

96. As a product designer, how can you manipulate the alignment of the fibrous macrostructure?

97. List some guidelines regarding the location of the parting line between the upper and lower halves of a die set.

98. What is meant by the term *draft* in forging?

99. A die was designed to forge an aluminum part. Can the same design be used to forge a similar part made of titanium? Why?

100. Explain the meaning of *matching draft,* using sketches.

101. Differentiate between a *web* and a *rib* in a forging.

102. What is the difference between a corner radius and a fillet radius? Use sketches.

103. What are *punchout holes* in a forging?

104. List some advantages of including punchout holes in a forging design.

105. Why are recesses sometimes included in a forging design?

106. List the different cold forming processes and use sketches to illustrate how they differ.

Problems

1. In hot rolling, determine the load on each roll of a two-high rolling mill, given the following:

Diameter of the roll:	20 inches (500 mm)
Stock width:	48 inches (1020 mm)
Initial thickness:	0.08 inch (2 mm)
Final thickness:	0.04 inch (1 mm)
Flow stress of rolled material:	14,200 lb/in.^{2} (100 MN/m^2)

2. In hot rolling low-carbon-steel plate 48 inches (1200 mm) in width, given the roll diameter as 20 inches (500 mm), initial thickness as 1.5 inches (37.5 mm), final thickness as 0.4 inch (10 mm), and the flow stress of steel as 28,400 lb/in.^{2} (200 MN/m^2), calculate the number of rolling passes if the maximum load on the roll in each pass is not to exceed 225,000 pounds force (1.0 MN).

3. Write a computer program to solve Problem 2, assuming that all the data are variables to be given for each design.

4. Calculate the maximum achievable reduction in a single drawing of a lead wire.

5. Estimate the largest possible extrusion ratio of 2.0-inch (50-mm) aluminum bar having mean flow stress of 21,900 lb/in.^{2} (150 MN/m^2) if the press available has a capacity of only 45,000 pounds force (200 kN).

6. Plot a curve indicating the efficiency of a drop hammer versus the ratio between the weights of the anvil and the moving parts if the value of K that represents the elasticity of the billet is taken as 0.1. What ratio do you suggest? Why should it not be justified to take large ratios?

Design Example

PROBLEM

Design a simple wrench that measures 1/2 inch (12.5 mm) across bolt-head flats and is used for loosening nuts and bolts. The torque required to loosen (or tighten) a bolt (or a nut) is 1 lb ft (6.8 N·m). The production volume is 25,000 pieces per year. Forging is recommended as a manufacturing process.

Solution

Because the wrench is going to be short, it cannot be held by the full hand but probably by only three fingers. The force that can be exerted is to be taken, therefore, as 4 pounds. The arm of the lever is equal to (1 × 12)/4, or 3 inches (75mm). Add on allowance for the holding fingers. The shape of the wrench will be as shown in Figure 5.70.

Now, let us select the materials. A suitable material would be AISI 1045 CD steel to facilitate machining (sawing) of the stock material. Closed-die forging of the billets

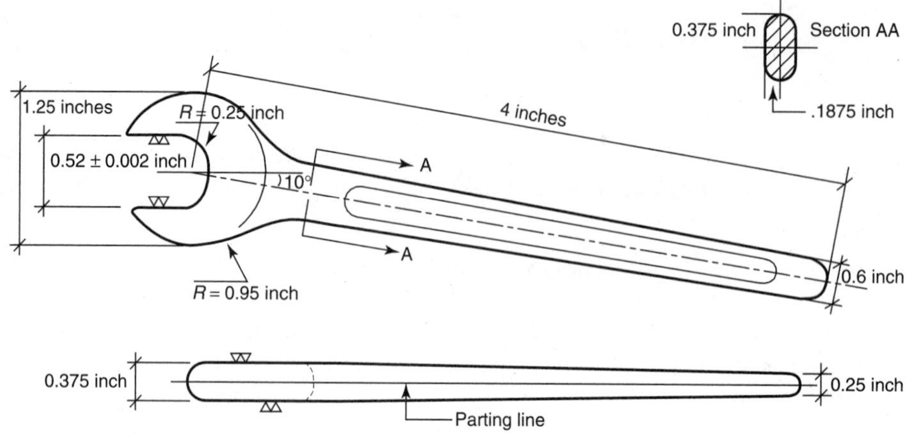

FIGURE 5.70
A wrench manufactured by forging

is recommended, as well as employing drop-forging hammers. To facilitate withdrawal of the part, the cross section of the handle should be elliptical (see Figure 5.70). The parting line should coincide with the major axis of the ellipse.

Let us check the stress due to bending:

$$I = \frac{1}{4}\pi\, a^3 b = \frac{1}{4}\,(\pi)(0.375)^3(0.1875) = 7.7 \times 10^{-3} \text{ in.}^4$$

where: a is half the major axis

b is half the minor axis

$$\text{stress} = \frac{My}{I} = \frac{Ma}{I} = \frac{5 \times 12 \times 0.375}{7.7 \times 10^{-3}} = 2922 \text{ lb/in.}^2$$

It is less than the allowable stress for 1045 CD steel, which is

$$\frac{60{,}000}{2} = 30{,}000 \text{ lb/in.}^2$$

In order to check the bearing stress, let us assume a shift of 0.25 inch between the forces acting on the faces of the nut to form a couple (this assumption can be verified if we draw the nut and the wrench to scale):

$$\text{each force} = \frac{60}{0.25} = 240 \text{ pounds}$$

Further assume that the bearing area is 0.375 by 0.25 inch. The bearing stress is, therefore,

$$\frac{240}{0.375 \times 0.25} = 2560 \text{ lb/in.}^2$$

It is less than the allowable stress of the 1045 CD steel.

The forged wrench finally has to be trimmed and then machined on the surfaces indicated in Figure 5.70. An allowance of 1/64 inch should be provided between the wrench open-head and the nut. Now, our design is complete and ready to be released to the workshop.

Design Projects

1. A clock frame 3 by 5 inches (75 by 125 mm) is manufactured by machining an aluminum-alloy stock. Make a design and a preliminary feasibility study so that it can be produced by extrusion. Assume the production volume is 20,000 pieces per year.
2. A motor frame that has a 6-inch (150-mm) internal diameter and that is 10 inches (250 mm) long is currently produced by casting. That process yields a high percentage of rejects, and the production cost is relatively high. Knowing that the production volume is 20,000 pieces per year, redesign the part so that it will be lighter and can be easily produced by an appropriate metal forming operation that has a high efficiency of material utilization.
3. A pulley transmits a torque of 600 lb ft (816 N·m) to a shaft that is 1¼ inches (31 mm) in diameter. It is to be driven by a flat belt that is 2 inches (50 mm) in width. Provide a detailed design for the pulley if the production volume is 10,000 pieces per year and the pulley is manufactured by forging.
4. A connecting lever is to be manufactured by forging. The estimated production volume is 50,000 pieces per year. The lever has two short bosses, each at one of its ends, and each has a vertical hole 3/4 inch (19 mm) in diameter. The horizontal distance between the centers of the two holes is 12 inches (300 mm), and the vertical difference in levels is 3 inches (75 mm). The lever during its functioning is subjected to a bending moment of 200 lb ft (272 N·m). Make a detailed design for this lever.
5. If the lever in Problem 4 is to be used in a space vehicle, would you use the same material? What are the necessary design changes? Make a design appropriate for this new situation.
6. Design a gear blank that transmits a torque of 200 lb ft (272 N·m) to a shaft that is 3/4 inch (19 mm) in diameter. The pitch diameter of the gear is 8 inches (200 mm), and 40 teeth are to be cut in that blank by machining. Assume the production volume is 10,000 pieces per year.
7. A straight-toothed spur-gear wheel transmits a torque of 1200 lb ft (1632 N·m) to a steel shaft (AISI 1045 CD steel) that is 2 inches (50 mm) in diameter. The pitch

diameter of the gear is 16 inches (400 mm), its width is 4 inches (100 mm), and the base diameter is 15 inches (375 mm). Make a complete design for this gear's blank (i.e., before teeth are cut) when it is to be manufactured by forging. Assume the production volume is 10,000 pieces per year.

8. A shaft has a minimum diameter of 1 inch (25 mm) at both its ends, where it is to be mounted in two ball bearings. The total length of the shaft is 12 inches (300 mm). The shaft is to have a gear at its middle, with 40 teeth and a pitch-circle diameter of 1.9 inches (47.5 mm). The width of the gear is 2 inches (50 mm). Make a design for this assembly if the production volume is 50,000 per year.

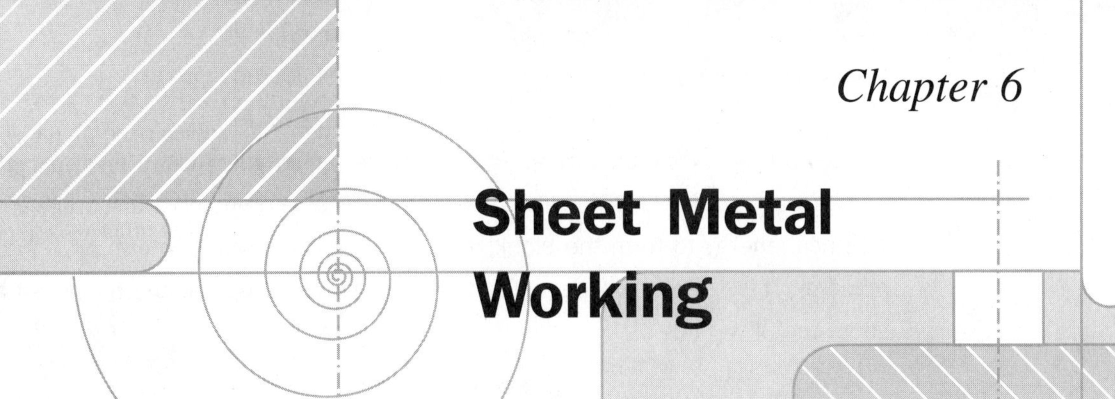

Chapter 6

Sheet Metal Working

INTRODUCTION

The processes of *sheet metal working* have recently gained widespread industrial application. Their main advantages are their high productivity and the close tolerances and excellent surface finish of the products (which usually require no further machining). The range of products manufactured by these processes is vast, but, in general, all of these products have thin walls (relative to their surface area) and relatively intricate shapes. Sheets made from a variety of metals (e.g., low-carbon steel, high-ductility alloy steel, copper and some of its alloys, and aluminum and some of its alloys) can be successfully worked into useful products. Therefore, these processes are continually becoming more attractive to the automotive, aerospace, electrical, and consumer goods industries. Products that had in the past always been manufactured by processes like casting and forging have been redesigned so that they can be produced by sheet metal working. Components like pulleys, connecting rods for sewing machines, and even large gears are now within the range of sheet metal products.

Sheet metals are usually worked while in their cold state. However, when processing thick sheets, which are at least 0.25 inch (6 mm) and are referred to as *plates,* thermal cutting is employed to obtain the required blank shape, and the blank is then hot-worked in a hydraulic or friction screw press. Thus, fabrication of boilers, tanks, ship hulls, and the like would certainly require hot working of thick plates.

By far, the most commonly used operations in sheet metal working are those performed in a press. For this reason, they are usually referred to as *press working,* or simply *stamping,* operations. Other techniques involve *high-energy-rate forming* (HERF), like using explosives or impulsive discharges of electrical energy to form the blank, and *spinning* of the sheet metal on a form mandrel. This chapter will describe each of the various operations employed in sheet metal working.

6.1 PRESS WORKING OPERATIONS

All press working operations of sheet metals can be divided into two main groups: *cutting* operations and *shape-forming* operations. Cutting operations involve separating a part of the blank, whereas forming operations involve nondestructive plastic deformation, which causes relative motion of parts of the blank with respect to each other. Cutting operations include shearing, cutoff, parting, blanking, punching, and notching. Shape-forming operations include various bending operations, deep drawing, embossing, and stretch-forming.

Cutting Operations

The mechanics of separating the metal are the same in all sheet metal cutting operations. Therefore, the operations are identified according to the shape of the curve along which cutting takes place. When the sheet metal is cut along a straight line, the operation is called *shearing* and is usually performed using inclined blades or guillotine shears in order to reduce the force required (see Figure 6.1). Cutting takes place gradually, not all at once, over the width of the sheet metal because the upper blade is inclined. The angle of inclination of the upper blade usually falls between 4° and 8° and must not exceed 15° so that the sheet metal is not pushed out by the horizontal component of the reaction.

When cutting takes place along an open curve (or on an open corrugated line), the operation is referred to as *cutoff,* provided that the blanks match each other or can be fully nested, as shown in Figure 6.2. The cutoff operation results in almost no waste of stock and is, therefore, considered to be very efficient with respect to material utilization. This operation is usually performed in a die that is mounted on a crank press. If the blanks do not match each other, it is necessary for cutting to take place along two open curves (or lines), as shown in Figure 6.3. In this case, the operation is called *parting.* It

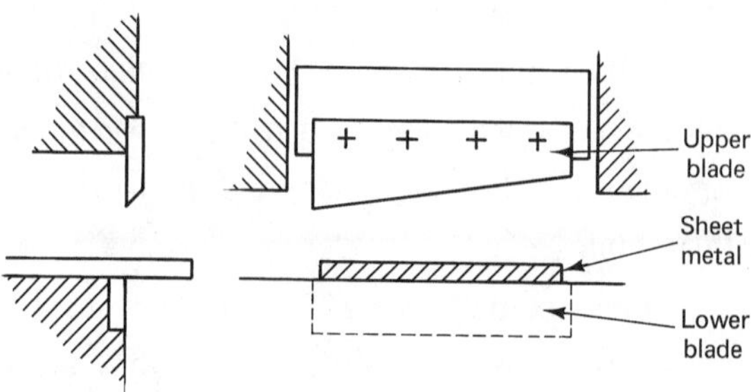

FIGURE 6.1
Shearing operation with inclined blades

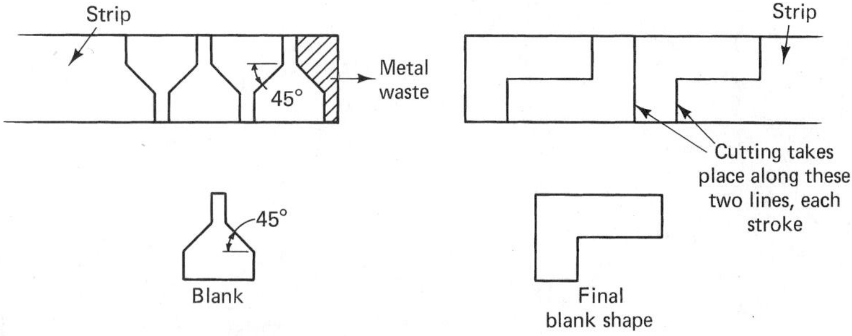

FIGURE 6.2 Examples of cutoff operations

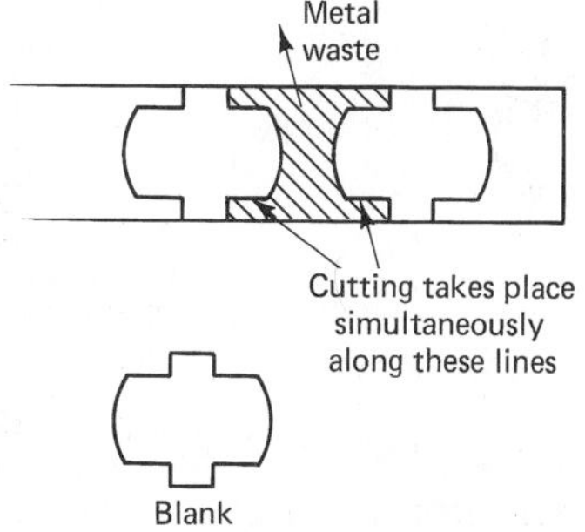

FIGURE 6.3 An example of a parting operation

is clear from the figure that a parting operation results in some waste of stock and is, therefore, less efficient than shearing and cutoff operations.

In *blanking* operations, cutting occurs along a closed contour and results in a relatively high percentage of waste in stock metal, a fact that makes blanking operations less efficient than other cutting operations. Nevertheless, this process is used for mass production of blanks that cannot be manufactured by any of the preceding operations. An efficient layout of blanks on the strip of sheet metal can result in an appreciable saving of material. An example of a good layout is shown in Figure 6.4a, where circular blanks are staggered. The in-line arrangement shown in Figure 6.4b is less efficient in terms of material utilization. Because a blanking operation is performed in a die, there is a limit to the minimum distance between two adjacent blanks. It is always advantageous to keep this minimum distance larger than 70 percent of the thickness of the sheet metal. In blanking, the part separated from the sheet metal is the product, and it is usually further processed. But if the remaining

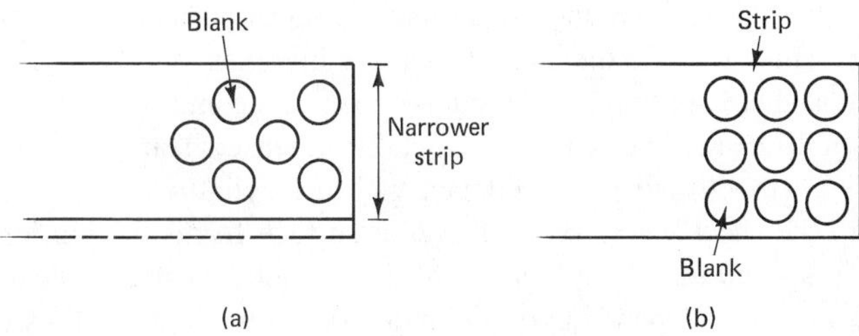

FIGURE 6.4 Two methods for laying out circular blanks for blanking operations: (a) staggered layout; (b) in-line arrangement

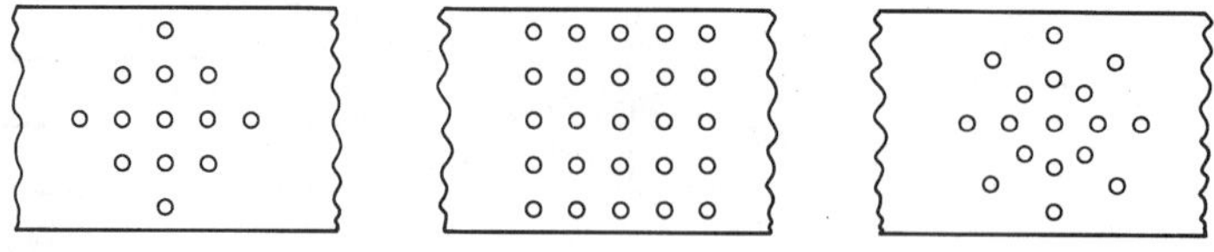

FIGURE 6.5 Different patterns of holes produced by perforating operations

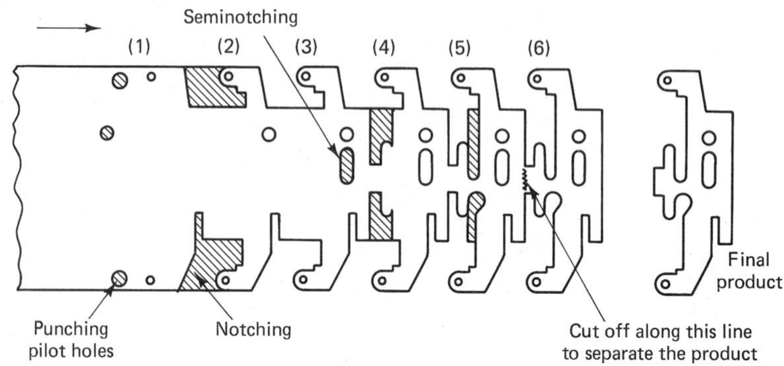

FIGURE 6.6 Progressive working operations

part of the sheet is required as a product, the operation is then termed *punching*. Sometimes, it is required to simultaneously punch a pattern of small holes as an ornament, for light distribution, or for ventilation; the operation is then referred to as *perforating*. Figure 6.5 illustrates some patterns of perforated holes.

A *notching* operation is actually a special case of punching, where the removed part is adjacent to the edge of the strip. It is clear that any required shape can be obtained by carrying out several notching operations. For this reason, notching is usually employed in progressive dies. A similar operation, called *seminotching,* in which the separated part is not attached to the side of the strip, is also used in progressive working of sheet metals. In Figure 6.6, we can see both of these operations and how they can be employed progressively to produce a blank with an intricate shape.

Mechanics of sheet metal cutting. Let us now look further at the process of cutting sheet metal. For simplicity, consider the simple case where a circular punch, together with a matching die, are employed to punch a hole. Figure 6.7 shows the punch, die, and sheet metal during a punching operation. When a load is applied through the punch, the upper surface of the metal is elastically bent over the edge of the punch, while the lower surface is bent over the edge of the die. With further increase in the punch load, the elastic curvature becomes permanent or plastic and is referred to as the *rollover.* Next, the punch sinks into the upper surface of the sheet, while the lower surface sinks into the die hole. This stage involves mainly plastic flow of metal by shearing as there are two forces equal in magnitude and opposite in direction, subjecting the cylindrical surface within the metal to intense shear stress. The result will be a cylindrical smooth surface in contact with the cylindrical surface of the punch as it sinks into the sheet metal. Also, a similar surface forms the border of the part of the metal sinking into the die hole. Each of these smooth surfaces is called a *burnish.* The extent of a burnish depends upon the metal of the sheet as well as on the design features of

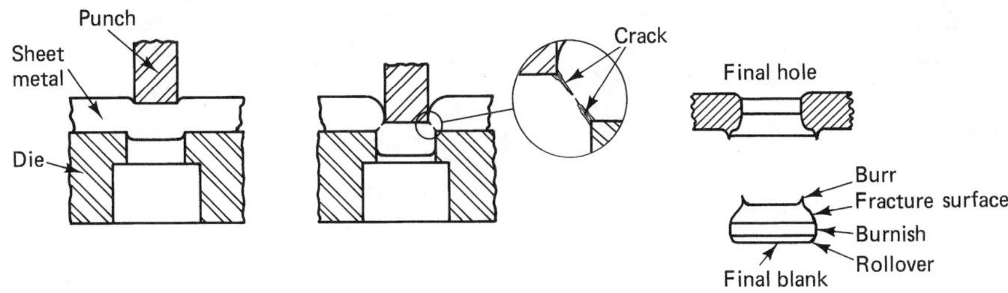

FIGURE 6.7
Stages of a blanking operation

the die. The burnish ranges approximately between 40 and 60 percent of the stock thickness, the higher values being for soft ductile materials like lead and aluminum. At this stage, two cracks initiate simultaneously in the sheet metal, one at the edge of the punch and the other at the edge of the die. These two cracks propagate and finally meet each other to allow separation of the blank from the sheet metal. This zone has a rough surface and is called the *fracture surface* (break area). Finally, when the newly formed blank is about to be completely separated from the stock, a *burr* is formed all around its upper edge. Thus, the profile of the edge of a blank involves four zones: a rollover, a burnish, a fracture surface, and a burr. In fact, the profile of the edge of the generated hole consists of the same four zones, but in reverse order.

We are now in a position to discuss the effects of some process parameters, such as the punch-die clearance. Figure 6.8a illustrates the case where the punch-die clearance is excessive and is almost equal to the thickness of the sheet. Initially, the metal is bent onto the round edges of the punch and the die, and it then forms a short circular wall connecting the flat bottom and the bulk of the sheet. With further increase in the applied load, the wall elongates under the tensile stress, and tearing eventually occurs. As can be seen in Figure 6.8a, the blank resulting in this case has a bent, torn edge all around and, therefore, has no value. On the other hand, if the punch-die clearance is too tight, as shown in Figure 6.8b, the two cracks that initiate toward the end of the operation do not meet, and another shearing must take place so that the blank can be separated. This operation is referred to as the *secondary shear.* As can be seen, the obtained blank has an extremely rough side. In addition, the elastically recovering

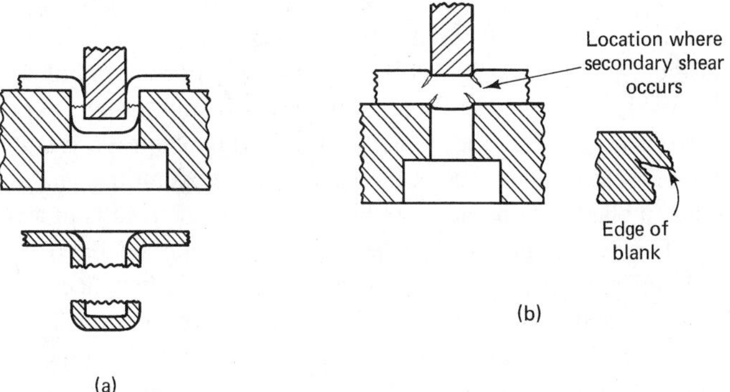

FIGURE 6.8
Blanking operations where the punch-die clearance is:
(a) excessive; (b) too tight

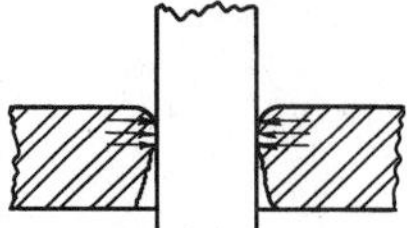

FIGURE 6.9
Elastic recovery of the metal around the hole gripping the punch

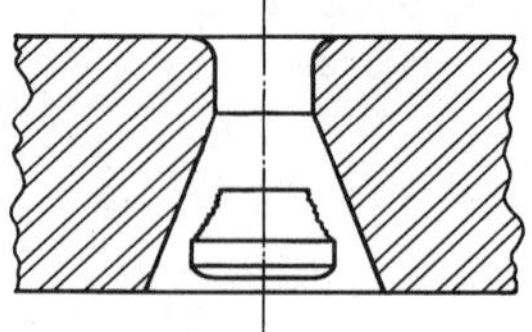

FIGURE 6.10
Elastic recovery of the blank necessitating die relief

sheet stock tends to grip the punch, as shown in Figure 6.9, thus increasing the force required to withdraw the punch from the hole, which is usually called the *stripping force*. This results in excessive punch wear and shorter tool life. On the other hand, the blank undergoes elastic recovery, and it is, therefore, necessary to provide relief by enlarging the lower part of the die hole, as shown in Figure 6.10.

Between these two extremes for the punch-die clearance, there exists an optimum value that reduces or minimizes the stripping force and the tool wear and also gives a blank with a larger burnish and smaller fracture surface. This recommended value for the punch-die clearance is usually taken as about 10 to 15 percent of the thickness of the sheet metal, depending upon the kind of metal being punched.

Forces required. Based on the preceding discussion, the force required for cutting sheet metal is equal to the area subjected to shear stress (the product of the perimeter of the blank multiplied by the thickness of the sheet metal) multiplied by the ultimate shear strength of the metal being cut. The blanking force can be expressed by the following equation:

$$F = K \times Q \times t \times \tau_{\text{ultimate}} \tag{6.1}$$

where: Q is the perimeter

t is the thickness

τ_{ultimate} is the ultimate shear strength

Note that K is an experimentally determined factor to account for the deviation of the stress state from that of pure shear and is taken as about 1.3. The ultimate shear stress can either be obtained from handbooks or be taken as approximately 0.8 of the ultimate tensile strength of the same metal.

We can now see that one of the tasks of a manufacturing engineer is to calculate the required force for blanking (or punching) and to make sure that it is below the capacity of the available press. This is particularly important in industries that involve blanking relatively thick plates. There is, however, a solution to the problem when the required force is higher than the capacity of the available press. It is usually achieved by beveling (or shearing) the punch face in punching operations and the upper surface of the die steel in blanking operations. Shearing the punch results in a perfect hole but a distorted blank, whereas shearing the die yields a perfect blank but a distorted hole. Nevertheless, in both cases, cutting takes place gradually, not all at once, along the contour of the hole (or the blank), with the final outcome being a reduction in the required blanking force. The shear angle is usually taken proportional to the thickness of the sheet metal and ranges between 2° and 8°. Double-sheared punches are quite common and are employed to avoid the possibility of horizontal displacement of sheet metals during punching. Figure 6.11 illustrates the basic concept of punch and die shearing. It also provides a sketch of a double-sheared punch.

Another important aspect of the punching (or blanking) operation is the stripping force (i.e., the force required to pull the punch out of the hole). It is usually taken as 10 percent of the cutting force, although it depends upon some process parameters, such as the elasticity and plasticity of the sheet metals, the punch-die clearance, and the kind of lubricant used.

Bar cropping. *Bar cropping* is similar to sheet metal cutting. Although bars, not sheets, are cut, the mechanics of the process are similar to those of sheet metal cutting, and separation of the cropped part is due to plastic flow caused by intense shear stress. The process is used for mass production of billets for hot forging and cold forming processes. Nevertheless, the distortion and work-hardening at the sheared cross section limit the application of bar cropping when the billets are to be cold formed. Therefore, a modified version of the cropping operation has to be used. It involves completely confining the cropped billet and applying an axial stress of approximately 20 percent of the tensile strength of the bar material. This bar-cropping technique, which is shown in Figure 6.12, yields a very smooth cropped surface and distortion-free billets.

Fine blanking. As we saw previously, the profile of the edge of a blank is not smooth but consists of four zones: the rollover, the burnish, the fracture surface (break area), and the burr. Sometimes, however, the blank must have a straight, smooth side for some functional reasons. In this case, an operation called *fine blanking* is employed, as

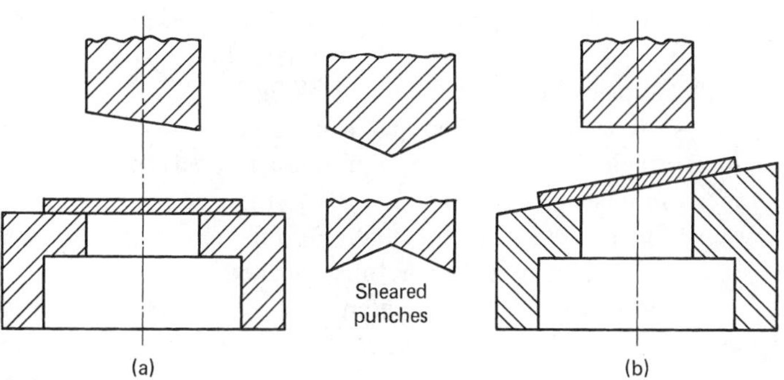

FIGURE 6.11 Shearing of the punch and the die: (a) sheared punch resulting in distorted blanks; (b) sheared die resulting in distorted holes

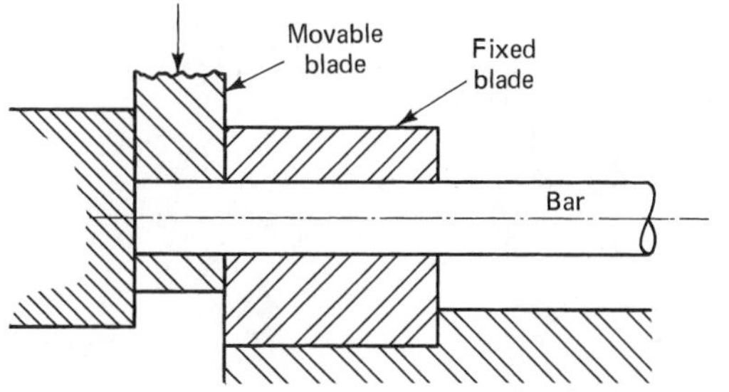

FIGURE 6.12 Bar cropping with workpiece totally confined

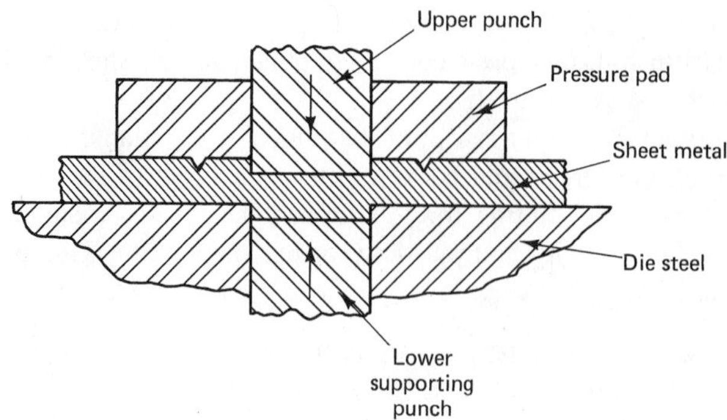

FIGURE 6.13
Fine-blanking operation

Figure 6.13 shows. This operation necessitates the use of a triple-action press and a special die with a very small punch-die clearance. As can be seen in the figure, the metal is squeezed and restrained from moving in the lateral directions in order to control the shear flow along a straight vertical direction. A variety of shapes can be produced by this method. They can have any irregular outer contour and a number of holes as well. The fine-blanking operation has found widespread application in precision industries.

Miscellaneous cutting operations. The primary operation that is used for preparing strips for blanking is needed because the available sheets vary in width between 32 and 80 inches (800 to 2000 mm), a range that is usually not suitable because of the dimensions of the die and the press. Therefore, coils having a suitable width have to be obtained first. The operation performed is called *slitting,* and it employs two circular cutters for each straight cut. Sometimes, slitting is carried out in a rolling plant, and coils are then shipped ready for blanking.

A secondary operation that is sometimes carried out on blanks (or holes) to eliminate rough sides and/or to adjust dimensions is the *shaving* operation. The excess metal in this case is removed in the form of chips. As can be seen in Figure 6.14, the punch-die clearance is very small. For this reason, the die must be rigid, and matching of its two halves must be carefully checked.

Sometimes, punching operations are mistakenly called *piercing.* In fact, the mechanics of sheet metal cutting in the two operations are completely different. We can see in Figure 6.15 that piercing involves a tearing action. We can also see the pointed

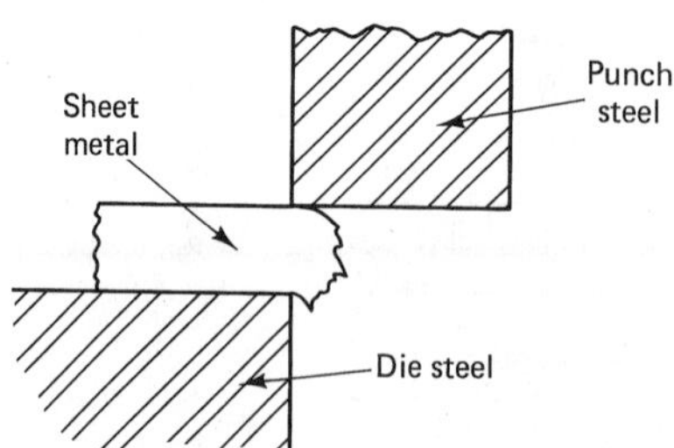

FIGURE 6.14
The shaving operation

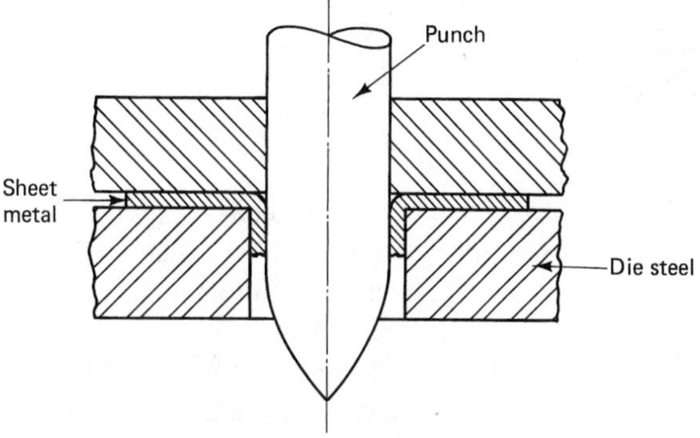

FIGURE 6.15
The piercing operation

shape of the punch. Neither blanks nor metal waste result from the piercing operation. Instead, a short sleeve is generated around the hole, which sometimes has functional application in toy construction and the like.

Cutting-die construction. The construction of cutting dies may take various forms. The simplest one is the *drop-through die,* which is shown in Figure 6.16. In addition to the punch and die steels, the die includes the upper and lower shoes, the guideposts, and some other auxiliary components for guiding and holding the metal strip. The *stripper plate* touches the strip first and holds it firmly during the blanking operation; it then continues to press it until the punch is totally withdrawn from the hole made in the strip. The generated blanks fall through the die hole, which has a relief for this reason, and are collected in a container located below the bed of the press.

Consequently, this die construction is applicable only if the bed of the press has a hole. On the other hand, if the diameter of the required blanks is too large, the use of a drop-through die may result in a defect called *dishing.* As shown in Figure 6.17, this defect involves slackening of the middle of the blank in such a manner that it becomes curved and not flat. The answer to this problem lies in employing a *return-type die.*

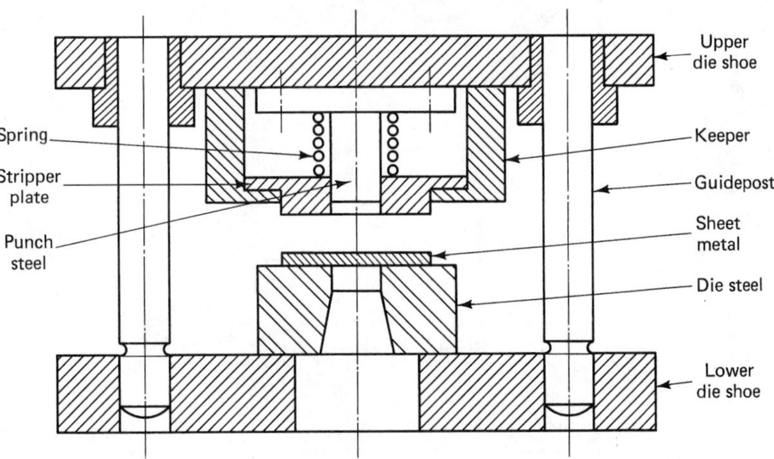

FIGURE 6.16
Die construction for simple drop-through blanking die

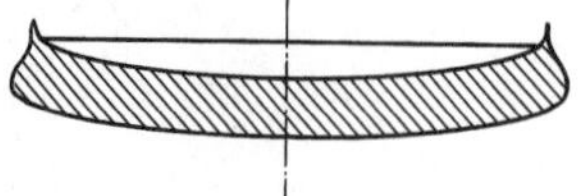

FIGURE 6.17
A vertical section through a blank with the dishing defect

Figure 6.18 shows that in this type of die construction, the blank is supported throughout the operation by a spring-actuated block that finally pushes the blank upward above the surface of the strip, where it is automatically collected. A more complicated die construction, like that shown in Figure 6.19, can be used to perform two operations simultaneously. This is usually referred to as a *compound die.* As can be seen in Figure 6.19, the hollow blanking punch is also a hole-punching die. This allows blanking and punching to be carried out simultaneously. The product, which is a washer, and the central scrap are removed by return blocks.

Bending Operations

Bending is the simplest operation of sheet metal working. It can, therefore, be carried out by employing simple hand tools. As opposed to cutting operations, there is always a clear displacement between the forces acting during a bending operation. The generated bending moment forces a part of the sheet to be bent with respect to the rest of it through local plastic deformation. Therefore, all straight unbent surfaces are not subjected to bending stresses and do not undergo any deformation. Figure 6.20 illustrates the most commonly used types of bending dies: the V-type, the wiping, and the channel (U-type) dies. We can see that the displacement between forces is maximum in the

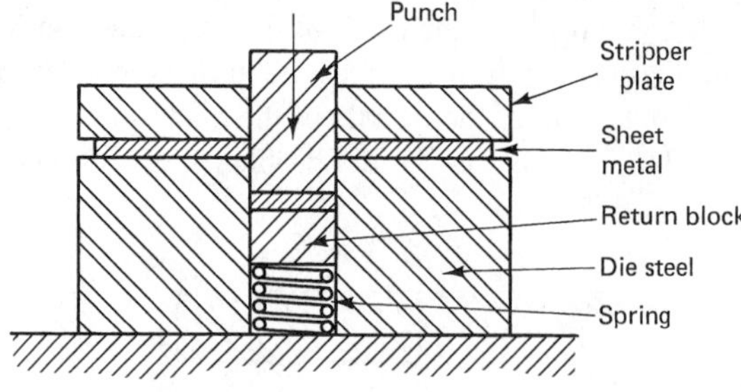

FIGURE 6.18
A return-type die

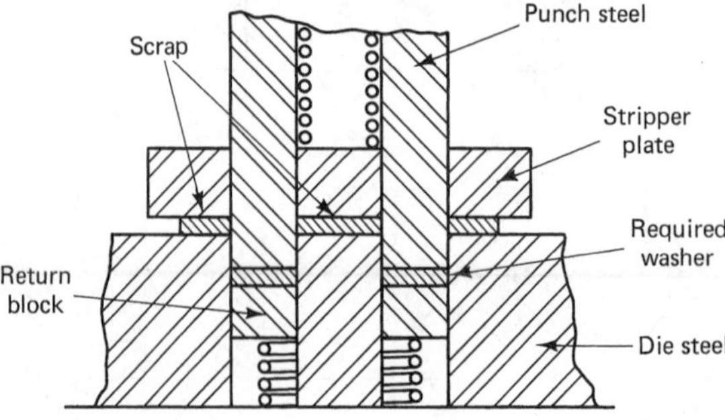

FIGURE 6.19
A compound die for producing a washer

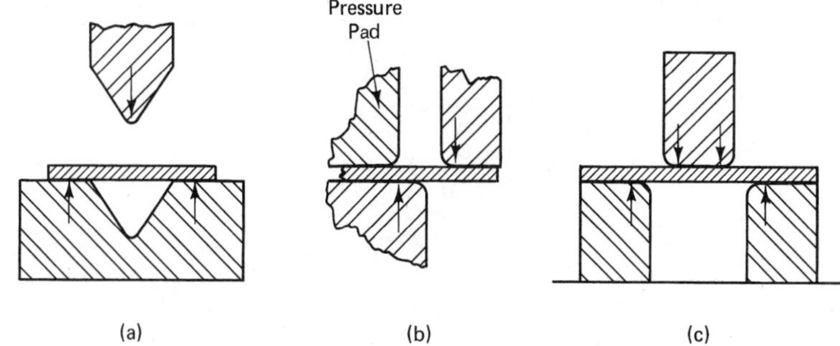

FIGURE 6.20 The three common types of bending dies: (a) V-type die; (b) wiping die; (c) channel (U-type) die

case of the V-type die, and, therefore, lower forces are required to bend sheet metal when using this kind of die.

Mechanics of bending. The bending of sheet metal resembles the case of a beam with a very high width-to-height ratio. When the load is applied, the bend zone undergoes elastic deformation; then plastic deformation occurs with a further increase in the applied load. During the elastic deformation phase, the external fibers in the bend zone are subjected to tension, whereas the internal fibers are subjected to compression. The distribution of stresses is shown in Figure 6.21a. Note that there is a neutral plane that is free of stresses at the middle of the thickness of the sheet. The length of the neutral axis remains constant and does not undergo either elongation or contraction. Next, when the plastic phase starts, the neutral plane approaches the inner surface of the bend, as can be seen in Figure 6.21b. The location of the neutral plane is dependent upon many factors, such as the thickness of the sheet metal, the radius, and the degree of bend. Nevertheless, the distance between the neutral plane and the inner surface of the bend is taken as equal to 40 percent of the thickness of the sheet metal as a first approximation for blank-development calculations.

Let us now consider a very important phenomenon—namely, *springback,* which is an elastic recovery of the sheet metal after the removal of the bending load. As Figure 6.22 indicates, for bending by an angle of 90°, the springback amounts to a few degrees. Consequently, the obtained angle of bend is larger than the required one. Even

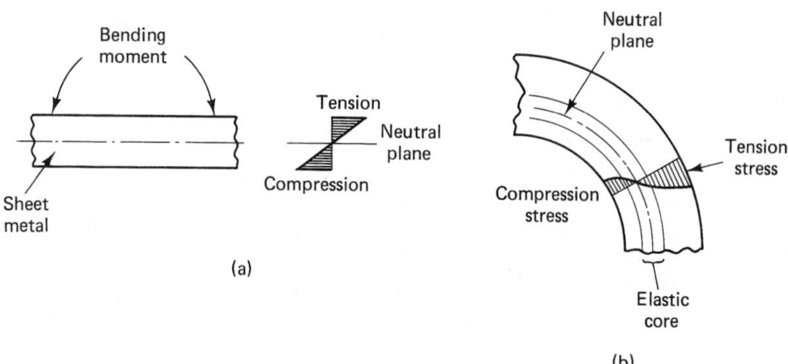

FIGURE 6.21 Distribution of stress across the sheet thickness: (a) in the early stage of bending; (b) toward the end of a bending operation

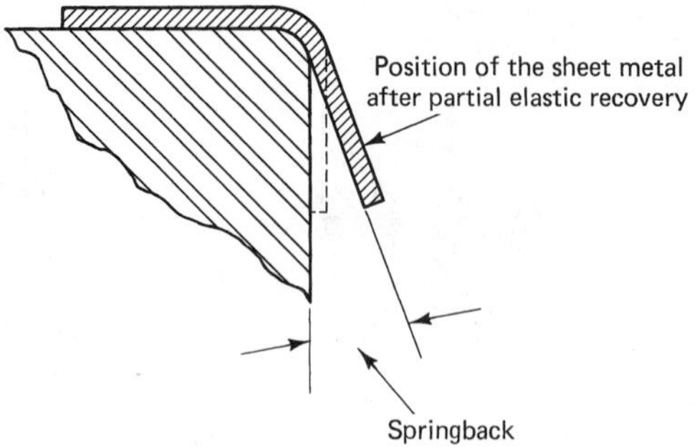

FIGURE 6.22 The springback phenomenon

toward the end of the bending operation, the zone around the neutral plane is subjected to elastic stresses and, therefore, undergoes elastic deformation (see Figure 6.21b). As a result, the elastic *core* tries to return to its initial flat position as soon as the load is removed. When doing so, it is impeded by the plastically deformed zones. The final outcome is, therefore, an elastic recovery of just a few degrees. Consequently, the way to eliminate springback involves forcing this elastic core to undergo plastic deformation. This can be achieved through either of the techniques shown in Figure 6.23a and b. In the first case, the punch is made so that a projection squeezes the metal locally; in the second case, high tensile stress is superimposed upon bending. A third solution is overbending, as shown in Figure 6.23c. In this case, the amount of overbending should be equal to the springback so that the exact required angle is obtained after the elastic recovery.

Blank development. We have previously referred to the fact that the neutral plane does not undergo any deformation during the bending operation and that its length, therefore, remains unchanged. Accordingly, the length of the blank before bending can be obtained by determining the length of the neutral plane within the final product. The lengths of the straight sections remain unchanged and are added together. The following equation can be applied to any general bending product, such as the one shown in Figure 6.24:

$$L = \text{total length of blank before bending}$$

$$= \ell_1 + \ell_2 + \ell_3 + \ell_4 + \frac{\pi\alpha_1}{180}R_1 + \frac{\pi\alpha_2}{180}R_2 + \frac{\pi\alpha_3}{180}R_3 \tag{6.2}$$

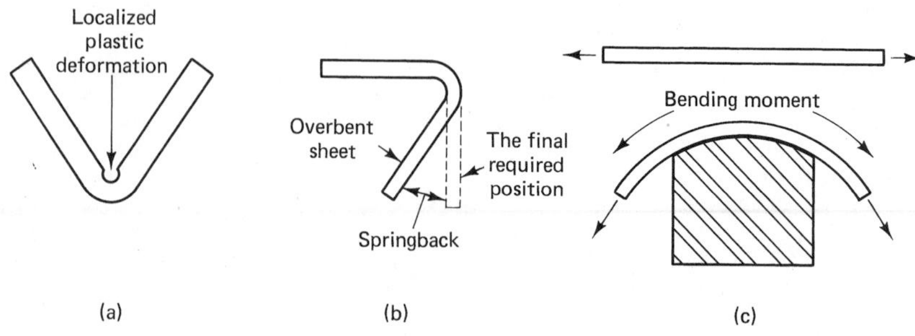

FIGURE 6.23 Methods used to eliminate springback: (a) bottoming; (b) stretch-forming; (c) overbending

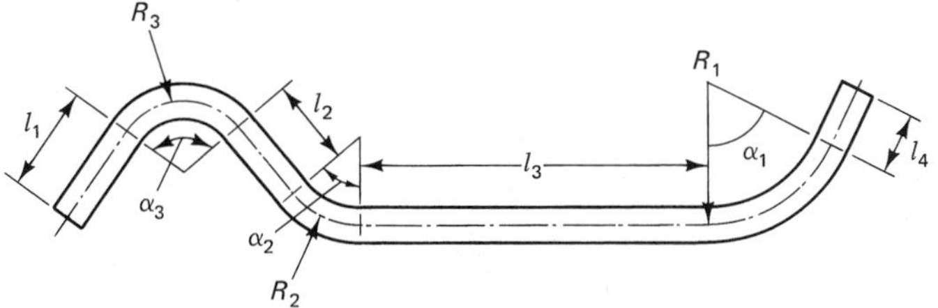

FIGURE 6.24 A bending product divided into straight and circular sections for blank development

where: R is equal to $r + 0.4t$

r is the inner radius of a bend

t is the thickness of the sheet metal

R is the radius of the neutral axis

Classification of bending operations. Various operations can be classified as bending, although each one has its own industrial name. They include, for example, conventional bending, flanging, hemming, wiring, and corrugating. The *flanging* operation is quite similar to conventional bending, except that the ratio of the lengths of the bent part to that of the sheet metal is small. Flanging is usually employed to avoid a sharp edge, thus eliminating the possibility of injury. It is also used to add stiffness to the edges of sheet metal and for assembly purposes.

Among the bending operations, *hemming* used to be a very important one, before the recent developments in welding and can-forming technologies. A hem is a flange that is bent by 180°; it is used now to get rid of a sharp edge and to add stiffness to sheet metal. A few decades ago, hems were widely employed for seaming sheet metals. Figure 6.25 shows four different kinds of hems. A similar operation is *wiring,* which is shown in Figure 6.26. *True* wiring involves bending the edge of the sheet metal around a wire. Sometimes, the operation is performed without a wire, and it is then referred to as *false* wiring.

Corrugating is another operation that involves bending sheet metal. Different shapes, like those shown in Figure 6.27, are obtained by this operation. These shapes possess better rigidity and can resist bending moments normal to the corrugated cross

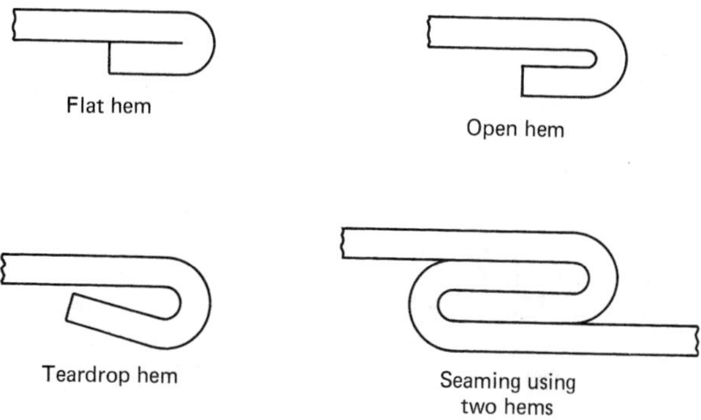

FIGURE 6.25 Different kinds of hems

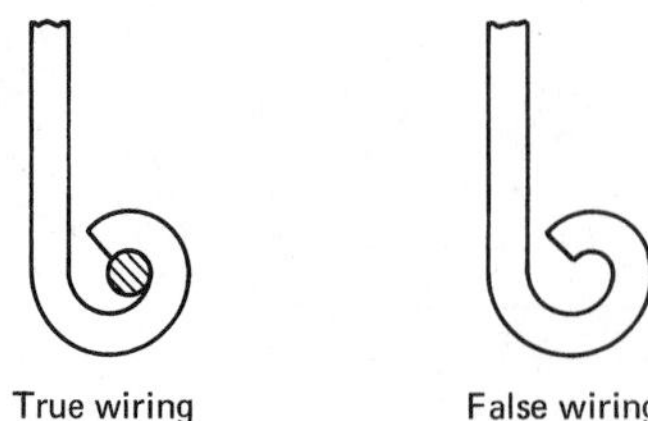

FIGURE 6.26
Wiring operation

FIGURE 6.27
Different shapes of corrugated sheet metal

section mainly because of the increase in the moment of inertia of the section due to corrugation and because of the work-hardened zones resulting from bending.

Miscellaneous bending operations. Conventional bending operations are usually carried out on a press brake. However, with the developments in metal forming theories and machine tool design and construction, new techniques have evolved that are employed in bending not only sheet metal but also iron angles, structural beams, and tubes. Figure 6.28 illustrates the working principles and the stages involved in *roll bending*. As can be seen in the figure, the rolls form a pyramid-type arrangement. Two rolls are used to feed the material, whereas the third (roll B) gradually bends it (see Figure 6.28a and b). The direction of feed is then reversed, and roll A now gradually bends the beam (see Figure 6.28c and d).

Another bending operation that recently emerged and that is gaining industrial application is *rotary bending*. Figure 6.29 illustrates the working principles of this oper-

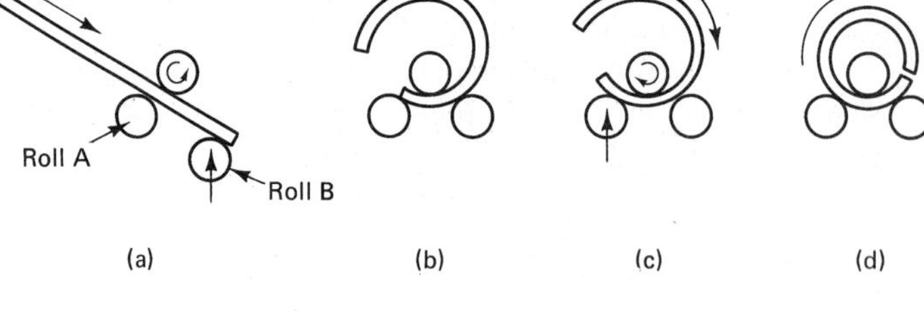

FIGURE 6.28
Stages involved in roll bending a structural beam: (a) feeding; (b) initial bending; (c) further bending; (d) reversing the direction of feed

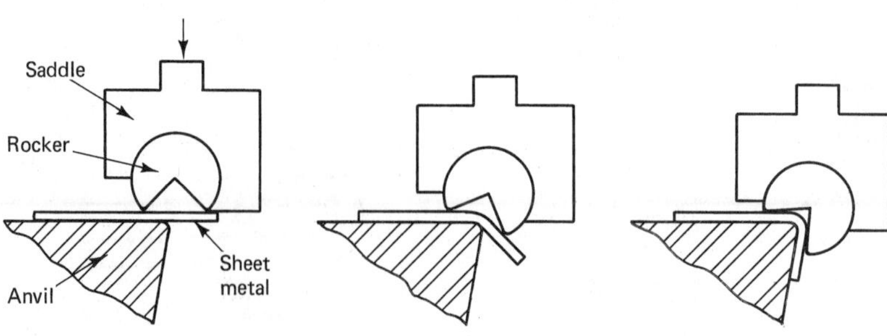

FIGURE 6.29
Working principles of rotary bending

ation. As can be seen, the rotary bender includes three main components: the saddle, the rocker, and the die anvil. The rocker is actually a cylinder with a V-notch along its length. The rocker is completely secured inside the saddle (i.e., the saddle acts like a housing) and can rotate but cannot fall out. The rotary bender can be mounted on a press brake. The rocker acts as both a pressure pad and a bending punch. Among the advantages claimed for rotary bending are the elimination of the pressure pad and its springs (or nitrogen cylinders), lower required tonnage, and the possibility of overbending without the need for any horizontal cams. This new method has been patented by the Accurate Manufacturing Association and is nicknamed by industrial personnel as the "Pac Man" bending operation.

A bending process that is usually mistakenly mentioned among the rolling processes is the manufacture of thin-walled welded pipes. Although rolls are the forming tools, the operation is actually a gradual and continuous bending of a strip that is not accompanied by any variation in the thickness of that strip. Figure 6.30 indicates the basic principles of this process. Notice that the width of the strip is gradually bent to take the form of a circle. Strip edges must be descaled and mechanically processed before the process is performed to improve weldability. Either butt or high-frequency induction welding is employed to weld the edges together after the required circular cross section is obtained. This process is more economical and more productive than seamless tube rolling. Poor strength and corrosion resistance of seams are considered as its main disadvantages.

Deep Drawing Operation

Deep drawing involves the manufacture of deep, cuplike products from thin sheet metal. As can be seen in Figure 6.31, the tooling basically involves a punch with a round corner and a die with a large edge radius. It can also be seen that the punch-die clearance is slightly larger than the thickness of the sheet metal. When load is applied through the punch, the metal is forced to flow radially and sink into the die hole to form a cup. This is an oversimplification of a rather complex problem. For the proper design of deep-drawn products as well as the tooling required, we have to gain a deeper insight into the process and understand its mechanics.

Mechanics of deep drawing. Consider what happens during the early stages of applying the load. As Figure 6.32a shows, the blank is first bent onto the round edge of the die hole. With further increase in the applied load, the part of the blank that was bent is straightened in order to sink into the annular clearance between the punch and

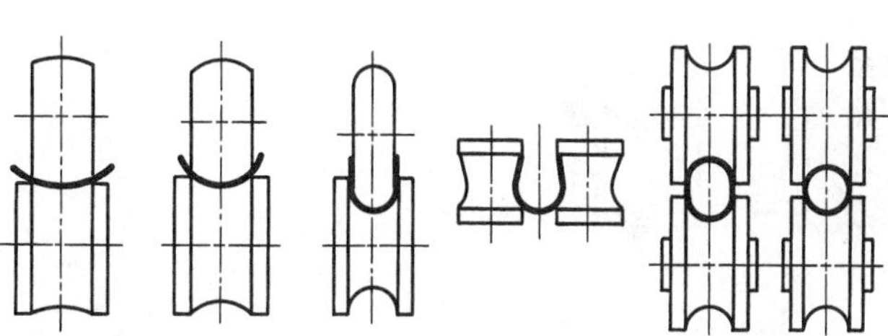

FIGURE 6.30 Roll bending as employed in the manufacture of seamed tubes

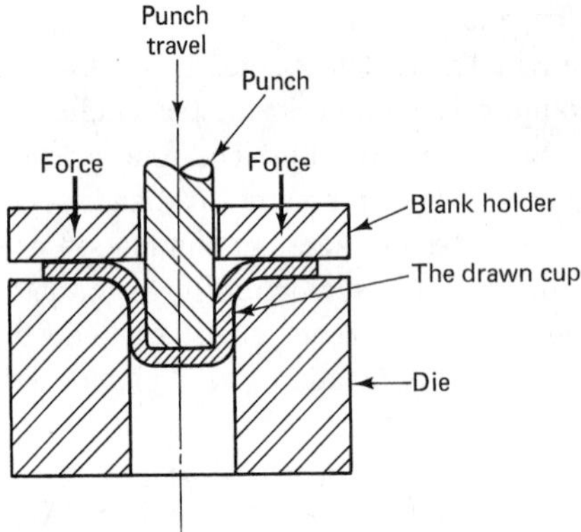

FIGURE 6.31
Basic concept of deep drawing

the die, thus forming a short, straight, vertical wall. Next, the rest of the blank starts to flow radially and to sink into the die hole, but because the lower surface of the blank is in contact with the upper flat surface of the die steel, frictional forces try to impede that flow. These forces are a result of static friction; their magnitude drops as the blank metal starts to move. Now consider what happens to a sector of the blank, such as that shown in Figure 6.32b, when its metal flows radially. It is clear that the width of the sector shrinks so that the large peripheral perimeter of the blank can fit into the smaller perimeter of the die hole. This is caused by circumferential compressive stresses acting within the plane of the blank. With further increase in the applied load, most of the blank sinks into the die hole, forming a long vertical wall, while the remaining part of the blank takes the form of a small annular flange (see Figure 6.31). The vertical wall is subjected to uniaxial tension whose magnitude is increasing when going toward the bottom of the cup.

We can see from the preceding discussion that the deep drawing process involves five stages: bending, straightening, friction, compression, and tension. Different parts of the blank being drawn are subjected to different states of stress. As a result, the deformation is not even throughout the blank, as is clear in Figure 6.33, which shows an exaggerated longitudinal section of a drawn cup. While the flange gets thicker because of the circumferential compressive stress, the vertical wall gets thinner, and thinning is

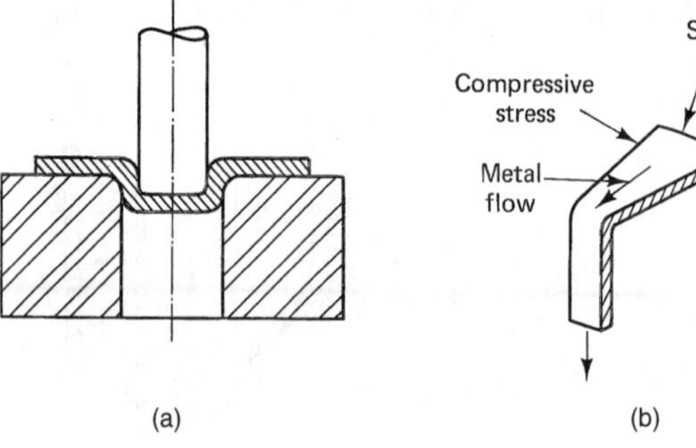

FIGURE 6.32
Mechanics of deep drawing: (a) first stage of deep drawing (i.e., bending);
(b) compression stage in deep drawing

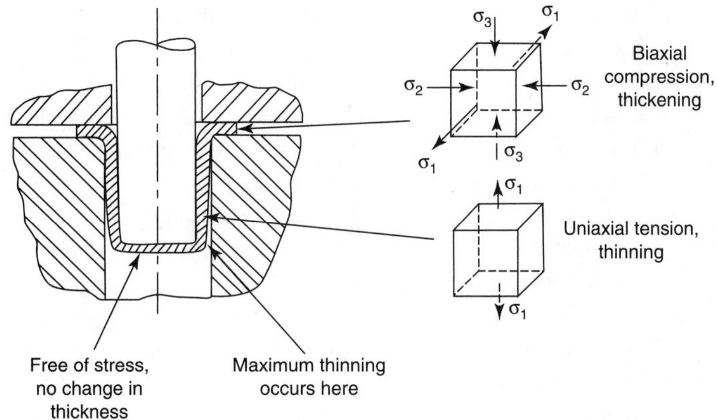

FIGURE 6.33 An exaggerated longitudinal section of a drawn cup, with the states of stress at different locations

maximum at the lowest part of the wall adjacent to the bottom of the cup. Accordingly, if the cup is broken during the drawing process, failure is expected to occur at the location of maximum thinning. An upper bound for the maximum drawing force can, therefore, be given by the following equation:

$$F = \pi \times (d + t)t\sigma_T \qquad \textbf{(6.3)}$$

where: F is the maximum required drawing force
d is the diameter of the punch
t is the thickness of the blank
σ_T is the ultimate tensile strength of the blank material

The blank holder. As previously mentioned, the thin blank is subjected to compressive stresses within its plane. This is similar to the case of a slender column subjected to compression, where buckling is expected to occur if the slenderness ratio (i.e., length/thickness) is higher than a certain value. Therefore, by virtue of similarity, if the ratio of the diameter of the blank to its thickness exceeds a certain value, buckling occurs. Actually, if $(D_o - d)/t \geq 18$, where D is the blank diameter, d is the punch diameter, and t is the thickness, the annular flange will buckle and crimple. This is a product defect referred to as *wrinkling.*

One way to eliminate wrinkling (buckling) of the thin blank is to support it over its entire area. This is done by sandwiching the blank between the upper surface of the die steel and the lower surface of an annular ring that exerts pressure upon the blank, as shown in Figure 6.31. This supporting ring is called the *blank holder,* and the force exerted on it can be generated by die springs or a compressed gas like nitrogen. On the other hand, higher frictional forces will initiate at both the upper and lower surfaces of the blank as a result of the blank-holding force. For this reason, lubricants like soap in water, waxes, mineral oil, and graphite are applied to both surfaces of the blank. Moreover, the upper surface of the die steel as well as the lower surface of the blank holder must be very smooth (ground and lapped). As a rule of thumb, the blank-holding force is taken as 1/3 the force required for drawing.

Variables affecting deep drawing. Now that we understand the mechanics of the process, we can identify and predict the effect of each of the process variables. For example, we can see that poor lubrication results in higher friction forces, and, accordingly, a higher drawing force is required. In fact, in most cases of poor lubrication, the cup cross section does not withstand the high tensile force, and failure of the wall at the bottom takes place during the process. A small die corner radius would increase the bending and straightening forces, thus increasing the drawing force, and the final outcome would be a result similar to that caused by poor lubrication.

In addition to these process variables, the geometry of the blank has a marked effect not only on the process but also on the attributes of the final product. An appropriate quantitative way of expressing the geometry is the number indicating the thickness as a percentage of the diameter, or $(t/D) \times 100$. For smaller values of this percentage (e.g., 0.5), excessive wrinkling should be expected, unless a high blank-holding force is used. If the percentage is higher than 3, no wrinkling occurs, and a blank holder is not necessary.

Another important variable is the drawing ratio, which is given by the following equation:

$$R = \frac{D}{d} \tag{6.4}$$

where: R is the drawing ratio

D is diameter of the blank

d is the diameter of the punch

It has been experimentally found that the deep drawing operation does not yield a sound cup when the drawing ratio is higher than 2 (i.e., for successful drawing, R must be less than 2).

Another number that is commonly used to characterize drawing operations is the percentage reduction. It can be given by the following equation:

$$r = \frac{D - d}{D} \times 100 \tag{6.5}$$

where: r is the percentage reduction

D is the diameter of the blank

d is the diameter of the punch

It is a common industrial practice to take the value of r as less than 50 percent in order to have a sound product without any tearing. When the final product is long and necessitates a value of r higher than 50 percent, an intermediate cup must be obtained first, as shown in Figure 6.34. The intermediate cup must have dimensions that keep the percentage reduction below 50. It can then be redrawn, as illustrated in Figure 6.35, once or several times until the final required dimensions are achieved. The maximum permissible percentage reduction in the redrawing operations is always far less than 50 percent. It is usually taken as 30 percent, 20 percent, and 13 percent, in the first, second, and third redraws, respectively. If several redrawing operations are required, the product should

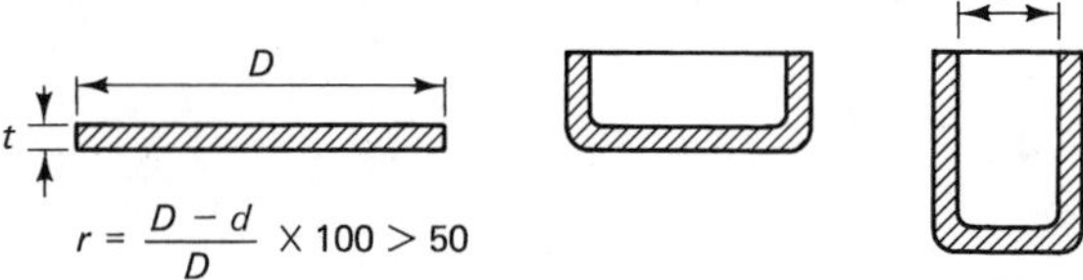

FIGURE 6.34 The use of an intermediate cup when the total required reduction ratio is high

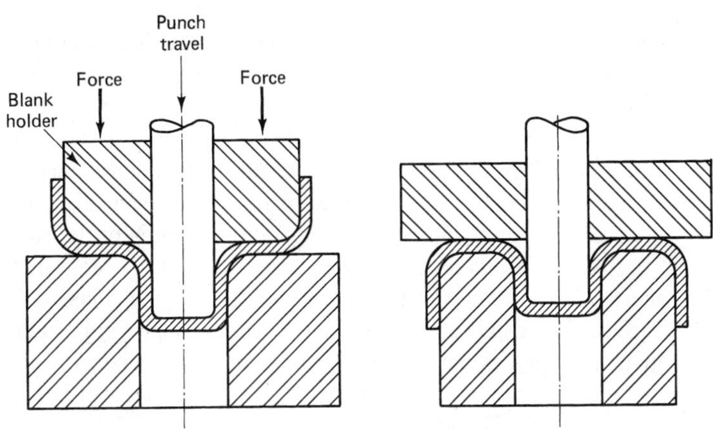

FIGURE 6.35 Redrawing an intermediate cup

then be annealed after every two operations in order to eliminate work-hardening and thus avoid cracking and failure of the product.

Blank-development calculations. For the sake of simplicity, it is always assumed that the thickness of the blank remains unchanged after the drawing operation. Because the total volume of the metal is constant, it can then be concluded that the surface area of the final product is equal to the surface area of the original blank. This rule forms the basis for the blank-development calculations. Consider the simple example shown in Figure 6.36. The surface area of the cup is the area of its bottom plus the area of the wall:

$$\text{surface area of cup} = \frac{\pi}{4}d^2 + \pi dh$$

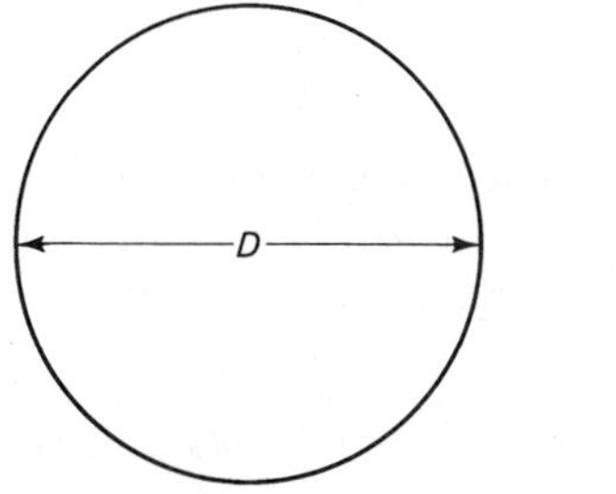

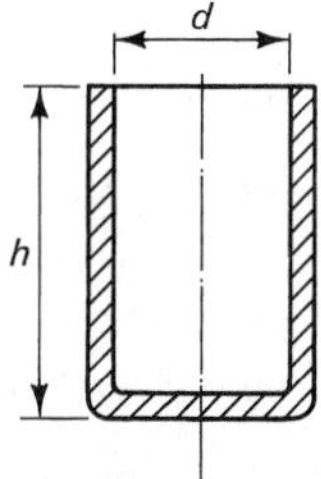

Surface area of blank $\frac{\pi}{4}D^2 = \frac{\pi}{4}d^2 + \pi dh$, i.e., surface area of the cup

FIGURE 6.36 A simple example of blank development

This is equal to the surface area of the original blank; therefore, we can state that

$$\frac{\pi}{4}D^2 = \frac{\pi}{4}d^2 + \pi dh$$

or

$$D^2 = d^2 + 4dh$$

Therefore, the original diameter of the blank, which is unknown, can be given by the following equation:

$$D = \sqrt{d^2 + 4dh} \tag{6.6}$$

Equation 6.6 gives an approximate result because it assumes the cup has sharp corners, which is not the case in industrial practice. However, this equation can be modified to take round corners into account by adding the area of the surface of revolution resulting from the rotation of the round corner around the centerline of the cup, when equating the area of the product to that of the original blank. Note that the area of any surface of revolution can be determined by employing Pappus's first theorem, which gives that area as the product of the path of the center of gravity of the curve around the axis of rotation multiplied by the length of that curve.

Planning for deep drawing. The process engineer usually receives a blueprint of the required cup from the product designer. His or her job is to determine the dimensions of the blank and the number of drawing operations needed, together with the dimensions of intermediate cups, so that the tool designer can start designing the blanking and the deep drawing dies. That job requires experience as well as close contact between the product designer and the process engineer. The following steps can be of great help to beginners:

1. Allow for a small flange around the top of the cup after the operation is completed. This flange is trimmed at a later stage and is referred to as the *trimming allowance.* It is appropriate to take an allowance equal to 10 to 15 percent of the diameter of the cup.

2. Calculate the total surface area of the product and the trimming allowance. Then, equate it to the area of the original blank with an unknown diameter. Next, solve for the diameter of the original blank.

3. Calculate the thickness as a percentage of the diameter or $(t/D) \times 100$, in order to get a rough idea of the degree of wrinkling to be expected (see the preceding discussion on process variables).

4. Calculate the required percentage reduction. If it is less than or equal to 50, then the required cup can be obtained in a single drawing. But if the required r is greater than 50, then a few redrawing operations are required; the procedure to be followed is given in the next steps.

5. For the first draw, assume r to be equal to 50 and calculate the dimensions of the intermediate cup. Then, calculate r required for the first redraw. If $r \leq 30$, only a single redraw is required.

6. If $r > 30$ for the first redraw, take it as equal to 30 and calculate the dimensions of a second intermediate cup. The percentage reduction for the second redraw should be less than 20; otherwise, a third redraw is required, and so on.

Ironing. We can see from the mechanics of the deep drawing operation that there is reasonable variation in the thickness of the drawn cup. In most cases, such thickness variation does not have any negative effect on the proper functioning of the product, and, therefore, the drawn cups are used as is. However, close control of the dimensions of the cups is sometimes necessary. In this case, cups are subjected to an *ironing* operation, in which the wall of the cup is squeezed in the annular space between a punch and its corresponding die. As can be seen in Figure 6.37, the punch-die clearance is smaller than the thickness of the cup and is equal to the final required thickness. Large reductions in thickness should be avoided in order to obtain a sound product. It is good industrial practice to take the value of the punch-die clearance in the range between 30 and 80 percent of the thickness of the cup. Also, the percentage reduction in thickness, which is given next, should fall between 40 and 60 in a single ironing operation. This is a safeguard against fracture of the product during the operation. Following is the equation to be applied:

$$\text{percentage reduction in thickness} = \frac{t_o - t_f}{t_o} \times 100 \qquad \textbf{(6.7)}$$

where: t_o is the original thickness of the cup

t_f is the final thickness of the cup after ironing

Drawing of stepped, conical, and domed cups. Stepped cups are those with two (or more) shell diameters (see Figure 6.38a). They are produced in two (or more) stages. First, a cup is drawn to have the large diameter, and, second, a redrawing operation is performed on only the lower portion of the cup. Tapered or conical cups (see Figure

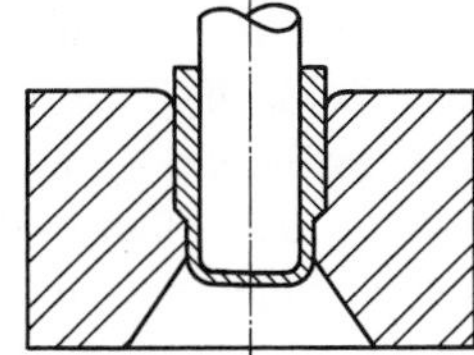

FIGURE 6.37
The ironing operation

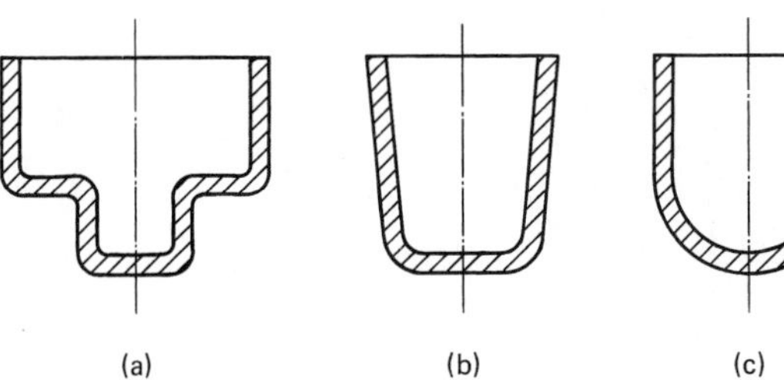

FIGURE 6.38
Deep-drawn cups:
(a) stepped; (b) conical;
(c) domed

6.38b) cannot be drawn directly. They first have to be made into stepped cups, which are then smoothed and stretched out to give the required tapered cups. A complex deep drawing operation is used for producing domed cups (see Figure 6.38c). So that the sheet metal stretches properly over the punch nose, higher blank-holding forces are required. Therefore, the process actually involves stretch-forming, and its variables should be adjusted to eliminate either wrinkling or tearing.

Drawing of box-shaped cups. When all press working operations of sheet metal are reviewed, there would be almost no doubt that the box drawing process is the most complex and difficult to control. Nevertheless, in an attempt to simplify the problem, we can divide a box into four round corners and four straight sides. Each of these round corners represents 1/4 of a circular cup, and, therefore, the previous analysis holds true for it. On the other hand, no lateral compression is needed to allow the blank metal to flow toward the die edge at each of the straight sides. Accordingly, the process in these zones is not drawing at all; it is just bending and straightening. For this reason, the metal in these zones flows faster than in the round corners, and a square blank takes the form shown in Figure 6.39 after drawing. Note that there is excess metal at each of the four round corners, which impedes the drawing operations at those locations. It also results in localized higher stresses and tears almost always beginning at one (or more) of the corners during box drawing, as can be seen in Figure 6.40.

Several variables affect this complex operation as well as the quality of the products obtained. They include the die bending radius, the die corner radius, and the shape of the original blank. These process variables have been investigated by research workers, and it has been found that in order to obtain sound box-shaped cups, it is very important to ensure easy, unobstructed flow of metal during the drawing operation. The absence of this condition results in the initiation of high tensile stresses in the vertical walls of the box, especially at the round corners, and results in considerable thinning, which is followed by fracture. Among the factors that can cause obstruction to the metal flow are smaller die radii, higher reduction ratios (at the corners), and poor lubrication. These are added to the presence of excess metal at the corners, which causes an appreciable increase in the transverse compressive stresses. Therefore, an optimum blank shape without excess metal at the corners is necessary for achieving successful drawing operations of box-shaped cups. A simple method for optimizing the shape of the blank is shown in Figure 6.41. It involves printing a square grid on the surface of the blank and determining the borders of the undeformed zone on the flanges at each corner (by observing the

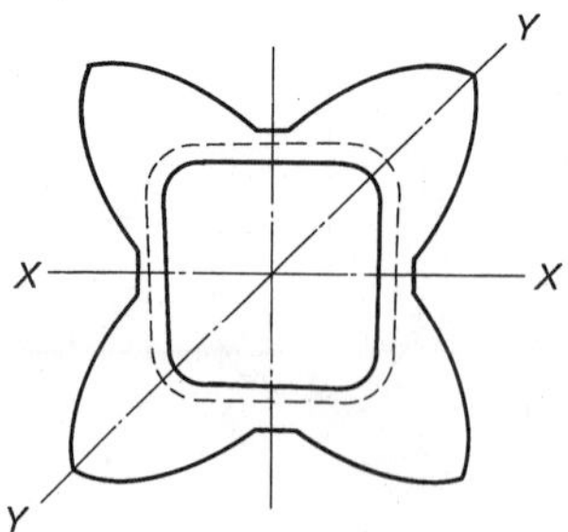

FIGURE 6.39
Final shape of a box-shaped cup, obtained by deep drawing a square blank

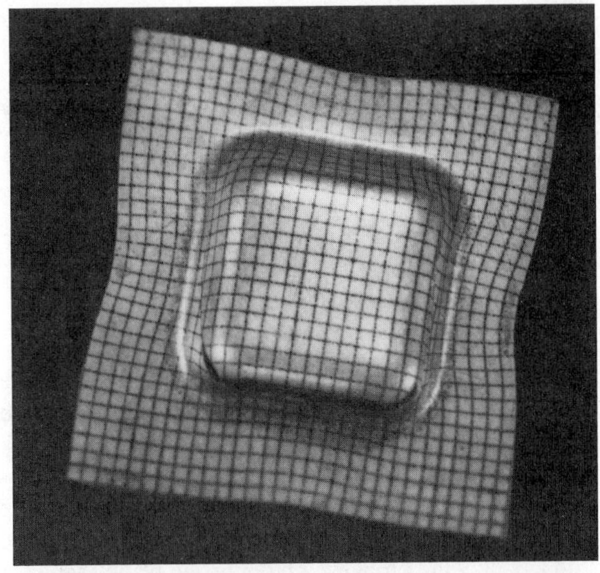

FIGURE 6.40
Tears occurring in box drawing

FIGURE 6.41
Optimized blank shape for drawing box-shaped cups

undistorted grid) so that it can be taken off the original blank. It has been found that the optimum shape is a circle with four cuts corresponding to the four corners. Also, the blank-holding force has been found to play a very important role. Better products are obtained by using a rubber-actuated blank holder that exerts low forces during the first third of the drawing stroke, followed by a marked increase in those forces during the rest of the drawing stroke to eliminate wrinkling and stretch out the product.

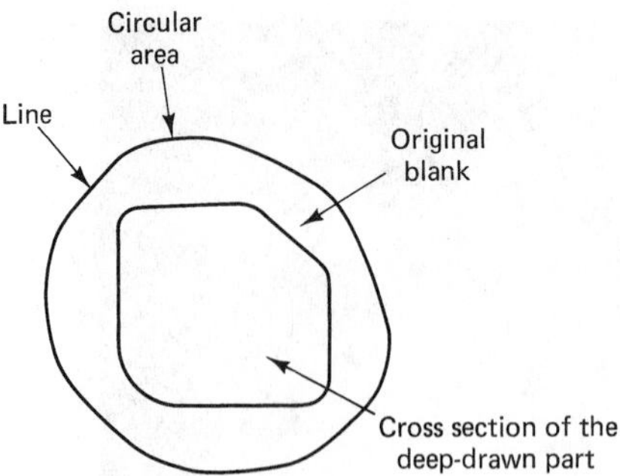

FIGURE 6.42 Optimized blank shape for drawing cups with an irregular cross section

The preceding discussion can be generalized to include the drawing of a cup with an irregular cross section. This can be achieved by dividing the perimeter into straight sides and circular arcs. Professor Kurt Lange and his coworkers (Institute Für Umformstechnik, Stuttgart Universitäte) have developed a technique for obtaining the optimum blank shape in this case by employing the slip-line field theory. The technique was included in an interactive computer expert system that is capable of giving direct answers to any drawing problem. An optimized blank shape obtained by that system is shown in Figure 6.42.

Recent developments in deep drawing. A recent development in deep drawing involves cup drawing without a blank holder. Cupping of a thick blank has been accomplished by pushing the blank through a die having a special profile, as shown in Figure 6.43, without any need for a blank holder. This process has the advantages of reducing the number of processing stages, eliminating the blank holder, and using considerably simpler tool construction. A further advantage is that the operation can be performed on a single-acting press, resulting in an appreciable reduction in the initial capital cost required.

Another new development is the employment of ultrasonics to aid the deep drawing operation. The function of the ultrasonic waves is to enlarge the die bore and then leave it to return elastically to its original dimension in a pulsating manner. This reduces the friction forces appreciably, resulting in a marked reduction in the required drawing force and in a clear improvement of the quality of the drawn cup. In many cases, the cup can be drawn by the force exerted by the human hand without the need

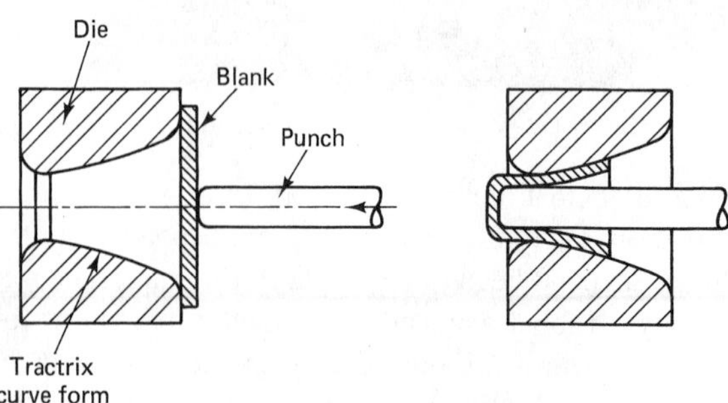

FIGURE 6.43 Drawing cups without a blank holder

for any mechanical force-generating device. It is, therefore, obvious that low-tonnage, high-production-rate presses can be used, which makes the process economically attractive.

Defects in deep-drawn parts. These defects differ in shape and cause, depending upon the prevailing conditions and also on the initial dimensions of the blank. Following is a brief description of the most common defects, some of which are shown in Figure 6.44:

1. **Wrinkling.** Wrinkling is the buckling of the undrawn part of the blank under compressive stresses; it may also occur in the vertical walls (see Figure 6.44a and b). If it takes place on the punch nose when drawing a domed cup, it is referred to as *puckering*.
2. **Tearing.** Tearing, which always occurs in the vicinity of the radius connecting the cup bottom and the wall, is caused by high tensile stresses due to the obstruction of the flow of the metal in the flange.
3. **Earing.** Earing is the formation of ears at the free edges of a deep-drawn cylindrical cup (see Figure 6.44c). It is caused by the anisotropy of the sheet metal. Ears are trimmed after a drawing operation, resulting in a waste of material.
4. **Surface irregularities.** Surface irregularities are caused by nonuniform yielding, like the orange-peel effect of Luder's lines.
5. **Surface marks.** Surface marks are caused by improper punch-die clearance or poor lubrication. These include draw marks, step rings, and burnishing.

Forming Operations

In this section, we will discuss the various forming operations performed on sheet metals—not just flat sheets, but tubular sheets (i.e., thin-walled tubes) as well. Therefore, not only will operations like embossing and offsetting be discussed, but also tube bulging, expanding, and necking will be considered.

Forming of sheets. True forming involves shaping the blank into a three-dimensional (or sculptured) surface by sandwiching it between a punch and a die. The strain is not uniform, and the operation is complex. The nonhomogeneity (or complexity) depends upon the nature and the unevenness of the required shape. Experience and trial and error were employed in the past to obtain an optimum blank shape and to avoid thinning the blank or tearing.

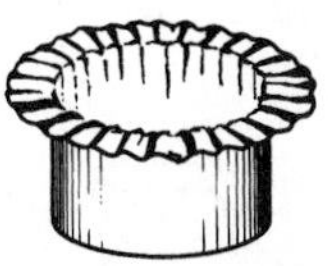

(a)

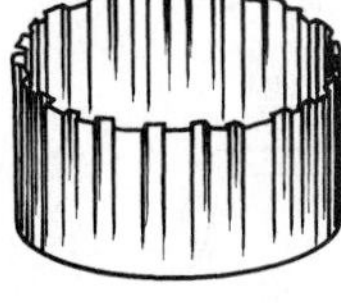

(b)

(c)

FIGURE 6.44 Some defects occurring in deep drawing operations: (a) wrinkling in the flange; (b) wrinkling in the wall; (c) earing

A printed grid on the original blank helps to detect the locations of overstraining where tearing is expected. It also helps in optimizing the shape of the original blank. With recent advances in computer graphics and simulation of metal deformation, rational design of the blank can be performed by the computer, without any need for trial and error. In fact, a successful software package has been prepared by the Mechanical Engineering Department of Michigan Technological University.

Embossing operations. Embossing operations involve localized deflection of a flat sheet to create depressions in the form of beads and offsets. This is sometimes called *oil canning*. Beads and offsets are usually employed to add stiffness to thin sheets, whether flat or tubular (e.g., barrels), as well as for other functional reasons. A typical example of a part that is subjected to embossing is the license plate of an automobile. The cross section of a bead can take different forms, such as those shown in Figure 6.45. Because this operation involves stretching the sheet, the achieved localized percentage elongation within the bead cross section must be lower than that allowable for the metal of the sheet. On the other hand, Figure 6.46 shows two kinds of offsets, where it is common practice to take the maximum permissible depth as three times the thickness of the sheet metal.

Rubber forming of flat sheets. Rubber forming is not new and actually dates back to the nineteenth century, when a technique for shearing and cutting paper and foil was patented by Adolph Delkescamp in 1872. Another rubber forming technique, called the *Guerin process,* was widely used during World War II for forming aircraft panels. It involved employing a confined rubber pad on the upper platen of the press and a steel form block on the lower platen, as shown in Figure 6.47a. This method is still sometimes used. As can be seen in the figure, when a block of elastomer (usually incompressible artificial rubber) is confined in a rigid box, the only way it can flow when the punch sinks into it is up, thus forming the blank around the punch under uniform pressure over the whole surface. It is also common industrial practice to place spacers on the base of the metal box in order to provide a relief for the elastomer block, which, in turn, helps to avoid the initiation of high localized strains in the blank area directly beneath the punch. Rubber forming has real potential when the number of parts required is relatively small and does not justify designing and constructing a forming die.

A modified version of this process, called the *hydroform* process, involves employing a pressurized fluid above the rubber membrane, as shown in Figure 6.47b.

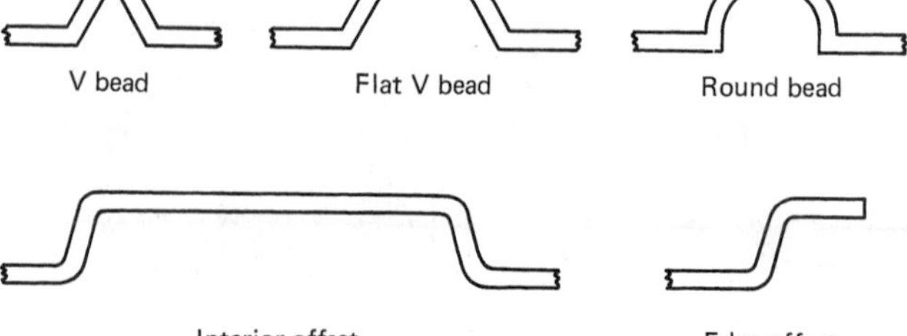

FIGURE 6.45 Different kinds of beads

FIGURE 6.46 Offsetting operations

FIGURE 6.47
Rubber forming of flat sheets: (a) conventional rubber forming; (b) hydroform process

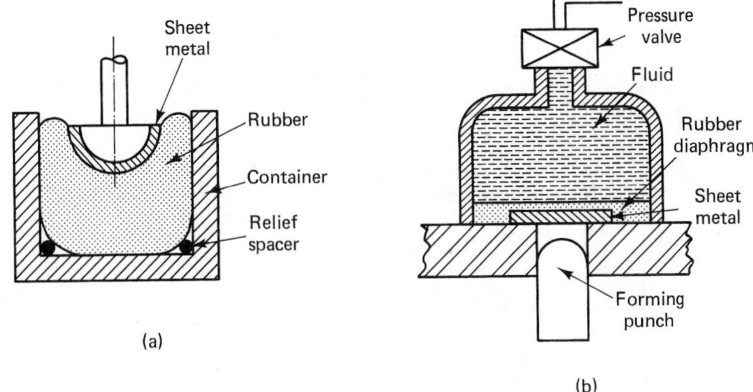

This is similar in effect to drawing the cup into a high-pressure container, as previously mentioned. Therefore, percentage reductions higher than those obtained in conventional deep drawing can be achieved.

Forming of tubular sheets. Figure 6.48a through d indicates tubular parts after they were subjected to beading, flattening, expanding, and necking operations, respectively.

Tube bulging is another forming operation, in which the diameter of the tube, in its middle part, is expanded and then restrained by a split die and forced to conform to the details of the internal surface of the die. This can be achieved by internal hydraulic pressure or by employing an elastomer (polyurethane) rod as the pressure-transmitting medium, causing expansion of the tube. A schematic of this operation is

FIGURE 6.48
Different tubular parts after forming operations: (a) beading; (b) flattening; (c) expanding; (d) necking

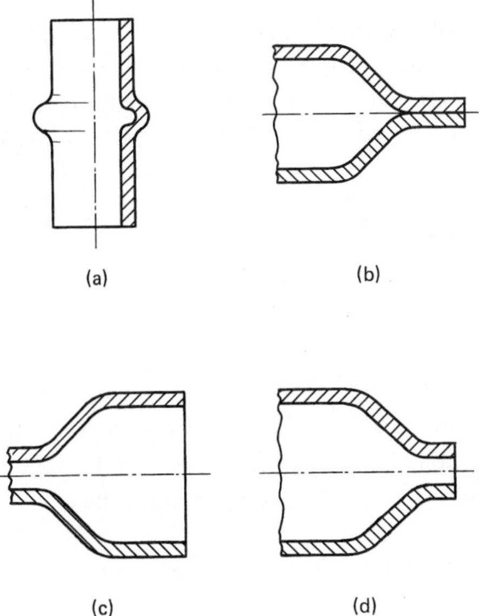

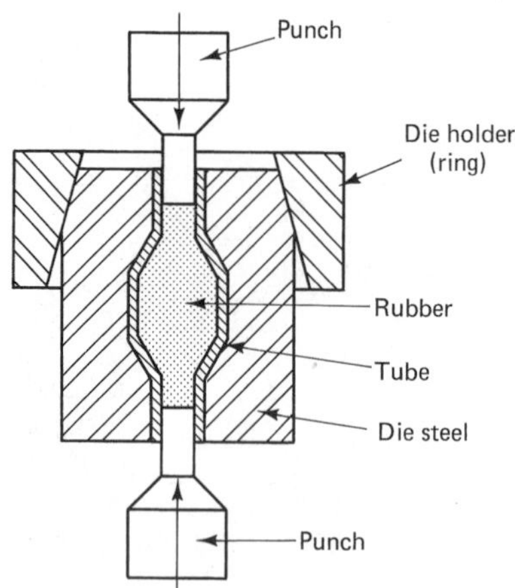

FIGURE 6.49 The tube-bulging operation with an elastomer rod

given in Figure 6.49. At the beginning of the operation, the elastomer rod fits freely inside the tube and has the same length. Compressive forces are then applied to both the rod and the tube simultaneously so that the tube bulges outward in the middle and the frictional forces at the tube-rod interface draw more metal into the die space, thus decreasing the length of the tube. The method of using a polyurethane rod is simpler and cleaner, and there is no need for using oil seals or complicated tooling construction. A further advantage of rubber bulging is that it can be used for simultaneous forming, piercing, and shearing of thin tubular sheets.

6.2 HIGH-ENERGY-RATE FORMING (HERF)

In HERF, the energy of deformation is delivered within a very short period of time—on the order of milliseconds or even microseconds. HERF methods include *explosive, electrohydraulic,* and *electromagnetic* forming techniques. These techniques are usually employed when short-run products or large parts are required. HERF is also recommended for manufacturing prototype components and new shapes in order to avoid the unjustifiable cost of dies. Rocket domes and other aerospace structural panels are typical examples. During a HERF process, the sheet metal is given an extremely high acceleration in a very short period of time and is thus formed as a result of consuming its own kinetic energy to cause deformation.

Explosive Forming

Explosive forming of sheet metal received some attention during the past decade. The various explosive forming techniques fall under one or the other of two basic systems: confined and unconfined. In a *confined* system, which is shown in Figure 6.50a, a charge of low explosives is detonated and yields a large amount of high-pressure gas, thus forcing the sheet metal to take the desired shape. This system is mainly used for

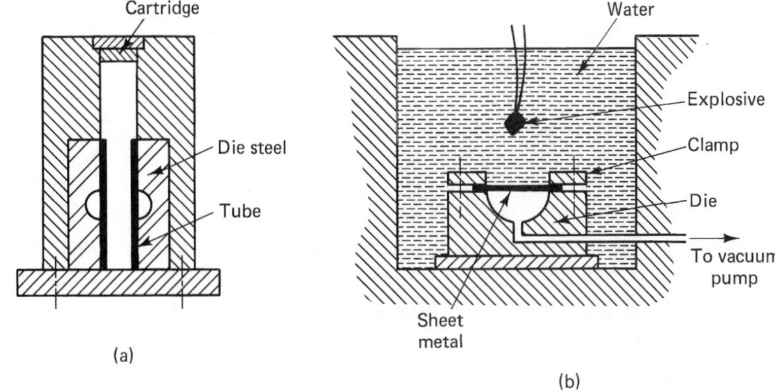

FIGURE 6.50 Explosive forming of sheet metal: (a) confined system; (b) standoff system

bulging and flaring of small tubular parts. Its main disadvantage is the hazard of die failure because of the high pressure generated.

In an *unconfined,* or *standoff,* system, which is shown in Figure 6.50b, the charge is maintained at a distance from the sheet blank (the standoff distance), and both the blank and the charge are kept immersed in water. When the charge is detonated, shock waves are generated, thus forming a large blank into the desired shape. It is obvious that the efficiency of the standoff system is less than that of the confined system because only a portion of the surface over which the shock waves act is utilized (actually, shock waves act in all directions, forming a spherical front). However, the standoff system has the advantages of a lower noise level and of largely reducing the hazard of damaging the workpiece by particles resulting from the explosion. In a simple standoff system, the distance from the explosive charge to the water surface is usually taken as twice the standoff distance. The latter depends upon the size of the blank and is taken as equal to D (the blank diameter) for D less than 2 feet (60 cm) and is taken as equal to $0.5D$ for D greater than that. Best results are obtained when the blank is clamped lightly around its periphery and when a material with a low modulus of elasticity, like plastic, is used as a die material. This eliminates springback, thus obtaining closer tolerances. A modified version of this method is illustrated in Figure 6.51, where a reflector is used to collect and

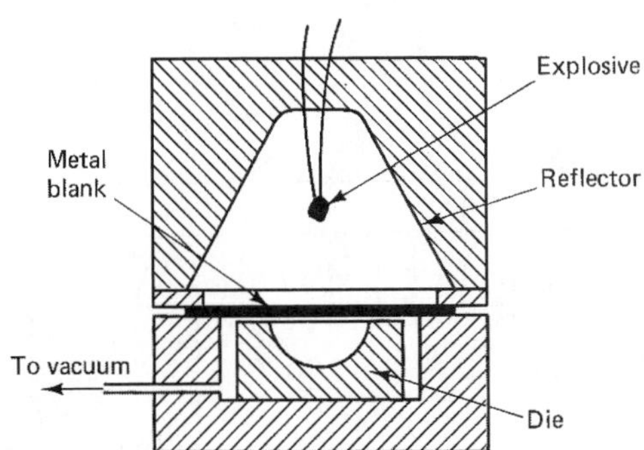

FIGURE 6.51 Increasing the efficiency of explosive forming by using a reflector

reflect the explosion energy that does not fall directly onto the blank surface. This leads to improved efficiency over the standoff system because a smaller amount of charge is needed for the same job.

Electrohydraulic Forming

The basic idea for the process of electrohydraulic forming, which has been known for some time, is based on discharging a large amount of electrical energy across a small gap between two electrodes immersed in water, as shown in Figure 6.52. The high-amperage current resulting from suddenly discharging the electrical energy from the condensers melts the thin wire between the electrodes and generates a shock wave. The shock wave lasts for a few microseconds; it travels through water to hit the blank and forces it to take the shape of the die cavity. The use of a thin wire between the electrodes has the advantages of initiating and guiding the path of the spark, enabling the use of nonconductive liquids; also, the wire can be shaped to suit the geometry of the required product. The method is also safer than explosive forming and can be used for simultaneous operations like piercing and bulging. Nevertheless, it is not suitable for continuous production runs because the wire has to be replaced after each operation. Moreover, the level of energy generated is lower than that of explosive forming. Therefore, the products are generally smaller than those produced by explosive forming.

Electromagnetic Forming

Electromagnetic forming is another technique based on the sudden discharge of electrical energy. As we know from electricity and magnetism in physics, when an electric current passes through a coil, it initiates a magnetic field whose magnitude is a function of the current. We also know that when a magnetic field is interrupted by a conductive material (workpiece), a current is induced in that material that is proportional to the rate of change of the flux. This is called *eddy current* and produces its own magnetic field that opposes the initial one. As a result, repulsive forces between the coil and the workpiece force the workpiece to conform to the die cavity. This technique can be used to form flat as well as tubular sheets. As can be seen in Figure 6.53, it is employed in expanding as well as compressing tubes. It has proven to be very effective when forming relatively thin materials.

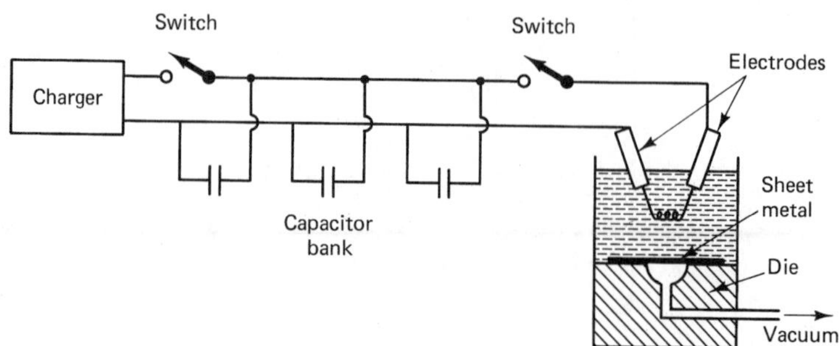

FIGURE 6.52
Electrohydraulic forming

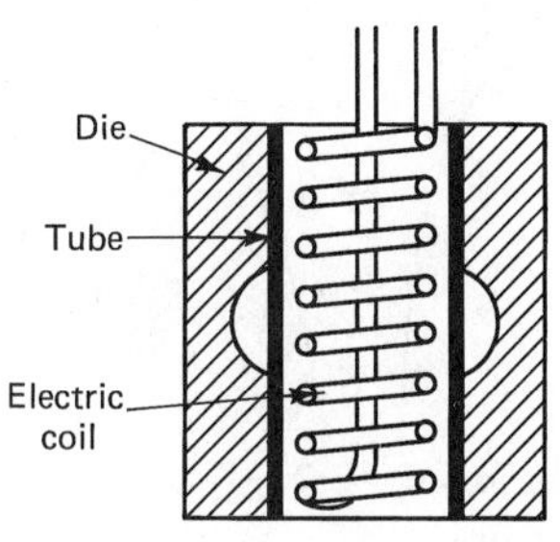

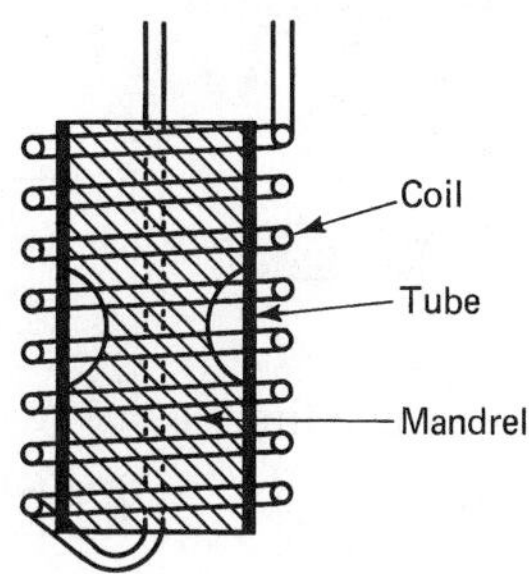

FIGURE 6.53 Examples of electromagnetic forming of tubes

6.3 SPINNING OF SHEET METAL

Spinning is the forming of axisymmetric hollow shells over a rotating-form mandrel by using special rollers. Generally, the shapes produced by spinning can also be manufactured by drawing, compressing, or flanging. However, spinning is usually used for forming large parts that require very large drawing presses or when there is a diversity in the products (i.e., when various shapes are needed but only a small number of each shape is required).

A schematic of the spinning operation is shown in Figure 6.54. At the beginning, the semifinished product (circular blank) is pushed by the tail stock against the front of the form mandrel (usually a wooden one) that is fixed on the rotating faceplate of the spinning machine (like a lathe). A pressing tool is pushed by the operator onto the external surface of the blank. The blank slips under the pressing devices, which causes localized deformation. Finally, the blank takes the exact shape of the form mandrel. This technique can also be used to obtain hollow products with a diameter at the end (neck) smaller than that at the middle. In this case, it is necessary to use a collapsible-form mandrel, which is composed of individual smaller parts that can be extracted from the neck of the final product after the process is completed. Figure 6.55 shows a group of parts produced by spinning.

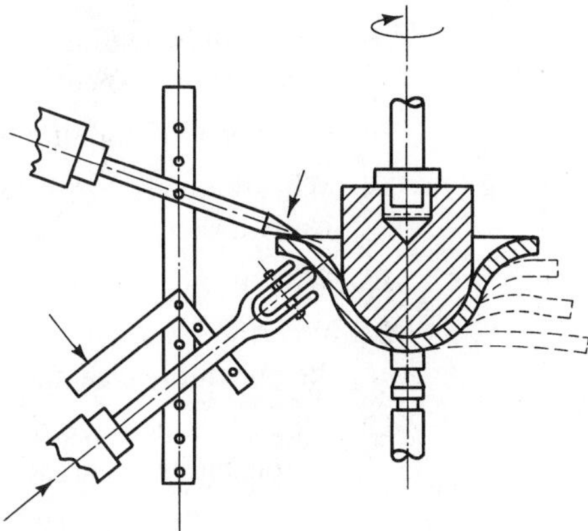

FIGURE 6.54 A schematic of the spinning operation

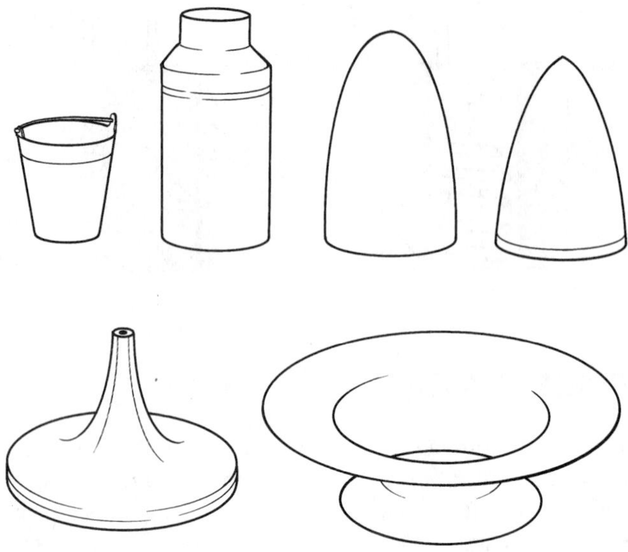

FIGURE 6.55
A group of parts produced by spinning

A modified version of this method involves replacing the operator by a numerically controlled (NC) tool. Auxiliary operations, like removing the excess metal, are also carried out on the same machine. Better surface quality and more uniform thickness are the advantages of NC spinning over the conventional techniques.

Review Questions

1. What main design feature characterizes sheet metal products?
2. List some of the advantages of press working sheet metals.
3. When are sheet metals formed in their hot state? Give examples.
4. What are the two main groups of press working operations?
5. What main condition must be fulfilled so that cutting of sheet metal (and not any other operation) takes place?
6. Use sketches to explain why the angle of inclination of the upper blade of guillotine shears must not exceed 15°.
7. Use sketches to differentiate between the following operations: shearing, cutoff, parting, blanking, and punching.
8. Why must attention be given to careful layout of blanks on a sheet metal strip?
9. Describe a perforating operation.
10. What does an edge of a blank usually look like? Draw a sketch.
11. What is meant by the *percentage penetration?*
12. What does an edge of a blank look like when the punch-die clearance is too large?
13. What does an edge of a blank look like when the punch-die clearance is too tight?

14. When are punches sheared and why?

15. When are dies sheared and why?

16. In what aspect is fine blanking different from conventional blanking?

17. Use sketches to explain each of the following operations: shaving, piercing, and cropping.

18. Can a drop-through die be used on any press? Why not?

19. What is the function of a stripper plate?

20. List two types of die constructions for blanking operations.

21. How can a washer be produced in a single stroke?

22. What condition must be fulfilled so that bending of sheet metal takes place?

23. Sketch the common types of bending dies.

24. Which die requires the minimum force for the same thickness of sheet metal?

25. Where is tearing expected to occur and where is wrinkling expected to occur when a sheet metal is subjected to bending?

26. What is *springback?* Why does it occur?

27. List three methods for eliminating the effects of springback.

28. On what assumption is blank development based?

29. List some operations that can be classified as bending. Use sketches and explain design functions of the products.

30. How can structural angles be bent?

31. Explain rotary bending, using sketches, and list some of the advantages of this operation.

32. Explain how a seamed tube can be produced by continuous bending.

33. What are the disadvantages of seamed tubes?

34. Explain deep drawing, using sketches.

35. What are the stages involved in deep drawing a circular cup? Explain, using sketches.

36. Indicate the states of stress at different locations in a cup toward the end of a drawing operation.

37. Where is thickening expected to occur?

38. At what location is thinning maximum? To what would this lead?

39. Why is a blank holder sometimes needed?

40. List some of the variables affecting the deep drawing operation.

41. What is *wrinkling?* Why does it occur?

42. Describe an ironing operation.

43. Is there any limitation on ironing?

44. Why are conical cups not drawn directly?

45. What is actually taking place when drawing domed cups?

46. Is it feasible to take any blank shape for box drawing operations and then trim the excess metal? Why?

47. What are the mechanics of deformation in the straight-sides areas?

48. What is the advantage of ultrasonic deep drawing?

49. How can plates be drawn without a blank holder?

50. List some of the defects experienced in deep-drawn products.

51. As a product designer, how can you make use of the embossing operation when designing sheet metal parts?

52. When would you recommend using rubber forming techniques?

53. What is meant by *high-energy-rate forming?*

54. When would you recommend using explosive forming?

55. Should the dies used in explosive forming be made of a hard material, like alloy steel, or a softer one, like plastic? Why?

56. What happens if you make the hydraulic head very small in explosive forming?

57. What are the advantages of electrohydraulic forming? What are the disadvantages?

58. Use a sketch to explain the electromagnetic forming operation.

59. Describe spinning. When is it recommended?

60. Can products with a diameter at the neck smaller than at the middle be produced by spinning? How?

Problems

1. The blank shown in Figure 6.56 is to be produced from a sheet metal strip 0.0625 inch (1.6 mm) in thickness. Material is low-carbon steel AISI 1020. Estimate the required blanking force.

2. The products shown in Figure 6.57a, b, and c are produced by bending. Obtain the length of the original blank to the nearest 0.01 inch (0.25 mm). Take t as 0.0625 inch (1.6 mm).

3. A cup is drawn from a sheet of 1020 steel. The thickness is 0.03 inch (0.8 mm), and the inner diameter is 1 inch (25 mm). Estimate the maximum force required for drawing. If the material is aluminum, what would the force be?

4. A cup with a height of 0.75 inch (18.75 mm) and an inner diameter of 1 inch (25 mm) is to be drawn from a steel strip 0.0625 inch (1.6 mm) in thickness. Plan for the drawing process by carrying out blank development, determining the number of drawings, and looking at the draw severity analysis.

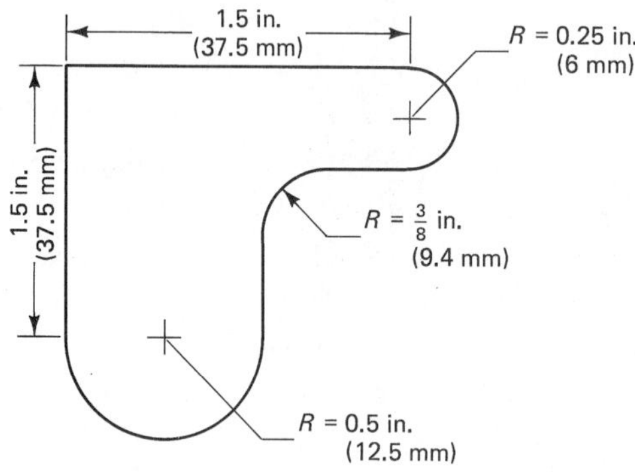

FIGURE 6.56
The blank shape required in Problem 1

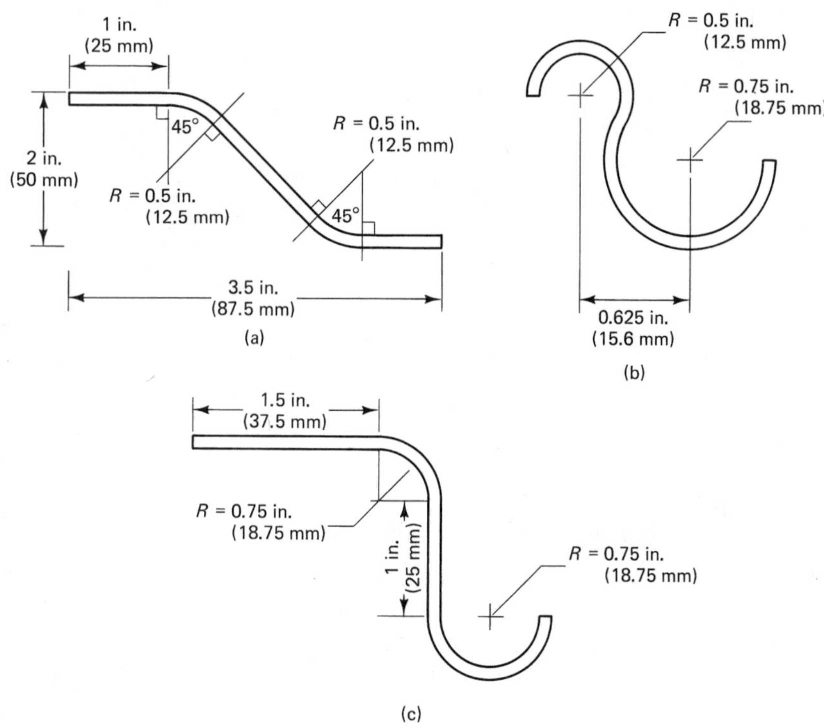

FIGURE 6.57 Products produced by bending in Problem 2

Design Example

PROBLEM

Design a simple wrench to loosen (or tighten) a 1/2-inch (12.5-mm) nut (or bolt head). The 1/2 inch (12.5 mm) measures across the nut flats. The torque is 1 lb ft (1.356 N·m), and 50,000 pieces are required annually. The wrench is to be produced by press working.

Solution

A suitable method for production is fine blanking as there will be no need for any further machining operations. We cannot select a steel that has a high carbon content because it will create problems during the fine-blanking operation. An appropriate choice is AISI 1035 CD steel. The dimensions of the wrench are the same as those given in the examples on forging and casting, although the tolerances can be kept much tighter. A detailed design is given in Figure 6.58.

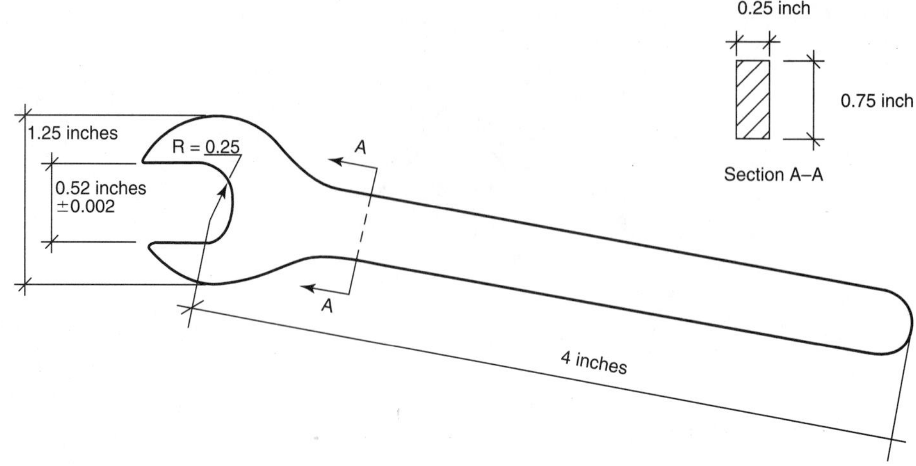

FIGURE 6.58 Detailed design of a wrench produced by stamping

Now it is time to check the maximum tensile stress due to bending:

$$I = \frac{1}{12}bh^3 = \frac{1}{2}(0.25)(0.75)^3 = 0.10546 \times 10^{-2} \text{ in.}^4$$

$$\sigma = \frac{My}{I} = \frac{1 \times 12 \times 0.5}{0.10546} = 5689.0 \text{ lb/in.}^2$$

It is less than the allowable stress for 1035 steel, which is about 20,000 lb/in.^{2}.

Design Projects

1. A pulley (for a V-belt) that has 4-inch (100-mm) outer diameter and is mounted on a shaft that is 3/4 inch (19 mm) in diameter was manufactured by casting. The process was slow, and the rejects formed a noticeable percentage of the production. As a product designer, you are required to redesign this pulley so that it can be produced by sheet metal working and welding.

2. Design a connecting rod for a sewing machine so that it can be produced by sheet metal working, given that the diameter of each of the two holes is 0.5 inch (12.5 mm) and the distance between the centers of the holes is 4 inches (100 mm).

3. If a connecting rod four times smaller than the one of Design Project 2 is to be used in a little toy, how would the design change?

4. Design a table for the machine shop. The table should be 4 feet in height, with a surface area of 3 by 3 feet (900 by 900 mm), and should be able to carry a load

of half a ton. Assume that 4000 pieces are required annually and that different parts will be produced by sheet metal working and then joined together by nuts and bolts.

5. A trash container having a capacity of 1 cubic foot (0.02833 m^3) is to be designed for manufacturing by sheet metal working. Assume that it is required to withstand an axial compression load of 200 pounds (890 N) and that the production rate is 50,000 pieces per year. Provide a detailed design for this trash container.

6. A connecting lever is produced by forging. The lever has two short bosses, each at one of its ends and each with a vertical hole that is 3/4 inch (19 mm) in diameter. The horizontal distance between the centers of the holes is 12 inches (300 mm), and the vertical distance is 3 inches (75 mm). The lever during functioning is subjected to a bending moment of 200 lb ft (271 N·m). Because of the high percentage of rejects and low production rate, this connecting lever is to be produced by sheet metal working. Provide a detailed design so that it can be produced by this manufacturing method.

Chapter 7

Powder Metallurgy

INTRODUCTION

Powder metallurgy is the technology of producing useful components shaped from metal powders by pressing and simultaneous or subsequent heating to produce a coherent mass. The heating operation is usually performed in a controlled-atmosphere furnace and is referred to as *sintering*. The sintering temperature must be kept below the melting point of the powder material or the melting point of the major constituent if a mixture of metal powders is used. Therefore, sintering involves a solid-state diffusion process that allows the compacted powder particles to bond together without going through the molten state. This, in fact, is the fundamental principle of powder metallurgy.

Historical background. Although powder metallurgy is becoming increasingly important in modern industry, the basic techniques of this process are very old indeed. The ancient Egyptians used a crude form of powder metallurgy as early as 3000 B.C. to manufacture iron implements. The technique involved reducing the ore with charcoal to obtain a spongy mass of metal that was formed by frequent heating and hammering to eject the slag and consolidate the iron particles together into a mass of wrought iron. This process was used because the primitive ovens then available were not capable of melting iron. The same technique was used later by smiths in India about A.D. 300 to manufacture the well-known Delhi pillar weighing 6.5 tons. This method was superseded when more advanced ovens capable of melting ferrous metals came into being.

At the beginning of the nineteenth century, powder metallurgy had its first truly scientific enunciation, in England, when Wallaston published details of the preparation of malleable platinum. As had happened in the past, Wallaston's technique was superseded by melting. However, the need for the powder metallurgy process arose again to satisfy the industrial demand for high-melting-point metals. An important application was the production of ductile tungsten in 1909 for manufacturing electric lamp filaments.

Why powder metallurgy? As a result of the development of furnaces and melting techniques, the powder-consolidation process is now usually used when melting metal is undesirable or uneconomical. Fusion is not suitable when it is required to produce parts with controlled, unique structures, such as porous bearings, filters, metallic frictional materials, and cemented carbides. Also, it has been found that powder metallurgy can produce certain complicated shapes more economically and conveniently than other known manufacturing processes. For this reason, the process currently enjoys widespread industrial application. As the price of labor and the cost of materials continue to rise, the powder-consolidation technique is becoming more and more economical because it eliminates the need for further machining operations, offers more efficient utilization of materials, and allows components to be produced in massive numbers with good surface finish and close tolerances.

7.1 METAL POWDERS

The Manufacture of Metal Powders

Different methods are used for producing metal powders. They include reduction of metal oxides, atomization of molten metals, electrolytic deposition, thermal decomposition of carbonyls, condensation of metal vapor, and mechanical processing of solid metals.

Reduction. In reduction, the raw material is usually an oxide that is subjected to a sequence of concentration and purification operations before it is reduced. Carbon, carbon monoxide, and hydrogen are used as reducing agents. Following is the chemical formula indicating the reaction between carbon and iron oxide:

$$2Fe_2O_3 + 3C \xrightarrow{\text{heat}} 4Fe + 3CO_2 \uparrow \tag{7.1}$$

Because the reaction takes place at a high temperature, the resulting metal particles sinter together and form sponges that are subsequently crushed and milled to a powder suitable for consolidation. Such powders have low apparent densities and often contain impurities and inclusions, but they are cheap. Metal powders produced by this method include iron, cobalt, nickel, tungsten, and molybdenum.

Atomization. Atomization is frequently used for producing powders from low-melting-point metals such as tin, lead, zinc, aluminum, and cadmium. Iron powder can also be produced by atomization. The process involves forcing a molten metal through a small orifice to yield a stream that is disintegrated by a jet of high-pressure fluid. When compressed gas is used as the atomizing medium, the resulting powder particles will be spherical. The reason is that complete solidification takes a relatively long period, during which surface tension forces have the chance to spheroidize the molten metal droplets. However, when water is used, the droplets solidify very quickly and have a ragged or irregular form. Figure 7.1 illustrates the atomization technique.

Electrolytic deposition. Electrolytic deposition involves obtaining metal powders from solutions by electrolysis. Process parameters such as current density and solution concentration are controlled to give a loose deposit instead of the coherent layer acquired in electroplating. The electrolytically deposited powders are then carefully

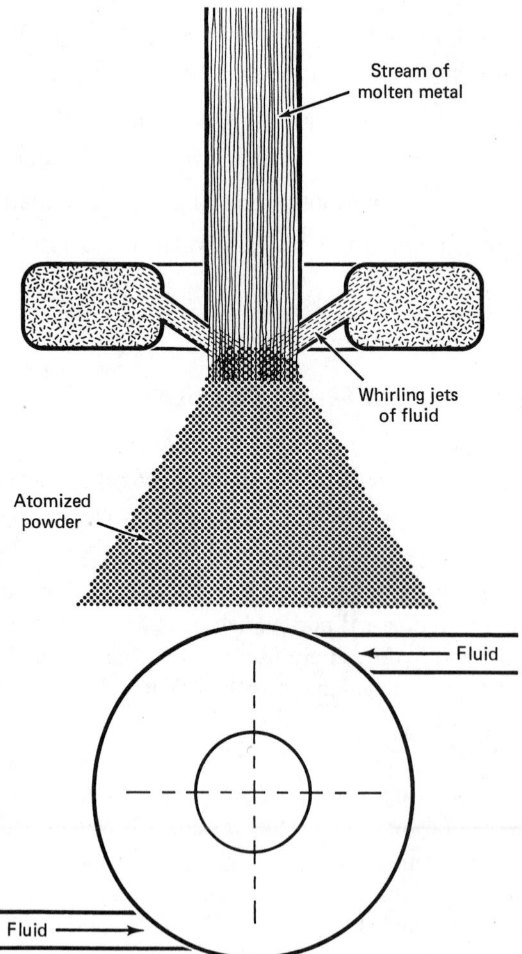

FIGURE 7.1 Production of metal powders by atomization

washed, dried, and annealed. Such powders are relatively expensive, but their important advantage is their high purity and freedom from nonmetallic inclusions.

Thermal decomposition of carbonyls. Nickel and iron carbonyls are volatile liquids having low boiling points of 110°F and 227°F (43°C and 107°C), respectively. They decompose at temperatures below 572°F (300°C), and the metal is precipitated in the form of a very fine powder.

Condensation of metal vapor. Condensation is employed only with some low-melting-point metals. For example, zinc powder can be obtained directly by condensation of the zinc vapor.

Mechanical processing of solid metals. Production of metal powders by comminuation of solid metals is accomplished by either machining, crushing, milling, or any combination of these. This method is limited to the production of beryllium and magnesium powders because of the expenses involved.

Properties of Metal Powders

The particular method used for producing a metal powder controls its particle and bulk properties, which, in turn, affect the processing characteristics of that powder. Therefore, comprehensive testing of all the physical and chemical properties of powders is essential prior to use in order to avoid variations in the final properties of the compacts. Following are the important characteristics of metal powders.

Chemical composition. In order to determine the chemical composition, conventional chemical analysis is used in addition to some special tests that are applicable only to metal powders, such as weight loss after reduction in a stream of hydrogen, which is an indirect indication of the amount of oxide present. For example, in the case of iron powder, the following equation is used:

$$\%\text{ iron oxide} = \%\text{ weight loss} \times \frac{159.7}{48} \tag{7.2}$$

$$= \%\text{ weight loss} \times 3.33 \tag{7.3}$$

In Equation 7.2, the fraction on the right-hand side is the ratio of the total weight of iron oxide to the weight of the combined oxygen in it, or $(Fe_2O_3)/(O_3)$, which can be calculated by summing up the atomic weights of each element in the numerator and denominator.

It is also important to mention that the percentages of nonmetallic inclusions will affect the maximum achievable density of the compacted powder (i.e., the full theoretical density). For example, if an iron powder (density of iron is 7.87 g/cm^3) consists of a percent Fe_2O_3, b percent carbon, and c percent sulfur, the following equation can be applied:

$$\text{max. achievable density} = \frac{100}{\dfrac{100-(a+b+c)}{7.87} + \dfrac{a}{\rho_{\text{oxide}}} + \dfrac{b}{\rho_{\text{carbon}}} + \dfrac{c}{\rho_{\text{sulfur}}}} \tag{7.4}$$

where ρ_{oxide}, ρ_{carbon}, and ρ_{sulfur} are the densities of oxide, carbon, and sulfur, respectively. Equation 7.4 can also be used in calculating the maximum achievable density for a mixture of powders.

Particle shape. The particle shape is influenced by the method of powder production and significantly affects the apparent density of the powder, its pressing properties, and its sintering ability.

Particle size. The flow properties and the apparent density of a metal powder are markedly influenced by the particle size, which can be directly determined by measurement on a microscope, by sieving, or by sedimentation tests.

Particle-size distribution. The particle-size distribution has a considerable effect on the physical properties of the powder. Sieve testing is the standard method used for the determination of the particle-size distribution in a quantitative manner. The apparatus used involves a shaking machine on which a series of standard sieves are stacked with the coarsest at the top and the finest at the bottom. The particle-size distribution is obtained from the percentage (by weight) of the sample that passes through one sieve but is retained on the next finer sieve. These sieves are defined by the mesh size, which indicates the number of apertures per linear inch. After the test is performed, the results are stated in a suitable form, such as a table of weight percentages, graphs of frequency distribution, or cumulative oversize and undersize curves where the cumulative size is the total weight percentage above or below a particular mesh size.

Specific surface. Specific surface is the total surface area of the particles per unit weight of powder, usually expressed in square centimeters per gram (cm^2/g). The specific surface has a considerable influence on the sintering process. The higher the specific surface, the higher the activity during sintering because the driving force for bonding during the sintering operation is the excess energy due to the large area (high specific surface).

Flowability. Flowability is the ease with which a powder will flow under gravity through an orifice. A quantitative expression of the flowability of a powder is its flow rate, which is determined using a Hall *flowmeter.* As illustrated in Figure 7.2, this apparatus involves a polished conical funnel made of brass having a half-cone angle of 30° and an orifice of 0.125 inch (3.175 mm). The funnel is filled with 50 grams of the powder, and the time taken for the powder to flow from the funnel is determined, the flow rate being expressed in seconds. The flow properties are dependent mainly upon the particle shape, particle size, and particle-size distribution. They are also affected by the presence of lubricants and moisture. Good flow properties are required if high production rates are to be achieved in pressing operations because the die is filled with powder flowing under gravity and because a shorter die-filling time necessitates a high powder-flow rate.

Bulk (or apparent) density. The bulk (or apparent) density is the density of the bulk of a powder mass. It can be easily determined by filling a container of known volume with the powder and then determining the weight of the powder. The bulk density is the quotient of the powder mass divided by its volume and is usually expressed in grams per cubic centimeter (g/cm^3). The apparent density is influenced by the same

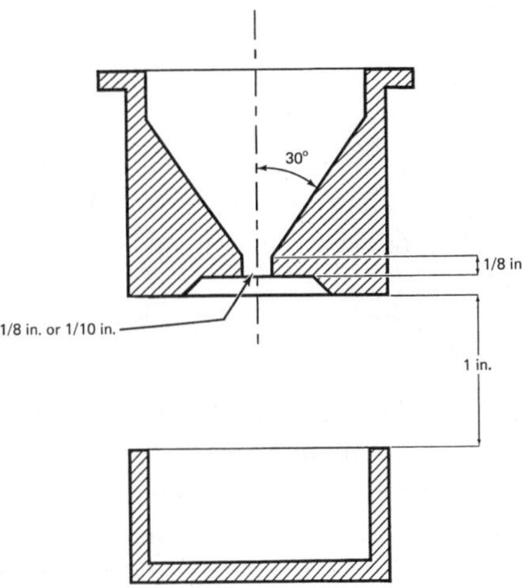

FIGURE 7.2
A sketch of the Hall flowmeter

factors as the flowability—namely, the particle configuration and the particle-size distribution.

Compressibility and compactibility. Compressibility and compactibility are very important terms that indicate and describe the behavior of a metal powder when compacted in a die. *Compressibility* indicates the densification ability of a powder, whereas *compactibility* is the structural stability of the produced as-pressed compact at a given pressure. A generalized interpretation of these terms involves graphs indicating the as-pressed density versus pressure (for compressibility) and the as-pressed strength versus pressure (for compactibility). It must be noted that these two terms are *not* interchangeable: A brittle powder may have good compressibility but usually has a weak as-pressed compactibility.

Sintering ability. Sintering ability is the ability of the adjacent surfaces of particles in an as-pressed compact to bond together when heated during the sintering operation. Sintering ability is influenced mainly by the specific surface of the powder used and is the factor responsible for imparting strength to the compact.

Factors Affecting the Selection of Metal Powders

Probably all metallic elements can be made in powderous form by the previously discussed manufacturing methods. However, the powder characteristics will differ in each case and will depend mainly upon the method of manufacture. The task of the manufacturing engineer is to select the type of powder appropriate for the required job. The decision generally depends upon the following factors:

1. Economic considerations
2. Purity demands
3. Desired physical, electrical, or magnetic characteristics of the compact

These considerations will be discussed in a later section.

7.2 POWDER METALLURGY: THE BASIC PROCESS

The conventional powder metallurgy process normally consists of three operations: powder blending and mixing, powder pressing, and compact sintering.

Blending and Mixing

Blending and mixing the powders properly is essential for uniformity of the finished product. Desired particle-size distribution is obtained by blending in advance the types of powders used. These can be either elemental powders, including alloying powders to produce a homogeneous mixture of ingredients, or prealloyed powders. In both cases, dry lubricants are added to the blending powders before mixing. The commonly used lubricants include zinc stearate, lithium stearate, calcium stearate, stearic acid, paraffin, acra wax, and molybdenum disulfide. The amount of lubricant added usually ranges between 0.5 and 1.0 percent of the metal powder by weight. The function of the lubricant is to minimize the die wear, to reduce the friction that is initiated between the die surface and powder particles during the compaction operation, and, hence, to obtain more even density distribution throughout the compact. Nevertheless, it is not recommended that the just-mentioned limits of the percentage of lubricant be exceeded, as this will result in extruding the lubricant from the surfaces of the particles during compaction to fill the voids, preventing proper densification of the powder particles and impeding the compaction operation.

The time for mixing may vary from a few minutes to days, depending upon operator experience and the results desired. However, it is usually recommended that the powders be mixed for 45 minutes to an hour. Overmixing should always be avoided because it may decrease particle size and work-harden the particles.

Pressing

Pressing consists of filling a die cavity with a controlled amount of blended powder, applying the required pressure, and then ejecting the as-pressed compact, usually called the *green compact,* by the lower punch. The pressing operation is usually performed at room temperature, with pressures ranging from 10 tons/in.^{2} (138 MPa) to 60 tons/in.^{2} (828 MPa), depending upon the material, the characteristics of the powder used, and the density of the compact to be achieved.

Tooling is usually made of hardened, ground, and lapped tool steels. The final hardness of the die walls that will come in contact with the powder particles during compaction should be around 60 R_c in order to keep the die wear minimal. The die cavity is designed to allow a powder fill about three times the volume (or height) of the green compact. The ratio between the initial height of the loose powder fill and the final height of the green compact is called the *compression ratio* and can be determined from the following equation:

$$\text{compression ratio} = \frac{\text{height of loose powder fill}}{\text{height of green compact}} = \frac{\text{density of green compact}}{\text{apparent density of loose powder}} \tag{7.5}$$

When pressure is first applied to metal powders, they will undergo repacking or restacking to reduce their bulk volume and to attain better packing density. The extent to which this occurs depends largely on the physical characteristics of the powder particles. The movement of the powder particles relative to one another will cause the oxide films covering their surfaces to be rubbed off. These oxide films will also collapse at the initial areas of contact between particles because these areas are small and the magnitude of the localized pressures are, therefore, extremely high. This leads to metal-to-metal contact and, consequently, to *cold-pressure welding* between the powder particles at the points of contact. When the pressure is further increased, interlocking and plastic deformation of the particles take place, extending the areas of contact between the individual particles and increasing the strength and density of the coherent compacted powder. Plasticity of the metal-powder particles plays a major role during the second stage of the pressing operation. As the compaction pressure increases, further densification is increasingly retarded by work-hardening of the particle material and by friction. Figure 7.3 shows a typical plot of the relationship between the achieved density and the compaction pressure. As can be seen, the density first goes up at a high rate, and then the rate of increase in density decreases with increasing pressure. Consequently, it is very difficult to achieve the full density because prohibitive pressure is required.

Frictional forces between the powder and the die wall always oppose the transmission of the applied pressure in its vicinity. Therefore, the applied pressure diminishes with depth in the case of single-ended pressing (i.e., when the compaction pressure is applied on only one side). This is accompanied by an uneven density distribution throughout the compact. The density always decreases with increasing distance from the pressing punch face. Figure 7.4 indicates the variation of pressure with depth along

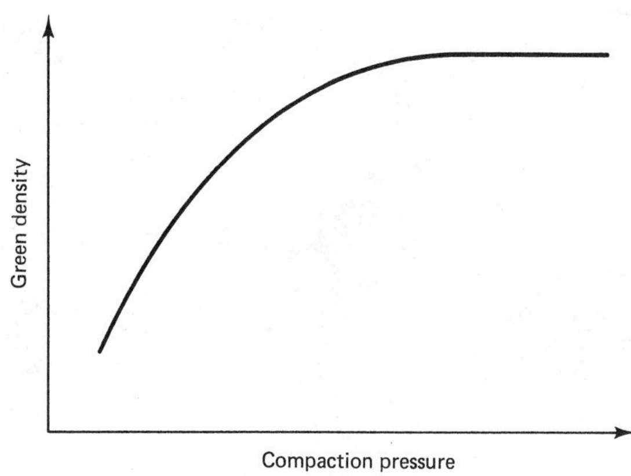

FIGURE 7.3 A typical plot of the relationship between achieved density and compaction pressure

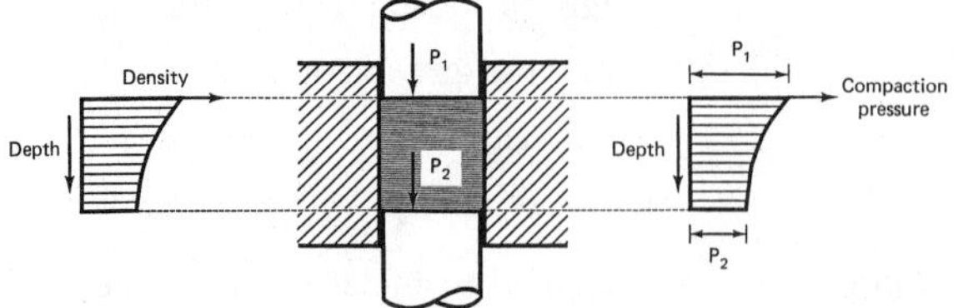

FIGURE 7.4
The variation of pressure with depth along the compact

the compact as well as the resulting variation in density. It is always recommended that the value of the length-to-diameter ratio of the compact be kept lower than 2.0 in order to avoid considerable density variations.

In order to improve pressure transmission and to obtain more even density distribution, lubricants are either admixed with the powder or applied to the die walls. Other techniques are also used to achieve uniform density distribution, such as compacting from both ends and suspending the die on springs or withdrawing it to reduce the effects of die-wall friction.

During the pressing of a metal powder in a die, elastic deformation of the die occurs in radial directions, leading to bulging of the die wall. Meanwhile, the compact deforms both elastically and plastically. When the compaction pressure is released, the elastic deformation tries to recover. But because some of the compact expansion is due to plastic deformation, the die tightly grips the compact, which hinders the die from returning to its original shape. Accordingly, a definite load, called the *ejection load,* has to be exerted on the compact in order to push it out of the die. Figure 7.5 illustrates the sequence of steps in a pressing operation.

Sintering

Sintering involves heating the green compact in a controlled-atmosphere furnace to a temperature that is slightly below the melting point of the powder metal. When the compact is composed of mixed elemental powders (e.g., iron and copper), the sinter-

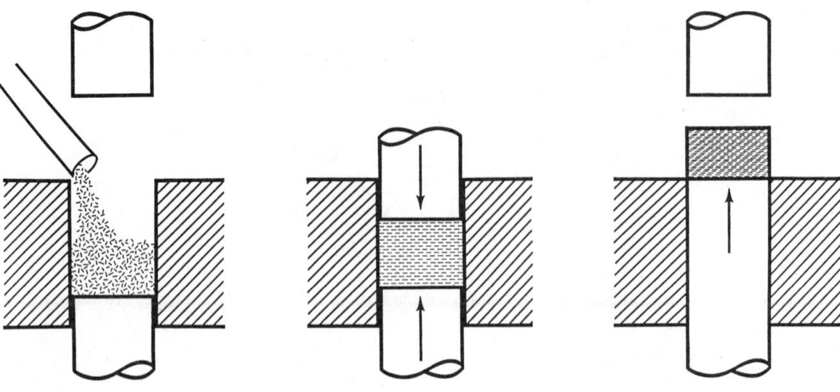

FIGURE 7.5
Sequence of steps in a pressing operation

ing temperature will then have to be below the melting point of at least one major constituent. The sintering operation will result in the following:

1. Strong bonding between powder particles
2. Chemical, dimensional, or phase changes
3. Alloying, in the case of mixed elemental powders

Such effects of the sintering operation are influenced by process variables such as sintering temperature, time, and atmosphere.

The amount, size, shape, and even nature of the pores are changed during sintering. There are two kinds of porosity: open, or interconnected, porosity (connected to the compact surface) and closed, or isolated, porosity. In a green compact, most of the porosity is interconnected and is characterized by extremely irregular pores. After sintering, interconnected porosity becomes isolated, and pore spheroidization takes place because of the surface tension forces. Also, the oxide films covering the particle surfaces of a green compact can be reduced by using the appropriate sintering atmosphere.

The most important atmospheres used in industrial sintering are carbon monoxide, hydrogen, and cracked ammonia. The latter is commonly used and is obtained by catalytic dissociation of ammonia, which gives a gas consisting of 25 percent nitrogen and 75 percent hydrogen by volume. Inert gases like argon and helium are occasionally used as sintering atmospheres, but cost is a decisive factor in limiting their use. Vacuum sintering is also finding some industrial application in recent years; nevertheless, the production rate is the main limitation of this method.

There are two main types of sintering furnaces: continuous and batch-operated. In *continuous* furnaces, the charge is usually conveyed through the furnace on mesh belts. These furnaces are made in the form of tunnels or long tubes having a diameter of not more than 18 inches (45 cm). Heating elements are arranged to provide two heating zones: a relatively low-temperature zone, called a *dewaxing* zone, in which lubricants are removed so that they will not cause harmful reactions in the next zone, and a uniform heating zone, which has the required high temperature where sintering actually takes place. A third zone of the furnace tube is surrounded by cooling coils in order to cool the compacts to ambient temperature in the controlled atmosphere of the furnace, thus avoiding oxidation of the compacts. Flame curtains (burning gases like hydrogen) are provided at both ends of the furnace tube to prevent air from entering into the furnace. Figure 7.6 is a sketch of a continuous sintering furnace. This type of furnace is suitable for mass production because of its low sintering cost per piece and its ability to give more consistent products. When small quantities of compacts must be sintered, however, *batch-operated* furnaces are used. These furnaces (e.g., vacuum furnaces) are also more suitable when high-purity products are required.

The sintering time varies with the metal powder and ranges between 30 minutes and several hours. However, 40 minutes to an hour is the most commonly used sintering time in industry.

FIGURE 7.6
A sketch of a continuous sintering furnace

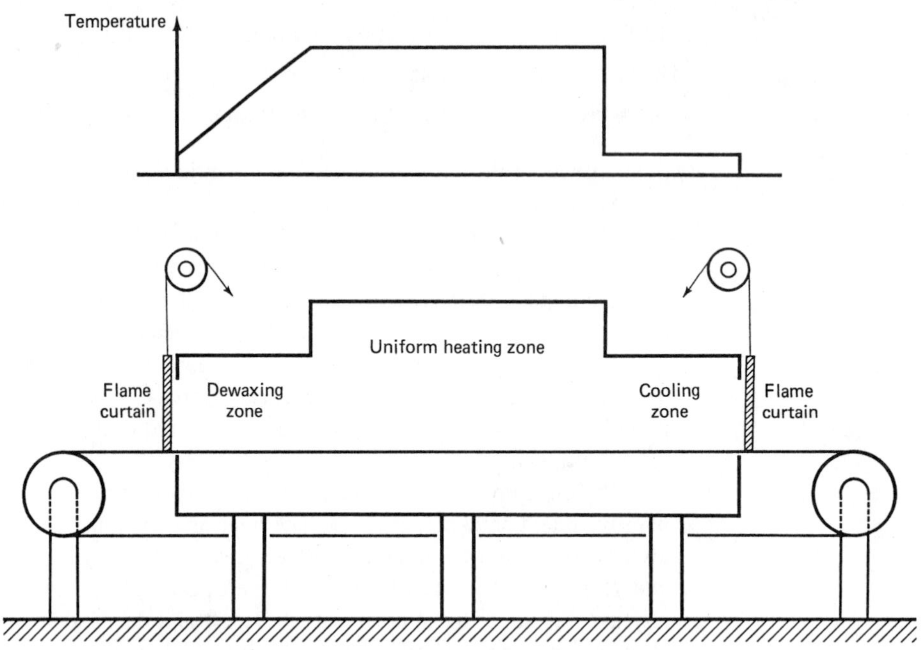

7.3 OPERATIONAL FLOWCHART

Because of the wide variety of powder metallurgy operations, it may be difficult for a person who is not familiar with this process to pursue the proper sequence of operations. The flowchart in Figure 7.7 is intended to clearly show the relationship between the various powder metallurgy operations (which will be discussed later) and to give a bird's-eye view of the flow of material to yield the final required product. Nevertheless, it must be remembered that there are exceptions and that some operations cannot be shown on the flowchart because they would make it overly detailed and complicated.

7.4 ALTERNATIVE CONSOLIDATION TECHNIQUES

There are many techniques of consolidating metal powders. They are classified, as shown in Figure 7.8, into two main groups: pressureless and pressure forming. The *pressureless* methods are those in which no external pressure is required. This group includes loose sintering, slip casting, and slurry casting. The *pressure forming* methods include conventional compaction, vibratory compaction, powder extrusion, powder

FIGURE 7.7
A flowchart showing the relationship between the various powder metallurgy operations

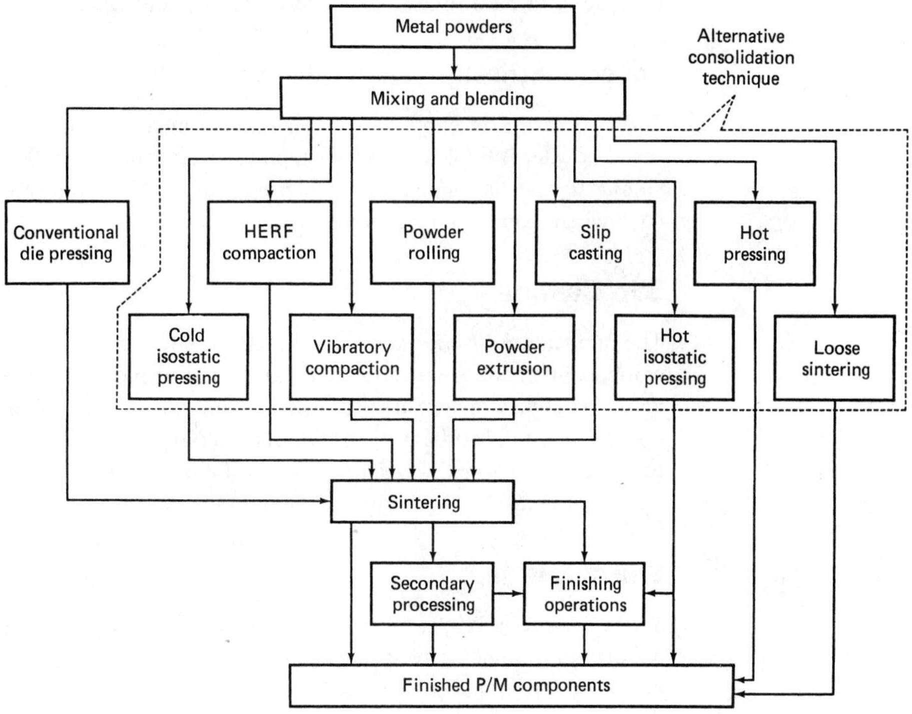

FIGURE 7.8
Classification of the techniques for consolidating metal powders

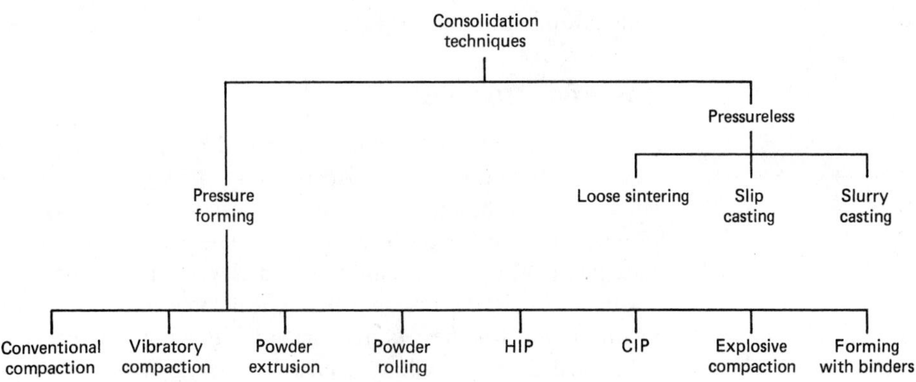

rolling, hot and cold isostatic pressing, explosive forming, and forming with binders. A detailed account of conventional powder metallurgy has been given; following is a brief description of these other consolidation techniques.

Loose Sintering

Loose sintering is employed in manufacturing filters. It involves sintering of loose metal powder in molds made of graphite or ceramic material. The temperature used is similar to that of conventional sintering, but the time involved is usually longer (two days when manufacturing stainless steel filters).

Slip Casting

The application of *slip casting* is usually limited to the production of large, intricate components made from refractory metals and cermets (mixtures of metals and ceramics). The slip, which is a suspension of fine powder particles in a viscous liquid, is poured into an absorbent plaster-of-paris mold. Both solid and hollow articles can be produced by this method. When making hollow objects, excess slip is poured out after a layer of metal has been formed on the mold surface.

Slurry Casting

Slurry casting is very similar to slip casting, except that the mixture takes the form of a slurry and binders are usually added. Also, because the slurry contains less water, nonabsorbent molds can be used.

Vibratory Compaction

Vibratory compaction involves superimposing mechanical vibration on the pressing load during the compaction operation. The advantages of this process include the considerable reduction in the pressure required and the ability to compact brittle particles that cannot be pressed by conventional techniques because the high compaction load required would result in fragmentation rather than consolidation of the powder particles. The main application involves the consolidation of stainless steel and uranium oxide powders for nuclear fuel elements.

Isostatic Pressing

In *isostatic pressing* (IP), equal all-around pressure is applied directly to the powder mass via a pressurized fluid. Accordingly, die-wall friction is completely eliminated, which explains the potential of the process to produce large, dense parts having uniform density distribution. The process can be performed at room temperature (cold isostatic pressing) or can be carried out at elevated temperatures (hot isostatic pressing).

In *cold isostatic pressing* (CIP), a flexible envelope (usually made of rubber or polymers) that has the required shape is filled with the packed powder. The envelope is then sealed and placed into a chamber that is, in turn, closed and pressurized to con-

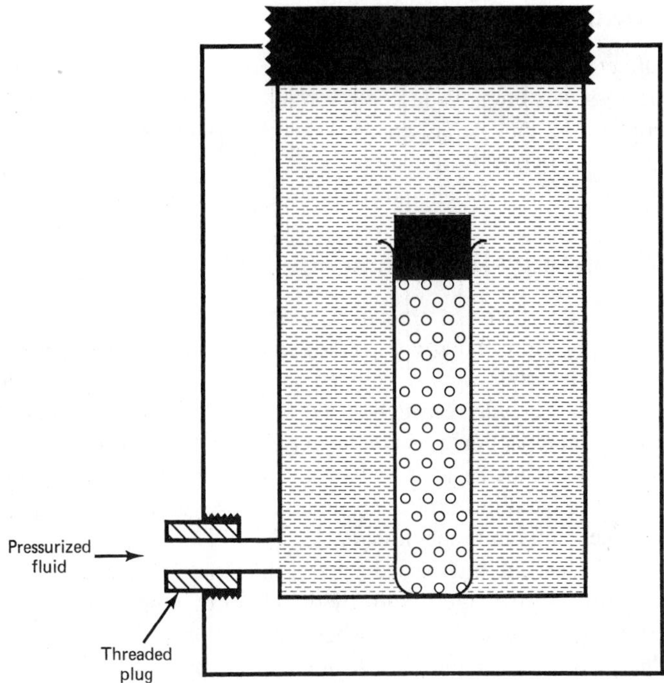

FIGURE 7.9
The isostatic pressing operation

solidate the powder. The lack of rigidity of the flexible envelope is countered by using a mesh or perforated container as a support (see Figure 7.9). The main disadvantage of this process is the low dimensional accuracy due to the flexibility of the mold.

In *hot isostatic pressing* (HIP), both isostatic pressing and sintering are combined. Powder is canned in order to separate it from the pressurized fluid, which is usually argon. The can is then heated in an autoclave, with pressure applied isostatically. Complete densification and particle bonding occur. The elevated temperature at which the powder is consolidated results in a softening of the particles. For this reason, the process is used to compact hard-to-work materials such as tool steels, beryllium, nickel-base superalloys, and refractory metals. A good example is the manufacture of jet-engine turbine blades, where a near-net shape is made from nickel-base superalloys. A main disadvantage of this method is the long processing time.

Powder Extrusion

Powder extrusion is a continuous compaction process and can be performed hot or cold. It is employed in producing semifinished products having a high length-to-diameter ratio, a geometry that makes producing them by conventional powder metallurgy impossible. The conventional technique involves packing metal powder into a thin container that is, in turn, evacuated, sealed, and then extruded. An emerging technique involves the extrusion of suitable mixtures of metal (or ceramic) powders and binders such as dextrin and sugars. It has been successfully employed in the production of highly porous materials used as filters or fuel cells in batteries.

Powder Rolling

Direct *powder rolling,* or roll compacting, is another type of continuous compaction process. It is employed mainly for producing porous sheets of nonferrous powders like copper and nickel. This process involves feeding the metal powder into the gap between the two rolls of a simple mill, where it is squeezed and pushed forward to form a sheet that is sintered and further rolled to control its density and thickness.

High-Energy-Rate Compaction

The various *HERF compaction* techniques are based on the same principle, which is the application of the compaction energy within an extremely short period of time. Several methods were developed for compacting metal powders at high speeds. Examples are explosives, high-speed presses, and spark sintering. It is believed that explosive compaction is suitable only when the size of the compact and the density required cannot be achieved by the isostatic compaction process. Nevertheless, the danger of handling explosives and the low cycling times impose serious limitations on this technique in production.

The use of high-speed presses like the Dynapak (built by General Dynamics) and the Petro-Forge (built by Mechanical Engineering Department, Birmingham University, England) for powder compaction is, in practicality, an extension of the die-pressing technique. These high-speed presses are particularly advantageous for pressing hard-to-compact powders and large components.

There are also some other powder-consolidation methods that can be classified as high-speed techniques. These include electrodynamic pressing, electromagnetic pressing, and spark sintering. Electrodynamic pressing involves utilizing the high pressure produced by the sudden discharge of electrical energy to compact powders at high speeds. Electromagnetic pressing is based upon the phenomenon that a strong magnetic field is generated when electric current is suddenly discharged through an inductance. This strong magnetic field is used for pressing a thin-walled metallic tube that contains the powder. Spark sintering involves the sudden discharge of electrical energy into the powder mass to puncture the oxide films that cover each individual powder particle and to build up pure metallic contacts between the particles. After about 10 seconds of impulsive discharging, the current is shut off, and a pressure of about 14,500 lb/in^2 (100 MPa) is applied to compact the powder to the final required form.

Injection Molding

Although *injection molding* is an emerging process, it can be considered as a version of forming with binders, which is a rather old method. The process involves injection molding metal powders that are precoated with a thermoplastic polymer into a part similar in shape to the final required component but having larger dimensions. After removing the polymer by an organic solvent, the porous compact is then sintered for a long time in order to allow for volume shrinkage and, consequently, an increase in density. The main advantage of this process is that it offers promise in the forming of intricate shapes.

Hot Pressing

Hot pressing is a combination of both the compaction and the sintering operations. It is basically similar to the conventional powder metallurgy process, except that powders are induction heated during pressing, and, consequently, a protective atmosphere is necessary. For most metal powders, the temperatures used are moderate (above recrystallization temperature), and dies made of superalloys are used. The hot pressing of refractory metals (e.g., tungsten and beryllium), however, necessitates the use of graphite dies. The difficulties encountered in this technique limit its application to laboratory research.

7.5 SECONDARY CONSOLIDATION OPERATIONS

In most engineering applications, the physical and mechanical properties of the as-sintered compact are adequate enough to make it ready for use. However, secondary processing is sometimes required to increase the density and enhance the mechanical properties of the sintered component, thus making it suitable for heavy-duty engineering applications. The operations involved are similar to those used in forming fully dense metals, though certain precautions are required to account for the porous nature of the sintered compacts. Following is a survey of the common secondary operations.

Coining (Repressing)

Coining involves the pressing of a previously consolidated and sintered compact in order to increase its density. This operation is performed at room temperature, and considerable pressures are thus required. It is often possible to obtain significant improvement in strength not only because of the increased densification but also because of the work-hardening that occurs during the operation. A further advantage of this process is that it can be employed to alter shape and dimensions slightly. *Repressing* is a special case of coining where no shape alteration is required.

Extrusion, Swaging, or Rolling

Sintered powder compacts, whether in their cold or hot state, can be subjected to any forming operation (extrusion, swaging, or rolling). When processing at elevated temperatures, either a protective atmosphere or canning of the compacts has to be employed. Such techniques are applied to canned sintered compacts of refractory metals, beryllium, and composite materials.

Forging of Powder Preforms

Repressing and coining of sintered compacts cannot reduce porosity below 5 percent of the volume of the compact. Therefore, if porosity is to be completely eliminated, *hot forging* of powder preforms must be employed. Sintered powder compacts having medium densities (80 to 85 percent of the full theoretical density) are heated, lubricated, and fed into a die cavity. The preform is then forged with a single stroke, as opposed to conventional forging of fully dense materials, where several blows and manual transfer of a billet through a series of dies are required. This advantage is a consequence of using a preform that has a shape quite close to that of the final forged product. The tooling used involves a precision flashless closed die; therefore, the trimming operation performed after conventional forging is eliminated.

The forging of powder preforms combines the advantages of both the basic powder metallurgy and the conventional hot forging processes while eliminating their shortcomings. For this reason, the process is extensively used in the automotive industry in producing transmission and differential-gear components. Examples of some forged powder metallurgy parts are shown in Figure 7.10.

7.6 FINISHING OPERATIONS

Many powder metallurgy products are ready for use in their as-sintered state; however, finishing processes are frequently used to impart some physical properties or geometrical characteristics to them. Following are some examples of the finishing operations employed in the powder metallurgy industry.

Sizing

Sizing is the pressing of a sintered compact at room temperature to secure the desired shape and dimensions by correcting distortion and change in dimensions that may have occurred during the sintering operation. Consequently, this operation involves only

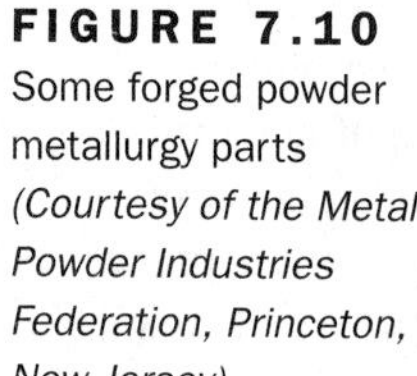

FIGURE 7.10
Some forged powder metallurgy parts *(Courtesy of the Metal Powder Industries Federation, Princeton, New Jersey)*

(a)

FIGURE 7.10 (Cont.) Some forged powder metallurgy parts *(Courtesy of the Metal Powder Industries Federation, Princeton, New Jersey)*

(b)

limited deformation and slight density changes and has almost no effect on the mechanical properties of the sintered compact.

Machining

Features like side holes, slots, or grooves cannot be formed during pressing, and, therefore, either one or two machining operations are required. Because cooling liquids can be retained in the pores, sintered components should be machined dry whenever possible. An air blast is usually used instead of coolants to remove chips and cool the tool.

Oil Impregnation

Oil impregnation serves to provide either protection against corrosion or a degree of self-lubrication or both. It is usually carried out by immersing the sintered porous compact in hot oil and then allowing the oil to cool. Oil impregnation is mainly used in the manufacturing of self-lubricating bearings made of bronze or iron.

Infiltration

Infiltration is permeation of a porous metal skeleton with a molten metal of a lower melting point by capillary action. Infiltration is performed in order to fill the pores and give two-phase structures with better mechanical properties. The widely used application of this process is the infiltration of porous iron compacts with copper. The process is then referred to as *copper infiltration* and involves placing a green compact of copper under (or above) the sintered iron compact and heating them to a temperature above the melting point of copper.

Heat Treatment

Conventional heat treatment operations can be applied to sintered porous materials, provided that the inherent porosity is taken into consideration. Pores reduce the thermal conductivity of the porous parts and thus reduce their rate of cooling. For sintered porous steels, this means poorer hardenability. Also, cyanide salts, which are very poisonous and are used in heat treatment salt baths, are retained in the pores, resulting in extreme hazards when using such heat-treated compacts. Therefore, it is not advisable to use salt baths for surface treatment of porous materials.

Steam Oxidizing

A protective layer of magnetite (Fe_3O_4) can be achieved by heating the sintered ferrous parts and exposing them to superheated steam. This will increase the corrosion resistance of the powder metallurgy parts, especially if it is followed by oil impregnation.

Plating

Metallic coatings can be satisfactorily electroplated directly onto high-density and copper-infiltrated sintered compacts. For relatively low-density compacts, electroplating must be preceded by an operation to seal the pores and render the compacts suitable for electroplating.

7.7 POROSITY IN POWDER METALLURGY PARTS

The structure of a powder metallurgy part consists of a matrix material with a microstructure identical to that of a conventional fully dense metal and pores that are a unique and controllable feature of sintered porous materials. For this reason, powder metallurgy materials are grouped according to their porosity, which is quantitatively expressed as the percentage of voids in a part. Those materials having less than 10 percent porosity are considered to be high density; those with porosity more than 25 percent, low density. There is a relationship between porosity and density (both being expressed as fractions of the full theoretical density), and it can be expressed by the following equation:

$$\text{porosity} = 1 - \text{density} \tag{7.6}$$

As previously explained, the theoretical density is not that of the fully dense pure metal but is the mean value of the densities of all constituents. These include not only alloying additives but also impurities. When considering green densities, the effect of lubricants must be taken into consideration.

Pores are classified with respect to their percentage, type, size, shape, and distribution. The type can be either interconnected or isolated. The volume of interconnected porosity can be determined by measuring the amount of a known liquid needed to saturate the porous powder metallurgy sample. The interconnected porosity is essential for successful oil impregnation and thus is very important for the proper functioning of self-lubricating bearings.

At this stage, it is appropriate to differentiate between the following three technical terms used to describe density:

$$\text{bulk density} = \frac{\text{mass of compact}}{\text{bulk volume of compact}} \tag{7.7}$$

$$\begin{aligned} \text{apparent density} &= \frac{\text{mass of compact}}{\text{apparent volume}} \\ &= \frac{\text{mass of compact}}{\text{bulk volume of compact} - \text{volume of open pores}} \end{aligned} \tag{7.8}$$

$$\begin{aligned} \text{true density} &= \frac{\text{mass of compact}}{\text{true volume}} \\ &= \frac{\text{mass of compact}}{\text{bulk volume of compact} - (\text{volume of open pores} + \text{volume of closed pores})} \end{aligned} \tag{7.9}$$

For a green compact produced by admixed lubrication, these densities are misleading and do not indicate the true state of densification due to the presence of lubricant within the space between metal particles. Therefore, the bulk density must be readjusted to give the true metal density (TMD) as follows:

$$\text{TMD} = \text{actual bulk density} \times \frac{\%\text{ of metal}}{100} \tag{7.10}$$

7.8 DESIGN CONSIDERATIONS FOR POWDER METALLURGY PARTS

The design of a powder metallurgy part and the design of the tooling required to produce it cannot be separated. A part design that needs either long, thin tubular punches, tooling with sharp corners, or lateral movement of punches cannot be executed. For this reason, the design of powder metallurgy parts is often different from that of parts produced by machining, casting, or forging, and a component that is being produced by these methods has to be redesigned before being considered for manufacture by powder metallurgy. Following are various tooling and pressing considerations, some of which are illustrated in Figure 7.11.

Holes

Holes in the pressing direction can be produced by using core rods. In this case, there is almost no limitation on the general shape of the hole. But side holes and side slots are very difficult to achieve during pressing and must be made by secondary machining operations (see Figure 7.11a).

Wall Thickness

It is not desirable to have a wall thickness less than 1/16 inch (1.6 mm) because the punch required to produce the thickness will not be rigid enough to withstand the high stresses encountered during the pressing operation.

Fillets

It is recommended that sharp corners be avoided whenever possible. Fillets with generous radii are desirable, provided that they do not necessitate the use of punches with featherlike edges (see Figure 7.11b).

Tapers

Tapers are not always required. However, it is desirable to have them on flange-type sections and bosses to facilitate the ejection of the green compact.

Chamfers

As mentioned earlier, it is sometimes not desirable to use radii on part edges. Chamfers are the proper alternative in preventing burrs.

Flanges

A small flange, or overhang, can be easily produced. However, for a large overhang, ejection without breaking the flange is very difficult (see Figure 7.11c).

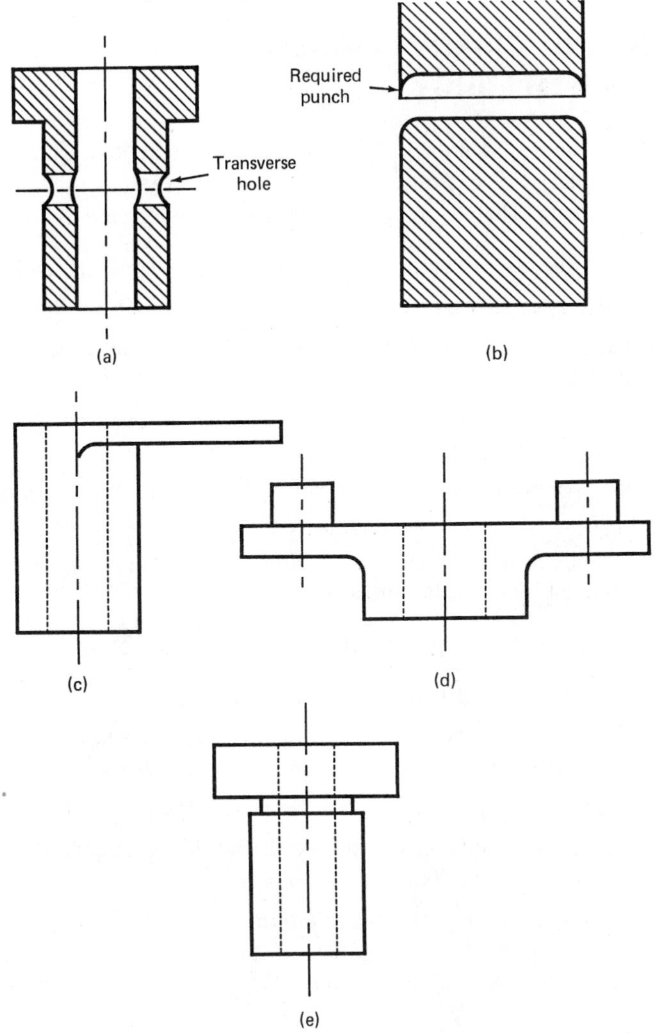

FIGURE 7.11 Design considerations for powder metallurgy parts: (a) holes; (b) fillets; (c) flanges; (d) bosses; (e) undercuts

Bosses

Bosses can be made, provided that they are round in shape (or almost round) and that the height does not exceed 15 percent of the overall height of the component (see Figure 7.11d).

Undercuts

Undercuts that are perpendicular to the pressing direction cannot be made because they prevent ejection of the part. If required, they can be produced by a secondary machining operation (see Figure 7.11e).

7.9 ADVANTAGES AND DISADVANTAGES OF POWDER METALLURGY

Like any other manufacturing process, powder metallurgy has advantages as well as disadvantages. The decision about whether to use this process or not must be based on these factors. The advantages of powder metallurgy are as follows:

1. Components can be produced with good surface finish and close tolerances.
2. There is usually no need for subsequent machining or finishing operations.
3. The process offers a high efficiency of material utilization because it virtually eliminates scrap loss.
4. Because all steps of the process are simple and can be automated, only a minimum of skilled labor is required.
5. Massive numbers of components with intricate shapes can be produced at high rates.
6. The possibility exists for producing controlled, unique structures that cannot be obtained by any other process.

The main disadvantages of the process are as follows:

1. Powders are relatively high in cost compared with solid metals.
2. Sintering furnaces and special presses, which are more complicated in principle and construction than conventional presses, are necessary.
3. Tooling is very expensive as several punches or die movements are often used.
4. High initial capital cost is involved, and the process is generally uneconomical unless very large numbers of components are to be manufactured.
5. Powder metallurgy parts have inferior mechanical properties due to porosity (this does not apply to forged powder metallurgy parts), and the process is thus primarily suitable for the production of a large number of small, lightly stressed parts.

7.10 APPLICATIONS OF POWDER METALLURGY PARTS

The applications of powder metallurgy parts fall into two main groups. The first group consists of those applications in which the part is used as a structural component that can also be produced by alternative competing manufacturing methods, powder metallurgy being used because of the low manufacture cost and high production rate. The second group includes those applications in which the part usually has a controlled, unique structure and cannot be made by any other manufacturing method. Examples are porous bearings, filters, and composite materials. Following is a quick review of the various applications.

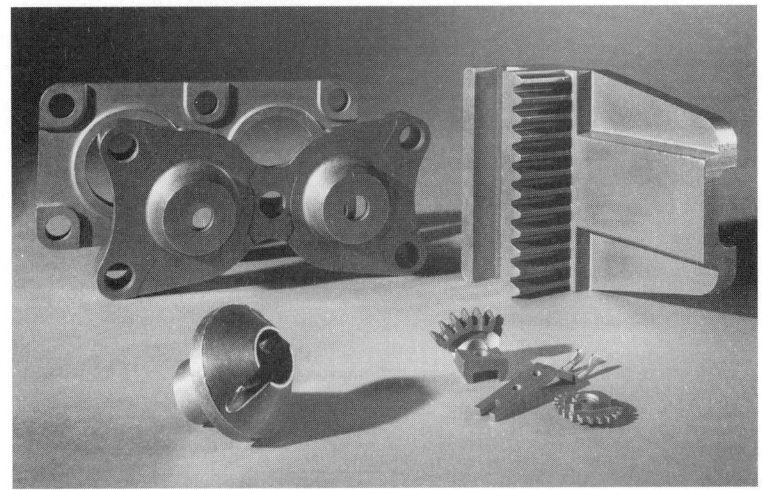

FIGURE 7.12 Some powder metallurgy products *(Courtesy of the Metal Powder Industries Federation, Princeton, New Jersey)*

Structural Components

Powder metallurgy used to be limited to the production of small, lightly stressed parts. However, with the recent development in forging powder preforms, the process is commonly used in producing high-density components with superior mechanical properties. Cams, gears, and structural parts of the transmission system are some applications of the powder metallurgy process in the automotive, agricultural machinery, and domestic appliance industries. Figures 7.12 and 7.13 show some examples of powder metallurgy products.

The structural powder metallurgy components are usually made of iron-base powders, with or without additions of carbon, copper, and other alloying elements like

FIGURE 7.13 More powder metallurgy products *(Courtesy of the Metal Powder Industries Federation, Princeton, New Jersey)*

nickel. Prealloyed powders are also employed, although they are less common than the mixed elemental powders.

Self-Lubricating Bearings

Self-lubricating bearings are usually made by the conventional die-pressing technique, in which a porosity level between 20 and 40 percent is achieved. A sizing operation is performed for dimensional accuracy and in order to obtain smooth surfaces. The bearings are oil impregnated either before or after sizing. Bronze powders are used in the manufacturing of porous bearings, but iron-base powders are also employed to give higher strength and hardness.

Filters

In manufacturing filters, the appropriate metal powder (e.g., bronze) is screened in order to obtain uniform particle size. The powder is then poured into a ceramic or graphite mold. The mold is put into a sintering furnace at the appropriate sintering temperature so that loose sintering can take place. The products must have generous tolerances, especially on their outer diameters, where 3 percent is typical.

Friction Materials

Clutch liners and brake bands are examples of friction materials. They are best manufactured by powder metallurgy. The composition includes copper as a matrix, with additions of tin, zinc, lead, and iron. Nonmetallic constituents like graphite, silica, emery, or asbestos are also added. The mixture is then formed to shape by cold pressing. After sintering, some finishing operations like bending, drilling, and cutting are usually required. It must be noted that friction materials are always joined to a solid plate, which gives adequate support to these weak parts.

Electrical Contact Materials

Electrical contact materials include two main kinds: sliding contacts and switching contacts. It is not possible to produce any of these contact materials except by powder metallurgy as both involve duplex structures.

Sliding contacts are components of electrical machinery employed when current is transferred between sliding parts (e.g., brushes in electric motors). The two main characteristics needed are a low coefficient of friction and good electrical conductivity. Compacts of mixtures of graphite and metal powder can fulfill such conditions. Powders of metals having high electrical conductivity, such as brass, copper, or silver, are used. These graphite-metal contacts are produced by conventional pressing and sintering processes.

Switching contacts are used in high-power circuit breakers. The three characteristics needed are good electrical conductivity, resistance to mechanical wear, and less tendency of the contact surfaces to weld together. A combination of copper, silver, and a refractory metal like tungsten provides the required characteristics. These contacts are produced either by conventional pressing and sintering or by infiltrating a porous refractory material with molten copper or silver.

Magnets

Magnets include soft magnets and permanent magnets. Soft magnets are used in dc motors or generators as armatures, as well as in measuring instruments. They are made of iron, iron-silicon, and iron-nickel alloys. Electrolytic iron powder is usually used because of its high purity and its good compressibility, which allows the high compact densities required for maximum permeability to be attained.

Permanent magnets produced by powder metallurgy have the commonly known name *Alnico.* This alloy consists mainly of nickel (30 percent), aluminum (12 percent), and iron (58 percent) and possesses outstanding permanent magnetic properties. Some other additives are often used, including cobalt, copper, titanium, and niobium.

Cores

The cores produced by powder metallurgy are used with ac high-frequency inductors in wireless communication systems. Such cores must possess high constant permeability for various frequencies as well as high electrical resistivity. Carbonyl iron powder is mixed with a binder containing insulators (to insulate the powder particles from one another and thus increase electrical resistivity) and then compacted using extremely high pressures, followed by sintering.

Powder Metallurgy Tool Steels

The production of tool steels by powder metallurgy eliminates the defects encountered in conventionally produced tool steels—namely, segregation and uneven distribution of carbides. Such defects create problems during tool fabrication and result in shorter tool life. The technique used involves compacting prealloyed tool-steel powders by hot isostatic pressing to obtain preforms that are further processed by hot working.

Superalloys

Superalloys are nickel- and cobalt-base alloys, which exhibit high strength at elevated temperatures. They are advantageous in manufacturing jet-engine parts like turbine blades. The techniques used in consolidating these powders include HIP, hot extrusion, and powder metallurgical forging.

Refractory Metals

The word *refractory* means "difficult to fuse." Therefore, metals with high melting points are considered to be refractory metals. These basically include four metals: tungsten, molybdenum, tantalum, and niobium. Some other metals can also be considered to belong to this group. Examples are platinum, zirconium, thorium, and titanium. Refractory metals, as well as their alloys, are best fabricated by powder metallurgy. The technique used usually involves pressing and sintering, followed by working at high temperatures. The applications are not limited to incandescent lamp filaments and heating elements; they also include space technology materials, the heavy metal used in radioactive shielding, and cores for armor-piercing projectiles. Titanium is gaining an expanding role in the aerospace industry because of its excellent strength-to-specific-weight ratio and its good fatigue and corrosion resistance.

Cemented Carbides

Cemented carbides are typical composite materials that possess the superior properties of both constituents. Cemented carbides consist of hard wear-resistant particles of tungsten or titanium carbides embedded in a tough strong matrix of cobalt or steel. They are mainly used as cutting and forming tools; however, there are other applications, including gages, guides, rock drills, and armor-piercing projectiles. They possess excellent red hardness and have an extremely long service life as tools. Cemented carbides are manufactured by ball-milling carbides with fine cobalt (or iron) powder, followed by mixing with a lubricant and pressing. The green compact is then presintered at a low temperature, machined to the required shape, and sintered at an elevated temperature. A new dimension in cemented carbides is Ferro-Tic, involving titanium carbide particles embedded in a steel matrix. This material can be heat treated and thus can be easily machined or shaped.

Review Questions

1. Define each of the following technical terms:
   a. compressibility
   b. compactibility
   c. green density
   d. impregnation
   e. infiltration
   f. flowability
   g. particle-size distribution
2. List five advantages of the powder metallurgy process.
3. List four disadvantages of the powder metallurgy process.
4. What are the important characteristics of a metal powder?
5. Describe three methods for producing metal powders.
6. Explain briefly the mechanics of pressing.
7. Why are lubricants added to metal powders before pressing?
8. Is it possible to eliminate all voids by conventional die pressing? Why?
9. Explain briefly the mechanics of sintering.
10. Why is it necessary to have controlled atmospheres for sintering furnaces?
11. Explain why it is not possible to use the conventional pressing techniques as a substitute for each of the following operations: isostatic pressing, slip casting, HERF compaction.
12. Differentiate between the following: coining, repressing, sizing.
13. How is copper infiltration accomplished and what are its advantages?

14. Can powder metallurgical forging be replaced by conventional forging? Why?

15. How can machining of some powder metallurgy components be inevitable?

16. How is plating of powder metallurgy components carried out?

17. Name five products that can only be produced by powder metallurgy.

18. Why are cemented carbides presintered?

19. Why is electrolytic iron powder used in manufacturing soft magnets?

20. Discuss four design limitations in connection with powder metallurgy components.

Problems

1. Following are the experimentally determined characteristics of three kinds of iron powder:

Sponge Iron Powder (I)					
Screen Analysis		**Chemical**	**Composition**	**Apparent Density**	**Flow Rate**
+70	0%	H_2 loss	0.2%		
−70 to +100	1%	C	0.02%	2.4 g/cm^3	35 s/50 g
−100 to +325	74%	SiO_2	0.2%		
−325	25%	P	0.015%		
		S	0.015%		

Sponge Iron Powder (II)					
Screen Analysis		**Chemical**	**Composition**	**Apparent Density**	**Flow Rate**
+40	2%	H_2 loss	0.2%		
−40 to +60	40%	C	0.1%	2.4 g/cm^3	35 s/50 g
−70 to +100	30%	SiO_2	0.3%		
−100 to +200	20%	S	0.015%		
−200	8%	P	0.015%		

Atomized Iron Powder						
Screen Analysis		Chemical	Composition	Apparent Density	Flow Rate
+100	1%	H_2 loss	0.2%		
+140	45%	C	0.01%	2.9 g/cm^3	24 s/50 g
+200	25%				
+230	15%				
+325	10%				
−325	4%				

Plot the following for each powder:

a. Cumulative oversize graph
b. Cumulative undersize graph
c. Frequency distribution curve (obtain median particle size for the powder)
d. Histogram of particle-size distribution

The axis usually indicates the particle size in microns and not mesh size. Use the following table:

Sieve	40	60	70	100	240	200	230	325
Microns	420	250	210	149	105	74	63	44

2. Calculate the full theoretical density for a compact made of atomized iron powder, knowing that the density of carbon equals 2.2 g/cm^3 and the density of iron oxide equals 2.9 g/cm^3.

3. Determine the approximate height of the powder fill for each kind of iron powder given in Problem 1 if the green density of the compact is 6.8 g/cm^3 and its height is 2.1 cm.

4. Plot a graph indicating the maximum achievable green density versus the percentage of admixed zinc stearate for atomized iron powder (density of zinc stearate is 1.1 g/cm^3). What can you deduce from the curve?

5. Which powder in Problem 1 would fill the die cavity faster?

6. Calculate the maximum achievable green density of a mixture of atomized iron powder plus 1 percent zinc stearate and 10 percent pure copper (density of copper is 8.9 g/cm^3).

7. Following is the relationship between density and pressure for atomized iron powder containing 1 percent zinc stearate:

Green density, g/cm^3	5.35	6.15	6.58	6.75	6.9
Pressure, MN/m^2	157.5	315	472.5	629.9	787.4

If it is required to manufacture a gear wheel having a green density of 6.8 g/cm^3 using a press with a capacity of 1 MN, calculate the diameter of the largest gear wheel that can be manufactured. How can you produce a larger gear by modifying the design?

8. A cylindrical compact of atomized iron powder plus 1 percent zinc stearate had a green bulk density of 7.0 g/cm^3, a diameter equal to 2 cm, and a height equal to 3 cm. After sintering, its bulk density increased to 7.05 g/cm^3. Calculate its new dimensions.

9. The sintered density of atomized iron compact containing 10 percent copper was 7.2 g/cm^3. What is the porosity?

Design Project

Figure 7.14 shows a part that is currently produced by forging and subsequent machining. Because the part is not subjected to high stresses during its actual service conditions, the producing company is considering the idea of manufacturing it by powder metallurgy in order to increase the production rate. Redesign this component so that it can be manufactured by the conventional die-pressing technique.

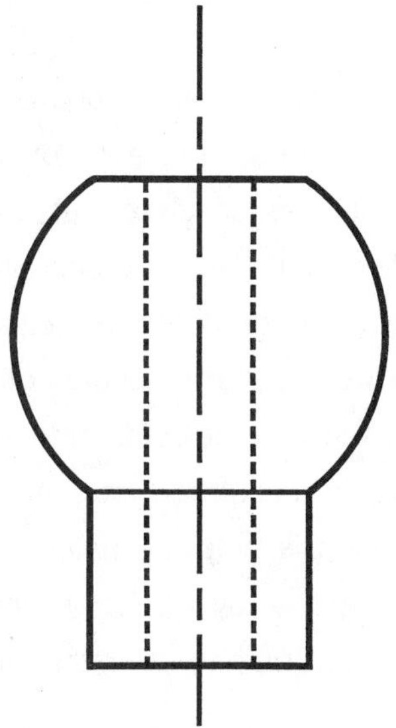

FIGURE 7.14
A part to be redesigned for production by powder metallurgy

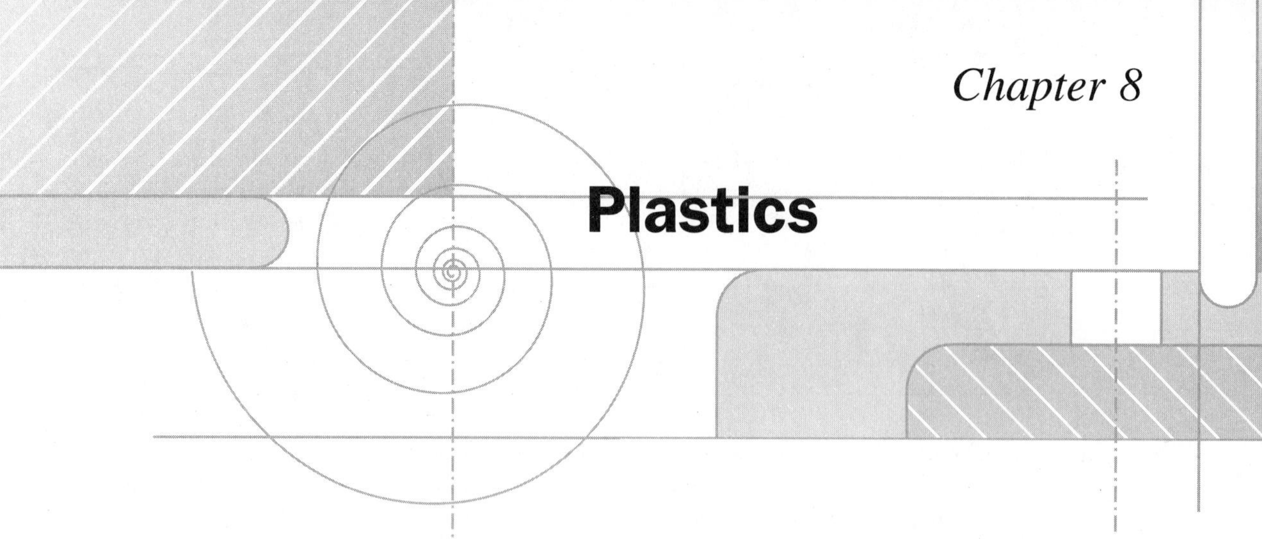

Chapter 8

Plastics

INTRODUCTION

Plastics, which are more correctly called *polymers,* are products of macromolecular chemistry. In fact, the term *polymer* is composed of the two Greek words *poly* and *meres,* which mean "many parts." This is, indeed, an accurate description of the molecule of a polymer, which is made up of a number of identical smaller molecules that are repeatedly linked together to form a long chain. As an example, consider the commonly used polymer polyethylene, which is composed of many ethylene molecules (C_2H_4) that are joined together, as shown in Figure 8.1. These repeated molecules are always organic compounds, and, therefore, carbon usually forms the backbone of the chain. The organic compound whose molecules are linked together (like ethylene) is referred to as the *monomer.*

Now, let us examine why the molecules of a monomer tend to link together. We know from chemistry that carbon has a valence of 4. Therefore, each carbon atom in an ethylene molecule has an unsaturated valence bond. Consequently, if two ethylene molecules attach, each to one side of a third molecule, the valence bonds on the two carbon atoms of the center molecule will be satisfied (see Figure 8.1). In other words, the molecules of the monomer tend to attach to one another to satisfy the valence requirement of the carbon atoms.

The molecules of a monomer in a chain are strongly bonded together. Nevertheless, the long chains forming the polymer molecules tend to be more or less amorphous and are held together by weaker secondary forces that are known as the *van der Waals forces* (named after the Dutch physicist). There-

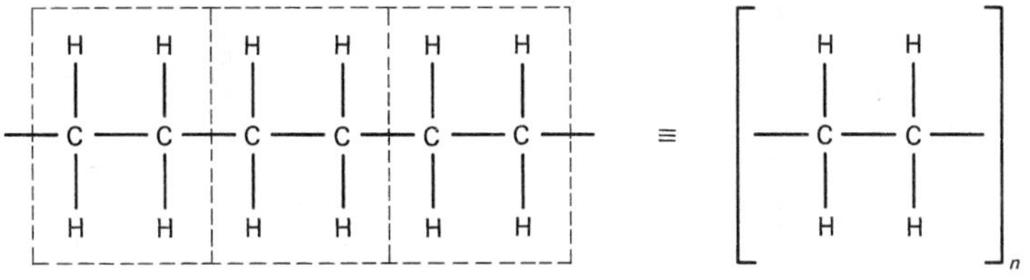

FIGURE 8.1
The molecular chain of polyethylene

fore, polymers are generally not as strong as metals or ceramics. It is also obvious that properties of a polymer such as the strength, elasticity, and relaxation are dependent mainly upon the shape and size of the long chainlike molecules, as well as upon the mutual interaction between them.

A common, but not accurate, meaning of the term *polymer* involves synthetic organic materials that are capable of being molded. Actually, polymers form the building blocks of animal life; proteins, resins, shellac, and natural rubber are some examples of natural polymers that have been in use for a long time. On the other hand, the synthetic, or manufactured, polymers have come into existence fairly recently. The first synthetic polymer, cellulose nitrate (celluloid), was prepared in 1869. It was followed in 1909 by the phenolics, which were used as insulating materials in light switches. The evolution of new polymers was accelerated during World War II due to the scarcity of natural materials. Today, there are thousands of polymers that find application in all aspects of our lives.

8.1 CLASSIFICATION OF POLYMERS

There are, generally, two methods for classifying polymers. The first method involves grouping all polymers based on their elevated-temperature characteristics, which actually dictate the manufacturing method to be used. The second method of classification groups polymers into *chemical families,* each of which has the same monomer. As an example, the ethenic family is based on ethylene as the monomer, and different polymers (members of this family such as polyvinyl alcohol or polystyrene) can be made by changing substituent groups on the basic monomer, as shown in Figure 8.2. As we will see later, this enables us to study most polymeric materials by covering just a limited number of families instead of considering thousands of polymers individually. But before reviewing the commonly used chemical families of polymers, let us discuss in depth their elevated-temperature behavior. Based on this behavior, polymers can be split into two groups: *thermoplastics* and *thermosets.*

FIGURE 8.2
Structural formula of some polymers of the ethenic group

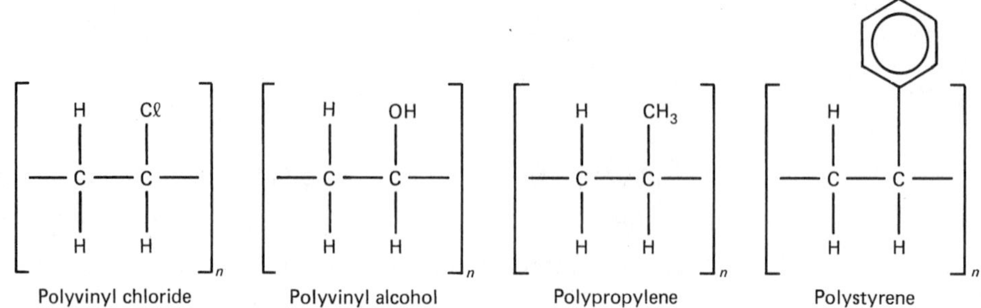

Thermoplastics

Thermoplastics generally have *linear* structures, meaning that their molecules look like linear chains having little breadth but significant length. This structure, as shown in Figure 8.3, is analogous to a bowl of spaghetti. Bonds between the various molecular chains are mainly of the van der Waals type (i.e., secondary forces). Therefore, this type of polymer softens by heating and can then flow viscously to take a desired shape because elevated temperatures tend to decrease the intermolecular coherence of the linear chains. When the solidified polymers are reheated and melted again, they can be given a different shape. This characteristic enables plastics fabricators to recycle thermoplastic scrap, thus increasing the efficiency of raw-material utilization.

Usually, a thermoplastic polymer consists of a mixture of molecular chains having different lengths. Therefore, each structure has a different melting point, and, consequently, the whole polymer melts, not at a definite temperature, but within a range whose limits are referred to as the *softening point* and the *flow point.* It has been observed that when a thermoplastic is given a shape at a temperature between the soft-

FIGURE 8.3
The molecular chains of a thermoplastic polymer

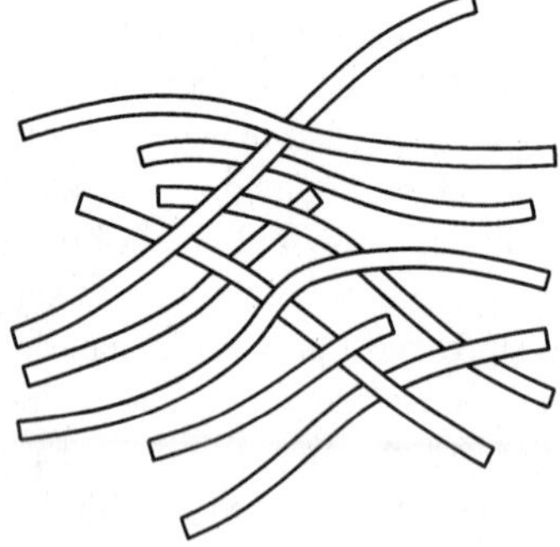

ening and the flow points, the intermolecular tension is retained after the thermoplastic cools down. Therefore, if the part is reheated to a temperature above the softening point, it will return to its original shape because of this intermolecular tension. This phenomenon, which characterizes most thermoplastic polymers, is known as *shaping memory.*

Many thermoplastic polymers are soluble in various solvents. Consequently, any one of these polymers can be given any desired shape by dissolving it into an appropriate solvent and then casting the viscous solution in molds. When the solvent completely evaporates, it leaves the rigid resin with the desired shape.

Several chemical families of polymeric materials can be categorized as thermoplastic. These include the ethenics, the polyamides, the cellulosics, the acetals, and the polycarbonates. Their characteristics, methods of manufacture, and applications are discussed in detail later in this chapter.

Thermosets

The molecules of thermosets usually take the form of a three-dimensional network structure that is mostly *cross-linked,* as shown in Figure 8.4. When raw (uncured) thermosetting polymers are heated to elevated temperatures, they are set, cross-linked, or polymerized. If reheated after this curing operation, thermosets will not melt again but will char or burn. Therefore, for producing complex shapes of thermosetting polymers, powders (or grains) of the polymers are subjected to heat and pressure until they are cured as finished products. Such polymers are referred to as *heat-convertible resins.*

Some raw thermosets can take the form of liquids at room temperature. When required, they are converted into solids by curing as a result of heating and/or additives (hardeners). This characteristic enables fabricators to produce parts by casting mixtures of liquid polymers and hardeners into molds. Therefore, these polymers are referred to as *casting resins.*

The cured thermosets are insoluble in solvents and do not soften at high temperatures. Thus, products made of thermosets can retain their shape under combined load and high temperatures, conditions that thermoplastics cannot withstand.

FIGURE 8.4 The molecular chains of a thermosetting polymer

8.2 PROPERTIES CHARACTERIZING PLASTICS AND THEIR EFFECT ON PRODUCT DESIGN

Properties of plastics differ significantly from those of metals, and they play a very important role in determining the form of the product. In other words, the form is dictated not only by the function but also by the properties of the material used and the method of manufacture, as we will see later. Following is a discussion of the effect of the properties characterizing plastics on the design of plastic products.

Mechanical Properties

The mechanical properties of polymers are significantly inferior to those of metals. Strength and rigidity values for plastics are very low compared with the lowest values of these properties for metals. Therefore, larger sections must be provided for plastic products if they are to have a similar strength and/or rigidity as metal products. Unfortunately, these properties get even worse when plastic parts are heated above moderate temperatures. In addition, some plastics are extremely brittle and notch-sensitive. Accordingly, any stress raisers like sharp edges or threads must be avoided in such cases.

A further undesirable characteristic of plastics is that they tend to deform continually under mechanical load even at room temperature. This phenomenon is accelerated at higher temperatures. Consequently, structural components made of plastics should be designed based on their creep strength rather than on their yield strength. This dictates a temperature range in which only a plastic product can be used. It is obvious that such a range is dependent principally upon the kind of polymer employed.

In spite of these limitations, the strength-to-weight ratio as well as the stiffness-to-weight ratio of plastics can generally meet the requirements for many engineering applications. In fact, the stiffness-to-weight ratio of reinforced polymers is comparable to that of metals like steel or aluminum.

Physical Properties

Three main physical properties detrimentally affect the widespread industrial application of polymers and are not shared by metals. First, plastics usually have a very high coefficient of thermal expansion, which is about ten times that of steel. This has to be taken into consideration when designing products involving a combination of plastics and metals. If plastics are tightly fastened to metals, severe distortion will occur whenever a significant temperature rise takes place. Second, some plastics are inflammable (i.e., not self-extinguishing) and keep burning even after the removal of the heat source. Third, some plastics have the ability to absorb large amounts of moisture from the surrounding atmosphere. This moisture absorption is unfortunately accompanied by a change in the size of the plastic part. Nylons are a typical example of this kind of polymer.

8.3 POLYMERIC SYSTEMS

This section surveys the commonly used polymeric materials and discusses their manufacturing properties and applications. Also discussed are the different additives that are used to impart certain properties to the various polymers.

Commonly Used Polymers

Following are some polymeric materials that are grouped into chemical families according to their common monomer.

Ethenic group. The monomer is ethylene. This group includes the following polymers:

1. Polyethylene.

$$\left[\begin{array}{c} \ \ \mathrm{H}\ \ \ \ \mathrm{H}\ \ \\ \ \ |\ \ \ \ \ |\ \ \\ -\mathrm{C}-\mathrm{C}- \\ \ \ |\ \ \ \ \ |\ \ \\ \ \ \mathrm{H}\ \ \ \ \mathrm{H}\ \ \end{array}\right]_n$$

The properties of polyethylene depend upon factors like degree of crystallinity, density, molecular weight, and molecular weight distribution. This thermoplastic polymer is characterized by its chemical resistance to solvents, acids, and alkalies, as well as by its toughness and good wear properties. Polyethylenes also have the advantage of being adaptable to many processing techniques, such as injection molding, blow molding, pipe extrusion, wire and cable extrusion, and rotational molding.

The applications of polyethylene are dependent upon the properties, which, in turn, depend upon the density and molecular weight. Low-density polyethylene is used in manufacturing films, coatings, trash bags, and throwaway products. High-density polyethylene is used for making injection-molded parts, tubes, sheets, and tanks that are used for keeping chemicals. The applications of the ultrahigh molecular weight (UHMW) polyethylene include wear plates and guide rails for filling and packaging equipment.

2. Polypropylene.

$$\left[\begin{array}{c} \ \ \mathrm{H}\ \ \ \ \mathrm{H}\ \ \\ \ \ |\ \ \ \ \ |\ \ \\ -\mathrm{C}-\mathrm{C}- \\ \ \ |\ \ \ \ \ |\ \ \\ \ \ \mathrm{H}\ \ \ \ \mathrm{CH_3} \end{array}\right]_n$$

Polypropylene is a thermoplastic material. A molecule of this polymer has all substituent groups (i.e., CH_3) on only one of its sides. This promotes crystallinity and, therefore, leads to strength higher than that of polyethylene. The resistance of polypropylene to chemicals is also good.

Polypropylene is mainly used for making consumer goods that are subjected to loads during their service life, such as ropes, bottles, and parts of appliances. This polymer is also used in tanks and conduits because of its superior resistance to chemicals.

3. Polybutylene.

$$\left[-\overset{H}{\underset{H}{\overset{|}{\underset{|}{C}}}} - \overset{CH_3}{\underset{CH_3}{\overset{|}{\underset{|}{C}}}} - \right]_n$$

Polybutylene is a thermoplastic polymer that has high tear, impact, and creep resistances. It also possesses good wear properties and is not affected by chemicals. Polybutylene resins are available in many grades, giving a wide range of properties and, therefore, applications.

The properties of polybutylene have made it an appropriate material for piping applications. These pipes can be joined together by heat fusion welding or by mechanical compression. Some grades of polybutylene are used as high-performance films for food packaging and industrial sheeting.

4. Polyvinyl chloride.

$$\left[-\overset{H}{\underset{H}{\overset{|}{\underset{|}{C}}}} - \overset{H}{\underset{Cl}{\overset{|}{\underset{|}{C}}}} - \right]_n$$

Polyvinyl chloride (PVC) is a thermoplastic polymer that can be processed by a variety of techniques like injection molding, extrusion, blow molding, and compression molding. It is fairly weak and extremely notch-sensitive but has excellent resistance to chemicals. When plasticized (i.e., additives are used to lubricate the molecules), it is capable of withstanding large strains.

The applications of rigid PVC include low-cost piping, siding, and related profiles, toys, dinnerware, and credit cards. Plasticized PVC is used in upholstery, imitation leather for seat covers and rainwear, and as insulating coatings on wires.

5. Polyvinyleidene chloride.

$$\left[-CH_2-C(Cl)_2-\right]_n$$

Polyvinyleidene chloride (PVDC) is nonpermeable to moisture and oxygen. It also possesses good creep properties. It is a preferred food-packaging material (e.g., saran wrap). Rigid grades are used for hot piping.

6. Polystyrene.

$$\left[-CH_2-CH(C_6H_5)-\right]_n$$

This thermoplastic polymer is known as "the cheap plastic." It has poor mechanical properties, can tolerate very little deflection, and breaks easily. Because of its cost, polystyrene is used for cheap toys and throwaway articles. It is also made in the form of foam (Styrofoam) for sound attenuation and thermal insulation.

7. Polymethyl methacrylate (Plexiglas acrylics).

$$\left[-CH_2-C(CH_3)(C(=O)OCH_3)-\right]_n$$

This polymer has reasonably good toughness, good stiffness, and exceptional resistance to weather. In addition, it is very clear and has a white-light transmission equal to that of clear glass. Consequently, this polymer finds application in safety glazing and in the manufacture of guard and safety glasses. It is also used in making automotive and industrial lighting lenses. Some grades are used as coatings and lacquers on decorative parts.

8. Fluorocarbons like polytetrafluoroethylene (Teflon).

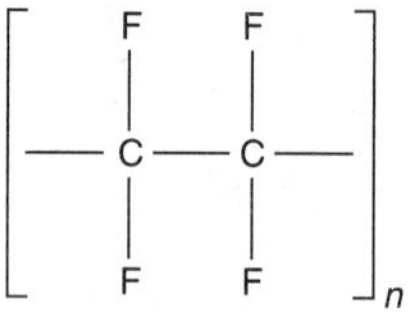

$$\left[\begin{array}{c} \quad F \quad F \\ \quad | \quad\;\; | \\ -C-C- \\ \quad | \quad\;\; | \\ \quad F \quad F \end{array}\right]_n$$

Teflon is characterized by its very low coefficient of friction and by the fact that even sticky substances cannot adhere to it easily. It is also the most chemically inert polymer. Nevertheless, it has some disadvantages, including low strength and poor processability. Because of its low coefficient of friction, Teflon is commonly used as a dry film lubricant. It is also used as lining for chemical and food-processing containers and conduits.

Polycarbonate group. These are actually polyesters. They are thermoplastic and have linear molecular chains. Polycarbonate exhibits good toughness, good creep resistance, and low moisture absorption. It also has good chemical resistance. It is widely used in automotive and medical and food packaging because of its cost effectiveness. It is also considered to be a high-performance polymer and has found application in the form of solar collectors, helmets, and face shields.

Polyacetal group. Included in this group is formaldehyde, with ending groups.

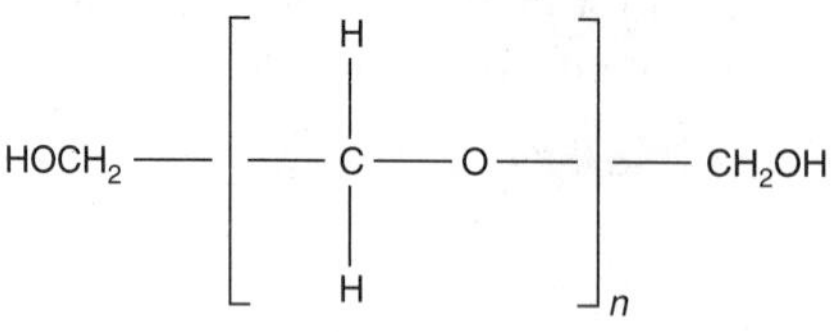

$$HOCH_2-\left[\begin{array}{c} H \\ | \\ -C-O- \\ | \\ H \end{array}\right]_n-CH_2OH$$

Formaldehyde is a thermoplastic polymer that can be easily processed by injection molding and extrusion. It has a tendency to be highly crystalline, and, as a result, this polymer possesses good mechanical properties. It also has good wear properties and a good resistance to moisture absorption.

Its applications involve parts that were made of nonferrous metals (like zinc, brass, or aluminum) by casting or stamping. These applications are exemplified by shower heads, shower-mixing valves, handles, good-quality toys, and lawn sprinklers.

Cellulosic group. The monomer is cellulose.

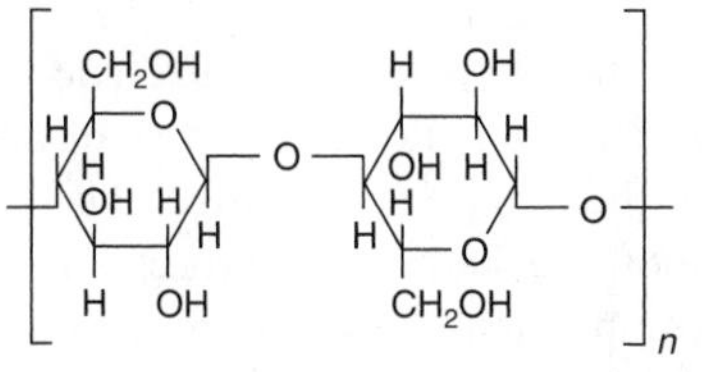

Cellulose itself is not a thermoplastic polymer. It can be produced by the viscous regeneration process to take the form of a fiber as in rayon, or a thin film, as in cellophane. Cellophane applications involve mainly decoration. Nevertheless, cellulose can be chemically modified to produce the following thermoplastics:

1. **Cellulose nitrate.** Good dimensional stability and low water absorption are the positive characteristics of this polymer. The major disadvantage that limits its widespread use is its inflammability. Cellulose nitrate is used in making table-tennis balls, fashion accessories, and decorative articles. It is also used as a base for lacquer paints.
2. **Cellulose acetate.** This polymer has good optical clarity, good dimensional stability, and resistance to moisture absorption. The uses of cellulose acetate include transparent sheets and films for graphic art, visual aids, and a base for photographic films. It is also used in making domestic articles.
3. **Cellulose acetate butyrate.** This thermoplastic polymer is tough, has good surface quality and color stability, and can readily be vacuum formed. It finds popular use in laminating with thin aluminum foil.
4. **Cellulose acetate propionate.** This thermoplastic polymer has reasonably good mechanical properties and can be injection molded or vacuum formed. It is used for blister packages, lighting fixtures, brush handles, and other domestic articles.

Polyamide group. This family includes high-performance melt-processable thermoplastics.

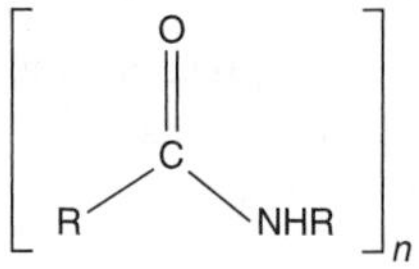

$$\left[R - \overset{O}{\overset{\|}{C}} - NHR \right]_n$$

R is a chemical group that differs for different members of this family.

One group of common polyamides is the nylons. These are characterized by their endurance and retention of their good mechanical properties even at relatively high temperatures. They also possess good lubricity and resistance to wear. The chief limitation is their tendency to absorb moisture and change size.

These polymers find use in virtually every market (e.g., automotive, electrical, wire, packaging, and appliances). Typical applications include structural components up to 10 pounds (4 kg), bushings, gears, cams, and the like.

ABS. The three monomers are acrylonitrile, butadiene, and styrene. Based on this three-monomer system (similar to an alloy in the case of metals), the properties of this group vary depending upon the components. Fifteen different types are commercially used. They possess both good mechanical properties and processability. Applications of the ABS group include pipes and fittings, appliances and automotive uses, telephones, and components for the electronics industry.

Polyesters. These polymers result from a condensation reaction of an acid and an alcohol. The type and nature of the polymer obtained depend upon the acid and alcohol used. This multitude of polymers are mostly thermoplastic and can be injection molded and formed into films and fibers. Their uses include bases for coatings and paints, ropes, fabrics, outdoor applications, construction, appliances, and electrical and electronic components. Polyester is also used as a matrix resin for fiberglass to yield the composite fiber-reinforced polymer.

Phenolic group. The monomer is phenol formaldehyde.

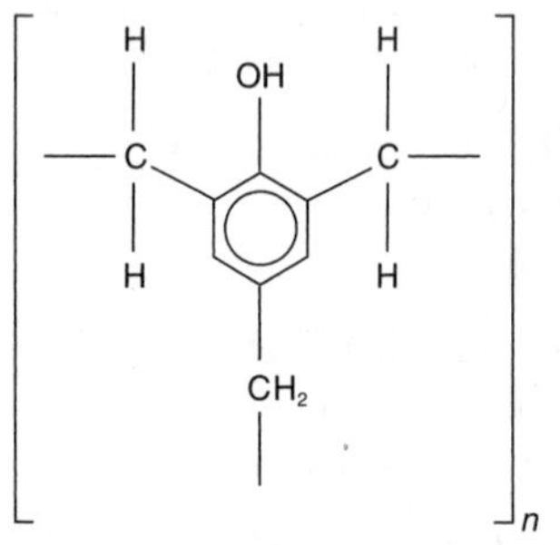

As previously mentioned, phenolics are actually the oldest manufactured thermosetting polymers. They are processed by compression molding, where a product with a highly cross-linked chain structure is finally obtained. Phenolics are characterized by their high strength and their ability to tolerate temperatures far higher than their molding temperature.

Phenolics are recommended for use in hostile environments that cannot be tolerated by other polymers. They are used in electrical switchplates, electrical boxes, and similar applications. Nevertheless, the chief field of application is as bonding agents for laminates, plywood-grinding wheels, and friction materials for brake lining.

Polyimides. Polyimides are mostly thermosetting and have very complex structures. They are considered to be one of the most heat-resisting polymers. They do not melt and flow at elevated temperatures and are, therefore, manufactured by powder metallurgy techniques.

The polyimides are good substitutes for ceramics. Applications include jet-engine and turbine parts, gears, coil bobbins, cages for ball bearings, bushings and bearings, and parts that require good electrical and thermal insulation.

Epoxies. Epoxies and epoxy resins are a group of polymers that become highly cross-linked by reaction with curing agents or hardeners. These polymers have low molecular weight and got their name from the epoxide group at the ends of the molecular chains. Epoxy resins are thermosetting and have good temperature resistance. They adhere very well to a variety of substrates. Another beneficial characteristic is their stability of dimensions upon curing.

The common application of epoxy resins is as adhesives. With the addition of fibers and reinforcements, laminates and fiber-reinforced epoxy resins can be obtained and are used for structural applications.

Polyurethanes. Polyurethanes involve a wide spectrum of polymers ranging from soft thermoplastic elastomers to rigid thermosetting foams. While all polyurethanes are products of a chemical reaction of an isocyanate and an alcohol, different polymers are apparently obtained by different reacting materials.

Elastomers are used as die springs, forming-die pads, and elastomer-covered rolls. Some of these elastomers are castable at room temperature and find popular application in rubber dies for the forming of sheet metals. Flexible foam has actually replaced latex rubber in home and auto seating and interior padding. The rigid thermosetting foam is used as a good insulating material and for structural parts. Other applications of polyurethanes include coating, varnishes, and the like.

Silicones. In this group of polymers, silicon forms the backbone of the molecular chain and plays the same role as that of carbon in other polymers.

$$\left[-\underset{R}{\overset{R}{\overset{|}{\underset{|}{Si}}}} - O - \right]_n$$

Silicones can be oils, elastomers, thermoplastics, or thermosets, depending upon the molecular weight and the functional group. Nevertheless, they are all characterized by their ability to withstand elevated temperatures and their water-repellent properties.

Silicones in all forms are mainly used for high-temperature applications. These include binders for high-temperature paints and oven and good-handling tubing gaskets. Silicone oils are used as high-temperature lubricants, mold release agents, and damping or dielectric fluids.

Elastomers. These polymeric materials possess a percentage elongation of greater than 100 percent together with significantly high resilience. This rubberlike behavior is attributed to the branching of the molecular chains. Elastomers mainly include five polymers: natural rubber, neoprene, silicone rubber, polyurethane, and fluoroelastomers. Natural rubber is extracted as thick, milky liquid from a tropical tree. Next, moisture is removed, additives (coloring, curing agents, and fillers) are blended with it, and the mixture is then vulcanized. The latter operation involves heating up to a temperature of 300°F (150°C) to start cross-linking and branching reactions.

The application of elastomers includes seals, gaskets, oil rings, and parts that possess rubberlike behavior such as tires, automotive and aircraft parts, and parts in forming dies.

Additives

Additives are materials that are compounded with polymers in order to impart and/or enhance certain physical, chemical, manufacturing, or mechanical properties. They are also sometimes added just for the sake of reducing the cost of products. Commonly used additives include fillers, plasticizers, lubricants, colorants, antioxidants, and stabilizers.

Fillers. Fillers involve wood flour, talc, calcium carbonate, silica, mica flour, cloth, and short fibers of glass or asbestos. Fillers have recently gained widespread industrial use as a result of the continued price increase and short supply of resin stocks. An expensive or unavailable polymer can sometimes be substituted by another *filled polymer,* provided that an appropriate filler material is chosen.

The addition of inorganic fillers usually tends to increase the strength because this kind of additive inhibits the mobility of the polymers' molecular chains. Nevertheless, if too much filler is added, it may create enclaves or weak spots and cause problems during processing, especially if injection molding is employed.

Plasticizers. Plasticizers are organic chemicals (high-boiling-temperature solvents) that are admixed with polymers in order to enhance resilience and flexibility. This is a result of facilitating the mobility of the molecular chains, thus enabling them to move easily relative to one another. On the other hand, plasticizers reduce the strength. Therefore, a polymer that meets requirements without the addition of plasticizers is the one to use.

Lubricants. Lubricants are chemical substances that are added in small quantities to the polymer to improve processing and flowability. They include fatty acids, fatty alcohols, fatty esters, metallic stearates, paraffin wax, and silicones. Lubricants are classified as external (applied externally to the polymer), internal, or internal-external. The last group includes most commercially used lubricants.

Colorants. Colorants may be either dyes or pigments. Dyes have smaller molecules and are transparent when dissolved. Pigment particles are relatively large (over 1 μm) and are, therefore, either translucent or opaque. Pigments are more widely used than dyes because dyes tend to extrude from the polymers.

Antioxidants. The use of antioxidants is aimed at enhancing the resistance to oxidation and degradation of polymers, thus extending their useful temperature range and service life. These substances retard the chemical reactions that are caused by the presence of oxygen.

Stabilizers. Stabilizers are substances that are added to polymers to prevent degradation as a result of heat or ultraviolet rays. The mechanism of inhibiting degradation of polymers differs for different stabilizers. However, ultraviolet stabilizers usually function by absorbing ultraviolet radiation.

8.4 PROCESSING OF PLASTICS

A variety of processing methods can be employed in manufacturing plastic products. However, it must be kept in mind that no single processing method can successfully be employed in shaping all kinds of plastics. Each process has its own set of advantages and disadvantages that influence product design. Following is a survey of the common methods for plastic processing.

Casting

Casting is a fairly simple process that requires no external force or pressure. It is usually performed at room temperature and involves filling the mold cavity with monomers or partially polymerized syrups and then heating to cure. After amorphous solidification, the material becomes *isotropic,* with uniform properties in all directions. Nevertheless, a high degree of shrinkage is experienced during solidification and must be taken into consideration when designing the mold. Sheets, rods, and tubes can be manufactured by casting, although the typical application is in trial jigs and fixtures as well as in insulating electrical components. Acrylics, epoxies, polyesters, polypropylene, nylon, and PVC can be processed by casting. The casting method employed is sometimes modified to suit the kind of polymer to be processed. Whereas nylon is cast in its hot state after adding a suitable catalyst, PVC film is produced by solution casting. This process involves dissolving the PVC into an appropriate solvent, pouring the solution on a substrate, and allowing the solvent to evaporate in order to finally obtain the required film.

Blow Molding

Blow molding is a fast, efficient method for producing hollow containers of thermoplastic polymers. The hollow products manufactured by this method usually have thin walls and range in shape and size from small, fancy bottles to automobile fuel tanks.

Although there are different versions of the blow molding process, they basically involve blowing a tubular shape (parison) of heated polymer in a cavity of a split mold. As can be seen in Figure 8.5, air is injected through a needle into the parison, which expands in a fairly uniform thickness and finally conforms to the shape of the cavity.

Injection Molding

Injection molding is the most commonly used method for mass production of plastic articles because of its high production rates and the good control over the dimensions of the products. The process is used for producing thermoplastic articles, but it can also be applied to thermosets. The main limitation of injection molding is the required high

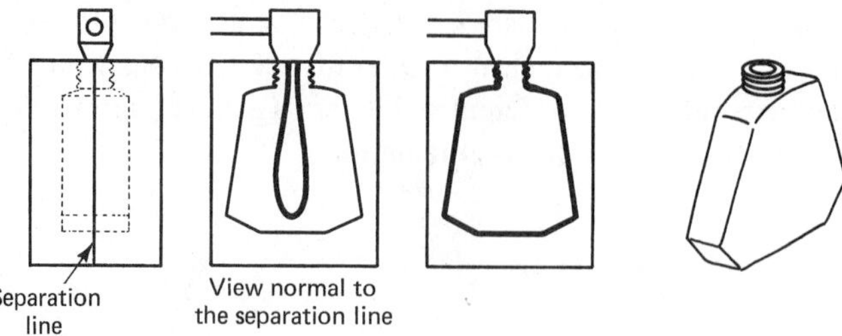

FIGURE 8.5
The blow molding process

initial capital cost, which is due to the expensive machines and molds employed in the process.

The process basically involves heating the polymer, which is fed from a hopper in granular pellet or powdered forms, to a viscous melted state and then forcing it into a split-mold cavity, where it hardens under pressure. Next, the mold is opened, and the product is ejected by a special mechanism. Molds are usually made of tool steel and may have more than a single cavity.

Figure 8.6 shows a modern screw-preplasticator injection unit employed in injection molding of thermoplastics. As can be seen, the diverter valve allows the viscous polymer to flow either from the plasticating screw to the pressure cylinder or from the cylinder to the cooled mold. When thermosets are to be injection molded, a machine with a different design has to be used. Also, the molds must be hot so that the polymer can cure.

Once the decision has been made to manufacture a plastic product by injection molding, the product designer should make a design that facilitates and favors this process. Following are some design considerations and guidelines.

Make the thickness of a product uniform and as small as possible. Injection molding of thermoplastics produces net-shaped parts by going from a liquid state to a solid state. (These net-shaped parts are used as manufactured; they do not require further processing or machining.) This requires time to allow the heat to dissipate so that the

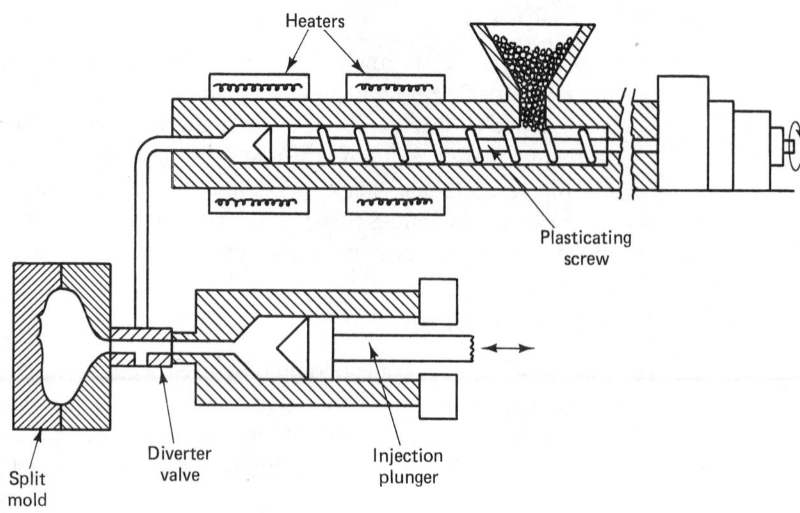

FIGURE 8.6
The injection molding process

polymer melt can solidify. The thicker the walls of a product, the longer the product cycle, and the higher its cost. Consequently, a designer has to keep the thickness of a product to a minimum without jeopardizing the strength and stiffness considerations. Also, thickness must always be kept uniform; if change in thickness is unavoidable, it should be made gradually. It is better to use ribs rather than increase the wall thickness of a product. Figure 8.7 shows examples of poor design and how they can be modified (by slight changes in constructional details) so that sound parts are produced.

Provide generous fillet radii. Plastics are generally notch-sensitive. The designer should, therefore, avoid sharp corners for fillets and provide generous radii instead. The ratio of the fillet radius to the thickness should be at least 1.4.

Ensure that holes will not require complex tooling. Holes are produced by using core pins. It is, therefore, clear that through holes are easier to make than blind holes. Also, when blind holes are normal to the flow, they require retractable core pins or split tools, thus increasing the production cost.

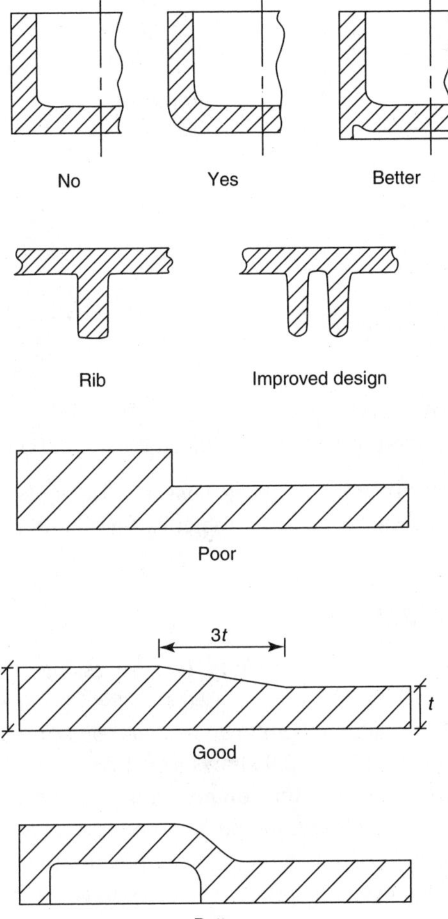

FIGURE 8.7 Examples of poor and good designs of walls of plastic products

FIGURE 8.8
Examples of poor and good designs of bosses in injection-molded parts

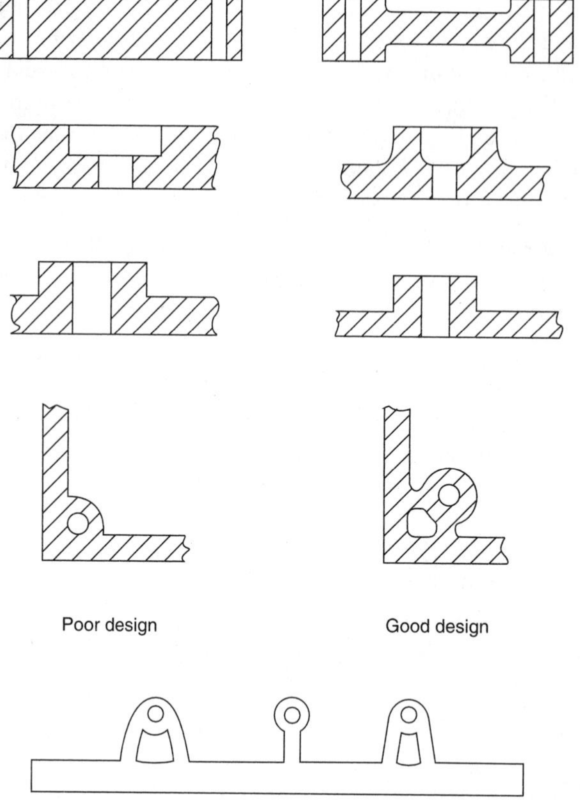

Provide appropriate draft. As is the case with forging, it is important to provide a draft of 1° so that the product can be injected from the mold.

Avoid heavy sections when designing bosses. Heavy sections around bosses lead to wrappage and dimensional control problems. Figure 8.8 shows poor and good designs of bosses.

Compression Molding

Compression molding is used mainly for processing thermosetting polymers. The process involves enclosing a premeasured charge of polymer within a closed mold and then subjecting that charge to combined heat and pressure until it takes the shape of the mold cavity and cures. Figure 8.9 shows a part being produced by this process.

Although the cycle time for compression molding is very long when compared with that for injection molding, the process has several advantages. These include low capital cost (because the tooling and the equipment used are simpler and cheaper) and the elimination of the need for sprues or runners, thus reducing the material waste. There

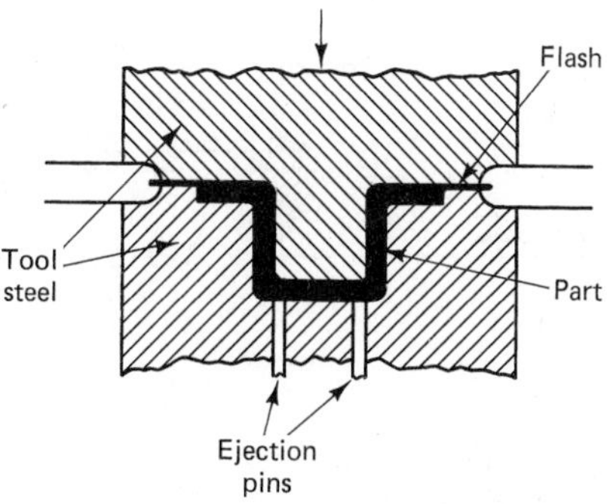

FIGURE 8.9
The compression molding process

are, however, limitations upon the shape and size of the products manufactured by this method. It is generally difficult to produce complex shapes or large parts as a result of the poor flowability and long curing times of the thermosetting polymers.

Transfer Molding

Transfer molding is a modified version of the compression molding process, and it is aimed at increasing the productivity by accelerating the production rate. As can be seen in Figure 8.10, the process involves placing the charge in an open, separate "pot," where the thermosetting polymer is heated and forced through sprues and runners to fill several closed cavities. The surfaces of the sprues, runners, and cavities are kept at a temperature of 280 to 300°F (140 to 200°C) to promote curing of the polymer. Next, the entire shot (i.e., sprues, runners, product, and the excess polymer in the pot) is ejected.

Rotational Molding

Rotational molding is a process by which hollow objects can be manufactured from thermoplastics and sometimes thermosets. It is based upon placing a charge of solid or liquid polymer in a mold. The mold is heated while being rotated simultaneously around two perpendicular axes. As a result, the centrifugal force pushes the polymer against the walls of the mold, thus forming a homogeneous layer of uniform thickness

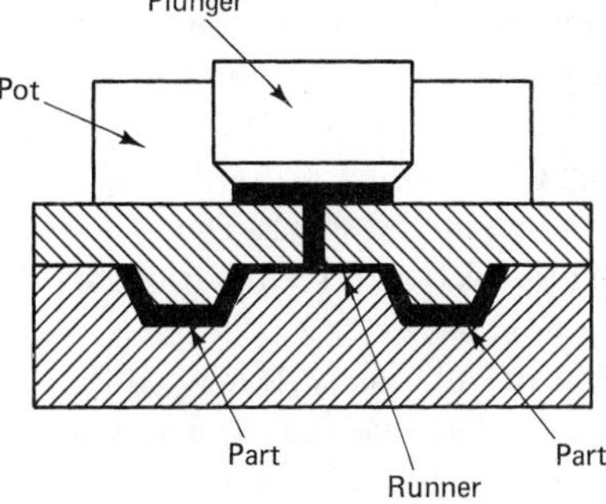

FIGURE 8.10
The transfer molding process

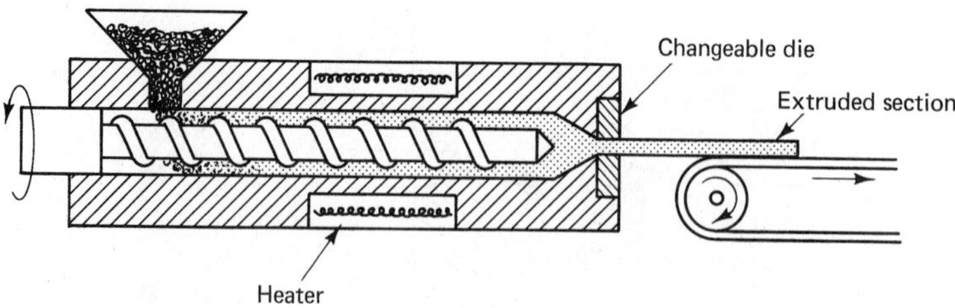

FIGURE 8.11
The extrusion process

that conforms to the shape of the mold, which is then cooled before the product is ejected. The process, which has a relatively long cycle time, has the advantage of offering almost unlimited product design freedom. Complex parts can be molded by employing low-cost machinery and tooling.

Extrusion

In *extrusion,* a thermoplastic polymer in powdered or granular form is fed from a hopper into a heated barrel, where the polymer melts and is then extruded out of a die. Figure 8.11 shows that plastics extrusion is a continuous process capable of forming an endless product that has to be cooled by spraying water and then cut to the desired lengths. The process is employed to produce a wide variety of structural shapes, such as profiles, channels, sheets, pipes, bars, angles, films, and fibers. Extrusions like bars, sheets, and pipes can also be further processed by other plastic manufacturing methods until the desired final product is obtained.

A modification of conventional extrusion is a process known as *coextrusion.* It involves extruding two or more different polymers simultaneously in such a manner that one polymer flows over and adheres to the other polymer. This process is used in industry to obtain combinations of polymers, each contributing some desired property. Examples of coextrusion include refrigerator liners, foamed-core solid-sheath telephone wires, and profiles involving both dense material and foam, which are usually used as gasketing in automotive and appliance applications.

Thermoforming

Thermoforming involves a variety of processes that are employed to manufacture cuplike products from thermoplastic sheets by a sequence of heating, forming, cooling, and trimming. First, the sheet is clamped all around and heated to the appropriate temperature by electric heaters located above it. Next, the sheet is stretched under the action of pressure, vacuum, or male tooling and is forced to take the shape of a mold. The polymer is then cooled to retain the shape. This is followed by removing the part from the mold and trimming the web surrounding it. Figure 8.12a through d illustrates the different thermoforming processes.

Although thermoforming was originally developed for the low-volume production of containers, the process can be automated and made suitable for high-volume applications. In this case, molds are usually made of aluminum because of its high thermal

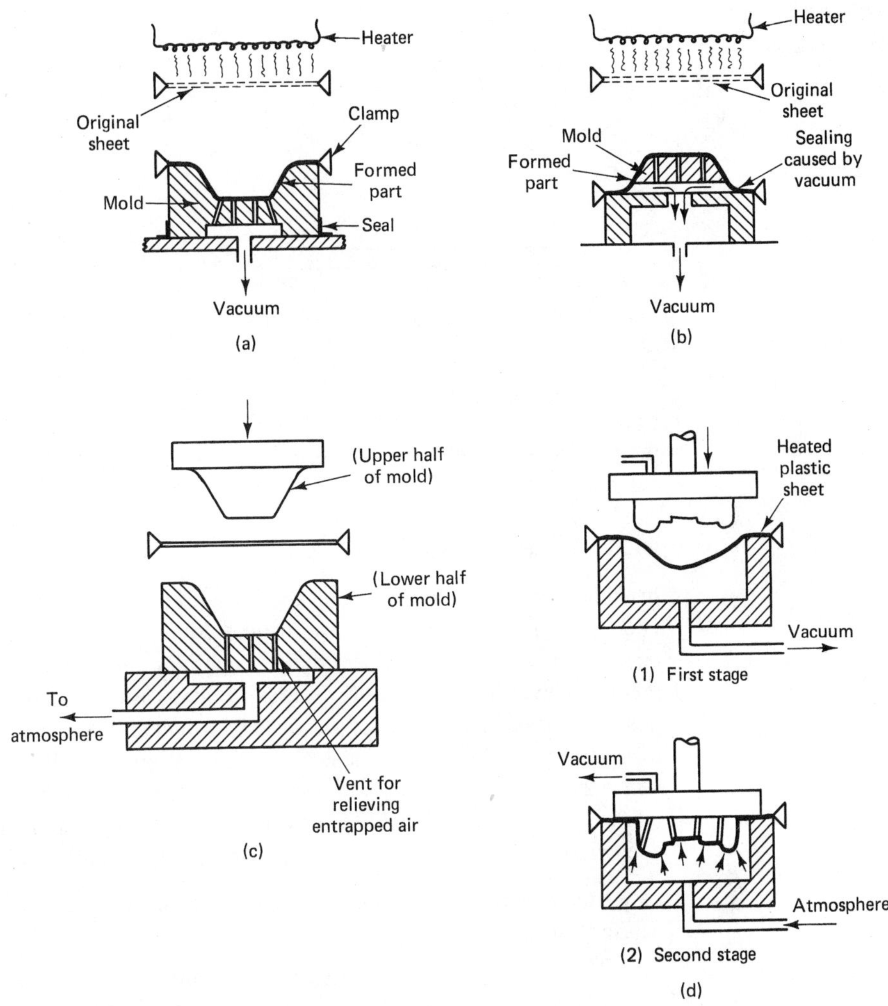

FIGURE 8.12 Different thermoforming processes: (a) straight vacuum forming; (b) drape forming; (c) matched-mold forming; (d) vacuum snapback

conductivity. For low-volume or trial production, molds are made of wood or even plaster of paris.

Examples of the parts produced by thermoforming include containers, panels, housings, machine guards, and the like. The only limitation on the shape of the product is that it should not contain holes. If holes are absolutely required, they should be made by machining at a later stage.

Calendering

Calendering is the process employed in manufacturing thermoplastic sheets and films. This process is similar to rolling with a four-high rolling mill, except that the rolls that squeeze the polymer are heated. The thermoplastic sheet is fed and metered in the first and second roll gaps, whereas the third roll gap is devoted to gaging and finishing.

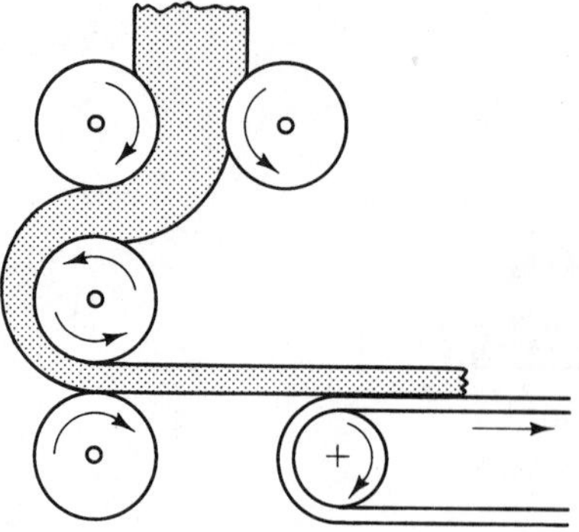

FIGURE 8.13 The calendering process

Most of the calendering products are flexible or rubberlike sheets and films, although the process is sometimes applied to ABS and polyethylene. Figure 8.13 illustrates the calendering process.

Machining of Plastics

In some cases, thermoplastic and thermosetting polymers are subjected to machining operations like sawing, drilling, or turning. Some configurations and small lot sizes can be more economically achieved by machining than by any other plastic-molding method. Nevertheless, there are several problems associated with the machining of plastics. For instance, each type of plastic has its own unique machining characteristics, and they are very different from those of the conventional metallic materials. A further problem is the excessive tool wear experienced when machining plastics, which results in the interruption of production as well as additional tooling cost. Although much research is needed to provide solutions for these problems, there are some general guidelines:

1. Reduce friction at the tool-workpiece interface by using tools with honed or polished surfaces.
2. Select tool geometry so as to generate continuous-type chips. Recent research has revealed that there exists a critical rake angle (see Chapter 9) that depends upon the polymer, depth of cut, and cutting speed.
3. Use twist drills that have wide, polished flutes, low helix angles, and tool-point angles of about 70° and 120°.

Recently, lasers have been employed in cutting plastics. Because a laser acts as a materials eliminator, its logical application is cutting and hole drilling. High-pressure water jets also currently find some application in the cutting of polymers and composites.

Welding of Plastics

There are several ways for assembling plastic components. The commonly used methods include mechanical fastening, adhesive bonding, thermal bonding, and ultrasonic welding. Only thermal bonding and ultrasonic welding are discussed next because the first two operations are similar to those used with metals.

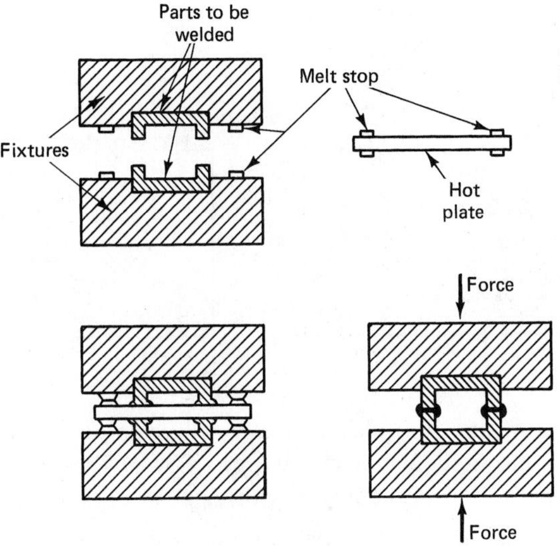

FIGURE 8.14 Steps involved in hot-plate joining

Thermal bonding of plastics. Thermal bonding, which is also known as *fusion bonding,* involves the melting of the weld spots in the two plastic parts to be joined and then pressing them together to form a strong joint. Figure 8.14 illustrates the steps involved in the widely used thermal bonding method known as *hot-plate joining*. As can be seen in the figure, a hot plate is inserted between the edges to be mated in order to melt the plastic parts; melting stops when the plate comes in contact with the holding fixture. Next, the plate is withdrawn, and the parts are pressed together and left to cool to yield a strong joint. The cycle time usually ranges from 15 to 20 seconds, depending upon the relationship between the melt time and the temperature (of the hot plate) for the type of plastic to be bonded. Also, this process is applied only to thermoplastics.

Figure 8.15 illustrates different types of joint design. The one to select is dependent upon both the desired strength and the appearance of the joint. The product designer must keep in mind that a small amount of material is displaced from each side to form the weld bead. This must be taken into account when dimensional tolerance is critical, such as when fusion-bonded parts are to be assembled together.

Another thermal bonding process, which is equivalent to riveting in the case of metals, is referred to as *thermostaking*. As can be seen in Figure 8.16, the process

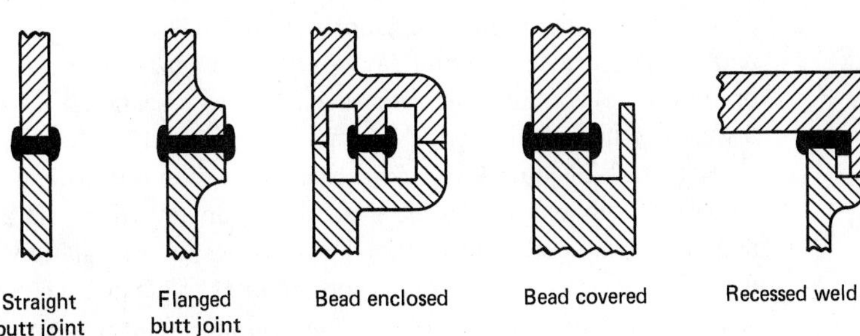

FIGURE 8.15 Different joint designs for fusion bonding

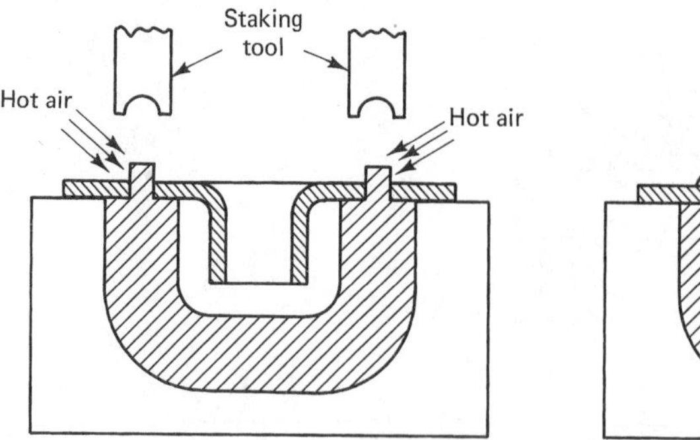

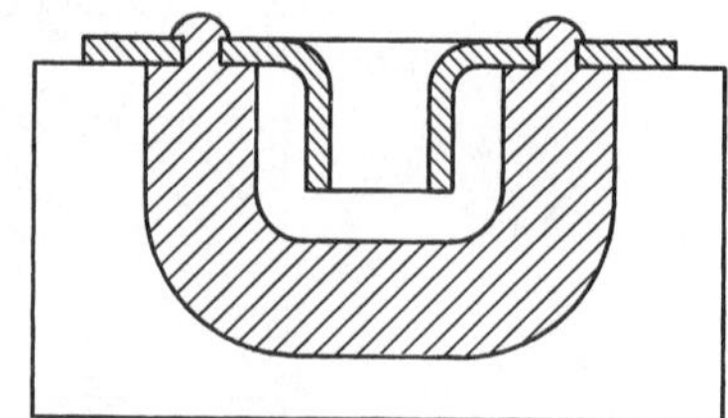

FIGURE 8.16 The thermostaking process

involves the softening of a plastic stud by a stream of hot air and then forming the softened stud and holding it while it cools down. Thermal bonding processes find widespread application in the automotive, appliance, battery, and medical industries.

Ultrasonic welding of plastics. Ultrasonic welding is gaining popularity in industry because of its low cycle time of about 0.5 second and the strong, tight joints that are easily obtainable. The process is used for thermoplastics and involves conversion of high-frequency electrical energy to high-frequency mechanical vibrations that are, in turn, employed to generate highly localized frictional heating at the interface of the mating parts. This frictional heat melts the thermoplastic polymer, allowing the two surfaces to be joined together.

The product designer must bear in mind that not all thermoplastics render themselves suitable for ultrasonic welding. Whereas amorphous thermoplastics are good candidates, crystalline polymers are not suitable for this process because they tend to attenuate the vibrations. Hydroscopic plastics (humidity-absorbing polymers, such as nylons) can also create problems and must, therefore, be dried before they are ultrasonically welded. In addition, the presence of external release agents or lubricants reduces the coefficient of friction, thus making ultrasonic welding more difficult.

The equipment used involves a power supply, a transducer, and a horn. The power supply converts the conventional 115-V, 60-Hz (or 220-V, 50-Hz) current into a high-frequency current (20,000 Hz). The transducer is usually a piezoelectric device that converts the electrical energy into high-frequency, axial-mechanical vibrations. The horn is the part of the system that is responsible for amplifying and transmitting the mechanical vibrations to the plastic workpiece. Horns may be made of aluminum, alloy steel, or titanium. The latter material possesses superior mechanical properties and is, therefore, used with heavy-duty systems. The horns amplify the mechanical vibration via a continuous decrease in the cross-sectional area and may take different forms to achieve that goal, as shown in Figure 8.17.

The task of joint design for ultrasonic welding is critical because it affects the design of the molded parts to be welded. Fortunately, there are a variety of joint designs, and each has its specific features, advantages, and limitations. The type of joint to be used should obviously depend upon the kind of plastic, the part geometry, the strength required, and the desired cosmetic appearance. Following is a discussion of the commonly used joint designs, which are illustrated in Figure 8.18.

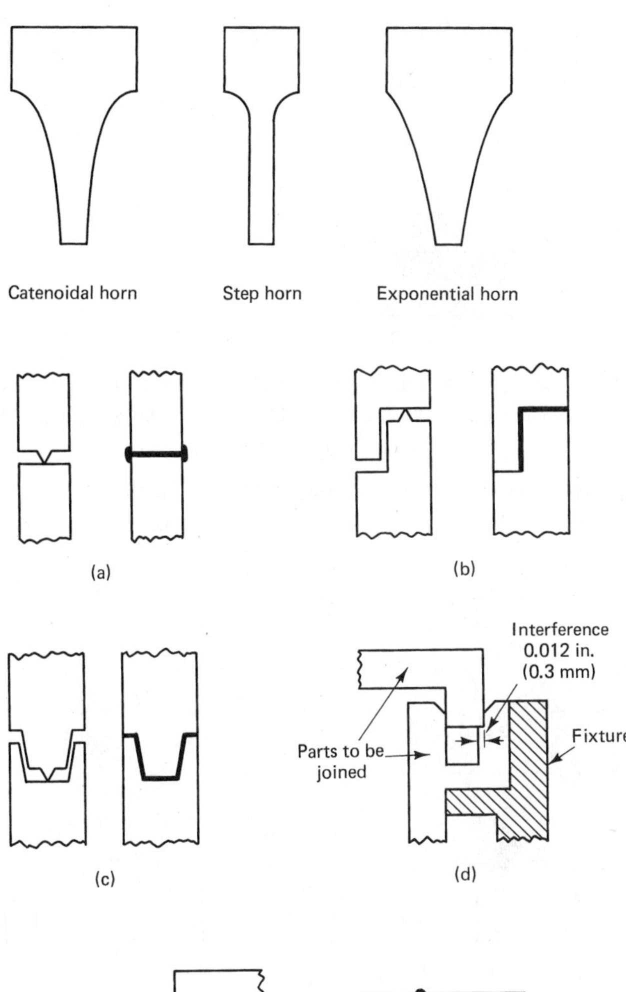

FIGURE 8.17 Different horn shapes employed in ultrasonic welding of plastics

FIGURE 8.18 Different joint designs for ultrasonic welding: (a) butt joint; (b) step joint; (c) tongue-and-groove joint; (d) interference joint; (e) scarf joint

1. **Butt joint with energy director.** The butt joint (see Figure 8.18a) is the most commonly used joint design in ultrasonic welding. As can be seen in the figure, one of the mating parts has a triangular-shaped projection. This projection is known as an *energy director* because it helps to limit the initial contact to a very small area, thus increasing the intensity of energy at that spot. This causes the projection to melt and flow and cover the whole area of the joint. This type of joint is considered to be the easiest to produce because it is not difficult to mold into a part.

2. **Step joint with energy director.** The step joint (see Figure 8.18b) is stronger than the butt joint and is recommended when cosmetic appearance is desired.

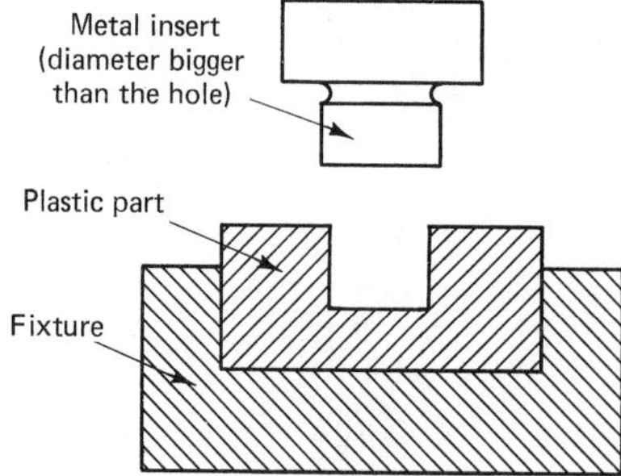

FIGURE 8.19
Ultrasonic installation of metal insert into plastic part

3. **Tongue-and-groove joint with energy director.** The tongue-and-groove joint (see Figure 8.18c) promotes the self-locating of parts and prevents flash. It is stronger than both of the previously mentioned methods.
4. **Interference joint.** The interference joint (see Figure 8.18d) is a high-strength joint and is usually recommended for square corners or rectangular-shaped parts.

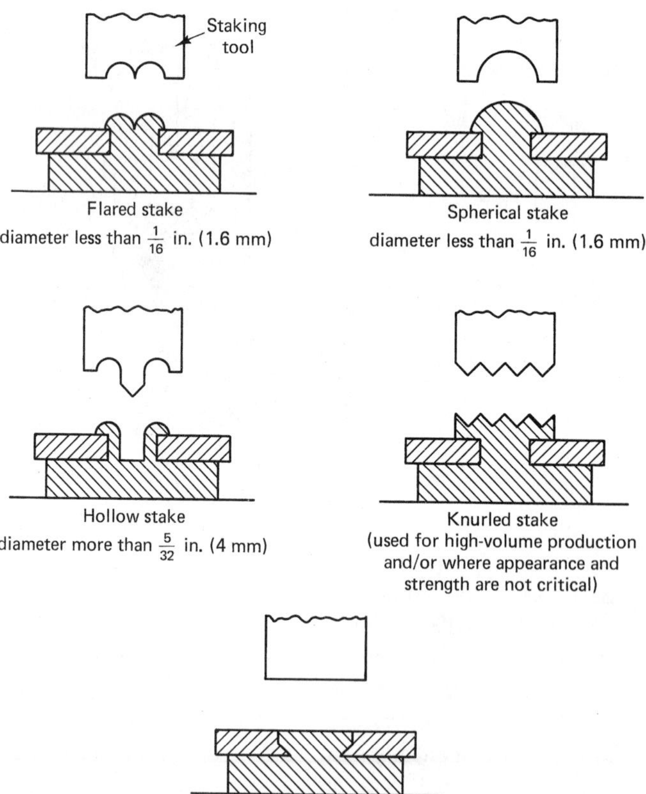

FIGURE 8.20
Ultrasonic staking

5. **Scarf joint.** The scarf joint (see Figure 8.18e) is another high-strength joint and is recommended for components with circular or oval shapes.

In addition to welding, ultrasonics are employed in inserting metallic parts into thermoplastic components. Figure 8.19 illustrates an arrangement for the ultrasonic installation of a metal insert into a plastic part.

Another useful application of these systems is ultrasonic staking, which is equivalent to riveting or heading. Figure 8.20 indicates the different types of stakes, as well as their recommended applications. Notice that these stakes can be flared, spherical, hollow, knurled, or flush.

8.5 FIBER-REINFORCED POLYMERIC COMPOSITES*

In this present age of new materials, at the forefront of advancing developments are materials based on the combination of organic polymer resins and high-strength, high-stiffness synthetic fibers. This section addresses the materials, processing, and design methodology of *fiber-reinforced polymeric composites.*

Historical Background

Although the merits of fiber-reinforced materials have been known for centuries, (straw-reinforced clay was reportedly used as a building material by the Egyptians in 600 B.C.), it is only in the past 40 years that fiber-reinforced polymers have become important engineering materials. New synthetic high-strength, high-modulus fibers and new resins and matrix materials have elevated fiber-reinforced composites into the material of choice for innovative lightweight, high-strength engineered products. These developments along with established engineering design criteria and special processing technology have advanced fiber-reinforced composites close to the realm of a commodity material of construction. In the areas of automobile bodies, recreational boat hulls, and bathroom fixtures (bathtubs and shower stalls), fiberglass-reinforced organic polymer resins have indeed become the material of choice. In more advanced applications, the first completely fiber-reinforced polymeric resin composite aircraft came into existence in the 1980s. For the 1990s, some important nonaerospace applications are emerging, such as sports equipment (sailboat spars) and, more recently, wind turbine blades.

The utilization of composite materials in functional engineering applications continues to grow. It is, therefore, important for engineering students to know about and understand these materials so that new uses may be developed and propagated. Consequently, a brief review of organic polymer engineering composites is presented next. A general description of these materials, their unique properties, processing techniques, and engineering design features will put into perspective present and future uses of fiber-reinforced polymer (FRP) engineering materials.

---

* Section 8.5 was written by Dr. Armand F. Lewis, Lecturer at the University of Massachusetts Dartmouth.

Nature of Composites

A *composite* may be defined as a material made up of several identifiable phases, combined in an ordered fashion to provide specific properties different from or superior to those of the individual materials. Many types of composites exist, including laminated materials, filamentary-wound or -layered and particulate-filled compositions, and multiphase alloys and ceramics. Most naturally occurring structural materials are composites (wood, stone, bone, and tendon).

Overall, composite materials can be classified according to Table 8.1. We will focus on fiber/resin composite materials composed of higher-strength, higher-modulus fibers embedded in an organic polymer/resin matrix. Table 8.2 lists some of the common resin and fiber materials employed. These composite materials are generally referred to as *fiber-reinforced polymers* (FRP). Currently, polyester and epoxy resins are the most common commercially used matrix resin polymers, while glass fibers are the most widely used reinforcing fiber. Resin matrix composites containing high-strength, high-elasticity-modulus carbon (graphite), polyaramid (Kevlar, a DuPont trade name), and boron fibers are also in use for specialty (advanced) composite material applications.

The integral combination of high-strength, high-elasticity-modulus fibers and relatively low-strength, low-rigidity polymer matrices forms some unique engineering materials. FRP composites possess the material processing and fabrication properties of polymeric materials yet, due to their fiber reinforcement, can be designed to possess directional stiffness and strength properties comparable to those of metals. These mechanical properties can be achieved at a very light weight. This feature can be illustrated by comparing the specific strength (tensile strength/density) to the specific elastic modulus (tensile elasticity-modulus/density) of various fiber-reinforced composite materials with plastics and metals. Figure 8.21 compares the specific strengths and specific elastic moduli of these materials. Notice that commodity elastomers, plas-

TABLE 8.1
Classifications of composite materials

Classification	Typical Example(s)
Fiber/resin composites Continuous Discontinuous	Glass fabric/mat reinforced polyester resin molded into sport boat hulls
Heterophase polymer mixtures Random particulate filled Flake or shaped particles	Aluminum and/or graphite powder blended into nylon plastic to form a machine gear
Interstitial polymeric materials Interpenetrating polymer networks Skeletal composites	"Marbleized" decorative plastic for wall panels
Laminar and linear composites Material hybrids Polymer-polymer	High-pressure laminates used in kitchen countertops and polyurethane rubber-impregnated polyaramid rope/cable

TABLE 8.2
Some materials used in organic polymer engineering composites

Matrix Resin	Maximum Service Temperature
Epoxy	Up to 121°C (250°F)
Polyester	Up to 62°C (non HT)
Vinyl ester	Up to 145°C (HT type)
Phenolic	Up to 149°C (300°F)
Polyimide	Up to 260°C (500°F)
Thermoplastics	
Nylon	Up to 80°C (175°F)
Polyphenylene sulfide	Up to 149°C (300°F)
Polyetheretherktone (PEEK)	Up to 200°C (392°F)
Fiber (Continuous Yarn/Filament, Woven Fabric, Nonwoven Mat, Chopped Fiber)	
Glass (especially E and S glass)	
Polyaramid organic fiber (Kevlar)® (Dupont trademark)	
Carbon	
Boron	
Form for Processing	
Liquid casting resin	
B-stage resin mixture	
Preimpregnated (prepreg) B-stage resin/fiber/fabric combination	

tics, and metals occupy only a very small portion of this *structural materials map.* Fibers and fiber-reinforced resin composites occupy the outer regions. Fiber-reinforced composites can have specific strengths and moduli up to six times those of common structural materials. For a given weight, fiber-reinforced composites far outperform other engineering materials in their strength and stiffness. These specific strengths and moduli approach the mechanical properties of theoretically perfect, ordered polymer crystals. This property makes composite materials unique among engineering structural materials and opens new horizons for novel engineering designs. For example, composite materials are widely used in aircraft and aerospace applications: The FRP property of high specific strength with high elasticity modulus made possible the design, construction, and functional deployment of the U.S. Air Force all-carbon fiber-reinforced epoxy resin composite Stealth reconnaissance aircraft.

The observed high strength and stiffness-to-weight ratio of fiber-reinforced composites can be easily explained. Various material properties of composites can be estimated by a *rule of mixtures* approach. Micromechanic properties such as modulus (stiffness), strength, Poisson's ratio, and thermal expansion of fiber-reinforced polymer composites can be estimated by the following equation:

$$M_c = V_f M_f + V_m M_m \tag{8.1}$$

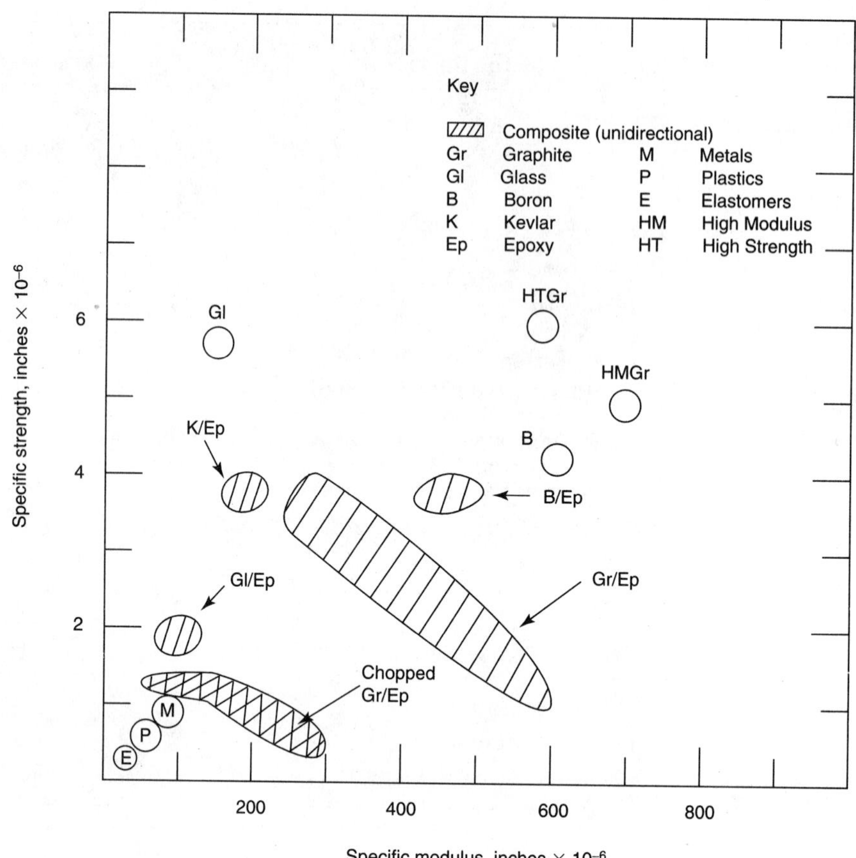

FIGURE 8.21 Specific strengths and specific elastic moduli of materials

where: M is the particular material property

V is the volume fraction of the fiber (f) or matrix (m)

M_c is the material property of the composite "mixture"

The individual material component properties, therefore, contribute by a volume fraction ratio to the properties of the combined composite materials. For this rule of mixtures equation to apply, several basic assumptions and limitations are involved:

1. The fiber/polymer matrix composites as well as the polymer matrix are assumed to be linearly elastic and homogeneous.
2. There are no voids in the composite, and there is good adhesion between the reinforcing fibers and the polymer matrix.
3. The proximity of the fiber and polymer does not alter the properties of the individual components.
4. The rule of mixtures has some directional limitations as many FRPs are not isotropic materials.

For example, if we are dealing with a continuous fiber-reinforced polymer resin composite, the modulus and strength properties of the composite will be very different in the direction longitudinal to the fiber length compared to the properties across or perpendicular to the fibers. For strength and modulus, Equation 8.1 is most appropriate for composites being tested in the longitudinal (fiber) direction. The mechanical contribution of the fibers are directly in line with the direction of pull. The fibers are strong and stiff in this longitudinal direction, and the polymer matrix is relatively weak and much less rigid. Note that the strength and stiffness of materials in fiber form are always much higher than bulk materials (e.g., bar, rod, plate) because the fiber form of a material has a more atomically ordered internal structure. Fibers have an internal crystalline structure that favorably alters the stiffness and fracture behavior of this form of material. The presence of fibers makes composites stiffer and stronger in the longitudinal (fiber) direction than the polymer matrix by itself. The term *fiber-reinforced polymer* is thus appropriate. Property directionality effects are very important to consider in the use of fiber-reinforced composites in engineering designs.

Fiber Reinforcement

Generally, reinforcement in FRPs can be either fibers, whiskers, or particles. In composite materials of the most commercial interest, fibers are the most important and have the most influence on composite properties. Table 8.3 presents a comparison of the most common reinforcement fibers used in preparing organic polymer engineering composites. Nylon fiber is included here as a reference fiber. All the materials listed in Table 8.3 are textile fibers and can, for the most part, be processed into manufactured products in the same manner as textile fibers (e.g., continuous yarn, wound filaments, woven and knitted fabrics, nonwoven mats). The high-strength and high-stiffness properties of the glass (S-2), carbon, and polyaramid fibers are evident. These reinforcing fibers, when used in composite material fabrication, can take several forms, such as <0.1 inch (3–4 mm) fiber "whiskers," 0.1–0.3 inch (3–10 mm) chopped fibers, 0.1–2.0 inch (3–50 mm) (nonwoven) matted fiber sheets, woven fabric (continuous) fiber with plain weave, and unidirectional/longitudinal (continuous) fiber ribbons. These fiber reinforcement forms are illustrated in Figure 8.22. When using fiber reinforcement in polymer composites, the surface of the fibers or yarns is pre-

TABLE 8.3
Comparison properties of various fibers

	S-2 Glass	Carbon T-300	Aramid-49	Nylon 6/6
Tensile Strength, lb/in.^{2} (MPa)	665,700 (4,590)	390,100 (2,690)	400,000 (2,758)	143,000 (2,758)
Modulus of Elasticity, lb/in.^{2} (MPa)	13,000,000 (87,000)	33,000,000 (229,000)	18,000,000 (124,110)	800,000 (5,516)
Elongation, %	5.4	1.2	2.5	18.3
Density, lb/in.^{3} (g/cm^3)	0.090 (2.49)	0.062 (1.73)	0.052 (1.44)	0.041 (1.14)

FIGURE 8.22
Comparison of fiber reinforcement forms

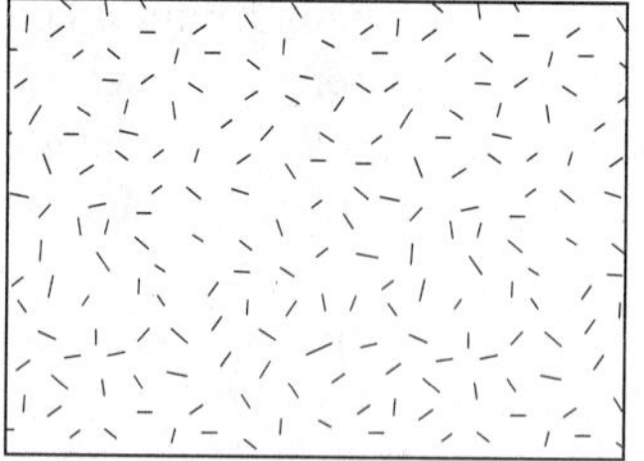

Whiskers
<3–4 mm

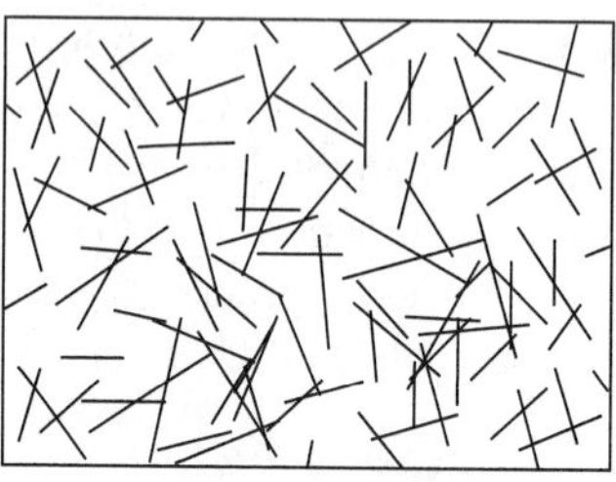

Chopped fibers
3–10 mm

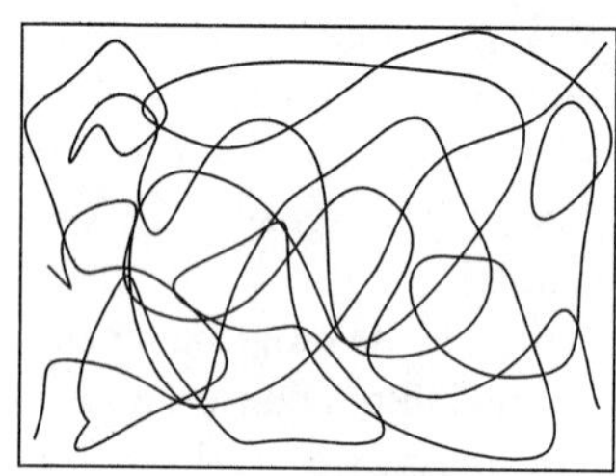

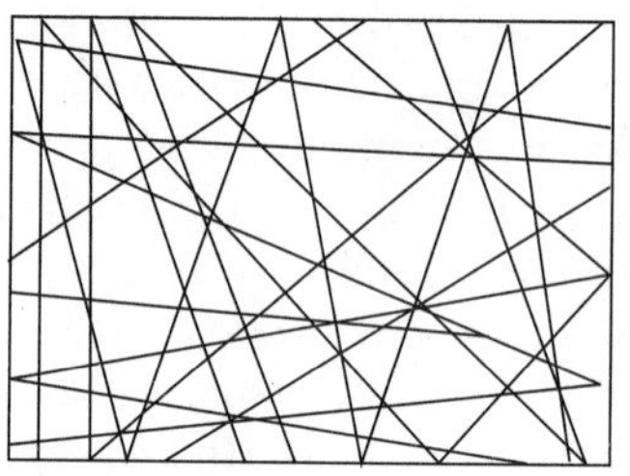

Fiber (non-woven) mat
>3–50 mm

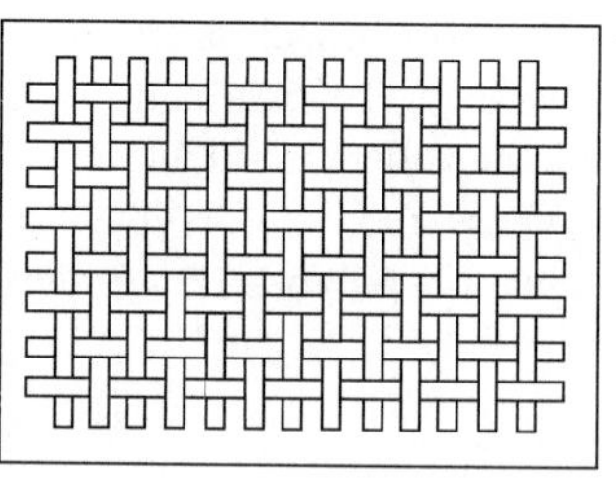

Woven fabric
continuous

Parallel aligned yarns
continuous

treated with a chemical coupling agent to enhance wetting and adhesion of the matrix resin to the fibers. Here, the chemical coupling agents are made specific to the chemical nature of the matrix resin being used. It is important that the fiber supplier be consulted for the proper type of fiber/resin coupling agent when fiber reinforcement materials are purchased.

As Table 8.3 shows, the most commonly used reinforcing fiber material is glass. In particular, S-2 glass is used in most high-performance applications. There exists an extensive applications, manufacturing, and processing history involving the use of glass fiber in polymer composite applications. Various forms of carbon fiber are also used for high-performance applications. The processing of carbon-fiber-reinforced polymer composites follows similar procedures as glass fibers. However, in the continuous-yarn processing of carbon fibers, precautions must be taken to protect electrical processing equipment from damage. Airborne, electrically conducting graphite dust is generated when the carbon fibers or yarns are processed through guide rings and rollers. This can occur before the fibers are wetted by the matrix resin during material fabrication. The dust can ruin electrical equipment if it is allowed to penetrate an instrument's enclosure. Sometimes, explosion proof electrical equipment is used when processing carbon fibers. Another approach is to fit the electrical instrument housing with a positive pressure differential of clean air (or nitrogen gas).

Matrix Resins

Classification of polymer matrices. Many types of polymers and resins can be reinforced by fibers to create FRP composite materials. Polymer matrices can be classified into two basic categories: thermoplastic and thermosetting.

1. **Thermoplastic.** Many of the polymers previously discussed can be reinforced with fibers to form composites. The most common types are chopped-fiber-reinforced thermoplastics. These materials can be processed in the same way as nonfiber-reinforced plastics. Generally, chopped fibers are blended and mixed into a molten mass of the engineering thermoplastic (e.g., nylon, polycarbonate, acetal) in a melt-extruder type of plastics-compounding machine. The fiber containing plastic is extruded into a thin rod and cut into molding powder or pellets. This thermoplastic molding powder is then used for injection molding or extrusion of engineered parts similar to the unreinforced plastics discussed in the preceding sections.

   Continuous fibers such as glass, carbon, or polyaramid have also been prepared with thermoplastic resin matrices. The concept here is to first coat thermoplastic resins onto continuous-fiber yarn by a hot melt or a polymer solution-solvent-dip process. These thermoplastic polymer-coated yarns can then be fabricated into shaped structures by a (hot press) matched-die compression molding technique or other techniques for affecting molten-polymer controlled consolidation. At this time, discontinuous chopped-fiber thermoplastic composites are much more widely used than continuous fiber-reinforced composites. The main advantages of thermoplastic matrix fiber composites is that they can be processed, for the most part, in conventional thermoplastic polymer fabrication equipment. Furthermore, any scrap or off-quality material can be recycled back into the injection molding or extruding machine. However, care must be taken during this thermoplastic processing not to overly "work" these materials in the molten state. Excessive processing in the molten state severely shortens the overall reinforcing fiber length, which can diminish the reinforcement effect of the fiber in the polymer matrix.

2. **Thermosetting.** Reinforced composites are traditionally associated with thermosetting polymers such as the unsaturated polyester and epoxy resins. In their cured state, thermosetting resins are composed of long polymer chains that are joined together through cross-bridges that link together all the molecules in the resin mass. The final, hardened, tough, and glassy state of the cured resin is the terminal condition of the polymer resin matrix. In this state, the resin serves the all-important role of structurally consolidating, supporting, and cohesively tying together the fiber reinforcement in the composite. However, during initial processing, it is important that thermosetting resins undergo a gradual liquid-to-solid conversion. It is this feature that renders thermosetting resins of the unsaturated polyester and epoxy type most readily adaptable to fiber-reinforced composite component fabrication.

Sequence of FRP fabrication with respect to the resin system involved. At first, the resin is in a liquid state as it is received from the supplier. It may be more or less fluid depending on its viscosity (from a flowable waterlike consistency to a high-viscosity syrup). At this stage, *rheological thickeners* to increase resin viscosity or *reactive diluents* to decrease resin viscosity may be added to the resin formulation. Frequently, the curative part of the resin system is much more fluid than the resin part. Here, the viscosity of the final mixed resin and curative system is low enough to accommodate proper flow in processing. Sometimes, the fluidity of the resin may be lowered by increasing the temperature of the resin upon its application to the fiber. In all, it is important that the viscosity of the liquid resin be adjusted so that it has the proper fluidity to wet, impregnate, and saturate the reinforcing fiber yarns, fabric, or mat.

The next consideration is the need to chemically catalyze the resin so that it properly cross-links and cures the resin under the prescribed conditions. It is also necessary to have the catalyzed resin react very slowly at ambient temperature so that it remains fluid while it is in the process of wetting the reinforcing fibers. This resin-system fluid time is referred to as the *pot life* or *open time* of the resin. This fluid-time feature is controlled by the nature of the catalyst, the ambient temperature, and the bulk volume of resin in the resin container. Note that a bulk of catalyzed resin is a resin undergoing a heat-generating exothermic reaction. If the bulk volume of the resin is too large, heat cannot be easily dissipated. The reaction in the fiber/resin dip tank will automatically accelerate, the resin will cure, or, worse, the heat of the reaction may cause a serious fire as well as noxious fumes. Most often, however, the processing equipment will contain dual-component pumps and a mixing head that will continuously meter and mix the proper components of the resin system (resin: part A; curative: part B) at the appropriate moment and position for wetting the fibers.

Once the resin part and the curative part of the resin system have been mixed, the liquid-to-solid cure reaction of the resin begins. The curing resin system will undergo several stages: liquid/fluid, gel stage, rubbery stage, and tough/glassy solid. Depending upon the processing temperature, the liquid-to-gel-to-rubber transition may occur from hours (for room temperature cures), to minutes, to seconds. The *gel point* in a thermosetting resin system is the point in the cure-time sequence when the resin undergoes a sharp rise in viscosity and ceases to be a fluid. Theoretically, the gel point is defined as the time in the cure when each polymer molecule in the system is tied together by at least one cross-link. Therefore, at the gel point, the polymer molecules in the resin system have combined and have reached an infinite molecular weight. After the gel point, the number of cross-links in the polymer system continues to increase, the cross-link network gets tighter and tighter, and the resin becomes a solid. It is at the gel stage that the influence of cross-linking takes hold. The rubbery stage is intermediate in cross-linking. In the solid glassy state of the resin, the ultimate number of cross-links in the resin system exists. Figure 8.23 illustrates the nature of the polymer resin and curative molecules during the curing sequence. Note that it is only in the solid-state stage that the fabricated composite part retains its shape and may be moved for additional processing or given a postcure if necessary. Let us now examine the specific resin chemistry of the unsaturated polyester and epoxy resin systems.

Chemistry of the unsaturated polyester resin system. Unsaturated organic polymers are polymer systems containing *double bonds,* or C=C. Double bonds can react with each other by an addition reaction that can be initiated by a free-radical catalyst. With the help of free-radical catalysts, unsaturated organic compounds can react with each other to form high-molecular-weight polymers:

$$n\left[\begin{array}{cc} H & H \\ | & | \\ C & = C \\ | & | \\ R & R' \end{array}\right] \xrightarrow[\text{bond opening}]{\text{catalyst}} \left[\begin{array}{c} H \quad H \\ | \quad\ | \\ -C-C- \\ | \quad\ | \\ R \quad R' \end{array}\right]_n$$

unsaturated monomer — polymer

In unsaturated polyester resins, the resin part of the mixture is represented by high-molecular-weight polymer molecules having unsaturated groups in their chain. These unsaturated polyesters are readily soluble in the unsaturated organic liquid compound called *styrene.* Styrene (known here as a monomer) can easily react with itself (using a free-radical catalyst initiator) to form a styrene polymer, or *polystyrene.* Because the monomeric styrene can readily react with unsaturated groups, when liquid styrene is mixed with unsaturated polyester resin, it serves the dual role of a reactive diluent and cross-linking agent. If a free-radical catalyst is added to a solution mixture of unsaturated polyester resin and styrene, the styrene simultaneously reacts with both the unsaturation in the backbone of the polyester chain and with itself. With free-radical catalysis, the polymerization reaction involving the growing polystyrene chains that react with the one polyester chain can also react with itself. When this reaction, in turn, connects with another polyester chain, a cross-link is formed between the two chains. In the molecular mixture mass of styrene, growing polystyrene chains, and unsaturated polyester molecules, an array of cross-links are formed between the multitude of polyester molecules (see Figure 8.24). As the polymer system reacts, from its initial mixing of the catalyst, the resin system will change from a liquid, to a gel-rubber when cross-linking starts to

FIGURE 8.23
Nature of molecules at various stages of thermosetting resin cure

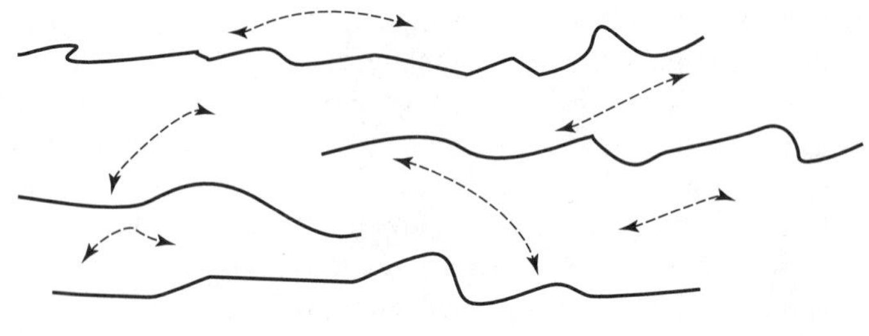

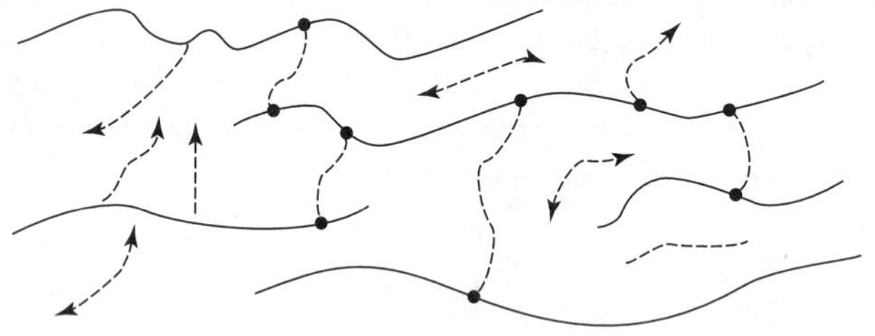

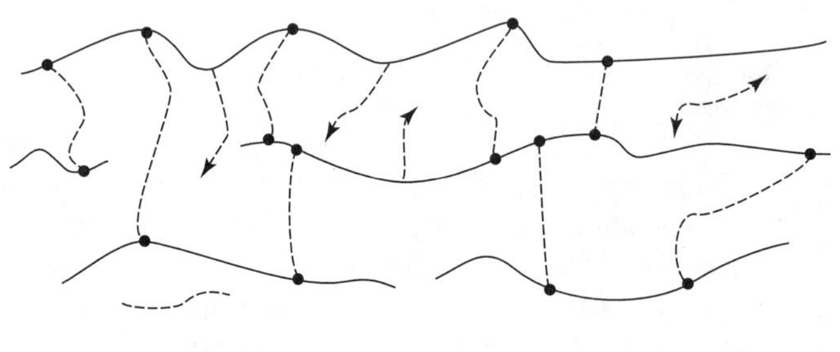

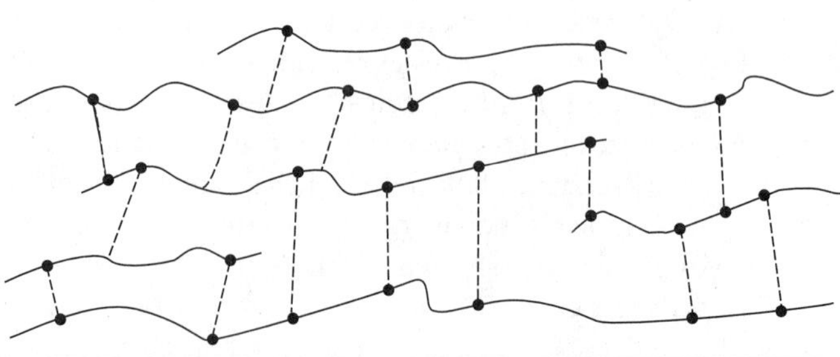

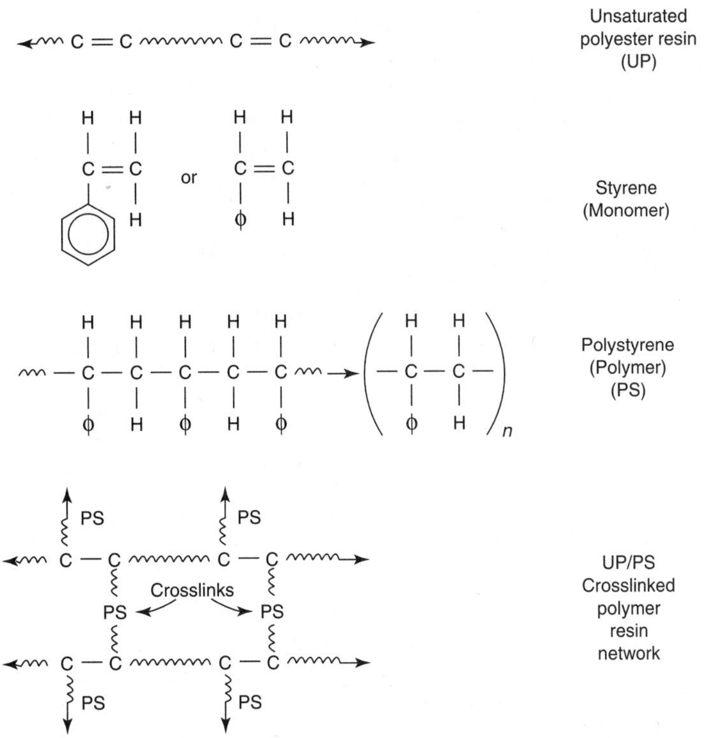

FIGURE 8.24 Chemical structure of unsaturated polyester/styrene resin

occur, and, finally, to a solid glassy vitreous state when numerous cross-links form and tie together the polyester molecules in the resin system. This, in essence, is the chemical mechanism that characterizes the cure of a typical polyester resin.

In the commercial formulation of unsaturated polyester/styrene thermosetting FRP resins, to make the resin more sag resistant when applied to vertical and more contoured surfaces, fumed silica is added to alter the rheology of the liquid resin. Another additive involves using a wax material that serves as a surface active agent that allows the resin to cure more evenly at its surface. Unsaturated polyester resin systems are by far the most widely used FRP matrix resins because of their low cost and availability. However, their use is being questioned because of environmental concerns. Styrene monomer is quite odiferous, and questions are being raised regarding its human toxicity after long-term process-operator exposure.

Chemistry of the epoxy resin system. Because of their inherent good adhesion to all types of surfaces, epoxy resins are generally more difficult to work with than polyesters. However, epoxies have much better thermal properties and exhibit very low shrinkage during cure. Their adhesive properties, while adding process difficulties, serve to enhance the structural integrity of the fiber/resin composite material system. Epoxies provide good adhesion of the resin matrix to the reinforcing fibers. The major hardeners for epoxy resins are amines and anhydrides. The chemistry of these hardener/curative systems is discussed next.

Epoxy resins are characterized by the reaction of the epoxy group $\overset{\;\;O}{C \!-\!-\!-\! C}$ known as the *oxirane ring*. Polymerization reactions proceed by the opening of this oxirane ring to form a difunctional chemical-reacting specie similar to the unsaturated C=C group in polyesters. Epoxy resins are low-molecular-weight polymers containing oxirane rings at each end of the chain. They are cured by adding a multifunctional chemical to the mixture that serves to cross-link the system by an addition reaction with the oxirane ring. The most common cross-linking agents for epoxies are the amines. Many of the amines used to cure epoxies are liquids, which makes the amines serve as reactive diluents. Such liquid material systems are also easily adaptable to dual-component pumps and the mixing of resin during dispensing for processing. The basic reaction between (primary) amine groups and the epoxy group is as follows:

$$RNH_2 + \overset{\quad O}{CH_2 - CH} \longrightarrow RNH - CH_2 - \overset{OH}{\overset{|}{CH}}$$

$$RNH - CH_2 - \overset{OH}{\overset{|}{CH}} + \overset{\quad O}{CH_2 - CH} \longrightarrow RN\,(CH_2 - CHOH)_2$$

As shown, each of the two hydrogen atoms of the primary amine, RNH_2, where R is a generic unspecified organic grouping, is capable of reacting with one epoxide group. In this chemical process, as with polyester resins, the epoxy polymer passes from a liquid to a gel-rubber to the solid glassy state as the cross-linking reaction proceeds. During the later stages of the reaction, the resultant OH groups that are formed in the amine reaction can also react with epoxy groups and further increase the cross-link density of the polymer.

One advantage of amine-cured epoxy resins is that they can harden or cure at room temperature. However, room temperature curing leads to polymers with low temperature stability. Also, the moisture resistance of these epoxy resins is generally low. Both temperature and moisture resistance can be improved by postcuring the resins above 212°F (100°C). Here, the chemical cross-links of the resin are maximized as complete reaction of all the epoxy groups is approached.

The reaction of anhydride curing agents with epoxy resins is more complex than that of amine cures. With anhydrides, amine catalysts are required along with cures at high temperature. During reaction, several competing reactions can take place. The most important reactions are as follows:

1. Opening of the anhydride ring with the OH groups from the catalytically reacted epoxy groups to form a carboxyl group:

$$\underset{\text{anhydride}}{\left(-\overset{|}{C}-\overset{O}{\overset{\|}{C}}-O-\overset{O}{\overset{\|}{C}}-\overset{|}{C}-\right)} + \underset{\text{epoxy resin moiety}}{HO-CH} \longrightarrow \underset{\text{reaction product}}{-\overset{|}{C}-\overset{O}{\overset{\|}{C}}-O-CH \quad -\overset{|}{C}-\underset{O}{\underset{\|}{C}}-OH}$$

2. Subsequent reaction of the carboxyl group with the epoxy group:

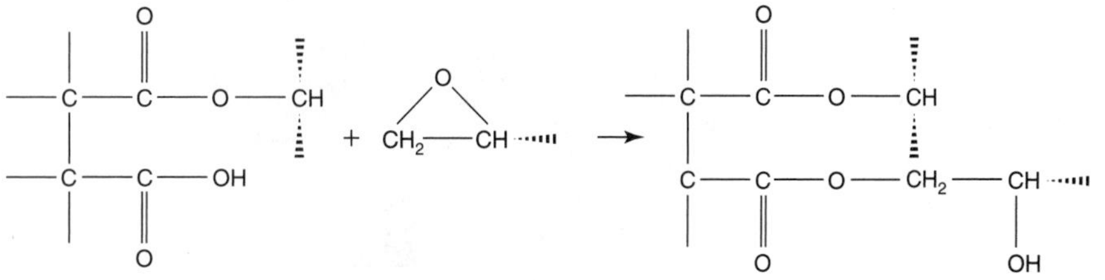

3. Epoxy groups, in turn, reacting with the formed OH groups:

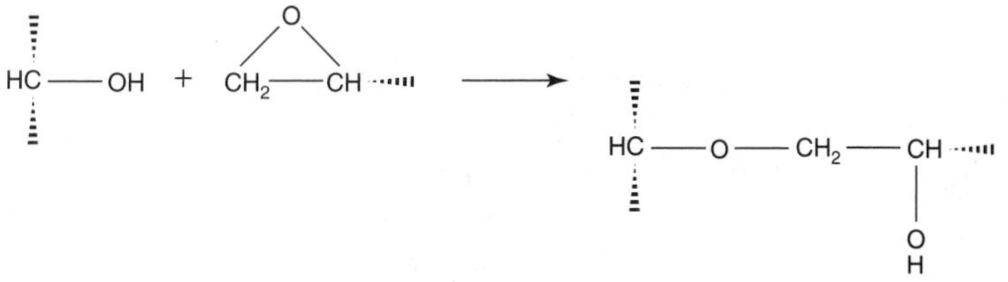

Although all three reactions can occur, which of the three reactions predominates depends on the reaction temperature.

Compared to amine cures, the pot life of anhydride cures is long, and the reaction produces a low exotherm. Long-time, elevated-temperature cures up to 392°F (200°C) are necessary if ultimate properties are desired. Overall, compared to amine-cured systems, anhydride cures result in much better chemical resistance for the final cured resin product.

From a processing standpoint, the environmental and industrial hygiene aspects of amine- or anhydride-cured epoxy resins are much better than the hygiene problems associated with unsaturated polyester resin processing. In all cases, proper protective clothing (coat, gloves, and goggles) must be worn while working with these resins. Amine and anhydride chemicals are generally quite corrosive to the skin and may cause dermatitis.

Forms of Composite Materials and Fabrication Techniques

Discontinuous fiber reinforcement. The reaction injection molding (RIM) process involves bringing together two components of a thermosetting polymeric resin system in a mixing head and injecting the reacting mixture into a closed mold before reaction is complete, as illustrated in Figure 8.25. The resin system then cures in the mold at a relatively low pressure of 50 psi (345 kPa). The timing of the curing reaction is very important because the reaction must occur at the moment the mold cavity is filled. Close process control is required. Because the process involves low-viscosity intermediates, complex parts can be fabricated using the RIM method.

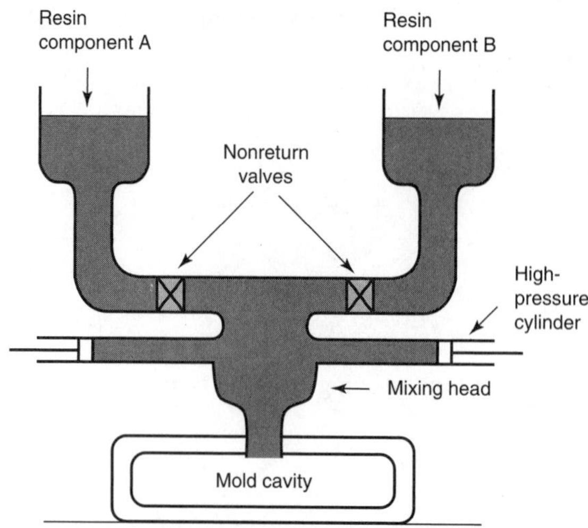

FIGURE 8.25 The reaction injection molding (RIM) process

Reinforcement (glass, fiber, or flake) can be added to one of the resin components prior to mixing if increased flexural modulus, thermal stability, and, in some instances, a special surface finish is desired in the final molded product. This process, reinforced reaction injection molding (RRIM), is shown in Figure 8.26.

Structural reaction injection molding (SRIM) and resin transfer molding (RTM) are similar to RRIM, except that the reinforcement is placed directly into the mold prior to the injection of the resin. In SRIM, the reinforcement is typically a *preform* of reinforcement fibers or mat of nonwoven fibers. In RTM, as shown in Figure 8.27, a catalyzed resin is pumped directly into the mold cavity containing the reinforcement.

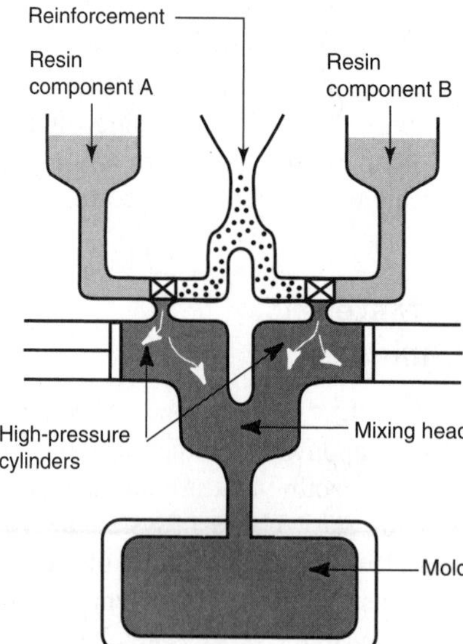

FIGURE 8.26 The reinforced reaction injection molding (RRIM) process

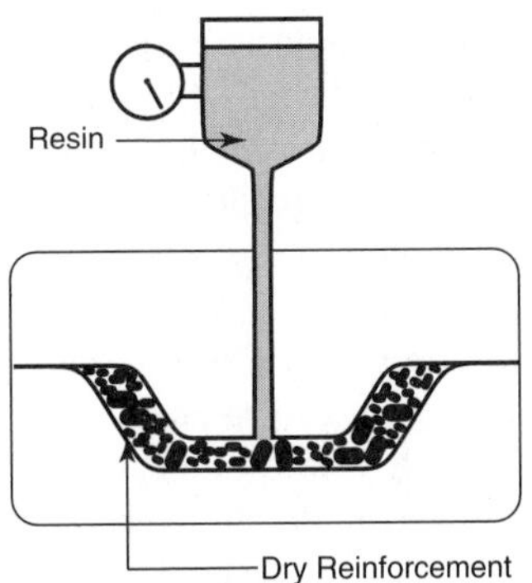

FIGURE 8.27 The resin transfer molding (RTM) process

The resin system is such that it cures without heat. The advantages of RTM are that, because no mixing head is involved, a relatively low investment is needed for equipment and tooling. Furthermore, large FRP parts and parts containing inserts and cores can be fabricated using the RTM process. RIM, RRIM, SRIM, and RTM processing are widely used in the automotive and aerospace industries.

Wet lay-up and vacuum bagging. Imbedding plies of glass, carbon, and/or polyaramid plain-weave fabric or fibrous mat into an uncured liquid resin and allowing the liquid resin to solidify (cure) while being constrained by a mold or form is a common processing technique used in the pleasure boat building industry. A typical arrangement of the plies used in this technique, called the *wet lay-up* process, is shown in Figure 8.28.

Related to this wet lay-up process is the *vacuum bagging* method of fabricating composite parts and shapes. The principle of vacuum bagging is quite simple. The shape to be fabricated is prepared by a room temperature wet lay-up procedure as just described. The part to be fabricated is usually assembled over a form or shape of the desired (complex and/or contoured) part. The assembly, like the lay-up arrangement shown in Figure 8.28, is then placed in an airtight disposable plastic "bag" fitted with a vacuum tube fitting or stem. If the air is sealed off and then evacuated from it, the bag will automatically close in on the wet laid-up plies of fiber and liquid (uncured) resin and consolidate these plies by

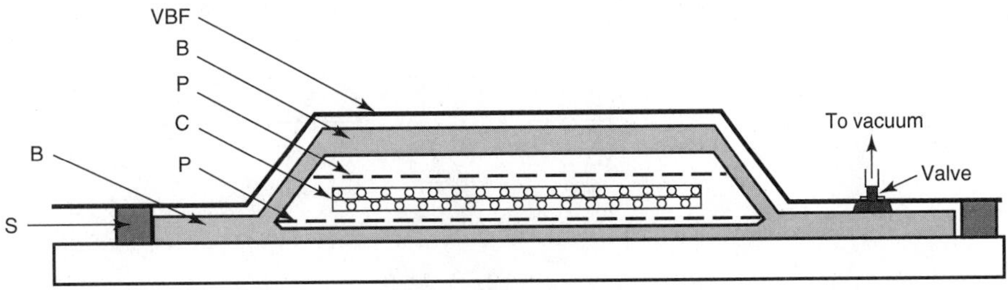

FIGURE 8.28 Arrangement of plies in wet lay-up assembly

the action of atmospheric pressure. This composite assembly is then allowed to solidify (cure) at room or elevated temperature. After this cure time, the vacuum bag, bleeder ply, and resin-absorber material are removed from the assembly and discarded, leaving the fabricated composite part ready for subsequent finishing or treatment.

A variation of the wet lay-up method is the *spray-up* process, where a spray gun simultaneously sprays catalyzed resin and chops continuous glass yarn into specific lengths. As shown in Figure 8.29, chopped fibers enter the spray nozzel of the spray gun, and the materials are comixed and sprayed onto an open-cavity mold. The mold usually is faced with a smooth coating of already cured resin called a *gel-coat* or a thermoplastic shell. This forms the outer surface of the structure being fabricated. When the sprayed-on fiber-reinforced resin cures, the part is removed from the mold. The laminar structure formed is composed of an aesthetically acceptable or otherwise finished outer skin. Adhered to and backing up this skin is the cured fiber-reinforced resin. Open-mold processing of this type is used extensively in bathtub and shower stall applications.

Unidirectional-fiber resin prepregs. Fiber-reinforced composite materials are commonly used in the form of a *prepreg.* Prepregs are typically side-by-side aligned fiber yarns that have been impregnated by a B-staged resin matrix (meaning that it has been deliberately partially cured). Unidirectional-fiber composite prepregs are commercially available in the form of rolls, tapes, and sheets. One drawback is that these prepregs must be kept frozen, below 32°F (0°C), for shipping and storage before use. They also have a relatively short shelf life. If not properly stored, the B-stage resins will cure slowly at room temperature, and their function will be destroyed.

Prepreg material is used to fabricate structures by plying together lay-ups of these resin-impregnated unidirectional fibers. The lay-ups can be designed to have different desired mechanical properties depending upon the geometrical arrangement or assembly of the reinforcing fibers in the cured lay-up. Some typical unidirectional-fiber ply arrangements are shown in Figure 8.30. Mechanically, these unidirectional (0°, 0°), cross-ply (0°, 90°), and quasi-isotropic (0°, +45°, 90°, –45°, 0°) plied laminates will

FIGURE 8.29
The spray-up process

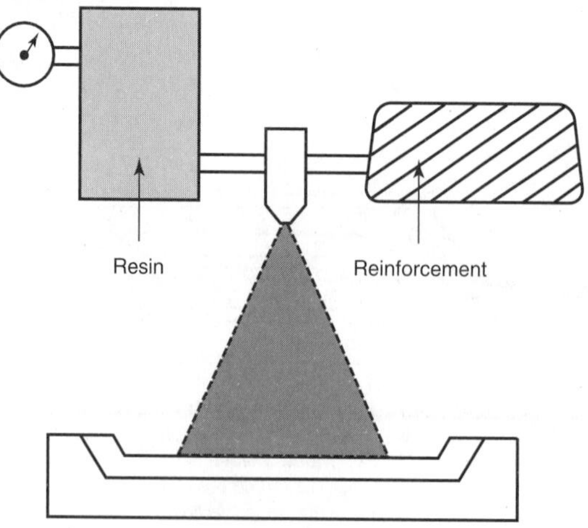

FIGURE 8.30 Various arrangements of unidirectional-fiber ply laminates

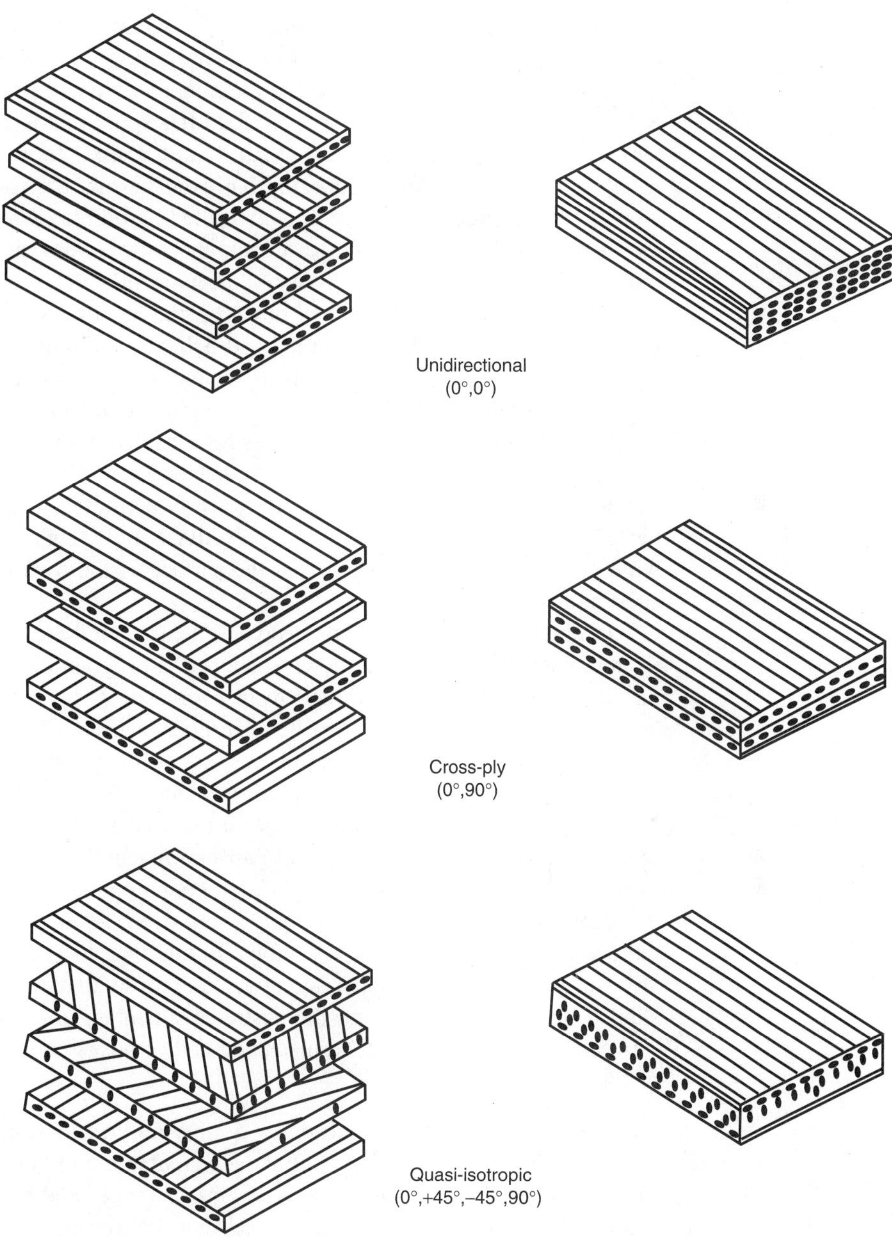

have planar anisotropic properties. Their flexural stiffness will always be higher in the longitudinal direction of the fibers. Other forms of B-stage resin-impregnated fiber forms are commercially available (e.g., fabrics and fibrous mats). The numerous B-stage precomposite forms and types of fiber are all available to the composite materials design engineer in the construction of a fiber-reinforced composite structure.

Filament winding. *Filament winding* is a fiber-reinforced composite processing procedure commonly used to fabricate tubular (hollow) and cylindrical tank or bottle-like structures. The apparatus used in the filament winding process is shown in Figure 8.31a. Basically, filamentary yarns are fed off a spool that is mounted on a creel. The yarn is immersed in a catalyzed, but still liquid, resin bath, where the yarn is impregnated with the resin. After squeezing out excess resin, the resin-impregnated yarn is wound onto a rotating mandrel in a controlled and directed manner. A computer system and control arm guide the yarn back and forth across the mandrel in a predetermined pattern. The computer controls the type of wind pattern and the number of layers of yarn filaments to be laid down on the mandrel surface. Two types of wind patterns are possible: circumferential and helical, (as Figure 8.31b shows). In the circumferential or hoop wind, the yarn is wound in a continuous manner in close proximity alongside itself. No crossover of the yarn occurs during the lay-down of a given layer, and the lay-down pattern can thus be considered to be at a zero wind angle. The wind proceeds back and forth across the mandrel until the desired number of layers is accomplished. In the helical wind, the yarn is permitted to cross over itself and traverses the length of the mandrel at a prescribed angle (e.g., 10°, 30°, 45°). Again, the wind proceeds back and forth across the surface of the rotating mandrel until the desired number of layers is formed.

In practice, combinations of hoop and helical wind are usually performed to fabricate a part. The desired lay-down sequence is programmed on the computer. While the desired (yarn) filament-wound resin composite is being formed on the mandrel, heating lamps can be focused on the resin/fiber mass to affect partial cure of the resin during this lay-down step. Once the desired winding pattern is completed, the mandrel with its wound fiber/resin composite outer surface is left rotating. Rotation and heat-lamp curing continue until the resin material is in a rigid enough state that the rotation can stop and the cylindrical part and mandrel can be removed from the filament winding machine. Postcuring of the wound composite and mandrel can then be accomplished by placing the assembly in an oven. After final curing, the mandrel is removed from the core of the assembly. To facilitate this, the mandrel form is generally made with a slight taper along its length so that the mandrel can easily be slipped out of an end, leaving the desired filamentary composite cylindrical "shell." The composite part can then be machined and/or post-treated to the desired condition or form.

Pultrusion processing. *Pultrusion* is a fiber-reinforced resin processing technique that is readily adaptable to the continuous manufacture of constant cross-sectional linear composite shapes. Rods, I beams, angles, channels, and hollow tubes and pipes are commonly produced by pultrusion processing. Pultrusion is a linear-oriented processing method whereby yarns of reinforcing fiber are continuously immersed in and impregnated with a catalyzed fluid resin. As the term *pultrusion* indicates, these resin-impregnated continuous-fiber yarns are concurrently *pulled* through an elongated heated die designed so that the fiber/resin composite mass exiting the die is sufficiently cured and retains the cross-sectional shape of the die. The apparatus used in the pultrusion process is shown in Figure 8.32a. In practice, prescribed lengths of the formed

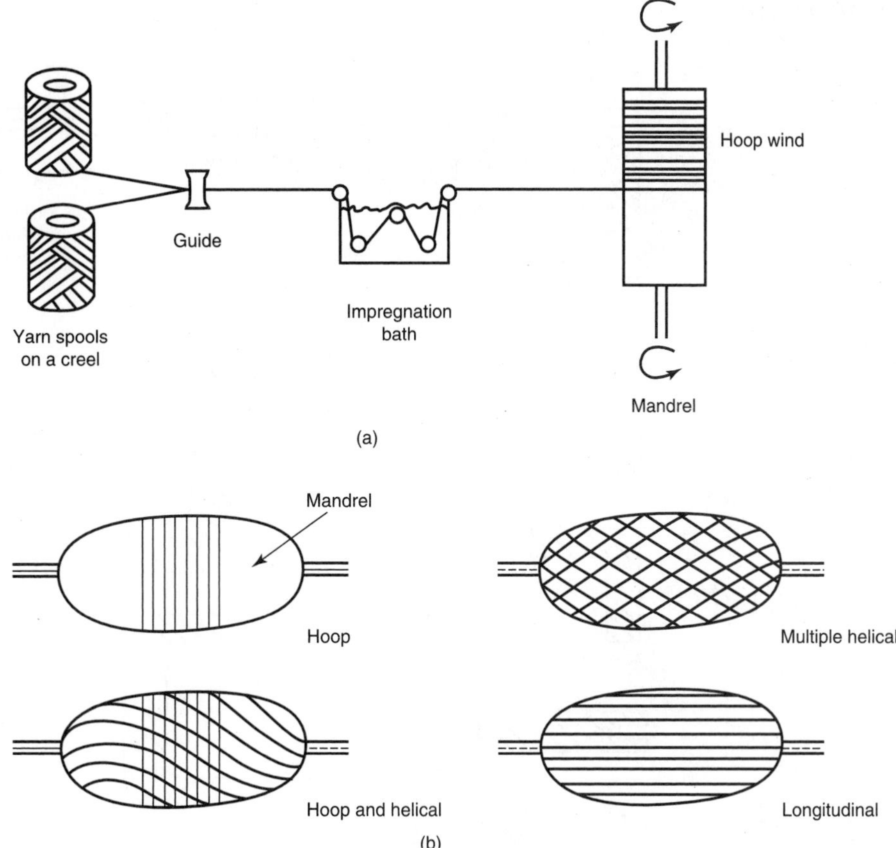

FIGURE 8.31
The filament winding process: (a) apparatus; (b) wind patterns

piece can be cut using an in-line cutoff wheel. Pultrusion is, therefore, adaptable to low-cost, continuous production of constant cross-sectional composite shapes. The process of pultrusion is critically controlled by the resin system used (e.g., unsaturated polyester, epoxy, and vinyl ester resins), the temperature and temperature profile of the heated die, and the rate of pulling through the die.

In the manufacture of pultruded shapes, such as those shown in Figure 8.32b, although the core cross section of the composite is linear oriented, there is often a need to wrap the outer surface of the composite with a webbing (nonwoven or woven tape) of fibrous material. This serves to consolidate the pultruded shape and gives a much more durable outer surface to the finished part. In this instance, thin veils of nonwoven or woven fabric tapes are fed into the entrance of the die along with the resin-impregnated continuous-fiber yarns. This assembled mass of fibers and resin proceeds to be pulled through the die as just described. The manufacture of hollow pultruded shapes is common, and a special die is then required. A shaped insert or "torpedo" is fitted at the die entrance and extends partway into it. The fluid resin-impregnated fibers entering the die are now constrained by this center-core obstruction. With the proper

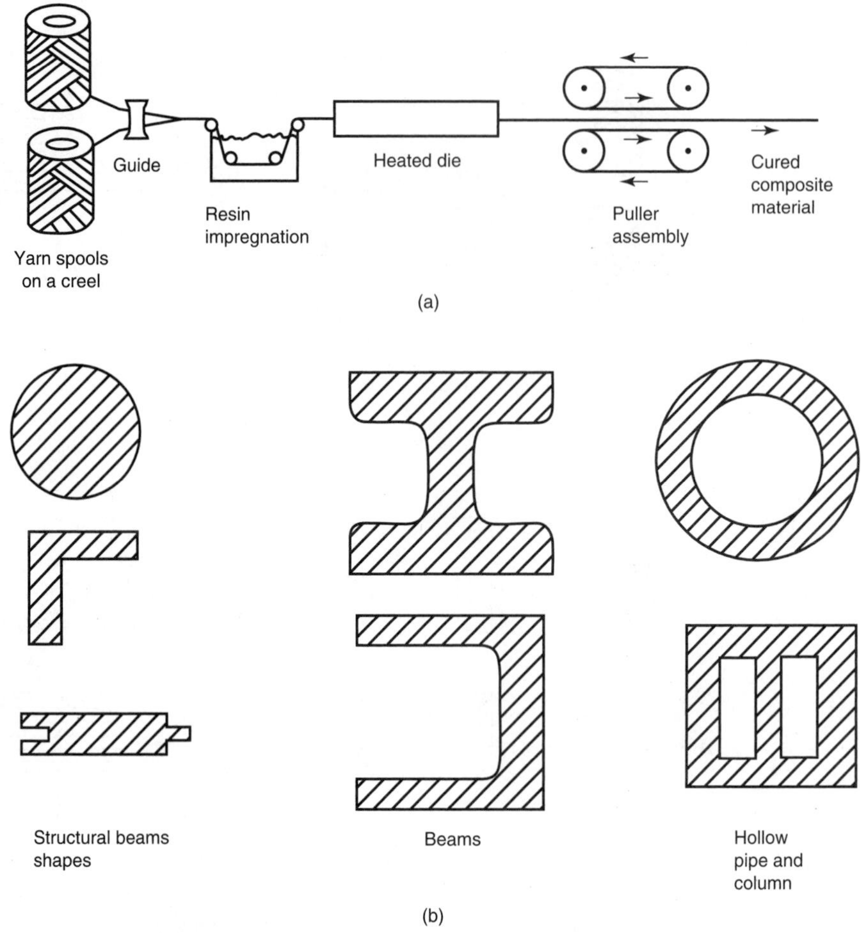

FIGURE 8.32 The pultrusion process: (a) apparatus; (b) cross-sectional designs

pulling speed, die temperature profile, and catalyzed resin formulation, the shape of the insert is retained as the desired hollow cross section of the part exits the die.

Engineering Design with Composite Materials

In the development of commercial products, there are many considerations. The particular field of organic polymer engineering composites is no exception. It is imperative for the engineer to have an integrated understanding of the design, materials behavior, processing, and service performance behavior of composite materials in order to develop a successful product. This integrated approach is diagrammed in Figure 8.33. Thus far, this section has reviewed some of the materials and processing aspects of fiber-reinforced organic polymer composites. The engineering design and final application aspects of composite materials are covered next.

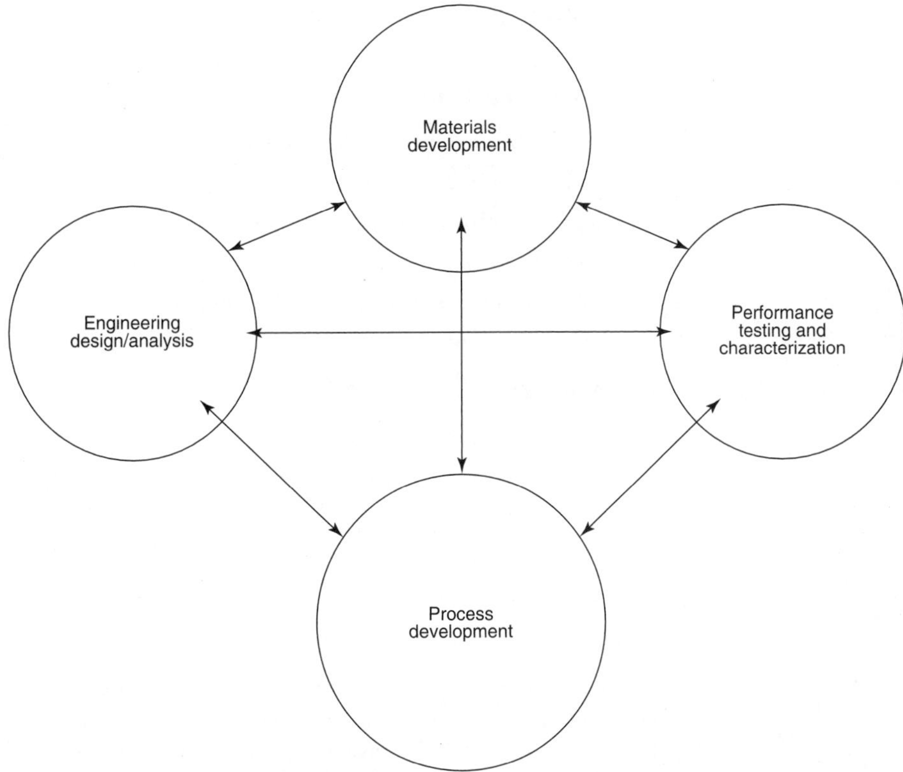

FIGURE 8.33
Model of technical base for engineered composite materials product development

First, however, in order to carry out an engineering design with organic polymer composites, the engineer must recognize and understand their advantages and limitations. Some of the advantages and disadvantages of carbon and polyaramid fiber-reinforced polymeric composites are as follows.

Advantages of carbon fibers.

1. High stiffness-to-weight and strength-to-weight ratios
2. High compressive strength
3. Excellent fatigue resistance
4. Good wear resistance (self-lubricating) and low friction coefficient
5. Mechanical vibration damping ability better than metals
6. Excellent creep resistance
7. Corrosion resistance (when not in contact with metals)
8. Some (directional) electrical and thermal conductivity
9. Very low (to slightly negative) directional thermal expansion coefficient
10. Very broad engineering design versatility
11. Broad processing versatility

12. Less energy required to manufacture engineering composite structures than to fabricate with metals

Advantages of polyaramid fibers.

1. High stiffness-to-weight and strength-to-weight ratios
2. Excellent fatigue resistance
3. Excellent corrosion resistance
4. Good vibration damping properties
5. Better impact resistance than carbon fiber composites
6. Electrically insulating

Disadvantages.

1. Limited service temperature
2. Moisture sensitivity/swelling/distortion
3. Anisotropic properties
4. Low compression strength (polyaramid fiber)
5. Bimetallic corrosion (carbon fiber)
6. Relatively high cost of advanced fibers

With these features and limitations in mind, the design engineer can proceed to create unique products. In the composites field, it is not appropriate to think only of using composite materials as a materials replacement for existing products. New products that take advantage of the unique properties of composite materials can also be conceived. Many of these new product concepts involve exploiting the remarkable specific strengths and specific elastic moduli of the "advanced" fiber-reinforced composites. The design engineer can choose from a multitude of reinforcing fiber types and fiber geometry arrangements, as well as from a variety of matrix materials. He or she has the freedom to mix in the design specification two or more diverse fiber types, as well as the freedom to directionally place the reinforcing fibers. All these degrees of freedom of choice are available so that the desired final component can be designed and fabricated. Fiber-reinforced organic polymer engineering composites are, therefore, capable of being used to create what can be referred to as *integral design engineering material structures* (IDEMS). Through computer-aided design (CAD) and finite-element stress analysis (FEM) techniques, new products are developed in computer-model form. In creating the actual fabricated product, the other facets of the integrated materials system manufacturing operation come into play (see Figure 8.33). Some specific areas in the design of organic polymer composites are discussed next.

Cutting, hole drilling, and machining. Although composite parts and structures are process molded to the near-finished state, machining, drilling, and trimming are often required as final steps. Therefore, the assembly and the finishing of the fabricated part are important in the creation of a final commercial product. There is always the possibility of damaging the composite material in these finishing post-treatments. Delamination, edge fraying, matrix cracking, or crazing leading to weak spots in the composite material structure are all possible. Great care must be taken to maintain the composite's structural integrity and appearance.

Post-treatment of fiber-reinforced composites involves different tooling and procedures compared to what is done for metal or plastics. The abrasiveness of the fiber and the possibility of fragmentation of the matrix resin are two factors to consider. Composites are machined, cut, and trimmed more easily using processes similar to grinding or abrasive cutting rather than conventional metal-cutting techniques. Also, the method used is dictated by the type of fiber reinforcement. Glass fiber, carbon fiber, and especially polyaramid fiber composites all require their own procedures. For example, the cutting of polyaramid-fiber-reinforced composites is difficult because the fiber is so tough and does not cleave or cut in a brittle, fracture mode. Polyaramid fibers undergo a process called *fibrillation* when "damaged" by the drilling, cutting, or machining tool. Fuzzy edge cuts or fiber-filled drill holes are produced when conventional machining and drilling tools are used. For polyaramid and for other fiber-reinforced composite materials for that matter, water-jet cutting, laser cutting, and diamond wire cutting are often used to achieve an acceptable edge profile to the final machine-finished parts. For carbon-fiber-reinforced composites, the thermal effects due to laser cutting, machining, and drilling can be a deterrent because the carbon fibers are thermally conductive. A weakened, charred, heat-damaged zone may surround the laser-cut edge. In summary, great care must be taken in the finishing post-treatments of fiber-reinforced composite materials.

Adhesive and mechanical joining. Adhesives are the principal means of joining composite materials to themselves and other materials of construction (metals, plastics, wood). The reasons for this are numerous. Most importantly, adhesive bonds are uniquely capable of distributing stress and can easily be joined into contoured shapes. In mechanical joining, hole drilling is required, which can lead to delamination of the composite and a stress concentration at the point of joining. The transfer of load from one material to another without creating large stress concentrations is the ultimate goal of materials joining. This can be achieved better by adhesive joining. Adhesives can often be incorporated into the structural laminar shape being fabricated as a one-step manufacturing process. Metal strips, layers, and/or fittings can easily be adhesively "molded" in the manufactured structure during the composite processing stage (e.g., wet lay-up, filament winding, RIM, RRIM, and so on). Adhesive joining techniques lend themselves to the creation of integrally designed structures as described previously. The various adhesive joint designs (lap shear, butt tensile, scarf joints, and so on) were discussed in Chapter 4.

Structural adhesives are available in various forms and types. Most common are the *two-package* epoxy resins. These formulated products are very similar to the epoxy matrix resins used to create the fiber-reinforced composite materials themselves. Usually, these two-package products consist of part A, the epoxy resin prepolymer, and part B, the curative (such as a primary amine or a polyamide/amine). Fillers, thickeners, reactive diluents, tackifiers, and other processing aids such as silicone compounds to improve the moisture durability of the adhesive are added to the final formulation.

These two-part adhesives are mixed just before being applied to the surfaces of the parts to be joined. The assembly is then placed in a compression mold, platen press, or vacuum bagging arrangement, where heat may be applied to consolidate the layers being joined and cure the adhesive. There are also some *one-package* paste adhesives that are formulated with a latent curative; the curative reacts only at high temperature.

Another useful form of adhesive is the *film* adhesive. Film adhesives are used extensively in the aerospace industry. Here, adhesives exist in the form of sheets. These sheets are malleable, are drapable, and can be cut using shears to the desired size and shape. These films are then placed between the surfaces to be joined and are cured under consolidation pressure and elevated temperature. Like the one-package adhesives, these adhesives are formulated with a high-temperature-reacting latent curative. Film adhesives, like the fiber-reinforced epoxy prepregs described earlier, must be stored at low temperature and kept frozen until ready to use.

Also used in bonding composite materials are the *acrylic* adhesives. Acrylic adhesives having different flexibilities are available. They cure at room temperature by a free-radical polymerization reaction. One feature of acrylic adhesives is that cure can be achieved by first coating the free-radical catalyst on the surfaces to be bonded. This "catalyst-primed" surface can then be stored until it is ready for bonding. An uncatalyzed acrylic adhesive is then coated onto the catalyst-primed surface. The surfaces to be joined are then mated under contact pressure and allowed to cure, undisturbed, at room temperature. Acrylic adhesives can produce bonds that are very oil resistant.

Finally, it is important that the surfaces to be joined be clean and free of oils, greases, and loose surface material layers. This is especially necessary when joining composite materials to metals. Vapor degreasing, followed by a chemically alkaline cleaning bath, is normally used for surface treating metals prior to adhesive bonding.

Sandwich-panel construction. Structural sandwich-panel construction consists of face sheets made up of fiber-reinforced laminar composite material (or metal sheet) adhesively bonded to both sides of a core material. This concept is illustrated in Figure 8.34. The principle behind sandwich construction is that the core material spaces the facings away from the symmetric center of the panel. Therefore, in flexure, the faces or outer skins of the panel are in tension or compression. This construction leads to the reinforcement in the faces, which resists the bending of the panel. The columnar strength of the honeycomb core material then provides the shear and compression strength of this unique panel structure. Above all, the adhesive must be strong and have a high enough shear and peel strength to withstand these shear stresses.

Sandwich construction leads to the use of panels that give the highest stiffness-to-weight ratio of any material design. Sandwich-panel construction is used extensively in aircraft and aerospace applications, where the core materials are generally honeycombed in geometric shape. Honeycomb cores can be made of thin metal (aluminum or titanium) or of fiber-reinforced resin sheet (e.g., thin sheet of resin-impregnated glass, carbon, or polyaramid mat). The manufacture of honeycomb core by the expansion process is shown in Figure 8.35. Manufacturing honeycomb core involves coating discrete strips of adhesive onto sheets of core material. The specially coated core material is then cured under compression to form a "log" or block of core material. The

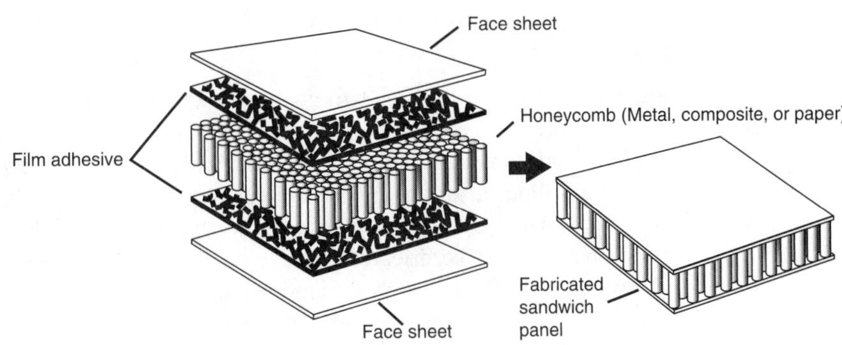

FIGURE 8.34 Structural sandwich panel construction *(Courtesy Strong, A. B.* Fundamentals of Composites Manufacturing: Materials, Methods, and Applications. *Dearborn, Michigan: Society of Manufacturing Engineers, 1989)*

log must then be cut to the desired core height and subsequently expanded to form the final core material. In some instances, the core material is dipped into a resin solution so that the core structure can be consolidated or stiffened. Another method of making honeycomb is the direct corrugation process. In some less demanding stiffness and compression applications, a rigid foam core material can be used. Rigid foam and

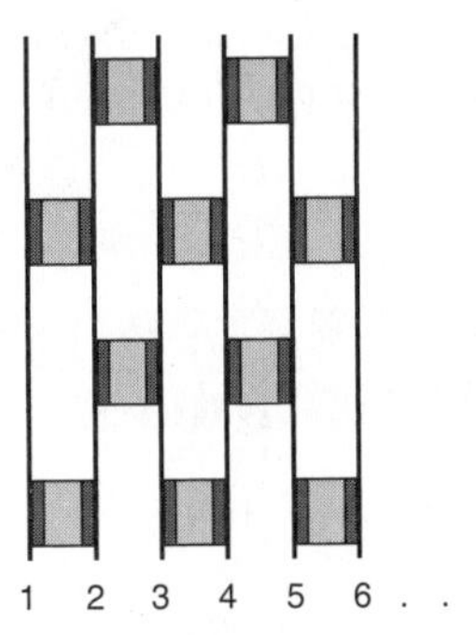

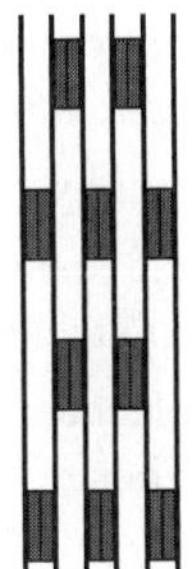

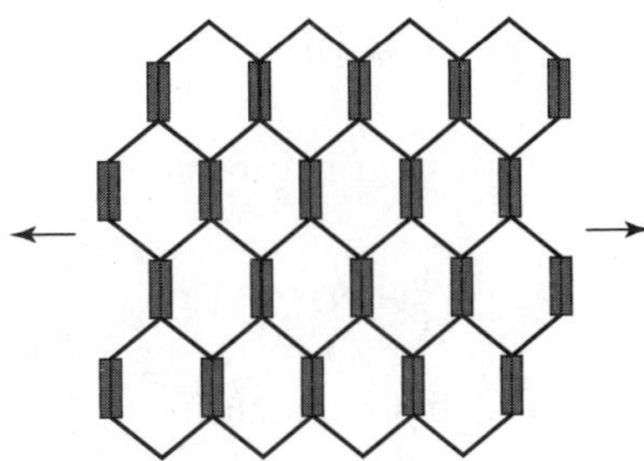

FIGURE 8.35 Manufacture of honeycomb core by the expansion process

Kraft-paper-based honeycomb core panels are often used in truck cargo bed panels and in door panels.

Painting and coating. Standard coating methods can be used for painting or coating fiber-reinforced composite structures. In all cases, the surface of the composite must be thoroughly prepared before the final coating is applied. Surface cleaning, sanding, abrading, filling in surface grooves/blemishes, and a solvent wipe must be carried out before the paint sealer and final paint finish are applied. Paint sealers and the final paint coating must be dried/cured at temperatures below the cure temperature of the composite part. Drying with infrared heaters can be troublesome as the heat location and temperature cannot be properly controlled using this technique. Epoxy and polyurethane-based surface coatings are especially useful in the painting of composite structures.

REFERENCES

Kaverman, R. D. "Reinforced Plastics and Composites." In Michael L. Berins, ed., *SPI Plastics Engineering Handbook.* New York: Van Nostrand Reinhold, 1991.

Mayer, Rayner M. *Design with Reinforced Plastics.* Design Council, K128 Haymarket, London SWIY 450. Bournemouth, England: Bourne Press Ltd., 1995.

Schwartz, Mel M., ed. *Composite Materials Handbook,* 2nd ed. New York: McGraw-Hill, 1992.

Strong, A. B. *Fundamentals of Composites Manufacturing: Materials, Methods, and Applications.* Dearborn, Michigan: Society of Manufacturing Engineers, 1989.

Review Questions

1. What are *plastics* and why are they called *polymers*?
2. What is a *monomer*?
3. Are all polymers artificial? Give examples.
4. Why is the strength of polymers lower than that of metals?
5. Why is the electrical conductivity of polymers lower than that of metals?
6. When did polymers start to gain widespread application and why?
7. How are polymers classified based on their temperature characteristics?
8. What is meant by *chemical families* of polymers? Give examples.
9. What are the main characteristics of a thermoplastic polymer?
10. Does a thermoplastic polymer have a fixed melting temperature? Why?
11. What is meant by *shaping memory*?
12. What are the main characteristics of a thermosetting polymer?
13. How do molecules of a thermosetting polymer differ from those of a thermoplastic polymer?

14. Compare the properties of plastics with those of metals. How do the differences affect the design of plastic products?
15. How can we have different polymers starting from the same monomer?
16. List four polymers that belong to the ethenic group. Discuss their properties and applications.
17. What are the main applications of polyacetals?
18. What is cellophane and how is it produced?
19. What is the major disadvantage of cellulose nitrate?
20. What are the major applications for cellulose acetate?
21. What is the chief limitation of nylons?
22. What are the major characteristics of phenolics?
23. How are polyimides manufactured?
24. List the common applications for epoxies.
25. Discuss the properties of polyurethanes and list some of their applications.
26. What property characterizes silicones? Suggest suitable applications to make use of that property.
27. Explain how natural rubber is processed.
28. Why are additives compounded with polymers?
29. List some fillers. Why are they added to polymers?
30. What happens when too much filler is added?
31. How does the addition of plasticizers affect the properties of a polymer?
32. List some of the lubricants used when processing polymers.
33. What are the mechanisms for coloring polymers?
34. Are all polymers cast in the same manner?
35. What are the design features of parts produced by blow molding?
36. Using sketches, explain the injection molding process.
37. What is the chief limitation of injection molding?
38. What kinds of polymers are usually processed by compression molding?
39. List some advantages of the compression molding process.
40. What is the main difference between compression molding and transfer molding?
41. Explain briefly the operating principles of rotational molding.
42. List examples of plastic products that are manufactured by extrusion.
43. What is the coextrusion process? Why is it used in industry?
44. What are the design features of parts produced by thermoforming? Give examples.
45. What are the products of the calendering process?
46. What is the major problem experienced when machining plastics?
47. Using sketches, explain the process of hot-plate joining.
48. Describe thermal staking.
49. Explain how ultrasonics are employed in welding and assembling plastic parts.
50. Do all plastics render themselves suitable for ultrasonic welding? Explain.
51. What are the basic components of ultrasonic welding equipment?
52. Using sketches, show some designs of ultrasonic-welded joints. List the characteristics of each.
53. Explain the sequence of operations involved in open-mold processing of reinforced polymers.
54. What are the similarities and differences between extrusion and pultrusion of polymers?
55. What are the design features of parts manufactured by filament winding?

56. Explain briefly the nature of FRP composites.

57. How can you predict the properties of a composite? Provide a quantitative equation.

58. List some of the fibers used as inforcement in FRP composites.

59. Briefly discuss the various matrix resins for FRP composites indicating their advantages, disadvantages, and limitations.

60. Why is vacuum bagging used in the modified version of the wet lay-up method?

61. What should we be careful about when using fiber resin prepregs?

62. What are the advantages of sandwich panels?

Design Projects

1. The current products of a company involve different fruit preserves in tin cans, each containing 8 ounces (about 250 g). The company uses 250,000 tin cans annually, and each costs 13 cents. Because their machines are almost obsolete and the cost of tin is rising every year, the company is considering replacing the tin cans with plastic containers. Design plastic containers to serve this goal, taking into account the plastic-processing method to be used. Also, make a feasibility study for the project.

2. Design a plastic cup that has a capacity of 8 ounces (about 250 g) of water. Assume the annual production volume is 20,000 pieces.

3. Design a high-quality plastic pitcher that has a capacity of 32 ounces (about 1 kg) of liquid. Assume the annual production volume is 15,000 pieces.

4. Design a wheel for a bicycle so that it can be produced by injection molding instead of sheet metal forming. The diameter is 24 inches (600 mm), and a load of 100 pounds (about 45 kg) is applied, through the axle, at its center. Assume the annual production volume is 100,000 wheels.

5. A trash container that has a capacity of 1 cubic foot (0.027 m^3) is made of sheet metal and can withstand an axial compressive load of 110 pounds (50 kg). Redesign it so that it can be made of plastic. Assume the annual production volume is 20,000 pieces.

Chapter 9

Physics of Metal Cutting

INTRODUCTION

Metal cutting can be defined as a process during which the shape and dimensions of a workpiece are changed by removing some of its material in the form of chips. The chips are separated from the workpiece by means of a cutting tool that possesses a very high hardness compared with that of the workpiece, as well as certain geometrical characteristics that depend upon the conditions of the cutting operation. Among all of the manufacturing methods, metal cutting, commonly called *machining*, is perhaps the most important. Forgings and castings are subjected to subsequent machining operations to acquire the precise dimensions and surface finish required. Also, products can sometimes be manufactured by machining stock materials like bars, plates, or structural sections.

Machining comprises a group of operations that involve seven basic chip-producing processes: shaping, turning, milling, drilling, sawing, broaching, and grinding. Although one or more of these metal-removal processes are performed at some stage in the manufacture of the vast majority of industrial products, the basis for all these processes (i.e., the mechanics of metal cutting) is yet not fully or perfectly understood. This is certainly not due to the lack of research but rather is caused by the extreme complexity of the problem. A wide variety of factors contribute to this complexity, including the large plastic strains and high strain rates involved, the heat generated and high rise in temperature during machining, and, finally, the effect of variations in tool geometry and tool material. It seems, therefore, realistic to try to simplify the cutting operation by eliminating

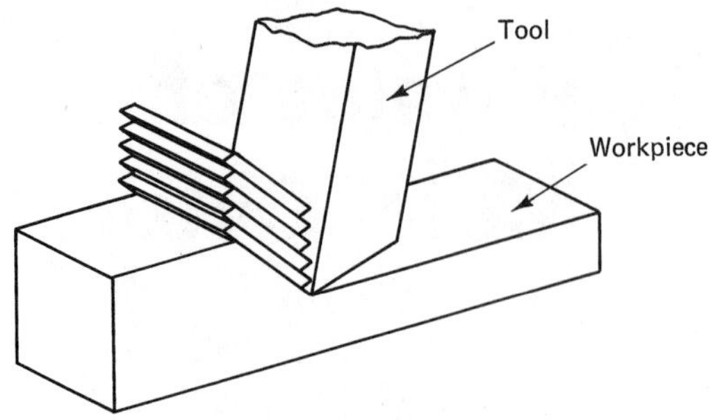

FIGURE 9.1 Two-dimensional cutting using a prismatic, wedge-shaped tool

as many of the independent variables as possible and making appropriately implicit assumptions if an insight into this complicated process is to be gained. In fact, we are going to take this approach in discussing the cutting tools and the mechanics of chip formation. We are going to consider two-dimensional cutting, in which a prismatic, wedge-shaped tool with a straight cutting edge is employed, as shown in Figure 9.1, and the direction of motion of the tool (relative to the workpiece) is perpendicular to its straight cutting edge. In reality, such conditions resemble the case of machining a plate or the edge of a thin tube and are referred to as *orthogonal cutting*.

9.1 CUTTING ANGLES

Figure 9.2 clearly illustrates that the lower surface of the tool, called the *flank,* makes an angle ψ with the newly machined surface of the workpiece. This *clearance angle* is essential for the elimination of friction between the flank and the newly machined surface. As can also be seen in Figure 9.2, there is an angle α between the upper surface, or *face,* of the tool along which chips flow and the plane perpendicular to the machined

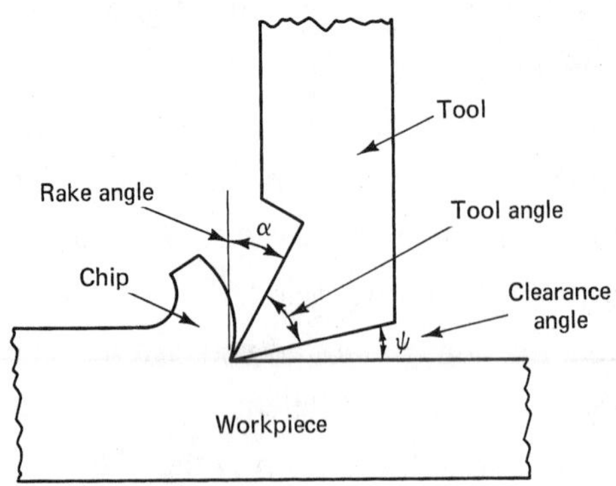

FIGURE 9.2 Tool angles in two-dimensional cutting

surface of the workpiece. It is easy to realize that the angle α indirectly specifies the slope of the tool face. This angle is known as the *rake angle* and is necessary for shoveling the chips formed during machining operations. The resistance to the flow of the removed chips depends mainly upon the value of the rake angle. As a consequence, the quality of the machined surface also depends on the value of the rake angle. In addition to these two angles, there is the *tool angle* (or wedge angle), which is the angle confined between the face and the flank of the tool. Note that the algebraic sum of the rake, tool, and clearance angles is always equal to 90°. Therefore, it is sufficient to define only two of these three angles. In metal-cutting practice, the rake and clearance angles are the ones that are defined.

As you may expect, the recommended values for the rake and clearance angles are dependent upon the nature of the metal-cutting operation and the material of the workpiece to be machined. The choice of proper values for these two angles results in the following gains:

1. Improved quality of the machined surface
2. A decrease in the energy consumed during the machining operation (most of which is converted into heat)
3. Longer tool life as a result of a decrease in the rate of tool wear because the elapsed heat is reduced to minimum

Let us now consider how the mechanical properties of the workpiece material affect the optimum value of the rake and clearance angles. Generally, soft, ductile metals require tools with larger positive rake angles to allow easy flow of the removed chips on the tool face, as shown in Figure 9.3. In addition, the higher the ductility of the workpiece material, the larger the tool clearance angle that is needed in order to reduce the part of the tool that will sink into the workpiece (i.e., reduce the area of contact between the tool flank and the machined workpiece surface). On the other hand, hard, brittle materials require tools with smaller or even negative rake angles in order to increase the section of the tool subjected to the loading, thus enabling the tool to withstand the high cutting forces that result. Figure 9.4 illustrates tools having zero and negative rake angles required when machining hard, brittle alloys. In this case, the clearance angle is usually taken as smaller than that recommended when machining soft, ductile materials.

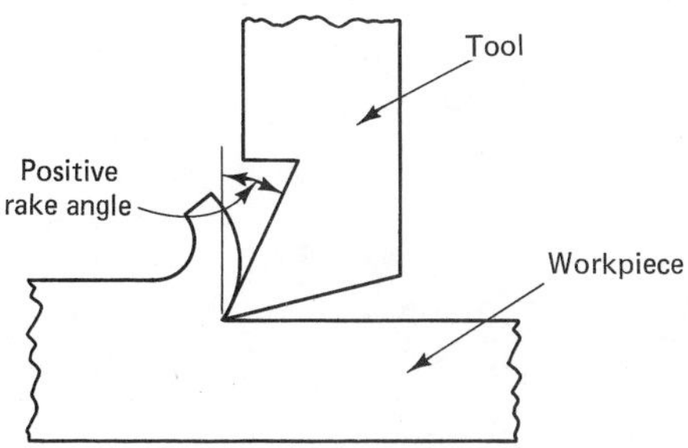

FIGURE 9.3
Positive rake angle required when machining soft, ductile metals

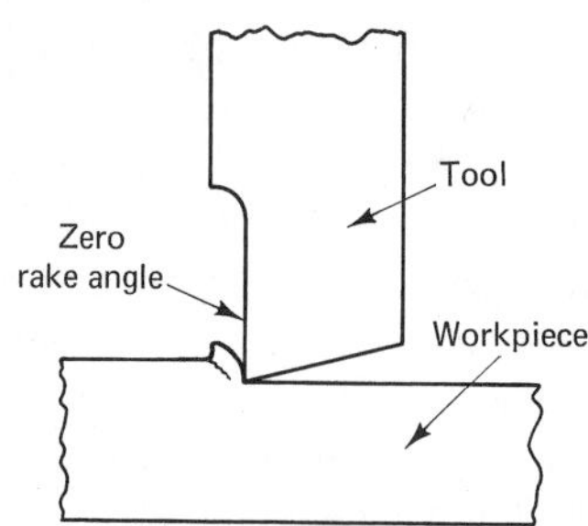

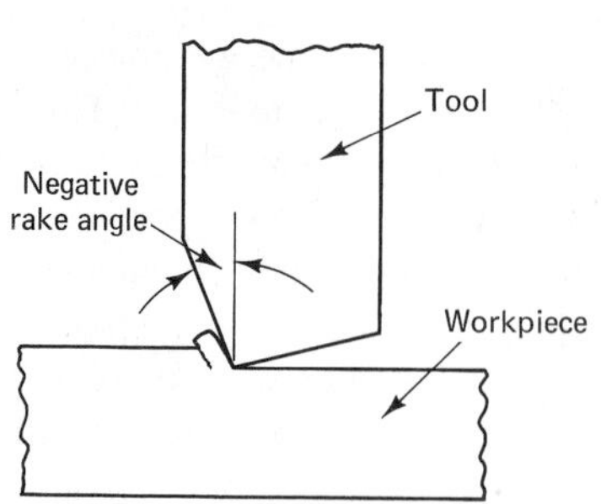

FIGURE 9.4
Zero and negative rake angles required when machining hard, brittle materials

9.2 CHIP FORMATION

Mechanics of Chip Formation

There was an early attempt by Reuleaux at the beginning of the twentieth century to explain the mechanics of chip formation. He established a theory that gained popularity for many years; it was based on assuming that a crack would be initiated ahead of the cutting edge and would propagate in a fashion similar to that of the splitting of wood fibers, as shown in Figure 9.5. Thanks to modern research that employed high-speed photography and quick stopping devices capable of freezing the cutting action, it was possible to gain a deeper insight into the process of chip formation. As a result, Reuleaux's theory collapsed and proved to be a misconception; it has been found that the operation of chip formation basically involves shearing of the workpiece material. Let us now see, step by step, how that operation takes place.

The stages involved in chip removal are shown in Figure 9.6. When the tool is set at a certain depth of cut (see Figure 9.6a) and is then pushed against the workpiece, the cutting edge of the tool and the face start to penetrate the workpiece material. The surface layer of the material is compressed; then pressure builds up and eventually exceeds the elastic limit of the material. As a result of the intense shear stress along the plane N–N, called the *shear plane,* plastic deformation takes place, and the material of the surface layer has no option but to flow along the face of the tool without being separated from the rest of the workpiece (see Figure 9.6b). With further pushing of the tool, the ultimate tensile strength is exceeded, and a little piece of material (a chip) is

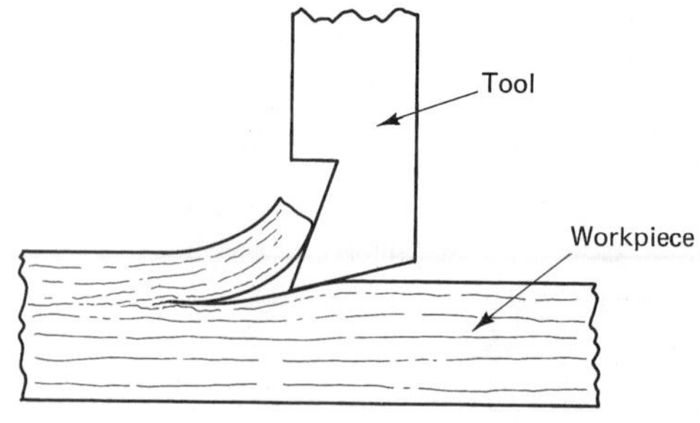

FIGURE 9.5
Reuleaux's misconception of the mechanics of chip removal

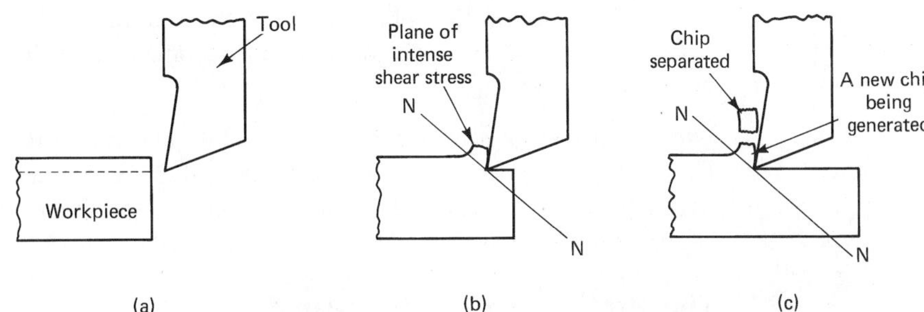

FIGURE 9.6 Stages in chip removal: (a) tool set at a certain depth of cut set; (b) workpiece penetration; (c) chip separation

separated from the workpiece by slipping along the shear plane (see Figure 9.6c). This sequence is repeated as long as the tool continues to be pushed against the workpiece, and the second, third, and subsequent chips are accordingly separated.

Types of Chips

The type of chip produced during metal cutting depends upon the following factors:

1. The mechanical properties (mainly ductility) of the material being machined
2. The geometry of the cutting tool
3. The cutting conditions used (e.g., cutting speed) and the cross-sectional area of the chip

Based on these factors, the generated chips may take one of the forms shown in Figure 9.7. Following is a discussion of each type of chip.

Continuous chip. When machining soft, ductile metals such as low-carbon steel, copper, and aluminum at the recommended cutting speeds (which are high), plastic flow predominates over shearing (i.e., plastic flow continues, and shearing of the chip never takes place). Consequently, the chip takes the form of a continuous, twisted ribbon (see Figure 9.7a). Because the energy consumed in plastically deforming the metal is eventually converted into heat, coolants and lubricants must be used to remove the generated heat and to reduce friction between the tool face and the hot, soft chip.

Discontinuous chips. When machining hard, brittle materials such as cast iron or bronze, brittle failure takes place along the shear plane before any tangible plastic flow occurs. Consequently, the chips take the form of discontinuous segments with irregular shape (see Figure 9.7b). As no plastic deformation is involved, there is no energy to be converted into heat. Also, the period of time during which a chip remains in con-

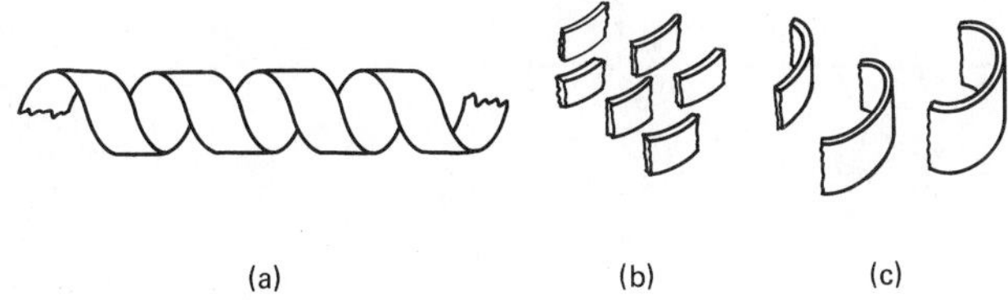

FIGURE 9.7 Types of machining chips: (a) continuous, twisted ribbon; (b) discontinuous, irregular segments; (c) sheared, short ribbons

tact with the face of the tool is short, and, therefore, the heat generated due to friction is very small. As a result, the tool does not become hot, and lubricants and coolants are not required.

Sheared chips. When machining semiductile materials with heavy cuts and at relatively low cutting speeds, the resulting sheared chips have a shape that is midway between the segmented and the continuous chips (see Figure 9.7c). They are usually short, twisted ribbons that break every now and then.

The Problem of the Built-Up Edge

When machining highly plastic, tough metals at high cutting speeds, the amount of heat generated as a result of plastic deformation and friction between the chip and the tool is large and results in the formation of a built-up edge, as shown in Figure 9.8. The combination of the resulting elevated temperature with the high pressure at the tool face causes localized welding of some of the chip material to the tool face (see Figure 9.8a). The welded material (chip segment) becomes an integral part of the cutting tool, thus changing the values of the cutting angles. This certainly increases friction, leading to the buildup of layer upon layer of chip material. This newly formed false cutting edge (see Figure 9.8b) is referred to as the *built-up edge.* The cutting forces also increase, the built-up edge breaks down, and the fractured edges adhere to the machined surface (see Figure 9.8c). The harmful effects of the built-up edge are increased tool wear and a very poorly machined workpiece. The manufacturing engineer must choose the proper cutting conditions to avoid the formation of a continuous chip with a built-up edge.

The Cutting Ratio

As can be seen in Figure 9.9, during a cutting operation, the workpiece material just ahead of the tool is subjected to compression, and, therefore, the chip thickness becomes greater than the depth of cut. The ratio of t_o/t is called the *cutting ratio* (r_c) and

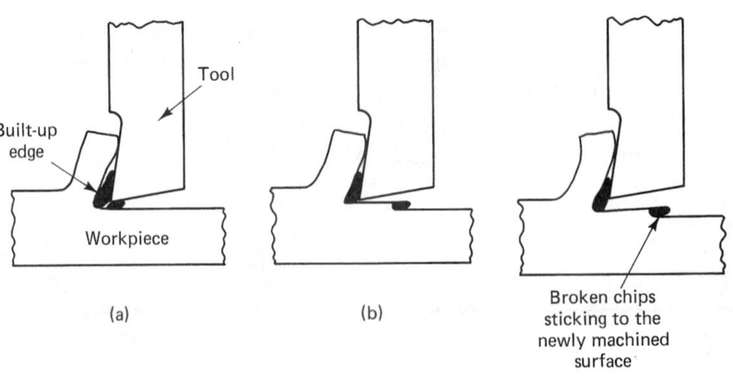

FIGURE 9.8
Stages in the formation of the built-up edge: (a) localized welding; (b) false cutting edge; (c) flawed surface

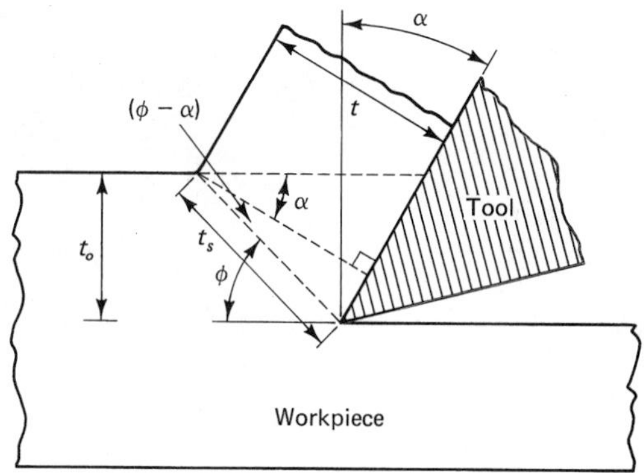

FIGURE 9.9
Geometry of a chip with respect to depth of cut

can be obtained as follows:

$$r_c = \frac{t_o}{t} = \frac{t_s \sin \phi}{t_s \cos (\phi - \alpha)} = \frac{\sin \phi}{\cos (\phi - \alpha)} \tag{9.1}$$

By employing trigonometry and carrying out simple mathematical manipulation, we can obtain the following equation:

$$\tan \phi = \frac{r_c \cos \alpha}{1 - r_c \sin \alpha} \tag{9.2}$$

Equation 9.2 is employed in obtaining the value of the *shear angle* ϕ when the rake angle α, the depth of cut, and the final thickness of the chips are known. In experimental work, the chip thickness is either measured directly with the help of a ball-ended micrometer or obtained from the weight of a known length of chip (of course, the density and the width of the chip must also be known).

Let us now study the relationship between velocities. Considering the constancy of mass and assuming the width of the chip to remain constant, it is easy to see that

$$V \times t_o = V_c \times t$$

or

$$\frac{V_c}{V} = \frac{t_o}{t} = r_c$$

In other words,

$$V_c = Vr_c = \frac{V \sin \phi}{\cos (\phi - \alpha)} \tag{9.3}$$

We can now draw the velocity triangle because we know the magnitudes and directions of two velocities, V and V_c. The shear velocity, V_s, which is the velocity with which the metal slides along the shear plane, can then be determined. Based on the velocity triangle shown in Figure 9.10 and applying the sine rule, the following can be stated:

$$\frac{V}{\sin (90 - \phi + \alpha)} = \frac{V_s}{\sin (90 - \alpha)} = \frac{V_c}{\sin \phi}$$

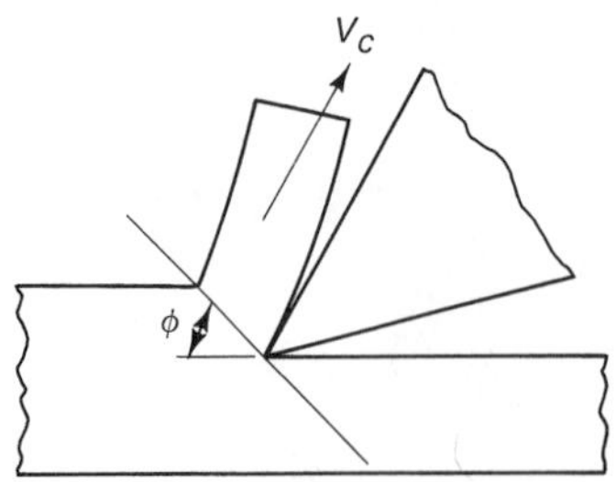

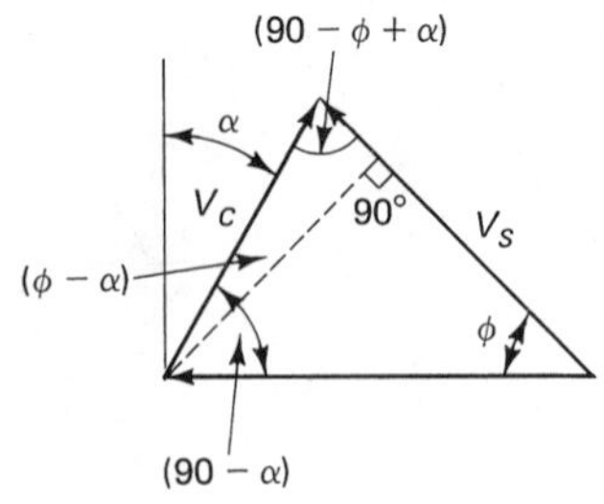

FIGURE 9.10 Velocity triangle and kinematics of the chip-removal process

This equation can also take the form

$$\frac{V}{\cos(\phi - \alpha)} = \frac{V_s}{\cos\alpha} = \frac{V_c}{\sin\phi}$$

Therefore,

$$V_s = V\frac{\cos\alpha}{\cos(\phi - \alpha)} \tag{9.4}$$

Shear Strain During Chip Formation

The value of the shear strain is an indication of the amount of deformation that the metal undergoes during the process of chip formation. As can be seen in Figure 9.11, the parallelogram *abda′* will take the shape *abcd′* due to shearing. The shear strain can be expressed as follows:

$$\gamma = \frac{a'n}{an} + \frac{d'n}{an} = \cot\phi + \tan(\phi - \alpha) \tag{9.5}$$

The shear strain rate can be obtained from Equation 9.5 as follows:

$$\dot{\gamma} = \frac{a'n}{an} \times \frac{1}{\Delta t} + \frac{d'n}{an} \times \frac{1}{\Delta t} = \frac{a'd'}{an} \times \frac{1}{\Delta t}$$

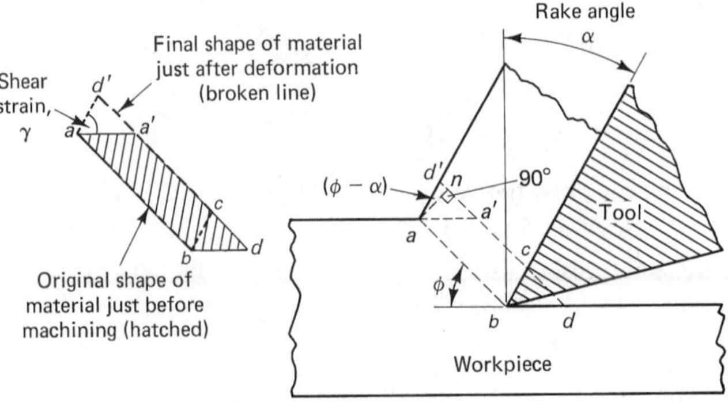

FIGURE 9.11 Shear strain during chip formation

But,

$$\frac{a'd'}{\Delta t} = V_s$$

Therefore,

$$\dot{\gamma} = \frac{V_s}{an}$$

where *an* is the thickness of the shear zone. Experimental results have indicated that the thickness of the shear zone is very small. Consequently, it can easily be concluded that the process of chip formation takes place at an extremely high strain rate. This finding is very important, especially for strain-rate-sensitive materials, where the strength and ductility of the material are markedly affected.

9.3 CUTTING FORCES

Theory of Ernst and Merchant

In order to simplify the problem, let us consider the two-dimensional, idealized cutting model of continuous chip formation. In this case, all the forces lie in the same plane and, therefore, form a coplanar system of forces. Walter Ernst and Eugene M. Merchant, both eminent American manufacturing scientists, based their analysis of this system of forces on the assumption that a chip acts as a rigid body in equilibrium under the forces acting across the chip-tool interface and the shear plane. As Figure 9.12 shows, the cutting edge exerts a certain force upon the workpiece. The magnitude of that force is dependent upon many factors, such as the workpiece material, the conditions of cutting, and the values of the cutting angles.

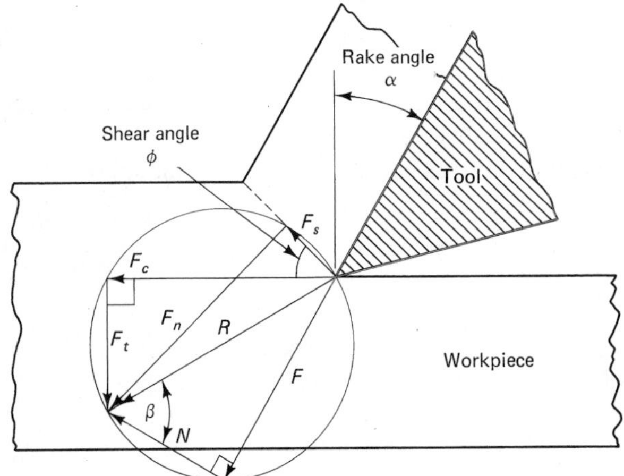

FIGURE 9.12
Cutting force diagram according to Ernst and Merchant

By employing simple mechanics, the force can be resolved into two perpendicular components, F_c and F_t. As can be seen in Figure 9.12, F_c acts in the direction of tool travel and is referred to as the *cutting component,* whereas F_t acts normal to that direction and is known as the *thrust component.* The resultant tool force can alternatively be resolved into another two perpendicular components, F_s and F_n. The first component, F_s, acts along the shear plane and is referred to as the *shearing force;* the second component, F_n, acts normal to it and causes compressive stress to act on the shear plane. Again, at the chip-tool interface, the components of the resultant force that acts on the chip are F and N. Notice from the figure that F represents the friction force that resists the movement of the chip as it slides over the face of the tool, while N is the normal force. The ratio between F and N is actually the coefficient of friction at the chip-tool interface. Because each two components are perpendicular, it is clear from Euclidean geometry that the point of intersection of each two components must lie on the circumference of the circle that has the resultant force as a diameter. The cutting force diagram of Figure 9.12 lets us express F_s, F_n, F, and N in terms F_c and F_t as follows:

$$F_s = F_c \cos\phi - F_t \sin\phi \tag{9.6}$$

$$F_n = F_c \sin\phi + F_t \cos\phi \tag{9.7}$$

$$F = F_c \sin\alpha + F_t \cos\alpha \tag{9.8}$$

$$N = F_c \cos\alpha - F_t \sin\alpha \tag{9.9}$$

The preceding equations can be used to determine different unknown parameters that affect the cutting operations. For instance, the coefficient of friction at the chip-tool interface can be obtained as follows:

$$\mu = \frac{F}{N} = \frac{F_c \sin\alpha + F_t \cos\alpha}{F_c \cos\alpha - F_t \sin\alpha} = \tan^{-1}\beta$$

Dividing both the numerator and denominator by $\cos\alpha$, we obtain

$$\mu = \frac{F_t + F_c \tan\alpha}{F_c - F_t \tan\alpha} \tag{9.10}$$

The shear force F_s is of particular importance as it is used for obtaining the magnitude of the mean shear strength of the material along the shear plane and during the cutting operation. This is equal to the mean shear stress acting through the shear plane and can be computed as follows:

$$\tau_s = \frac{F_s}{A_s}$$

where A_s, the area of the shear plane, equals $A_{\text{chip}}/\sin\phi$, where A_{chip} is the cross-sectional area of the chip. Therefore,

$$\tau_s = \frac{[F_c \cos\phi - F_t \sin\phi]\sin\phi}{A_{\text{chip}}} \tag{9.11}$$

Experimental work has indicated that the mean shear stress, calculated from Equation 9.11, is constant for a given metal over a wide variation in the cutting conditions. This can be explained by the fact that the strain rate at which metal cutting occurs is sufficiently high to be the only factor that affects the shear strength for a given material. Therefore, the cutting speed, amount of strain, or temperature do not have any appreciable effect on the value of the mean shear stress of the metal being machined.

Ernst and Merchant extended their analysis and studied the relationship between the shear angle and the cutting conditions. They suggested that the shear angle always takes the value that reduces the total energy consumed in cutting to a minimum. Because the total work done in cutting is dependent upon and is a direct function of the component F_c of the cutting force, they developed an expression for F_c in terms of ϕ and the constant properties of the workpiece material. Next, that expression was differentiated with respect to ϕ and then equated to zero in order to obtain the value ϕ for which F_c and, therefore, the energy consumed in cutting is a minimum. Following is the mathematical treatment of this problem.

From Figure 9.12, we can see that

$$F_s = R\cos(\phi + \beta - \alpha) \tag{9.12}$$

Therefore,

$$R = \frac{F_s}{\cos(\phi + \beta - \alpha)}$$

But,

$$F_s = \tau_s A_s = \tau_s \frac{A_{\text{chip}}}{\sin\phi}$$

Therefore,

$$R = \frac{\tau_s A_{\text{chip}}}{\sin\phi} \times \frac{1}{\cos(\phi + \beta - \alpha)} \tag{9.13}$$

Again, it can be seen from Figure 9.12 that

$$F_c = R\cos(\beta - \alpha) \tag{9.14}$$

Hence, from Equations 9.13 and 9.14,

$$F_c = \frac{\tau_s A_{\text{chip}}}{\sin\phi} \times \frac{\cos(\beta - \alpha)}{\cos(\phi + \beta - \alpha)} \tag{9.15}$$

Differentiating Equation 9.15 with respect to ϕ and equating the outcome to zero, we obtain the condition that will make F_c minimal. This condition is given by the following equation:

$$2\phi + \beta - \alpha = \frac{\pi}{2} \tag{9.16}$$

It was found that the theoretical value of ϕ obtained from Equation 9.16 agreed well with the experimental results when cutting polymers, but this was not the case when machining aluminum, copper, or steels.

Theory of Lee and Shaffer

The theory of American manufacturing scientists E. Lee and Bernard W. Shaffer is based on applying the slip-line field theory to the two-dimensional metal-cutting problem. A further assumption is that the material behaves in a rigid, perfectly plastic manner and obeys the von Mises yield criterion and its associated flow rule. After constructing the slip-line field for that problem, it was not difficult for Lee and Shaffer to obtain the relationship between the cutting parameters and the shear angle. The result can be given by the following equation:

$$\phi + \beta - \alpha = \frac{\pi}{4} \tag{9.17}$$

In fact, neither of the preceding theories quantitatively agrees with experimental results. However, the theories yield linear relationships between ϕ and $(\beta - \alpha)$, which is qualitatively in agreement with the experimental results.

Cutting Energy

We can see from the previous discussion that it is the component F_c that determines the energy consumed during machining because it acts along the direction of relative tool travel. The power consumption P_m (i.e., the rate of energy consumption during machining) can be obtained from the following equation:

$$P_m = F_c \times V \tag{9.18}$$

where V is the cutting speed.

The rate of metal removal during machining Z_m is also proportional to the cutting speed and can be given by

$$Z_m = A_o \times V \tag{9.19}$$

where A_o, the cross-sectional area of the uncut chip, equals t_o times the width of the chip. Now, the energy consumed in removing a unit volume of metal can be obtained from Equations 9.18 and 9.19 as follows:

$$P_c = \frac{P_m}{Z_m} = \frac{F_c \times V}{A_o \times V} = \frac{F_c}{A_o} \tag{9.20}$$

In Equation 9.20, P_c, a parameter that indicates the efficiency of the process, is commonly known as the *specific cutting energy* and also sometimes is called the *unit horsepower*. Unfortunately, the specific cutting energy for a given metal is not constant but rather varies considerably with the cutting conditions, as we will see later.

9.4 OBLIQUE VERSUS ORTHOGONAL CUTTING

Until now, we have simplified the metal-cutting process by considering only *orthogonal* cutting. In this type of cutting, the cutting edge of the tool is normal to the direction of relative tool movement, as shown in Figure 9.13a. It is actually a two-dimensional process in which each longitudinal section (i.e., parallel to the tool travel) of the tool and chip is identical to any other longitudinal section of the tool and chip. The cutting force is, therefore, also two-dimensional and can be resolved into two components, both lying within the plane of the drawing. Although this approach facilitated the analysis of chip formation and the mechanics of metal cutting, it is seldom used in practice because it applies only when turning the end face of a thin tube in a direction parallel to its axis.

The more common type (or model) of cutting used in the various machining operations is *oblique cutting*. In this case, the cutting edge of the tool is inclined to (i.e., not normal to) the relative tool travel, as can be seen in Figure 9.13b. It is a three-dimensional problem in which the cutting force can be resolved into three perpendicular components, as indicated in Figure 9.14. The magnitudes of these components can be measured by means of a special apparatus that is mounted either in the workholder or toolholder and is known as a *dynamometer*. As you may expect, the tool geometry

FIGURE 9.13 Types of cutting: (a) orthogonal; (b) oblique

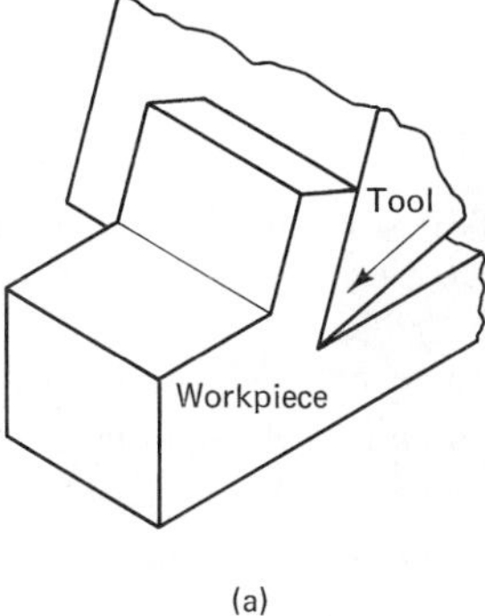

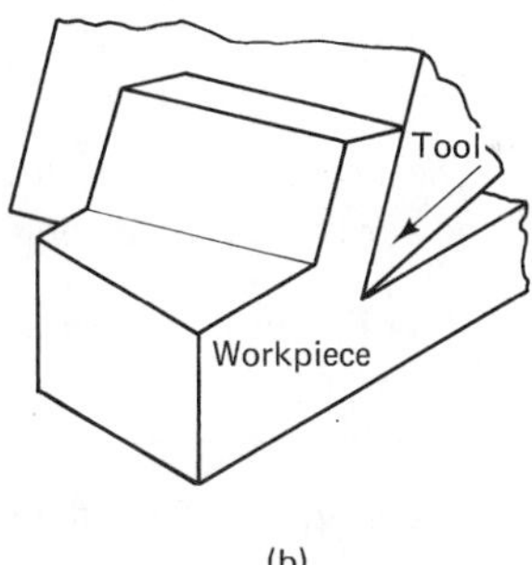

FIGURE 9.14 Components of the three-dimensional cutting force

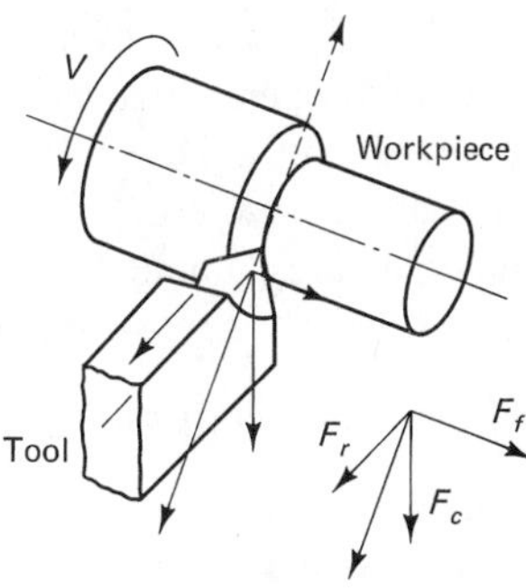

is rather complicated and will be discussed later. For now, let us see the effect of each of the cutting force components on the oblique cutting operations.

Forces in Oblique Cutting

Following is a discussion of the three components referred to as F_c, F_f, and F_r in Figure 9.14:

1. F_c is the *cutting force* and acts in the direction where the cutting action takes place. It is the highest of the three components and results in 99 percent of the energy consumed during the process. The horsepower due to this force, hp_c, can be given by the following equation:

$$hp_c = \frac{F_c \times V_c}{550 \times 60} \quad \textbf{(9.21)}$$

In Equation 9.21, F_c is in pounds and V_c is in feet per minute. Consequently, the appropriate conversion factors must be used if the horsepower is to be obtained in SI units.

2. F_f is the *feed force* (or longitudinal force in turning). The term *feed* means the movement of the tool to regenerate the cutting path in order to obtain the machined surface. This force amounts to only about 40 percent of the cutting force. The horsepower required to feed the tool, hp_f, can be given as follows:

$$hp_f = \frac{F_f \times V_f}{550 \times 60} \quad \textbf{(9.22)}$$

The horsepower given by Equation 9.22 amounts to only 1 percent of the total power consumed during cutting.

3. F_r is the *thrust force* (or radial force in turning) and acts in the direction of the depth of cutting. This force is the smallest of the three components and amounts to only 20 percent of the cutting force or, in other words, 50 percent of the feed force. This component does not result in any power consumption as there is no tool movement along the direction of the depth of cut.

These components of the cutting force are measured only in scientific metal-cutting research. The manufacturing engineer is, however, interested in determining beforehand the motor horsepower required to perform a certain job in order to be able to choose the right machine for that job. Therefore, use is made of the concept of *unit horsepower,* which was mentioned previously. Experimentally obtained values of unit horsepower for various common materials are compiled in tables ready for use. The total cutting horsepower can be obtained from the following equation:

$$hp_c = \text{unit hp} \times \text{rate of metal removal} \times \text{correction factor} \quad \textbf{(9.23)}$$

where the rate of metal removal is in cubic inches per minute and the correction factor is introduced to account for the tool geometry and and the variation in feed.

Table 9.1 indicates the unit horsepower values for various ferrous metals and alloys having different hardness numbers. Table 9.2 provides the unit horsepower values for nonferrous metals and alloys. Figure 9.15a through c indicates the different correction factors for the unit horsepower to account for variations in the cutting conditions.

The cutting horsepower is not of practical importance by itself. Its significance is that it is used in computing the motor horsepower. Obviously, the motor horsepower

TABLE 9.1
Unit horsepower values for ferrous metals and alloys

	Brinnel Hardness Number					
Ferrous Metals and Alloys	**150–175**	**176–200**	**201–250**	**251–300**	**301–350**	**351–400**
ANSI						
1010–1025	.58	.67	—	—	—	—
1030–1055	.58	.67	.80	.96	—	—
1060–1095	—	—	.75	.88	1.0	—
1112–1120	.50	—	—	—	—	—
1314–1340	.42	.46	.50	—	—	—
1330–1350	—	.67	.75	.92	1.1	—
2015–2115	.67	—	—	—	—	—
2315–2335	.54	.58	.62	.75	.92	1.0
2340–2350	—	.50	.58	.70	.83	—
2512–2515	.50	.58	.67	.80	.92	—
3115–3130	.50	.58	.70	.83	1.0	1.0
3160–3450	—	.50	.62	.75	.87	1.0
4130–4345	—	.46	.58	.70	.83	1.0
4615–4820	.46	.50	.58	.70	.83	.87
5120–5150	.46	.50	.62	.75	.87	1.0
52100	—	.58	.67	.83	1.0	—
6115–6140	.46	.54	.67	.83	1.0	—
6145–6195	—	.70	.83	1.0	1.2	1.3
Plain cast iron	.30	.33	.42	.50	—	—
Alloy cast iron	.30	.42	.54	—	—	—
Malleable iron	.42	—	—	—	—	—
Cast steel	.62	.67	.80	—	—	—

Source: Turning Handbook of High-Efficiency Metal Cutting, *1980, courtesy Carboloy Inc., a Seco Tools Company.*

TABLE 9.2
Unit horsepower values for nonferrous metals and alloys

Nonferrous Metals and Alloys	Properties	Unit Horsepower
Brass	Hard	0.83
	Medium	0.50
	Soft	0.33
	Free machining	0.25
Bronze	Hard	0.83
	Medium	0.50
	Soft	0.33
Copper	Pure	0.90
Aluminum	Cast	0.25
	Hard (rolled)	0.33
Monel	(rolled)	1.0
Zinc alloy	(die cast)	0.25

Source: Turning Handbook of High-Efficiency Metal Cutting, *1980, courtesy Carboloy Inc., A Seco Tools Company.*

has to be higher than the cutting horsepower as some power is lost in overcoming friction and inertia of the moving parts. The following equation can be used for calculating the motor horsepower:

$$hp_m = hp_c \times \frac{1}{\eta} \tag{9.24}$$

where η is the machine efficiency, which can be taken from Table 9.3.

The cutting horsepower is used not only in calculating the motor horsepower but also for giving a fair estimate of the cutting force component F_c by using Equation 9.21. This force is very important when studying the vibrations associated with metal cutting, as we will see later. The following example illustrates how to estimate the cutting force component.

Example of Estimating Cutting Force Component

During a turning operation, the metal-removal rate (M.R.R.) was found to be 3.6 cubic inches per minute. Following are other data of the process:

Material:	ANSI 1055, HB 250
Cutting speed:	300 feet per minute
Undeformed chip thickness:	0.01 inch
Tool character:	0-7-7-7-15-15-1/32 (see Section 9.5)

Estimate the cutting force component F_c.

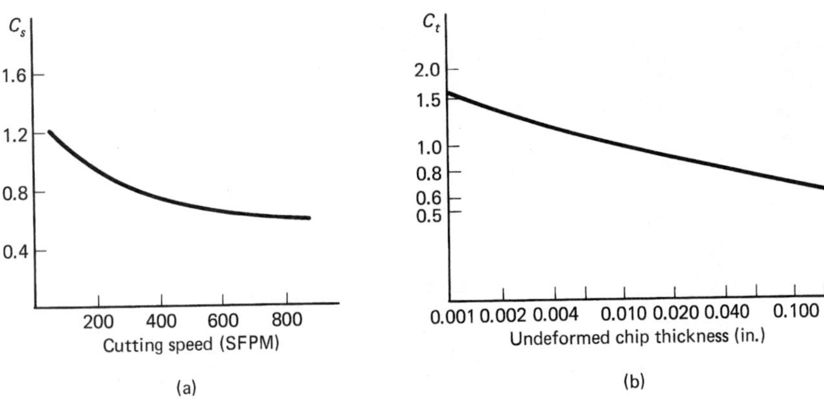

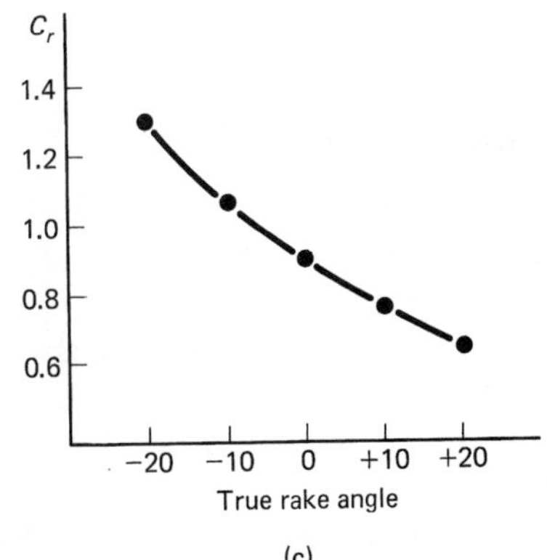

FIGURE 9.15 Different correction factors to account for variations in the cutting conditions: (a) cutting speed; (b) chip thickness; (c) rake angle *(Source:* Turning Handbook of High-Efficiency Metal Cutting, *1980, courtesy Carboloy Inc., A Seco Tools Company)*

Here is the solution to this example problem:

$$\text{spindle hp} = \text{M.R.R.} \times \text{unit hp} \times \text{correction factor}$$

The correction factor because of the cutting speed is 0.8, and the correction factor because of the undeformed chip thickness is 1. The true rake angle is

$$\tan \alpha_{\text{true}} = \cos 15° \tan 7° + \sin 15° \tan 0$$

TABLE 9.3 Typical overall machine tool efficiencies (except milling machines)

Type	Efficiency (%)
Direct-spindle drive	90
One-belt drive	85
Two-belt drive	70
Geared head	70

Source: Turning Handbook of High-Efficiency Metal Cutting, *1980, courtesy Carboloy Inc., A Seco Tools Company.*

where $\alpha_{\text{true}} \simeq 6°$ (see Section 9.5). The correction factor because of the true rake angle is 0.83, and the unit hp is 0.8 from Table 9.1. Thus,

$$\text{spindle hp} = 3.6 \times 0.8 \times 0.8 \times 1 \times 0.83 = 1.9 \text{ hp}$$

$$\text{hp}_c = \frac{F_c \times 300 \text{ ft/min.}}{550 \times 60}$$

Therefore,

$$F_c = \frac{1.9 \times 550 \times 60}{300} = 209 \text{ pounds}$$

Note that the undeformed chip thickness equals feed (inches per revolution) times the cosine of the side cutting-edge angle.

9.5 CUTTING TOOLS

Basic Geometry

In order for a tool to cut a material, it must have two important characteristics: First, it must be harder than that material, and, second, it must possess certain geometrical characteristics. The cutting tool geometry differs for different machining operations. Nevertheless, it is always a matter of rake and clearance angles. Therefore, we are going to limit our discussion, at the moment, to single-point tools for the sake of simplicity. Other types of tools will be considered when we cover the various machining operations.

As can be seen in Figure 9.16, the geometry of a single-point cutting tool can be adequately described by six cutting angles. These can be shown more clearly by projecting them on three perpendicular planes using orthogonal projection, as is done in Figure 9.17. Let us now consider the definition of each of the six angles.

Side cutting-edge angle. The side cutting-edge angle (SCEA) is usually referred to as the *lead angle*. It is the angle enclosed between the side cutting edge and the longitudinal direction of the tool. The value of this angle varies between 0° and 90°, depending upon the machinability, rigidity, and, sometimes, the shape of the workpiece (e.g.,

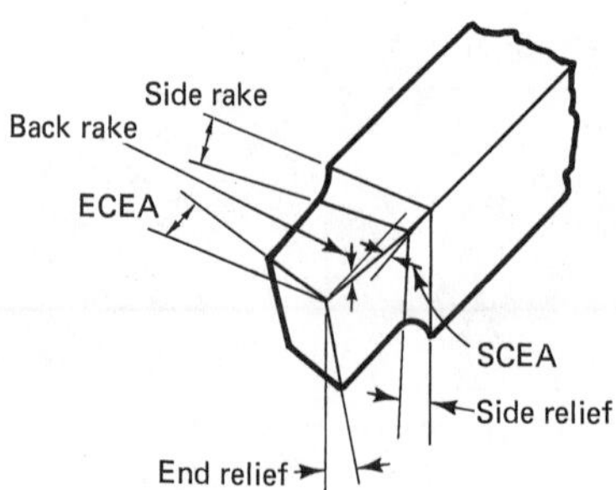

FIGURE 9.16
Geometry of a single-point cutting tool

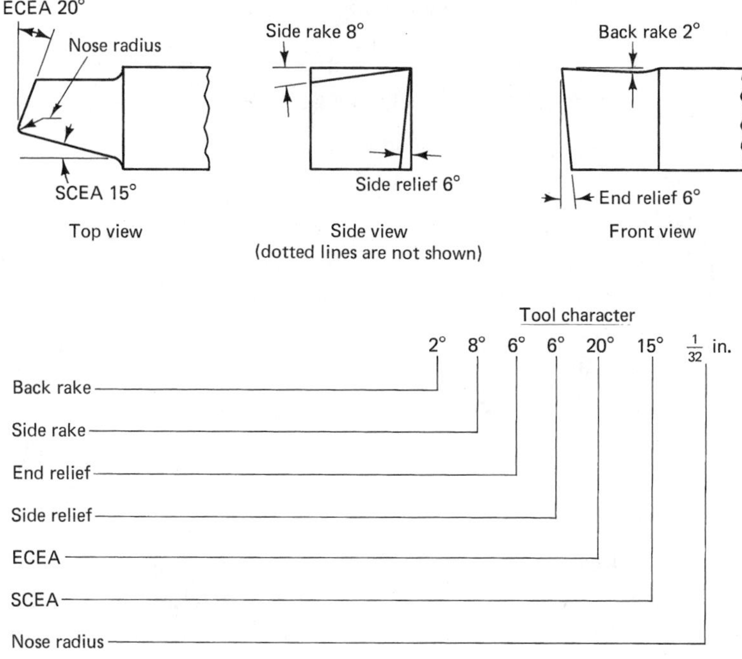

FIGURE 9.17 Orthogonal projection of the cutting angles of a single-point tool and tool character

a 90° shoulder must be produced by a 0° SCEA). As this angle increases from 0° to 15°, the power consumption during cutting decreases. However, there is a limit for increasing the SCEA, beyond which excessive vibrations take place because of the large tool-workpiece interface. On the other hand, if the angle were taken as 0°, the full cutting edge would start to cut the workpiece at once, causing an initial shock. Usually, the recommended value for the lead angle should range between 15° and 30°.

End cutting-edge angle. The end cutting-edge angle (ECEA) serves to eliminate rubbing between the end cutting edge and the machined surface of the workpiece. Although this angle takes values in the range of 5° to 30°, commonly recommended values are 8° to 15°.

Side relief and end relief angles. Side and end relief angles serve to eliminate rubbing between the workpiece and the side and end flank, respectively. Usually, the value of each of these angles ranges between 5° and 15°.

Back and side rake angles. Back and side rake angles determine the direction of flow of the chips onto the face of the tool. Rake angles can be positive, negative, or zero. It is the side rake angle that has the dominant influence on cutting. Its value usually varies between 0° and 15°, whereas the back rake angle is usually taken as 0°.

Another useful term in metal cutting is the *true rake angle,* which is confined between the line of major inclination within the face of the tool and a horizontal plane. It determines the actual flow of chips across the face of the tool and can be obtained from the following equation:

$$\text{true rake angle} = \tan^{-1}(\tan \alpha \sin \lambda + \tan \beta \cos \lambda) \quad \textbf{(9.25)}$$

where: α is the back rake angle
β is the side rake angle
λ is the lead angle (SCEA)

As previously mentioned, the true rake angle has a marked effect on the unit horsepower for a given workpiece material, and a correction factor has to be used when calculating the power in order to account for variations in the true rake angle.

Tool character. The tool angles are usually specified by a standard abbreviation system called the *tool character,* or the *tool signature.* As also illustrated in Figure 9.17, the tool angles are always given in a certain order: back rake, side rake, end relief, side relief, ECEA, and SCEA, followed by the nose radius of the tool.

Cutting Tool Materials

Cutting tools must possess certain mechanical properties in order to function adequately during the cutting operations. These properties include high hardness and the ability to retain it even at the elevated temperatures generated during cutting, as well as toughness, creep and abrasion resistance, and the ability to withstand high bearing pressures. Cutting materials differ in the degree to which they possess each of these mechanical properties. Therefore, a cutting material is selected to suit the cutting conditions (i.e., the workpiece material, cutting speed or production rate, coolants used, and so on). Following is a survey of the commonly used cutting tool materials.

Plain-carbon steel. Plain-carbon steel contains from 0.8 to 1.4 percent carbon, has no additives, and is subjected to heat treatment to increase its hardness. Plain-carbon steel is suitable only when making hand tools or when soft metals are machined at low cutting speeds as it cannot retain its hardness at temperatures above 600°F (300°C) due to tempering action.

Alloy steel. The carbon content of alloy steel is similar to that of plain-carbon steel, but it contains alloying elements (in limited amounts). Tools made of alloy steel must be heat treated and are used only when machining is carried out at low cutting speeds. The temperature generated as a result of cutting should not exceed 600°F (300°C) to avoid any tempering action.

High-speed steel. High-speed steel (HSS) is a kind of alloy steel that contains a certain percentage of alloying elements, such as tungsten (18 percent), chromium (4 percent), molybdenum, vanadium, and cobalt. High-speed steel is heat treated by heating (at two stages), cooling by employing a stream of air, and then tempering it. Tools made of HSS can retain their hardness at elevated temperatures up to 1100°F (600°C). These tools are used when relatively high cutting speeds are required. Single-point tools, twist drills, and milling cutters are generally made of high-speed steel, except when these tools are required for high-production machining.

Cast hard alloys. Cast hard alloys can be either ferrous or nonferrous and contain about 3 percent carbon, which, in turn, reacts with the metals to form very hard carbides. The carbides retain their hardness even at a temperature of about 1650°F (900°C). Because such a material cannot be worked or machined, it is cast in ceramic molds to take the form of tips that are mounted onto holders by brazing or by being mechanically fastened.

Sintered cemented-carbide tips. Sintered cemented carbide was developed to eliminate the main disadvantage of the hard cast alloys: brittleness. Originally, the composition of this material involved about 82 percent very hard tungsten carbide particles and 18 percent cobalt as a binder. Sintered cemented carbides are always molded to shape by the powder metallurgy technique (i.e., pressing and sintering, as was explained in Chapter 7). As it is impossible to manufacture the entire tool out of cemented carbide because of the strength consideration, only tips are made of this material; these tips are brazed or mechanically fastened to steel shanks that have the required cutting angles.

Cemented carbides used to be referred to as Widia, taken from the German expression "Wie Diamant," meaning diamondlike, because they possess extremely high hardness, reaching about 90 Rc, and they retain such hardness even at temperatures of up to 1850°F (1000°C). Recent developments involve employing combinations of tungsten, titanium, and tantalum carbides with cobalt or nickel alloy as binders. The result is characterized by its low coefficient of friction and high abrasion resistance. Tools with cemented-carbide tips are recommended whenever the cutting speeds required or the feed rates are high and are, therefore, commonly used in mass production. Recently, carbide tips have been coated with nitrites or oxides to increase their wear resistance and service life.

Ceramic tips. Ceramic tips consist basically of very fine alumina powder, Al_2O_3, which is molded by pressing and sintering. Ceramics have almost the same hardness as cemented carbides, but they can retain that hardness up to a temperature of 2200°F (1100°C) and have a very low coefficient of thermal conductivity. Such properties allow for cutting to be performed at speeds that range from two to three times the cutting speed used when carbide tips are employed. Ceramic tips are also characterized by their superior resistance to wear and to the formation of crater cavities. They require no coolants. Their toughness and bending strength are low, which must be added to their sensitivity to creep loading and vibration. Therefore, ceramic tips are recommended only for finishing operations (small depth of cut) at extremely high cutting speeds of up to 180 feet per minute (600 m/min.). Following are the three common types of ceramic tips:

1. Oxide tips, consisting mainly of aluminum oxide, have a white color with some pink or yellow tint.
2. Cermet tips, including alumina and some metals such as titanium or molybdenum, are dark gray in color.
3. Tips that consist of both oxides as well as carbides are black in color.

Ceramic tips should not be used for machining aluminum because of their affinity to oxygen.

Diamond. Diamond pieces are fixed to steel shanks and are used in precision cutting operations. They are recommended for machining aluminum, magnesium, titanium, bronze, rubber, and polymer. When machining metallic materials, a mirror finish can be obtained.

Tool Wear

There are two interrelated causes for tool wear: *mechanical abrasion* and *thermal erosion.* Although these two actions take place simultaneously, the role of each varies for various cutting conditions. Mechanical wear is dominant when low cutting speeds are used or when the workpiece possesses high machinability. Thermal wear prevails when high cutting speeds are used with workpieces having low machinability. Thermal wear is due to diffusion, oxidation, and the fact that the mechanical properties of the tool change as a result of the high temperature generated during the cutting operation.

The face of the cutting tool is subjected to friction caused by the fast relative motion of the generated chips onto its surface. Similarly, the flanks are also subjected to friction as a result of rubbing by the workpiece. Although the tool is harder than the workpiece, friction and wear will take place and will not be evenly distributed over the face of the tool. Wear is localized in the vicinity of the cutting edge and results in the formation of a crater. There are different kinds of tool wear:

1. Flank wear
2. Wear of the face that comes in contact with the removed chip
3. Wear of the cutting edge itself
4. Wear of the nose
5. Wear and formation of a crater
6. Cracks in the cutting edges occurring during interrupted machining operations such as millings

Tool Life

Tool life is defined as the length of actual machining time beginning at the moment when a just-ground tool is used and ending at the moment when the machining operation is stopped because of the poor performance of that tool. Different criteria can be used to judge the moment at which the machining operation should be stopped. It is common to consider the tool life as over when the flank wear reaches a certain amount (measured as the length along the surface generated due to abrasion starting from the tip). This maximum permissible flank wear is taken as 0.062 inch (1.58 mm) in the case of high-speed steel tools and 0.03 inch (0.76 mm) for carbide tools.

The tool life is affected by several variables, the important ones being cutting speed, feed, and the coolants used. The effect of these variables can be determined experimentally and then represented graphically for practical use. It was found by Frederick W. Taylor that the relationship between tool life and cutting speed is exponential. It can, therefore, be plotted on a logarithmic scale so that it takes the form of a straight

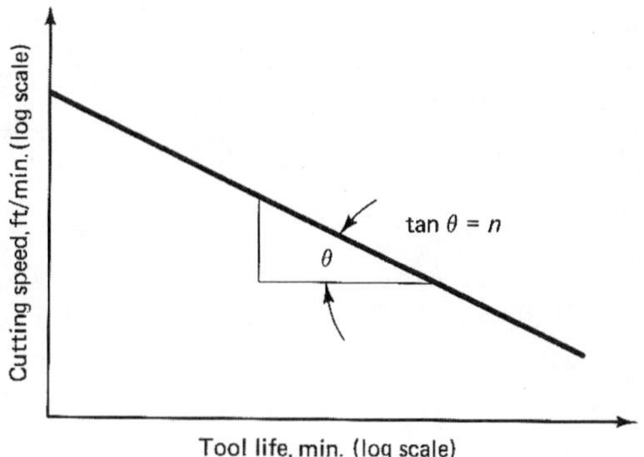

FIGURE 9.18 Relationship between tool life and cutting speed on a log-log scale

line, as shown in Figure 9.18. In fact, this was the basis for establishing an empirical formula that correlates tool life with cutting speed. A correction factor is also introduced into the formula to account for the effects of other variables. The original formula had the following form:

$$VT^n = c \tag{9.26}$$

where: n is a constant that depends upon the tool material (0.1 for HSS, 0.20 for carbides, and 0.5 for ceramic tools)

c is a constant that depends upon the cutting conditions (e.g., feed)

T is the tool life measured in minutes

V is the cutting speed in feet per minute

Equation 9.26 is very useful in obtaining the tool life for any cutting speed if the tool life is known at any other cutting speed.

9.6 MACHINABILITY

Machinability Defined

Machinability is a property characterizing the material of the workpiece: It is the ease with which that material can be machined. In order to express machinability in a quantitative manner, one of the following methods is used:

1. The maximum possible rate of chip removal
2. Surface finish of the machined workpiece
3. Tool life
4. Energy required to accomplish the cutting operation

It is clear that the tool life is the most important of these criteria as it plays an important role in maximizing the production while minimizing the production cost. Moreover, criteria such as surface finish and machining precision depend upon many factors, such as the sharpness of the cutting edge, the rigidity of the tool, and the pos-

TABLE 9.4
Machinability indices for some metals and alloys (using carbide tools)

Metal or Alloy	Machinability Index (%)
Steel SAE 1020 (annealed)	65
Steel SAE A2340	45
Cast iron	70–80
Stainless steel 18-8 (austenitic)	25
Tool steel (low tungsten, chrome, and carbon)	30
Copper	70
Brass	180
Aluminum alloy	300 and above

sibility of formation of a built-up edge. As a consequence, it is the tool life that is most suitable as a criterion of machinability.

Machinability Index

Because machinability cannot be expressed in an absolute manner, it is appropriate to take a highly machinable metal as a reference and express the machinability of any other ferrous metal as a percentage of that of the reference metal. The reference metal chosen was steel SAE-AISI 1112 because of its superior machinability, which exceeds that for any other steel. Such steel is usually referred to as *free cutting steel.* The machinability index can now be given:

$$\text{machinability index} = \frac{\text{cutting speed of metal for tool life of 20 minutes}}{\text{cutting speed of steel SAE 1112 for tool life of 20 minutes}} \times 100 \qquad \textbf{(9.27)}$$

Table 9.4 indicates the machinability index for some commonly used metals and alloys.

9.7 CUTTING FLUIDS

Necessary Characteristics

As previously mentioned, the process of metal cutting results in the generation of a large amount of heat and a localized increase in the temperature of the cutting tool. This effect is particularly evident when machining ductile metals. Accordingly, coolants are required to remove any generated heat, to lower the temperature of the cutting tool, and, consequently, to increase the tool service life. In order to fulfill such conditions and function properly, a cutting fluid must possess certain characteristics:

1. The cutting fluid must possess suitable chemical properties (i.e., to be appropriate from the point of view of chemistry), must not react with the workpiece material or cause corrosion in any component of the machine tool, and should not promote

the formation of rust or spoil the lubricating oil of the machine bearing and slides whenever it comes in contact with that oil.

2. The cutting fluid must be chemically stable (i.e., must not change its properties with time).
3. No poisonous gases or fumes should evolve during machining so that there is no possibility of problems regarding the safety or health of the workers.
4. The lubricating and cooling properties of the cutting fluid must be superior.
5. The fluid used should be cheap and should be recycled by a simple filtration process.

Types of Cutting Fluids

The following discussion involves the different kinds of cutting fluids that are used in industry to satisfy the preceding requirements.

Pure oils. Mineral oils such as kerosene or polar organic oils such as sperm oil, linseed oil, or turpentine can be used as cutting fluids. The application of pure mineral oils is permissible only when machining metals with high machinability, such as free cutting steel, brass, and aluminum. This is a consequence of their poor lubricating and cooling properties. Although the polar organic oils possess good lubricating and cooling properties, they are prone to oxidation, give off unpleasant odors, and tend to gum.

Mixed oils. Mineral oils are mixed with polar organic oils to obtain the advantages of both constituents. In some cases, sulfur or chlorine is added to enable the lubricant to adhere to the tool face, giving a film of lubricant that is tougher and more stable. The oils are then referred to as *sulfurized* or *chlorinated* oils. The chlorinated oils have the disadvantage of the possible emission of chlorine gas during the machining operation.

Soluble oils. Soluble oils are sometimes called *water-miscible fluids* or *emulsifiable oils.* By blending oil with water and some emulsifying agents, soapy or milky mixtures can be obtained. These liquids have superior cooling properties and are recommended for machining operations requiring high speeds and low pressures. Sometimes, extreme-pressure additives are blended with the mixtures to produce emulsions with superior lubricating properties.

Water solutions. A solution of sodium nitrate and trinolamine in water can be employed as a cutting fluid. Caustic soda is also used, provided that the concentration does not exceed 5 percent. If the concentration of the solution exceeds this limit, the paint of the machine and the lubricating oil of the slides may be affected.

Synthetic fluids. Synthetic fluids can be diluted with water to give a mixture that varies in appearance from clear to translucent. Extreme-pressure additives like sulfur or chlorine can be added to the mixture so that it can be used for difficult machining operations.

9.8 CHATTER PHENOMENON

When we feel cold in winter, our jaws and teeth may start to chatter. A similar phenomenon occurs when the cutting tool and workpiece are exposed to certain unfavorable cutting conditions and dynamic characteristics of the machine tool structure. The analysis of this chatter phenomenon is an extremely complex task. However, thanks to the work of the late Professor Stephen A. Tobias of the University of Birmingham in England, we are able to understand how vibrations of the cutting tool initiate and how they can be minimized. Left without remedy, these vibrations result in breakage of the cutting tool (especially if it is ceramic or carbide) and poor surface quality. They may also cause breakage of the entire machine tool. Two basic types of vibrations are generated during machining: forced vibrations and self-excited vibrations.

Forced vibrations take place as a result of periodic force applied within the machine tool structure. This force can be due to an imbalance in any of the machine tool components or interrupted cutting action, such as milling, in which there is a periodic engagement and disengagement between the cutting edges and the workpiece. The frequency of these forced vibrations must not be allowed to come close to the natural frequency of the machine tool system or any of its components; otherwise, resonance (vibrations with extremely high amplitude) takes place. The remedy in this case is to try to identify any possible source for the imbalance of the machine tool components and eliminate it. In milling machines, the stiffness and the damping characteristics of the machine tool are controlled so as to keep the forcing frequency away from the natural frequency of any component and/or the natural frequency of the system.

Self-excited vibrations, or *chatter,* occur when an unexpected disturbing force, such as a hard spot in the workpiece material or sticking friction at the chip-tool interface, causes the cutting tool to vibrate at a frequency near the natural frequency of the machine tool. As a result, resonance takes place, and a minimum excitation produces an extremely large amplitude. Such conditions drastically reduce tool life, result in poor surface quality, and may cause damage to either the workpiece or the machine tool or both. This unfavorable condition can be eliminated, or at least reduced, by controlling the stiffness and the damping characteristics of the system. This is usually achieved by selecting the proper material for the machine bed (cast iron has better damping characteristics than steel), by employing dry-bolted joints as energy dissipators where the vibration energy is absorbed in friction, or by using external dampers or absorbers. Advanced research carried out at the University of Birmingham in England indicated the potentials of employing layers of composites as a means to safeguard against the occurrence of chatter.

9.9 ECONOMICS OF METAL CUTTING

Our goal now is to find out the operating conditions (mainly the cutting speed) that maximize the metal-removal rate or the tool life. These two variables are in opposition to each other; a higher metal-removal rate results in a shorter tool life. Therefore, some

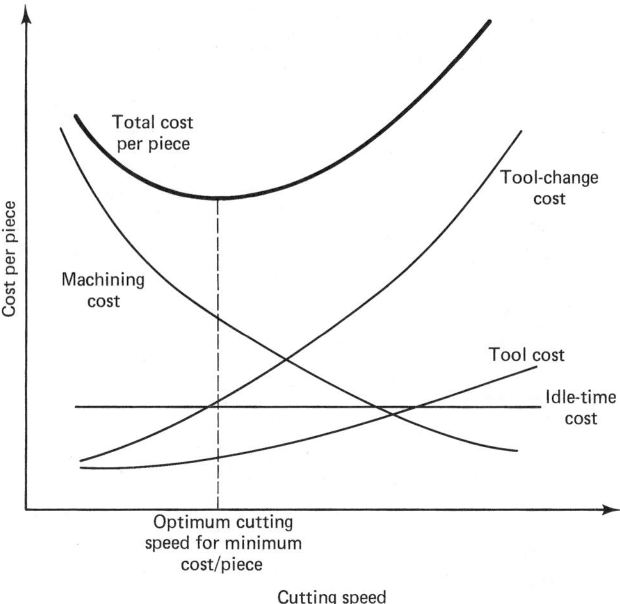

FIGURE 9.19 Relationship between cost per piece and cutting speed

trade-off or balance must be made in order to achieve either minimum machining cost per piece or maximum production rate, whichever is necessitated by the production requirements.

Figure 9.19 indicates how to construct the relationship between the cutting speed and the total cost per piece for a simple turning operation. The total cost is composed of four components: machining cost, idle-time (nonproductive) cost, tool cost, and tool-change cost. An increase in cutting speed obviously results in a reduction in machining time and, therefore, lower machining cost. This is accompanied by a reduction in tool life, thus increasing tool and tool-change costs. As can be seen in Figure 9.19, the curve of the cost per piece versus the cutting speed has a minimum that corresponds to the optimum cutting speed for the minimum cost per piece.

The relationship between the production time per piece and the cutting speed can be constructed in the same manner, as shown in Figure 9.20. There is also a minimum for this curve that corresponds to the optimum cutting speed for the maximum productivity (minimum time per piece). Usually, this value is higher than the maximum economy speed given in Figure 9.19. Obviously, a cutting speed between these two limits (and depending upon the goals to be achieved) is recommended.

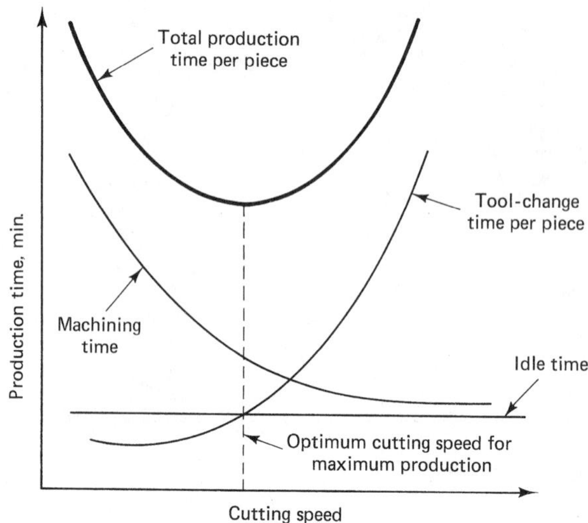

FIGURE 9.20
Relationship between production time per piece and cutting speed

Review Questions

1. How can the complex process of metal cutting be approached?
2. Define the *rake angle* and the *clearance angle* in two-dimensional cutting.
3. Why are the angles in Question 2 required?
4. What is the upper surface of the tool called?
5. What is the lower surface of the tool called?
6. What are the cutting variables that affect the values of the rake and clearance angles?
7. List some drawbacks if the cutting angles are not properly chosen.
8. When should the rake angle be taken as a positive value?
9. When should the rake angle be taken as a negative value?
10. Can orthogonal cutting actually take place? Explain.
11. Use sketches to explain the stages involved in the formation of chips during machining.
12. Use sketches to illustrate the different types of machining chips and explain when and why we can expect to have each of these types.
13. Explain the stages involved in the formation of the built-up edge.
14. Does the built-up edge have useful or harmful effects?
15. What is meant by the *shear angle?*
16. What is meant by the *cutting ratio?*
17. Derive an expression for the shear strain that takes place during orthogonal cutting.
18. Draw a sketch of the cutting force diagram proposed by Ernst and Merchant.
19. How can the relationship between the shear and rake angles be expressed according to Ernst and Merchant?
20. On what basis have Lee and Shaffer developed their theory?

21. Derive an expression for the specific energy during two-dimensional cutting.

22. Illustrate the difference between orthogonal and oblique cutting.

23. What are the components of the cutting force in oblique cutting? How do you compare their magnitudes with each other?

24. Define the *unit horsepower.*

25. Describe fully the geometry of single-point cutting tools.

26. Explain the effect of each of the cutting angles in oblique cutting on the mechanics of the process.

27. List the different cutting tool materials and enumerate the advantages, disadvantages, and applications of each.

28. What are the two main causes for tool wear?

29. List the different kinds of tool wear.

30. Define *tool life.*

31. What is the relationship between tool life and cutting speed?

32. Define *machinability* and explain how it is quantitatively expressed by the machinability index.

33. What are the necessary characteristics of cutting fluids?

34. List the different types of cutting fluids and provide the advantages and limitations of each.

35. What are the causes for forced vibrations during machining?

36. How can forced vibrations be minimized?

37. What is *chatter* and why does it occur?

38. How can we eliminate chatter?

39. What trouble can vibrations cause during machining?

40. Use sketches to explain how the value of the optimum cutting speed can be obtained for maximum economy and for maximum productivity.

Problems

1. In a turning operation, the diameter of the workpiece is 2 inches (50 mm), and it rotates at 360 revolutions per minute. How long will a carbide tool last (n = 0.3) under such conditions if an identical carbide tool lasted for 1 minute when used at 1000 feet per minute (305.0 m/min.)?

2. Determine the increase in the tool life of a carbide tip as a result of a decrease in the cutting speed of 25, 50, and 75 percent.

3. When turning a thin tube at its edge, the following conditions were observed:

Depth of cut:	0.125 inch
Chip thickness:	0.15 inch
Back rake angle:	8°
Cutting speed:	300 ft/min.

Calculate the

a. Cutting ratio
b. Shear angle
c. Chip velocity

4. A geared-head lathe is employed for machining steel AISI 1055, BHN 250. The cutting speed is 400 feet per minute, and the rate of metal removal is 2.4 cubic inches per minute. If the tool used has the character 0-7-7-7-15-15-1/32, estimate the following:

a. The energy consumed in machining per unit time
b. The power required at the motor
c. The tangential component of the cutting force

Neglect the correction factor for the undeformed chip thickness.

5. A 5-hp, 2-V, belt-driven lathe is to be used for machining brass under the following conditions:

Cutting speed:	600 ft/min.
Rate of metal removal:	7.2 in.^{3}/min.
SCEA of the tool:	30°

Neglect the effect of chip thickness. Does this lathe have enough power for the required job?

Design Project

Prepare a computer program that determines the optimum cutting speed that results in maximum productivity. The program should be interactive, the input being workpiece material, tool material, and depth of cut. Assume the time for changing the tool is 60 seconds and the time to return the tool to the beginning of the cut is 20 seconds. Take the workpiece material to be

a. Steel 1020
b. Brass
c. Aluminum
d. Stainless steel

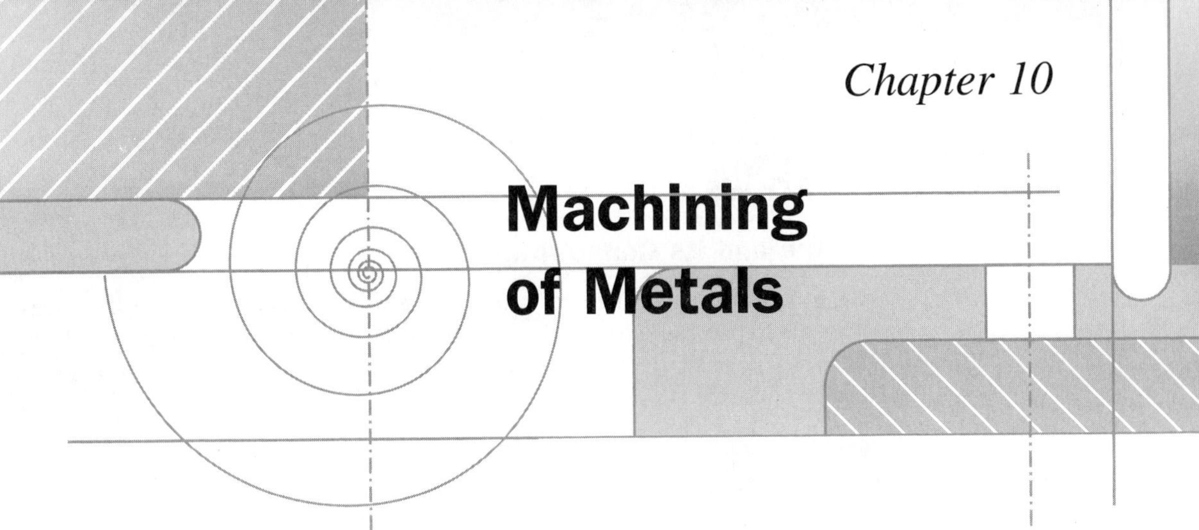

Chapter 10

Machining of Metals

INTRODUCTION

This chapter will focus on the technological aspects of the different machining operations, as well as the design features of the various machine tools employed to perform those operations. In addition, the different shapes and geometries produced by each operation, the tools used, and the work-holding devices will be covered. Special attention will be given to the required workshop calculations that are aimed at estimating machining parameters such as cutting speeds and feeds, metal-removal rate, and machining time.

Machine tools are designed to drive the cutting tool in order to produce the desired machined surface. For such a goal to be accomplished, a machine tool must include appropriate elements and mechanisms capable of generating the following motions:

1. A relative motion between the cutting tool and the workpiece in the direction of cutting
2. A motion that enables the cutting tool to penetrate into the workpiece until the desired depth of cut is achieved
3. A feed motion that repeats the cutting action every round or every stroke to ensure continuation of the cutting operation

10.1 TURNING OPERATIONS

The Lathe and Its Construction

A *lathe* is a machine tool used for producing surfaces of revolution and flat edges. Based on their purpose, construction, number of tools that can simultaneously be mounted, and degree of automation, lathes, or more accurately, lathe-type machine tools, can be classified as follows:

1. Engine lathes
2. Toolroom lathes
3. Turret lathes
4. Vertical turning and boring mills
5. Automatic lathes
6. Special-purpose lathes

In spite of the diversity of lathe-type machine tools, there are common features with respect to construction and principles of operation. These features can be illustrated by considering the commonly used representative type, the *engine lathe,* which is shown in Figure 10.1. Following is a description of each of the main elements of an engine lathe.

Lathe bed. The *lathe bed* is the main frame, a horizontal beam on two vertical supports. It is usually made of gray or nodular cast iron to damp vibrations and is made by casting. It has guideways that allow the carriage to slide easily lengthwise. The height of the lathe bed should be such that the technician can do his or her job easily and comfortably.

Headstock. The *headstock* assembly is fixed at the left-hand side of the lathe bed and includes the *spindle,* whose axis is parallel to the guideways (the slide surface of the bed). The spindle is driven through the gearbox, which is housed within the headstock. The function of the gearbox is to provide a number of different spindle speeds (usually 6 to 18 speeds). Some modern lathes have headstocks with infinitely variable spindle speeds and that employ frictional, electrical, or hydraulic drives.

The spindle is always hollow (i.e., it has a through hole extending lengthwise). Bar stocks can be fed through the hole if continuous production is adopted. Also, the hole has a tapered surface to allow the mounting of a plain lathe center, such as the one shown in Figure 10.2. It is made of hardened tool steel. The part of the lathe center that fits into the spindle hole has a Morse taper, while the other part of the center is conical with a 60° apex angle. As explained later, lathe centers are used for mounting long workpieces. The outer surface of the spindle is threaded to allow the mounting of a chuck, a faceplate, or the like.

Tailstock. The *tailstock* assembly consists basically of three parts: its lower base, an intermediate part, and the quill. The lower base is a casting that can slide on the lathe

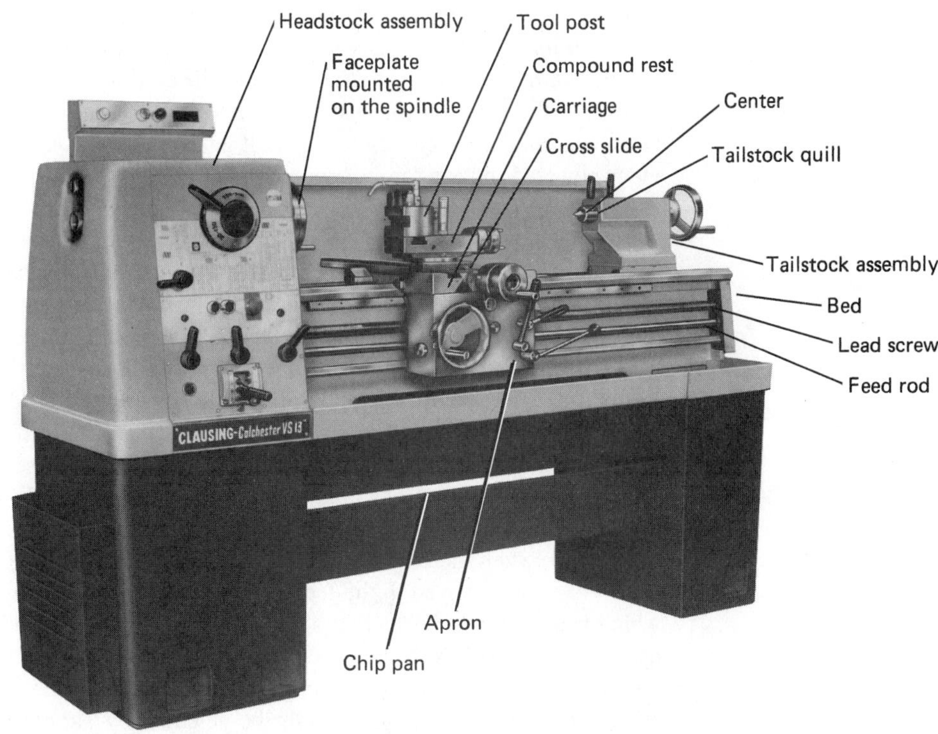

FIGURE 10.1
An engine lathe
(Courtesy of Clausing Industrial, Inc., Kalamazoo, Michigan)

bed along the guideways, and it has a clamping device so that the entire tailstock can be locked at any desired location, depending upon the length of the workpiece. The intermediate part is a casting that can be moved transversely so that the axis of the tailstock can be aligned with that of the headstock. The third part, called the *quill,* is a hardened steel tube that can be moved longitudinally in and out of the intermediate part as required. This is achieved through the use of a handwheel and a screw, around which a nut fixed to the quill is engaged. The hole in the open side of the quill is tapered to allow the mounting of lathe centers or other tools like twist drills or boring bars. The quill can be locked at any point along its travel path by means of a clamping device.

Carriage. The main function of the *carriage* is to mount the cutting tools and generate longitudinal and/or cross feeds. It is actually an H-shaped block that slides on the lathe bed between the headstock and tailstock while being guided by the V-shaped

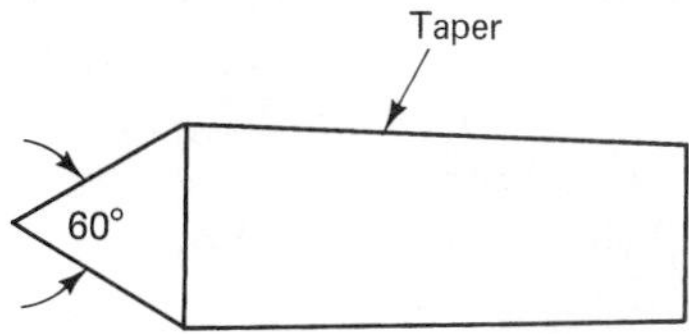

FIGURE 10.2
A plain lathe center

guideways of the bed. The carriage can be moved either manually or mechanically by means of the apron and either the feed rod or the lead screw.

The *apron* is attached to the saddle of the carriage and serves to convert the rotary motion of the feed rod (or lead screw) into linear longitudinal motion of the carriage and, accordingly, the cutting tool (i.e., it generates the axial feed). The apron also provides powered motion for the cross slide located on the carriage. Usually, the tool post is mounted on the compound rest, which is, in turn, mounted on the cross slide. The compound rest is pivoted around a vertical axis so that the tools can be set at any desired angle with respect to the axis of the lathe (and that of the workpiece). These various components of the carriage form a system that provides motion for the cutting tool in two perpendicular directions during turning operations.

When cutting screw threads, power is provided from the gearbox to the apron by the lead screw. In all other turning operations, it is the feed rod that drives the carriage. The lead screw goes through a pair of half nuts that are fixed to the rear of the apron. When actuating a certain lever, the half nuts are clamped together and engage with the rotating lead screw as a single nut that is fed, together with the carriage, along the bed. When the lever is disengaged, the half nuts are released and the carriage stops. On the other hand, when the feed rod is used, it supplies power to the apron through a worm gear. This gear is keyed to the feed rod and travels with the apron along the feed rod, which has a keyway extending along its entire length. A modern lathe usually has a quick-change gearbox located under the headstock and driven from the spindle through a train of gears. It is connected to both the feed rod and the lead screw so that a variety of feeds can easily and rapidly be selected by simply shifting the appropriate levers. The quick-change gearbox is employed in plain turning, facing, and thread-cutting operations. Because the gearbox is linked to the spindle, the distance that the apron (and the cutting tool) travels for each revolution of the spindle can be controlled and is referred to as the *feed.*

The Turret Lathe

A *turret lathe* is similar to an engine lathe, except that the conventional tool post is replaced with a hexagonal (or octagonal) turret that can be rotated around a vertical axis as required. Appropriate tools are mounted on the six (or eight) sides of the turret. The length of each tool is adjusted so that, by simply indexing the turret, any tool can be brought into the exactly desired operating position. These cutting tools can, therefore, be employed successively without the need for dismounting the tool and mounting a new one each time, as is the case with conventional engine lathes. This results in an appreciable saving in the time required for setting up the tools. Also, on a turret lathe, a skilled machinist is required only initially to set up the tools. A laborer with limited training can operate the turret lathe thereafter and produce parts identical to those that can be manufactured when a skilled machinist operates the lathe. Figure 10.3 illustrates a top view of a hexagonal turret with six different tools mounted on its sides. Sometimes, the turret replaces the tailstock and can be either vertical (i.e., with a horizontal axis) or horizontal (i.e., with a vertical axis). In this case, four additional tools can be mounted on the square tool post, sometimes called a *square turret,* thus allow-

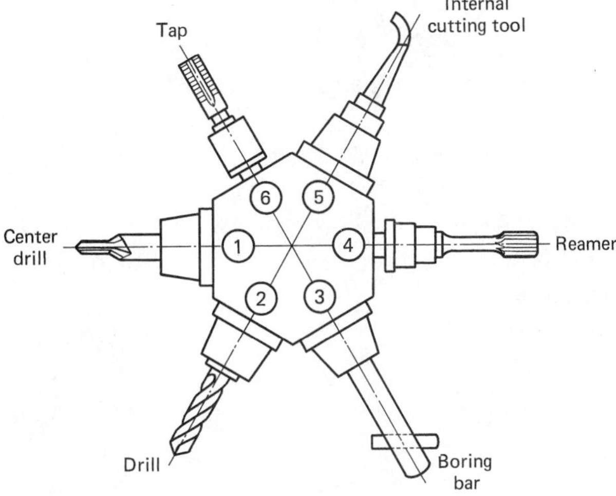

FIGURE 10.3
Top view of a hexagonal turret with six different tools

ing twelve machining operations to be performed successively. Turret lathes always have work-holding devices with quick-release (and quick-tightening) mechanisms.

Specifying a Lathe

It is important for a manufacturing engineer to be able to specify a lathe in order to place an order or to compare and examine contract bids. The specifications of a lathe should involve data that reveal the dimensions of the largest workpiece to be machined on that lathe. They also must include the power consumption, as well as information that is needed for shipping and handling. Table 10.1 indicates an example of how to specify a lathe.

Tool Holding

Tools for turning operations are mounted in a toolholder (tool post). On an engine lathe, it is located on the compound rest. More than one cutting tool (up to four) can be mounted in the toolholder in order to save the time required for changing and setting up each tool should only one tool be mounted at a time. In all turning operations, the following conditions for holding the tools must be fulfilled:

1. The tip of the cutting edge must fully coincide with the level of the lathe axis. This can be achieved by using the pointed edge of the lathe center as a basis for adjustment, as shown in Figure 10.4. Failure to meet this condition results in a change in the values of the cutting angles from the desired ones.
2. The centerline of the cutting tool must be horizontal.
3. The tool must be fixed tightly along its length and not just on two points.
4. A long tool-overhang should be avoided in order to eliminate any possibility for elastic strains and vibrations.

TABLE 10.1
Example of specifications of a lathe

Model	Example
Maximum swing over bed (largest diameter of workpiece)	12 in. (300 mm)
Maximum swing over carriage (largest diameter over carriage)	8 in. (200 mm)
Hole through spindle	0.75 in. (19 mm)
Height of centers	6 in. (150 mm)
Turning length	24 in. (600 mm)
Thread on spindle nose	
Taper in spindle and tailstock sleeves	3 Morse
21 spindle speeds	20–2000 rev/min.
Metric threads	2–6 mm
Whitworth	4–28 teeth
Feeds per revolution	0.0002–0.008 in. (0.05–0.2 mm)
Power required	1.6 kW
Net weight	1 ton
Floor space requirement (length/width/height)	64/36/56 in. (1600/900/1400 mm)

Lathe Cutting Tools

The shape and geometry of lathe cutting tools depend upon the purpose for which they are employed. Turning tools can be classified into two main groups: *external* cutting tools and *internal* cutting tools.

Types of tools. Each of these groups includes the following types of tools:

1. **Turning tools.** Turning tools can be either *finishing* or *rough* turning tools. Rough turning tools have small nose radii and are employed when deep cuts are made. Finishing tools have larger nose radii and are used when shallower cuts are made in order to obtain the final required dimensions with good surface finish. Rough turning tools can be right-hand or left-hand tools, depending upon the direction of feed. They can have straight, bent, or offset shanks. Figure 10.5 illustrates the different kinds of turning tools.

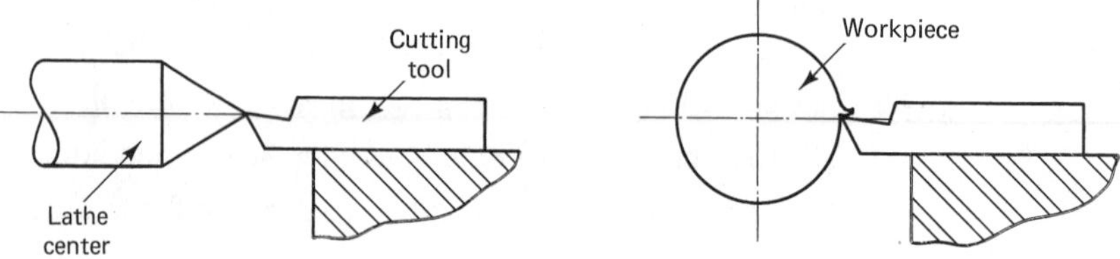

FIGURE 10.4
A simple method for tool setup

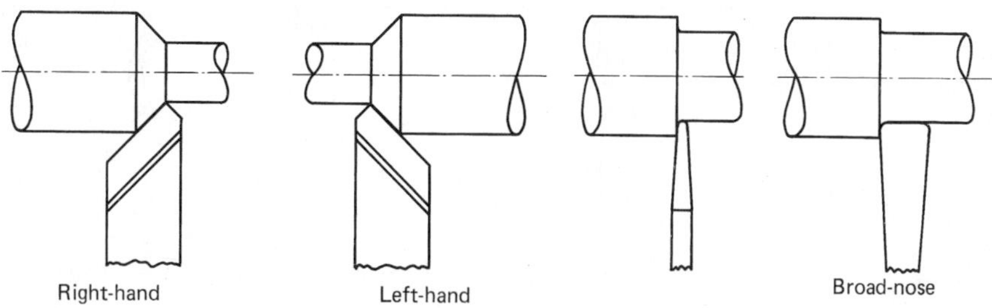

FIGURE 10.5 Different kinds of turning tools

2. **Facing tools.** Facing tools are employed in facing operations for machining flat side or end surfaces. As can be seen in Figure 10.6, there are tools for machining both left and right side surfaces. These side surfaces are generated through the use of cross feed, contrary to turning operations, where longitudinal feed is used.

3. **Cutoff tools.** Cutoff tools, which are sometimes called *parting tools,* serve to separate the workpiece into parts and/or machine external annular grooves, as shown in Figure 10.7.

4. **Thread-cutting tools.** Thread-cutting tools have either triangular, square, or trapezoidal cutting edges, depending upon the cross section of the desired thread. Also, the plane angles of these tools must always be identical to those of the thread forms. Thread-cutting tools have straight shanks for external thread cutting and bent shanks for internal thread cutting. Figure 10.8 illustrates the different shapes of thread-cutting tools.

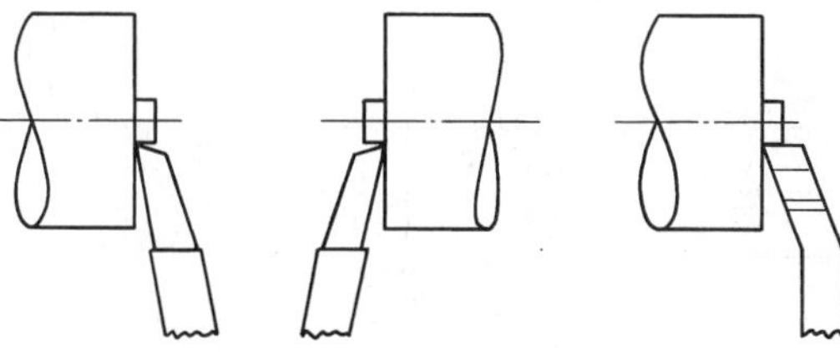

FIGURE 10.6 Different kinds of facing tools

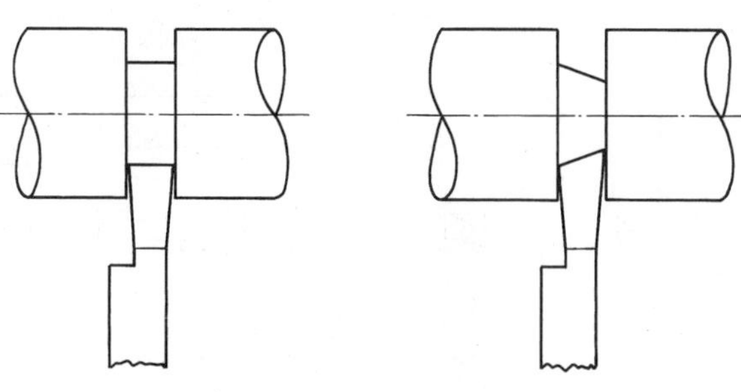

FIGURE 10.7 Cutoff tools

FIGURE 10.8
Different shapes of thread-cutting tools

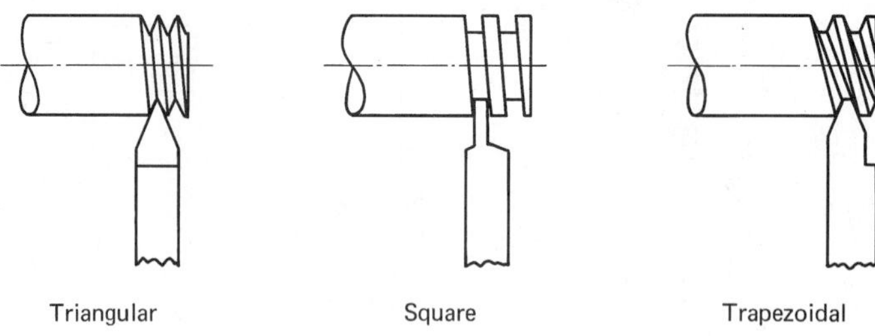

5. **Form tools.** As shown in Figure 10.9, form tools have edges specially manufactured to take a form that is opposite to the desired shape of the machined workpiece.

Internal and external tools. The types of internal cutting tools are similar to those of the external cutting tools. They include tools for rough turning, finish turning, thread cutting, and recess machining. Figure 10.10 illustrates the different types of internal cutting tools.

Carbide tips. As previously mentioned, a high-speed steel tool is usually made in the form of a single piece, contrary to cemented carbides or ceramics, which are made in the form of tips. The tips are brazed or mechanically fastened to steel shanks. Figure 10.11 shows an arrangement that includes a carbide tip, a chip breaker, a seat, a clamping screw (with a washer and a nut), and a shank. As its name suggests, the function of a *chip breaker* is to break long chips every now and then, thus preventing the formation of very long, twisted ribbons that may cause problems during the machining operation. As shown in Figure 10.12, the carbide tips (or ceramic tips) have different shapes, depending upon the machining operations for which they are to be employed. The tips can either be solid or have a central through hole, depending upon whether brazing or mechanical clamping is employed for mounting the tip on the shank.

FIGURE 10.9
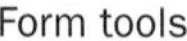
Form tools

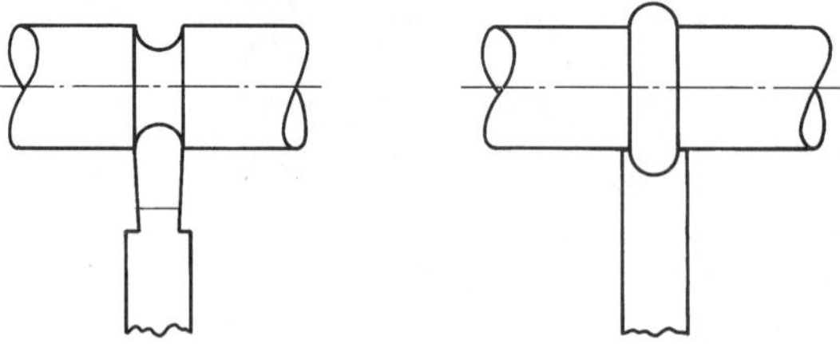

FIGURE 10.10
Different types of internal cutting tools

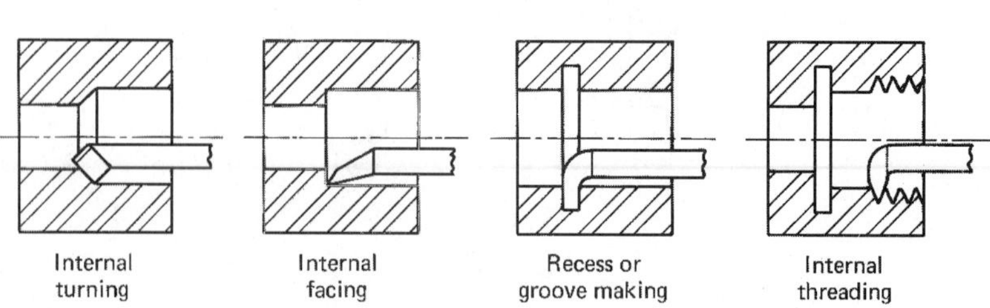

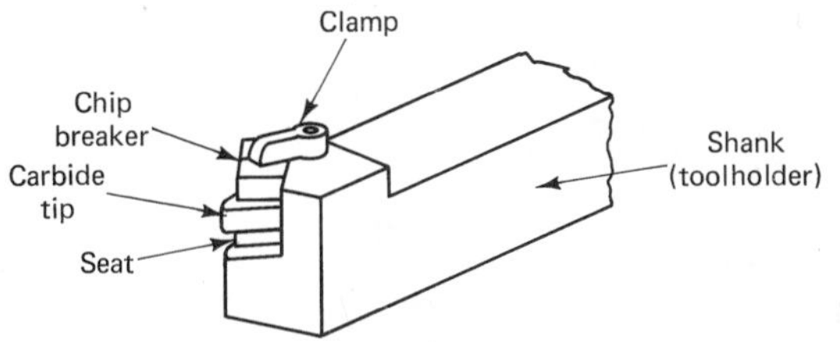

FIGURE 10.11
A carbide tip fastened to a toolholder

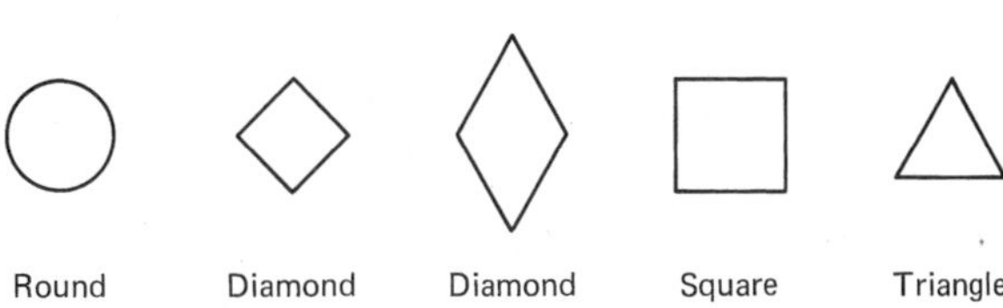

FIGURE 10.12
Different shapes of carbide tips

Methods of Supporting Workpieces in Lathe Operations

Some precautions must be taken when mounting workpieces on a lathe to ensure trouble-free machining. They can be summarized as follows:

1. It is recommended that an appropriate gripping force that is neither too high nor too low be used. A high gripping force may result in distortion of the workpiece after the turning operation, whereas a low gripping force causes either vibration of the workpiece or slip between the workpiece and the spindle (i.e., the rotational speed, or rpm, of the workpiece will be lower than that of the spindle).

2. The workpiece must be fully balanced, both statically and dynamically, by employing counterweights and the like if necessary.

3. The cutting force should not affect the shape of the workpiece or cause any permanent deformation. A manufacturing engineer should calculate the cutting force using his or her knowledge of metal cutting (Chapter 9) and then check whether or not such a force will cause permanent deformation by using stress analysis. Such calculations are very important when machining slender workpieces (i.e., those with high length-to-diameter ratios). Whenever it becomes evident that the cutting force will cause permanent deformation, the machining parameters must be changed to reduce the magnitude of the force (e.g., use a smaller depth of cut or lower feed).

Following is a brief discussion of each of the work-holding methods employed in lathe operations.

Holding the workpiece between two centers. The workpiece is held between two centers when turning long workpieces like shafts and axles having length-to-diameter ratios higher than 3 or 4. Before a workpiece is held, each of its flat ends must be prepared by drilling a 60° center hole. The pointed edges of the *live center* (mounted in the tailstock so that its conical part rotates freely with the workpiece) and the *dead center* (mounted in the spindle hole) are inserted in the previously drilled center holes.

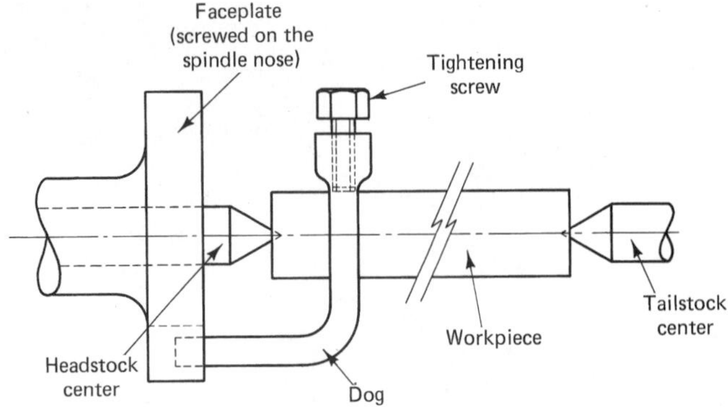

FIGURE 10.13
Holding the workpiece between two centers during turning

As shown in Figure 10.13, a driving dog is clamped on the left end of the workpiece by means of a tightening screw. The tail of the lathe dog enters a slot in the driving-dog plate (or faceplate), which is screwed on the spindle nose.

When very long workpieces having length-to-diameter ratios of 10 or more are turned between centers, rests must be used to provide support and prevent sagging of the workpiece at its middle. *Steady rests* are clamped on the lathe bed and thus do not move during the machining operation; *follower rests* are bolted to and travel with the carriage. A steady rest employs three adjustable fingers to support the workpiece. However, in high-speed turning, the steady rest should involve balls and rollers at the end of the fingers where the workpiece is supported. A follower rest has only two fingers and supports the workpiece against the cutting tool. A steady rest can be used as an alternative to the tailstock for supporting the right-hand end of the workpiece. Figure 10.14 illustrates a steady rest used to support a very long workpiece.

Holding the workpiece in a chuck. When turning short workpieces and/or when performing facing operations, the workpiece is held in a *chuck,* which is screwed on the spindle nose. A universal, self-centering chuck has three jaws that can be moved sep-

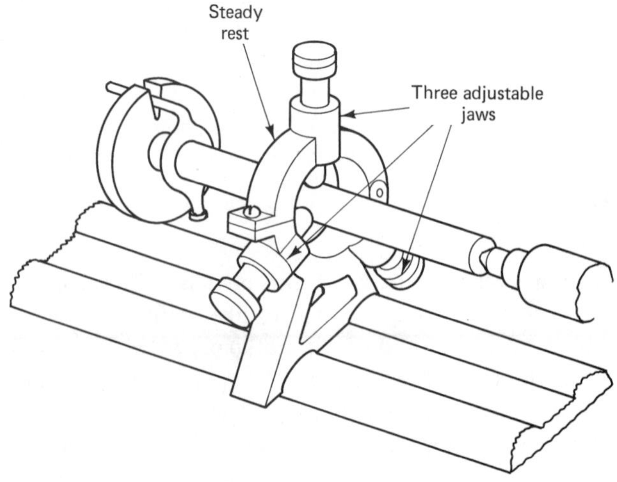

FIGURE 10.14
A steady rest used to support a very long workpiece

arately or simultaneously in radial slots toward its center to grip the workpiece or away from its center to release the workpiece. This movement is achieved by inserting a chuck wrench into a square socket and then turning it as required. Four-jaw chucks are also employed; these are popular when turning complex workpieces and those having asymmetric shapes. Magnetic chucks (without jaws) are used to hold thin, flat workpieces for facing operations. There are also pneumatic and hydraulic chucks, and they are utilized for speeding up the processes of loading and unloading the workpieces. Figure 10.15 shows how a workpiece is held in a chuck.

Mounting the workpiece on a faceplate. A *faceplate* is a large circular disk with radial plain slots and T-slots in its face. The workpiece can be mounted on it with the help of bolts, T-nuts, and other means of clamping. The faceplate is usually employed when the workpiece to be gripped is large or noncircular or has an irregular shape and cannot, therefore, be held in a chuck. Before any machining operation, the faceplate and the workpiece must be balanced by a counterweight mounted opposite to the workpiece on the faceplate, as shown in Figure 10.16.

Using a mandrel. Disklike workpieces or those that have to be machined on both ends are mounted on *mandrels,* which are held between the lathe centers. In this case, the mandrel acts like a fixture and can take different forms. As Figure 10.17 shows, a

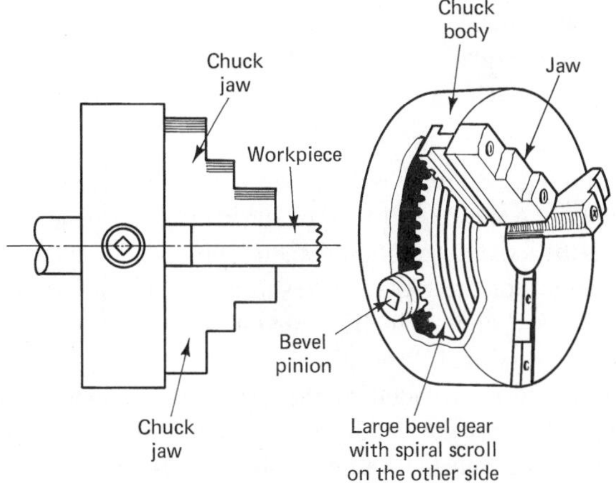

FIGURE 10.15 Holding the workpiece in a chuck

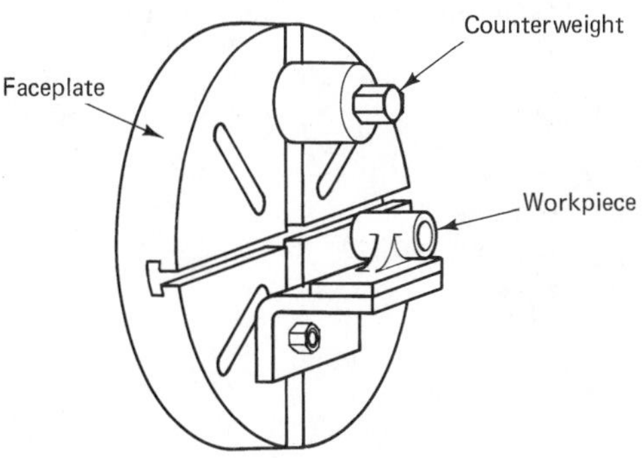

FIGURE 10.16 Mounting the workpiece on a faceplate

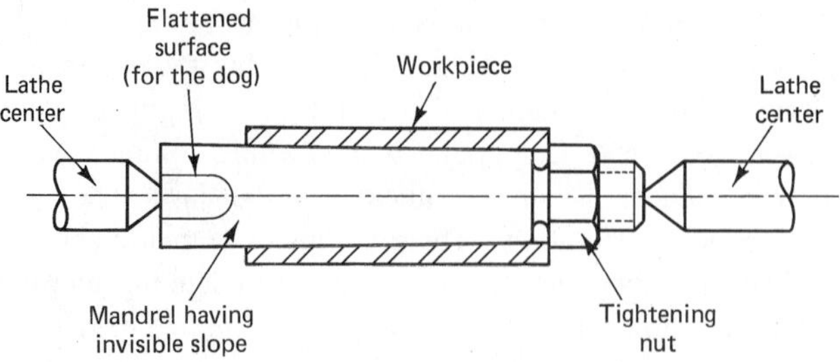

FIGURE 10.17
Mounting the workpiece on a mandrel

mandrel can be a truncated conical rod with an intangible slope on which the workpiece is held by the wedge action. A split sleeve that is forced against a conical rod is also employed. There are also some other designs for mandrels.

Holding the workpiece in a chuck collet. A *chuck collet* consists of a three-segment split sleeve with an external tapered surface. The collet can grip a smooth bar placed between these segments when a collet sleeve, which is internally tapered, is pushed against the external tapered surface of the split sleeve, as shown in Figure 10.18.

Lathe Operations

The following sections focus on the various machining operations that can be performed on a conventional engine lathe. It must be born in mind, however, that modern computerized numerically controlled (CNC) lathes have more capabilities and can do other operations, such as contouring, for example. Following are the conventional lathe operations.

Cylindrical turning. *Cylindrical turning* is the simplest and the most common of all lathe operations. A single full turn of the workpiece generates a circle whose center falls on the lathe axis; this motion is then reproduced numerous times as a result of the axial feed motion of the tool. The resulting machining marks are, therefore, a helix having a very small pitch, which is equal to the feed. Consequently, the machined surface is always cylindrical.

The axial feed is provided by the carriage or compound rest, either manually or automatically, whereas the depth of cut is controlled by the cross slide. In roughing cuts, it is recommended that large depths of cuts, up to 1/4 inch (6 mm) depending upon the workpiece material, and smaller feeds be used. On the other hand, very fine feeds, smaller depths of cut, less than 0.05 inch (0.4 mm), and high cutting speeds are

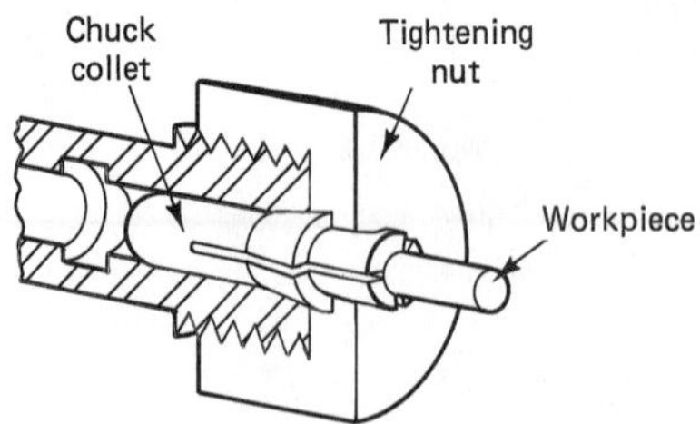

FIGURE 10.18
Holding the workpiece in a chuck collet

FIGURE 10.19
Equations applicable to lathe operations

Operation		Cutting Speed	Machining Time	Material-removal Rate
Turning (external)	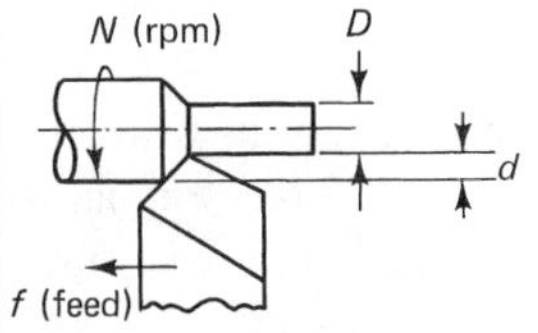 N (rpm), D, d, f (feed)	$V = \pi(D + 2d)N$	$T = \frac{L}{fN}$ where $L = L_{workpiece}$ + allowance i.e., length of the workpiece plus allowance	$MRR = \pi(D + d)N \cdot f \cdot d$
Boring	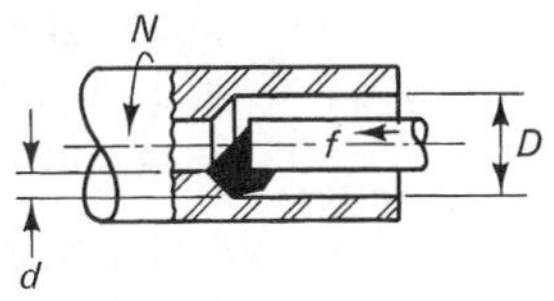 N, f, D, d	$V = \pi DN$	$T = \frac{L}{fN}$	$MRR = \pi(D - d)N \cdot f \cdot d$
Facing	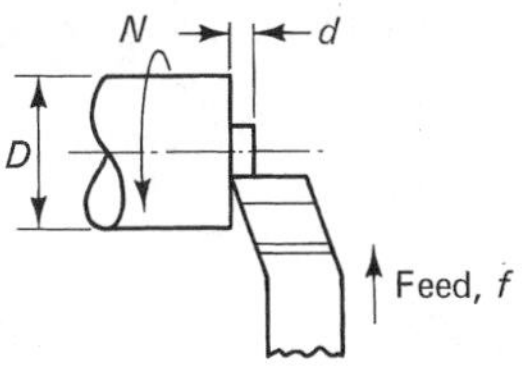 N, d, D, Feed, f	max. $V = \pi DN$ min. $V = 0$ mean $V = \frac{\pi DN}{2}$	$T = \frac{D + \text{allowance}}{2fN}$	max. $MRR = \pi DN \cdot f \cdot d$ mean $MRR = \frac{\pi DN \cdot f \cdot d}{2}$
Parting	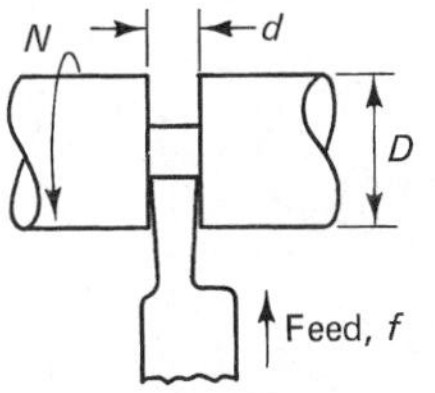 N, d, D, Feed, f	max. $V = \pi DN$ min. $V = 0$ mean $V = \frac{\pi DN}{2}$	$T = \frac{D + \text{allowance}}{2fN}$	max. $MRR = \pi DN \cdot f \cdot d$ mean $MRR = \frac{\pi DN \cdot f \cdot d}{2}$

preferred for finishing cuts. Figure 10.19 indicates the equations used to estimate the different machining parameters in cylindrical turning.

Facing. The result of a *facing* operation is a flat surface that is either the entire end surface of the workpiece or an annular intermediate surface like a shoulder. During a facing operation, feed is provided by the cross slide, whereas the depth of cut is controlled by the carriage or compound rest. Facing can be carried out either from the periphery inward or from the center of the workpiece outward. It is obvious that the machining marks in both cases take the form of a spiral. Usually, it is preferred to clamp the carriage during a facing operation as the cutting force tends to push the tool (and, of course, the whole carriage) away from the workpiece. In most facing operations, the workpiece is held in a chuck or on a faceplate. Figure 10.19 also indicates the equations applicable to facing operations.

Groove cutting. In *cutoff* and *groove-cutting* operations, only cross feed of the tool is employed. The cutoff and grooving tools that were previously discussed are employed.

Boring and internal turning. *Boring* and *internal turning* are performed on the internal surfaces by a boring bar or suitable internal cutting tool. If the initial workpiece is solid, a drilling operation must be performed first. The drilling tool is held in the tailstock, which is then fed against the workpiece.

Taper turning. *Taper turning* is achieved by driving the tool in a direction that is not parallel to the lathe axis but inclined to it with an angle that is equal to the desired angle of the taper. Following are the different methods used in taper turning:

1. One method is to rotate the disk of the compound rest with an angle equal to half the apex angle of the cone, as is shown in Figure 10.20. Feed is manually provided by cranking the handle of the compound rest. This method is recommended for the taper turning of external and internal surfaces when the taper angle is relatively large.
2. Special form tools can be used for external, very short, conical surfaces, as shown in Figure 10.21. The width of the workpiece must be slightly smaller than that of the tool, and the workpiece is usually held in a chuck or clamped on a faceplate. In

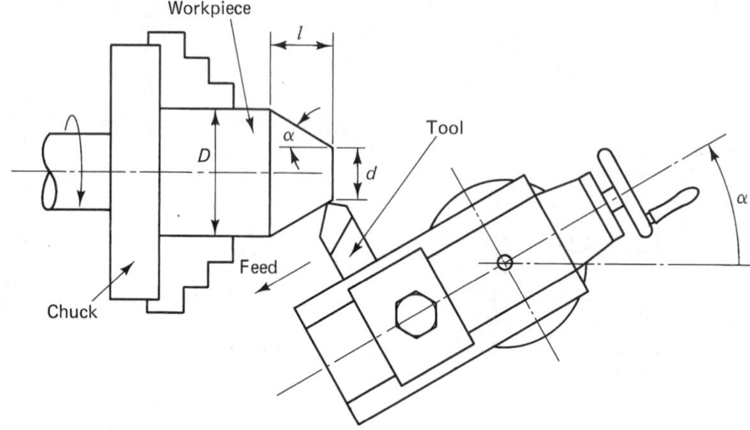

FIGURE 10.20 Taper turning by rotating the disk of the compound rest

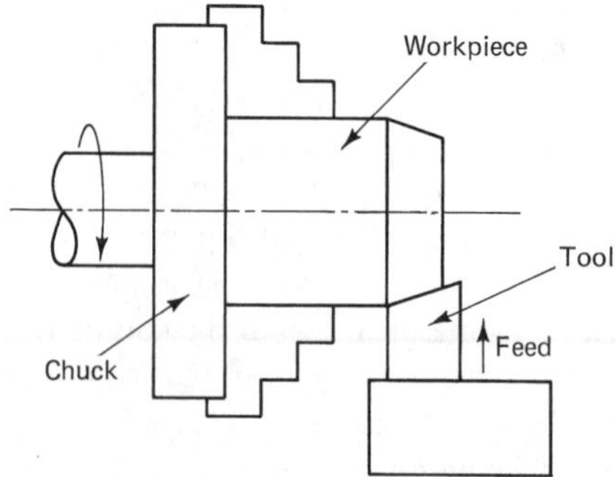

FIGURE 10.21 Taper turning by employing a form tool

this case, only the cross feed is used during the machining process, and the carriage is clamped to the machine bed.

3. The method of offsetting the tailstock center, as shown in Figure 10.22, is employed for the external taper turning of long workpieces that are required to have small taper angles (less than 8°). The workpiece is mounted between the two centers; then the tailstock center is shifted a distance S in the direction normal to the lathe axis. This distance can be obtained from the following equation:

$$S = \frac{L(D - d)}{2\ell} \tag{10.1}$$

where: L is the full length of the workpiece
D is the largest diameter of the workpiece
d is the smallest diameter of the workpiece
ℓ is the length of the tapered surface

4. A special taper-turning attachment, such as the one shown in Figure 10.23, is used for turning very long workpieces, when the length is larger than the full stroke of the compound rest. The procedure followed in such cases involves complete disengagement of the cross slide from the carriage, which is then guided by the taper-turning attachment. During this process, the automatic axial feed can be used as usual. This method is recommended for very long workpieces with a small cone angle (8° through 10°).

Thread cutting. For *thread cutting,* the axial feed must be kept at a constant rate, which is dependent upon the rotational speed (rpm) of the workpiece. The relationship between both is determined primarily by the desired pitch of the thread to be cut.

As previously mentioned, the axial feed is automatically generated when cutting a thread by means of the lead screw, which drives the carriage. When the lead screw rotates

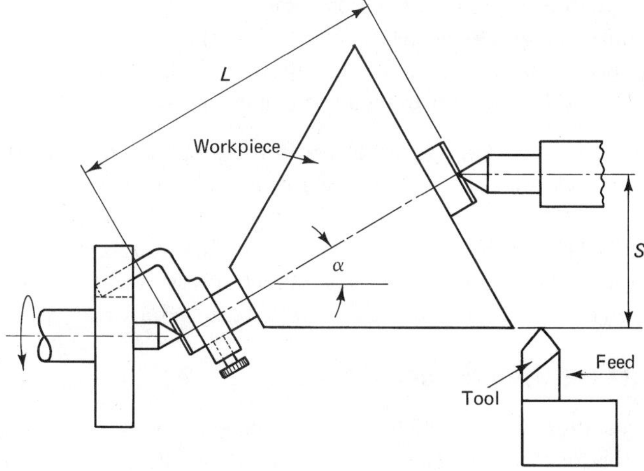

FIGURE 10.22
Taper turning by offsetting the tailstock center

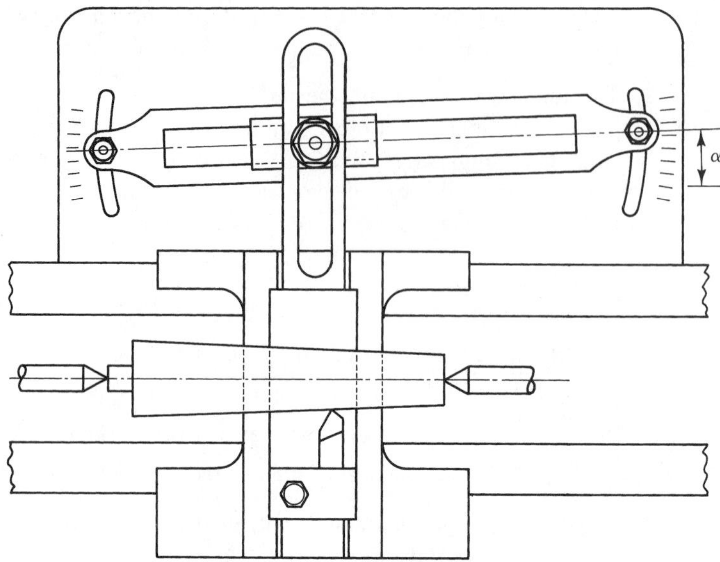

FIGURE 10.23 Taper turning by employing a special attachment

a single revolution, the carriage travels a distance equal to the pitch of the lead screw. Consequently, if the rotational speed of the lead screw is equal to that of the spindle (i.e., that of the workpiece), the pitch of the resulting cut thread is exactly equal to that of the lead screw. The pitch of the resulting thread being cut, therefore, always depends upon the ratio of the rotational speeds of the lead screw and the spindle:

$$\frac{\text{pitch of lead screw}}{\text{desired pitch of workpiece}} = \frac{\text{rpm of workpiece}}{\text{rpm of lead screw}} = \text{spindle-to-carriage gearing ratio} \quad \textbf{(10.2)}$$

This equation is useful in determining the kinematic linkage between the lathe spindle and the lead screw and enables proper selection of the gear train between them.

In thread-cutting operations, the workpiece can be either held in a chuck or mounted between two lathe centers for relatively long workpieces. The form of the tool used must exactly coincide with the profile of the thread to be cut (i.e., triangular tools must be used for triangular threads, and so on).

Knurling. *Knurling* is basically a forming operation in which no chips are produced. It involves pressing two hardened rolls with rough filelike surfaces against the rotating workpiece to cause plastic deformation of the workpiece metal, as shown in Figure 10.24. Knurling is carried out to produce rough, cylindrical (or conical) surfaces that are usually used as handles. Sometimes, surfaces are knurled just for the sake of decoration, in which case there are different knurl patterns to choose from.

Cutting Speeds and Feeds

The cutting speed, which is usually given in surface feet per minute (SFM), is the number of feet traveled in the circumferential direction by a given point on the surface (being cut) of the workpiece in one minute. The relationship between the surface speed

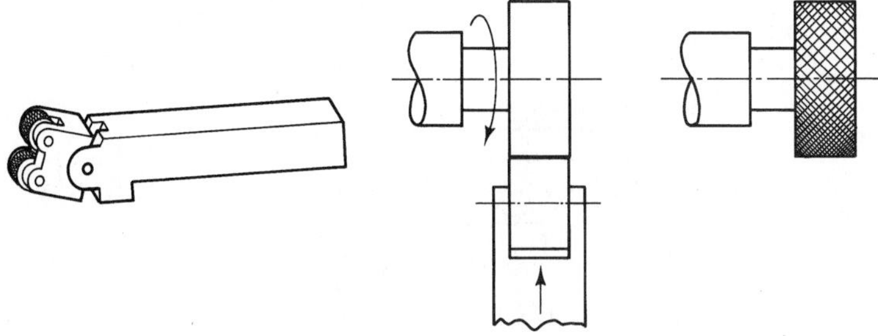

FIGURE 10.24
The knurling operation

and the rpm can be given by the following equation:

$$\text{SFM} = \pi DN \qquad \text{(see Table 10.1)}$$

where: D is the diameter of the workpiece in feet
N is the rpm

The surface cutting speed is dependent upon the material being machined as well as the material of the cutting tool and can be obtained from handbooks and information provided by cutting-tool manufacturers. Generally, the SFM is taken as 100 when machining cold-rolled or mild steel, as 50 when machining tougher metals, and as 200 when machining softer materials. For aluminum, the SFM is usually taken as 400 or above. There are also other variables that affect the optimal value of the surface cutting speed. These include the tool geometry, the type of lubricant or coolant, the feed, and the depth of cut. As soon as the cutting speed is decided upon, the rotational speed (rpm) of the spindle can be obtained as follows:

$$N = \frac{\text{SFM}}{\pi D} \tag{10.3}$$

The selection of a suitable feed depends upon many factors, such as the required surface finish, the depth of cut, and the geometry of the tool used. Finer feeds will produce better surface finish, whereas higher feeds reduce the machining time during which the tool is in direct contact with the workpiece. Therefore, it is generally recommended to use high feeds for roughing operations and finer feeds for finishing operations. Again, recommended values for feeds, which can be taken as guidelines, are found in handbooks and in information booklets provided by cutting-tool manufacturers.

Design Considerations for Turning

When designing parts to be produced by turning, the product designer must consider the possibilities and limitations of the turning operation as well as the machining cost. The cost increases with the quality of the surface finish, with the tightness of the tolerances, and with the area of the surface to be machined. Therefore, it is not recommended that high-quality surface finishes or tighter tolerances be used in the product design unless they are required for the proper functioning of the product. Figure 10.25

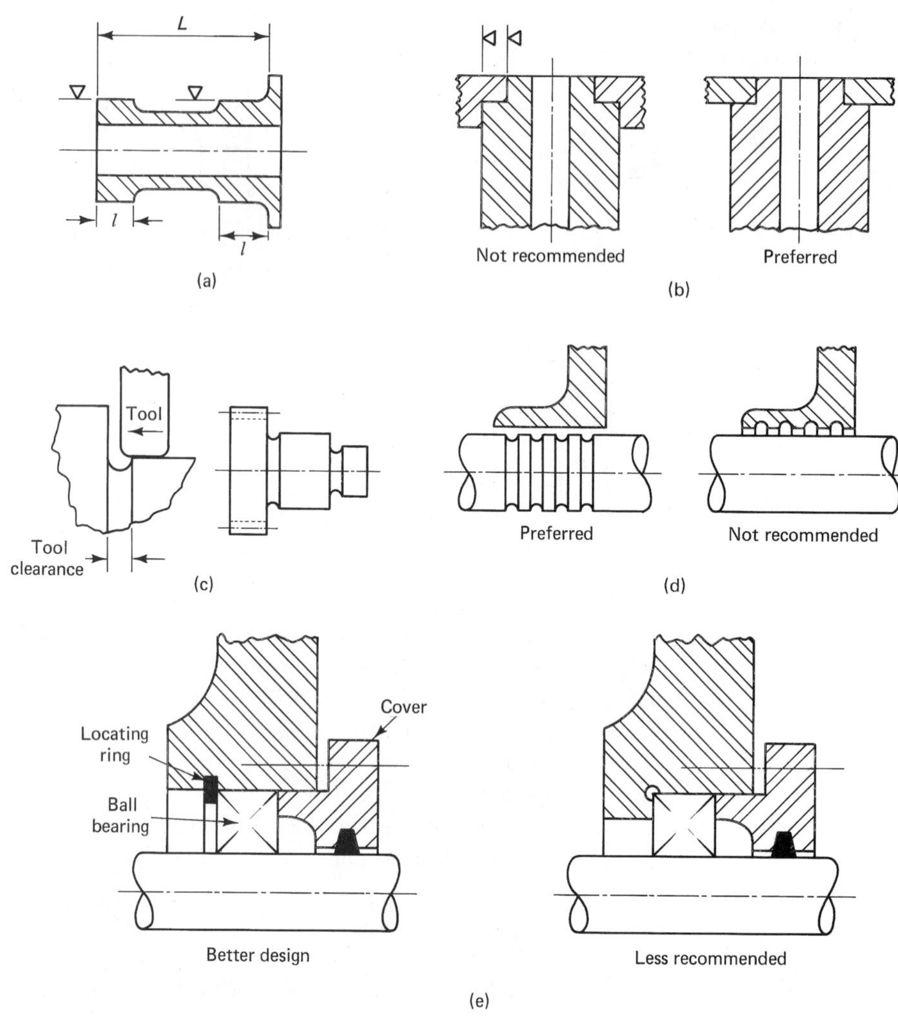

FIGURE 10.25 Design considerations for turning: (a) reduce area of surface to be machined; (b) reduce number of operations required; (c) provide allowance for tool clearance; (d) opt for machining external over internal surfaces; (e) opt for through boring over alternatives

graphically depicts some design considerations for turning. Here are the guidelines to be followed:

1. Try to reduce the area of the surfaces to be machined, especially when a large number of parts is required or when the surfaces are to mate with other parts (see Figure 10.25a).

2. Try to reduce the number of operations required by appropriate changes in the design (see Figure 10.25b).

3. Provide an allowance for tool clearance between different sections of a product (see Figure 10.25c).

4. Always keep in mind that machining of exposed surfaces is easier and less expensive than machining of internal surfaces (see Figure 10.25d).

5. Remember that through boring is easier and cheaper than other alternatives (see Figure 10.25e).

10.2 SHAPING AND PLANING OPERATIONS

Planing, shaping, and *slotting* are processes for machining horizontal, vertical, and inclined flat surfaces, slots, or grooves by means of a lathe-type cutting tool. In all these processes, the cutting action takes place along a straight line. In planing, the workpiece (and the machine bed) is reciprocated, and the tool is fed across the workpiece to reproduce another straight line, thus generating a flat surface. In shaping and slotting, the cutting tool is reciprocated, and the workpiece is fed normal to the direction of tool motion. The difference between the latter two processes is that the tool path is horizontal in shaping and it is vertical in slotting. Shapers and slotters can be employed in cutting external and internal keyways, gear racks, dovetails, and T-slots. Shapers and planers have become virtually obsolete because most shaping and planing operations have been replaced by more productive processes such as milling, broaching, and abrasive machining. The use of shapers and planers is now limited to the machining of large beds of machine tools and the like.

In all three processes, there are successive alternating cutting and idle return strokes. The cutting speed is, therefore, the speed of the tool (or the workpiece) in the direction of cutting during the working stroke. The cutting speed may be either constant throughout the working stroke or variable, depending upon the design of the shaper or planer. Let us now discuss the construction and operation of the most common types of shapers and planers.

Horizontal Push-Cut Shaper

Construction. As can be seen in Figure 10.26, a *horizontal push-cut shaper* consists of a frame that houses the speed gearbox and the quick-return mechanism that transmits power from the motor to the ram and the table. The ram travel is the primary motion that produces a straight-line cut in the working stroke, whereas the intermittent cross travel of the table is responsible for the cross feed. The tool head is mounted at the front end of the ram and carries the clapper box toolholder. The toolholder is pivoted at its upper end to allow the tool to rise during the idle return stroke in order not to ruin the newly machined surface. The tool head can be swiveled to permit the machining of inclined surfaces.

The workpiece can be either bolted directly to the machine table or held in a vise or other suitable fixture. The cross feed of the table is generated by a ratchet and pawl mechanism that is driven through the quick-return mechanism (i.e., the crank and the slotted arm). The machine table can be raised or lowered by means of a power screw and a crank handle. It can also be swiveled in a universal shaper.

Quick-return mechanism. As can be seen in Figure 10.27, the *quick-return mechanism* involves a rotating crank that is driven at a uniform angular speed and an oscillating slotted arm that is connected to the crank by a sliding block. The working stroke takes up an angle (of the crank revolution) that is larger than that of the return stroke.

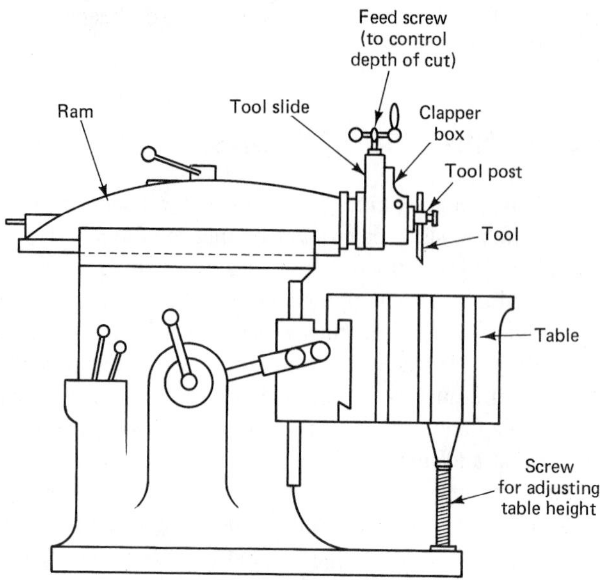

FIGURE 10.26
Design features of a horizontal push-cut shaper

Because the angular speed of the crank is constant, it is obvious that the time taken by the idle return stroke is less than that taken by the cutting stroke. In fact, it is the main function of the quick-return mechanism to reduce the idle time during the machining operation to a minimum.

Now, let us consider the average speed (s) of the tool during the cutting stroke. It can be determined as a function of the length of the stroke and the number of strokes per minute as follows:

$$s = \frac{2LN}{C} \text{ in ft/min. (m/min.)} \tag{10.4}$$

where: L is the length of stroke in feet (m)
N is the number of strokes per minute
C is the cutting time ratio

Note that the cutting time ratio is

$$C = \frac{\text{cutting time}}{\text{total time for one crank revolution}} = \frac{\text{angle corresponding to cutting stroke}}{2\pi}$$

It is also obvious that the total number of strokes required to machine a given surface can be given by the following equation:

$$n = \frac{W}{f} \tag{10.5}$$

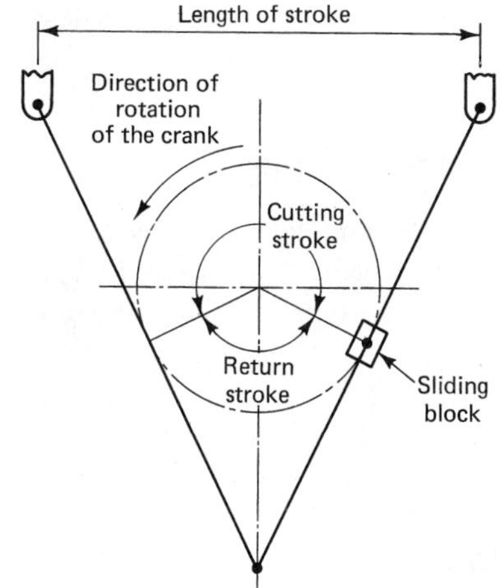

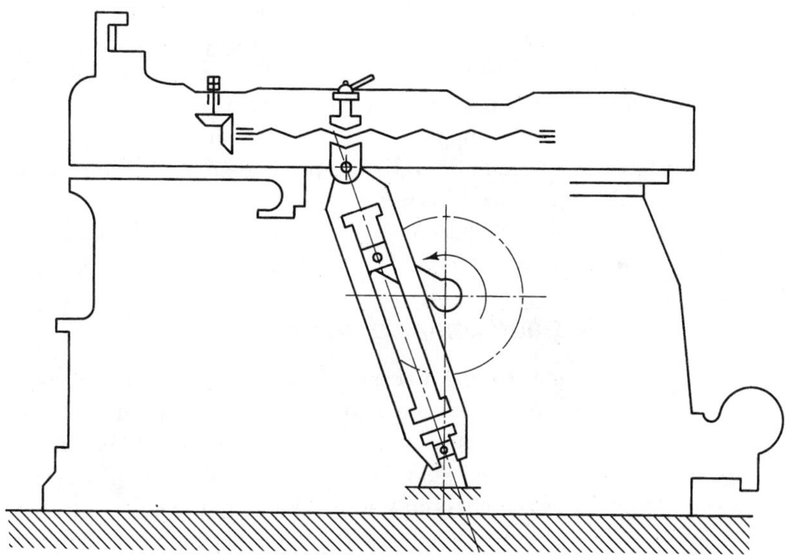

FIGURE 10.27 Details and working principles of the quick-return mechanism

where: W is the total width of the workpiece

f is the cross feed (e.g., inches per stroke)

Therefore, the machining time T is n/N. After mathematical manipulation, it can be given as follows:

$$T = \frac{2LN}{s \times C} \tag{10.6}$$

Next, the metal-removal rate (MRR) can be given by the following equation:

$$\text{MRR} = T \times f \times L \times N \text{ (in.}^3\text{/min.)} \tag{10.7}$$

Vertical Shaper

The *vertical shaper* is similar in construction and operation to the push-cut shaper, the difference being that the ram and the tool head travel vertically instead of horizontally. Also, in this type of shaper, the workpiece is mounted on a round table that can have a rotary feed whenever desired to allow the machining of curved surfaces (e.g., spiral grooves). Vertical shapers, which are sometimes referred to as *slotters,* are used in internal cutting. Another type of vertical shaper is known as a *keyseater* because it is specially designed for cutting keyways in gears, cams, pulleys, and the like.

Planer

A *planer* is a machine tool that does the same work as the horizontal shaper but on workpieces that are much larger than those machined on a shaper. Although the designs of planers vary, most common are the double-housing and open-side constructions. In a *double-housing* planer, two vertical housings are mounted at the sides of the long, heavy bed. A cross rail that is supported at the top of these housings carries the cutting tools. The machine table (while in operation) reciprocates along the guideways of the bed and has T-slots in its upper surface for clamping the workpiece. In this type of planer, the table is powered by a variable-speed dc motor through a gear drive. The cross rail can be raised or lowered as required, and the inclination of the tools can be adjusted as well. In an *open-side* planer, there is only one upright housing at one side of the bed. This construction provides more flexibility when wider workpieces are to be machined.

Planing and Shaping Tools

Planing and shaping processes employ single-point tools of the lathe type, but heavier in construction. They are made of either high-speed steel or carbon tool steel with carbide tips. In the latter case, the machine tool should be equipped with an automatic lifting device to keep the tool from rubbing the workpiece during the return stroke, thus eliminating the possibility of breaking or chipping the carbide tips.

The cutting angles for these tools depend upon the purpose for which the tool is to be used and the material being cut. The end relief angle does not usually exceed 4°, whereas the side relief varies between 6° and 14°. The side rake angle also varies between 5° (for cast iron) and 15° (for medium-carbon steel).

10.3 DRILLING OPERATIONS

Drilling involves producing through or blind holes in a workpiece by forcing a tool that rotates around its axis against the workpiece. Consequently, the range of cutting from this axis of rotation is equal to the radius of the required hole. In practice, two symmetrical cutting edges that rotate about the same axis are employed.

Drilling operations can be carried out by using either hand drills or drilling machines. The latter differ in size and construction. Nevertheless, the tool always rotates around its axis while the workpiece is kept firmly fixed. This is contrary to drilling on a lathe.

Cutting Tools for Drilling Operations

In drilling operations, a cylindrical rotary-end cutting tool, called a *drill,* is employed. The drill can have one or more cutting edges and corresponding flutes that are straight or helical. The function of the flutes is to provide outlet passages for the chips generated during the drilling operation and also to allow lubricants and coolants to reach the cutting edges and the surface being machined. Following is a survey of the commonly used types of drills.

Twist drill. The *twist drill* is the most common type of drill. It has two cutting edges and two helical flutes that continue over the length of the drill body, as shown in Figure 10.28. The drill also consists of a neck and a shank that can be either straight or tapered. A tapered shank is fitted by the wedge action into the tapered socket of the spindle and has a tang that goes into a slot in the spindle socket, thus acting as a solid means for transmitting rotation. Straight-shank drills are held in a drill chuck that is, in turn, fitted into the spindle socket in the same way as tapered-shank drills.

As can be seen in Figure 10.28, the two cutting edges are referred to as the *lips* and are connected together by a *wedge,* which is a chisel-like edge. The twist drill also has two margins that allow the drill to be properly located and guided while it is in operation. The *tool point angle* (TPA) is formed by the two lips and is chosen based on the properties of the material to be cut. The usual TPA for commercial drills is 118°, which is appropriate for drilling low-carbon steels and cast irons. For harder and tougher metals, such as hardened steel, brass, and bronze, larger TPAs (130° or 140°)

FIGURE 10.28
A twist drill

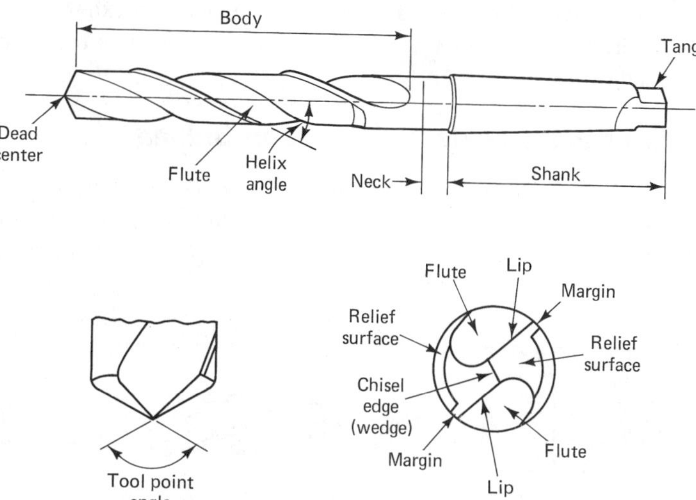

give better performance. The helix angle of the flutes of a twist drill ranges between 24° and 30°. When drilling copper or soft plastics, higher values for the helix angle are recommended (between 35° and 45°). Twist drills are usually made of high-speed steel, although carbide-tipped drills are also available. The sizes of twist drills used in industrial practice range from 0.01 inch to 3½ inches (0.25 up to 80 mm).

Core drill. A *core drill* consists of the chamfer, body, neck, and shank, as shown in Figure 10.29. This type of drill may have three or four flutes and an equal number of margins, which ensures superior guidance, thus resulting in high machining accuracy. The figure also shows that a core drill has a flat end. The chamfer can have three or four cutting edges, or lips, and the lip angle may vary between 90° and 120°. Core drills are employed for enlarging previously made holes and not for originating holes. This type of drill promotes greater productivity, high machining accuracy, and superior quality of the drilled surfaces.

Gun drill. A *gun drill* is used for drilling deep holes. All gun drills are straight-fluted, and each has a single cutting edge. A hole in the body acts as a conduit to transmit coolant under considerable pressure to the tip of the drill. As can be seen in Figure 10.30, there are two kinds of gun drills: the *center-cut* gun drill used for drilling blind holes and the *trepanning* drill. The latter has a cylindrical groove at its center, thus generating a solid core that guides the tool as it proceeds during the drilling operation.

Spade drill. A *spade drill* is used for drilling large holes of 3½ inches (90 mm) or more. The design of this type of drill results in a marked saving in tool cost as well as in a tangible reduction in tool weight that facilitates its ease of handling. Moreover, this drill is easy to grind. Figure 10.31 shows a spade drill.

Saw-type cutter. A *saw-type cutter,* like the one illustrated in Figure 10.32, is used for cutting large holes in thin metal.

Drills made in combination with other tools. An example is a tool that involves both a drill and a tap. Step drills and drill and countersink tools are also sometimes used in industrial practice.

Cutting Speeds and Feeds in Drilling

We can easily see that the cutting speed varies along the cutting edge. It is always maximum at the periphery of the tool and is equal to zero on the tool axis. Nevertheless, we consider the maximum speed because it is the one that affects the tool

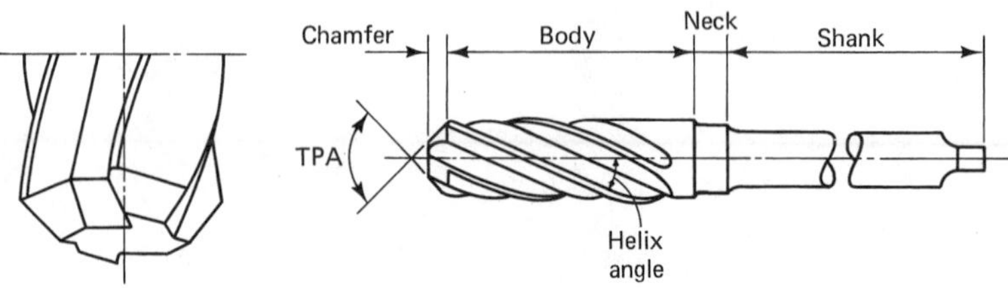

FIGURE 10.29
A core drill

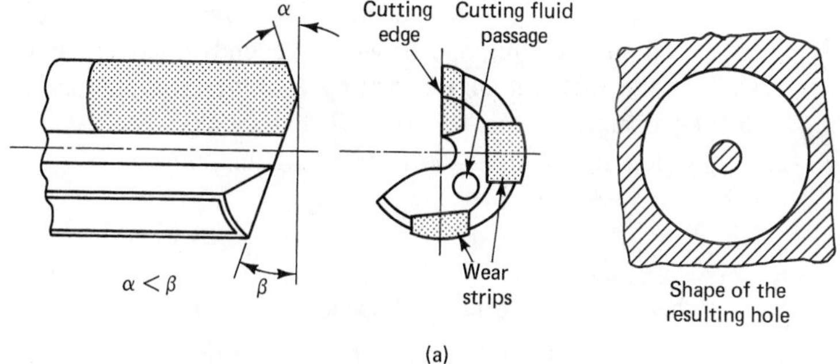

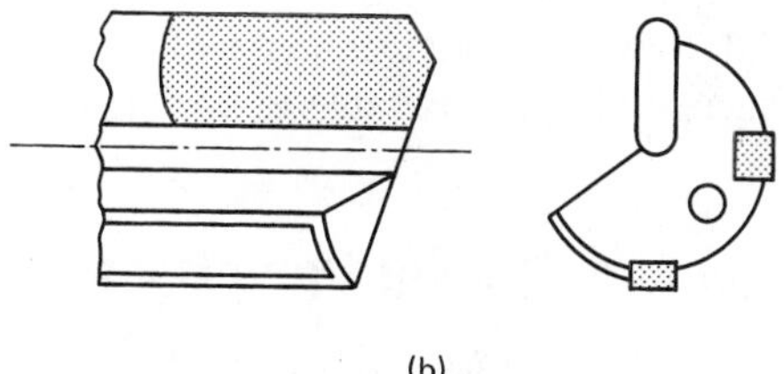

FIGURE 10.30
Gun drills: (a) trepanning gun drill; (b) center-cut gun drill

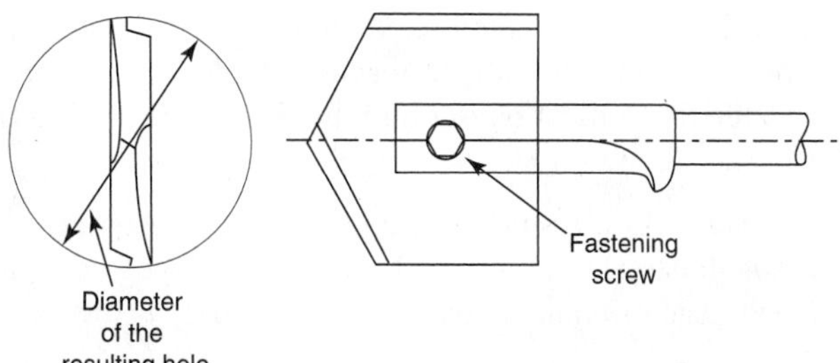

FIGURE 10.31
A spade drill

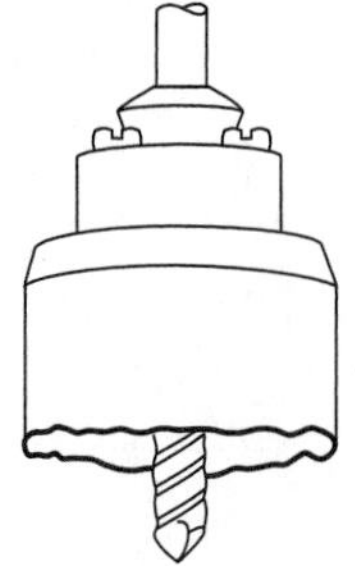

FIGURE 10.32
A saw-type cutter

wear and the quality of the machined surface. The maximum speed must not exceed the permissible cutting speed, which depends upon the material of the workpiece as well as the material of the cutting tool. Data about permissible cutting speeds in drilling operations can be found in handbooks. The rotational speed of the spindle can be determined from the following equation:

$$N = \frac{\text{CS}}{\pi D} \tag{10.8}$$

where: N is the rotational speed of the spindle (rpm)

D is the diameter of the drill in feet (m)

CS is the permissible cutting speed in ft/min. (m/min.)

In drilling operations, feeds are expressed in inches or millimeters per revolution. Again, the appropriate value of feed to be used depends upon the metal of the workpiece and drill material and can be found in handbooks. Whenever the production rate must be increased, it is advisable to use higher feeds rather than increase the cutting speed.

Other Types of Drilling Operations

In addition to conventional drilling, there are other operations that are involved in the production of holes in industrial practice. Following is a brief description of each of these operations.

Boring. *Boring* involves enlarging a hole that has already been drilled. It is similar to internal turning and can, therefore, be performed on a lathe, as previously mentioned. There are also some specialized machine tools for carrying out boring operations. These include the vertical boring mill, the jig boring machine, and the horizontal boring machine.

Counterboring. As a result of *counterboring,* only one end of a drilled hole is enlarged, as is illustrated in Figure 10.33a. This enlarged hole provides a space in which to set a bolt head or a nut so that it will be entirely below the surface of the part.

Spot facing. *Spot facing* is performed to finish off a small surface area around the opening of a hole. As can be seen in Figure 10.33b, this process involves removing a minimal depth of cut and is usually performed on castings or forgings.

Countersinking. As shown in Figure 10.33c, *countersinking* is done to accommodate the conical seat of a flathead screw so that the screw does not appear above the surface of the part.

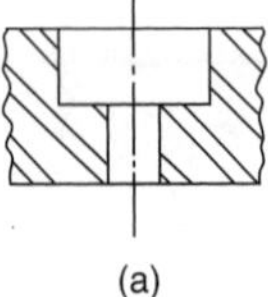
(a)

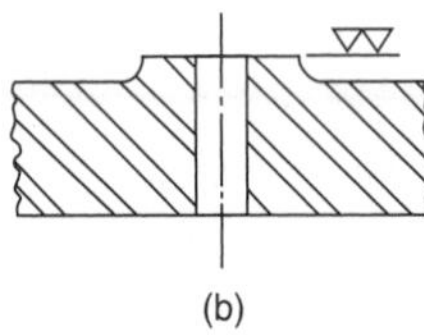
(b)

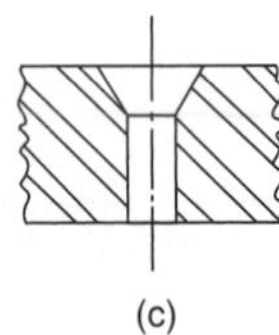
(c)

FIGURE 10.33 Operations related to drilling: (a) counterboring; (b) spot facing; (c) countersinking

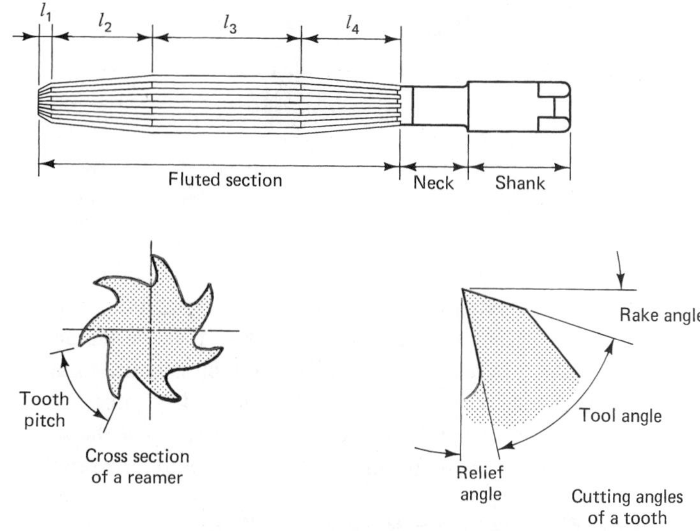

FIGURE 10.34
Details of a reamer

Reaming. *Reaming* is actually a "sizing" process, by which an already drilled hole is slightly enlarged to the desired size. As a result of a reaming operation, a hole has a very smooth surface. The cutting tool used in this operation is known as a *reamer.* As shown in Figure 10.34, a reamer has a fluted section, a neck, and a shank. The fluted section includes four zones: the chamfer, the starting taper, the sizing zone, and the back taper. The chamfer or bevel encloses an angle that depends upon the method of reaming and the material being cut. This is a consequence of the fact that this angle affects the magnitude of the axial reaming force. The larger the chamfer angle, the larger the required reaming force. Table 10.2 indicates some recommended values of the chamfer angle for different reaming conditions. The starting taper is the part of the reamer that actually removes chips. Figure 10.34 also shows that each tooth of that part of the reamer has a cutting edge as well as rake, relief, and tool (or lip) angles. The sizing zone guides the reamer and smooths the surface of the hole. Finally, the back taper serves to reduce friction between the reamer and the newly machined surface.

Reamers are usually made of hardened tool steel. Nevertheless, reamers that are used in mass production are tipped with cemented carbides in order to increase the tool life and the production rate.

Tapping. *Tapping* is the process of cutting internal threads. The tool used is called a *tap.* As shown in Figure 10.35, it has a boltlike shape with four longitudinal flutes. Made of hardened tool steel, taps can be used for either manual or machine cutting of

TABLE 10.2
Recommended values of the chamfer angle of reamers

Metal to Be Reamed	Steel	Cast Iron	Soft Metals
Manual reaming	1°–3°	1°–3°	1°–3°
Machining reaming	8°–10°	20°–30°	3°–5°

FIGURE 10.35
A tap

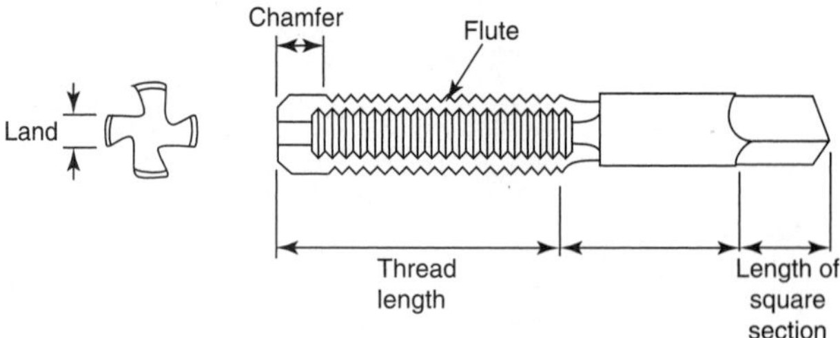

threads. In the latter case, the spindle of the machine tool must reverse its direction of rotation at the end of the cutting stroke so that the tap can be withdrawn without destroying the newly cut thread. When tapping is carried out by hand, a set of three taps is used for each desired threaded hole size. The three taps differ slightly in size, and two of them are actually undersized. The first tap of the set to be used is always a *tapper tap;* it reduces the torque (and, consequently, the power) required for tapping.

Design Considerations for Drilling

Figure 10.36 graphically depicts some design considerations for drilling. Here are the guidelines to be followed:

1. Make sure the centerline of the hole to be drilled is normal to the surface of the part. This is to avoid bending and breaking the tool during the drilling operation. As previously mentioned, the twist drill has a chisel edge and not a pointed edge at its center. This, although it facilitates the process of grinding the tool, causes the tool to shift from the desired location and makes it liable to breakage, especially if it is not normal to the surface to be drilled. (See Figure 10.36a for examples of poor and proper design practice for drilled holes.)
2. When tapping through holes, ensure that the tap will be in the clear when it appears from the other side of the part (see Figure 10.36b).
3. Remember that it is impossible to tap the entire length of a blind or counterbored hole without providing special tool allowance (see Figure 10.36c).

Classification of Drilling Machines

Drilling operations can be carried out by employing small portable machines or by using the appropriate machine tools. These machine tools differ in shape and size, but they have common features. For instance, they all involve one or more twist drills, each rotating around its own axis while the workpiece is kept firmly fixed. This is contrary to the drilling operation on a lathe, where the workpiece is held in and rotates with the chuck. Following is a survey of the commonly used types of drilling machines.

Bench-type drilling machines. *Bench-type* drilling machines are general-purpose, small machine tools that are usually placed on benches. This type of drilling machine includes an electric motor as the source of motion, which is transmitted via pulleys and belts to the spindle, where the tool is mounted. The feed is manually generated by low-

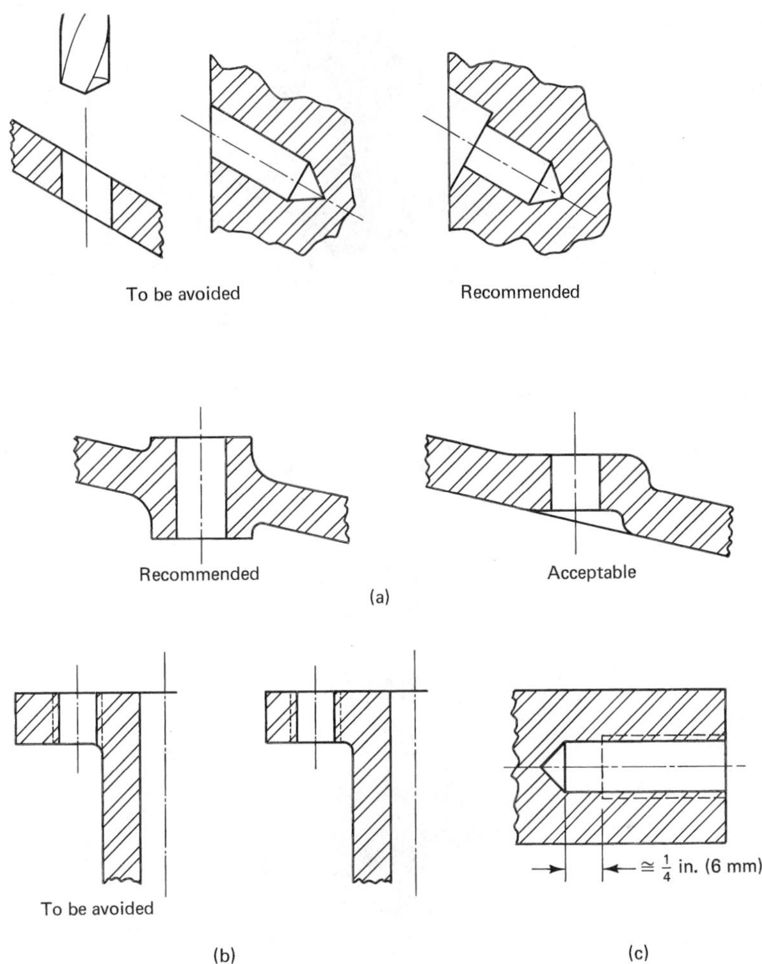

FIGURE 10.36 Design considerations for drilling: (a) set centerline of tool normal to surface to be drilled; (b) ensure tap is clear when it appears from other side; (c) provide allowance when tapping a blind hole

ering a lever handle that is designed to lower (or raise) the spindle. The spindle rotates freely inside a sleeve (which is actuated by the lever through a rack-and-pinion system) but does not rotate with the spindle.

The workpiece is mounted on the machine table, although a special vise is sometimes used to hold the workpiece. The maximum height of a workpiece to be machined is limited by the maximum gap between the spindle and the machine table.

Upright drilling machines. Depending upon the size, *upright* drilling machines can be used for light, medium, and even relatively heavy jobs. A light-duty upright drilling machine is shown in Figure 10.37. It is basically similar to a bench-type machine, the main difference being a longer cylindrical column fixed to the base. Along the column is an additional sliding table for fixing the workpiece that can be locked in position at any desired height. The power required for this type of machine is greater than that for a bench-type drilling machine as this type is employed in performing medium-duty jobs.

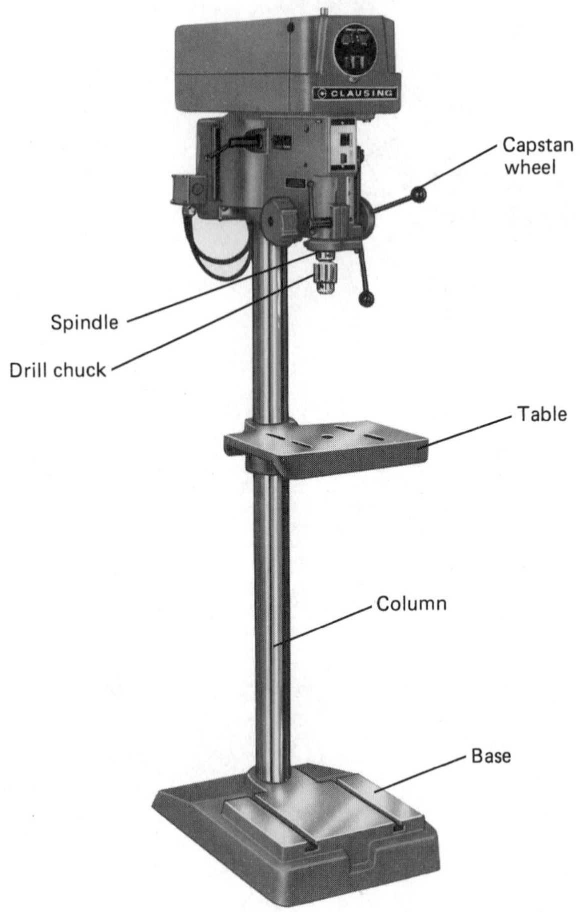

FIGURE 10.37 An upright drilling machine *(Courtesy of Clausing Industrial, Inc., Kalamazoo, Michigan)*

There are also large drilling machines of the upright type. In this case, the machine has a box column and a higher power to deal with large jobs. Moreover, gearboxes are employed to provide different rotational spindle speeds as well as axial feed motion, which can be preset at any desired rate.

Multispindle drilling machines. *Multispindle* drilling machines are sturdily constructed and require high power; each is capable of drilling many holes simultaneously. The positions of the different tools (spindles) can be adjusted as desired. Also, the entire head (which carries the spindles and the tools) can be tilted if necessary. This type of drilling machine is used mainly for mass production in jobs having many holes, such as cylinder blocks.

Gang drilling machines. When several separate heads (each with a single spindle) are arranged on a single common table, the machine tool is then referred to as a *gang* drilling machine. This type of machine tool is particularly suitable where several operations are to be performed in succession.

Radial drills. *Radial* drills are particularly suitable for drilling holes in large and heavy workpieces that are inconvenient to mount on the table of an upright drilling machine. As shown in Figure 10.38, a radial drilling machine has a main column that is fixed to the base. The cantileverd guide arm, which carries the drilling head spindle and tool, can be raised or lowered along the column and clamped at any desired position. The drilling head slides along the arm and provides rotary motion and axial feed motion. The cantilevered guide arm can be swung, thus allowing the tool to be moved in all directions according to a cylindrical coordinate system.

Turret drilling machines. Machine tools that belong in the *turret* drilling machine category are either semiautomatic or fully automatic. A common design feature is that the main spindle is replaced by a turret that carries several drilling, boring, reaming, and threading tools. Consequently, several successive operations can be carried out with only a single initial setup and without the need for setting up the workpiece again between operations.

Automatic turret drilling machines that are operated by NC or CNC systems (see Chapter 14) are quite common. In this case, the human role is limited to the initial setup and monitoring. This type of machine tool has advantages over the gang-type drilling machine with respect to the space required (physical size of the machine tool) and the number of workpiece setups.

Deep-hole drilling machines. *Deep-hole* drilling machines are special machines employed for drilling long holes like those of rifle barrels. Usually, gun-type drills are used and are fed slowly against the workpiece. In this type of machine tool, it is the

FIGURE 10.38
A radial drill

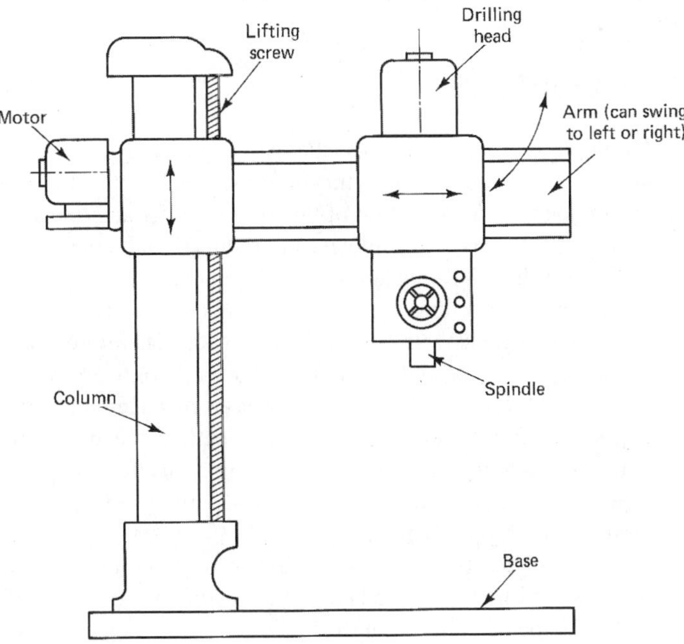

workpiece that is rotated, while the drill is kept from rotary motion. A deep-hole drilling machine may have either a vertical or a horizontal construction. However, in both cases, the common feature is the precise guidance and positive support of the workpiece during the drilling operation.

Jig-boring machines. *Jig-boring* machines are specially designed to possess high precision and accuracy. A machine of this type not only drills the holes but also locates them because the table movements are monitored by electronic measuring devices. Jig-boring machines are usually employed in the manufacture of forming and molding dies, gages, and work-holding devices like jigs and fixtures.

Work-Holding Devices in Drilling

During conventional drilling operations, the workpiece must be held firmly on the machine table. The type of work-holding device used depends upon the shape and the size of the workpiece, the desired accuracy, and the production rate. For low production when the accuracy is not very important, conventional machine vises or vises with V-blocks (for round work) are used. For moderate production and when accuracy is of some importance, jigs are usually employed. A *jig* is a work-holding device that is designed to hold a particular workpiece (i.e., it cannot be used for workpieces having different shapes) and to guide the cutting tool during the drilling operation. This eliminates the need for laying out the workpiece prior to machining, thus saving the time spent in blueing and scribing when no jigs are employed. The design of jigs and fixtures is a separate topic and is beyond the scope of this text. Interested readers are referred to the books dealing with tool design and with jig and fixture design that are given at the end of this text.

10.4 MILLING OPERATIONS

Milling is a machining process that is carried out by means of a multiedge rotating tool known as a *milling cutter.* In this process, metal removal is achieved by simultaneously combining the rotary motion of the milling cutter and linear motions of the workpiece. Milling operations are employed in producing flat, contoured, and helical surfaces, as well as for thread- and gear-cutting operations.

Each of the cutting edges of a milling cutter acts as an individual single-point cutter when it engages with the workpiece metal. Therefore, each of the cutting edges has the appropriate rake and relief angles. Because only a few of the cutting edges are engaged with the workpiece at a time, heavy cuts can be taken without adversely affecting the tool life. In fact, the permissible cutting speeds and feeds for milling are three to four times higher than those for turning or drilling. Moreover, the quality of the surfaces machined by milling is generally superior to the quality of surfaces machined by turning, shaping, or drilling.

A wide variety of milling cutters is available in industry. This, together with the fact that a milling machine is a very versatile machine tool, make the milling machine the backbone of a machining workshop.

Milling Methods

As far as the direction of cutter rotation and workpiece feed are concerned, milling is performed by either of the following two methods.

Up milling (conventional milling). In *up milling,* the workpiece is fed against the direction of cutter rotation, as shown in Figure 10.39a. The depth of the cut (and, consequently, the load) gradually increases on the successively engaged cutting edges. Therefore, the machining process involves no impact loading, thus ensuring smoother operation of the machine tool and longer tool life. The quality of the machined surface obtained by up milling is not very high. Nevertheless, up milling is commonly used in industry, especially for rough cuts.

Down milling (climb milling). In *down milling,* the cutter rotation coincides with the direction of feed at the contact point between the tool and the workpiece, as shown in Figure 10.39b. The maximum depth of cut is achieved directly as the cutter engages with the workpiece. This results in a kind of impact, or sudden loading. Therefore, this method cannot be used unless the milling machine is equipped with a backlash eliminator on the feed screw. The advantages of this method include higher quality of the machined surface and easier clamping of workpieces as the cutting forces act downward.

Types of Milling Cutters

Milling cutters come in a wide variety of shapes, each designed to effectively perform a specific milling operation. Generally, a milling cutter can be described as a multiedge cutting tool having the shape of a solid of revolution, with the cutting teeth arranged either on the periphery or on an end face or on both. Following is a survey of the commonly used types of milling cutters.

Plain milling cutter. A *plain* milling cutter, as shown in Figure 10.40a, is a disk-shaped cutting tool that may have straight or helical teeth. This type of cutter is always mounted on horizontal milling machines and is used for machining flat surfaces.

Face milling cutter. A *face* milling cutter, like the one in Figure 10.40b, is also used for machining flat surfaces. It is bolted at the end of a short arbor that is, in turn, mounted on a vertical milling machine.

FIGURE 10.39
Milling methods: (a) up milling; (b) down milling

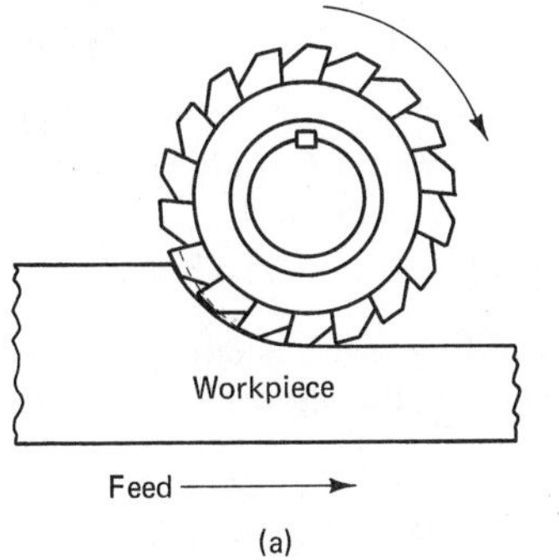

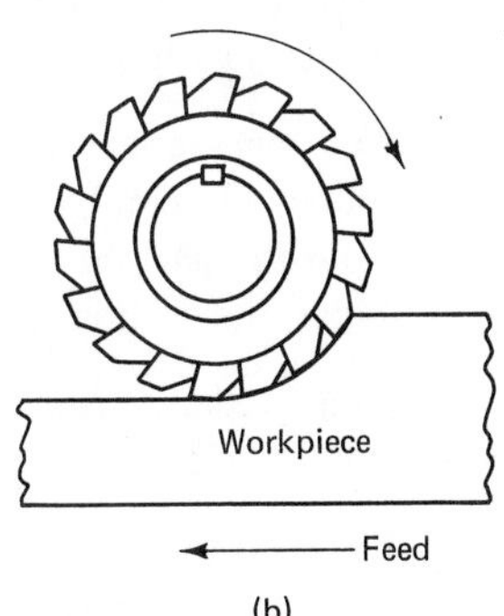

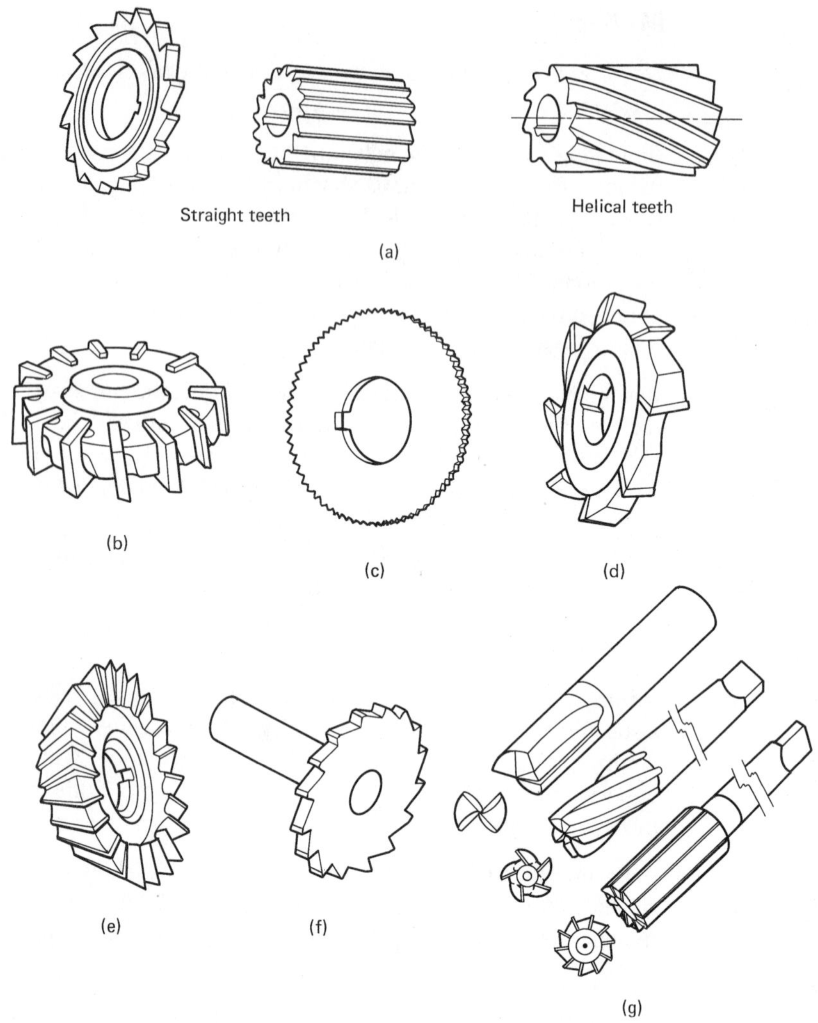

FIGURE 10.40 Types of milling cutters: (a) plain milling cutter; (b) face milling cutter with inserted teeth; (c) plain metal-slitting saw cutter; (d) side milling cutter; (e) angle milling cutter; (f) T-slot cutter; (g) end mill cutter

Plain metal-slitting saw. Figure 10.40c illustrates a plain *metal-slitting* saw cutter. Notice that it actually involves a very thin plain milling cutter.

Side milling cutter. A *side* milling cutter is used for cutting slots, grooves, and splines. As can be seen in Figure 10.40d, it is quite similar to the plain milling cutter, the difference being that this type has teeth on the sides. As is the case with the plain cutter, the cutting teeth can be straight or helical.

Angle milling cutter. An *angle* milling cutter is employed in cutting dovetail grooves, ratchet wheels, and the like. Figure 10.40e shows a milling cutter of this type.

T-slot cutter. As shown in Figure 10.40f, a *T-slot* cutter involves a plain milling cutter with an integral shaft normal to it. As the name suggests, this type of cutter is used for milling T-slots.

End mill cutter. An *end* mill cutter finds common application in cutting slots, grooves, flutes, splines, pocketing work, and the like. As Figure 10.40g indicates, an end mill cutter is always mounted on a vertical milling machine and can have two or four flutes, which may be straight or helical.

Form milling cutter. The teeth of a *form* milling cutter have a shape that is identical to the section of the metal to be removed during the milling operation. Examples of this type of cutter include gear cutters, gear hobs, and convex and concave cutters. Form milling cutters are mounted on horizontal milling machines, as is explained later when we discuss gear cutting.

Materials of Milling Cutters

The commonly used milling cutters are made of high-speed steel, which is generally adequate for most jobs. Milling cutters tipped with sintered carbides or cast nonferrous alloys as cutting teeth are usually employed for mass production, where heavier cuts and/or high cutting speeds are required.

Cutting Speeds and Feeds in Milling

Figure 10.41 indicates methods of estimating the different machining parameters during milling operations. These parameters include the cutting speed, the feed, and the metal-removal rate. The cutting speed is the peripheral velocity at any point on the circumference of the cutter. The allowable value for the cutting speed in milling is dependent upon many factors, including the cutter material, material of the workpiece, diameter and life of the cutter, feed, depth of cut, width of cut, number of teeth on the cutter, and the type of coolant used. The feed in milling operations is the rate of movement of the cutter axis relative to the workpiece. It is expressed in inches (or mm) per revolution or inches (or mm) per minute. It can also be expressed in inches (or mm) per tooth, especially for plain and face milling cutters.

The depth of cut is the thickness of the metal layer that is to be removed in one cut. The maximum allowable depth of cut depends upon the material being machined and is commonly taken up to 5/16 inch (8 mm) in roughing operations and up to 1/16 inch (about 1.5 mm) in finishing operations. Another parameter that affects milling operations is the width of cut, which is the width of the workpiece in contact with the cutter in a direction normal to the feed. The width of cut should decrease with increasing depth of cut to keep the load and power requirement below those that can be met by the cutter and the machine tool, respectively.

Cutting Angles of Milling Cutters

As previously mentioned, the geometry of any tool is basically a matter of rake and relief angles. Figure 10.42 shows the cutting angles of a plain, straight-tooth milling cutter (for simplicity). The radial rake angle facilitates the removal of chips and ranges from 10° to 20°, depending upon the workpiece material to be cut. When machining hard metals with carbide-tipped cutters, a negative rake angle of 10° is usually employed. The relief angle also depends upon the workpiece material and varies between 12° and 25°.

FIGURE 10.41
Equations applicable to milling operations

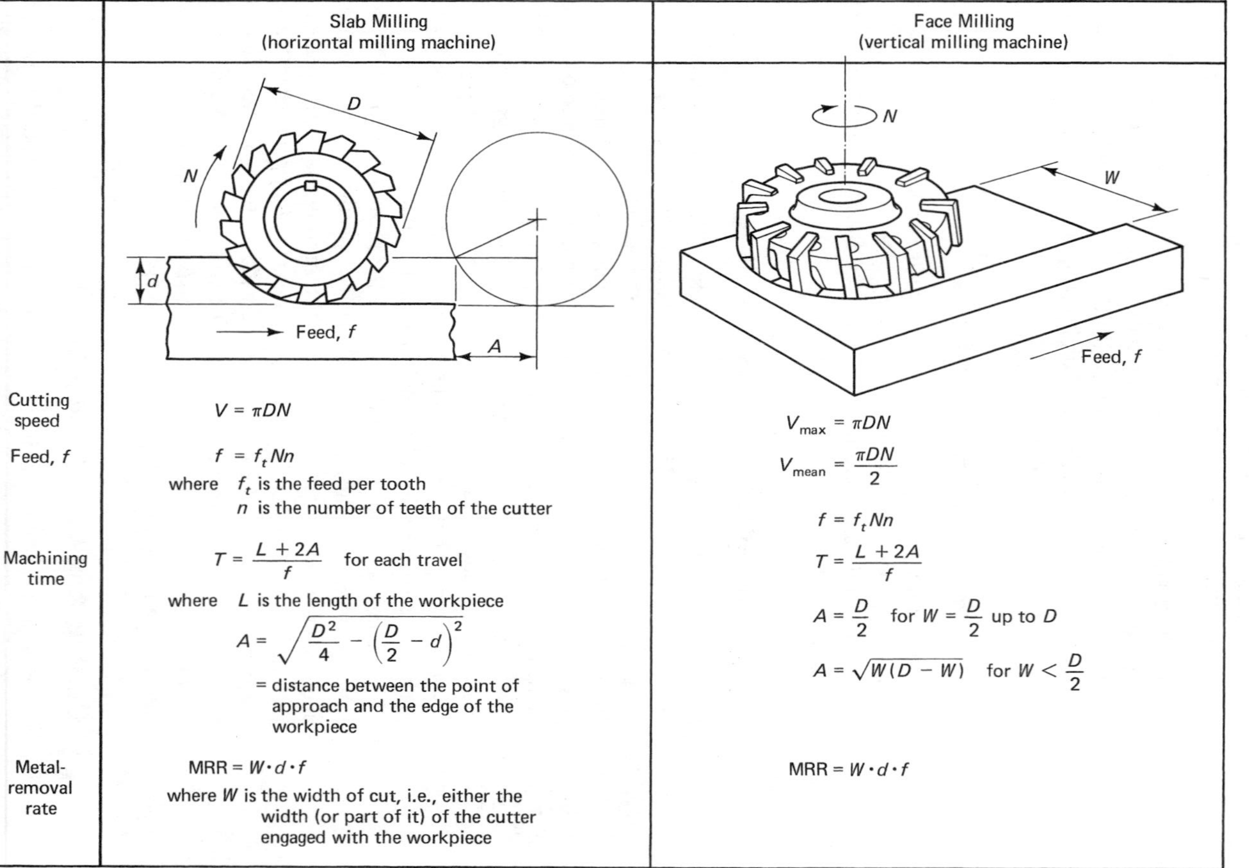

	Slab Milling (horizontal milling machine)	Face Milling (vertical milling machine)
Cutting speed	$V = \pi DN$	$V_{max} = \pi DN$ $V_{mean} = \frac{\pi DN}{2}$
Feed, f	$f = f_t Nn$ where f_t is the feed per tooth n is the number of teeth of the cutter	$f = f_t Nn$
Machining time	$T = \frac{L + 2A}{f}$ for each travel where L is the length of the workpiece $A = \sqrt{\frac{D^2}{4} - \left(\frac{D}{2} - d\right)^2}$ = distance between the point of approach and the edge of the workpiece	$T = \frac{L + 2A}{f}$ $A = \frac{D}{2}$ for $W = \frac{D}{2}$ up to D $A = \sqrt{W(D - W)}$ for $W < \frac{D}{2}$
Metal-removal rate	$MRR = W \cdot d \cdot f$ where W is the width of cut, i.e., either the width (or part of it) of the cutter engaged with the workpiece	$MRR = W \cdot d \cdot f$

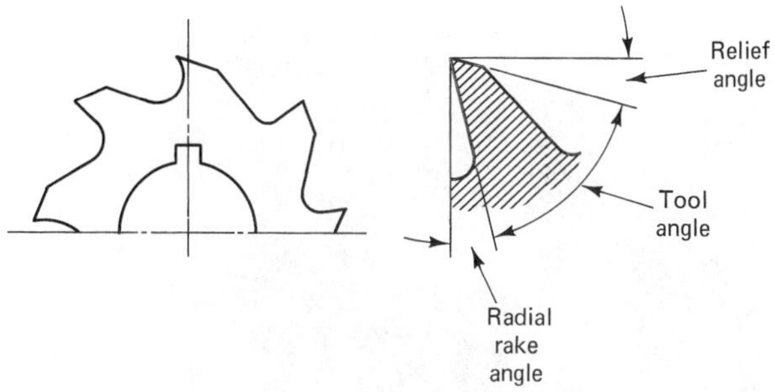

FIGURE 10.42
Cutting angles of a plain, straight-tooth milling cutter

Types of Milling Machines

Several types of milling machines are employed in industry. They are generally classified by their construction and design features. They vary from the common general-purpose types to duplicators and machining centers that involve a tool magazine and are capable of carrying out many machining operations with a single workpiece setup. Following is a survey of the types of milling machines commonly used in industry.

Plain horizontal milling machine. The construction of the *plain horizontal milling machine* is very similar to that of a universal milling machine (see discussion that follows), except that the machine table cannot be swiveled. Plain milling machines usually have a column-and knee type of construction and three table motions (i.e., longitudinal, transverse, and vertical). The milling cutter is mounted on a short arbor that is, in turn, rigidly supported by the overarm of the milling machine.

Universal milling machine. The construction of a *universal milling machine* is similar to that of the plain milling machine, except that it is more accurate and has a sturdier frame and its table can be swiveled with an angle up to 50°. Universal milling machines are usually equipped with an index or dividing head that allows for the cutting of gears and cams, as is discussed later. Figure 10.43 shows a machine tool of this type.

Vertical milling machine. As the name *vertical milling machine* suggests, the axis of the spindle that holds the milling cutter is vertical. Table movements are generally similar to those of plain horizontal milling machines; however, an additional rotary motion is sometimes provided for the table when helical and circular grooves are to be machined. The cutters used with vertical milling machines are almost always of the end-mill type. Figure 10.44 shows a vertical milling machine.

Duplicator. A *duplicator* is sometimes referred to as a *copy milling machine* because it is capable of reproducing an exact replica of a model. The machine has a stylus that scans the model, at which time counterpoints on the part are successively machined. Duplicators were used for the production of large forming dies for the automotive industry, where models made of wood, plaster of paris, or wax were employed. Duplicators are not commonly used in industry now because they have been superseded by CAD/CAM systems.

FIGURE 10.43
A universal milling machine *(Courtesy of Manuel Pereira, photography specialist, University of Massachusetts, Dartmouth)*

Machining center. A *machining center* is comprised of a multipurpose CNC machine (see Chapter 14) that is capable of performing a number of different machining processes. A machining center has a tool magazine in which many tools are held. Tool changes are automatically carried out, and so are functions such as coolant turn-on/off. Machining centers are, therefore, highly versatile and can perform a number of machining operations on a workpiece with a single setup. Parts having intricate shapes can easily be produced with high accuracy and excellent repeatability.

Universal dividing head. The *universal dividing head* is an attachment mounted on the worktable of a universal milling machine that is employed for cutting gears. The function of the dividing head is to index the gear blank through the desired angle each time the metal between two successive teeth is removed. Therefore, this attachment is sometimes known as an *index head.*

Figure 10.45 shows a universal dividing head, which consists of the body, the swivel block, the work spindle and its center, the index plate, and the index crank with a latch pin. The workpiece (with one of its ends supported by the center of the work

FIGURE 10.44
A vertical milling machine *(Courtesy of Manuel Pereira, photography specialist, University of Massachusetts, Dartmouth)*

FIGURE 10.45
A universal dividing head *(Courtesy of Manuel Pereira, photography specialist, University of Massachusetts, Dartmouth)*

spindle) is rotated through the desired angle by rotating the index crank through an angle that is dependent upon the desired angle. The index crank is fixed to a shaft that is, in turn, attached to a worm-gear reducer with a ratio of 40 to 1. Consequently, 40 turns of the index crank result in only one full turn of the workpiece. This index plate has six concentric circles of equally spaced holes to assist in measuring and controlling any fraction of revolution in order to crank the correct angle. The following equation is used to determine the angle through which the crank is to be rotated in gear cutting:

$$\text{number of turns of index crank} = \frac{40}{\text{number of teeth of desired gear}} \tag{10.9}$$

We can see from Equation 10.9 that if the gear to be cut has 20 teeth, the index crank should be rotated two full turns each time a tooth space is to be produced. As a consequence, the workpiece will be rotated each time through an angle equal to 18°. Similarly, if the desired gear has 30 teeth, the index crank must be rotated 1¼ turns each time.

10.5 GRINDING OPERATIONS

Grinding is a manufacturing process that involves the removal of metal by employing a rotating abrasive wheel. The wheel simulates a milling cutter with an extremely large number of miniature cutting edges. Generally, grinding is considered to be a finishing process and is used for obtaining high-dimensional accuracy and superior surface finish. Grinding can be performed on flat, cylindrical, or even internal surfaces by employing specialized machine tools, referred to as *grinding machines.* Obviously, grinding machines differ in construction as well as capabilities, and the type to be employed is determined mainly by the geometrical shape and nature of the surface to be ground (e.g., cylindrical surfaces are ground on cylindrical grinding machines).

Types of Grinding Operations

Surface grinding. As the name *surface grinding* suggests, this operation involves the grinding of flat or plane surfaces. Figure 10.46 indicates the two possible variations: either a horizontal or a vertical machine spindle. With a horizontal spindle (see Figure 10.46a), the machine usually has a planer-type reciprocating table on which the workpiece is held. However, grinding machines with vertical spindles can have either a planer-type table like that of the horizontal-spindle machine or a rotating worktable. Also, the grinding action in this case is achieved by the end face of the grinding wheel (see Figure 10.46b), contrary to the case of horizontal-spindle machines, where the workpiece is ground by the periphery of the grinding wheel. Figure 10.46 also indicates the equations used to estimate the different parameters of the grinding operation, such as the machining time and the metal-removal rate. During the surface grinding operations, heavy workpieces are either held in fixtures or clamped on the machine table by strap clamps and the like, whereas smaller workpieces are usually held by magnetic chucks.

FIGURE 10.46
Surface grinding: (a) horizontal spindle; (b) vertical spindle

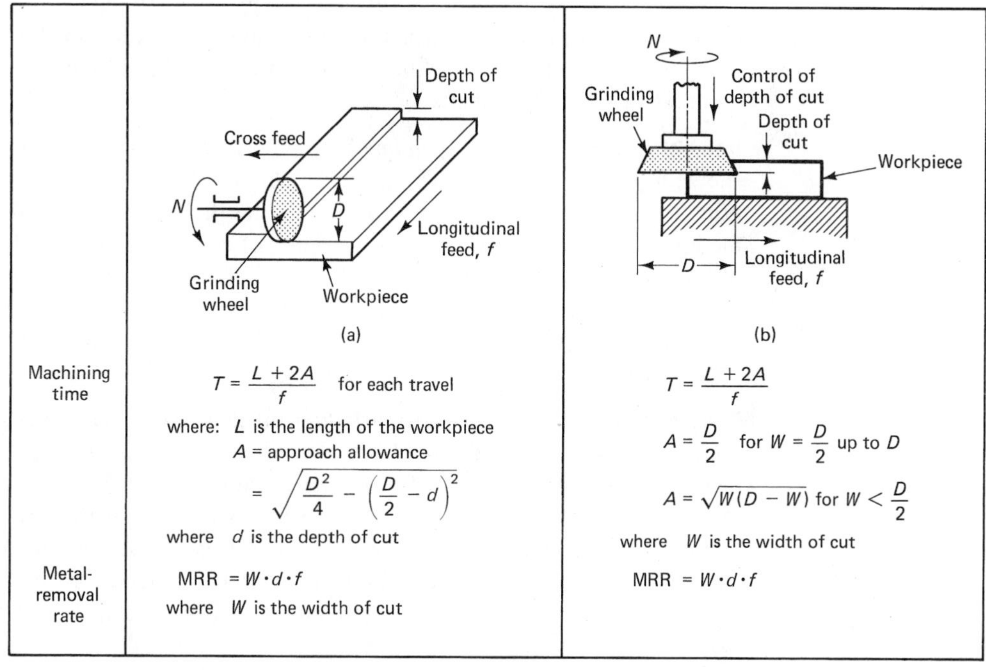

	(a)	(b)
Machining time	$T = \frac{L + 2A}{f}$ for each travel where: L is the length of the workpiece A = approach allowance $= \sqrt{\frac{D^2}{4} - \left(\frac{D}{2} - d\right)^2}$ where d is the depth of cut	$T = \frac{L + 2A}{f}$ $A = \frac{D}{2}$ for $W = \frac{D}{2}$ up to D $A = \sqrt{W(D - W)}$ for $W < \frac{D}{2}$ where W is the width of cut
Metal-removal rate	MRR $= W \cdot d \cdot f$ where W is the width of cut	MRR $= W \cdot d \cdot f$

Cylindrical grinding. In *cylindrical grinding,* the workpiece is held between centers during the grinding operation, and the wheel rotation is the source and cause for the rotary cutting motion, as shown in Figure 10.47. Cylindrical grinding can be carried out by employing any of the following methods:

1. In the *transverse* method, both the grinding wheel and the workpiece rotate, and longitudinal linear feed is applied so that the entire length can be ground. The depth of cut is adjusted by the cross feed of the grinding wheel into the workpiece.

FIGURE 10.47
Cylindrical grinding

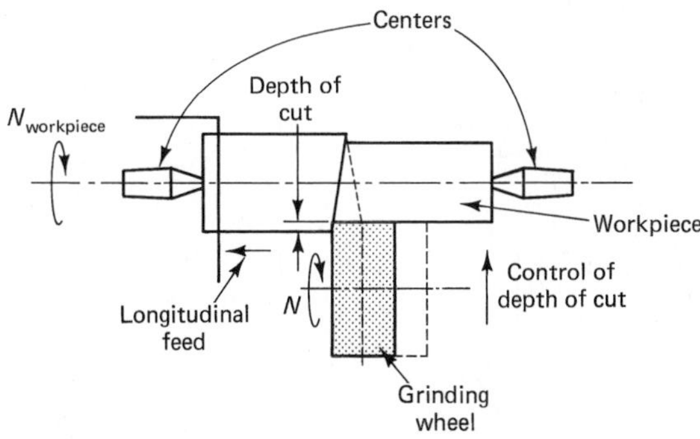

2. In the *plunge-cut* method, grinding is achieved through the cross feed of the grinding wheel, and no axial feed is applied. This method can be applied only when the surface to be ground is shorter than the width of the grinding wheel used.
3. In the *full-depth* method, which is similar to the transverse method, the grinding allowance is removed in a single pass. This method is usually recommended when grinding short, rigid shafts.

Internal grinding. *Internal grinding* is employed for grinding relatively short holes, as shown in Figure 10.48. The workpiece is held in a chuck or a special fixture. Both the grinding wheel and the workpiece rotate during the operation, and feed is applied in the longitudinal direction. Any desired depth of cut can be obtained by the cross feed of the grinding wheel. A variation of this type of grinding is *planatery* internal grinding, and it is recommended for heavy workpieces that cannot be held in chucks. In this case, the grinding wheel not only spins around its own axis but also rotates around the centerline of the hole that is being ground.

Centerless grinding. *Centerless grinding* involves passing a cylindrical workpiece, which is supported by a rest blade, between two wheels (i.e., the grinding wheel and the regulating or feed wheel). The grinding wheel does the actual grinding, while the regulating wheel is responsible for rotating the workpiece as well as generating the longitudinal feed. This is possible because of the frictional characteristics of this wheel, which is usually made of rubber-bonded abrasive. As can be seen in Figure 10.49, the axis of the regulating wheel is tilted at a slight angle with the axis of the grinding wheel. Consequently, the peripheral velocity of the regulating wheel can be resolved into two components: workpiece rotational speed and longitudinal feed. These can be given by the following equations:

$$V_{\text{workpiece}} = V_{\text{regulating wheel}} \times \cos \alpha \tag{10.10}$$

$$\text{axial feed} = V_{\text{regulating wheel}} \times C \times \sin \alpha \tag{10.11}$$

Note that C is a constant coefficient that accounts for the slip between the workpiece and the regulating wheel ($C = 0.94–0.98$).

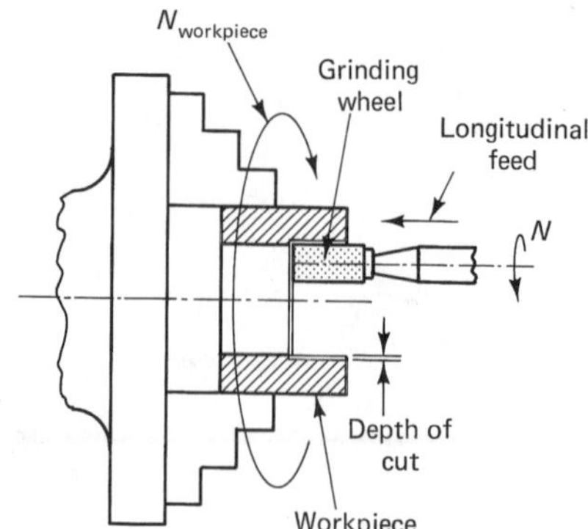

FIGURE 10.48
Internal grinding

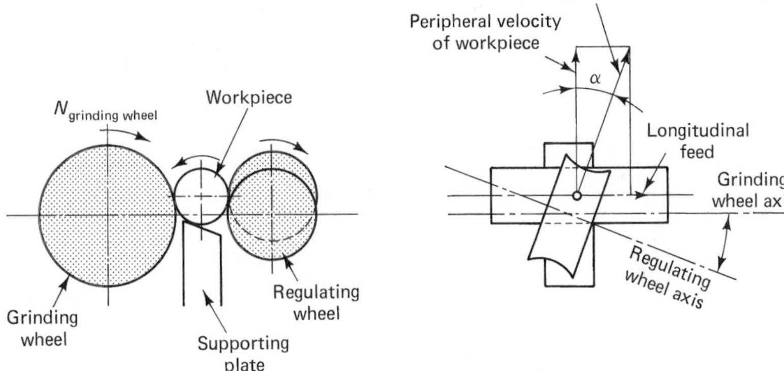

FIGURE 10.49
Centerless grinding

The velocity of the regulating wheel is controllable and is used to achieve any desired rotational speed of the workpiece. The angle α is usually taken from 1° to 5°; the larger the angle, the larger the longitudinal feed will be. When α is taken as 0° (i.e., the two axes of the grinding and regulating wheels are parallel), there is no longitudinal feed of the workpiece. Such a setting is used for grinding short shoulders or heads of workpieces having such features.

Grinding Wheels

Grinding wheels are composed of abrasive grains having similar size and a binder. The actual grinding process is performed by the abrasive grains. Pores between the grains within the binder enable the grains to act like separate single-point cutting tools. These pores also provide space for the generated chips, thus preventing the wheel from clogging. In addition, pores assist the easy flow of coolants so that heat generated during the grinding process is efficiently and promptly removed.

Grinding wheels are identified by their shape and size, kind of abrasive, grain size, binder, grade (hardness), and structure.

Shape and size of grinding wheels. Grinding wheels differ in shape and size, depending upon the purpose for which they are to be used. Various shapes are shown in Figure 10.50 and include the following types:

1. Straight wheels for surface, cylindrical, internal, and centerless grinding
2. Beveled-face or tapered wheels for grinding threads, gear teeth, and the like
3. Straight recessed wheels for cylindrical grinding and facing
4. Abrasive disks for cutoff and slotting operations when thickness is 0.02 to 0.2 inch (0.5 to 5 mm)
5. Cylindrical, straight, and flaring cups for surface grinding with the end of the wheel

FIGURE 10.50
Various shapes of grinding wheels

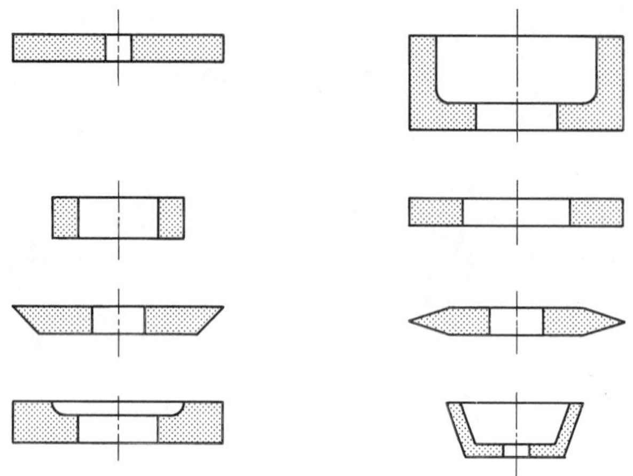

The main dimensions of a grinding wheel are the outside diameter *D*, the bore diameter *d*, and the height *H*. These dimensions vary widely, depending upon the grinding process for which the wheel is to be used.

Kind of abrasive. Grinding wheels can be made of natural abrasives such as quartz, emery, and corundum or of industrially prepared chemical compounds such as aluminum oxide or silicon carbide (carborundum). Generally, silicon-carbide grinding wheels are used when grinding low-tensile-strength materials like cast iron, whereas aluminum-oxide wheels are employed for grinding high-strength metals such as alloy steel and hardened steel.

Grain size of abrasive used. As you may expect, the grain size of the abrasive particles of the wheel plays a fundamental role in determining the quality of the ground surface obtained. The finer the grains, the smoother the ground surface is. Therefore, coarse-grained grinding wheels are used for roughing operations, whereas fine-grained wheels are employed in final finishing operations.

Grade of the bond. The grade of the bond is an indication of the resistance of the bond to the pulling off of the abrasive grains from the grinding wheel. Generally, wheels having hard grades are used for grinding soft materials, and vice versa. If a hard-grade wheel were to be used for grinding a hard material, the dull grains would not be pulled off from the bond quickly enough, thus impeding the self-dressing process of the surface of the wheel and finally resulting in a clogged wheel and a burnt ground surface. The cutting properties of all grinding wheels must be restored periodically by dressing with a cemented-carbide roller or a diamond tool to give the wheel the exact desired shape and remove all worn abrasive grains.

Structure. Structure refers to the amount of void space between the abrasive grains. When grinding soft metals, large void spaces are needed to facilitate the flow of the removed chips.

FIGURE 10.51
Standard marking system for grinding wheels

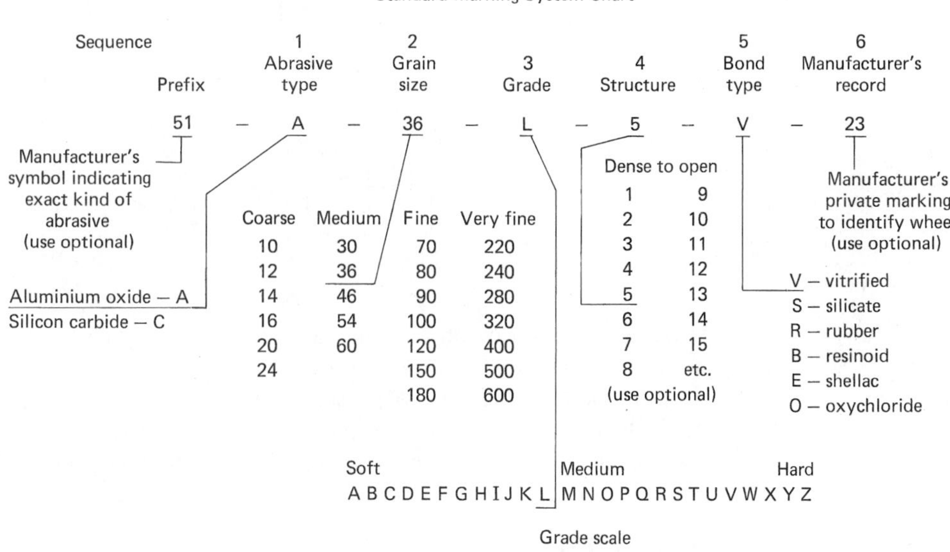

Binder. Abrasive particles are bonded together in many different ways. These include the use of vitrified bond, silicate, rubber, resinoid, shellac, or oxychloride. The vitrified bond is the most commonly used binder.

Standard marking system. The standard marking system shown in Figure 10.51 is employed for distinguishing grinding wheels by providing all the preceding parameters in a specific sequence.

10.6 SAWING OPERATIONS

Parting or cutoff operations can be performed on machine tools such as engine lathes, milling machines, and grinding machines. When cutting off is a basic operation in a large-volume production line, special sawing machines are required to cope with the production volume.

Types of Sawing Teeth

The cutting tool may take different forms, depending upon the type of sawing machine used. The tool can be a blade, a circular disk, or a continuous band. However, all these tools are multiedged with several cutting edges (i.e., teeth) per inch. As can be seen in Figure 10.52, teeth can be *straight, claw,* or *buttress.* Each tooth, irrespective of its form, must have a rake and a relief angle. Also, teeth are offset in order to make the kerf wider than the thickness of each individual tooth. This facilitates easy

FIGURE 10.52 Types of sawing teeth

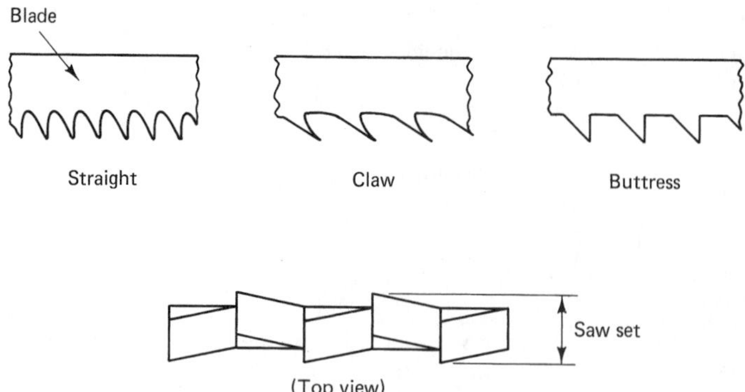

movement of the saw blade in the kerf, thus reducing the friction and the generated heat. The maximum thickness is usually referred to as the *saw set* and is equal to the width of the resulting kerf. When selecting the cutting speed and the number of teeth per inch, several factors have to be taken into consideration, such as the cutting tool material, the material of the workpiece, the tooth form, and the type of lubricant (coolant) used.

Sawing Machines

Sawing machines differ in shape, size, and construction, depending upon the purpose for which they are to be used. They can be classified into three main groups.

Reciprocating saw. In a *reciprocating saw,* a relatively large hacksaw blade is mechanically reciprocated. Depending upon the construction of the saw, the cutting blade may be either horizontal or vertical. Each cycle has a working (cutting) stroke as well as an idle stroke. Consequently, this type of saw is considered to be a low-productivity saw and is used only in small shops with low-to-moderate production volume.

FIGURE 10.53 Basic idea of a band saw

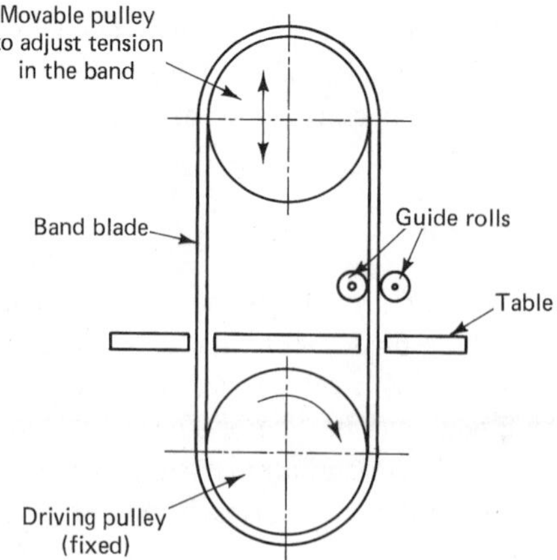

Circular saw. The cutting tool in a *circular saw* is a circular disk, with the cutting teeth uniformly arranged on its periphery. It looks like the slitting cutter used with milling machines. Although it is highly efficient, it can only be used for parting or cut-off operations of bar stocks or rolled sections.

Band saw. The highly flexible and versatile *band saw* employs a continuous-band sawing blade. As can be seen in Figure 10.53, the band-saw blade is mounted on two pulleys, one of which is the source of power and rotation. Each machine has a flash-welding attachment that is used to weld the edges of the band-saw blade together after adjusting the length, thus forming a closed band. Band saws can be used for contouring and for large-volume cutoff operations. Loading and unloading of the bar stock as well as length adjustment are done automatically (by special attachments in the latter case).

10.7 BROACHING OPERATIONS

Broaching is a metal-removing operation in which a multiedge cutting tool, like that shown in Figure 10.54, is used. In this operation, only a thin layer or limited amount of metal is removed. Broaching is commonly used to generate internal surfaces or slots, like those shown in Figure 10.55, that are very difficult to produce otherwise. However, it can also be used for producing intricate external surfaces that require tight tolerances.

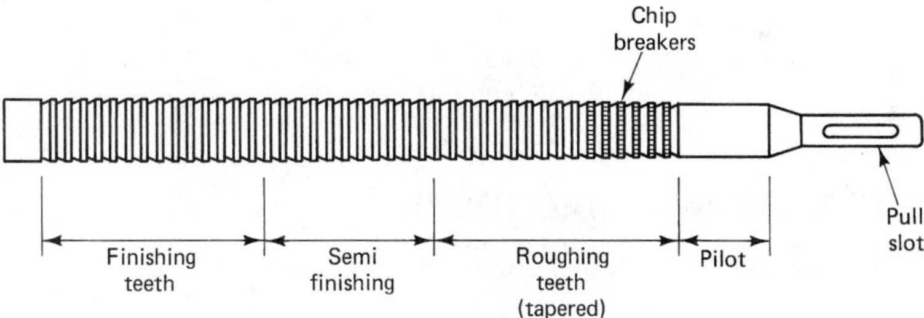

FIGURE 10.54
A broaching tool

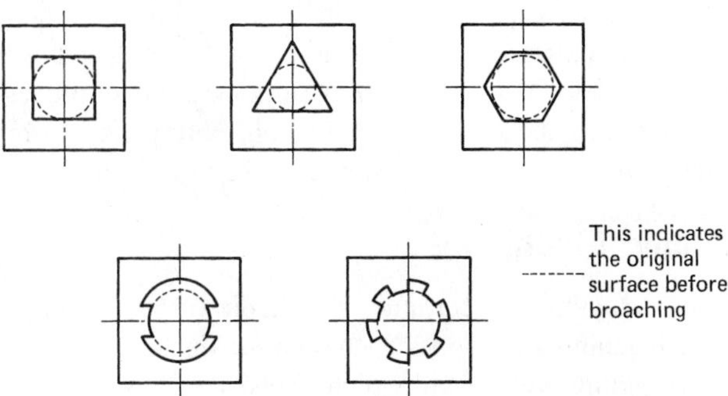

FIGURE 10.55
Different shapes produced by broaching

Broaching Machines

A broaching machine is simply comprised of a sturdy frame (or bed), a device for locating and clamping the workpiece, the cutting tool, and a means for moving the cutting tool (or the workpiece). The commonly used types of broaching machines are as follows:

1. *Pull-type* machines, in which the broaching tool is withdrawn through the initial hole in the tightly clamped workpiece
2. *Push-type* machines, in which the broaching tool is pushed to generate the required surface
3. *Surface-broaching* machines, where either the tool or the workpiece move to generate the desired surface
4. *Continuous-broaching* machines, where the workpiece moves continually over a fixed broaching tool in a straight or circular path

Advantages and Limitations of Broaching Operations

It is important to know the advantages and limitations of broaching in order to make full use of the potential of this operation. The advantages include the high cutting speed and high cycling time, the close tolerances and superior surface quality that can easily be achieved, and the fact that both roughing and finishing are combined in the same stroke of the broaching tool. Nevertheless, this operation can be performed only on through holes or external surfaces and cannot be carried out on blind holes, for example. Also, broaching involves only light cuts and, therefore, renders itself unsuitable for operations where the amount of metal to be removed is relatively large. Finally, the high cost of broaching tools and machines, together with the expensive fixturing, make this operation economically unjustifiable unless a large number of products are required.

10.8 NONTRADITIONAL MACHINING OPERATIONS

The need for nontraditional machining processes came as a result of the shortcomings and limitations of the conventional, mechanical, chip-generating processes. Whereas conventional processes can be applied only to soft and medium-hard materials, very fine features of extremely hard materials can be produced using the nontraditional machining processes. There are a variety of nontraditional processes, and each has its own set of advantages and fields of application. Following is a brief discussion of each of the nontraditional processes commonly used in industry.

Ultrasonic Machining

Ultrasonic machining is particularly suitable for machining hard, brittle materials because the machining tool does not come in contact with the workpiece. They are separated by a liquid (vehicle) in which abrasive grains are suspended. Equal volumes of water and very fine grains of boron oxide are mixed together to produce the desired suspension. Ultrasonic energy applied to the tool results in high-frequency mechanical vibrations (20 to 30 kHz). These vibrations impart kinetic energy to the abrasive grains, which, in turn, impact the workpiece and abrade it. The machining tool must be made of a tough ductile material such as copper, brass, or low-carbon steel so that it will not be liable to fretting wear or abrasion, as is the case with the workpiece. Ultrasonic machining is employed mainly in making holes with irregular cross sections. Both through and blind holes can be produced by this method.

Abrasive-Jet Machining

In *abrasive-jet machining,* liquids in which abrasive particles are suspended are pumped under extremely high pressure out from a nozzle. The resulting jet stream is then employed in processes like deburring, drilling, and cutting of thin sheets and sections. The process is particularly advantageous when cutting glass and sheets of composites. The shortcomings of this process involve the problems associated with using high-pressure pumps and the relatively slow feed rate employed.

Chemical Machining

Chemical machining involves attacking the surfaces of the workpiece to be machined with a chemical etch that reacts with the metal and dissolves the resulting chemical compound. The procedure consists of first covering the surfaces of the workpiece that are not to be machined with neoprene rubber or enamel and then dipping the workpiece into a basin of the appropriate chemical etch. Very fine details can be etched by this method, and the quality of the machined surface is high and free from any chips. A further advantage of this process is that it does not result in any work-hardening.

Electrochemical Machining

The mechanism with which *electrochemical machining* (ECM) takes place is reciprocal to that of the electroplating process, although similar equipment is used in both cases. In electrochemical machining, the workpiece is connected to the anode, while the cathode is connected to a copper ring that is used as the machining tool. Low-voltage, high-amperage direct current is used, and an electrolyte is pumped into the small gap between the workpiece and the copper ring. As is the case in electroplating, the amperage plays an important role in determining the rate of metal transfer from the anode to the cathode (i.e., the rate of metal removal during the electrochemical machining process). Electrochemical machining can be applied to all electrically conductive metals, including hardened alloy steel and tungsten, and it is particularly advantageous for machining thin sheets of nickel and titanium.

Electrodischarge Machining

Electrodischarge machining (EDM) is used for producing parts having intricate shapes, and it can be applied to all metallic materials, whether they are ductile or brittle. It cannot, however, be used with ceramics, plastics, or glass. Metal removal takes

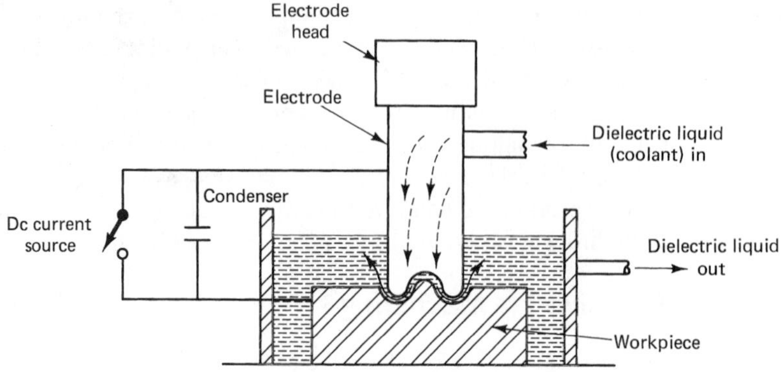

FIGURE 10.56
The EDM process

place as a result of an electric arc between the electrode and the workpiece, which are kept apart. A dielectric liquid like kerosene is pumped through the small gap of about 0.02 inch (0.5 mm) between the electrode and the workpiece, as shown in Figure 10.56. The dielectric liquid also acts as a coolant and a flushing medium to whip away the removed metallic dust.

The electrode is usually made of a material that can easily be shaped, such as copper, brass, graphite, or a copper-tungsten mixture. The electrode must be given a shape that fits exactly into the desired final cavity. Consequently, intricately shaped parts can easily be produced by this method, which has gained widespread industrial application in the manufacture of tools, metal-forming and forging dies, plastic, and die-casting

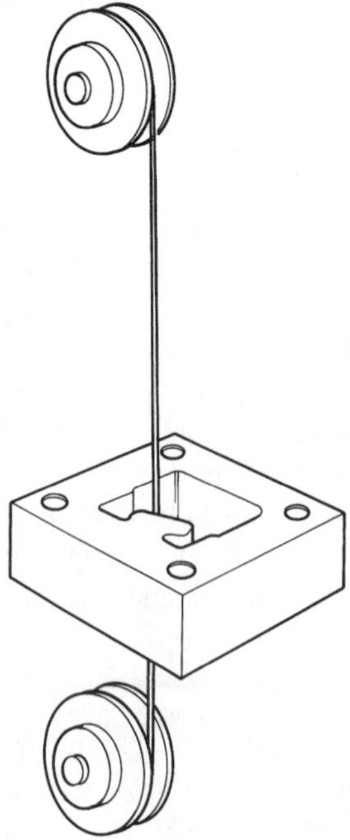

FIGURE 10.57
The concept of wire EDM

FIGURE 10.58 Some stamping dies cut by wire EDM *(Courtesy of Capitol Concept and Engineering, a Member of Synergis Technologies Group, Grand Rapids, Michigan)*

molds. Generally, it can be stated that the quality of the machined surface is dependent upon the number of sparks (electric arc sparks) per second (they range between 3000 and 10,000).

A new version of EDM is shown in Figure 10.57, wherein the conventional electrode is replaced by a tensioned wire of copper or tungsten that is guided by a CNC system to trace any desired contour. This process has revolutionized the tool and die-making industry. Whereas sharp corners are avoided in tools manufactured by conventional processes to prevent breakage or cracking during subsequent heat treatment, wire EDM can cut heat-treated steels directly to the desired shape. Therefore, large dies having intricate shapes and sharp corners can be produced by this technique. Figure 10.58 shows some stamping dies that were made by wire EDM.

Review Questions

1. What are the three motions necessary to generate a surface during machining operations?
2. List six different types of lathes.
3. What are the main elements of an engine lathe?
4. Why is the spindle of an engine lathe hollow and why does it have a Morse taper?
5. Discuss briefly the construction of the tailstock.
6. What is the main function of the carriage?
7. How does the carriage receive its motion?
8. Use sketches to explain the difference between a turret lathe and an engine lathe.
9. How do you specify a lathe?
10. What are the conditions for proper tool holding?

11. Use sketches to illustrate the following: turning tools, facing tools, cutoff tools, thread-cutting tools, form tools.

12. What precautions should be taken when supporting a workpiece during lathe operations?

13. When should a workpiece be held between two centers?

14. When is it necessary to hold a workpiece in a chuck?

15. When would a workpiece be mounted on a faceplate?

16. How would you hold a disklike workpiece that has to be machined on both sides?

17. What do the machining marks look like in cylindrical turning?

18. How is the axial feed provided and how is the depth of cut controlled in cylindrical turning?

19. What do the machining marks look like in facing operations?

20. What are the suitable work-holding devices for facing operations?

21. List three methods that can be employed to generate a tapered surface.

22. What provides the feed during thread cutting?

23. Describe knurling.

24. What are the variables that affect the optimal value of the surface cutting speed?

25. Discuss the considerations that must be taken into account when designing turned parts.

26. What is the difference between shaping and planing?

27. What kind of surfaces can be produced by shaping and planing operations?

28. Explain the working principles of the quick-return mechanism.

29. List the commonly used types of drills and discuss the applications of each.

30. What is meant by the *tool point angle?* How does the workpiece material affect the optimal value for this angle?

31. List some other hole-making operations and discuss the applications of each.

32. Discuss the considerations that must be taken into account when designing drilled parts.

33. List the various types of drilling machines and discuss the characteristics and fields of application of each.

34. What is a *jig?* When is the use of jigs recommended?

35. Define *milling.*

36. Why are the permissible cutting speeds in milling four times higher than those for turning?

37. Differentiate between up milling and down milling.

38. List the various types of milling cutters and discuss the applications of each.

39. List the various types of milling machines and discuss the applications of each.

40. What is the function of a universal dividing head?

41. Define *grinding.*

42. List the types of grinding operations and discuss the applications of each.

43. Of what are grinding wheels composed?

44. How can grinding wheels be identified?

45. List the different types of sawing machines and discuss the constructional features as well as the fields of application of each.

46. Use sketches to illustrate the types of teeth of saw blades.

47. Define *broaching.* When is the use of this process recommended? Discuss the advantages and limitations of broaching operations.

48. Why did the need arise for nontraditional machining operations?

49. How is ultrasonic energy employed to machine surfaces?

50. Explain the working principles of abrasive-jet machining.

51. When is the use of chemical machining recommended?

52. Discuss the advantages and limitations of electrochemical machining.

53. Explain the working principles of electrodischarge machining. Discuss the advantages and limitations of this process.

54. Discuss the advantages of wire EDM.

Problems

1. It is required to maintain a cutting speed of 120 feet per minute (37 m/min.) in a turning operation. If the initial workpiece diameter is 3.25 inches (82 mm) and the depth of cut is 0.1 inch (2.5 mm), calculate the rpm of the spindle during the third and sixth passes.

2. A 24-inch-diameter (600-mm) part with a 6-inch-diameter (150-mm) hole in the center is to be faced starting at the outside. The rotational speed of the spindle is 7 revolutions per second, the depth of cut is 0.15 inch (3.75 mm), and the feed is 0.01 inch per revolution (0.25 mm/rev). Calculate the machining time as well as the maximum and minimum rate of metal removal.

3. Two thousand bars that are 3.25 inches (81 mm) in diameter and 12 inches (300 mm) long are to be turned down to 2.75-inch (69-mm) diameters. Heavy cuts followed by a light finishing cut are to be used. For finishing, the feed is 0.005 inch (0.125 mm), the cutting speed is 300 feet per minute (90 m/min.), and the depth of cut is 0.07 inch (1.75 mm). Two roughing cuts (two passes) are required, where the cutting speed is only 200 feet per minute (60 m/min.) and the feed is 0.01 inch (0.25 mm). Calculate the overall production time when the time taken to return the tool to the beginning of cut is 15 seconds and the load/unload time is 2 minutes.

4. A bronze bushing is 2 inches (50 mm) in diameter, is 3 inches (75 mm) long, and has a central hole of 1.25 inches (31.25 mm). It is to be produced on a lathe, starting with a solid bar stock having a 2-inch (50-mm) outer diameter. Estimate the production time per piece. Take the feed as 0.01 inch per revolution (0.25 mm/rev) and the cutting speed as 200 feet per minute (60 m/min.). Assume any missing data.

5. A workpiece having a length of 3 inches (75 mm) is to be taper-turned by offsetting the tailstock. If the maximum diameter of the workpiece is 1⅛ inches (28.125 mm) and the minimum diameter is 1.0 inch (25 mm), calculate the amount of offset.

6. A 10-inch-diameter (250-mm) part having a 4-inch-diameter (100-mm) hole is to be bored for 4 inches (100 mm) of its length to a diameter of 4.4 inches (110 mm). A depth of cut of 0.08 inch (2 mm) is to be used with a feed of 0.004 inch (0.1 mm) and a cutting speed of 330 feet per minute (100 m/min.). If it takes 15 seconds to return the tool to the starting point and set the depth of cut, calculate the time required to complete this job.

7. A part is to be tapered in such a manner as to have the following dimensions:

Total length:	6 inches (150 mm)
Tapered length:	1½ inches (62.5 mm)
Large diameter:	1.0 inch (25 mm)
Small diameter:	0.625 inch (16 mm)

Calculate the tailstock offset.

8. How far must the tailstock be offset to cut a 0.5-inch-per-foot (41.7-mm/m) taper on an 8-inch-long (200-mm) workpiece?

9. In a drilling operation, the desired depth of the hole is 1 inch (25 mm), the drill size is 0.4 inch (10 mm), the rpm is 100, and the feed is 0.01 inch (0.25 mm). Calculate the cutting speed and estimate the drilling time.

10. A standard twist drill is used to drill a number of 3/8-inch (9.5-mm) through holes in a 5/8-inch-thick (16-mm) SAE 1020 steel plate. Cutting speed is 60 feet per minute (18.3 m/min.), and feed is 0.004 inch (0.1 mm). Calculate the time required for drilling each hole and the metal-removal rate.

11. It is required to drill a 1-inch-deep (25-mm) hole in each of 75,000 cast-iron blocks. If the rotational speed is 600 rpm and the feed is 0.002 inch (0.05 mm), estimate the required working hours. Assume that it takes 30 seconds to load and unload the part and that 15 seconds must be allowed each time the drill bit is changed. Take the number of bit changes as 10.

12. In a drilling operation, the feed rate is 1 inch per minute (25 mm/min.), the cutting speed is 37.2 feet per minute (12 m/min.), and the diameter of the hole to be drilled is 0.6 inch (15 mm). What is the feed?

13. In milling a step 1/8 by 1/8 inch (3.18 by 3.18 mm) in a 2-inch-long (50-mm) workpiece, a two-fluted 1/2-inch-diameter (12.5-mm) end mill is used. The rotational speed is 700 rpm, and the feed is 0.006 inch per tooth (0.15 mm per tooth). Estimate the milling time and the metal-removal rate.

14. In a face milling operation, the depth of cut is 1/4 inch (6 mm), and the table moves at 0.2 inch per second (10 mm/s). The width of the workpiece is 2.0 inches (50 mm), and the cutter has a diameter of 3.25 inches (81 mm). The rotational speed of the cutter is 120 rpm, and the feed is 0.01 inch per tooth (0.25 mm per tooth). If the length of the workpiece is 10 inches (250 mm), calculate the number of teeth of the cutter, the metal-removal rate, and the milling time.

15. An 18-tooth, 1-inch-wide (25-mm) HSS cutter having a diameter of 4 inches (100 mm) is to be used in slot milling a 10-inch-long (250-mm) workpiece. If the desired depth of slot is 0.24 inch (6 mm), the cutting speed is 93 feet per minute (30 m/min.), and the feed is 0.005 inch per tooth (0.125 mm per tooth), estimate the milling time.

16. Estimate the machining time in face milling given the following data:

Cutter:	20 teeth and 4 inches (100 mm) in diameter
Rotational speed:	300 rpm
Depth of cut:	0.25 inch (6 mm)
Feed:	0.001 inch per tooth (0.025 mm per tooth)
Length of workpiece:	20 inches (500 mm)
Cutter width:	larger than that of the workpiece

17. The recommended feed for milling a kind of steel is 0.01 inch per tooth (0.25 mm per tooth) when using a helical milling cutter with 20 teeth. If the cutting speed is 70 feet per minute (22.5 m/min.) and the cutter diameter is 4 inches (100 mm), calculate the feed rate of the table.

18. When gear-cutting processes are to be performed on a universal milling machine by using the indexing head, explain the procedure in each case when the gear has the following number of teeth:

a. 20 teeth
b. 32 teeth
c. 22 teeth
d. 15 teeth

Chapter 11

Product Cost Estimation

INTRODUCTION

As mentioned in Chapter 1, the production turn cannot continue unless the manufactured products are successfully marketed, the fixed and working capital is recovered, and a profit is made. Cost plays a vital role in the marketing process because it provides the information required to set up the selling price of a product. An overpriced product cannot penetrate the market and will eventually lose out to similar but more competitively priced products. Underestimation of the production cost may result in products sold at a loss and, consequently, financial problems for the manufacturing corporation.

Because our main concern is design for manufacturing and because design is an open-ended process that yields more than one workable solution, a logical criterion for evaluating these "solutions" or "designs" would certainly be the cost required to bring each design into being and manufacture the product. Therefore, it is fair to state that cost estimation is initiated by, linked to, and follows the product design in order to ensure the profitability of new products. Cost is also used to determine the most economical operation or sequence of operations for manufacturing a product, and it can be used as a means for establishing a cost-reduction program aimed at manufacturing the product so that it can be priced more competitively.

11.1 COSTS: CLASSIFICATION AND TERMINOLOGY

Costs can be classified in different ways based on their relationship to the production volume and the nature of the manufacturing operations. The first, and most logical, way to classify costs is to split them into two groups: *capital costs* and *operating costs.* As the name suggests, capital costs are incurred because of buildings, production machinery, and land. It is important to remember, when carrying out cost estimation, that buildings and machinery are depreciable (i.e., they tend to lose most of their value with time) whereas land is not. Operating costs are "running" costs that reoccur as long as the plant is in operation.

Another way to classify costs is to view them as belonging to one of two categories: *fixed costs,* which are independent of the production volume, and *variable costs,* which are dependent on it. Here are some examples of cost elements in the fixed-cost category:

- Depreciation on buildings, machinery, and equipment
- Insurance premiums (fire, theft, flooding, and occupational hazards)
- Property taxes (sometimes states and communities give tax breaks to industrial corporations to attract them to a region)
- Interest on investment (money borrowed from a bank, as explained in Chapter 1)
- Factory indirect-labor cost (wages of security, personnel, secretarial, clerical, janitorial, and financial staffs)
- Engineering cost (high-level engineering jobs and R & D expenses)
- Cost of rentals, if any (sometimes the building itself is rented, or some equipment may be rented for a short term)
- Cost of general supplies (supplies used by the factory indirect-labor force)
- Management and administrative expenses (salaries paid to corporate staff, plus legal expenses or salaries paid to legal staff)
- Marketing and sales expenses (salaries and wages paid to marketing and sales staffs, transportation and delivery expenses, rentals of warehouses, if any)

The following cost elements fall in the variable-cost category:

- Cost of materials
- Cost of labor (including production supervision)
- Cost of power (electricity, gas, or fuel oil) and utilities (water, sewer, etc.)
- Cost of maintenance of production equipment

The logical way to determine the total cost of a product is to add up all the cost elements, as is indicated by the bar diagram in Figure 11.1.

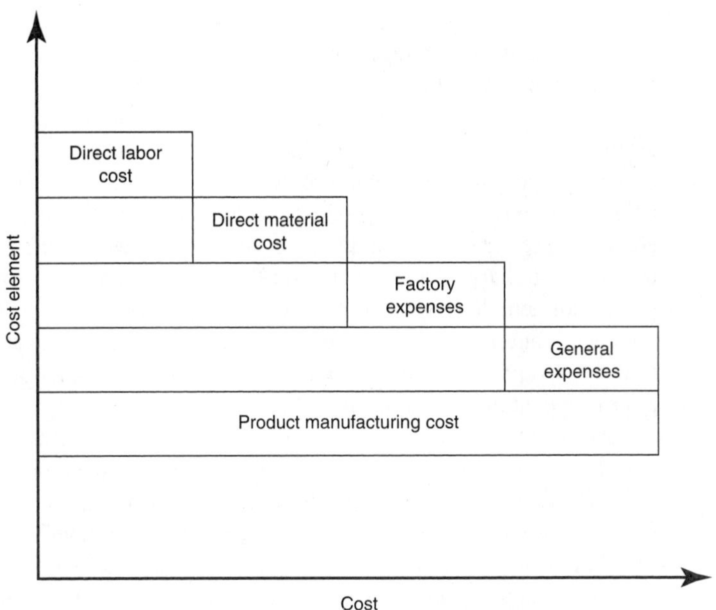

FIGURE 11.1
Elements contributing to the total cost of a product

In product cost estimation, the use of spreadsheets is extremely important, and the student is, therefore, encouraged to learn about and practice using them. The value of each cost element can be inserted in a spreadsheet cell, and, as an alternative, the formula for computing a cost element can be employed in the cell if the value of that element is not known. A further advantage of using spreadsheets is the ease with which alternative designs can be compared and evaluated from the point of view of cost. Figure 11.2 shows a spreadsheet where the specific cost elements and the total cost of

FIGURE 11.2
A spreadsheet that compares the cost of four alternative designs

Element of Product Cost / Design Alternatives	Direct labor cost ($)	Direct material cost ($)	Factory expenses ($)	General expenses ($)	Total manufacturing cost ($)
Design 1	2.4	37.0	8.0	7.0	54.4
Design 2	2.6	36.0	8.5	7.1	54.2
Design 3	2.5	36.5	9.0	7.3	55.3
Design 4	2.3	35.5	7.5	7.2	52.5

each design are shown in columns that make comparisons and conclusions easy. From a quick look at the figure, for example, it is not difficult to realize that design 4 is the optimal choice based on cost.

As easy as it may look, however, it is impossible to carry out the process of determining the total cost of a product unless rational procedures and analyses are employed to overcome two main problems. First, some costs cannot be directly assigned or traced to any particular product, but rather are spread over the entire factory; they are, therefore, labeled as "indirect" costs. In other words, the problem is how to calculate the cost share of a product from the salary of a receptionist or secretary. Second, we do not actually know the time taken to produce a design because that design has not been manufactured. It is the objective of this chapter to provide adequate answers to these two problems and to show the student how to independently carry out an engineering cost analysis for any desired design.

Now, with our stated goal as product cost estimation that is based on and begins after a detailed product design is available, we must develop highly accurate cost estimates that are suitable for submission on a bid or purchase order. This type of estimate is referred to as a *detailed estimate* and must have a level of accuracy of ±5 percent. The American Association of Cost Engineers came up with a list of five types of cost estimates, each having a certain level of accuracy, a different approach, and recommended applications. For example, the first type, a *rough estimate,* has an accuracy of ±40 percent and is based on indexing and modifying the cost of existing similar designs. It is, therefore, recommended for initial feasibility studies that are used to decide whether or not a probable profit justifies pursuing a project any further. Other types of cost estimates fall between the two extremes of rough to detailed and are, consequently, recommended for applications that depend upon their level of accuracy.

Before we attempt to gain a deeper insight into each of the elements that contribute to the total cost of a product, we must consider some factors that, if overlooked, may adversely affect the accuracy and validity of the estimate. For instance, the cost estimate cannot be held valid for more than a few months if the inflation rate in the country of production is noticeable. Further complications arise when the time taken to construct the plant and manufacture the products is so long that initial costs are affected by inflation (meaning that the money loses its purchasing power). Also, there are sometimes uncertain and unforeseen expenses, or contingency factors, a typical example being the escalation of R & D costs when developing new technology for manufacturing the products.

11.2 LABOR COST ANALYSIS

Labor can be either *direct* or *indirect:* Direct labor is explicitly related to the process of building the design, whereas indirect labor involves the work of foremen, stockroom keepers, and so on. We will be concerned here with the cost analysis of direct labor because the indirect-labor cost is generally covered by factory overhead costs in the form of a percentage of the cost of direct-labor hours. At this point, our goal is to estimate the labor time for building a design and then to multiply that time by the combined value of wages and fringe benefits, which is usually called *gross hourly cost.* Note, however, that wages are sometimes not based just on attendance, but also on performance (i.e., incentives are given when the hourly output exceeds a certain established goal).

Methods for Measurement of Time

Although there are quite a few approaches for the measurement and estimation of labor time, two methods are well accepted in industry and will, therefore, be covered here. The first method is based on *time and motion study,* a modern subject that was established by the eminent American engineer Frederick W. Taylor of Pennsylvania in the early twentieth century. This method, which is favored by industrial engineers, involves breaking down the manual work of an operation into individual simple motions. A typical motion is, for example, "reach and grab" (i.e., the worker stretches his or her hand to reach a tool and grab it). The operation is then converted into a tabular form that includes the entire sequence of basic motions that comprise the desired manual operation. Because these basic motions were thoroughly studied by industrial engineers and because time measurements were taken and standardized for each basic motion, our job is fairly easy. It is just to read, from published data that is readily available, the standard time unit for each motion included in the manual operation and insert it into a time and motion study table. By summing up all the time values, the total time required by an average worker to carry out the operation can be obtained. This time is modified by dividing it by the efficiency or the "rating" of the actual worker to account for interruptions and fatigue. This approach has the clear advantage of including a mechanism for rationalization of the operation by eliminating unnecessary and wasteful motions. The procedure described here is used to estimate the time for a single operation only and must be carried out for all operations required to produce a design. Consequently, our first step (after the design is available) is actually to prepare detailed process routing sheets indicating all operations included in the production of the part. It is clear that the time and motion study method requires a considerable amount of work, but it has usually been found that the effort and time spent are well worthwhile.

The second method is based on the historical value of time. Time cards for a similar design that has already been built are obtained and studied in order to determine the number of "man-hours" required to do the job (a man-hour is a unit indicating the output of one person working for one hour). Data analysis using spreadsheets is then employed to make the necessary adjustments, taking into consideration such factors as the skill level of the workers, the workplace environment, and cost escalation, if any.

The Learning Curve

It is a well-known fact that doing a job for the first time requires more time from a worker than when doing it for the fifth time, for example. This is evidently due to the phenomenon of self-teaching while performing the work, which, in turn, leads to a gain in work experience and thus a shorter time for doing that job. This is what is usually referred to as the *learning curve* or *product improvement curve.* As can be seen in Figure 11.3, the learning curve indicates the relationship between

FIGURE 11.3
The learning curve

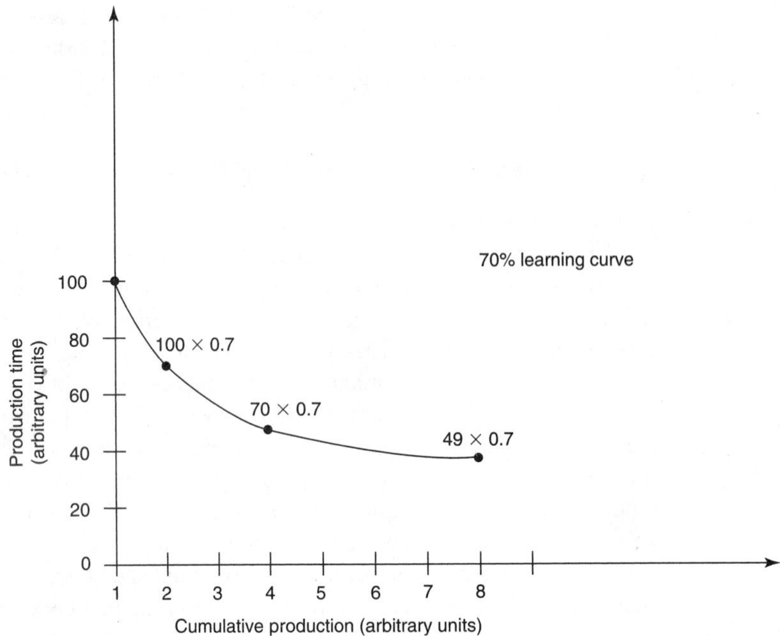

the production time (say, per unit) and the total cumulative production. Learning curves are usually labeled by percentages. In Figure 11.3, for example, we have a "70% learning curve." This simply means that whenever the total cumulative production doubles, the production time per unit decreases by 30 percent. This description suggests that the learning curve is exponential and can, therefore, be expressed by the following equation:

$$t = t_o \times p^n \tag{11.1}$$

where: t is the production time per unit after producing a number of units equal to p

t_o is the time taken to produce the first unit

p is the total cumulative production (i.e., total number of units produced)

n is a constant that depends on the constant percentage reduction characterizing a learning curve (e.g., –0.5146 for 70 percent curve, –0.322 for 80 percent curve, –.152 for 90 percent curve)

Taking the concept of the learning curve into consideration results in reducing the total production time (and cost) as estimated by the previously mentioned conventional methods, and adjustments have to be made. By simple mathematics, the total production time T is given by

$$T = t \times p \tag{11.2}$$

Substituting the value of t from Equation 11.1, we obtain

$$T = t_o \times p^n \times p = t_o \times p^{n+1} \tag{11.3}$$

Labor Laws

Some legal aspects must be considered when estimating the cost of labor. Federal and state laws regulate wages; for instance, minimum wage is $5.15 per hour. Also, the number of regular working hours per week is 40 and per day is limited to 8. If either (or both) of these is exceeded, a production worker must be paid at a rate equal to 150 percent of his or her regular hourly wage for the number of working hours that exceed the limits. Labor hourly wage rates in the United States, as well as other important relevant information, are compiled and published by the Bureau of Labor Statistics for various industrial sectors and can be obtained from the Department of Labor.

It is also important to remember that the labor cost is not limited to the money spent on wages but must include the fringe benefits paid to workers. Fringe benefits differ for different companies and may include any of the following:

- Health, dental, and life insurance premiums
- Expenses of insurance against job injury and hazards
- Holidays, paid vacations, and sick leave when actually taken
- The company's share in pension plans
- Payments to union stewards (if the company is unionized)
- Profit-sharing bonus money (that part of the company's profits paid to workers)

Fringe benefits can amount to as much as 30 percent or more of wages. It is, therefore, fairly common to combine wages with fringe benefits into the so-called gross hourly cost in order to avoid repetitive calculations.

11.3 MATERIAL COST ANALYSIS

Amount of Material Used

In order to carry out material cost analysis for a product, the amount of material used to manufacture that product first must be determined and then multiplied by the price of material, (or materials) in the form of dollars per unit weight or volume. Consequently, some documents must be available before this cost-estimating operation is initiated. These include, for example, the bill of materials, the engineering design documents, and printouts of inventory data. The engineer is then in a position to determine the bill of material required to build the design—a process that is sometimes referred to as the *quantity survey.* It is not difficult to see that the raw material required is more than the amount of material indicated in the design blueprints. This difference includes but is not limited to the waste during manufacturing, which, in turn, depends upon the specific production processes employed. For example, when the part is to be produced by machining, the amount of material removed from the stock in the form of chips must be added to the amount calculated from the design drawing. When the part is to be produced by casting, the material in the risers, sprues, and gating system must be added, as well as the material that is removed by machining (the "skin," drilled holes, slots, recesses, etc.). The same rule of adding the waste applies to the various manufacturing processes of forging, press working, and extrusion. Note that the waste is sometimes sold to junk dealers for recycling and the money paid for it must accordingly be subtracted from the cost of material. In some cases, however, the waste is valueless and is disposed at a cost that has to be added to the cost of material. In addition to waste, losses due to scrap (i.e., defective parts that are functionally obsolete) must be added. Again, scrap can be sold at a cost for recycling or may require disposal and additional expenditure. Losses may also include "shrinkage," which is a loss caused by environmental conditions (e.g., oxidation of steel or decrease in the volume of lumber when it dries).

The amount of raw material required to build the design and calculated according to the approach just described is a part of a cost analysis category known as *direct material.* This category can also include standard purchased items like nuts, bolts, springs, and washers. These items have no labor cost in our cost estimate, and the purchase price is considered as material cost (the company that makes these items has to consider the labor cost when estimating the selling price). Direct material also includes subcontracted items, which are assemblies or subassemblies manufactured outside the company and are supplied by an external subcontractor. Again, this does not include any labor cost in our estimate and must be categorized as material cost.

Determination of the amount of direct material required to build a design can sometimes be complicated. Consider, for example, the case when plastic is injection molded into a mold that has several dissimilar cavities for producing different parts. The sprues and runners form the waste in the injection molding operation. The question is, How do we determine the share of each part from that waste? Let us consider dividing the amount of material of the runners equally between the parts in order to get the amount of waste for each product. Unfortunately, the results may be totally misleading, especially when some of these products are very small while others are large. In fact, this case is referred to as one of *joint material cost* and arises whenever there is a multiple-product manufacturing process where the tracing of the raw material share of each individual product is difficult. A well-accepted approach in this case is to agree upon a *primary product* and attribute most of the untraceable common expenditures to it. In addition to direct material, there is material that is consumed during the process of transforming the raw stock into useful products. It is usually both necessary and untraceable to a particular design and is, therefore, referred to as *indirect material.* Typical examples include lubricating oils, soaps, and coolants. As it is clear that direct mathematics will not work here, a simpler method, which has gained acceptance in industrial cost-estimating practice, is to add this item to factory overhead costs.

Purchasing Price of Material

As previously mentioned, the purchasing price of material has to be known, in the form of dollars per unit weight or volume, in order to be able to estimate the total material cost to build the design. When the material used is contractual (i.e., purchased specifically for manufacturing a certain product), the actual purchase price can be directly employed in estimating the cost of material. However, when the material is taken from inventory, it is difficult to find its value and use it in the cost analysis because inventories usually contain various lots of the same material that have been purchased at different times at different prices. So, the question is, Which price do we use in cost estimation? In fact, nobody can provide a precise answer to this question, and, therefore, a number of approaches have been adopted by different schools of thought in industry. Following is a brief summary of some of the commonly used methods.

First-in-first-out method. The *first-in-first-out* method is based on following the rule of issuing first the material that was purchased first (i.e., having the longest time in stock) to the factory for processing and using its purchase price in the cost analysis. A clear drawback of this method arises when the time between purchasing and processing of the material becomes long. The original price may not be a true representation of the current value of the material, thus resulting in an inaccurate cost estimate.

Last-in-first-out method. In the *last-in-first-out* method, the material that was added last to the inventory is issued first to the factory. The material used can be from more than one purchased lot, with different costs for those lots—a fact that complicates the process of estimating the cost of materials.

Current-cost method. The approach taken by the *current-cost* method is to use the cost of materials corresponding to the time when the estimate is prepared. Once again, material issued from the inventory that would have been purchased earlier or later may have a cost different from the current cost.

Actual-price method. The *actual-price* method is based on calculating the amount of money originally spent to purchase the material used. If the material issued from the inventory belongs to the same lot, then the calculations are easy and simply take the original purchase price of material. However, when the material is taken from two (or more) lots having different purchase prices, an average or equivalent cost has to be used in the cost-estimating process. Here is the applicable equation:

$$C_{\text{equivalent}} = \frac{\sum_{i=1}^{n} c_i a_i}{\sum_{i=1}^{n} a_i} \tag{11.4}$$

where: c_i is the cost of material of lot i

a_i is the amount of material taken from lot i

i to n are used-lot serial numbers

11.4 EQUIPMENT COST ANALYSIS

The cost of equipment belongs to the fixed-cost category, and the depreciation of machinery as well as the interest on investment must be taken into account. It is often difficult, however, to get a quote for the cost of equipment, especially during the early phase of a feasibility study. Published data about the cost of equipment are, unfortunately, not directly applicable due to inflation and devaluation of the purchasing power of the dollar, as well as the difference in capacities or ratings given in the published data and those of the required equipment. Adjustments must, therefore, be made to account for factors that affect the validity of the published cost-of-equipment data. Following are some of the methods used.

Cost Indexing

A *cost index* is an indication of the buying power of money at a particular time for a certain category of equipment and machinery. Accordingly, if the cost of equipment and the corresponding index are known at some initial time, the current cost can be determined if the current index is obtained:

$$C_c = C_i\left(\frac{I_c}{I_i}\right) \tag{11.5}$$

where: C_c is the current cost

C_i is the cost at some initial time

I_c is the current index

I_i is the index at the same initial time as C_i

There are many published indexes, and each pertains to a certain area of application (e.g., chemical plants, consumer price). The Marshall and Swift cost index is the one to use for industrial equipment. It is readily available and has different values for the various kinds of industry. Care must, therefore, be taken not to use the index for the paper industry to calculate the cost of a steam turbine, for example. More importantly, indexing does not hold true when there is a radical change in the technology. This is evidenced by the fact that the prices of many electronic products have actually decreased as a result of technology change. Also, further adjustments are sometimes needed to account for regional conditions because published indexes are indications of national averages.

Size Effect

Sometimes, it is possible to only get hold of the cost of a machine similar to the required one but having a different size or rating. Corrections must, therefore, be made to that cost in order to obtain the cost of the desired machine. Consequently, a mathematical relationship between the cost and the capacity (size or rating) of capital equipment must first be established. As you may have guessed, the relationship is not linear due to the effect of the economy of scale on engineering design and production. The relationship can be given by the following empirical formula, which is usually referred to as the *six-tenths rule:*

$$C_2 = C_1\left(\frac{S_2}{S_1}\right)^{0.6} \tag{11.6}$$

where: C_2 is the cost of capital equipment 2

C_1 is the cost of capital equipment 1

S_2 is the capacity (size or rating) of capital equipment 2

S_1 is the capacity (size or rating) of capital equipment 1

Regression Analysis

Statistical techniques are used for collecting factual data that is, in turn, subjected to mathematical analysis and curve fitting in order to establish a relationship between cost and the various parameters that affect it. Here is the general equation:

$$C = C_o \sum_{i=1}^{i=n} P_i^{m_i} \tag{11.7}$$

where: C is the cost

C_o is a constant

P_i is the parameter i affecting the cost

m_i is a constant exponent for parameter i

C_o, P_i, and m_i are determined by mathematical and statistical methods. Although the formula is limited in scope to the specific equipment or system for which it is developed, it has proven to be very useful in cost models because it elaborates the elements and parameters that contribute most to the cost. It also leads to the ability to minimize and optimize cost using simple mathematical manipulations such as those of differential calculus.

11.5 ENGINEERING COST

Engineering cost includes salaries for high-level engineering jobs as well as expenditures (whether salaries or general expenses) for R & D. Usually, the engineering cost for a product is considered as part of the factory overheads (or even as part of the corporate overheads in some cases). Nevertheless, it is sometimes estimated separately based on previous records of existing similar products. In some cases (especially when the product is supplied to the federal government), a firm is hired to do the engineering on a contractual basis. The contract may specify a lump sum for the engineering cost or may involve the true engineering cost plus a negotiated fee or profit of, say, 15 percent of the cost. In this latter case, the engineering cost must be accurately determined.

11.6 OVERHEAD COSTS

Overhead costs are usually viewed by cost engineers as a burden because such costs cannot be directly or specifically related to the manufacturing of any particular product or even to a particular category of the company's production. Overhead costs can be divided into two main groups: factory overheads and corporate overheads.

Factory Overheads

Factory overheads include the previously mentioned engineering costs as well as other factory expenses that are not related to direct labor or material. An example would be the wages paid to personnel for security, safety, shipping and receiving, storage, and maintenance. The challenge here is how to calculate the "share" of each different product from these expenses. There are many approaches for charging these expenses to the cost of the various products. Following are three bases upon which factory overhead costs can be allocated:

1. The ratio between the direct-labor hours required to manufacture the product and the total number of direct-labor hours spent on the factory floor (this ratio, when multiplied by the total overhead expenses, yields the share of that product from the overhead cost)
2. The ratio between the material cost of the product and the total cost of material consumed on the factory floor (again, the share of a product from overhead cost is the product of multiplication of this ratio by the total overhead expenses)
3. The ratio between the space occupied by the production equipment (e.g., furnace or machine tool) and the total area of the factory floor

The direct-labor-hours method is by far the most commonly used approach for allocating factory overhead costs. As is clear, the production volume has a tangible effect on the factory overhead cost per product. If the production is reduced to half its normal level, for example, without reducing the total overhead expenses, the overhead share of a product will automatically double. Consequently, it is always a good idea to carry out cost analysis for any potential product at different production volumes (i.e., percentages of full capacity of production lines). Note that increasing productivity results in a reduced number of direct-labor hours. This is sometimes misinterpreted by management, and decisions may be made to reduce the budget for maintenance and other factory overhead items. It is the duty of the manufacturing engineer to eliminate any misinterpretations on the part of management. An alternative, in this case, is to use a different basis for allocating the overhead costs and requesting budgetary funds.

Corporate Overheads

Corporate overheads basically involve the cost of daily operation of the company beyond the factory floor throughout the year. These expenses include, for example, the salaries and fringe benefits of corporate executives as well as those of the business, administrative, and legal staffs. Again, the commonly adopted approach is to obtain an overhead rate that is the product of dividing the total corporate overhead expenses by the total cost of direct labor. Knowing the direct-labor time and cost for manufacturing a product, you can easily calculate the corporate overhead cost using this overhead rate. It is worth mentioning that corporations may operate more than one plant from the corporate headquarters—a fact that has to be taken into consideration when calculating both the corporate and the direct-labor costs in order to obtain the overhead rate.

11.7 DESIGN TO COST

The preceding discussions reflect the usual or conventional sequence of preparing the design and then costing the product based on the information provided in that design. With increasing global competition, however, cost is becoming more and more the driving force. Consequently, a need arises for costing a potential product before its design is completed or even made. This unusual approach is aimed at continuously improving the design in order to manufacture the desired product at a designated price that is equal to or less than the market price of the competitor's product. This "reverse" procedure is known as *design to cost* and is gaining popularity in industry, especially with newly emerging methodologies such as reengineering.

The process starts with benchmarking a given product, taking market price and quality as the judging criteria. By removing the retail profit, the manufacturing cost is obtained. Next, the various overhead rates that are well established in the company are employed to remove the different overhead cost items, yielding the prime cost. Then comes the difficult task of meticulously breaking down the prime cost among components, assemblies, and subassemblies. Favoring one component at the expense of another is a big mistake as setting a target cost below reasonable limits will make the design of the component virtually impossible. Once the target cost for a component is allocated, design begins using that target cost as an incentive for continuously improving the design. If the direct-labor cost, for example, is found to be less than the target, this will give some relief in the process of selecting materials. The same rule is applicable to subassemblies and assemblies (i.e., if the cost of a component is less than the target, this will give more flexibility when designing other components). When the design is finalized, it must be subjected to the conventional and accurate cost-estimating process.

Review Questions

1. Why is cost estimation of vital importance for a design engineer?
2. What role does cost play in process planning?
3. List two methods for classifying costs.
4. List some important elements of fixed cost.
5. List some important elements of variable cost.
6. What are the two main problems that complicate cost estimation?
7. Do all types of cost-estimating methods have the same accuracy? Explain why.
8. Can you rely upon a cost estimate that was done last year? Why not?
9. Assuming that the construction of the plant takes a long time, what effects would this have on the cost-estimating process?
10. What is meant by *direct labor?*
11. Is there any indirect labor? Explain.

12. What is the *gross hourly cost?*
13. Explain the difference between the two cases of wages based on attendance and wages based on performance.
14. How can we measure the direct-labor time before the product is actually manufactured?
15. What are the pros and cons of the Industrial Engineering approach for measuring the direct-labor time?
16. Explain the time-card method for estimating the direct-labor time.
17. What is the *learning curve?* What effects does it have on cost-estimating results?
18. List some important labor laws that must be considered when estimating the cost of a product.
19. List some common fringe benefits.
20. What is the *quantity survey?*
21. What are the sources of difference between the material in a product as indicated by the design drawing and the material actually consumed in the manufacture of that product?
22. Explain the term *indirect material.*
23. What is the *joint material cost?* Give examples.
24. Why is it difficult to get the cost of unit material when the material is issued from inventory?
25. Explain briefly the different methods used to obtain the cost of unit material when the material is issued from inventory. Give the pros and cons of each.
26. What is a *cost index?* Why is it important in cost estimation?
27. How can you calculate the cost of a machine with a known capacity if you know the cost and the capacity of another machine?
28. Show how regression analysis and statistics can be employed in cost estimation.
29. What is meant by *engineering cost?* How is it estimated?
30. What are the different types of overhead costs?
31. On what bases are factory overhead costs allocated?
32. In some cases, increasing productivity might have an adverse effect on budget allocations. Explain how.
33. Explain the concept of *design to cost.*
34. What is the driving force for design to cost?
35. What is the main problem encountered in the procedure of design to cost?

Problems

A number of stock bars, each 3.25 inches (81 mm) in diameter and 12 feet (3.6 m) in length, are to be used to produce 2000 bars, each 2.75 inches (69 mm) in diameter and 12 inches (300 mm) in length. The material cost is \$1.05 per pound (\$2.11/kg), and the density is 0.282 pound per cubic inch (789 kg/m^3). The total overhead and other expense is \$95,000. The total direct-labor expense for the plant is \$60,000. Estimate the production cost for a piece.

Solution

First, we have to calculate the production time per piece. Consequently, technical production data have to be either obtained or assumed. Following are some assumptions:

- The facing dimension necessary for a smooth end finish is 1/16 inch (1.6 mm).
- The width of the cutoff tool is 3/16 inch (4.76 mm).
- The collet requires 4 inches (100 mm) of length for last-part gripping.
- Heavy cuts are to be done followed by a light finishing cut: For two rough cuts, cutting speed is 200 feet per minute (60 m/min.) and feed is 0.01 inch (0.25 mm); for finishing, cutting speed is 300 feet per minute (90 m/min.) and feed is 0.005 inch (0.125 mm).
- The time taken to return the tool to the beginning of cut is 15 seconds.
- Load (setup) and unload time is 1 minute.

Machining Time

Position tool to perform cutoff: 15 seconds

Cutoff time:

$$\frac{D + a}{2} \times \frac{1}{\text{radial feed rate}}$$

$$= \frac{3.25 + 0.75}{2} \times \frac{\pi \times 3.25}{200 \times 12 \times 0.01} \times 60 = \quad 51 \text{ seconds}$$

Position tool to carry out facing operation: 15 seconds

Facing time:

$$\frac{D + a}{2} \times \frac{1}{\text{radial feed rate}} = \frac{2 \times \pi \times 3.25}{300 \times 12 \times 0.005} \times 60 = \quad 68 \text{ seconds}$$

Position tool to perform first rough cut: 15 seconds

First rough cut:

$$\frac{12 + 1/16 + 3/16 + 4/16}{\text{feed rate}} = \quad 320 \text{ seconds}$$

Position tool to perform second rough cut: 15 seconds

Second rough cut: 320 seconds

Position tool for finishing: 15 seconds

Finishing:

$$\frac{12.5}{\text{feed rate}} = \quad 360 \text{ seconds}$$

Cutoff: does not count with this piece; it is included in the time for next piece.

Load/unload time per piece (1 minute): 60 seconds

The total machining time per piece is 1254 seconds.

Cost of Labor/Piece

$$\text{number of pieces produced from a single bar} = \frac{12 \times 12}{12.25} = \frac{\text{total length}}{\text{length/piece}} = 11 \text{ pieces}$$

$$\text{number of stock bars} = \frac{2000}{11} = 181.8 = 182 \text{ bars}$$

$$\text{total loading time} = 2 \times 182 \times 60 \qquad \text{(assuming loading time/bar = 2 minutes)}$$

$$\text{share of each piece} = \frac{2 \times 182 \times 60}{2000} = 10.9 \simeq 11 \text{ seconds}$$

$$\text{total average production time/piece} = 1254 + 11 = 1265 \text{ seconds} \qquad \text{(direct labor)}$$

$$\text{cost of labor/piece} = \frac{1265}{3600} \times \$10/\text{hr} = \$3.52 \qquad \text{(assuming a CNC machine is used)}$$

Cost of Material/Piece

We have to consider the waste. Assume no scrap as the operation is simple:

$$\text{cost of material/piece} = \frac{182 \times \frac{\pi}{4}(3.25)^2 \times 12 \times 12 \times .282 \times 1.05}{2000} = \$32.17$$

Cost of Overhead/Piece

In this, all other costs are included:

$$\text{overhead rate} = \frac{95{,}000 \times 100}{60{,}000} = 158.33\%$$

$$\text{overhead cost/piece} = 3.52 \times \frac{158.33}{100} = \$5.57$$

Total Cost/Piece

$$\text{total cost/piece} = \text{direct labor cost} + \text{material cost} + \text{overhead cost} = 3.52 + 32.17 + 5.57 = \$41.26$$

Design Project

Select a few of the design projects that supplement Chapters 3 through 7, preferably a project for each manufacturing process, and then carry out cost estimation for the product. You are strongly advised to obtain real values for the different cost elements (e.g., material) by contacting industrial companies and obtaining quotations.

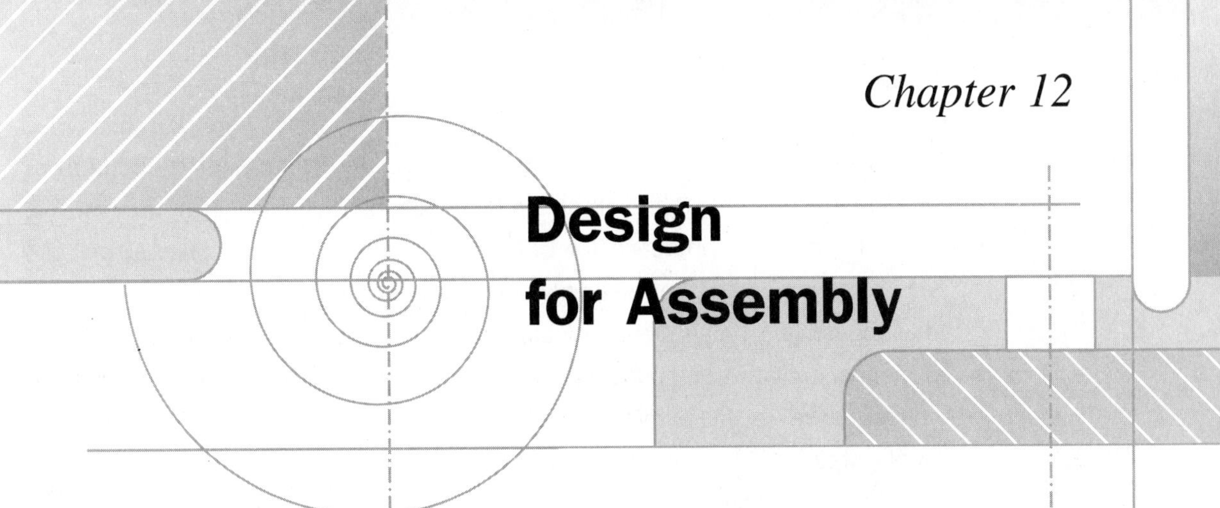

Chapter 12

Design for Assembly

INTRODUCTION

Modern societies are undergoing continuous development, which necessitates large-scale use of sophisticated products like appliances, automobiles, and health-care equipment. Each of these products involves a large number of individual components, assemblies, and subassemblies that must be brought together and assembled into a final product during the last step of the manufacturing sequence. A rational design should, therefore, be concerned with the ease and cost of assembly, especially when given the fact that 70 to 80 percent of the cost of manufacturing a product is determined during the design phase. It is for this reason that the concept of *design for assembly* (DFA) emerged. It is simply a process for improving the product design for easy and low-cost assembly. In other words, this assembly-conscious design approach not only focuses on functionality but also concurrently considers *assemblability.*

Although the use of the term *design for assembly* is fairly recent, several companies can claim, in good faith, that they have developed and have been using guidelines for assembly-conscious product design for a long time. For instance, the General Electric Company published in 1960, for internal use only, the *Manufacturing Producibility Handbook.* It included compiled manufacturing data that provided designers in the company with information necessary for sound and cost-efficient design. Later, in the 1970s, research institutions and research groups started to become more and more interested in the subject when the Conférence Internationale pour le Recherches de Production (CIRP) established a subcommittee for that purpose and Professor Geoffrey Boothroyd began his pioneering research at the University of Massachusetts Amherst.

The traditional approach for DFA has been to reduce the number of individual components in an assembly and to ensure an easy assembly for the remaining parts through design modifications. When a design is altered in such a manner that two components are replaced by just a single one, the logical consequence is the elimination of one operation in manual assembly or a whole station of an automatic assembly machine. Accordingly, many benefits have been credited to the DFA methodology, including simplification of products, lower assembly costs, reduced assembly (and manufacture) time, and reduced overheads. Recently, the DFA concept has been extended to incorporate process capacity and product mix considerations so that products can be designed to assist in balancing assembly flow, thus eliminating the problem of stressing one process too heavily while underutilizing others. Many people now are calling for extending DFA over the whole product life cycle, in which case environmental concerns would be addressed and designs would be developed that facilitate disassembly for service as well as for recycling at the end of the life cycle.

A first step toward a rational design for easy and low-cost assembly is the selection of the most appropriate method for assembling the product under consideration. The design guidelines for the selected method can then be applied to an assembly-conscious design for that product. The next step is the use of a quantitative measure to evaluate the design in terms of the ease of assembly and to pinpoint the sources of problems so that the design can then be subjected to improvement. As many iterations of this evaluation/improvement process as are necessary can be done in order to achieve an optimal design. It is, therefore, essential for us now to discuss the different assembly methods currently available.

12.1 TYPES AND CHARACTERISTICS OF ASSEMBLY METHODS

As you may expect, there is no single method that is always "better" than other methods under all conditions. In other words, each method has its own domain or range within which it can most successfully and economically be applied. Factors like the number of products assembled per year and the number of individual components in an assembly play a major role in determining the range of economical performance of an assembly system. Following is a description of the different assembly methods, as well as the characteristics of each.

Manual Assembly

In *manual assembly,* the operations are carried out manually with or without the aid of simple, general-purpose tools like screwdrivers and pliers. Individual components are transferred to the workbench either manually or by employing mechanical equipment such as parts feeds or transfer lines and then are manually assembled. This assembly method is characterized by its flexibility and adaptability—a direct consequence of the very nature of the key element of the system, the human brain. The assembly cost per product, however, is virtually constant and is independent of the production volume. There is an upper limit to the production volume above which the practicality and feasibility of the manual assembly method is, to say the least, questionable. This upper limit depends upon the number of individual components in an assembly and the number of different products assembled. Nevertheless, it is important to remember that the capital investment required for this type of assembly system is close to zero.

Automatic Assembly Using Special-Purpose Machines

In the type of assembly system referred to as *fixed automation* or the *Detroit type,* either synchronous indexing machines and automatic feeders or nonsynchronous machines where parts are handled by a free-transfer device are used. The system, in both cases, should be built to assemble only one specific product. Such is the case with the automotive assembly lines in Detroit, where each one is dedicated to the production of a specific model of car (and hence the reason this name is given to this type of assembly system). There is an inherent rigidity in this method of assembly, meaning that these systems lack any flexibility to accommodate tangible changes in the design of the product. Moreover, a system of this type requires a large-scale capital investment, as well as considerable time and engineering work before actual production can be started. Also, the individual components must be subjected to strict quality-control inspection before they can be assembled because any downtime due to defective parts will result in considerable production and, therefore, cash losses. Nevertheless, a real advantage of this assembly system is the decreasing assembly cost per product for increasing production volume. Naturally, when the production volume increases, the share of each product from the capital investment becomes smaller, which makes this assembly system particularly appropriate for mass production. It is worth mentioning that an underutilized assembly system will simply result in an increase in the assembly cost per product because the cost of equipment has to be divided between a smaller number of products, thus increasing the cost share of each product.

In order to come up with a more flexible version of the automatic assembly system that can tolerate some minor changes in the design of the product being assembled, the nonsynchronous machines are fitted with programmable workheads and parts magazines. Thus, the assembly sequence and characteristics can be tailored to match the attributes of the modified design. Although this system provides some flexibility, it is still considered to be most appropriate for mass production.

Automatic Assembly Using Robots

In *robotic assembly,* the production volume is higher than that of a manual assembly system but lower than that of an automatic assembly system that incorporates special-purpose machines. It, therefore, fills a gap, in production volume, between these other two assembly systems. Robotic assembly systems may take one or more of the following forms:

1. A one-arm, general-purpose robot operating at a single workstation that includes parts feeders, magazines, and so on. The end effector of the arm is tailored to suit the specific operation performed.

2. Two robotic arms operating at a single workstation. A programmable controller (PLC) is employed to simultaneously control and synchronize the motions of the two arms. This setup is referred to as a *robotic assembly cell* and is, in fact, very similar to a flexible manufacturing cell. Other supporting equipment like fixtures and feeders are also included in the cell.

3. Multistation robotic assembly system. This system is capable of performing several assembly operations simultaneously. It can also perform different assembly operations at each station. Accordingly, this robotic assembly system possesses extremely high flexibility and adaptability to design changes. On the other hand, a production volume that is quite close to that of the automatic assembly mass production system can be achieved using this type of system.

Comparison of Assembly Methods

Clearly, manual assembly requires the least capital investment followed by the two simplest forms of robotic assembly. On the other hand, compared to the automatic system with special-purpose machines, the multistation robotic assembly system requires more capital investment for a large production volume but less capital investment for a moderate production volume. A better way of illustrating this comparison is to plot a graph indicating the relationship between the assembly cost per product and the annual production volume for the three assembly methods. As shown in Figure 12.1, the assembly cost per product is constant for manual assembly and decreases linearly with increasing production volume for automatic assembly using special-purpose machines. In the case of robotic assembly, the assembly cost per product also decreases with increasing production volume but not linearly because the type of system used and its physical size depend upon the production volume as well. Figure 12.1 also helps to determine the range of production volume within which each of the assembly methods is cost effective. Consequently, such a graph is a valuable tool for selecting the appropriate assembly method for a specific project.

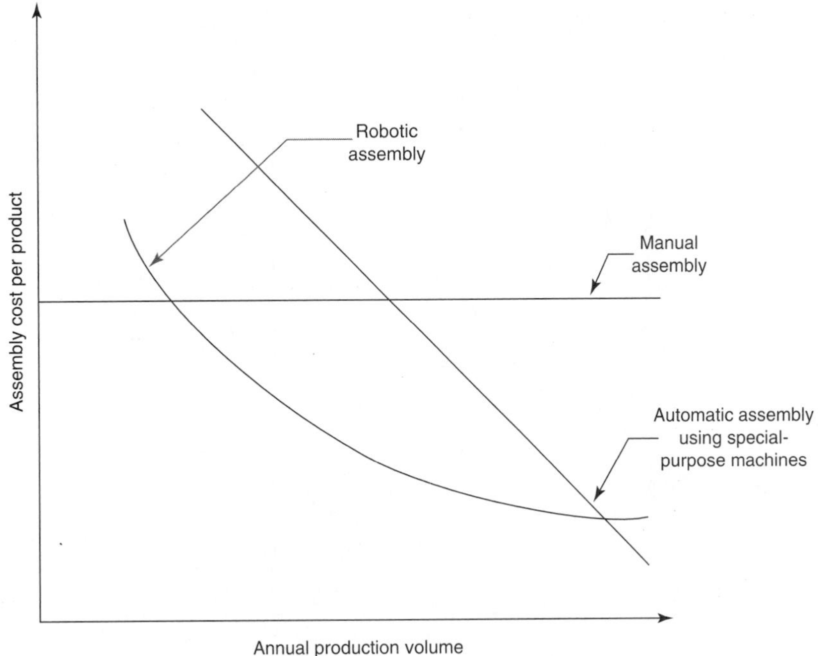

FIGURE 12.1 Assembly cost per product versus annual production volume for three assembly methods

12.2 SELECTION OF ASSEMBLY METHOD

Several factors must be taken into consideration by the product designer and the manufacturer when selecting an assembly method. These factors include the cost of assembly, the annual production volume (or production rate), the number of individual components to be assembled in a product, the number of different versions of a product or products, the availability of labor at a reasonable cost, and, last but not least, the payback period. The factors are interactive, and it is impossible to have a single mathematical relationship or a single graph that incorporates them all and indicates an appropriate range or domain for each assembly method. Usually, a two-variable chart is constructed based on fixed specific values for the other variables.

Figure 12.2 indicates the appropriate ranges of application for each of the various assembly methods when there is only one type (or version) of the product to be assembled. As can be seen, the two variables, which are pivotal in most cases, are the annual production volume and the number of individual components in an assembly. Notice that the manual assembly method is suitable for low production volumes and a limited number of individual components per assembly. Robotic assembly is recommended for moderate production, with the one-arm robot being more appropriate for assemblies that have less than eight individual components. When a large number of assemblies is to be produced, the use of assembly systems with special-purpose machines becomes a must. Remember that with an increasing

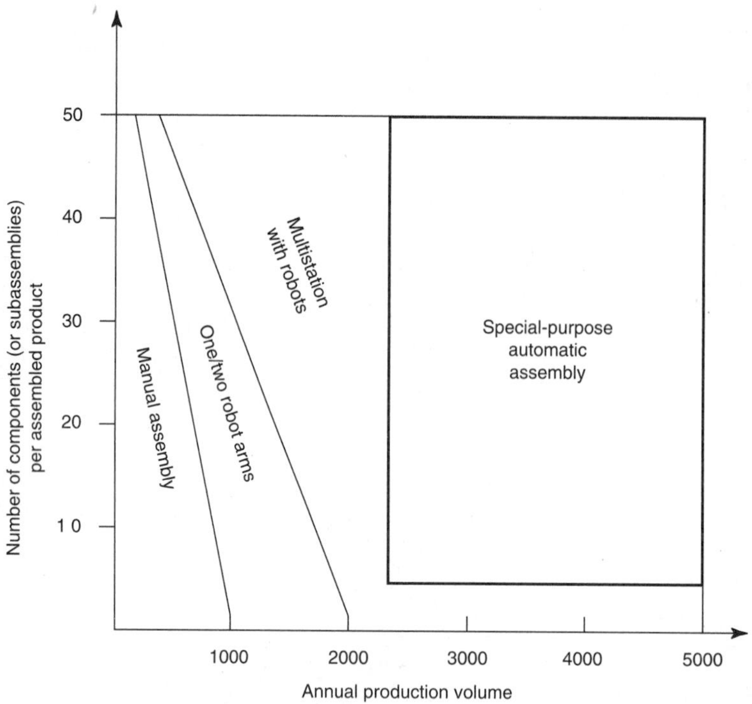

FIGURE 12.2 Appropriate ranges of application for various assembly methods

number of different types or versions of assemblies, the recommended ranges of application for each assembly method will differ from those shown in Figure 12.2. For instance, a multistation robotic assembly system would be more appropriate than an automatic assembly system with special-purpose machines for relatively high production volumes. The most important point here is that the assembly rate of the selected assembly method should not result in any bottleneck but rather should ensure trouble-free production. Also, it is always advisable to estimate the cost of assembly whenever more than one assembly method is under consideration. Assuming that all other factors are comparable, the method that gives the lowest assembly cost is the one to select.

12.3 PRODUCT DESIGN FOR MANUAL ASSEMBLY

We are now in a position to discuss the rules and guidelines to be followed when designing components for manual assembly. It is important here to emphasize that blind adherence to these rules is not recommended. In fact, this approach can result in very complex components that are difficult and expensive to manufacture. The use of good engineering sense, rational thinking, and accumulated knowledge will ensure that these rules are wisely applied. The strategy to adopt when designing products for manual assembly is to strive to reduce both the assembly time and the skills required of assembly workers. Here are the guidelines for product design for manual assembly:

1. Eliminate the need for any decision making by the assembly worker, including his or her having to make any final adjustments. Remember that assembly workers are usually unskilled and are paid at or close to the minimum wage and it is, therefore, not logical or fair to rely on them to make these adjustments.
2. Ensure accessibility and visibility. It is not logical or fair to require the worker, for example, to insert and tighten a bolt in a hole that is not visible or easily accessible.
3. Eliminate the need for assembly tools or special gages by designing the individual components to be self-aligning and self-locating. Parts that fit and snap together eliminate the need for fasteners, thus resulting in an appreciable reduction in both the assembly time and cost. Also, features like lips and chamfers can greatly aid in making parts self-locating, as is clearly demonstrated in Figure 12.3, where two pins, one having a chamfer and the other without, are being inserted into two identical holes during an assembly operation. Obviously, it is far easier and takes less time to insert the pin with the chamfer.
4. Minimize the types of parts by adopting the concept of standardization as a design philosophy. Expand the use of standard parts as well as multifunction and multipurpose components. Although more material may be consumed to manufacture multipurpose parts, the gains in reducing assembly time and cost will exceed that waste.
5. Minimize the number of individual parts in an assembly by eliminating excess parts and, whenever possible, integrating two or more parts together. Certainly, handling one part is far easier than handling two or more. The criteria for reducing the parts count per assembly, established by G. Boothroyd and P. Dewhurst (see the references at the end of this book), involve negative answers to the following questions:

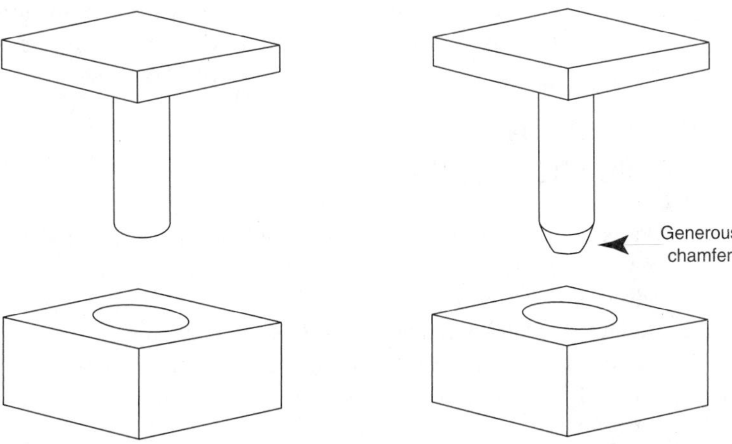

FIGURE 12.3 Using a chamfer to make a part self-locating

- Does the part move relative to all other parts already assembled?
- Must the part be of a different material or be isolated from other parts already assembled? (Only fundamental reasons concerned with material properties are acceptable.)
- Must the part be separate from all other parts already assembled because otherwise necessary assembly or disassembly of other parts would be impossible?

If the answer to each of these questions is no, then the part can be integrated or combined with another neighboring part in the assembly. When applying this rule, however, remember that combining two or more parts into a complicated one may result in making the part difficult to manufacture.

6. Avoid or minimize reorienting the parts during assembly. Try to make all motions simple by, for example, eliminating multimotion insertions. Avoid rotating or reorienting the assembly as well as releasing and regripping individual components. These are wasteful motions and result in increased assembly time and cost. The best time to eliminate them is during the design phase. The use of vertical insertion (along the Z axis) is ideal, especially when you take advantage of gravity.
7. Ensure ease of handling of parts from the bulk by eliminating the possibility of nesting or tangling them. This is achieved by simple modifications in the design. In addition, avoid the use of fragile or brittle materials, as well as flexible parts like cords and cables.
8. Design parts having maximum symmetry in order to facilitate easy orientation and handling during assembly. If symmetry is not achievable, the alternative is to design for asymmetry that is easily recognizable by the assembly worker.

Failure to observe the preceding rules may result in serious problems during assembly in terms of higher assembly costs or jams and delays. Consequently, many companies avoid manual assembly and sell their products unassembled. Examples in the United States include grills, furniture, and toys. As you may have experienced, some of these products are not properly designed for easy assembly, and it takes customers an extremely long time to assemble them. It is no surprise that such faulty designs do not pay off as they adversely affect the sales of the unassembled products.

12.4 PRODUCT DESIGN FOR AUTOMATIC ASSEMBLY

Parts that are designed to be assembled by automatic special-purpose machines must possess different geometric characteristics from those of parts to be assembled manually. Automatic assembly requires parts that are uniform, are of high quality, and have tighter geometric tolerances than those of manually assembled parts. These requirements are dictated by the need to eliminate any downtime of the assembly system due to parts mismatch or manufacturing defects. As a consequence, problems related to locating and inserting parts, though they need to be addressed, are not of primary importance. These problems require design changes to ease assembly; by revising the product design, each

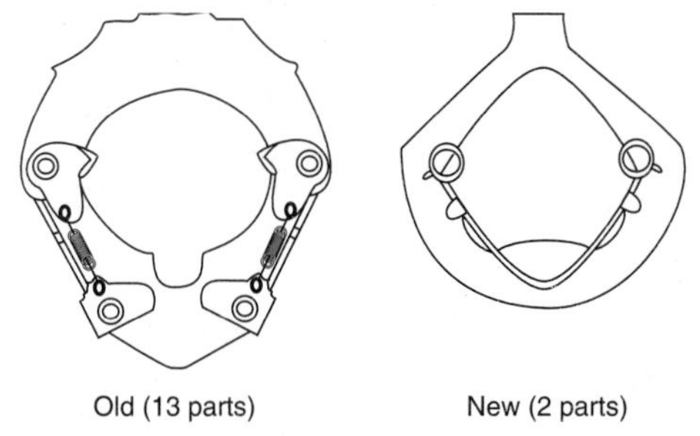

FIGURE 12.4
Facilitating assembly through reduction in parts count *(Redrawn after Iredale, R. "Automatic Assembly—Components and Products,"* Metal-working Production, *8 April 1964. Used by permission)*

assembly operation becomes simple enough to be performed by a machine rather than by a human being. The most important concerns to address involve the orientation, handling, and feeding of parts to the assembly machine. The efficiency of performing these tasks has a considerable effect on the efficiency and output of the assembly system and, of course, on the assembly cost. This approach is referred to as *design for ease of automation.* Here are the guidelines for product design for automatic assembly:

1. Reduce the number of different components in an assembly by using the three questions listed previously in the design guidelines for manual assembly. An appropriate approach is to use value analysis in identifying the required functions performed by each part and finding out the simplest and easiest way to achieve those functions. An example is shown in Figure 12.4, where two products are contrasted, one designed to facilitate assembly through a reduction in the parts count and the other designed without ease of assembly being taken into consideration.

   With the new developments in casting and plastics injection-molding technologies, complex components can replace entire subassemblies. Nevertheless, the designer has to be very careful when combining parts so as not to adversely affect the manufacturing cost. In fact, in order to reduce the parts count in assemblies, subcontractors and suppliers of electronics manufacturers have been continually asked to fabricate extremely complex parts. In short, the rule of reducing the number of parts should not be applied blindly because, in many cases, more efficient manufacturing can be achieved by breaking a single component into two or more parts, as shown in Figure 12.5, which indicates two methods for manufacturing a 2-foot axle shaft and flange.

FIGURE 12.5
Two methods for manufacturing an axle shaft and flange *(Redrawn after Lane, J. D., ed.* "Automated Assembly," *2nd ed., Society of Manufacturing Engineers, 1986. Used by permission)*

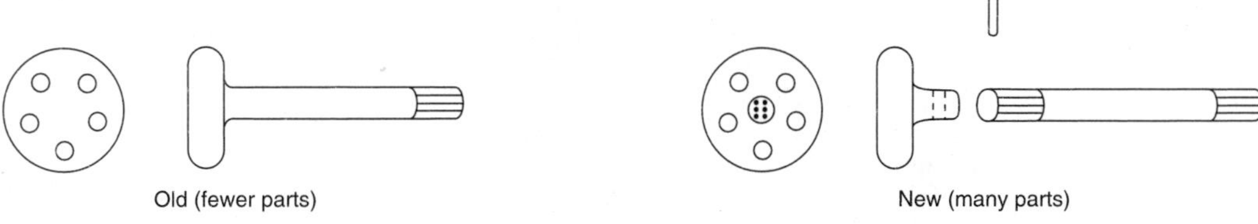

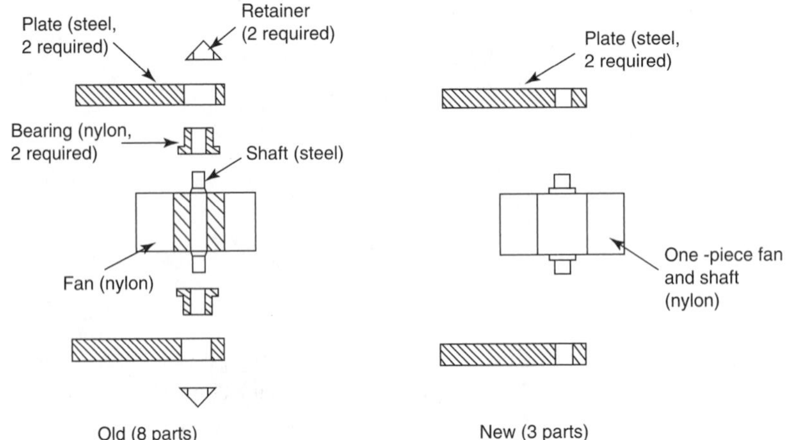

FIGURE 12.6 Facilitating assembly through simplification of design *(Redrawn after Iredale, 1964. Used by permission)*

2. Use self-aligning and self-locating features in parts to facilitate the process of their assembly. Considerable improvement can be achieved by using chamfers, guide-pins, dimples, molded-in locators, and certain types of screws (e.g., cone and oval screws). Figure 12.6 is an example of how to facilitate assembly through simple design modifications, while Figure 12.7 shows the types of screws that are suitable for assembly operations.

3. Avoid, whenever possible, fastening by screws because that process is both expensive and time-consuming. It is, therefore, recommended to design parts that will snap together or be joined together by a press fit. Tighter tolerances are then required, and problems may also be encountered in disassembly for maintenance,

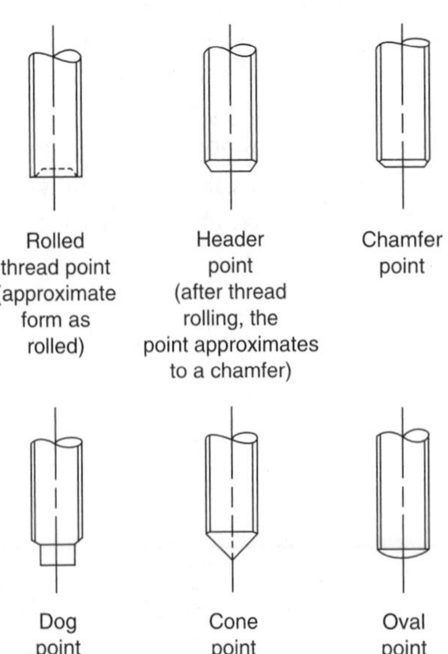

FIGURE 12.7 Types of screws suitable for assembly operations *(Redrawn after Tipping, W.V.* "Component and Product Design for Mechanized Assembly," *Conference on Assembly, Fastening, and Joining Techniques and Equipment, Production Engineering Research Association of Great Britian, 1965. Used by permission)*

repair, or recycling. If screws must be used, then unify their types and head shapes.

4. Make use of the largest and most rigid part of the assembly as a base or fixture where other parts are stack-assembled vertically in order to take advantage of gravity. This will eliminate the need for employing an assembly fixture, thus saving time and cost. Also, remember that the best assembly operation is one that is performed in a sandwichlike or layered fashion. If this is difficult or impossible to do, the alternative is to divide the assembly into a number of smaller subassemblies, apply the rule stated herein to each separately, and then plug all the subassemblies together.

5. Actively seek the use of standard components and/or materials. There should be a commitment, at all levels, to the goal of using a high percentage of standard parts in any new design. A very useful concept to be adopted in order to achieve this goal is *group technology.* Standardization should begin with fasteners, washers, springs, and other individual components. This translates into standardization of assembly motions and procedures. The next step is to use standard modules that are assembled separately and then plugged together as a final product. Each module can include a number of individual components that are self-contained in a subassembly having a specific performance in response to one or more inputs. This approach can lead to a considerable reduction in assembly cost, as well as in manufacturing and inventory costs.

6. Avoid the possibility of parts tangling, nesting, or shingling during feeding. A few changes in the geometric features may eliminate these problems without affecting the proper functioning of the component. Figure 12.8 shows some parts that tend to nest during feeding and the design modifications that eliminate this problem.

7. Avoid flexible, fragile, and abrasive parts and ensure that the parts have sufficient strength and rigidity to withstand the forces exerted on them during feeding and assembly.

8. Avoid reorienting assemblies because each reorientation may require a separate station or a machine, both of which cause an appreciable increase in cost.

9. Design parts to ease automation by presenting or admitting the parts to the assembly machine in the right orientation after the minimum possible time in the feeder. The process in the feeder consists of rejecting parts resting in any position but the one desired. Consequently, reducing the number of possible orientations of a part actually increases the odds of that part's going out of the feeder on its first try. Figure 12.9 shows the effect of the possible number of orientations on the efficiency of feeding. According to W. V. Tipping, two types of parts can easily be oriented: parts that are symmetrical in shape (e.g., a sphere or cube) and parts with clear asymmetry (preferably with marked polar properties either in shape or weight).

   Symmetrical parts are easily oriented and handled. Therefore, try to make parts symmetrical by adding nonfunctional design features like a hole or a projection.

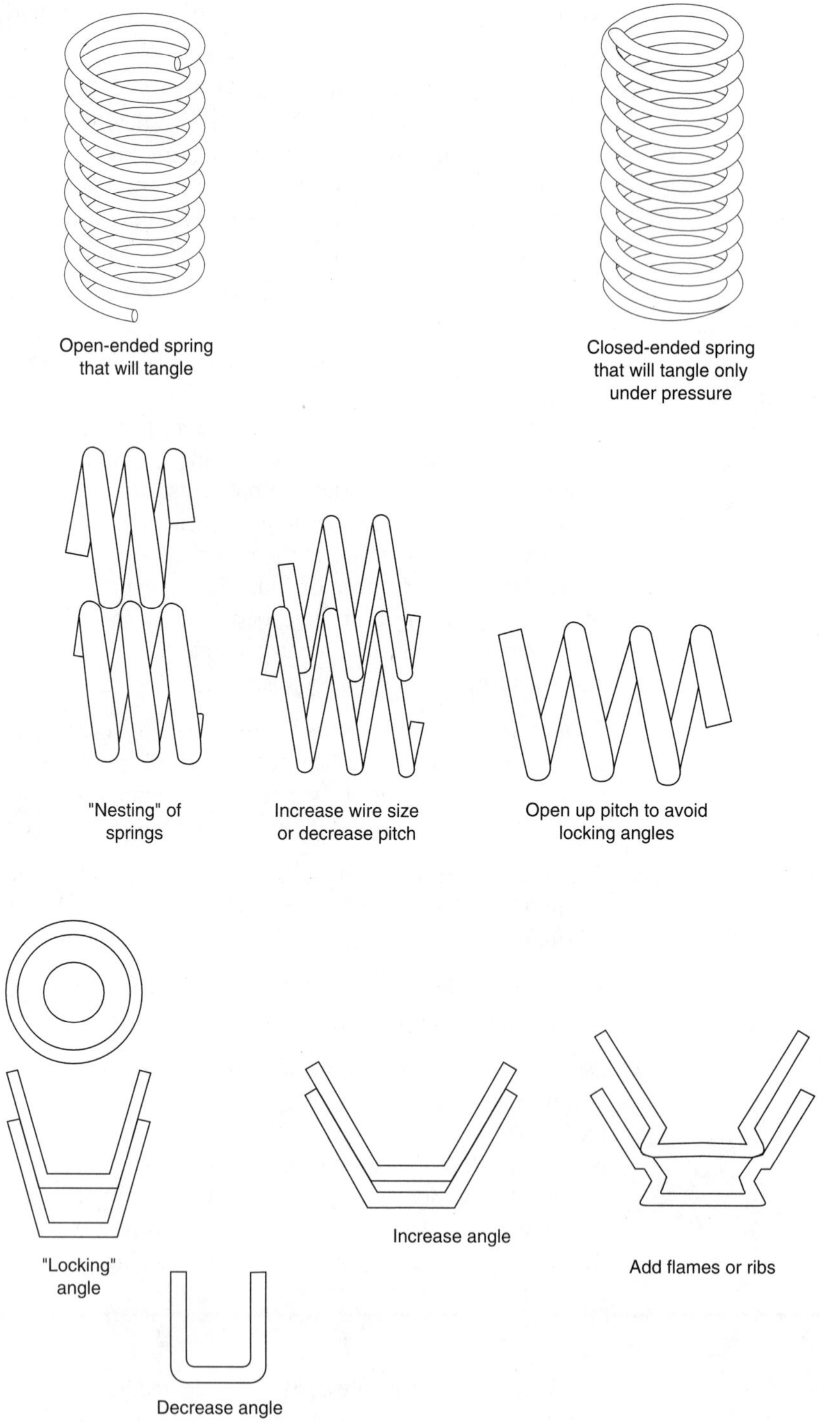

FIGURE 12.8
Parts that tend to nest during feeding and design modifications that eliminate the problem *(Redrawn after Lane 1986. Used by permission)*

Open-ended spring that will tangle

Closed-ended spring that will tangle only under pressure

"Nesting" of springs

Increase wire size or decrease pitch

Open up pitch to avoid locking angles

"Locking" angle

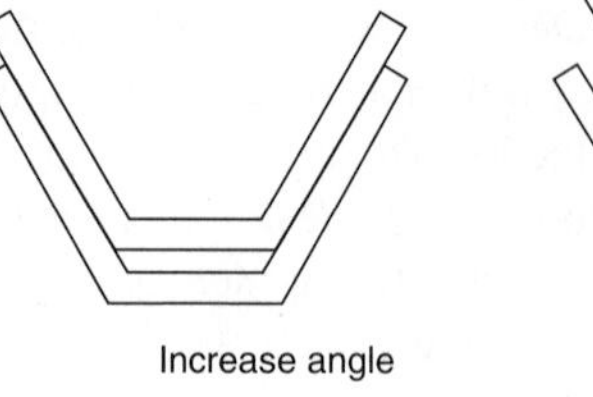

Increase angle

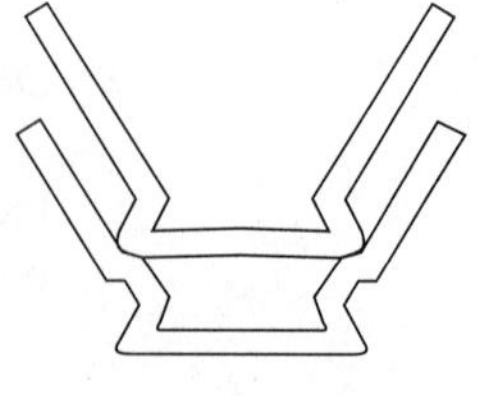

Add flames or ribs

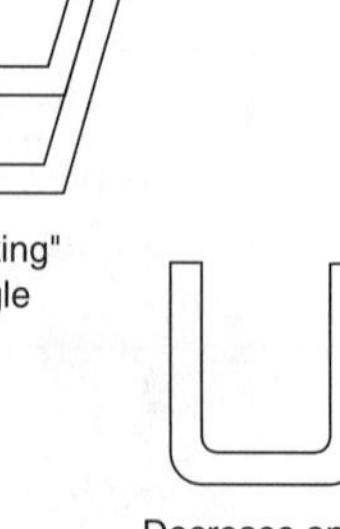

Decrease angle

FIGURE 12.9
Effect of possible number of orientations on efficiency of feeding *(Redrawn after Lane, 1986. Used by permission)*

Number of Orientations	Types of Parts	Required Number of Parts/Hour (out of the feeder)	Minimum Required Rate of Feeding Parts/Hour (into the feeder)
1	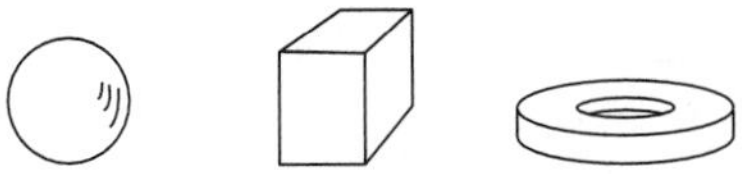 Sphere; Symmetrical cube; Symmetrical flat washer	600	600
2	 Tapered washer; Parts that naturally fall in one of two possible positions	600	1200
4	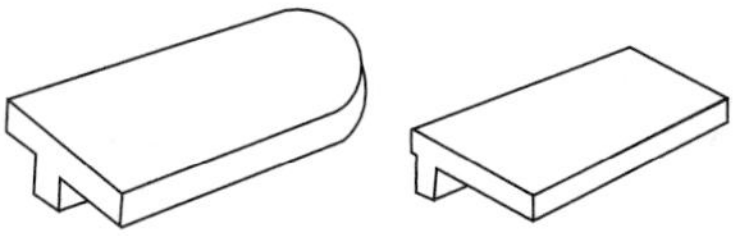 Parts having four natural positions	600	2400

Figure 12.10 shows some small changes in the design of parts that result in full symmetry. Generally, it is easy to achieve symmetry with sheet metal and injection-molded parts because the manufacturing cost of adding a feature is relatively low.

If it is too difficult or too expensive to achieve symmetry, nonfunctional features must then be added to make identification and grasping easier. This approach is also employed for parts for which orientation is based on hard-to-detect features like internal holes. In addition, components having similar shape and dimensions are difficult to identify and orient, and changes in dimensions or additions of design features must be made. Recent research work has come up with a concept, called *feedability,* that involves quantitative estimation of the odds of feeding a part having certain geometric characteristics to the assembly station in a specific orientation. Figure 12.11 shows some design changes that exaggerate asymmetry or indicate hidden features, while Figure 12.12 shows the effect of changing geometric features on the calculated values of feedability.

10. Try to design parts with a low center of gravity (i.e., it should not be far above the base). This gives the part a natural tendency to be fed in one particular orientation.

FIGURE 12.10
Examples of design changes that give full symmetry *(Redrawn after Lane, 1986. Used by permission)*

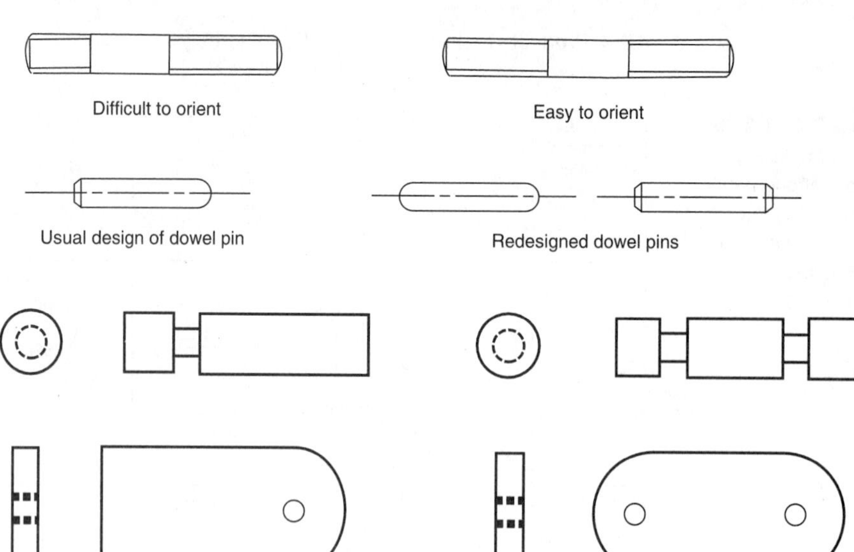

Before (2 natural orientation)

After (1 orientation required)

FIGURE 12.11
Examples of design changes that exaggerate asymmetry or indicate hidden features *(Redrawn after Iredale, 1964. Used by permission)*

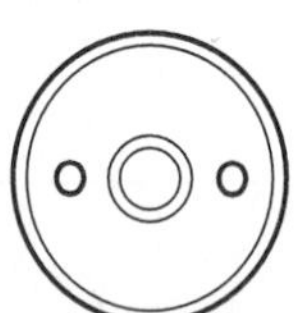

Difficult to orient with respect to small holes

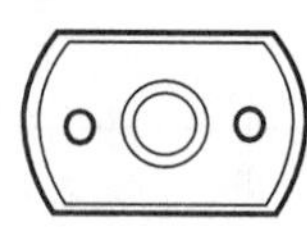

Flats on the sides make it much easier to orient with respect to the small holes

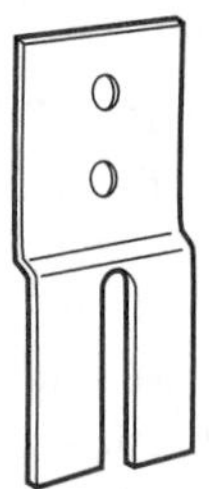

No feature sufficiently significant for orientation

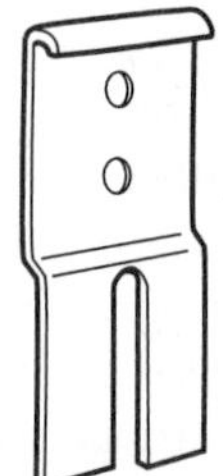

When correctly oriented will hang from rail

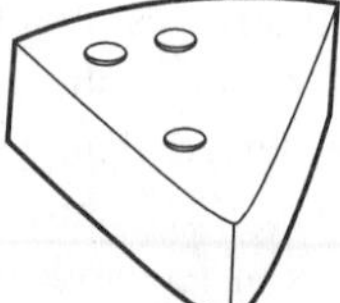

Triangular shape of part makes automatic hole orientation difficult

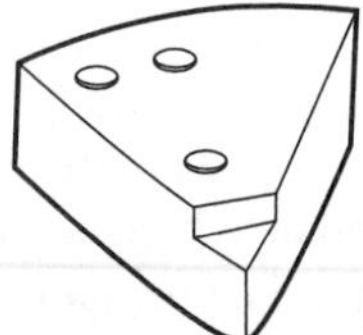

Nonfunctional shoulder permits proper orientation to be established in a vibratory feeder and maintained in transport rails

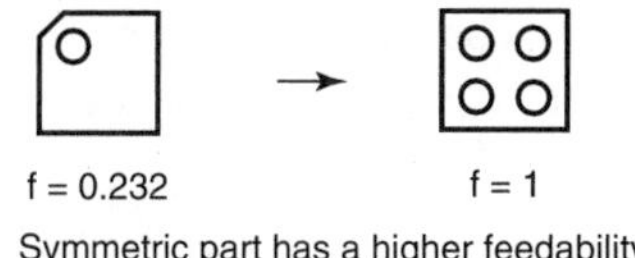

Symmetric part has a higher feedability

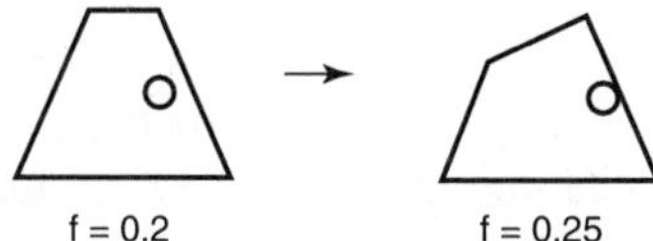

Asymmetric part has higher feedability than part with ambiguous symmetry

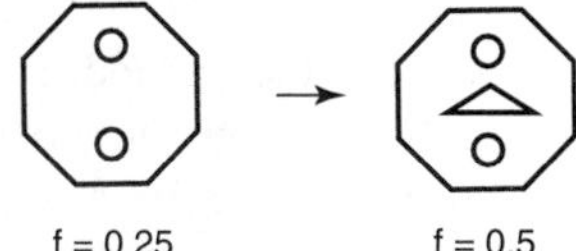

Gripping feature can increase the feedability

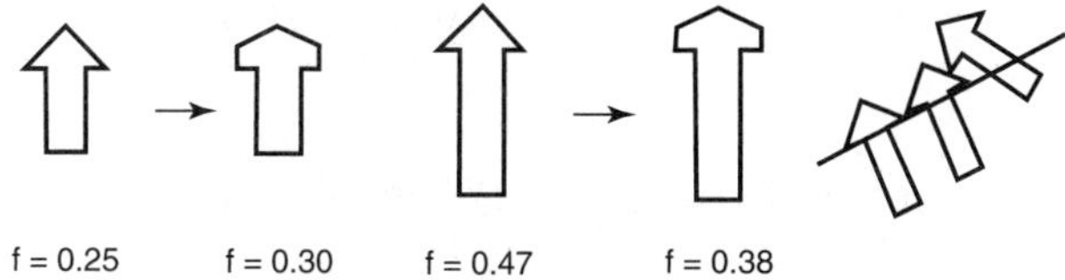

Avoiding toppling does not necessarily increase feedability

FIGURE 12.12 Effect of changing geometric features on calculated values of feedability *(Redrawn after Kim et al., "A Shape Metric For Design-for-Assembly," Proceedings of the International Conference on Robotics and Automation, 1992. Used by permission)*

Also, when such a part is transferred on a conveyor belt, it will not tip or be misoriented due to the force of inertia.

12.5 PRODUCT DESIGN FOR ROBOTIC ASSEMBLY

The product design rules for robotic assembly are basically the same as those for manual and/or automatic assembly. There are, however, two very important and crucial considerations that have to be taken into account when designing components for robotic assembly. They can be summed up as follows:

1. Design a component so that it can be grasped, oriented, and inserted by that robot's end effector. Failure to do so will result in the need for an additional robot and, consequently, higher assembly cost.

2. Design parts so that they can be presented to the robot's arm in an orientation appropriate for grasping. Also, eliminate the need for reorienting assemblies (or subassemblies) during the assembly operation. Ignoring this rule will cause an increase

in assembly time by consuming the robot's time for no valid reason. It also will cause an increase in the assembly cost per unit.

12.6 METHODS FOR EVALUATING AND IMPROVING PRODUCT DFA

At this point, let us review some of the methods currently used in industry, in America and abroad, for evaluating and improving product DFA. Because so many methods, systems, and software packages have recently been developed, the survey here will be limited to the most commonly known and used methods, for which substantial information and details have been published. There is no bias here for or against any method that has or has not been covered.

As you will soon see, most of the methods are based on measuring the ease or difficulty with which parts can be handled and assembled together into a given product. This does not mean that the components are physically brought together but rather that an analytical procedure is followed where the problems associated with the components' design are detected and quantitatively assessed. The right answer or optimal design comes from you, the engineer, when you use a particular DFA method as a tool in evaluating and comparing alternative design solutions. Following is a survey of each method.

The Boothroyd–Dewhurst DFA Method

The Boothroyd–Dewhurst DFA method was developed in the late 1970s by Professor Geoffrey Boothroyd, a pioneer in the area of DFA, at the University of Massachusetts Amherst in cooperation with Salford University of England. First, the appropriate assembly method is selected by means of charts. Then, the analytical procedure corresponding to the assembly method selected is used (i.e., there is a separate, though similar, procedure for each of the assembly methods). Figure 12.13 is a diagram of the stages of the Boothroyd–Dewhurst DFA method.

As an example, let us now examine the analytical procedure for manual assembly as the DFA analysis procedures for the other assembly methods are not much different. Note that the analysis cannot be employed to create a design from nothing but rather is used to evaluate and refine an existing design. In other words, the starting point is an assembly drawing of the product (either a prototype or an actual product). The first step in the analysis is to determine the assembly sequence (i.e., the part that is to be placed first and the parts that are to follow it in the order to be used for attaching them together). Boothroyd and Dewhurst proposed the worksheet shown in Figure 12.14 for effective bookkeeping of the assembly time and cost.

When more than one part is to be used in an operation, the assembly time for that operation is obtained by multiplying the assembly time for one part by the number of parts (see Figure 12.14). Required but nonassembly operations must also be included in the sequence. Each time the unfinished assembly is reoriented during the assembly process, the reorientation operation is entered into the worksheet and a time is allo-

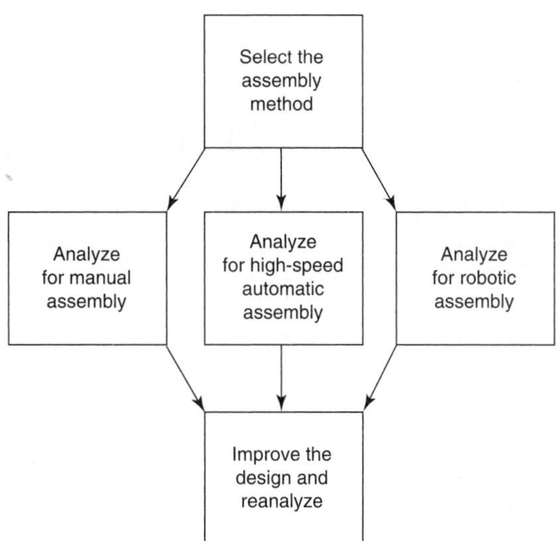

FIGURE 12.13 Stages of the Boothroyd–Dewhurst DFA method *(Redrawn after Miles, B.L. "Design for Assembly—A Key Element within Design for Manufacture," Proceedings of the Institution of Mechanical Engineers, 1989. Used by permission)*

cated for it. The assembly time for each component part is then obtained by adding the handling time of that part to its insertion time. These two times are extracted from charts that include assembly data. The data were compiled by Boothroyd, Dewhurst, and their coworkers based on practical observation over long periods of time and on research. In order to use the handling-time chart, a two-digit handling code must first be determined for each part based on its size, weight, and geometric attributes. A two-digit insertion code (and thus time) must also be obtained for each part based on accessibility, vision restriction, and resistance to insertion. Once the components and the assembly time for each are known, it is easy to estimate the total assembly time and assembly cost for the existing design.

The next step is aimed at reducing the parts count by totally eliminating some parts or combining them with neighboring parts. This is achieved by answering the previously listed three questions about the movement of the part relative to adjacent parts, its materials, and the need to have it separate for assembly and/or disassembly. Candidates for elimination can be identified and subtracted from the total number of parts to obtain the number of "theoretically needed" parts. Assuming that an ideal assembly operation of a component takes 3 seconds (1.5 seconds for handling and 1.5 seconds for inserting), the total ideal assembly time is given by the following equation:

$$\text{total ideal assembly time} = 3N_M \tag{12.1}$$

where N_M is the theoretical minimum number of parts.

Boothroyd and Dewhurst used a *design efficiency* index to evaluate the improvement in design in a quantitative manner. This index can be given by the following equation:

$$\text{design efficiency} = \frac{3N_M}{\text{calculated total assembly time}} \tag{12.2}$$

FIGURE 12.14 The Boothroyd–Dewhurst bookkeeping worksheet

1	2	3	4	5	6	7	8	9*	Name of Assembly ↓
Part (component) ID number	Number of times the operation is carried out consecutively	Two-digit manual handling code (obtained from charts)	Manual handling time per part	Two-digit manual insertion code (obtained from charts)	Manual insertion time per part	Operation time [(2) × (4) + (6)]	Operation cost	Figure for estimating the theoretical minimum number of parts	
4	1							1	
2	1							0	
3	2							1	
1	1							1	
Record totals here →						T_M	C_M	N_M	

* In column 9, if part is not essential, put 0; if required, put 1.

The mechanism for improving the design, according to this method, involves a review of the worksheet in order to pinpoint components that can be eliminated and that have relatively high handling and insertion times. The number of components or parts must then be reduced by eliminating some or most of the components so identified. This process is repeated until an optimal design (i.e., one having a design efficiency much higher than that of the initial design) is obtained.

Because it is rather time-consuming to perform the Boothroyd–Dewhurst procedure manually, a software package for DFA analysis based on their structured analysis has been developed. The latest commercially available version is very user friendly and runs in a Windows environment. Again, note that the system does not make any decisions for the designer; it is the designer who, with rational thinking and good engineering sense, ultimately decides what is right and appropriate.

One final note here: Although this DFA analysis would certainly decrease the parts count, it can often result in the manufacture and use of complex components. Bearing in mind that the assembly cost is only about 5 percent of the total production cost, the finalized "optimal design" may be easy to assemble but expensive (or difficult) to manufacture. In fact, the absence of a manufacturing-knowledge-based supporting system was the main shortcoming of the initial DFA techniques. Realizing that fact, Boothroyd and Dewhurst supplemented their DFA software with what they called *design for manufacture* software. This software is actually a product cost estimator for a few selected manufacturing processes and is used to estimate the manufacturing cost of the different alternative designs. The optimal design can then be selected based on both the assembly and the manufacturing costs.

The Hitachi Assembly Evaluation Method

Another method with a proven record of success is the Hitachi assembly evaluation method (AEM). It was employed to refine the designs of tape recorder mechanisms in order to develop an automatic assembly system for producing those subassemblies. That pioneering and original work by S. Hashizure (a research engineer at Hitachi) and his coworkers was awarded the Okochi Memorial Prize in 1980. Although this method does not explicitly distinguish between manual and automatic assembly, this difference is accounted for implicitly within the structured analysis. Also, the method was subjected to refinement in 1986 with improvements to its methodology, and a computer-based version is now available.

The Hitachi AEM approach is based on assessing the *assemblability* of a design by virtue of the following two indices:

1. An *assemblability evaluation score (E)* is used to assess design quality or difficulty of assembly operations. The procedure to compute E is based on considering the simple downward motion for inserting a part as the "ideal reference." For more complicated operations, penalty scores that depend upon the complexity and nature of each operation are assigned. The Hitachi method uses symbols to represent operations, and there are about 20 of them covering operations like the straight downward movement for part insertion and the operation of soldering, as shown in Figure 12.15.

FIGURE 12.15 Examples of Hitachi method symbols and penalty points (*Redrawn after Miyakawa, S., and T. Ohashi, "The Hitachi Assemblability Evaluation Method (AEM)," Proceedings of the International Conference on Product Design for Assembly, April 1986. Courtesy of Institute for Competitive Design)*

Elemental operation		AEM symbol	Penalty score
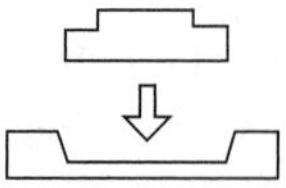	Downward movement	↓	0
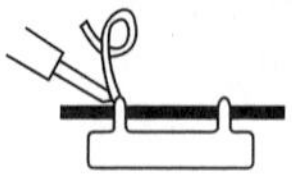	Soldering	S	20

After completing a worksheet in the same order as the anticipated assembly sequence, the penalty score for each part is manipulated to give the assemblability evaluation score for that part. The E values for all parts are then combined to produce an assemblability evaluation score for the whole assembly. Because a penalty score of zero corresponds to an E value of 100 percent, the higher the E score for an assembly, the lower the assembly time and cost. Accordingly, if each part of an assembly is to be added by a simple downward motion, the E score for each part and, therefore, for the whole assembly will be 100 percent. The E score is employed in simplifying the various operations and not explicitly in reducing the parts count.

2. An *estimated assembly cost ratio (K)* is an indication of the assembly cost improvements. As the name suggests, K is the ratio between the assembly cost of the new (modified) design divided by the assembly cost of the initial and/or standard design. It is clear that when K is 0.7, there is a 30-percent saving in the assembly cost as a result of modifying the design. The method of estimating the time (and cost) of an operation involves breaking it into its elemental components and allocating time for each elemental motion based on compiled practical observations. Any saving in the assembly cost can be achieved by reducing the parts count in a product and/or simplifying the assembly operations.

The Lucas DFA Method

The Lucas DFA method was developed in the 1980s as a result of collaborative work between the Lucas Corporation and the University of Hull (both in England). The motivation for developing this method, as stated by its creators, B. L. Miles and K. G. Swift, was to have the best features of the commercially available DFA software packages within a simple system and to aim its application at an early stage of the design process. Unlike the previous two methods, the Lucas DFA evaluation is not based on monetary costs, but on three indices that give a relative measure of assembling difficulty. The goal of reducing the parts count and the analysis of the insertion operations based on an encoded classification system, however, are shared with the previous two methods. Also, an easy-to-use computer version of this method is now commercially available.

Figure 12.16 shows an *assembly sequence flowchart* (ASF) of the Lucas DFA procedure. As can be seen, the analysis is carried out in three sequential stages: the *functional, feeding* (or handling), and *fitting* analyses. It can also be seen that the existence of a well-defined product design specification (PDS) is a must for carrying out the first stage of the DFA analysis.

Functional analysis. In the functional analysis, components are divided into two main groups. The first group includes components that perform a primary function and, therefore, exist for fundamental reasons. These components are considered to be essential, or "A," parts. The second group involves nonessential, or "B," components that perform only secondary functions like fastening and locating. The design

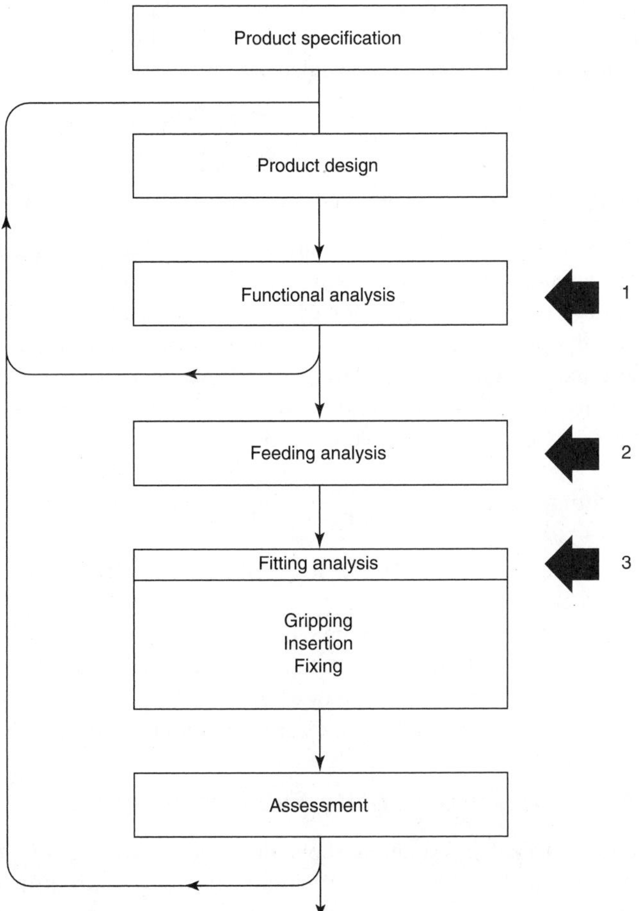

FIGURE 12.16 The Lucas DFA assembly sequence flowchart *(Redrawn after Miles, 1989. Used by permission)*

efficiency is the product of dividing the number of essential parts by the total number of parts and can be given by the following equation:

$$\text{design efficiency} = \frac{A}{A+B} \times 100 \tag{12.3}$$

According to the flowchart (see Figure 12.16), if the design efficiency is low, it should be improved through design modifications aimed at eliminating most of the nonessential parts. A clear advantage of the Lucas DFA method is that performing the functional analysis separately, before the other two analyses, acts as an initial "screening mechanism" that returns back poor designs before further effort is encountered in the detailed analysis. For this initial stage, the target objective is to achieve a design efficiency of 60 percent.

Feeding analysis. The feeding analysis is concerned with the problems associated with handling components (and subassemblies) until they are admitted to the assembly system. By answering a group of questions about the size, weight, handling difficulties, and orientation of a part, its feeding/handling index can be calculated. Next, the feeding/handling ratio can be calculated by using the following equation:

$$\text{feeding/handling ratio} = \frac{\text{feeding/handling index}}{\text{number of essential components}} \tag{12.4}$$

An ideal value for this ratio and one that is often taken as a target goal is 2.5.

Fitting analysis. The fitting analysis is divided into a number of subsystems including gripping, insertion, and fixing analyses. An index is given to each part based on its fixturing requirements, resistance to insertion, and whether or not there will be restricted vision during assembly. High individual values and/or a high total value of these indices means costly fitting operations, in which case the product should be redesigned with the goal of eliminating or at least reducing these operations. The fitting index is manipulated to yield the fitting ratio as given by the following equation:

$$\text{fitting ratio} = \frac{\text{fitting index}}{\text{number of essential components}} \tag{12.5}$$

Again, for the design to be acceptable, the value of the fitting ratio should be around 2.5.

Note that while the feeding/handling and fitting ratios can certainly be used as "measures of performance" to indicate the effectiveness of the design quality with respect to assembly, the absence of a mechanism to evaluate the effect of design changes on the manufacturing cost is a clear shortcoming of this method.

The Fujitsu Productivity Evaluation System

Some technical information about the Fujitsu productivity evaluation system (PES) was published in the *Fujitsu Scientific Technical Journal* in August of 1993. Unlike other DFA techniques, this method was developed not as a refinement procedure after the completion of the detailed design, but rather as a tool to aid in obtaining a detailed

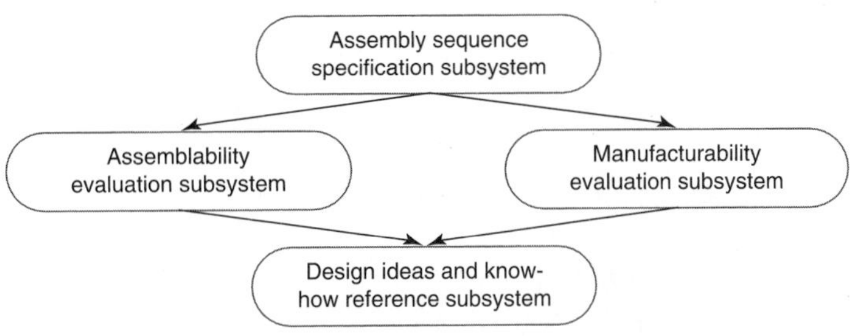

FIGURE 12.17 The Fujitsu PES *(Redrawn after Miyazawa, A. "Productivity Evaluation System," Fujitsu Science Technology Journal, December 1993. Used by permission)*

design that is easy to manufacture and assemble and also is cost effective. This method is, however, limited to bench-type manual assembly of relatively small parts, excluding, for example, products like automobiles and refrigerators. As can be seen in Figure 12.17, the Fujitsu PES (which is actually a software package) consists of four subsystems. It is based upon making full use of an expert system involving practical manufacturing and design data and rules of thumb that are gathered from the finest industry experts. The software addresses a problem by carrying out a rough evaluation that can be followed by detailed evaluations made concurrently with the product development process. The system is capable of performing absolute evaluation of assembly time and cost, as well as comparative evaluation as a percentage of that of a reference design. Figure 12.18 indicates the procedure for applying the productivity evaluation system throughout the product development cycle. Let us now discuss the function of each of the subsystems.

Assembly sequence specification subsystem. The function and the operation of the assembly sequence specification subsystem are shown in Figure 12.19. The designer selects parts similar to those envisioned to be used in the product, according to the conceptual design, and forms a library of parts and then specifies their assembly sequence. The system promptly retrieves previously stored values for assemblability and manufacturability that can be used by the evaluation subsystem to obtain assembly time and cost.

Assemblability evaluation subsystem. This tool is employed to estimate the assembly time and evaluate the ease of assembly. It is based upon a library of subassemblies (or "mechanisms") and their number of essential parts that are stored by functional module. The printing module, for example, includes dot printing (10 essential parts), thermal printing (8 essential parts), and laser printing (15 essential parts). As soon as the designer specifies the subassembly, a detailed drawing together with all the information appears on the screen of the monitor. The analysis addresses the handling and insertion of parts, specifies the target number of essential parts, and identifies high-cost processes and parts. Figure 12.20 shows the operation of the assemblability evaluation subsystem, as well as the types of input and output data. In fact, the system breaks down the assembly time of each part into handling time, insertion time, and so on, and displays it as a bar chart, as shown in Figure 12.21a. The system also shows the assemblability evaluation score for the whole product as well as for assembly and adjustment operations, as shown in Figure 12.21b.

FIGURE 12.18
Procedure for applying the Fujitsu PES *(Redrawn after Miyazawa, 1993. Used by permission)*

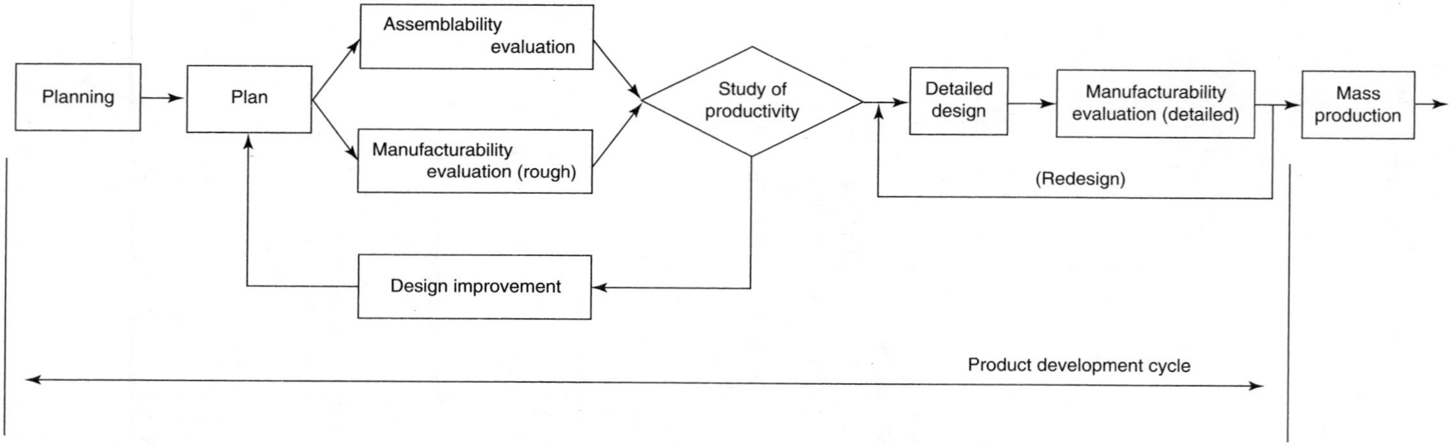

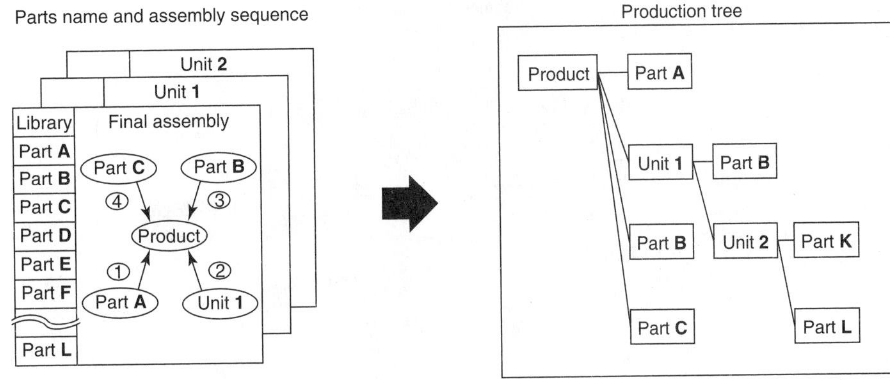

FIGURE 12.19 The assembly sequence specification subsystem *(Redrawn after Miyazawa, 1993. Used by permission)*

FIGURE 12.20 The assemblability evaluation subsystem *(Redrawn after Miyazawa, 1993. Used by permission)*

Manufacturability evaluation subsystem. As previously mentioned, the objective of reducing the parts count in a product can be achieved at the expense of the ease and cost of manufacturing some of the parts of that product. For this reason, the manufacturability evaluation subsystem was developed. As can be seen in Figure 12.22, it is used as a tool by the designer to estimate the manufacturing cost and evaluate manufacturability in a quantitative manner (on a score scale of 100). This can be done at two levels as desired by the designer: a rough evaluation and a detailed one.

Note that although the Fujitsu PES seems to address the requirement essential for carrying out comprehensive design for manufacturability and design for assembly analyses, it is based on retrieving previously compiled data. Because these data are gathered during the production, the PES software may only be successfully applied to products identical or similar to those falling within the range of products of that company. The system certainly has a proven record of success in the design of Fujitsu products, but its success when applied to other types of products has not yet been demonstrated. Figure 12.23 is an example of a product that was redesigned using the Fujitsu PES.

Other Methods

In addition to the DFA methods previously discussed, there are some software packages that are commercially available and/or being applied in-house in large corporations. These include packages by AT&T, Sony, and Sapphire. Unfortunately, technical details are not yet available, so coverage of these methods was not possible in this text.

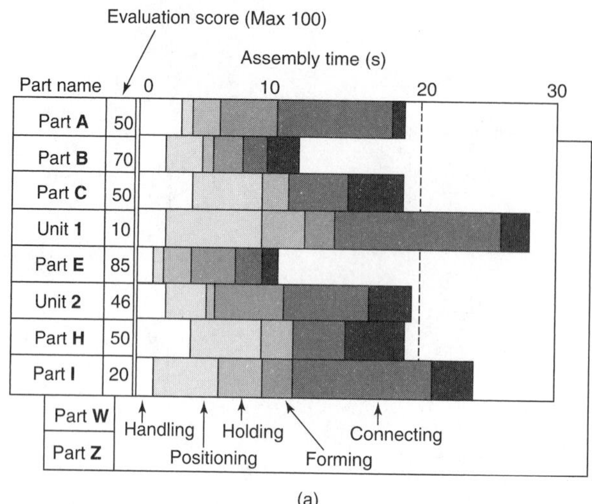

(a)

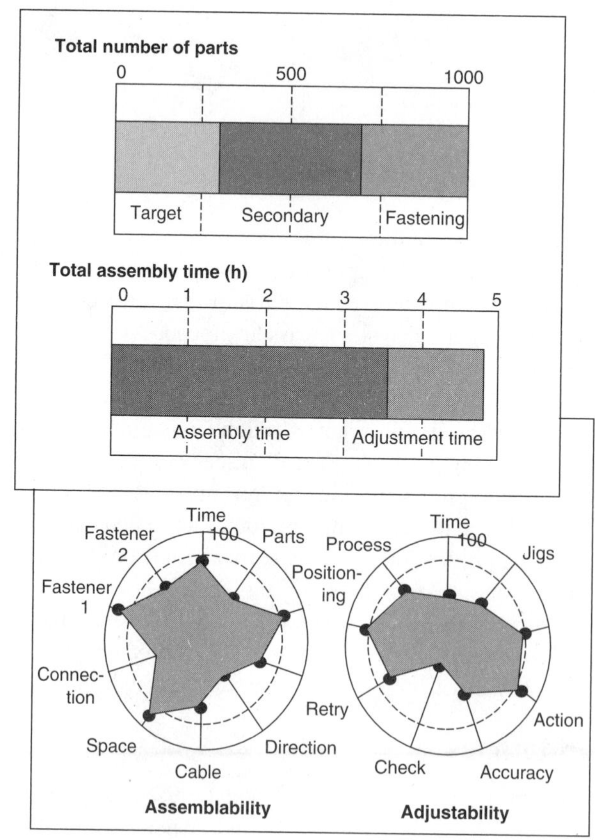

(b)

FIGURE 12.21 Assemblability evaluation results for a product: (a) assembly time for each part; (b) assemblability evaluation for the whole product *(Redrawn after Miyazawa, 1993. Used by permission)*

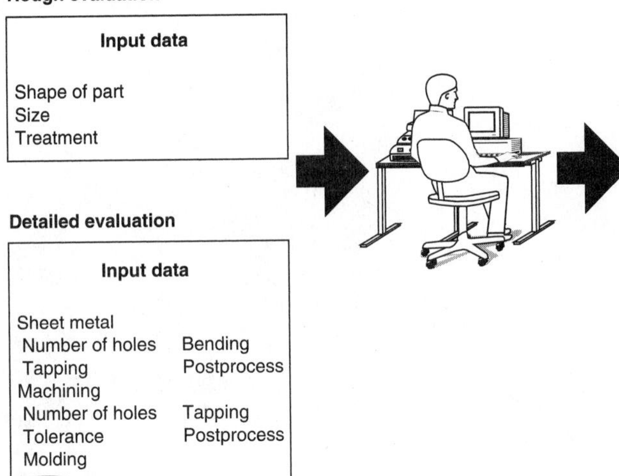

FIGURE 12.22 The manufacturability evaluation subsystem *(Redrawn after Miyazawa, 1993. Used by permission)*

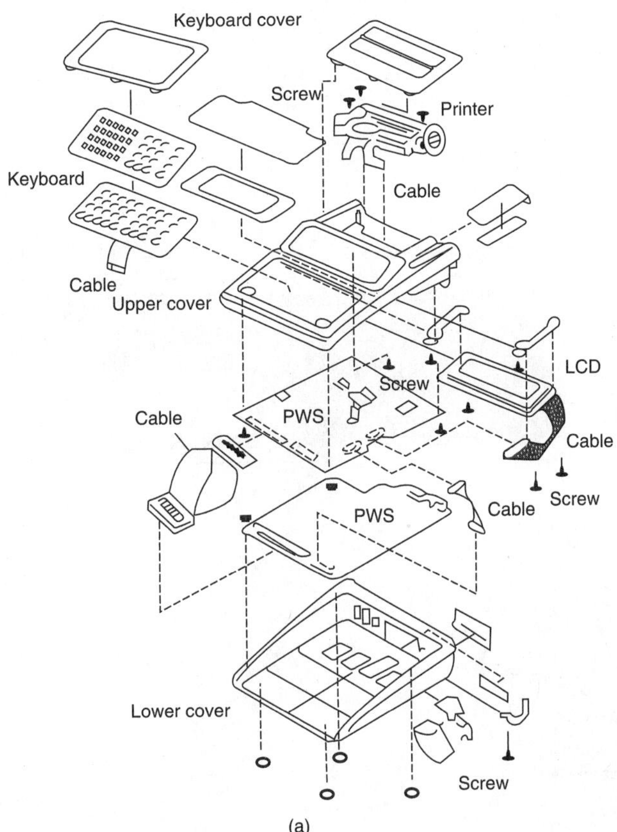

FIGURE 12.23 Example of a product redesigned using the Fujitsu PES: (a) before redesign *(Redrawn after Miyazawa, 1993. Used by permission)*

FIGURE 12.23 (CONT.)
Example of a product redesigned using the Fujitsu PES: (b) after redesign; (c) product cost; (d) assemblability estimation score *(Redrawn after Miyazawa, 1993. Used by permission)*

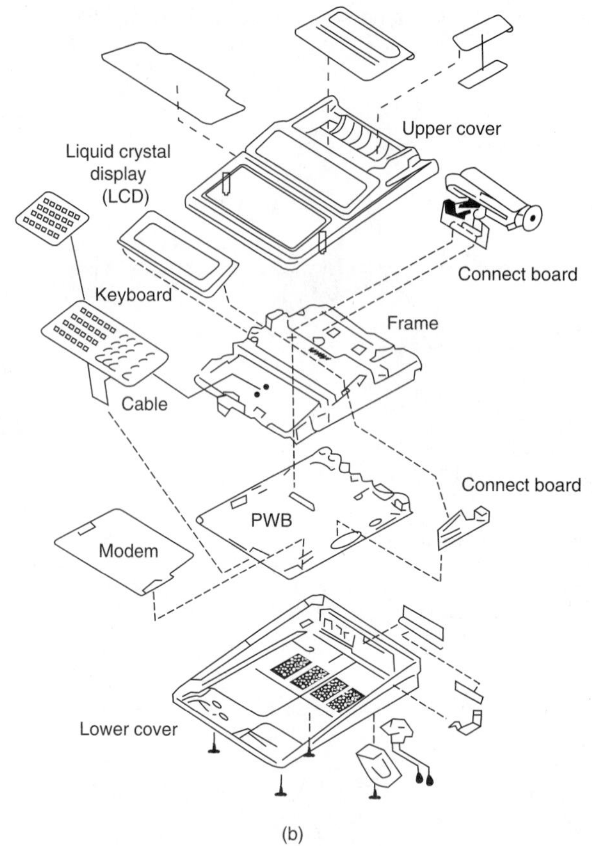

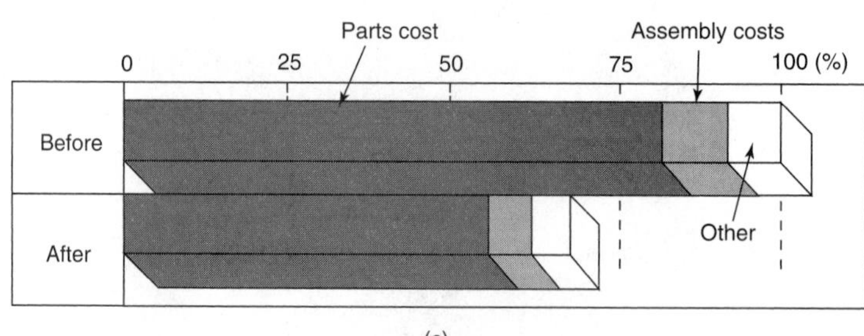

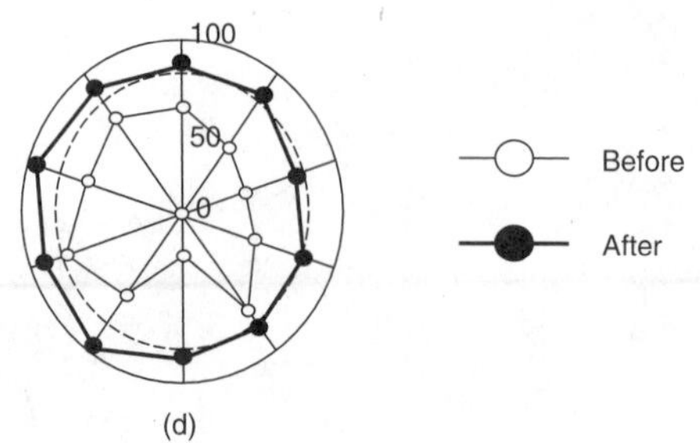

Review Questions

1. What options are available when selecting an assembly method?
2. What are the major characteristics of each of the available assembly methods?
3. Discuss some of the factors that affect the selection of an appropriate assembly method.
4. Define the term *design for assembly.*
5. What are the benefits of applying the concept of design for assembly?
6. What has always been the traditional approach for DFA?
7. What three questions form the criteria for eliminating a part from an assembly or combining it with its neighboring part?
8. List the guidelines for product design for manual assembly.
9. What is the ideal insertion motion? Why?
10. Why should you try to avoid reorienting parts during assembly?
11. What effect does the concept of standardization of parts have on the assembly process?
12. Does nesting (or tangling) of parts while in the bulk have any effect on the assembly operation?
13. Why should parts be symmetrical?
14. What is your advice if you cannot get the parts to be symmetrical?
15. List the guidelines for product design for automatic assembly.
16. How can the use of self-aligning and self-locating features facilitate automatic assembly?
17. Can screws be considered as essential parts in a product? Why?
18. What is the ideal fixturing method in automatic assembly?
19. Discuss the concept of *feedability.*
20. Why should parts with a low center of gravity be favored in automatic assembly?
21. List two rules for product design for robotic assembly.
22. What are the methods for performing DFA analysis?
23. List some of the advantages, characteristic features, and limitations of each of the DFA analysis methods.

Design Project

Choose a fairly simple product (e.g., a shower handle or coffeemaker), disassemble it, and make an assembly drawing or an exploded view of its parts. Next, study the function and material of each part, as well as the assembly sequence. Then, use the three questions (elimination criteria) of Boothroyd and Dewhurst to identify parts that are candidates for elimination or combining with other parts. Finally, modify your design in order to reduce the parts count and provide an assembly drawing of the new design, as well as a workshop drawing for each part.

Chapter 13

Environmentally Conscious Design and Manufacturing

INTRODUCTION

The increasing problems of landfill usage, the rising cost of energy and raw material, the greenhouse effect, and the decay of the ozone layer are among the major environmental concerns that prompted the second environmental revolution that is now taking place in the United States. Unlike the first environmental revolution in the 1970s, which was aimed at cleaning up hazardous waste from contaminated sites and natural resources, the second revolution is addressing waste reduction at the source. This goal can be achieved through the design of products that promote recycling as well as through the design of manufacturing processes that minimize waste, by-products, and emissions, and, therefore, utilize resources more efficiently. The magnitude of our current environmental problems is vast indeed. Consider, for example, the alarming trend of solid-waste generation. According to the Environmental Protection Agency (EPA), Washington, D.C., the United States generates 160 million tons of solid waste every year, and most of it goes to landfills that are nearly full. Furthermore, the EPA predicts that slightly less than half of the existing landfills will close before the turn of the century. The key to the solution of these problems lies in the policy of adopting environmentally friendly products and production operations—what environmentalists refer to as the concept of the *eco-factory.*

Currently, Europe seems to have a lead over the United States in solid-waste management and control as a result of government mandates that make both producer and consumer responsible for disposing of a product after its service life is over. Germany, for example, used to send 800,000 metric tons of appliances and computers to landfills every year, but as of January 1, 1994, producers had to take back and salvage their products and design new ones in a way that facilitated recycling. With 3 million metric tons of solid waste generated annually in Europe as a result of the disposal of vehicles, the automotive industry is expected to be targeted next.

Specialists believe that environmental problems are interrelated and, therefore, should be addressed at the same time (see *Manufacturing Engineering,* October 1993). In other words, progress has to be made on all fronts and not just in one specific area. Accordingly, reduction of wastes and pollution at the source should take the form of an overall process with the objective of meeting all of the following requirements:

- Design products for reusability and recycling.
- Design production processes to eliminate unusable waste, by-products, and emissions and to make efficient use of raw materials. Consider the waste, not as an unavoidable result of the process, but rather as a factor that adversely affects the efficiency.
- Design products to be serviced and maintained easily so as to ensure longer service life. This will eliminate one reason for obsolescence of a product (i.e., failure) and thus minimize the number of obsolete products dumped every year.
- Establish a material reclamation process based on waste management, recycling and recovery of materials, and minimal residues.

You may think that meeting these requirements would be very expensive and would, therefore, increase the production cost and make the products less competitive. But companies that have successfully adopted such policies claim direct and indirect benefits that surpass expenditure. It has been reported that over 50 percent of the activities of waste reduction at the source pay back in only six months. This actually means that money is saved after the initial payback period. In addition to meeting the expectations and demands of the public, which has been showing an ever-increasing environmental conscience, reducing solid waste at its source can yield the following benefits:

- Eliminating the cost of the disposal of used products in landfills and junkyards.
- Conserving natural resources as a consequence of reusing recovered and recycled materials in new products. (This would save the sources of raw materials and reduce energy consumption, especially in the aluminum industry.)
- Providing a cash return as a result of selling the recycled material to other companies.
- Improving yield and quality (as a consequence of reducing waste and scrap) and increasing the efficiency of material utilization.
- Reducing pollution and toxins.
- Providing safer workplaces where occupational health hazards are absent.

Before discussing the guidelines for environmentally conscious design and manufacture, let us first examine the sources of solid waste and the various methods of solid-waste management.

13.1 SOLID-WASTE SOURCES

In our modern societies, there is an abundance of products, such as appliances, electronic equipment, and transportation vehicles, that sooner or later have to be dumped in landfills. Because parts of the used products are either reused, recycled, or recovered, the term *solid waste* is used to describe the parts that remain in landfills. The extremely high and ever-increasing annual disposal rates of solid waste can be attributed to two main causes: The first is the huge amount of mass-produced appliances and electronic equipment sold to consumers every year, and the second is the high mortality of those items due to a relatively short service life. The service life is short not only because the products fail but also because they go out of style or become technologically obsolete. Currently, the service life ranges from 13 years for major appliances to less than 4 years for personal-care items like hair dryers (50 million of which are disposed of annually in landfills worldwide). The following discussion focuses on some major sources of solid waste.

Automotive Industry

About 30 million vehicles are scrapped every year worldwide, with the shares of Europe and the United States being 14 and 10 million, respectively. In the United States, more than 90 percent of the vehicles are sent to scrap dealers and then to shredders, where the various metals are easily separated and salvaged. Annually, about 11 million tons of ferrous metals and 800,000 tons of nonferrous metals are recovered. About 30 percent of each vehicle (by weight) is left unrecovered in a landfill. This solid waste, or "fluff," is comprised mainly of various types of plastics and rubber. Plastics are particularly hard to recycle because, although they may look the same, they have many different chemical structural formulas (see Chapter 8). Furthermore, scrap plastic may be coated with paint or other chemically dissimilar material. Unfortunately, this landfilled fluff amounts to 3 million tons every year in the United States alone, and it is increasing at a steady rate. This increase can be attributed to the current trend of using more plastics in cars to reduce the weight of the car, provide resistance to corrosion, improve noise-damping characteristics, and ensure excellent thermal insulation properties.

Appliances Industry

Examples of appliances that are disposed of in landfills at the end of their service life include refrigerators, stoves, dishwashers, and washing machines. These major appliances usually have a service life of 10 years or more. Small appliances for such uses as personal care, entertainment, and coffeemaking are also included under this category. About 350 million appliances, both small and major, were disposed of in landfills worldwide in 1993. As in the case with automobiles, plastic components are rapidly replacing metal components previously produced by stamping, die casting, or machining.

Business Equipment and Computers

As a result of adopting the philosophy of design for assembly, several metal components in an electronic unit can be replaced by a single, complex, injection-molded plastic component. Although this design would certainly facilitate assembly, it would create environmental problems at the end of the unit's service life because plastic is considered to be a major challenge for the recycling industry.

Housing and Construction Industry

Plastics and fiber-reinforced plastic composites are finding widespread application in the housing and construction industry. In 1995, it was estimated that about 9 percent of all plastic solid waste would come from construction. Examples of plastic parts currently used (and, of course, eventually requiring disposal) include pipes (water, drainage, and sewer), bathtubs, and floor tiles.

Consumer Goods

Consumer goods represent the third largest use of plastics after the packaging and construction industries. As of 1995, about 10 percent of all plastic solid waste was estimated as coming from scrapped consumer goods. Examples include disposable diapers and napkins and throwaway plasticware (utensils, trays, razors, lighters, pens, watches, and cameras).

Furniture Industry

Plastic furniture is replacing wood furniture, especially for use on beaches, in gardens, and in offices. In addition, synthetic carpets are becoming very popular in homes, offices, and public places. In 1995, the amount of plastic furniture and synthetic carpets disposed of in the United States was estimated to be about 3.1 billion pounds, or 7.2 percent of all plastic solid waste.

Packaging Industry

Packaging (e.g., for cosmetics and food) is currently the biggest market for plastics and the largest source of plastic solid waste as well (about 44 percent of plastic waste). This is due to the very low level of recycling of plastic packaging and is what makes paper sometimes more appealing than plastic in the packaging industry. In some cases, the paper recycling level goes as high as 50 percent. Nevertheless, paper amounts to 38 percent of total landfill volume in the United States, as opposed to plastic, which comprises only 18 percent of landfilled solid waste.

13.2 SOLID-WASTE MANAGEMENT

Before we discuss the various methods for solid-waste management, we need to define some important technical terms that are frequently used in the solid-waste management industry. The following definitions were originally given by M. Grayson (see the references at the end of this book):

- *Reuse* refers to further or repeated use of a waste in its original form. An example is the refilling of cigarette ignitors with fuel.
- *Recycling* is the use of a waste, or a waste-driven material, as a raw material for products or fuel that may or may not be similar to the original. A typical example is the shredding of soda bottles and then recycling the material into outdoor furniture and the like. Although, in this case, the final product after recycling is obviously different from the original one, this may not be the case when, for example, recycling paper.
- *Recovery* is the processing of a waste to prepare a usable material or fuel in a form in which and to a specification by which it can be recycled. An example is the processing of scrap iron into pig-iron ingots that can be used as a raw material and further processed into steel.

As we begin to discuss the main methods for managing solid waste, as well as the advantages and limitations of each method, it is important to remember that the choice of a particular method depends upon several factors, including public health and safety, cost, and the technology available. It is also important to first establish some cost models, to consider any demand for certain by-products, and to then make the necessary trade-offs so that the most appropriate method can be chosen. Following are the main methods for solid-waste management.

Disposal in Landfills

Disposal in landfills is certainly the easiest method. However, with the current rate of solid-waste generation, it is anticipated that about 50 percent of the landfills now open in the United States will close by the year 2000 (according to studies released by the EPA, Washington, D.C.). In addition, about 70 percent of all U.S. landfills have already closed since 1978. Moreover, this method has the disadvantage of wasting land and materials, which, when recycled, can result in a cash return as well as a saving of valuable resources.

Incineration

Although incineration eliminates the main disadvantage of landfilling (i.e., the limited finite capacity), it sometimes has serious drawbacks. For instance, there is the possibility of dangerous gases being emitted into the atmosphere, thus creating a public health hazard. Also, the mere existence of an incinerator may discourage the general public from recycling waste that would otherwise be recycled. Finally, incineration necessitates a very high initial capital investment. The cost of a modern incinerator can be as high as $200 million.

The Molten-Metal Process

The molten-metal process, which is also called *catalytic extraction processing* (CEP), has been developed for industrial and commercial applications by Molten Metal Technology, Inc., Waltham, Massachusetts. The company's process is believed to offer significant environmental and economic advantages over the other currently used methods of getting rid of solid waste, especially when hazardous or toxic materials (e.g., PCBs) are involved. Unlike burning, the molten-metal process is basically a recovery process that prepares useful by-products from waste in the form of usable raw materials and/or fuel. Figure 13.1 gives a general view of how the cycle works.

This revolutionary method involves injecting waste materials of all types (i.e., solids, liquids, sludges, and even gases) into a molten-metal bath that is usually molten iron heated to about 3000°F (1650°C). Highly reactive chemical catalysts are also injected into the bath. The waste first dissolves into the molten iron; it is then broken up into its elemental building blocks. By controlling the parameters of the process and adding reactants to the bath, useful metals like nickel as well as fuel like carbon monoxide and hydrogen can be obtained. Generally, the by-products (or, in reality, the useful products) fall into one (or more) of three classes: gases, specialty organics, and metals. Figure 13.2 sketches the CEP method in greater detail.

The CEP method has been successfully demonstrated in commercial-scale tests on a variety of waste materials at the pilot plant of Molten Metal Technology, Inc., Fall River, Massachusetts. These waste materials range from simple compounds like paraffins to complex materials containing heavy metals, halogens, cyanides, PCBs, and polyaromatic hydrocarbons.

This process not only destroys the hazardous waste but also yields useful raw materials and/or fuel—something that makes it particularly appealing in terms of

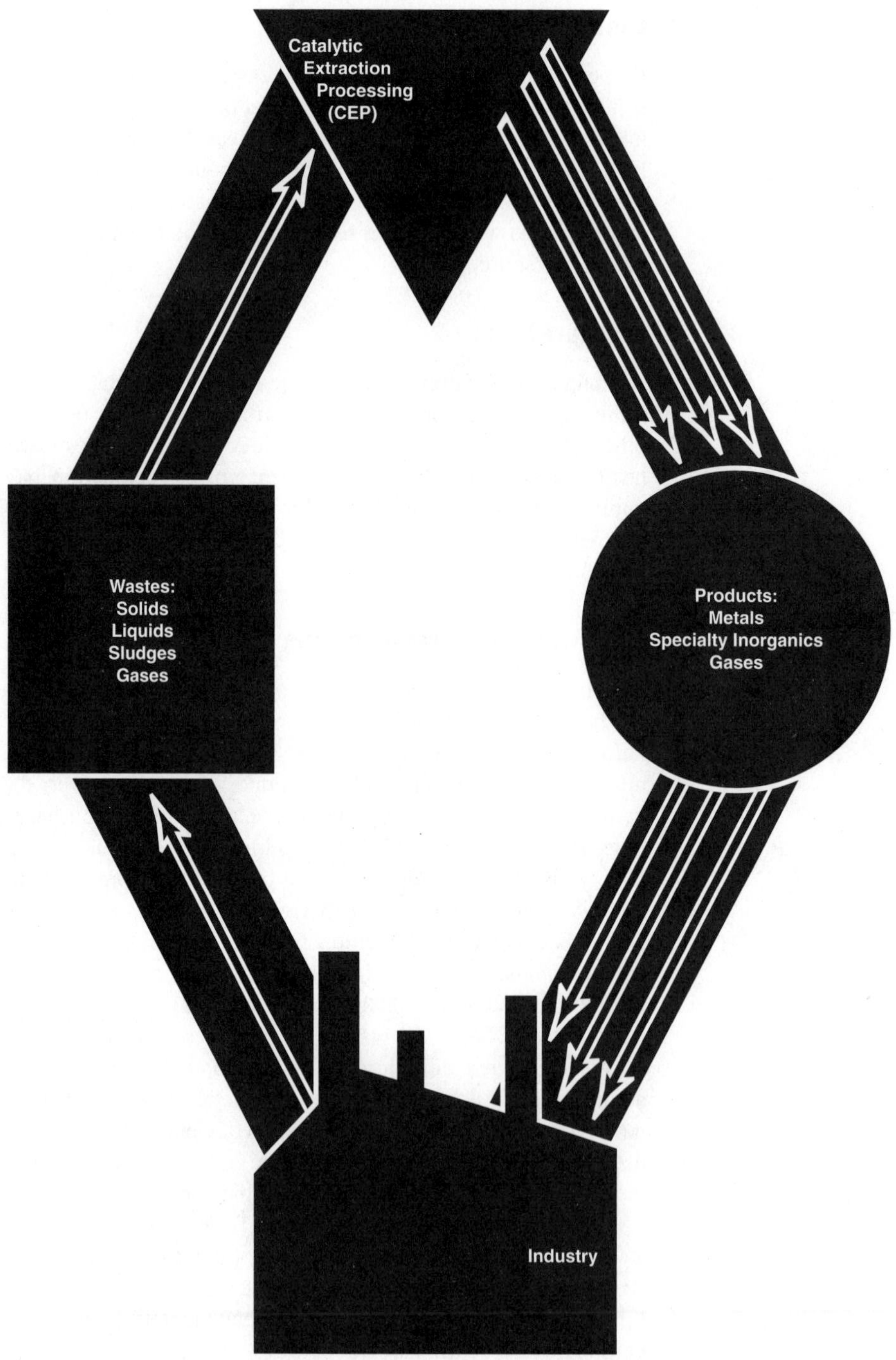

FIGURE 13.1 The waste material cycle in the CEP method *(Courtesy of Molten Metal Technology, Inc., Waltham, Massachusetts)*

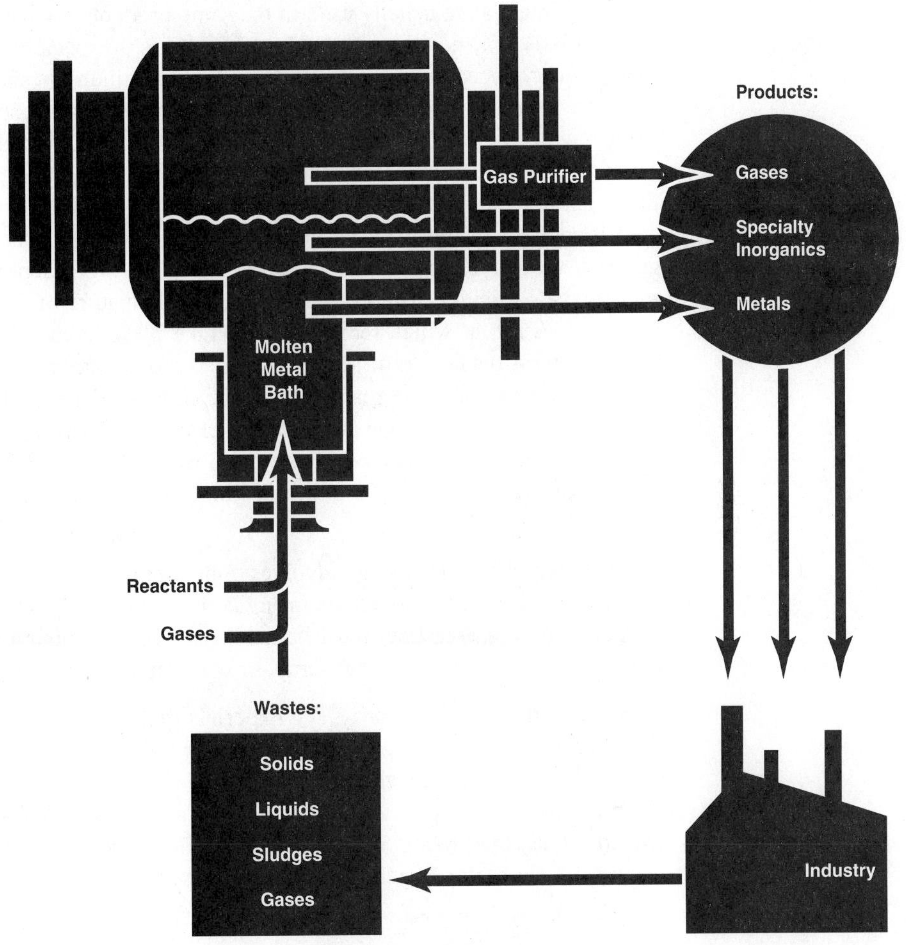

FIGURE 13.2 A sketch of the CEP method *(Courtesy of Molten Metal Technology, Inc., Waltham, Massachusetts)*

economics. In addition, the capital investment required is relatively moderate, especially when compared with the cost of modern incinerators. During 1995, Molten Metal Technology completed construction on six plants in the United States, each plant costing $25 million. The actual performance of these plants and their capacity or rate of waste destruction will determine the appropriate range of applications for the CEP method. It can then be decided whether the process will be limited to destroying just industrial hazardous waste or will be used on a larger scale for the management of solid waste generated in metropolitan areas.

Recycling

Recycling is a well-established method of solid-waste management and has a proven record of success. Whereas the vast majority of metal products in the United States are already recycled effectively after their service life, not as much paper, glass, and plastic is recycled. In fact, plastics currently pose the biggest recycling challenge. After recycling, metals are usually used in the same products for which they were originally used. This is not, however, the case with plastics (especially if they are used in the food industry) because of contamination and strict industry standards.

The adoption and use of recycling as a commercial solid-waste management process will always depend on how economically attractive that process can be. In other words, there must be a financial motivation for recycling to be successfully used. Evidently, this is the case when the cash value of the raw material (and/or fuel) recovered is more than the expense incurred during recycling. As you may have already guessed, the recycling expense or (cost) is not fixed but depends upon many factors, such as the particular material being recycled, the nature of the products, and the geographic location at which recycling is to take place. Accordingly, recycling must be preceded by a detailed cost analysis, carried out by an environmental manufacturing specialist, to evaluate the feasibility of the process. This approach has, in fact, been rigorously followed in Japan and, to some extent, in Germany, which have a slight lead in this respect over the United States, as shown in Figure 13.3.

To gain a deeper insight into recycling, let us now examine the different stages involved.

Material retrieval. This stage involves collecting commercially large amounts of materials and securing the source to ensure continuous supply. In addition, certain classes of materials (e.g., plastics) must be subjected to contamination inspection. Contaminants can be dirt, grease, metal scrap, or even another type of plastic.

Identification and separation. The materials that have been collected must be identified according to class (e.g., metals, glass, ceramics, plastics) and, in the case of plastics, by the exact chemical compound of which the waste is composed. As there are hundreds of kinds of plastics that look more or less the same but have different chemical formulas, the process of identification is extremely difficult and costly. In fact, it accounts for a high percentage of the recycling cost. After the identification operation

Solid Waste Management:

U.S., Japan & West Germany Compared–Percent of Total Waste

	Recycle or Reuse	Waste to Energy	Landfill or Other	Total
United States	11	6	83	100
Japan	50	23	27	100
West Germany	15	30	55	100

FIGURE 13.3 Solid waste management in the United States, Japan, and Germany *(Source: U.S. Environmental Protection Agency, Washington, D.C.)*

is complete, the various materials must be separated for recycling. Whenever two dissimilar materials or plastics are joined together by adhesives or ultrasonic welding, an appropriate means must be employed to separate them.

Reprocessing. When dealing with metals, this stage involves melting the scrap and then casting it into ingots that can be subjected to further processing as required. The properties of the recycled metal are predictable as well as controllable. This is not the case, however, when reprocessing plastics. The recycled plastic must be considered as a new material having properties that are different from those of the original material, and this consideration affects the selection of appropriate applications for a particular recycled plastic.

Marketing. Finding markets for a reprocessed plastic is crucial for the recycling process of that plastic to be successful. The selling price, the continuous availability of the product, and the properties of the recycled plastic are factors that affect the process of marketing it. These same factors determine the applications for which the recycled material can be employed. Finding the appropriate applications is an essential step in securing and expanding existing markets as well as in opening new markets for any recycled plastic. Marketing recycled metal does not present any problem because the properties of the reprocessed metal are predictable and controllable.

13.3 GUIDELINES FOR ENVIRONMENTALLY CONSCIOUS PRODUCT DESIGN

It should be clear from the preceding discussion that some slight changes in the design of a product can easily promote and facilitate its recyclability. As a consequence, several industries, including the automotive, electronics, and plastics industries, have developed guidelines for product design for recycling. When applying these guidelines, however, consideration must be given not only to the product to be recycled but also to the concept of reducing or eliminating the waste material generated during production.

Design for Disassembly

Dismantling a product for recycling has cost implications (e.g., the cost of labor required to take apart the different components). Accordingly, components and products have to be designed so that they can be disassembled with ease, thus reducing the cost of disassembly and making recycling economically attractive. This leads us to the concept of *design for disassembly* (DFD), which involves designing the product to be amenable to extremely rapid disassembly. Fortunately, most of the guidelines adopted in DFA hold true for DFD (e.g., minimizing the number of components and ensuring minimal handling of components). Other rules include making designs that facilitate reaching separation points for disassembly. Sometimes, the rules for DFD are not compatible with those of DFA. For example, joining two components with an adhesive may be an easy way to assemble them but would create problems when trying to disassemble them, whether for maintenance or at the end of the service life of the product. The analysis and selection of joining methods that promote recycling will be covered in more detail later in this chapter.

Because product servicing and recycling are both based on the same premise (i.e., ease of disassembly), adhering to the DFD rules would certainly have a positive effect on customers' satisfaction. This is a consequence of lower service and repair cost as a result of easier removal and replacement of components. Further advantages include reducing insurance premiums for products, as well as extending the service life of products and thus delaying their disposal in landfills. This is an indirect contribution to solving the problem of the accumulation of solid waste in landfills.

Material Selection

The ultimate goal of a designer, though unrealistic in most cases, is to come up with a one-material product. The practical alternative is to minimize the number of different materials used and to make them clearly identifiable. In order to realize the extent of this problem, you have to bear in mind that there are about a hundred different kinds of plastics in an automobile and that most of them look similar to one another. Following are some guidelines to be observed when selecting materials for product design for recycling:

1. Reduce the number of different materials employed in your design, especially when dealing with various kinds of plastics. This will sometimes result in overdesigning the properties of some parts so that they can use the same material as other parts that require these properties. Consider the personal computer as an example. The outer casing of a PC is made of polycarbonate (even though it is slightly expensive) because of its good impact resistance properties that meet the functional requirements of this component. The internal components, however, are usually made of ABS in order to reduce the cost. But if recyclability is to be taken into consideration, all components should be made of polycarbonate. This may result in an initial increase in the cost of the product. Nevertheless, some cash can be recovered at the end of the service life of the product by giving the product back to the manufacturer for recycling.

   Another good approach that is recommended for cutting down the variety of plastics used in a product is to design complex components out of only one kind of plastic that has different properties depending on the molecular weight or degree of polymerization. A plastic part with a label usually requires three different materials: the part, the label, and an adhesive. All these items, however, can be made from the same material but having a different degree of polymerization in each case so as to promote recycling. In fact, an appropriate candidate here would be polycarbonate.

2. When designing plastic components, try to employ thermoplastic materials rather than thermosets whenever the functional requirements permit it because thermoplastics are easier to recycle. It is also possible in many cases to use a thermoplastic elastomer as an alternative to difficult-to-recycle thermosets.

3. Use compatible materials and avoid secondary finishing operations like painting and coating. If the use of incompatible materials is inevitable, strive to make them easy to separate even though extra cost will be involved.

4. Label plastic components that each weigh slightly less than 1/4 pound (100 g) for identification and sorting. Use a molded-in material name or logo (or better, use an SAE standard material symbol) in multiple locations. An example of an SAE standard symbol is shown in Figure 13.4.

5. Factors such as ease and cost of recycling as well as the potential markets for recycled material must be considered when selecting materials for a new product design. In fact, the issue of marketing recycled plastic is particularly crucial because, without it, there would be no cash return from the recycling process. Therefore, there should be a clear "strategy" for marketing the recycled plastic during the early design phase. If the anticipated production volume is high, for example, the designer can then select a commodity material. Recycling companies would make a profit, in that case, from the large volume of recycled plastic as the price (and profit) per pound is relatively low. On the other hand, when working with a low production volume, it is advisable to employ a plastic of high market value in the design so as to make the price of the plastic waste (after the service life of the products) economically attractive to justify recycling.

Fastening and Joining Considerations

There are basically two methods of disassembling a product for recycling; the selection of one of them depends upon the method of combining the components together to form a product. The first method is reversed assembly, which involves following the same steps included in the assembly process in reverse order. For example, if two components are snap-fitted together during assembly, they would be similarly separated during dismantling for recycling. This, obviously, will not work for dissimilar plastic parts that are welded together. The second approach for disassembly involves crude methods of dismantling with brute force. This approach can make the recovery process profitable, especially when the dissimilarity between components is employed to facilitate separation. A good example would be plastics and metals, which can be shredded and then easily separated by means of a magnetic separator because of their physical dissimilarity. The designer has to decide upon the method of disassembly during the early design phase and to promote that method when preparing the design. It is also important to remember to strive to make fastening points accessible. Here are the pros and cons of the different fastening and joining methods from the viewpoint of design for environment:

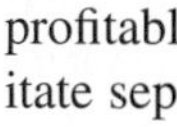

PVC

FIGURE 13.4
SAE standard symbol for polyvinyl chloride

1. *Welded parts* may or may not be easily recycled: Metals are recycled effectively after use, but this is not the case with plastics, where two dissimilar resins are joined together by stacking, ultrasonic welding, and so on. Brute force is usually required to separate the components but may create problems unless considered during design.

2. *Screws* are undesirable for both assembly and disassembly. You are advised to replace them with snap fits whenever possible. If these are not feasible, standardization of screw types, sizes, and head shapes is strongly recommended to facilitate disassembly.

3. *Adhesives,* although they facilitate assembly, are undesirable for disassembly because they create many problems. Although brute force is an efficient way of dismantling glued parts, adhesives are considered to be contaminants when used to join plastic parts. Solvents constitute another method for dismantling adhesive-bonded parts, but here again, the disposal of these solvents may create environmental problems.

4. *Snap-fit latches* are ideal for DFA and DFD. They certainly facilitate recycling as additional parts or dissimilar materials are not required. The design should, however, ensure that snaps are easy to "unsnap," can withstand the anticipated service conditions, and will not inadvertently unsnap during use. Note that snaps generally increase the complexity and cost of the mold, which must be considered in the economic analysis for recycling feasibility.

13.4 ENVIRONMENTALLY CONSCIOUS MANUFACTURING

The world is becoming more and more aware of the importance of considering the environmental impact of manufacturing. The consequences of ignoring environmental concerns are clearly evident in Eastern Europe and the former Soviet Union, where improper disposal of industrial hazardous waste has seriously affected the quality of life of the population. This new awareness has led to the outgrowth of programs for pollution prevention and the administration and processing of industrial wastes and by-products—hence, the birth of environmentally responsible activities like "zero avoidable pollution" and "green manufacturing." Industries in general, and the chemical industry in particular, are beginning to view waste, not as an unavoidable result of their processes, but as a measure of efficiency. In other words, the more waste a process generates, the less efficient it is considered, and the greater is the need for improvement. Here are some ideas that promote the concept of environmental consciousness:

1. Minimize material use and reduce and conserve energy used in manufacturing. The goal is to conserve natural resources.

2. Understand the side effects of processes and equipment emissions, such as paint vapor and abusive molding. This is the first step in preventing pollution at its source.

3. Whenever possible, avoid using or generating toxic materials, heavy metals, and the like by substituting nonpolluting chemicals in production processes. This is indeed the case for some heat treatment operations. Manufacturing processes should be reengineered to generate less waste and/or to improve their environmental impact. Also, minimize the use of substances and chemicals that have a short shelf life.

4. Use R & D and rational analysis to minimize the amount of coolants, lubricants, and cutting fluids in the different machining and forming processes. Only environmentally safe fluids should be used.

5. Whenever a new substance is used in daily production operations, make sure it will not result in a waste that requires off-site disposal. In fact, some solvents (e.g., those used in circuit-board fabrication) have higher disposal costs than purchase costs.

6. Do not allow solvents, catalysts, and reagents to cross the boundaries of the plant. They have to be recovered for recycling and/or treatment to eliminate their toxic or hazardous nature.

7. Always bear in mind the hierarchy put forward by the EPA that establishes the priorities for waste management as follows:
   - Waste reduction (highest priority)
   - Recycle and recovery
   - Treatment
   - Disposal (lowest priority)

8. Try to produce and use environmentally friendly packaging. Minimize the amount and select appropriate materials, and always remember that packaging amounts to 40 percent of the plastics in landfills.

13.5 ENVIRONMENTAL PROTECTION AND POLLUTION CONTROL LEGISLATION

It is important now to become aware of the various environmental protection and pollution control laws and regulations so as to take them into consideration when planning to build a manufacturing facility and/or when managing the daily operation of a production plant. These laws and regulations differ for different countries. Only the U.S. environmental protection legislation will be covered in this text, and the international reader is advised to consult with the environmental protection agency in his or her country to learn about the equivalent regulations. A summary of each of the commonly encountered federal environmental protection laws follows:

1. The *Clean Air Act* (CAA) is considered to be one of the most significant pieces of environmental legislation in years. Its main goal is to protect and enhance the quality of the nation's air. It establishes four general air quality goals:

   - Attainment and maintenance of national ambient air quality standards (NAAQS)
   - Prevention of significant deterioration (PSD) in clean air areas
   - Preservation of natural visibility in major national parks and wilderness areas
   - Avoidance of significant risks from hazardous air pollutants

   The amendments of the CAA contain a long list of regulatory requirements to increase pollution control through installation of more advanced pollution control equipment and to make changes in industrial operations that lead to reduction in emissions of air pollutants.

   The CAA has a significant impact on the manufacturing industry. It lists 189 chemicals, many used in manufacturing, that are under strict control. The list includes not only lead, nickel, and chromium but also many organic chemical compounds.

2. The *Clean Water Act* (CWA) is aimed at regulating the discharge of pollutants into navigable waters of the United States. It specifies the amount of pollutants that may be discharged into navigable waters. It also prohibits the discharge of oil in quantities determined to be harmful.

3. The *Comprehensive Environmental Response, Compensation, and Liability Act* (CERCLA), which is often referred to as the *Superfund,* provides sufficient resources, through the creation of an $8.5 billion fund, to facilitate the cleanup of hazardous substances in uncontrolled or abandoned waste sites. Under CERCLA, the EPA may compel the owner or operator of such a site to clean it up. The EPA can also go ahead and clean up the site and then compel the owner, operator, or other liable parties to pay the cleanup costs.

4. The *Resource Conservation and Recovery Act* (RCRA) requires the EPA to regulate the generation, transportation, storage, discharge, and final disposal of solid and hazardous waste. It also includes regulatory requirements for postdisposal monitoring care.

5. The *Toxic Substances Control Act* (TSCA) is designed to regulate commercially produced chemical substances through identification and control of the manufacture, processing, commercial distribution, use, or disposal of chemicals that pose an unreasonable risk to health or the environment. Although the TSCA is famous for regulating PCBs, it actually regulates about 60,000 chemical substances. It also empowers the EPA to gather information about a chemical's toxicity and exposure, to regulate existing chemical risks, and to identify and prevent future risks.

6. The *Underground Storage Tank Act* (USTA) regulates the detection, prevention, and correction of releases from underground storage tanks containing regulated substances.

7. The *Safe Drinking Water Act* (SDWA), unlike the CWA, focuses on the effects of industrial products and by-products on human and animal health rather than on environmental concerns. The SDWA covers two areas: public water systems and underground sources of drinking water. The EPA is authorized to publish regulations to prevent underground injection that endangers drinking water sources.

Review Questions

1. What are the motivations for environmentally conscious design and manufacturing?
2. What problems does the generation of industrial solid waste cause and what is the magnitude of these problems?
3. What are the requirements for the overall process of reduction of wastes and pollution at the source?
4. List and discuss the benefits of reducing solid waste at its source.
5. What are the major sources of solid waste? Discuss each of them briefly and rank them in descending order based on their impact on the environment.
6. Explain the following terms, showing the difference between them: *reuse, recycling, recovery.*
7. What are the main methods for solid-waste management? Discuss each of them briefly and elaborate on their advantages and limitations.
8. What are the factors that control the selection of a particular solid-waste management method?
9. Explain, in detail, the different stages involved in recycling and elaborate on the cost implications of each.
10. Show how slight changes in the product design can promote recycling.
11. Explain the term *design for disassembly* and show how it can facilitate recycling and promote environmental consciousness.
12. How can the designer, through rational selection of materials, improve the recyclability of his or her designs?
13. List some guidelines to be observed when selecting materials for product design for recycling. Discuss each briefly, showing its impact on the environment (especially when plastics are employed).
14. In what way does the method of fastening and joining the component of a product affect the feasibility and profitability of recycling?
15. Discuss, from the viewpoint of design for environment, the pros and cons of each of the different fastening and joining methods.
16. List some guidelines to be observed when planning a new manufacturing facility and/or when managing the daily operation of a production plant so as to make these activities environmentally friendly. Discuss each briefly.
17. List some of the important federal laws and regulations for environmental protection and pollution control. Explain how these can reduce and/or eliminate pollution at its source.

Chapter 14

Computer-Aided Manufacturing

INTRODUCTION

Computer-aided manufacturing (CAM) has been defined by Computer-Aided Manufacturing International (CAM-I) as "the effective utilization of computer technology in the management, control, and operations of the manufacturing facility through either direct or indirect computer interface with the physical and human resources of the company." Although this definition of CAM is broad and flexible and covers a variety of tasks, the dominant application of CAM is still in numerical control (NC) part programming. For this reason, many people subscribe to a narrower concept of CAM that involves mainly computer-assisted NC part programming. This concept of CAM may also stem from the fact that CAM has its roots in NC systems. Consequently, a logical step in studying CAM is to discuss NC systems as well as manual (unassisted by computer) NC part programming and then proceed to extensions of NC and the different methods of computer-assisted part programming.

14.1 NUMERICAL CONTROL (NC)

Overview

Before discussing NC systems, how they began, and how they evolved to their present status, let us consider a simple definition that will be a logical entry to the subject.

Definition. *Numerical control* (NC) can be defined as control of the operation of machine tools (or other sheet working and welding machines) by a series of coded instructions called a *program,* which consists mainly of alphanumeric characters (numbers and letters). It is obvious from this definition that the sequence of events is both preplanned and predictable. In other words, any desired sequence of events can be obtained by coding the appropriate instructions and can also be changed by changing those coded instructions. Therefore, NC systems are considered to be the typical form of programmable automation.

Historical background. The basic concept of numerical control is not new at all and dates back to the early years of the Industrial Revolution, when Joseph Jacquard developed a method to control textile looms by using punched cards. When he applied for a patent for his invention, however, he was denied that right by the Queen of England because she believed that it would displace poor workers (notice the similarity with the use of robots today). In fact, this old invention, together with the player piano, which is operated by a roll of punched paper tape, can be considered as simple, crude forms of mechanical NC.

A modern version of NC emerged in 1947 at the Parsons Engineering Company of Traverse City, Michigan, as a result of the need of John C. Parsons (the owner of the company) to manufacture helicopter rotor blades fast enough to meet his contracts. Later, Parsons Engineering was awarded a study contract by the U.S. Air Force Material Command to speed up production and develop continuous-path machining, with the subcontractor being the Massachusetts Institute of Technology. The job was later given in its entirety to MIT, and the machine they developed was successfully demonstrated in 1952. Between the years of 1953 and 1960, the rate of building and selling NC machines in the United States was very slow. This type of machine tool later gained widespread industrial application because of the need for consistency of dimensions and tighter tolerances.

Simplified idea of numerical control. To understand the basic idea behind numerical control, let us assume that a hole has to be drilled in a plate using a drill press and that, according to the blueprint, the hole is 8 inches to the right of the left edge and 5 inches above the lower edge. We start by clamping the workpiece on a positioning table on the drill press; we then crank the two handwheels of the two perpendicular slides to locate the corner of the plate (the point where the left and the lower edges meet) exactly under the center of the spindle. Now, if a single turn of the handwheel causes the table of the drill press to move by 0.10 inch, it is obvious that we need to move the table to the left 8 inches and forward 5 inches by cranking the appropriate handwheels 80 and 50 turns, respectively. In order to automate this operation, we replace the handwheels by electric servomotors that are operated by push buttons. Let us say for convenience that a single quick push of a button (like a dot in the telegram code) causes the attached servomotor to turn by 1/100 revolution. Consequently, we need to push the button of the first servomotor 8000 times and that of the second servomotor 5000 times in order to position the center of the spindle exactly above the desired location of the hole. After doing so, we will be in a position to perform the drilling operation and obtain a part that conforms precisely to the blueprint.

A closer look at this example shows that the machine is driven by numerical values (a number of pulses or button pushes for each direction) and responds by converting these values to meaningful physical quantities. This is, in fact, what is meant by numerical control, no matter how these numerical values are fed into the machine. We can, therefore, say that an NC system is a system that readily converts numerical values into physical quantities such as dimensions.

Advantages of numerical controls. The advantages of NC machine tools are felt not only on the factory floor but also in many other departments of a business corporation. Following are some advantages that can be used as justification for employing NC machine tools:

1. NC machine tools ensure positioning accuracy and repeatability. In other words, if the same program is employed to produce a number of parts, they will have identical dimensions.

2. NC machine tools can produce complex-shaped components automatically with closer tolerances and very high degrees of reliability, which provides the designer with a great degree of flexibility and freedom.

3. Because NC machine tools have high dimensional accuracy and repeatability, parts can be manufactured that require a long series of operations. Such parts are difficult to produce by conventional methods because accumulated errors result in completely unacceptable results.

4. Because, after being programmed, NC machine tools can perform any desired task (within their capability) without the need for a human operator on the shop floor, they can be employed to carry out operations in hostile environments, such as the machining of polymeric materials that emit poisonous gases.

5. NC systems transfer a substantial portion of the planning for the processing operation from the shop floor to the engineering offices, where specialists prepare NC part programs in comfortable surroundings and production is then directly monitored and controlled by management.

6. NC machine tools have the capability of performing more than one machining operation by automatically changing the tool used without changing the location of the workpiece. In other words, a sequence of machining operations can be performed in a single setup, which reduces the number of transfers of a workpiece between different machine tools or machining departments. This capability is considered to be one of the major advantages of NC machine tools because the nonproductive time used in setups and workpiece transfer amounts to a high percentage of the total production time, as evidenced by statistical data. In this respect, it is worth mentioning that actual machining time involves only 5 percent of the production life cycle of a typical component.

7. As a result of the preceding advantages and because of the minimal idle time involved, the use of NC machine tools is always accompanied by increased productivity.

8. The high dimensional accuracy and repeatability of NC machine tools provide a profound basis for the interchangeability of work between different production plants.

9. NC systems reduce part scrapping due to machining errors and lower the inspection and assembly cost as a result of the uniformity and reliability of the products produced by this technology.

Elements of an NC System

This section will focus on the elements of a tape-operated NC system. Although not many of these systems exist in the United States now, an understanding of this basic NC system will provide an adequate basis for comprehending the advanced NC systems currently used in industry. A sketch of the basic elements of an NC system that controls a machine table along a single direction is shown in Figure 14.1. Two systems like this one have to be used when the motion of the machine table must be controlled along two directions. Following is a brief discussion of the elements of an NC system.

Tape reader. As previously mentioned, the desired sequence of events is converted into a series of coded instructions (i.e., the program). The program is then recorded onto a tape. Next, the coded tape is read by the *tape reader* (a device, located in the machine control console, that has the function of winding and reading the tape). There are different types of tape readers (electromechanical, electronic, and optical); each has a different method of operation.

Machine control unit. The *machine control unit* (MCU) receives the coded instructions from the tape reader, decodes them by converting them into signals representing the preplanned commands, and then transmits the signals to the servomotors to generate the machine movements. In the early days of NC, the MCU was hard-wired; today, it consists mainly of a microcomputer.

Servomechanism. The function of the *servomechanism* is to amplify the signals received from the MCU and to provide power to produce the required tool (or machine table) movements. These signals generally take the form of pulses, whereas

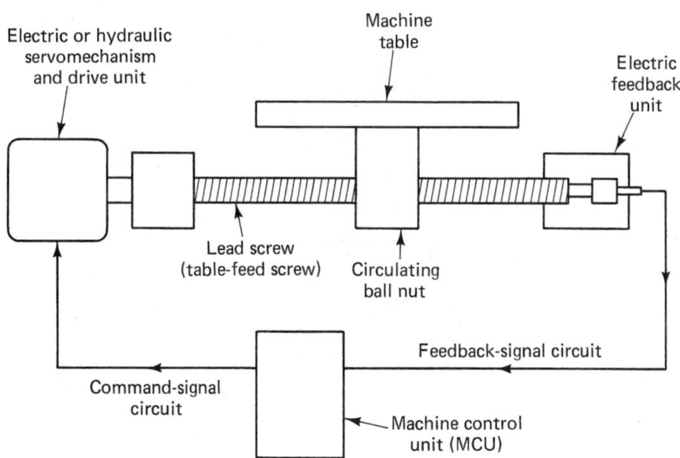

FIGURE 14.1
A sketch of the basic elements of an NC system

the servomotor is often a dc electric motor that drives the tool (or machine table) through a lead screw. Hydraulic systems are also in use.

Controlled element. A *controlled element* is any part (of the machine tool) that is numerically controlled. It can be a tool, a turret for an NC lathe, or the machine table for an NC drill press.

Feedback unit. The function of the *feedback unit* is to record the achieved movement of the tool (or machine bed) and then send a feedback signal to the MCU. The MCU compares the achieved position with the required or programmed one and automatically compensates for any discrepancy. Systems with feedback units are usually referred to as *closed-loop* systems.

The Coordinate System and Dimensioning Modes

As with any engineering application, NC programming is based on the Cartesian coordinate system (sometimes on the polar coordinate system as well). According to the Cartesian system, any point within a plane can be defined by its distances from the *X* axis and the *Y* axis (i.e., the *X* and the *Y* coordinates, respectively). Also, the point of intersection of these two perpendicular axes is called the *origin,* or *zero point.* The coordinates of a point can be both positive, both negative, or one negative and the other positive, depending upon the location of that point. The two perpendicular intersecting axes divide the plane into four quadrants, which are numbered counterclockwise, as shown in Figure 14.2. Notice that all values of *X* and *Y* are positive in the first quadrant and negative in the third quadrant. In the second quadrant, all values of *X* are negative, and all values of *Y* are positive; in the fourth quadrant, it is the other way around. In other words, when a point falls to the right of the *Y* axis, its *X* coordinate is positive, but when it is to the left of that axis, its *X* coordinate is negative. Similarly, the *Y* coordinate of a point is positive when the point is above the *X* axis, but it is negative when the point is below that axis.

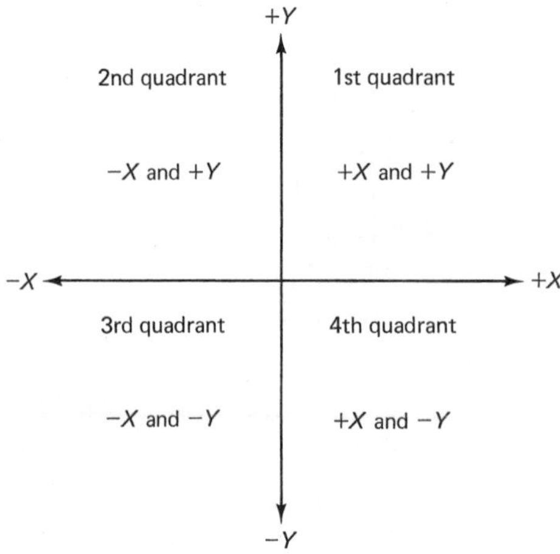

FIGURE 14.2
The coordinate system and quadrant notation

The Cartesian coordinate system can be extended to describe a point in space by adding a third dimension along the *Z* axis, which is perpendicular to the plane of the *X* and *Y* axes. As we will see later, this third dimension will enable us to deal with more complicated work.

We are now in a position to discuss the dimensioning modes, which include two types: *absolute* and *incremental.* The absolute mode is similar to the Cartesian coordinate system, where all coordinates of points are always given in reference to the origin or the machine zero point. When the incremental mode is used, reference is made to the latest position, which is actually equivalent to considering each location to be the zero point for the next location. In other words, incremental programming involves an *increment* from the present position to the new one, together with an associated sign indicating the direction. The following example will clarify the difference between the two programming modes. As can be seen in Figure 14.3, the centerline of the tool (e.g., a drill) coincides with the zero point, and we are required to write statements to move it first to point A, then to points B and C, and finally back to the origin. Here are the desired statements in both absolute and incremental modes:

Absolute Programming		
Step 1	*X* +2	*Y* +3
Step 2	*X* −2	*Y* +1
Step 3	*X* −3	*Y* −1
Step 4	*X* 0	*Y* 0

Incremental Programming		
Step 1	*X* +2	*Y* +3
Step 2	*X* −4	*Y* −2
Step 3	*X* −1	*Y* −2
Step 4	*X* +3	*Y* +1

Although absolute programming is more logical, it requires all drawings (i.e., blueprints of parts) to be done with all dimensions referenced to a single point in order to make the programmer's job easier. Incremental programming is advantageous for

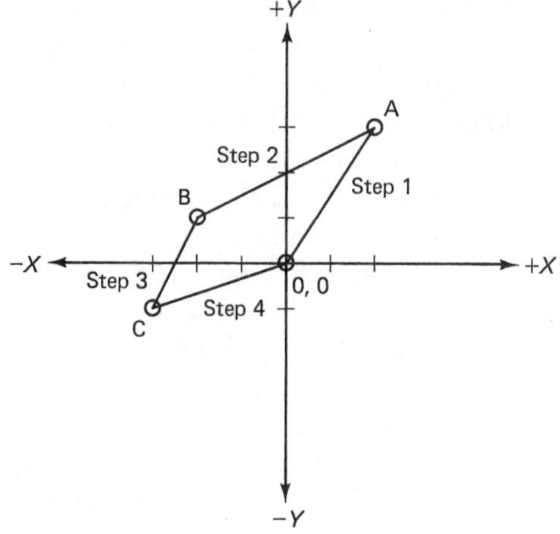

FIGURE 14.3 Absolute and incremental dimensioning modes

positioning work. NC programs are, therefore, usually a blend of both programming modes so that use can be made of the as-drafted dimensions of the part.

NC Machine Motions

The Electronic Industries Association (EIA) lists in its RS-267A standards the various types of NC machine motions or axes designations, whereas RS-267 indicates some 25 different NC machines. The single-spindle drilling machine is the simplest of all. It is generally a two-axis NC machine tool because it can be program-controlled on two axes: the *X* and *Y* axes. The *Z* motion of the spindle (raising and lowering) is controlled manually or by using a system of cams. In some NC machines, a tape command calls a preset depth, but this cannot be considered as an axis of motion. A *true axis of motion* is one along which an infinite number of locations for the tool (or machine bed) can be obtained. An axis of motion may be either *linear* or *rotational.* According to the EIA standards, the *X, Y,* and *Z* axes are the linear axes, whereas the rotational axes are the *a, b,* and *c* axes, which are used to indicate the rotary motion around the *X, Y,* and *Z* axes, respectively, as shown in Figure 14.4a. Positive direction of any of the rotary motions can be obtained by employing the *right-hand rule.* As can be seen in Figure 14.4c, this rule involves using three fingers of the right hand to indicate the linear axes and then using the thumb pointing out in the positive direction of the linear axis (the one that forms the center of rotation of the rotary motion under investigation) with the other fingers curved to indicate the positive direction of the rotary motion.

It is important to remember that the *X, Y,* and *Z* axes are neither arbitrary nor interchangeable. It has been agreed upon by the EIA and NC machine tool builders that the *Z* axis is always a line parallel to the spindle of the machine tool. Consequently, the *Z* axis can be either vertical or horizontal, depending upon the kind of machine tool. In the case of an NC lathe, the *Z* axis is horizontal, whereas it is vertical for a vertical milling machine or a drill press. The function of the spindle differs, depending upon the kind of machine tool. The spindle is employed for rotating the tool on milling and drilling machine, whereas it is the workpiece-rotating means on engine lathes and similar machines.

In addition to the primary linear axes *X, Y,* and *Z,* sometimes secondary linear axes are used that are parallel to the primary axes and are designated *U, V,* and *W,* respectively. NC systems that have both primary and secondary linear axes provide more flexibility when a program is to consist of both absolute and incremental dimensioning modes. In such a case, it is common to devote the *X, Y,* and *Z* coordinates to absolute dimensions and employ *U, V,* and *W* to indicate incremental motions.

It is also common to have the controlled element be the machine table (or, in other words, the workpiece) and not the cutting tool. The controlled element then responds to the tape command in an opposite direction; any movement of the workpiece in the established positive direction for the tool is considered to be negative. This is, in fact, equivalent to saying that it is the relative movement of the tool with respect to the workpiece that is actually considered.

NC machines can have two, three, four, or even five axes. In this respect, the word *axis* means any direction of linear or angular motion that is truly and fully controlled by the NC system. As previously mentioned, indexing or calling a preset dimension

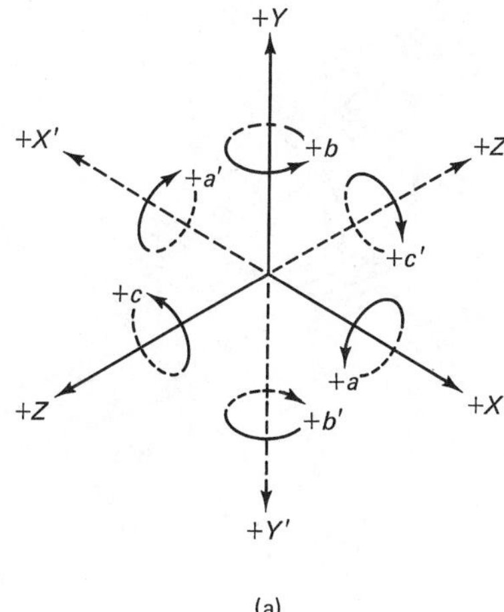

(a)

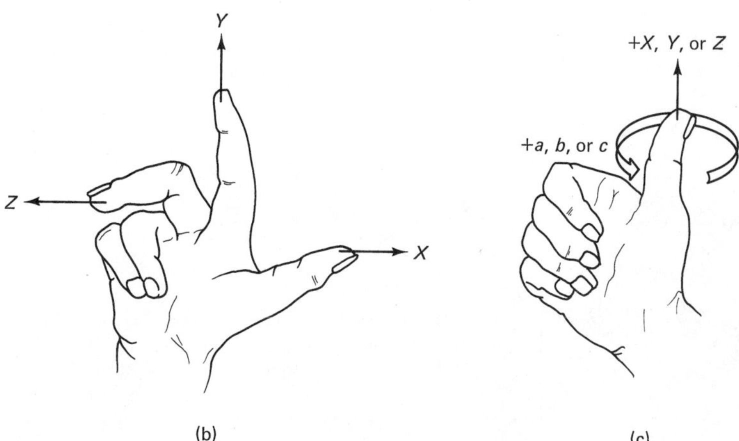

(b) (c)

FIGURE 14.4 The right-hand rule for the relative location of coordinate axes

does not fall under the definition of a true NC axis of motion. Following is a brief survey of each type of NC machine.

Two-axis NC machines. In two-axis NC machines, motions along only two axes (usually *X* and *Y* or *X* and *Z*) are fully controlled by tape commands. Figures 14.5 and 14.6 show an NC turret lathe and an NC drill press, respectively, that belong to this type of machine. Notice that, for the turret lathe, the positive direction of *X* is going away from the workpiece and the positive direction of *Z* is going away from the headstock. It is also important to note the difference between the positive direction of the axes and the machine table movements in Figure 14.6.

Three-axis NC machines. Vertical knee mills, drilling machines, and jig borers are examples of three-axis NC machines. In this type of machine, the motion is controlled along the *Z* axis as well as the *X* and the *Y* axes.

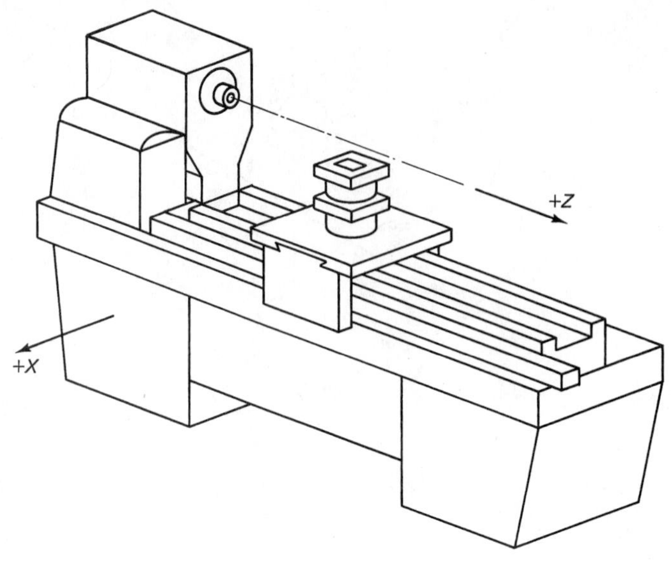

FIGURE 14.5
A numerically controlled turret lathe

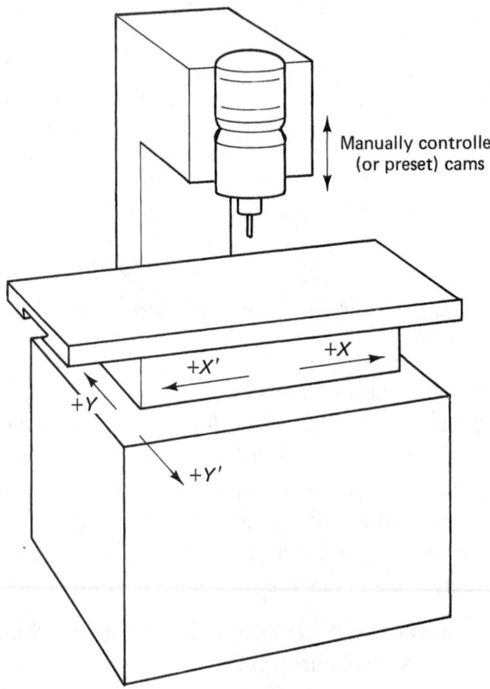

FIGURE 14.6
A numerically controlled drill press

Four-axis NC machines. In four-axis NC machines, in addition to motion along the *X, Y,* and *Z* axes, the machine table is rotated by command at a controlled rate during the machining operation. Again, note the difference between NC controlling and indexing (even when the latter involves a very large number of indexed positions).

Five-axis NC machines. Five-axis NC machines are used for producing sculptured surfaces because the machine head can swivel at a controlled rate (in addition to the previously mentioned four axes of motion). The tool can, therefore, be brought perpendicular to the desired surface. An example of this type of machine is the five-axis profile and contour mill with a tilting head.

Types of NC Systems

There are three basic types of control systems for NC machine tools: point-to-point, straight-cut, and contouring.

Point-to-point system. The point-to-point system is also referred to as *numerical positioning control* (NPC) and is usually used in NC drilling machines that are employed in drilling precise patterns of holes. The function of the NC system is, therefore, to move the spindle (or machine table) to the exact location, as given by a tape command, so that a hole can be drilled. As soon as the desired hole is drilled, the NPC system moves the spindle to the next programmed location to drill another hole, and so on. The spindle (or machine table) movement from one hole location to the next must be done as fast as possible to bring to a minimum the nonproductive time spent in movement. Accordingly, speeds of more than 100 inches per minute (2500 mm/min.) are quite common.

Provided that our main concern is positioning the spindle on each of the desired locations, it is of no importance to control the path along which the spindle moves from one location to the next. In fact, that path is not necessarily a straight line as it just covers the shortest amount of time. It usually involves two intersecting straight lines, as is explained later.

Straight-cut system. The straight-cut system is quite similar to the point-to-point system, except that the feed rate of the spindle along each machine axis is controlled so as to be suitable for machining (e.g., a milling operation on a vertical mill). Again, the spindle cannot be controlled so that it moves along a line inclined to the *X* and *Y* axes of the machine; the motion along one axis is independent from the motion along the other axis because it is controlled by a separate NC circuit (or subsystem). Nevertheless, motions along lines coinciding with or parallel to either the *X* or the *Y* axis can be accurately controlled. The spindle can move forward, to the right, backward, and to the left in a rectangular path, and, for this reason, the system is sometimes referred to as a *picture-frame* system. The sequence of motions may not necessarily yield a rectangular path. Figure 14.7 indicates some tool paths that can be produced by this NC system and that are employed in machining rectangular configurations, in face milling, and in pocketing.

NC machines fitted with straight-cut control are also capable of performing point-to-point positioning at very high speeds. They are, therefore, more versatile than NPC machines. However, their cost is also higher and must be justified by the kind of products required.

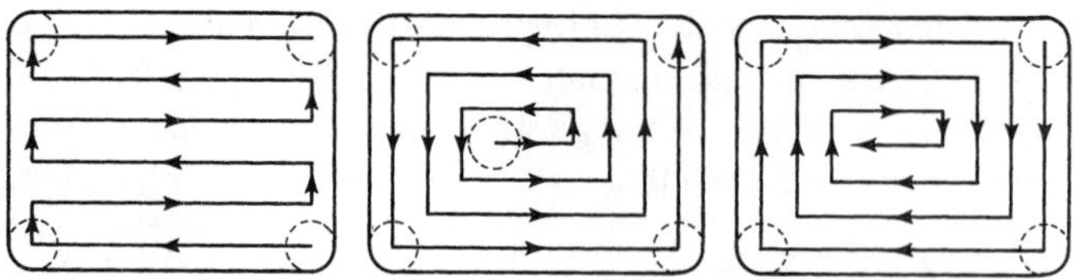

FIGURE 14.7 Some tool paths produced by a straight-cut NC system

Contouring system. In order to make angular cuts on a workpiece, the two driving servomotors (one for X-axis motion and the other for Y-axis motion) have to run at unequal speeds. In fact, the rate of travel in the Y direction divided by the rate of travel in the X direction must be equal to tan θ, where θ is the angle that the angular cut makes with the X direction. The capability of a control system to regulate the rate of spindle (or table) travel along two axes of motion at the same time is called *linear interpolation.*

In addition to point-to-point positioning and picture-frame cutting, a contouring control system can produce curves to very close tolerances. Therefore, it is sometimes referred to as the *continuous-path* system. The method of employing linear interpolation to produce curves involves breaking down a curve or an arc into a large number of straight lines in such a manner that the end of each line is the beginning of the next one (tip-to-tail fashion). Each and every line segment must, therefore, be programmed in order for the path to conform to the desired curve. Obviously, the larger the number of segments taken, the smaller each of the line segments becomes and the smoother the machined curve becomes. This concept is illustrated in Figure 14.8. Breaking down a curve into hundreds of straight lines and programming each line is very complex. This process, if carried out manually, would take much time and effort. Therefore, it is almost always performed using a computer, as we will see later when discussing computerized numerical control.

Punched Tape and Tape Coding

The NC tape. The NC tape is the oldest type of input media used for storing all the data (i.e., the NC program) needed to generate a desired part. When the tape is run on an NC machine, the prepared program is simply read, and the desired part can thus be machined. The same part can be produced several times by running the prepared tape

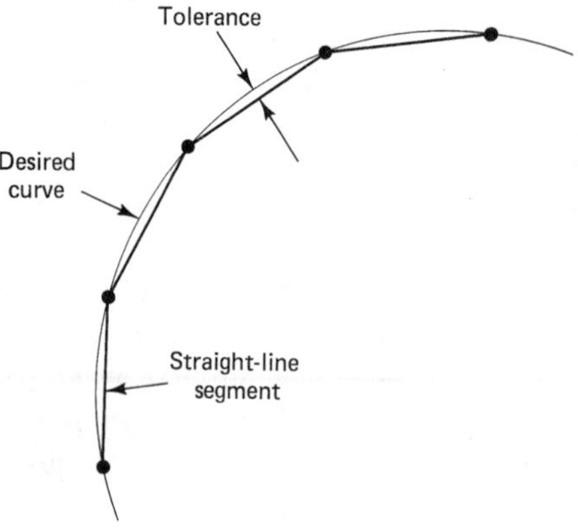

FIGURE 14.8 Approximation of a curve using straight-line segments

as many times as required. The identical part can also be made years after the tape is prepared, as long as the tape is kept in good condition. It is very important that the MCU and the tape always be compatible. In other words, they both must be based on the same coding system and the same coding format. Although punched tapes are not as commonly used as they used to be, a discussion of the coding and format of a punched tape will, nevertheless, provide an adequate and clear picture of how instructions are fed into NC machines.

The punched tape used for NC systems is standardized by the EIA to have a width of 1.000 ± 0.003 inch (25.4 ± 0.076 mm) and a thickness of 0.004 ± 0.0003 inch (0.1 ± 0.008 mm). The tape can be made of paper, a paper-mylar sandwich, or an aluminum-mylar laminate. Paper tapes are cheap and easy to damage, so their use is limited to short runs. For high production and frequent use, aluminum-mylar tapes are more suitable because of their durability, but they are more expensive. NC tapes are purchased in the form of rolls that are 8 inches (200 mm) in diameter, each having a tape length of up to 2000 feet (600 m). The tape is divided into eight main channels, or *tracks* (i.e., parallel to the edges), where holes can be punched. There is also a track of smaller holes to the right of the third main track. These smaller holes fit the tape-feeding sprocket in order to ensure positive drive of the tape. Letters of the English alphabet, digits from 0 to 9, and symbolic signals to the MCU each have a specific arrangement of punched holes in a line, or *row,* perpendicular to the edges of the NC tape. A single instruction given to the MCU usually consists of a set of letters and numbers; a set of rows of punched holes is referred to as a *word.* There are, however, some words that take only a single row. A number of words that are grouped together form a *data block.* The block is the smallest unit of a program that provides the NC system with complete information for an operation (or tool motion).

Punching the tape. First, the programmer prepares the manuscript of the part program, which is commonly called the *program sheet.* It involves a list of detailed instructions that describe the step-by-step operation of the NC system. The information on the program sheet is then transferred to a blank tape by punching holes into it that stand for the required codes. This is done by typing on a flexowriter or similar tape-punching piece of equipment. The result is not only the punched tape but also a printout of the program sheet that can be used to check for errors and make corrections.

Tape codes. NC tapes are coded in a *binary-coded decimal* (BCD) system, which is a further development of the binary coding system. This system is based on considering the presence of a hole as *on* and its absence as *off;* each is called a *bit.* The presence of a hole in the first track means 2^0 (i.e., two to the power zero), or 1, whereas its absence means 0. The second, third, and fourth tracks mean the number 2 raised to the powers 1, 2, and 3, respectively. In other words, the presence of a hole in the second track is equivalent to 2, the presence of a hole in the third track means 4, and the presence of a hole in the fourth track means 8. Thus, any digit (i.e., a number from 0 to 9) can be represented in one row of the tape by an arrangement of holes in the first four tracks (from left to right). Some examples will clarify this coding system. The digit 3 is designated by a combination of two holes, one in the first track and the other in the second track. The digit 7 is a combination of holes in the first, second, and third tracks.

In fact, each numerical digit, letter, or symbolic signal has its own designated combination of holes in a single row. When the tape reader reads a numerical value such as 4732, it reads a single digit (one row) at a time and, through its electronic circuit, places that digit in its proper decimal position with respect to the preceding and succeeding digits. In this way, the decimal value of any digit is determined by its relative position in a set of rows representing a numerical value.

Tape coding has to be standardized in order to facilitate interchangeability of tapes and communications between industrial firms. Two tape codes are commonly used: the EIA code and the ASCII (American Standard Code for Information Interchange) sponsored by the American National Standards Institute (ANSI). Figure 14.9 indicates both standard tape codes. As can be seen, the EIA code uses only six tracks of the eight available on a tape and always has an odd (uneven) number of holes in any row. The fifth track contains a hole whenever the number of holes representing a character is even. This method of detecting errors in a punched tape is called a *parity check.* An EIA-coded tape must have odd parity as an indication that no punching mistakes have been made. Also, as previously mentioned, each operation or movement is represented by a set of rows on the tape (i.e., a data block). Each block of information must be separated from the following one by a special character called the *end of block* (EOB). It is represented by a hole in the eighth track of the tape. The EIA code provides 63 different combinations of holes, which is both logical and sufficient for NC applications.

The ASCII code was introduced to more appropriately meet the needs of computer organizations, government, and the communications industry. It utilizes all eight tracks of a tape and, therefore, provides 128 characters (i.e., possible combinations of holes). It has even parity, contrary to the EIA code.

Tape formats. The term *format* refers to the way NC words are arranged in a data block. Although three formats gained some common use, two of them are rarely used now, and it is only the third one, the word-address format, that has gained widespread application in industry.

In the first type of format, called *fixed-block format,* words are arranged in the same sequence in all blocks throughout the program. In addition, not only blocks but also words within each block must have the same length (i.e., number of rows). If a word remains unchanged (e.g., in a motion parallel to the *Y* axis, the *X* coordinate remains the same) from one block to another, it must be repeated in the second block. This format is rarely used now because it lacks flexibility and results in lengthy and complicated programming.

In *TAB sequential format,* words are given in the same fixed order in all blocks and are separated by the TAB symbol. This symbol is represented in the EIA code by five holes in tracks 2 through 6. Although the order of words within the block is always the same, the length of blocks need not be the same. This is due to the fact that if a word remains unchanged, as in the preceding block, it need not be given again. This format is more flexible than the fixed-block format but, nevertheless, is not commonly used today.

In *word-address format,* which is currently the most commonly used format, a word is not identified by its location in the data block, but rather by a single letter, or *word code,* that precedes it. As an example, the value of the *X* coordinate of a point to which a tool is to move is preceded by the word code X. Similarly, the values of the *Y*

FIGURE 14.9
ASCII and EIA standard tape codes

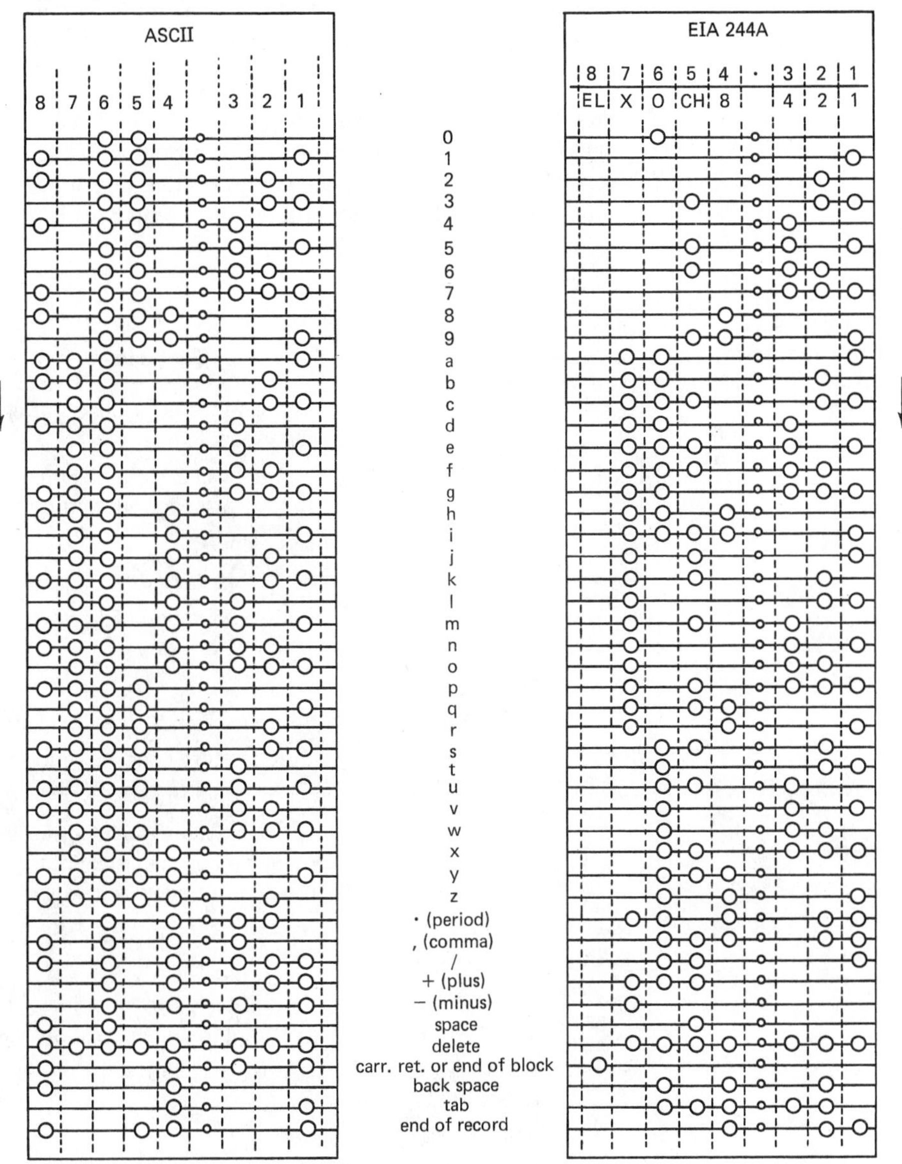

and *Z* coordinates are preceded by the word codes Y and Z, respectively. Therefore, words need not be presented in any special order. It is, however, a good idea to keep the order of each word the same in all blocks of a program for the sake of simplifying programming and checking even though neither the length of a block nor the order of words must be fixed. It is also important to remember that each data block must be followed by an EOB. Following are the word codes used in the word-address format:

1. Word code N stands for the *sequence number* and is a means of identifying each data block in a program. This word code is usually followed by three digits that indicate the order of the blocks in the program. It is quite common to number the blocks by tens (e.g., N010, N020, and so on) so that extra blocks (operations) can be inserted between existing ones whenever necessary.

2. Word code G stands for the *preparatory function* and is usually followed by two digits. This code specifies the mode of operation of the control (i.e., it commands the MCU, thus causing the spindle to operate in a specified manner). A list of the commonly used G codes according to EIA standards is given in Table 14.1.

TABLE 14.1
Preparatory function codes according to EIA Standards

Code	Function
G001	Rapid traverse, or straight-line travel at speed of about 200 inches per minute (5 m/min.) for tool positioning
G01	Linear interpolation, or straight-line interpolation between previously programmed point and the next at specified feed rate
G02	Circular interpolation clockwise
G03	Circular interpolation counterclockwise
G04	Establishment of a dwell, the duration of which is defined by the programmed value
G33	Threading
G70	Inch data input (first preparatory code in an English-system part program)
G71	Metric data input (first preparatory code in a metric-based part program)
G80	Cancel cycle (positioning at rapid traverse with spindle up)
G81	Drill cycle (rapid traverse positioning, spindle moving down rapidly to a point slightly above workpiece surface, spindle fed downward at preset feed rate, and rapid retraction of spindle upward)
G82	Drill cycle with dwell
G83	Pecking cycle
G84	Tapping cycle (similar to drill cycle except spindle reverses its rotation and rises at feed rate until tap disengages from workpiece)
G85	Boring cycle (similar to tapping cycle except spindle does not reverse its rotation)
G90	Absolute programming
G91	Incremental programming
G92	Preset absolute register
G94	Inches per minute (mm/min.)
G95	Inches per revolution (mm/rev.)
G96	Direct rpm programming

3. Word code M stands for the *miscellaneous function* and is followed by two digits. It is sometimes referred to as the *auxiliary function* and basically controls the on-off machine operations such as coolant on, tool change, and the like. A list of the commonly used M codes according to EIA standards is given in Table 14.2.
4. Word code X is for the *X* coordinate dimension.
5. Word code Y is for the *Y* coordinate dimension.
6. Word code Z is for the *Z* coordinate dimension.

Manual Part Programming

As previously mentioned, the job of the NC programmer involves manually (unassisted by computer) preparing step-by-step detailed instructions on a program sheet. This task requires that he or she be familiar with the NC machine on which the part is to be processed. The programmer should know, for example, the location of the *setup point* with respect to the machine *zero point.* Let us define these terms before we discuss manual part programming.

Zero point. The *zero point* is the point where all coordinate axes meet. Therefore, at the zero point location, each of the *X, Y,* and *Z* values is equal to zero. Also, as is the case in analytical geometry, the coordinate dimensions of any point are measured from that origin or zero point.

TABLE 14.2
Miscellaneous function codes according to EIA Standards

Code	Function
M00	Program stop (stops tape reader, spindle, and coolant)
M01	Optional stop
M02	End of program (stops tape reader, spindle, and coolant and resets control for next run)
M03	Spindle on forward
M04	Spindle on reverse
M05	Spindle off
M06	Tool change
M08	Coolant on
M09	Coolant off
M30	Tape rewind*
M40	Speed changes
M46	Gear changes (for spindle speeds)

*M30 is exactly like M02 with initiation of tape rewind. It is necessary that a tape rewind stop code be programmed prior to the first tape command so that no information is lost when rerunning the tape. The stop code is sometimes called the *end of record* (EOR). It is a percent sign (%) in ASCII.

Some NC machines have the zero point at a specific point (on the machine table) that cannot be changed. This is referred to as the *fixed zero point.* On the other hand, some machine control units allow the zero point to be established at any convenient spot selected by the programmer. This is referred to as the *floating zero point.* In this case, it is necessary to let the MCU know where the tool is located with reference to the selected zero point. This must be the first piece of information on the tape, directly after the units and dimensioning mode (i.e., inches or millimeters and absolute or incremental). It is usually achieved with the preparatory function code G92, followed by the coordinates of the location of the tool at the home position with reference to the selected program zero.

Setup point. Consider the simple case of an NC machine with a fixed zero point. It is not difficult to see that the programmer must know where the workpiece is to be located on the machine table with reference to the zero point so that he or she can refer all dimensions to that zero point and thus be able to write the program. There is, therefore, a need for an actual point on the workpiece whose location in relation to the fixed zero point must be known beforehand. This point is called the *setup point,* and it can be the intersection of two straight edges of the workpiece or a machined hole in the workpiece. It is also obvious that the setup point can be a defined point on the fixture holding the workpiece. Figure 14.10 shows how the absolute coordinates of any point on the workpiece can be obtained if the coordinates of the setup point are known.

Program preparation. In NC programming, it is not enough for the program to be capable only of producing the required part. The goal of the programmer should also be to reduce the time spent by the workpiece on the table of the NC machine. The task of eliminating wasteful and unnecessary movements of the tools as well as reducing the setup time is not easy; it requires a lot of experience and skill. Here are some guidelines that the beginning programmer should follow:

1. Check dimensions on the part blueprint to see whether they can be given in a way that makes programming easier.

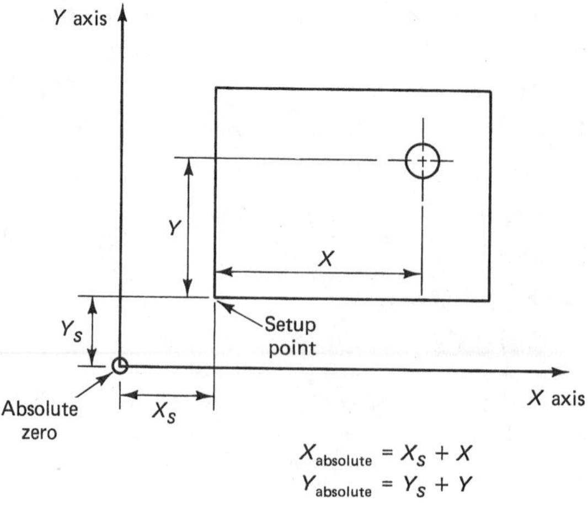

FIGURE 14.10
Using coordinates of the setup point to obtain absolute coordinates of any point on the workpiece

2. Study the part blueprint and prepare process sheets indicating the details and sequence of operations required to complete the job. Also, check the number of setups needed and try to divide the number of tools to be used between the setups.
3. Determine the most suitable fixturing method by studying the part configuration and correlate basic dimensions between the blueprint and the machine layout.
4. Prepare a tool layout, including the sizes and lengths of all tools to be used, in order to facilitate replacement of broken tools and to simplify setup.
5. Prepare the program sheet using the information gained in the preceding four planning steps.
6. Have a typist prepare the tape using a flexowriter. The printout of the program for that tape should be checked to make sure that there are no errors. (This step is not valid for modern CNC machine tools.)
7. Test the tape on the machine while operating in a single-block mode to ensure against collisions and to eliminate wasteful motion. Run the entire tape, bypassing errors but keeping a record of them for subsequent corrections.
8. Inspect the obtained part to make sure that its attributes fall within an acceptable range.
9. Correct the tape.
10. Prepare a folder containing the blueprint of the part, the fixture design, the program sheet, and a copy of an actual punched tape (or a magnetic tape) for the part program. Always write down the NC machine tool that can be used.

Program Example

A simple example that involves a drilling job will demonstrate the basic method of manual part programming. A drawing of the workpiece to be drilled using the fixed zero point of the machine tool as a dimensioning reference is shown in Figure 14.11.

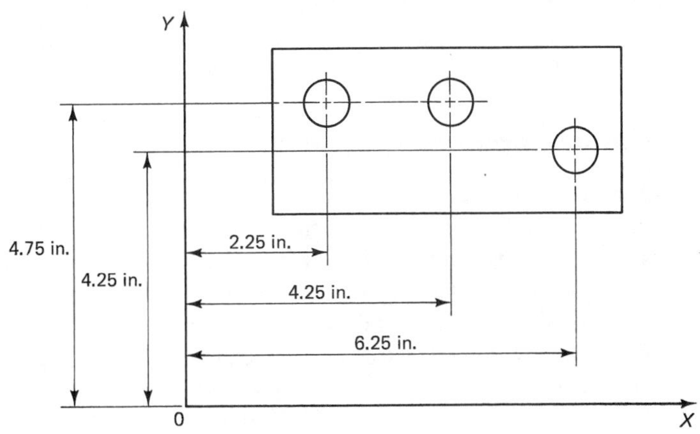

FIGURE 14.11 A workpiece to be drilled with an NC drill press and for which a program is to be prepared

In this example, the holes must be drilled only. Any further reaming or boring operation would require a tool-change command given by the preparatory function code for reaming or boring and the coordinate dimensions of the same points. Assuming that the feeds and speeds are manually set (that was the case with the early NC machines), the program can be given as follows:

```
EOR%  N001 G70 (EOB)
      N002 G80 X2250 Y4750 M51 (EOB)
      N003 G80 X4250 (EOB)
      N004 G80 X6250 Y4250 (EOB)
      N005 G80 X0000 Y0000 M02 (EOB)
```

In this program, take note of the following:

1. The length of the blocks is not always the same.
2. The EOB code does not appear on the printout of the program. It is only punched on the tape.
3. The X, Y, and M51 codes are modal (i.e., remain in effect until canceled by a new command).
4. M51 is a cam number command that determines the depth of the feed stroke for drilling and similar operations.
5. No decimal point is used with dimensions. Instead, they are expressed as multiples of the smallest possible movements of the machine table.

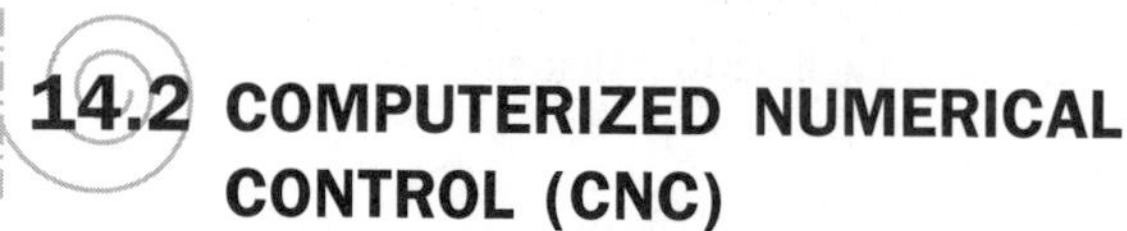

14.2 COMPUTERIZED NUMERICAL CONTROL (CNC)

In 1970, a new era for NC systems began with the emergence of *computerized numerical control* (CNC) technology, which involves replacing the hard-wired MCU of a conventional NC system by a microcomputer that, together with its software, accomplishes all the functions of a standard MCU. In addition to data decoding, feed rate control, buffering, and position loop control, a CNC system has many new features that are possible simply because a digital computer is used and improves the usefulness of the MCU.

Features of a CNC System

Ability to store programs. There is no need for the frequent use of a tape reader. Once a tape is run, it can be stored in the memory of the computer. It can be recalled later, as many times as required, directly from computer memory and without the need of re-running the tape. A tape reader fitted on a CNC system thus requires less maintenance than one used with a conventional NC system. Also, the computer is much faster in obtaining information (by retrieval from its memory) than the tape reader. Therefore, the use of CNC systems results in an appreciable saving in time.

Editing. It is very seldom that a satisfactory part program is obtained on the first attempt. Even experienced programmers need to make corrections, modifications, and improvements after running a program. The editing feature of CNC systems enables the programmer to make changes right on the factory floor. Also, all changes made go directly to computer memory without any reference to or use of the original punched tape. Consequently, a data-input device is needed in addition to a means of editing the program. Figure 14.12 shows a modern CNC lathe that has these features. The program is edited on a cathode-ray tube (CRT) that is similar to but smaller than that of a computer. The *manual data input* (MDI) device provides a means of entering programs into computer memory without any need for a tape reader.

Ability to produce tapes. After all necessary changes and improvements in the part program are made, a corrected punched (or magnetic) tape can easily be obtained by using an appropriate device that is plugged into the machine controller.

Expanded tool offsets. In CNC systems, the tool offsets (i.e., deviations in the lengths of the different tools from a reference value) are stored in the memory of the computer. Large numbers of offsets can be stored, which is not the case in conventional NC systems.

Expanded control of machine-sequence operations. CNC software usually handles machine operations for tool changes or control of the spindle or turret, thus making programming and operation of the machine much easier.

FIGURE 14.12
A modern CNC lathe *(Courtesy of Clausing Industrial, Inc., Kalamazoo, Michigan)*

Digitizing. Digitizing is usually provided as an option at extra cost. This feature allows a part program to be obtained directly from a model or an existing part. This is achieved by employing a stylus to scan the model while the CNC system monitors the movements and records the signals indicating the coordinate dimensions of points on the surface of the model. Again, a punched (or magnetic) tape can be obtained if needed. This feature eliminates the time-consuming and cumbersome operation of manual program preparation.

Circular interpolation. Very smooth arcs can be obtained because the computer of the CNC system has the capability to divide an arc into a large number of very small chord segments, calculate the coordinates of the endpoint of each segment, and establish a subprogram to generate the desired arc. This is achieved in programming by using either G02 or G03 as appropriate, together with the coordinates of the endpoint of the arc (say, X and Y) and the offsets of the center of the arc from its starting point along the X and Y axes, which are referred to as I and J, respectively. Following is a block of information that can drive a tool of a CNC lathe along the path required to machine an arc:

```
N030 G02 X3410 Z1606 I0400 K0325
```

Parametric programming. Parametric programming provides flexibility to a program by allowing several different-sized components having similar shapes to be machined using the same program. This is easily accomplished by changing the value of a few parameters, or *program variables.* Parametric programming can also be employed to obtain very smooth curves, provided that their mathematical equations are known.

Do loops. A do loop is useful in cutting the program short when it involves an operation that is to be repeated several times through incremental steps.

Roughing to a defined shape. An example of roughing is using the code G68, which involves a cycle to rough out a bar to a defined shape when cutting along the Z axis (turning). The shape is defined in a series of blocks called up in the G68 block, together with a parameter that defines the incremental depth of cut.

Subroutines. Subroutines allow the programmer to program, store, and repeat a pattern on different locations on the workpiece. Examples of applications include patterns of holes (bolt-hole circles), series of standard grooves on a shaft, and canned cycles that are created by the user.

Diagnostic capability. Diagnostic capability refers to the ability to detect faults when the CNC system goes down. This feature is a major advantage of CNC systems; it is due mainly to the diagnostic capabilities inherent in computer systems in general. In most cases, error messages are displayed to the operator. Also, a special diagnostic tape may be supplied with the CNC system, or the system may be connected through the telephone to the computer of the manufacturer's service department.

Program Example

A simple example that involves a machining job will demonstrate some of the capabilities of CNC technology. The part shown in Figure 14.13 is to be machined on a Clausing CNC lathe Hydro NC 540, fitted with GE 1050HL control, and it is necessary for a program to be prepared.

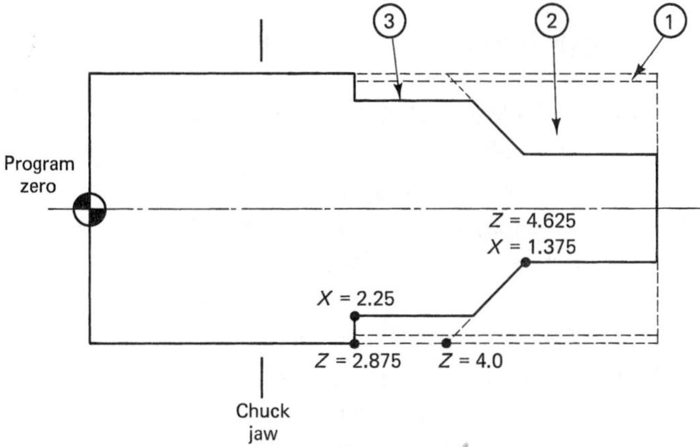

FIGURE 14.13
A part to be machined on a CNC lathe and for which a program is to be prepared

The program can be given as follows:

```
N010   G95   (in./rev.)
N020   G90   (absolute programming)
N030   G92   X13.33 Z13.485 S1500 T2100 M40
N040   G96   R13.33 S400 M03
N050   G00   X2.74 Z6.125 T2100 M08 (position for cut 1)
N060   G01   W -3.25 F.015 (cut 1)
N070   G00   X2.94 Z6.125 (position for cut 2)
N080   G81   X1.375 Z4.0 F.015 P1 .03 P2 4.625 P4 1 (cut 2)
N090   G00   X2.94 Z4.4375 T2100 (position for cut 3)
N100   G81   X2.25 Z2.875 F.015 P1 .03 P2 2.875 P4 1 (cut 3)
N110   G00   X13.33 Z13.485 M09 (coolant off, return home)
N120   M05   (spindle off)
```

In this program, take note of another important feature of CNC systems, decimal-point programming, which is not possible with conventional NC systems. Now, let us examine the program step-by-step:

- N010: The G95 code informs the MCU that it is going to receive English units (inches per revolution).
- N020: The G90 code informs the MCU that it is going to use the absolute dimensioning mode.
- N030: The function of the G92 code is to locate the program zero point; the numbers following the X and Z characters indicate the distance between the tool tip and the desired program zero point along the *X* and *Z* axes, respectively. Note that in programming NC lathes, the dimension *X* is taken as a diameter value (i.e., distance to the centerline multiplied by 2). The S character is the spindle-speed code; the value following it indicates the maximum spindle rpm required. The T code enables turret indexing as required and also indicates the tool offsets; it is followed by four digits, the first two indicating the required tool station and the second two indicating the tool offset. The M40 code calls up the lower speed range.

- N040: The G96 code is followed by an R word that always has the same value as the X word. The S word in this block applies to the required surface speed. The M03 code moves the spindle forward.
- N050: This block positions the tool for the first operation. The M08 code turns the coolant on.
- N060: This block involves a turning operation. Notice that the character W is used for incremental movement. The negative sign indicates that the motion is toward the chuck. The F stands for the feed rate word; in this case, it is in inches per revolution.
- N070: This block positions the tool for the second operation.
- N080: The G81 code indicates a canned cycle in which the tool cuts parallel to the *Z* axis. The X and Z words indicate the smallest values in this operation, whereas the parameter P1 indicates the incremental depth of cut and P2 indicates the value of *Z* at the root.
- N090: This block positions the tool for the third operation.
- N100: This block involves another canned cycle to carry out the third turning operation. Notice how the end of the cut is perpendicular to the centerline and not inclined as in the previous canned cycle.
- N110: The job is now finished. This block is employed to return the tool (turret) to the home position and to turn the coolant off.
- N120: The M05 code shuts the spindle off. It is now possible to release the chuck and obtain the desired product.

14.3 DIRECT NUMERICAL CONTROL (DNC)

As is the case with CNC, *direct numerical control* (DNC) is a hybrid of both NC and computer technologies. It involves bypassing the weakest and least reliable link in the system, the tape reader, by supplying part program data directly from the bulk storage device to the controller of the machine tool through telecommunication lines. Figure 14.14 shows how a DNC system functions by employing an additional hardware module as a connection between the mainframe computer and the machine tool controller. This piece of hardware is referred to as a *behind-the-tape-reader* (BTR). It is actually an additional source of data that does not depend on or make any use of the tape reader in its functioning. The tape reader is used, however, as a practical backup when the DNC system breaks down.

The main elements of a DNC system include a large, remote computer, a bulk storage device, telecommunication lines, and a number (as many as 100) of NC machine tools. The operation of a DNC system is based upon continuous flow of information from the bulk storage device to the NC machine tools, and vice versa. This takes place on a time-sharing basis and in real time. As a result, information is transmitted to the NC machine tool almost instantaneously when a signal from the machine indicating a need for instructions reaches the computer. This DNC configuration forms the backbone of today's flexible manufacturing systems.

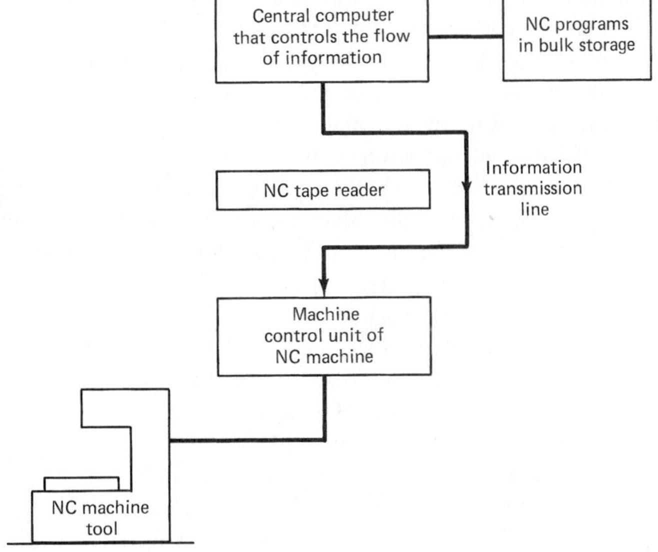

FIGURE 14.14
Structure and operation of a DNC system

Direct numerical control has several advantages. For instance, a 10 to 20 percent increase in productivity has been reported when DNC is employed. This is due to the monitoring abilities of the system, together with the increased machine run time. In addition to the elimination of the problems of the tape reader and the cost of tapes, DNC offers a step-by-step approach for establishing an integrated system that starts with a few NC machine tools and expands as required.

14.4 COMPUTER-AIDED PART PROGRAMMING

There are generally two methods of regenerating an NC part program: by *manual programming* or with *computer assistance.* In manual part programming, the programmer prepares a set of detailed instructions by which the desired sequence of operations is performed. He or she has to calculate manually the coordinates of all the various points along the required tool path before writing the program directly in a coded form and arranged in a format that can be understood by the MCU of the NC system. In other words, the programmer writes the part program in a *machine language* that can be directly read and processed by the NC system. Consequently, unless the part configuration is relatively simple and only a few different types of NC machines are employed, the task of manual part programming becomes time-consuming and cumbersome. When programs are to be prepared for complex parts requiring contouring or having complicated patterns of holes, it is almost impossible, or at least impractical, to do all the required geometric and trigonometric calculations by hand, and use must be made of that magic data-processing tool, the computer. Computer-aided part programming becomes a necessity when programming three-, four-, or five-axis NC machines that are used for generating sculptured surfaces or when complicated contouring is required. Also, when the plant includes several different types of NC machines, each having its own programming codes and format, computer-aided part programming is the right solution.

As can be expected from this discussion, computer-aided part programming is much easier and faster than manual part programming. All a programmer has to do is define the desired operations using the Englishlike words of the NC computer language that he or she is using. Once the part program is loaded into the computer, the computer takes care of all calculations and converts the input statements into a machine language compatible with the particular NC machine to be used. Typing errors can be corrected by means of the editing routines before a program is compiled. Programming errors are detected when compiling a program, thanks to the diagnostic error messages given by the computer. Nevertheless, the program may still contain some undetected errors that, if left uncorrected, will result in a part configuration different from the desired one. Therefore, the graphics capabilities of the NC computer system are always used to obtain a plot of the part geometry and the tool path in order to verify that the program will indeed produce the desired part. One of the most important advantages of computer-aided part programming is program verification because it results in fewer scrapped parts and saves most of the time used in debugging the program at the machine tool. This method of programming also has the advantages of simplicity, reducing the time needed for programming, and accuracy due to the elimination of any cumulative errors in calculations.

Internal Computer Operation

NC computer software systems can be divided into two distinct parts: the *general processor* and the *postprocessor*. The postprocessor is, in turn, composed of the postprocessor control system and the postprocessor machine segment. Let us now see what happens when a part program is loaded into the computer.

As mentioned, a part program consists of Englishlike statements that are used to define the desired operations. Therefore, the first step in processing a part program involves translating the input file (written in a general-purpose programming language such as APT and consisting of Englishlike words) into an equivalent computer machine-language program. This process, referred to as *compilation,* is necessary because the computer can understand only its own machine language. During the compilation process, if any syntax error is detected, further processing of the program is promptly stopped. Once compilation is complete, the processor handles all geometry and motion commands in the machine-language version of the program and carries out all the necessary calculations in the arithmetic-logic unit (ALU) of the central processing unit (CPU) of the computer. The output of these calculations adequately defines the tool path or cutter location (CL). The tool path indicates the center of the cutting tool and not the boundaries of the workpiece. Therefore, some of the calculations are concerned with offsetting the tool path from the desired part outline by a distance equal to the radius of the cutter (in milling operations).

Different NC systems use different control-tape codings and formats. Also, NC machine tools have different characteristics, depending upon the builder. Therefore, the output from the processor (which is in the form of CL data) must be reprocessed so that the precise output codes and format required for the given NC system and machine tool can be obtained. This is referred to as *postprocessing,* and it is carried out by a subprogram called the *postprocessor.* The output from the postprocessor is a tape image that can be converted to a punched (or magnetic) tape that can be employed to operate the given machine tool.

The postprocessor is a very specialized program that can reprocess data (the output from the processor) to operate only a particular combination of machine tool and NC controller. Because there are a large number of such combinations in industry, the postprocessor is divided into two distinct parts in order to facilitate the task of tailoring a complete postprocessor suitable for a given combination of machine tool and controller. The *control-system part* of the postprocessor is developed mainly to format the numerical data in a way to be received and understood by the NC controller. The *machine-segment part* of the postprocessor is in charge of processing statements dealing with coolant control (e.g., ON, OFF, FLOOD, MIST), automatic tool changing, spindle-speed selection, and the like. This part of the postprocessor is machine-dependent (i.e., it depends upon the features of the individual machine tool and changes from machine to machine). In order to better understand the relationship between the different parts of a computer-aided programming system, let us consider the block diagram shown in Figure 14.15. As can be seen in the figure, two identical types of controllers operating two different machines require two similar but not identical postprocessors. Also, two different NC controllers running two identical machines require two different postprocessors. We have to remember, however, that the output from the postprocessor in each case will always be exactly the same as the coded manual program prepared for the given machine tool.

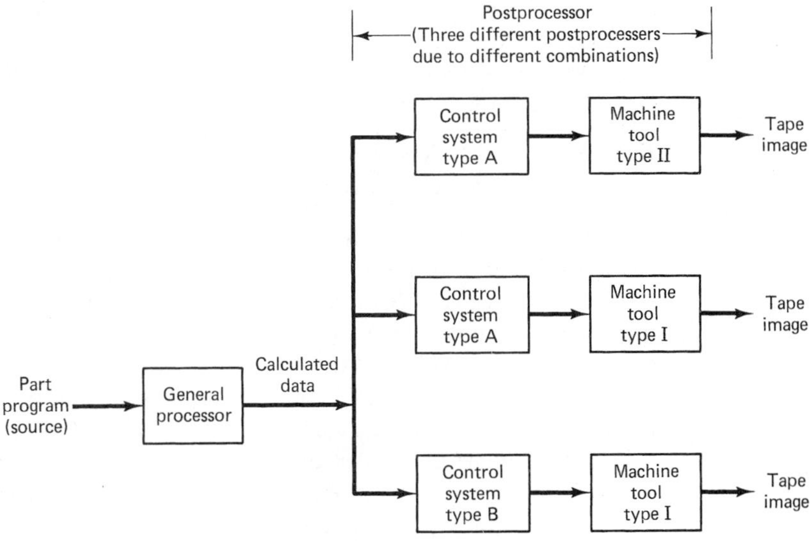

FIGURE 14.15
A computer-aided programming system involving various combinations of postprocessors and controllers

NC Programming Languages

Since the emergence of computer-aided part programming in the 1970s, numerous NC programming languages have been developed. Most of them have found limited use, and only a few are commonly used in industry. Therefore, the survey of NC programming languages is limited to the more or less general-purpose languages.

APT. APT stands for *Automatically Programmed Tools.* The language, which was originally developed at the Massachusetts Institute of Technology (MIT), is the most widely used and most comprehensive language available. An APT part program is written in Englishlike words and consists of a series of statements that define the part geometry, the desired operations, and the machine (and tool) characteristics. Each statement consists of a major word followed by a slash and some modifier words. There are 80 major words and 180 modifiers plus punctuation in the APT language. APT, although capable of producing sculptured surfaces, requires a large mainframe, which has limited the use of the language in the past. APT is the parent of two other NC programming languages that eliminate the need for a large computer. Recently, a simplified PC version of APT that is capable of driving a two-axis NC machine has become commercially available.

ADAPT. ADAPT stands for *Air Material Command-Developed APT.* This language is a simplified version of the APT language and can be run on a much smaller computer. ADAPT has 160 words (major and modifiers) plus punctuation, and it is limited mainly to applications that require plane contouring with a third axis of linear control.

UNIAPT. UNIAPT is another modified compatible version of the APT language. Although it can be used for programming three-axis and most of the four- and five-axis NC machines, UNIAPT was specifically developed to be run on minicomputers.

SPLIT. SPLIT is the acronym for *Sundstrand Processing Language Internally Translated.* SPLIT was developed to be used with Sundstrand machine tools. Therefore, its processor was dedicated and machine-dependent, and there was no need for a separate postprocessor. Accordingly, SPLIT could not be used with any type of machine tool other than Sundstrand, which had markedly limited its industrial use.

ACTION. ACTION is considered to be a child of SPLIT. It is a modified version that has a general-purpose processor and a machine-dependent postprocessor.

COMPACT II. COMPACT II is a child of ACTION and a grandchild of SPLIT. It has a general-purpose processor and a machine-dependent postprocessor. The COMPACT II language has the advantages of being simple and easy to learn, and it satisfies the vast majority of programming requirements. In addition to availability on a time-sharing basis from M.D.S.I./APPLICON in Ann Arbor, Michigan, there is now a version that can run on a VAX 785/11, as well as another new version that has been specially developed to run on a microcomputer. Many of the microcomputer CAM systems now available are similar to COMPACT II. Because microcomputers have found their way into manufacturing and computer-aided part programming, let us now study the COMPACT II language in more detail. The information provided here is published with special permission from M.D.S.I./APPLICON, Ann Arbor, Michigan. The cooperation of that firm is greatly appreciated.

Details of the COMPACT II Language

A COMPACT II program consists of a series of statements, each providing information or instructions to the system. A statement must always begin with a major word, followed by a set of associated minor words. Major words indicate the operation to be performed by the system, whereas minor words provide details about the location and manner in which that operation is to be accomplished. Minor words are also used to indicate tool description and to define cutting speeds, feeds, and the like.

Guidelines. In a COMPACT II program, statements must form a logical sequence of events. Following are some guidelines for building a program:

1. The initialization statements must come first in the program.
2. If the BASE (similar to the setup point in conventional NC) is to be used as a reference for the coordinate system, it must be defined prior to its use.
3. Points, lines, and circles must first be defined before being used in defining the part geometry.
4. A tool-change statement should precede the motion statements in which that tool is used.
5. The termination statement END must be the last statement in the program.

Syntax. Like any computer language, the COMPACT II language has a certain syntax that must be followed. Here are the rules for punctuation and arithmetic operations:

1. The comma is used to separate the units of information that form a statement.
2. The slash operator is usually used to modify the parameters associated with a geometric element.
3. Parentheses are employed to combine a set of information into a single unit. They must also be used to enclose a division operation.
4. The semicolon allows the programmer to avoid repeating the same major word in successive statements by acting as a substitute for the major word after the first statement.
5. The percent sign (%) is used to specify the opposite input mode for dimension. If the input mode is inches, a dimension followed by % is in millimeters.
6. The dollar sign is used in pairs to enclose comments that may be continued up to three lines.

Structure. The structure of a COMPACT II program involves five different groups of instructions, each group consisting of one or more statements. The five groups of instructions serve different purposes. The statements in the first group, which are always given at the beginning of a program, are the initialization statements. Other groups are used for defining the geometry of the workpiece, giving tool-change commands, initiating and defining tool motion, and terminating the program. A tool-change statement can be given as many times as required, depending upon the number of tools needed. In addition, each time a tool-change statement is given, it must be followed by the corresponding tool-motion statements. Following is a discussion of each of these kinds of statements:

1. Initialization sequence. The initialization sequence usually includes four statements:

a. The MACHIN statement is always the first statement of a COMPACT II program. It provides the name of the machine tool and consists of the major word MACHIN, followed by the name of the machine-tool link—for example,

```
MACHIN, MILL
```

b. The IDENT statement is the second statement in the program. It is used for identifying the program (or the machine-control tape) and consists of the major word IDENT, followed by the part name or number or any alphanumeric combination—for example,

```
IDENT, TEST 2 PROGRAM
```

c. The SETUP statement is used mainly to specify the home position of the gauge-length reference point (GLRP), relative to the absolute zero of the machine tool, at both the beginning and the end of the program. The GLRP is actually the point from which the tool gauge lengths are measured (e.g., for a milling machine tool, it is the center point at the surface of the quill). This statement is used to specify the program zero when a floating-zero machine is used. For a milling machine tool, this statement takes the following form:

```
SETUP, 3LX, 4.5LY, 10LZ
```

where 3 is the dimension from the absolute zero along the X axis to the GLRP, 4.5 is the dimension from the absolute zero along the Y axis to the GLRP, and 10 is the dimension from the absolute zero along the Z axis to the GLRP. The given numbers (3, 4.5, and 10) are arbitrary and differ from program to program.

You should always keep in mind that the home position is, at the same time, the load-unload position. Therefore, when using a lathe, for example, the home position of the turret should be selected such that the longest tool is clear of the maximum outer diameter of the workpiece. Sometimes, the SETUP statement is also used to establish the travel limits of the machine tool by specifying the LIMIT parameter, as follows:

```
SETUP, 10X, 20Z, LIMIT (X0/15,Z0/30)
```

For a lathe, the SETUP statement takes the following form:

```
SETUP, 5.75X, 7.5Z
```

where 5.75 is the dimension from the spindle centerline to the GRLP and 7.5 is similar to SETUP X but along the Z axis.

d. The BASE statement is used to define a secondary coordinate system shifted from but parallel to the original coordinate system, with the aim of facilitating the programming task. Here, BASE is a datum point located on the part blueprint and from which the part blueprint has been dimensioned (similar to the setup point in conventional NC). It is always advantageous to reference the BASE to the absolute zero, although it can also be referenced to other defined points. Following is a BASE statement where A means absolute (as opposed to XB, meaning with reference to the BASE):

```
BASE, 3XA, 4YA, 2ZA
```

2. Geometry definition. The shape of a workpiece can be precisely defined by defining its geometric elements (i.e., points, lines, circles, and planes):

a. A *point* is defined by using the major word DPT (define point). Any associated minor words describe how that point is specified. As is the case in analytical geometry, a point can be defined by providing a set of coordinate dimensions from absolute zero (or BASE) or by specifying the location of a point as lying at the intersection of two lines, a line and a circle, or two circles. In the latter two cases, a *selector* is required because the intersection will yield two points. This, as well as other methods for defining a point in COMPACT II, are given in the appendix at the end of this chapter. Following are some examples of statements used in defining points:

```
DPT1, 5XB, 3YB,  6ZB
DPT5, LN1, LN2,  5ZB
DPT6, LN2, CIR2, XL
```

In the last statement, XL is the selector, meaning that the point that has a larger coordinate dimension along the X axis is the required one.

b. A *line* is defined by using the major word DLN (define line). Any associated minor words describe how that line is specified. A line can be defined as passing through a point and making a certain angle with a reference axis, as passing through two defined points, or as an implied line perpendicular to one of the coordinate system axes. Following are some examples of statements used in defining lines:

```
DLN1, PT1, 30CW
DLN2, PT1, PT2
DLN3, 4XB
```

The various methods for defining lines in COMPACT II are given in the chapter appendix.

c. A *circle* is defined by using the major word DCIR (define circle), followed by any associated minor words. As is the case when defining points and lines, the methods used are adopted from analytical geometry. A circle can be defined by its center and radius, by three points through which it passes, by being concentric with an existing circle, or by being tangential to two existing lines. Following are some examples of statements used in defining circles:

```
DCIR1,  PT1,  1.5R
DCIR2,  LN2,  LN3,  2.5R
DCIR3,  PT1,  PT2,  PT3
DCIR4,  CIR3/5R
```

The various methods for defining circles in COMPACT II are given in the chapter appendix.

d. A *plane* is defined by using the major word DPLN (define plane), followed by any associated minor words. One of the methods for defining a plane involves specifying three points through which that plane passes. A plane can also be defined as perpendicular to an axis by programming its axis intercepts. Following are some examples of statements used in defining planes:

```
DPLN1,  PT1,  PT2,  PT3
DPLN2,  10ZA
```

3. Tool-change statements. The tool-change cycle is started by using one of the two major words ATCHG or MTCHG, followed by any associated minor words specifying the tool configuration, feed rate, spindle speed, and so on. The ATCHG (automatic tool-change) statement causes the spindle to stop and move to the tool-change position, where the current tool is returned to the magazine and replaced by the new tool whose number is given in the statement. The MTCHG (manual tool-change) statement serves to stop the machine function so that the operator can perform the tool-change operations manually. Following is a typical tool-change statement used in milling applications:

```
ATCHG,  TOOL4,  GL6,  .5TD,  300RPM,  .01IPR
```

The minor words in this statement have the following meanings:

a. TOOL4 is a command to get the tool in pocket 4 in the magazine and mount it in the spindle.

b. GL6 means that tool 4 has a gauge length (i.e., the length appearing beyond the GLRP) of 6 inches.

c. .5TD means that tool 4 has a tool diameter of 0.5 inch.

d. 300RPM indicates the rpm of the spindle after tool 4 is mounted. Sometimes, it is replaced by the cutting speed in feet per minute (e.g., 80FPM). In this case, the system automatically calculates the rpm using the tool diameter and the cutting speed.

e. .01IPR indicates the feed in inches per revolution. Sometimes, the feed rate in inches per minute is used instead (e.g., 1.5IPM).

In lathe applications, the minor words in a tool-change statement are slightly different from the preceding ones. For one thing, the tool gauge length has to be specified along both the X and the Z axes (i.e., the distances of the tool tip from the reference point at the center of the turret along the X and Z axes). Also, the radius of the tool nose has to be given. Following is a typical tool-change statement used in lathe applications:

```
ATCHG, 3GLX, 6GLZ, TOOL2, .05TLR, 100FPM, .015IPR
```

4. **Motion statements.** Major words are used to identify either linear or circular motion; these are followed by minor words that specify and terminate the path of the tool. The major and minor words used for linear motion are different from those used for circular motion:
   a. The two major words that generate linear motion are MOVE and CUT. MOVE generates rapid traverse motion and is used to position the tool prior to a cutting operation. (A clearance must be left between the final position of the tool and the workpiece surface to avoid accidents.) CUT generates feed rate motion for machining with the tool in contact with the part. Following are some examples of statements used to generate linear motion:

```
MOVE, OFFLN1/0.3XS, OFFLN2/YS
CUT, PARLN1, OFFLN3/XL
CUT, PARLN2, PASTLN4
CUT, PARLN8, TOLN7
CUT, PARLN11, ONLN12
```

   In the preceding statements, TOLN and PASTLN are determined relative to present tool position, and OFFLN must be followed by a modifier (XL, XS, YL, YS), which is sometimes accompanied by a tool offset (OFFLN1/.3XS).
   b. The three major words that generate circular motion are CONT, ICON, and OCON. They indicate the location of the path of the tool center with respect to the circular arc to be obtained after machining. As can be seen in Figure 14.16, CONT is used when the tool center always falls on the arc, while OCON (outside contour) and ICON (inside contour) are used to produce convex and concave surfaces, respectively. In all cases, the major words are followed by minor words indicating the direction of motion and its start and finish locations. An interesting feature of the COMPACT II language is that the linear motion from

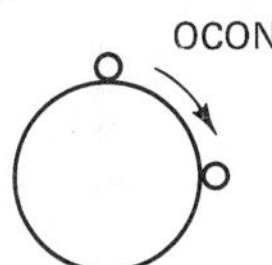

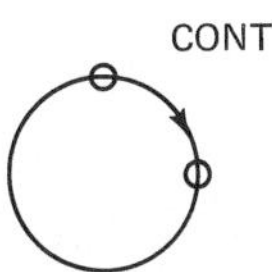

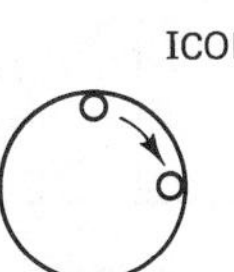

FIGURE 14.16
Major words for circular motion in COMPACT II

the current tool location to the start location of an arc need not be programmed and is automatically included in all circular motion statements. Following are some examples of statements used to generate circular motion:

```
ICON,  CIR2,  CCW,  S(TANLN3),  F(TANLN4)
OCON,  CIR4,  CW,   S(TANLN4),  F(TANLN6)
ICON,  CIR3,  CCW,  S(TANLN5),  F(90)
```

5. **Program termination.** The END statement is always the last statement of a COMPACT II program, and it contains only the major word END. As soon as this command is given, the resulting tape image will have blocks of information that return the machine tool axes to the home position, reset the control, and rewind the tape.

Program Example

Figure 14.17 indicates the required final shape of a workpiece after contouring. The original stock is an aluminum plate 3 by 6 by 0.5 inches that is located on a fixture whose surface is 1.5 inches above the absolute zero. Home position is located at 8YA, 5XA, 10ZA. The tool used is an end mill with a diameter of 0.875 inch and a gauge

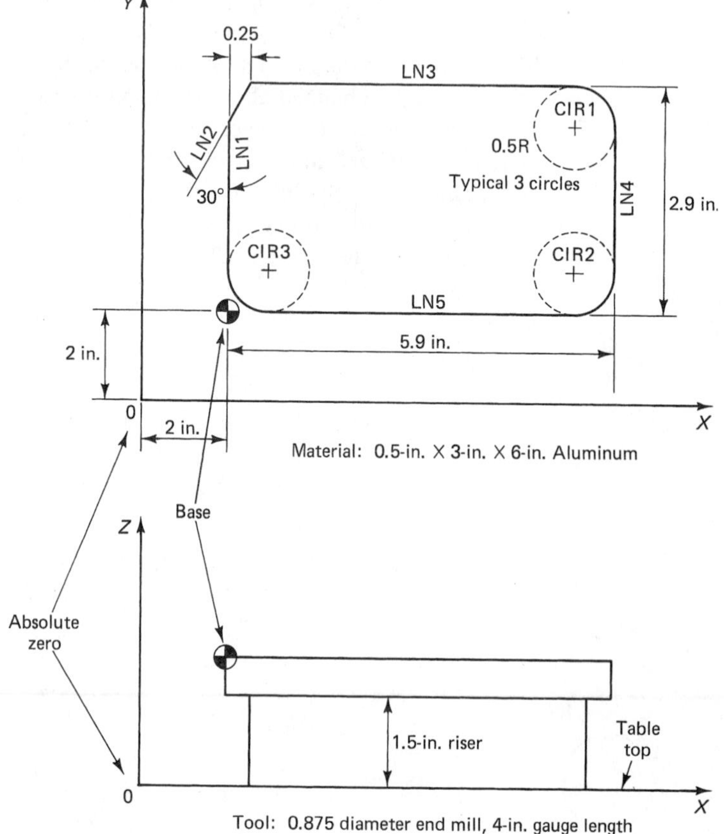

FIGURE 14.17 A workpiece to be contoured and for which a COMPACT II program is to be prepared

length of 4 inches. The following program in COMPACT II produces the desired shape:

```
MACHIN, MILL
IDENT, CONTOURING JOB
SETUP, 8LY, 5LX, 10LZ
BASE, 2XA, 2YA, 2ZA
DLN1, 0XB
DLN3, 2.9YB
DLN2, LN1/.25XL, LN3, 60CCW
DLN4, 5.9XB
DLN5, 0YB
DCIR1, LN3/.5YS, LN4/.5XS, .5R
DCIR2, LN4/.5XS, LN5/.5YL, .5R
DCIR3, LN1/.5XL, LN5/.5YL, .5R
ATCHG, TOOL1, 4GL, 600RPM, 2IPM, .875TD
MOVE, OFFLN5/.1YS, OFFLN1/XS, .1ZB
;-.60ZB
CUT, PARLN1, OFFLN2/XS
; PARLN2, OFFLN3/YL
OCON, CIR1, CW, S(TANLN3), F(TANLN4)
OCON, CIR2, CW, S(TANLN4), F(TANLN5)
OCON, CIR3, CW, S(TANLN5), F(TANLN1)
END
```

This program is self-explanatory. However, it is important to notice how circle 1 is defined:

```
DCIR1, LN3/.5YS, LN4/.5XS, .5R
```

The unit LN3/.5YS indicates a line parallel to LN3 and shifted by 0.5 inch below it. Actually, this line is the locus of the centers of the circles tangential to LN3. The unit LN4/.5XS represents the locus of the centers of the circles tangential to LN4. The intersection is, therefore, the center of the circle tangential to both lines, which is CIR1. This concept is frequently used in the COMPACT II language. Also notice the MOVE statement,

```
MOVE, OFFLN5/.1YS, OFFLN1/XS, .1ZB,
;-.60ZB
```

where the tool is brought to a point 0.1 inch above the surface of the workpiece and then lowered to the required length.

Note that only the basics of programming in COMPACT II have been covered here. This language contains many very useful features, including programming patterns of holes, do loops, and macro routines. Interested readers are advised to explore this advanced level in the *COMPACT II Programming Manual* published by M.D.S.I./APPLICON, Ann Arbor, Michigan.

Graphics NC Systems

Graphics NC systems, which have been developed for machining centers and lathes, represent the most up-to-date version of CNC systems. Figure 14.18 shows a machining center, and Figure 14.19 shows a lathe; both have graphics NC capabilities. When

FIGURE 14.18
A machining center having graphics NC capability *(Courtesy of MAZAK Corporation, Florence, Kentucky)*

FIGURE 14.19
A lathe having graphics NC capability *(Courtesy of MAZAK Corporation, Florence, Kentucky)*

programming these CNC systems, the programmer inputs data and communicates with the system in user-friendly language. A message and a menu corresponding to each position of the cursor are displayed in the lower part of the CRT screen. Depending upon the message, the programmer inputs data by pressing either the appropriate menu key (as shown in Figure 14.20, there are nine menu keys) or one or more numeric keys. Generally, the operational procedure for machine tools with graphics NC systems takes the following steps:

1. Register tools (input tools in the tool file picture).
2. Prepare desired program.
3. Allocate tools in the different pockets of the magazine or in the different stations of a turret in the case of a lathe (tool layout picture).

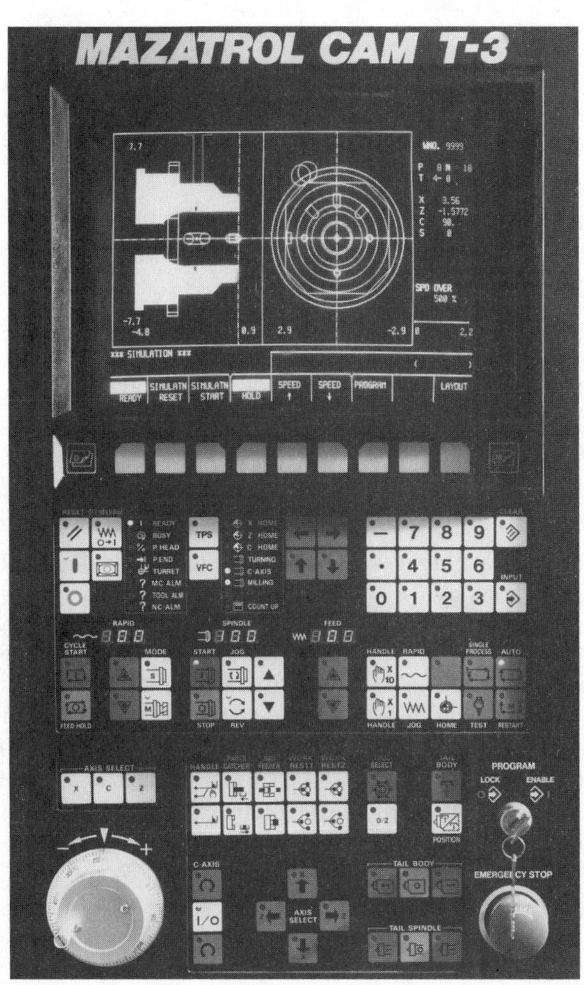

FIGURE 14.20
A controller for a graphics NC system
(Courtesy of MAZAK Corporation, Florence, Kentucky)

4. Input tool data, such as actual diameter and lengths. A tool gauge length can be accurately measured using a special unit attached to the machine.
5. Input coordinate dimensions for the program zero (this is similar to the BASE point in COMPACT II).
6. Check geometry of the part and the tool path on the graphic display (the CRT screen). If the desired part configuration and tool path are obtained, start automatic machining. However, if an error is observed in the product shape or in the tool path, modify the program until the error is eliminated.

Advantages of graphics NC systems. Following are some advantages of the graphics NC systems as exemplified by the MAZAK systems (i.e., these advantages are experienced in but may not be limited to the MAZAK systems):

1. Graphics NC systems involve a user-friendly NC data input method with respect to operation, routines, capabilities, and efficiency.
2. These systems have increased machine performance and high productivity.
3. They provide automatic programming routines, such as automatic optimum tool selection, automatic determination of tool paths, calculation of cutting conditions, insertion of chamfers and rounding corners, and calculation of the points of intersection of the geometric elements.
4. Programming and editing are possible while the computer is controlling the machine tool. The prepared program can be checked on the graphic display of the controller.
5. The tool path that is to be followed during machining can be checked on the CRT screen at a higher speed without the need for actually running the machine tool. This eliminates the danger of collision between the tools and any obstacle. It also allows the actual machining time to be obtained and then displayed on the CRT screen.
6. The systems have the capability of automatic machining-accuracy compensation for the tool wear that occurs during machining and that depends upon the workpiece machining time and the number of workpieces machined.
7. The systems are highly reliable as a result of full adoption of the latest microelectronics technology.

Programming. In order to successfully discuss the programming of these systems, let us consider one of the most commonly used graphics NC languages, the MAZATROL language. Only slight differences exist between the version of the language used for programming a machining center (CAM-M2) and that used for programming a lathe (CAM-T4). MAZATROL for CAM-T3 will serve as an example here.

As mentioned, programming is accomplished through interactive data entry using the controller. Questions from the computer are indicated on the CRT screen in the form of a message. The answer is then provided by the programmer by pressing the appropriate menu key or inputting numerical data using the numeric keys.

A MAZATROL program can be broken up into *units*. The first unit, which is given the serial number 0, is devoted entirely to describing the blank that will be machined, its material, its tolerances, and the way it is mounted on the lathe. Each of the succeeding units deals with a portion of the finished product and comprises two kinds of data, as follows:

1. **Process data.** These data define the kind of operation to be used when machining a portion of the product (e.g., external turning, internal turning, edge facing). This is achieved by selecting the menu keys that indicate first the type of MODE and then the kind of PART. If, for example, the MODE is chosen to be BAR and the PART to be OUT, the current program unit involves all the required conventional external turning to be performed on this portion of the workpiece (this does not include threading or grooving).

2. **Sequence data.** These data define the final shape of the preceding portion of the product after machining. The shape of this portion of the workpiece (or any other shape) can be broken up into a group of geometric entities such as lines or convex or concave curves. Radii, chamfers, and relief ways can also be included. Depending upon the kind of machining operation required, the sequence data may be given in a single block or (usually) in more than one block, as is clearly indicated in Figure 14.21.

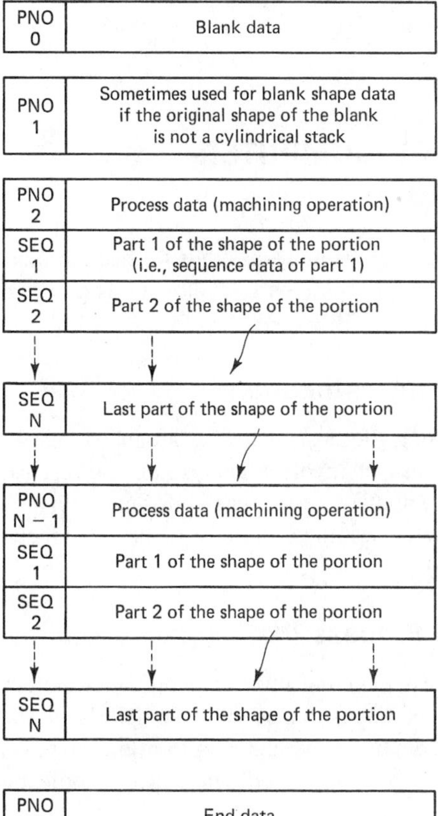

FIGURE 14.21 Structure of a MAZATROL program

Figure 14.21 also indicates that the last unit of a MAZATROL program is always the program end data (i.e., the number of pieces to be produced and whether or not the tool returns to the initial position).

In the past, machine tools with graphics NC systems could not be operated by tapes (or tape images) produced by manual or computer-assisted part programming. However, up-to-date models now have a port that accepts an electronic tape image in EIA code.

CAD/CAM Systems

The recent trend to establish a direct link, through electronic channeling, between the product design and manufacturing departments is aimed at eliminating the duplication of efforts by design and manufacturing personnel. When using such systems, interactive graphics software is employed to establish the geometry, dimensions, and tolerances for the desired part design, which can be displayed on the CRT screen. The geometry of the part design can be stored in the memory of the computer in the form of digital data that can, in turn, be adopted as a database for preparing an NC part program. By entering the tool data and employing an NC processor and post-processor, a tape image for the part program can be obtained. These CAD/CAM systems are particularly advantageous when the shop includes different types of NC machine tools.

14.5 OTHER APPLICATIONS OF COMPUTER-AIDED MANUFACTURING

The discussion of CAM applications has thus far been limited to the use of computers to drive tools in order to machine parts. Following is a brief discussion of some other applications of CAM.

Computerized Cost Estimation

The task of determining the cost of a new product is usually both time-consuming and cumbersome because it involves analysis of indirect expenses as well as overheads. Because the computer is efficient at information handling and processing, it is widely used to accurately estimate the cost of new products in the shortest possible time (see Chapter 11).

Computer-Aided Process Planning

Computer-aided process planning involves employing the computer to determine the optimal sequence of operations that should be employed to manufacture a desired part and also keep the production time and cost to a minimum. This application of CAM has recently been used in computerized automated manufacturing systems (see Chapter 16).

Computerized Machinability Data Systems

In computerized machinability data systems, the role of the computer is to provide the feed and speed that should be used to machine a given workpiece material by a given tool material. This is achieved either through retrieving the recommended values from a database created by experienced people from experimental observations or by employing mathematical modeling and Taylor's equation (see Chapter 9).

Computer-Aided Monitoring and Control of Manufacturing Processes

Computer-aided monitoring involves a variety of applications, ranging from data acquisition and computer process control to computerized numerical control and adaptive control. Adaptive control has special important applications in modern automated manufacturing systems, so let us discuss it here in some detail.

We previously came to the conclusion that although NC machines result in an appreciable reduction in the overall production time, the actual machining time remains virtually identical to that for conventional machine tools. This is a consequence of the fact that an NC system provides *preplanned control* without any feedback mechanism to account for real-time variations in the process parameters. For this reason, Bendix Research Laboratories developed an *adaptive control* system in the early 1960s to operate the machining process more efficiently. The main function of a practical adaptive control system is the real-time optimization of a performance index. This index is taken as the metal-removal rate or the cost per volume of metal removed in machining operations. The adaptive control system performs its function by detecting any variability in the condition of the workpiece being machined and adjusts the feed and the cutting speed to account for that variability and to maximize the performance index. The variability can take different forms, such as hard spots, which require reduction in feeds and cutting speeds, or the presence of air gaps in the workpiece, where the feed should be doubled or even tripled to minimize the idle time during which the tool travels across the air gap.

Figure 14.22 shows a typical adaptive control system used in industry for controlling on NC machine tools. In this system, the controlled variable is the feed, and the system monitors the spindle deflection (or the horsepower consumed) and keeps it below a certain predetermined value by controlling the feed. This type of adaptive control system is practical and is most commonly used in industry. It is referred to as *adaptive control constraint* because it limits, or constrains, the measured variable (e.g., spindle deflection or horsepower consumed) below a desired value.

In the 1970s, adaptive control was not widely used because it reduced only the in-process time, which usually accounted for less than 5 percent of the production cycle of a part. The system was used either when the actual machining time amounted to a high percentage of the total production time or when there were significant sources of variability in the workpiece. Now, adaptive control is gaining wider industrial application, especially in Computer Integrated Manufacturing (CIM) systems, where human intervention to compensate for variabilities is not required. Also, adaptive control has found application in chipless manufacturing processes like sheet metal working and welding.

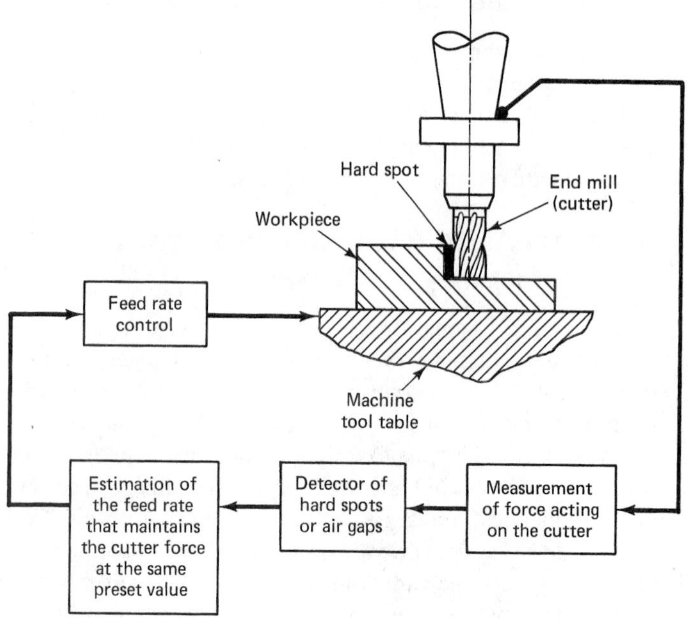

FIGURE 14.22
A typical adaptive control system used in industry

Review Questions

1. Define CAM.
2. What is the dominant application of CAM?
3. What is meant by *numerical control?*
4. When and from where did numerical control emerge in modern times?
5. Discuss some of the main advantages of NC machine tools.
6. What are the main elements of an NC system? Discuss the functions of each briefly.
7. What is the difference between the absolute and incremental dimensioning modes?
8. Explain the NC machine axes.
9. How many axes does an NC machine have?
10. An NC machine can be indexed in eight positions by tape command. Can this be considered as a machine axis?
11. How do you identify the Z axis of an NC machine tool?
12. What kind of NC machine tools can be employed for producing sculptured surfaces?
13. Define NPC. What is the main application for this kind of system?
14. What is the difference between the NPC and picture-frame systems?
15. What additional features must an NC system have so that it can perform contouring jobs? What do we call such an NC system?

16. What is the function of a punched tape?
17. What materials are NC tapes made of? Give an application for each of these materials.
18. What is meant by the *binary-coded decimal* system? Explain how numbers are coded on NC tapes.
19. Define the following terms: *row, bit, word, track,* and *block.*
20. How many tracks does an NC tape have?
21. What is a *parity check?* How is it performed in the EIA and the ASCII codes?
22. Why is the word EOB used?
23. Why is the word EOR used?
24. Describe some tape formats.
25. Explain the basic concept of word-address programming.
26. Explain the meaning of the N, G, and M codes.
27. Describe manual part programming.
28. What is the absolute zero?
29. Differentiate between the fixed zero and floating zero points.
30. What is the setup point?
31. List the steps included in a programming task.
32. Define CNC.
33. List some of the features of CNC systems and discuss each briefly.
34. Define DNC.
35. Explain what a BTR is.
36. What are the advantages of DNC?
37. In what way can a computer assist in preparing a part program?
38. What are the advantages of computer-assisted part programming? Explain each briefly.
39. Discuss briefly what you feed into a computer and what happens within it so that a tape image can be obtained to drive a machine tool.
40. What are the functions of the processor and the postprocessor?
41. Of what is a postprocessor usually composed?
42. List some NC computer programming languages.
43. Briefly describe each of the languages you listed in Question 42.
44. What is the relationship between SPLIT, ACTION, and COMPACT II?
45. Explain the function of each of the following in COMPACT II: comma, slash, parentheses, semicolon, percent sign, dollar sign.
46. What kinds of statements form a COMPACT II program? Explain each briefly.
47. Explain the meaning of BASE in COMPACT II.
48. How can we define the geometry of a part in COMPACT II?
49. What are ATCHG and MTCHG in COMPACT II?
50. What data does a tool-change statement in COMPACT II include?
51. What is the GLRP?
52. Describe the home position.
53. What statement in COMPACT II follows the tool-change statement? What major words can be used in that statement?
54. What is the last statement in a COMPACT II program?
55. What is graphics NC?
56. Describe the general procedure of preparing a graphics NC program.
57. List and discuss the main advantages of graphics NC systems.
58. Why is programming a graphics NC system considered to be very easy?
59. Explain briefly the structure of a MAZATROL program.
60. Differentiate between the process data and sequence data in a MAZATROL program.
61. What are the advantages of CAD/CAM systems? When would you recommend these systems?

62. What are the other applications of computer-aided manufacturing?

63. What is adaptive control and why is it now gaining industrial application?

Problems

1. Using a CNC lathe similar to the one mentioned earlier in the chapter in the program example on CNC machining, write a program to produce the part shown in Figure 14.23. Use the same tool length and data as in the program example.

2. Write a program in COMPACT II to contour around the workpiece shown in Figure 14.24. The tool is an end mill having a diameter of 0.5 inch and a gauge length of 3.5 inches. The rpm is to be taken as equal to 300, and the feed is equal to 0.015 inch per revolution. Here are the setup instructions:

a. Home position is located at 8LY, 5LX, 10LZ.

b. BASE is located at the lower left corner of the part and the top of the part.

c. Absolute zero is located at the lower left corner of the part and the table-top end view of the part and the fixture.

3. Design a chess piece (king, queen, pawn, or knight) and then select the appropriate tools, plan the tool path, and write a CNC program to machine it using the CNC lathe available in the lab.

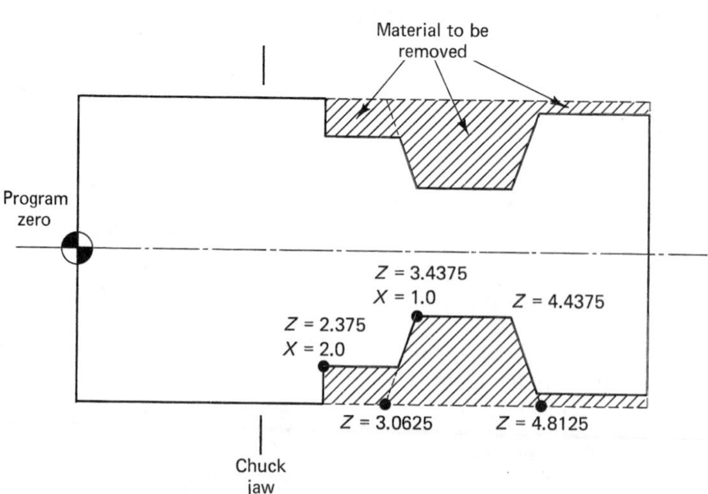

FIGURE 14.23
The part in Problem 1 for which a CNC program is to be prepared

FIGURE 14.24 The workpiece in Problem 2 for which a COMPACT II program is to be prepared

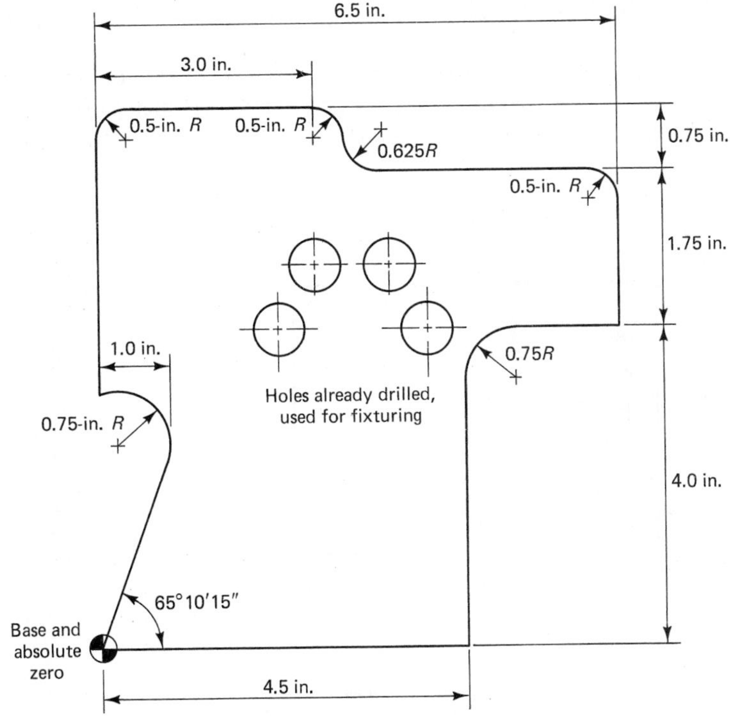

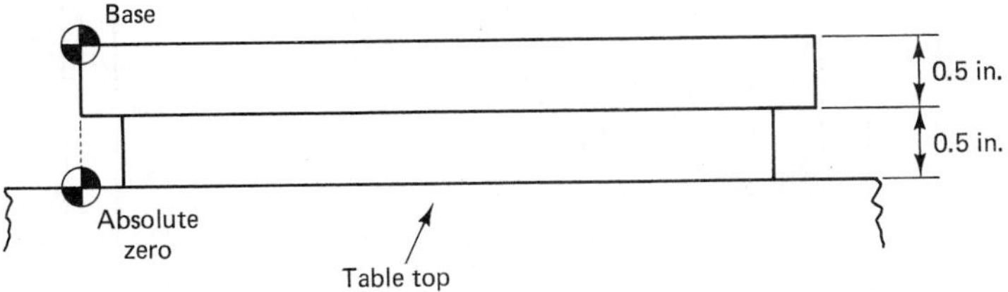

CHAPTER 14 APPENDIX

The following appendix depicts various ways of defining geometric entities, which can in turn be used to adequately describe the geometry of a workpiece. All of the figures are reproduced courtesy of M.D.S.I./Applicon, Ann Arbor, Michigan.

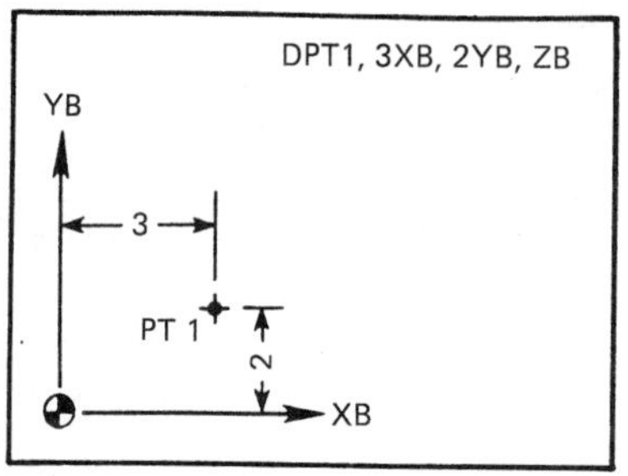

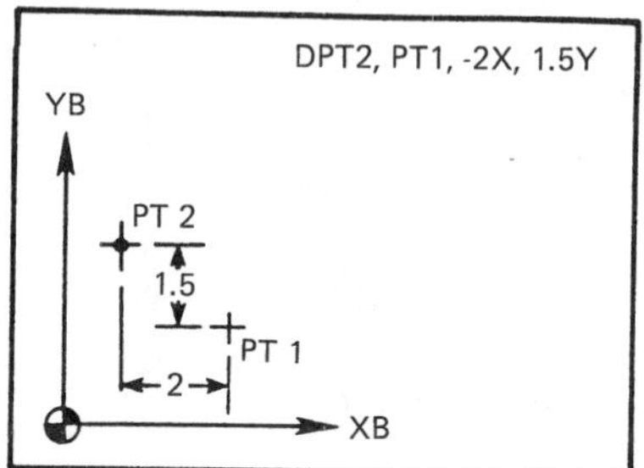

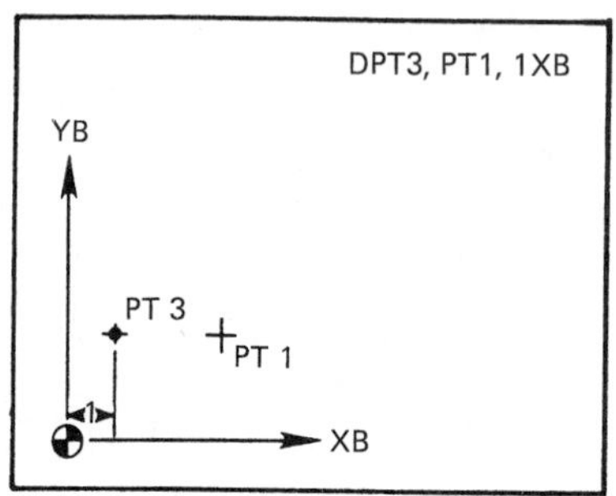

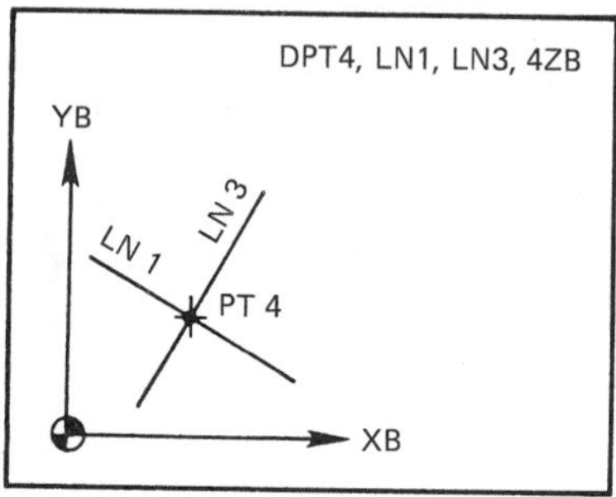

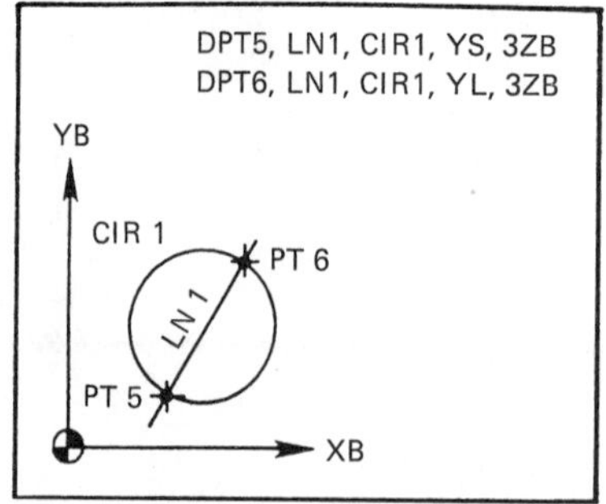

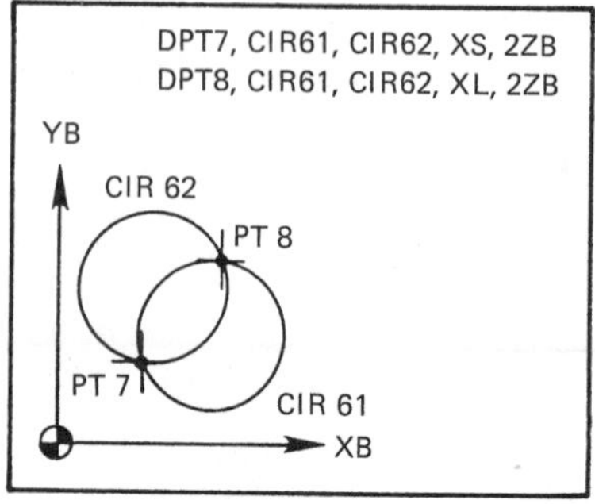

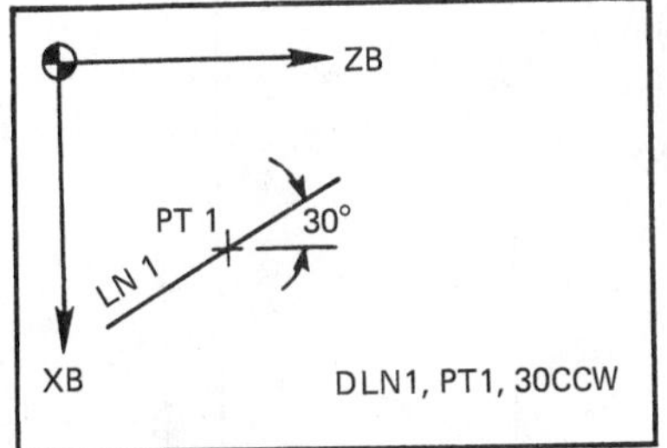
ZB
PT 1
30°
LN 1
XB
DLN1, PT1, 30CCW

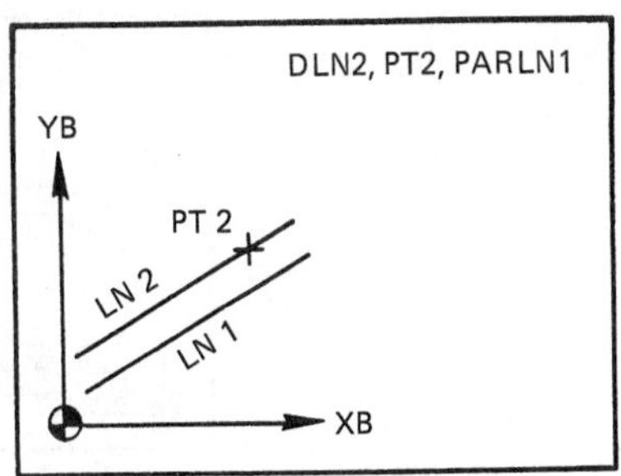
DLN2, PT2, PARLN1
YB
PT 2
LN 2
LN 1
XB

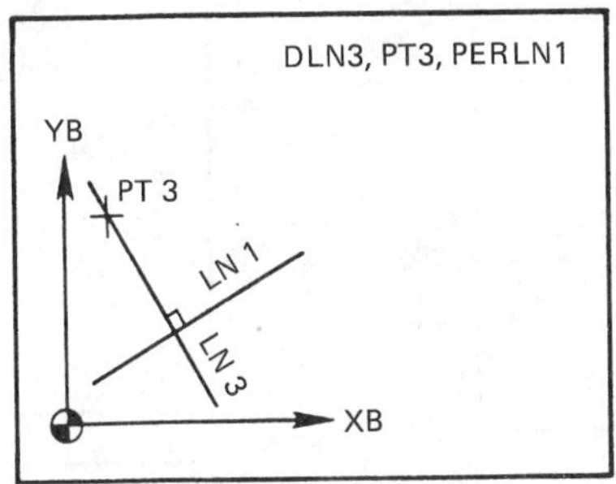
DLN3, PT3, PERLN1
YB
PT 3
LN 1
LN 3
XB

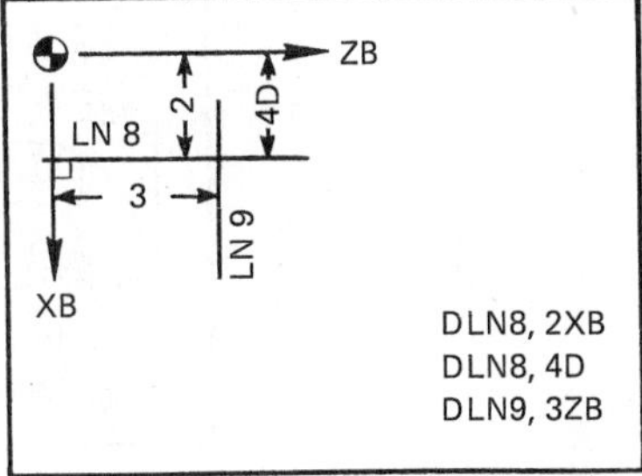
ZB
2
4D
LN 8
3
LN 9
XB
DLN8, 2XB
DLN8, 4D
DLN9, 3ZB

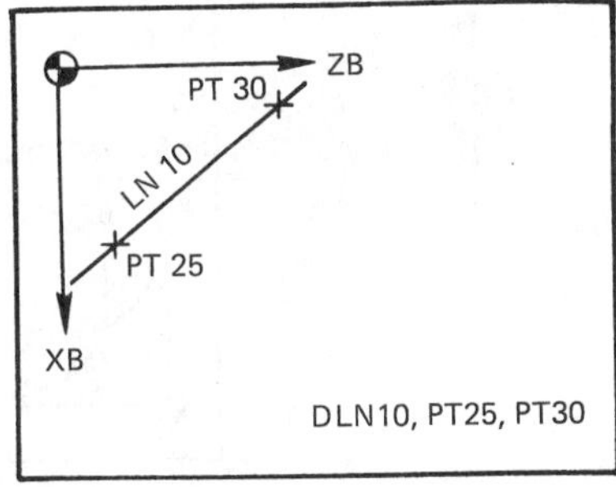
ZB
PT 30
LN 10
PT 25
XB
DLN10, PT25, PT30

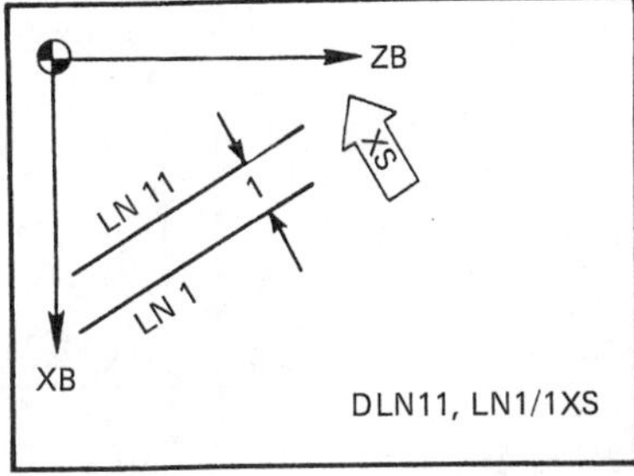
ZB
XS
LN 11
1
LN 1
XB
DLN11, LN1/1XS

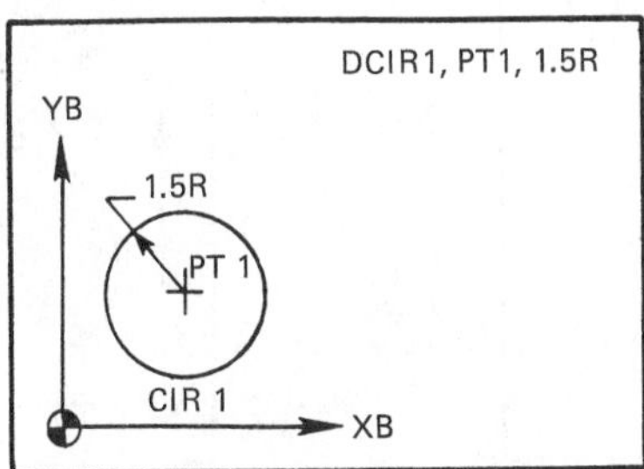
DCIR1, PT1, 1.5R
YB
1.5R
PT 1
CIR 1
XB

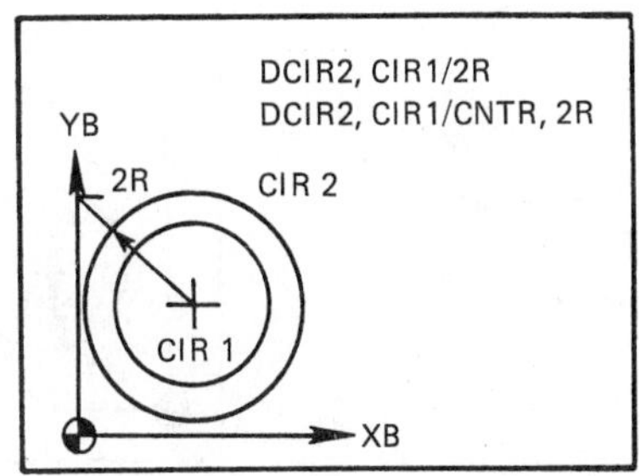
DCIR2, CIR1/2R
DCIR2, CIR1/CNTR, 2R
YB
2R
CIR 2
CIR 1
XB

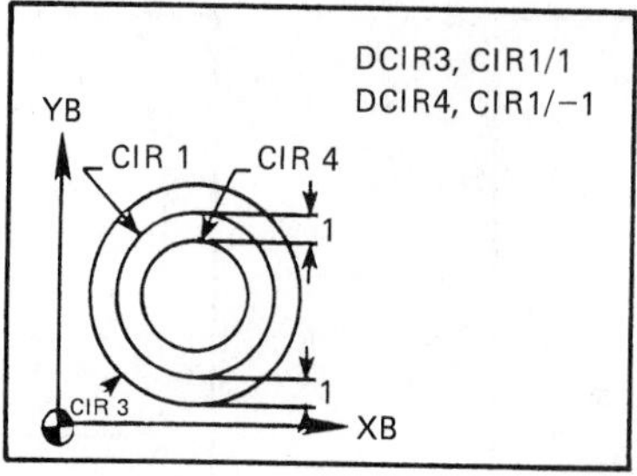
DCIR3, CIR1/1
DCIR4, CIR1/−1
YB
CIR 1
CIR 4
1
1
CIR 3
XB

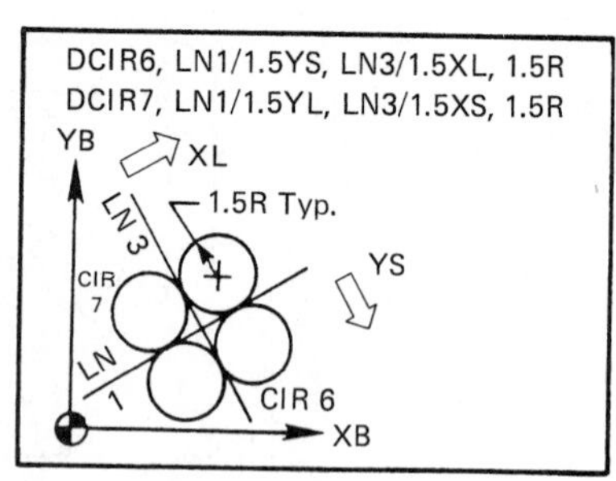
DCIR6, LN1/1.5YS, LN3/1.5XL, 1.5R
DCIR7, LN1/1.5YL, LN3/1.5XS, 1.5R
YB
XL
LN 3
1.5R Typ.
YS
CIR 7
LN 1
CIR 6
XB

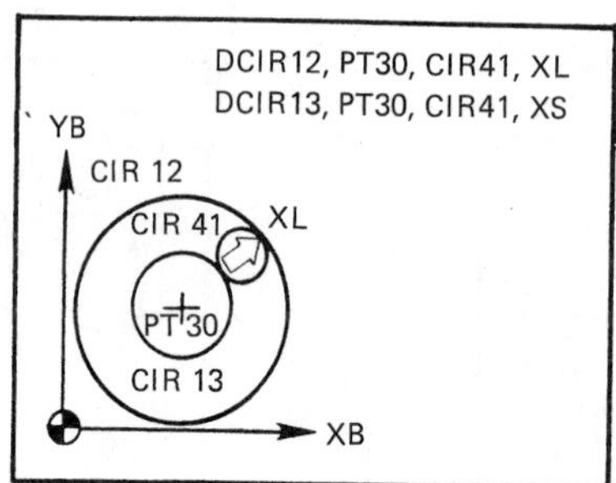
DCIR12, PT30, CIR41, XL
DCIR13, PT30, CIR41, XS
YB
CIR 12
CIR 41
XL
PT 30
CIR 13
XB

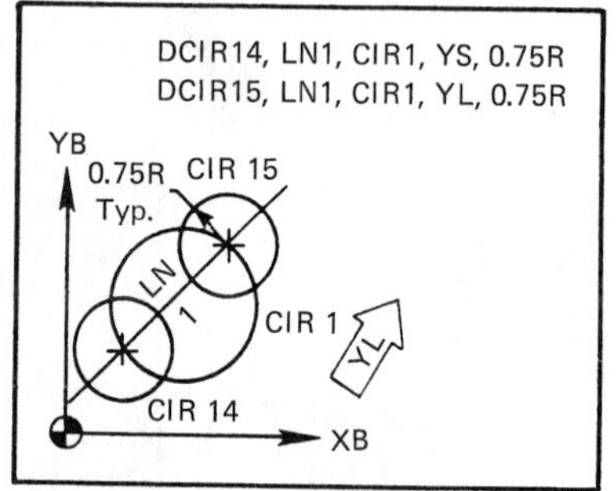
DCIR14, LN1, CIR1, YS, 0.75R
DCIR15, LN1, CIR1, YL, 0.75R
YB
0.75R Typ.
CIR 15
LN 1
CIR 1
YL
CIR 14
XB

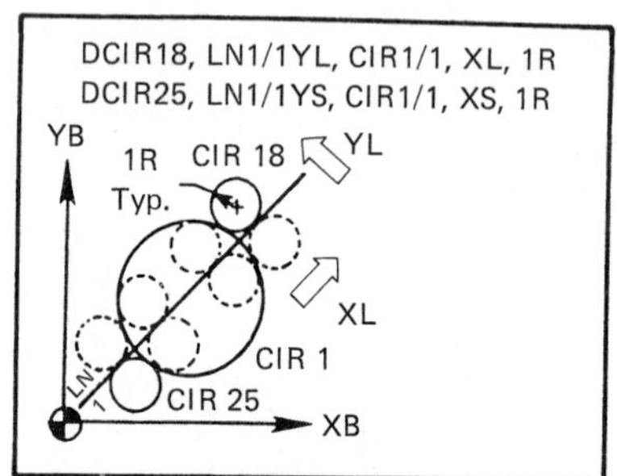
DCIR18, LN1/1YL, CIR1/1, XL, 1R
DCIR25, LN1/1YS, CIR1/1, XS, 1R
YB
1R Typ.
CIR 18
YL
XL
CIR 1
LN 1
CIR 25
XB

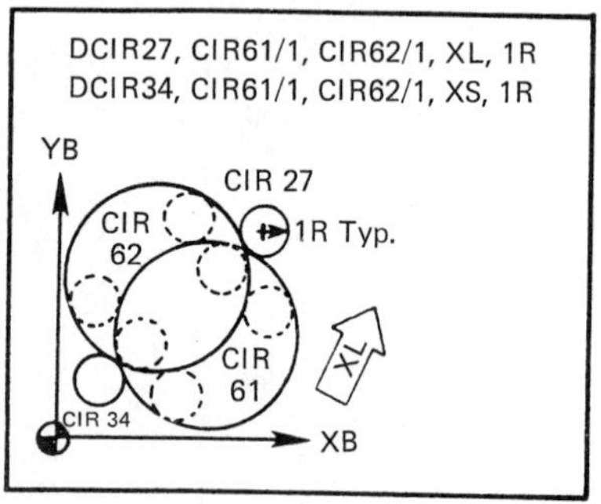
DCIR27, CIR61/1, CIR62/1, XL, 1R
DCIR34, CIR61/1, CIR62/1, XS, 1R
YB
CIR 27
CIR 62
1R Typ.
XL
CIR 61
CIR 34
XB

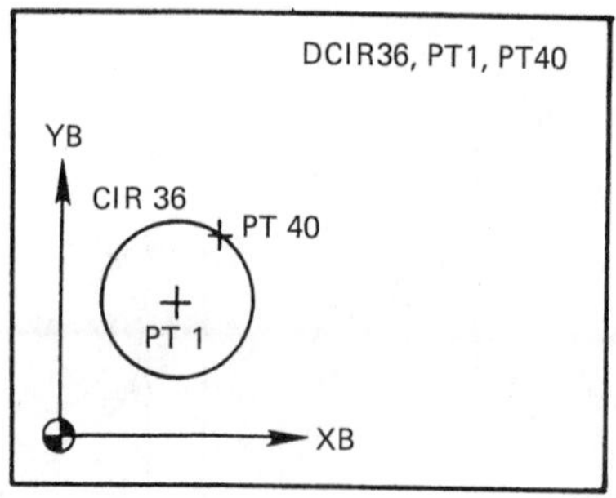
DCIR36, PT1, PT40
YB
CIR 36
PT 40
PT 1
XB

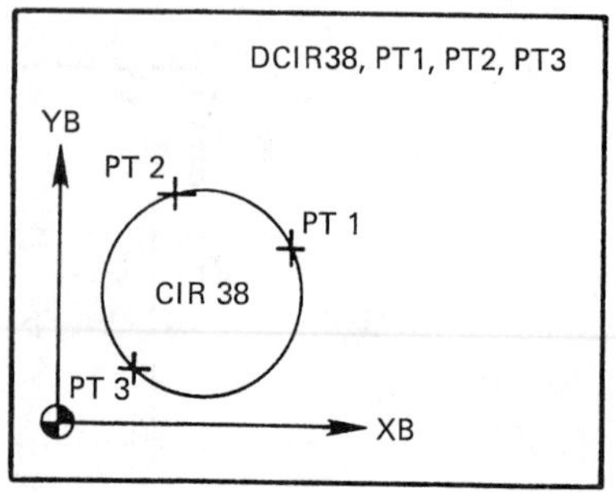
DCIR38, PT1, PT2, PT3
YB
PT 2
PT 1
CIR 38
PT 3
XB

Chapter 15

Industrial Robots

INTRODUCTION

This chapter will focus on industrial robots. It is aimed at providing the manufacturing engineer with an introduction to, yet complete picture of, this emerging kind of automation. Readers interested in further details are advised to consult any of the specialized books listed in the references at the end of this book.

Let us begin by defining the word *robot.* In fact, there are many definitions; some are general, whereas others include a variety of devices, some of which can hardly be called robots. In the early days of robots, the Japanese defined an industrial robot as an all-purpose machine, equipped with a memory and an appropriate mechanism to perform motions automatically, thus replacing human labor. Later, the Japanese Industrial Robots Association (JIRA), Tokyo, gave a definition of industrial robots that includes the following six categories:

1. *Manual manipulator,* a manipulator that is worked by an operator.
2. *Fixed-sequence robot,* a manipulator that repetitively performs successive steps of a given operation according to a predetermined sequence, condition, and position and whose set information *cannot* easily be changed.
3. *Variable-sequence robot,* a manipulator that repetitively performs successive steps of a given operation according to a predetermined sequence, condition, and position and whose set information *can* easily be changed.
4. *Playback robot,* a manipulator that can produce, from memory, operations originally executed under human control. (A human operator initially oper-

ates the robot in order to input instructions. All the information relevant to the operations—sequence, conditions, and positions—is put into memory. When needed, this information is recalled, or played back, and the operations are repetitively executed automatically from memory.)

5. *NC robot,* a manipulator that can perform a given task according to the sequence, conditions, and positions commanded via numerical data. (The software used for these robots includes punched tapes, cards, and digital switches. This robot has the same control mode as an NC machine.)
6. *Intelligent robot,* a manipulator with sensory perception (visual and/or tactile) that can detect changes by itself in the work environment or work condition and use its own decision-making facility to proceed with its operation accordingly.

A narrower definition is given by the Robot Institute of America (RI), Dearborn, Michigan, which defines a robot as "a reprogrammable, multifunctional manipulator designed to move materials, parts, tools, or specialized devices through variable programmed motions for the performance of a variety of tasks." The RI definition of robots excludes the first and second categories of the JIRA definition. This is, in fact, in agreement with recent trends that require robots to be easily programmed. By excluding manual manipulators and fixed-sequence robots, it is easy to see that robots are actually a kind of NC machine. In fact, the control units of NC machine tools and robots are virtually alike, and the servomotor and drive circuitry in each are exactly the same. We can, therefore, always refer to what we know about NC machines when discussing robots.

15.1 REASONS FOR USING ROBOTS

Robots are used in industry for different reasons, depending upon the kind of industry and its environment, and it is believed that robots will form the nucleus around which the factory of the future will be established.

To Work in Hostile Environments

Historically, robots were developed to carry out operations in hostile environments that create danger to human operators (e.g., when handling radioactive or toxic materials or when picking and placing hot components) or that are not appropriate for operators according to the Occupational Safety and Health Administration (OSHA).

To Perform Dull and Monotonous Jobs

At the beginning of this century, the technology of mass production became a reality, thanks to Henry Ford. Assembly lines resulted in an appreciable reduction in the cost of automobiles. Nevertheless, there were also some adverse social and psychological drawbacks. The workers in those mass-production plants were often assigned very simple, repetitive jobs for prolonged periods of time, such as tightening the same shape and size of nut or picking and placing the same components. Although the pay was good, many workers became alcoholics or suffered nervous breakdowns. A robot seemed to be the most suitable candidate to carry out such dull and monotonous jobs.

To Lift and Move Heavy Objects

Whenever heavy parts beyond the capability of a human must be lifted and moved, the "steel-collar" worker gets the job. In fact, modern robots can easily handle objects heavier than half a ton.

To Work During Unfavorable Hours

Holidays, second and third shifts, and the like are considered by many workers to be the least favorable hours for work. Obviously, this is not the case with robots, where all times and hours are the same.

To Provide Repeatability and Consistency

Like NC machine tools, robots possess excellent repeatability, which cannot be achieved by a human. Repeatability on the order of 0.002 inch (0.05 mm) is quite common for the most recent models of robots and ensures consistency of product dimensions and quality.

15.2 METHODS FOR CLASSIFYING ROBOTS

There are a number of methods for classifying robots: by the control system used, by their geometry and coordinate system, by their degrees of freedom (i.e., controlled axes), and by the method of programming. Two robots having different control systems can still have the same coordinate system and/or degrees of freedom, but they would not be able to perform similar tasks. It must be understood, therefore, that to classify robots means to group them based on similar characteristics that facilitate our choosing specific robots to meet certain demands for required applications.

Classification by Control System

Based on the control-system method of classification, industrial robots can be grouped into only two categories: *nonservo-controlled* and *servo-controlled* robots. Following is a description of the features and characteristics of each type.

Nonservo-controlled robots. Nonservo-controlled robots are the simplest type of robot, and the members of this group fall under the first and second categories of the

JIRA definition of robots. These robots are inexpensive, are simple to set up and program, have reasonable precision, and provide a "Start simple" approach to robotics.

Like other robots, a nonservo-controlled robot is composed of a mechanical unit (manipulator) and a control unit (controller). The mechanical unit has moving parts that perform the work, and the control unit directs the mechanical unit in carrying out its task. Nonservo-controlled robots are usually air-powered and operate by employing transporters along a Cartesian coordinate system on a point-to-point basis. The different points throughout a cycle are fixed through the use of mechanical or electromechanical stops, but the actual path between those points cannot be defined or controlled. Angular motion around one or more of the coordinate axes is provided by rotators and resembles wrist motion (i.e., roll, pitch, and yaw). The control of motion along a single axis is based on only one command, which is either on or off, because the stops determine the starting and ending points of the stroke. The command (on or off) is executed through the use of a four-way solenoid-actuated valve that controls the flow of pressurized air through the two inlets of the double-acting air cylinder of the transporter, as illustrated in Figure 15.1a and b.

Obviously, the control of a multimotion robot is not as easy as the preceding case because several on-off commands must be given in a certain order to perform a required task. Accordingly, programming nonservo-controlled multimotion robots involves preparing a table that includes the sequence as well as the duration of the on-off commands for the different axes of motion in order to produce the required cycle of events. The program plan is then fed into the control unit, which usually involves a solid-state programmable controller for nonservo-controlled robots. When the robot is operated, on-off commands are conveyed to the different solenoid-actuated valves according to the program plan. Simple cycles, like pick–place or pick–dip–place, can be easily programmed. Complex motion patterns can also be programmed, provided that the controller used has the appropriate programming capability.

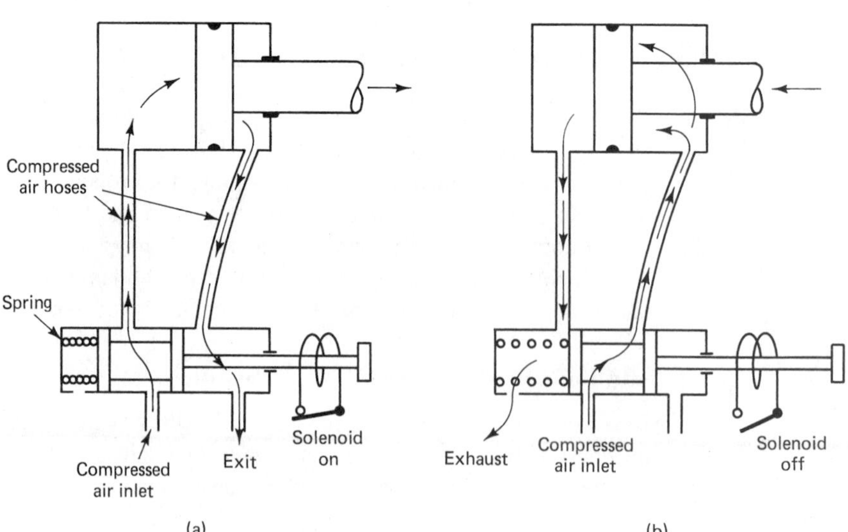

FIGURE 15.1 Working principles of a solenoid-actuated valve: (a) stretching of arm; (b) retraction of arm

Servo-controlled robots. Servo-controlled robots are basically NC or CNC machines and thus can be broken into two groups: *point-to-point* robots (types similar to numerical control positioning machines) and *continuous-path* robots (types similar to numerical control contouring machines). The path of the gripper of a point-to-point servo-controlled robot cannot be defined or controlled; nevertheless, the robot can be programmed to make the gripper stop at any one of an infinite number of points enclosed within the working envelope of the robot. This is not the case with a point-to-point nonservo-controlled robot, where positive stops must be used to define the desired points that are not included in the program. It should, therefore, be evident that the control unit of a servo-controlled robot and the MCU of an NC machine tool are virtually alike. Also, as is the case with NC machines, this type of robot is provided with a positional feedback system to detect any deviation of the achieved position from the programmed one. A hydraulic power source is usually used to drive the manipulator of the point-to-point servo-controlled robot because of the economy of this type of power source and the high power density that it provides. Consequently, this type of robot is characterized by its high load capacity and large working range. Moreover, its programming is fairly easy because it involves a word-address format similar to NC word-address programming.

Continuous-path robots have a wide range of capabilities, including driving the manipulator along a controlled path. Consequently, a continuous-path robot can be programmed so that its manipulator will avoid any obstacles. As is always the case with numerical control contouring systems, the control unit involves a microcomputer with software capable of providing linear and circular interpolation. In addition, the feedback system continually monitors not only position but also velocity after each short increment of time. Therefore, greater precision and repeatability are the main characterizing features of this type of robot, which generally tends to be smaller in physical size than the point-to-point type of robot and has lower load capacity as well. Recently, there has been a trend to use electric motors and drives in high-precision continuous-path robots to ensure accuracy and smoothness of motion and to completely eliminate the possibility of oil leakage, which can cause problems in some industries.

Classification by Geometry and Coordinate System

All commercially available industrial robots can be classified into four groups according to the coordinate system used in their operation. These include *Cartesian-, cylindrical-,* and *polar-coordinate* robots as well as *articulated-arm* robots.

Cartesian-coordinate robots. In Cartesian-coordinate robots, motion takes place along three linear perpendicular axes. That is, the robot's arm can move up or down, in or out, and in a motion perpendicular to the plane created by the previous two motions. The *work envelope* of a Cartesian-coordinate robot (i.e., the volume in space within which its grippers can move) is a parallelogram whose sides are the maximum strokes along the *X, Y,* and *Z* axes. Figure 15.2 shows a nonservo-controlled robot

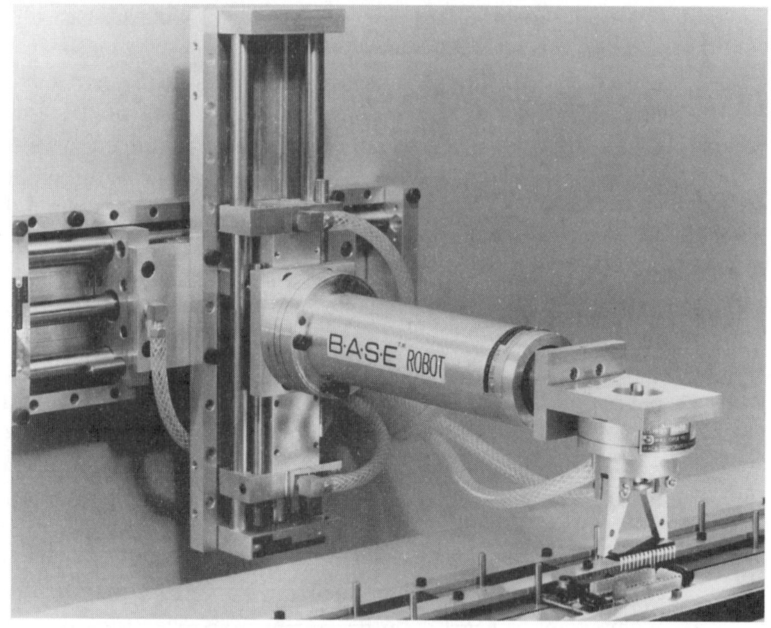

FIGURE 15.2
The mechanical unit of a nonservo-controlled robot having Cartesian coordinates *(Courtesy of MACK Corporation, Flagstaff, Arizona)*

FIGURE 15.3
A cylindrical-coordinate robot *(Courtesy of Prab Robots, Kalamazoo, Michigan)*

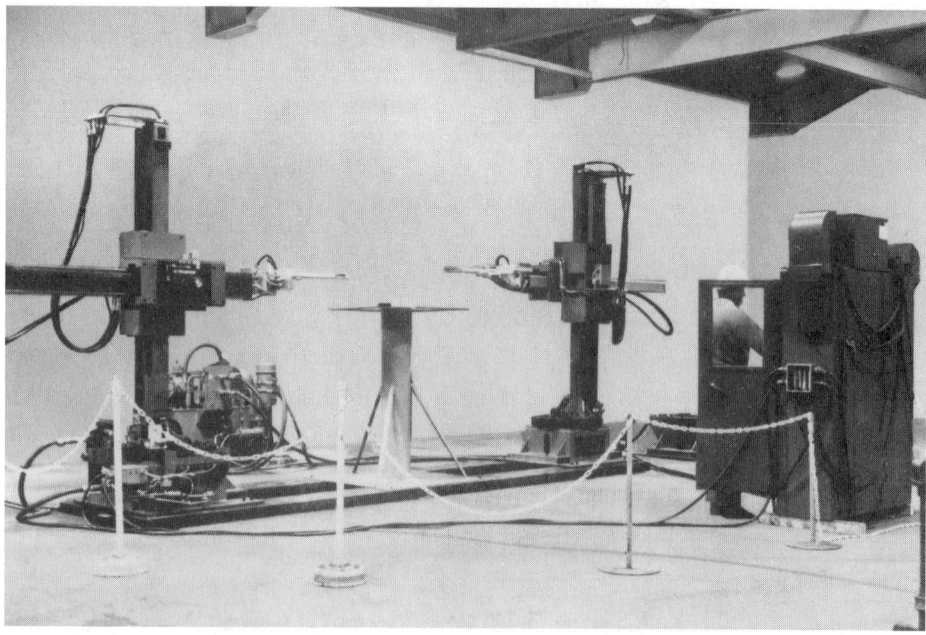

based on this coordinate system. Cartesian-coordinate robots that are servo-controlled have also been developed for precision assembly work.

Cylindrical-coordinate robots. In cylindrical-coordinate robots, the arm motion is similar to that of Cartesian-coordinate robots. The motion of the base, however, is rotary. That is, the robot's arm can move up or down and in or out, and it can also swing around the vertical axis. Figure 15.3 shows a cylindrical-coordinate robot, and Figure 15.4 illustrates its motion and working envelope. Notice that the working envelope is a sector of a cylinder; the external diameter is the maximum reach along the horizontal axis, and the internal diameter is the distance between the grippers and

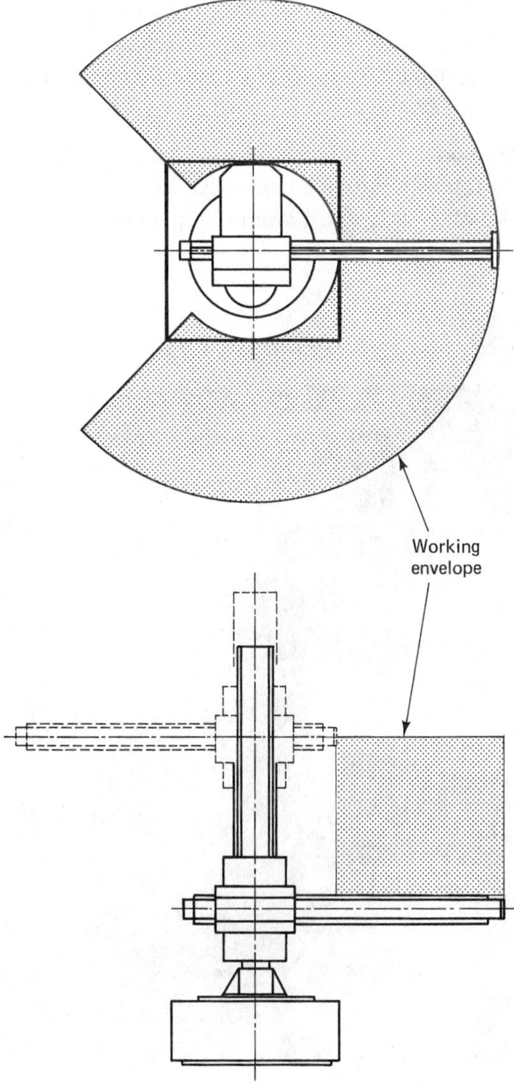

FIGURE 15.4
The motion and working envelope of a cylindrical-coordinate robot *(Courtesy of Prab Robots, Kalamazoo, Michigan)*

the vertical axis when the arm is in its retracted position. Because of their high structural rigidity, cylindrical-coordinate robots possess high load-carrying capacities and are, therefore, used in material handling of heavy objects.

Polar-coordinate robots. In polar-coordinate robots, which are usually referred to as *polar-spherical-coordinate* or *spherical-geometry* robots, there are three axes of motion. Two of them are rotary, and the third is linear. That is, the body of a polar-spherical-coordinate robot can pivot in either a vertical or a horizontal plane or both (creating a polar vector in space), and its arm can be extended out or retracted in. Figure 15.5 shows a polar-coordinate robot, and Figure 15.6 illustrates its motion and working envelope. Notice that the working envelope is a sector of the volume between two concentric spheres; the radius of the external sphere is the maximum reach along the linear axis, and the radius of the internal sphere is the shortest distance between the end of the arm (i.e., when it is retracted) and the vertical axis. The volume between the two spheres is truncated by the four extreme limits of the two rotary axes. Polar-spherical-coordinate robots are used in machine-tool loading and unloading.

Articulated-arm robots. In articulated-arm robots, a series of linkages and rotary motions enable the robots to provide humanlike motion. It is, therefore, quite common to refer to the parts of an articulated-arm robot (and sometimes to their motions) by their

FIGURE 15.5
A polar-coordinate robot *(Courtesy of Prab Robots, Kalamazoo, Michigan)*

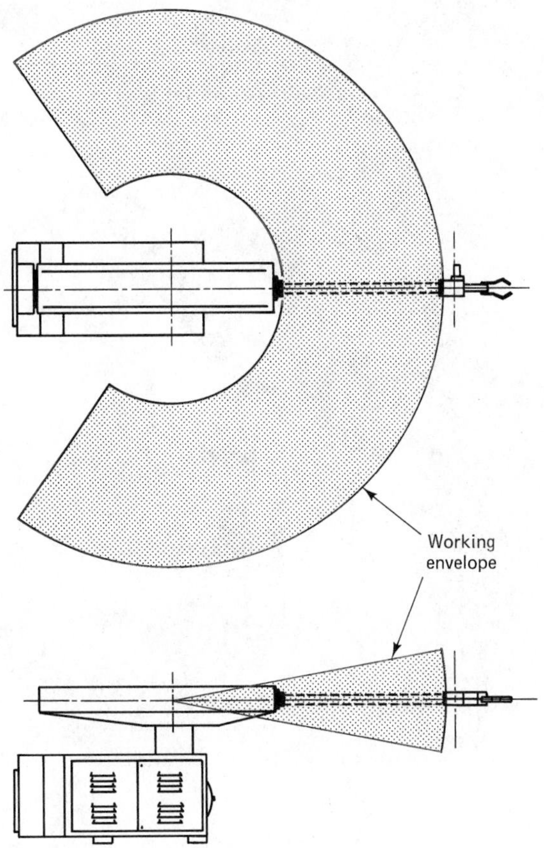

FIGURE 15.6
The motion and working envelope of a polar-coordinate robot *(Courtesy of Prab Robots, Kalamazoo, Michigan)*

human-body counterparts (e.g., shoulder and elbow). Figure 15.7 shows an articulated-arm robot, and Figures 15.8 and 15.9 illustrate the working envelope for two different models. Notice that the working envelope of an articulated-arm robot depends upon its design and model and that the geometry, or configuration, of this type of robot enhances its *reachability*. Therefore, articulated-arm robots are usually built to be high-precision, continuous-path manipulators and are used in applications such as MIG welding. Moreover, reachability can be further enhanced by mounting the robot overhead (upside down), as shown in Figure 15.10.

Classification by Degrees of Freedom

The term *degrees of freedom* refers to the number of linear and/or rotary axes of motion. However, the relative orientation of these axes must be taken into account when defining the degrees of freedom of a robot. Motions along two parallel linear axes cannot be considered as two degrees of freedom because one of these two axes of motion is redundant. Similarly, two rotary motions around the same centerline (or parallel centerlines) are also not counted as two degrees of freedom. Therefore, the maximum

FIGURE 15.7 An articulated-arm robot *(Courtesy of Prab Robots, Kalamazoo, Michigan)*

possible number of degrees of freedom is six (three mutually orthogonal linear axes and three rotary motions, each around one of these linear axes) but not all robots have six degrees of freedom. They generally have either three, four, five, or six degrees of freedom; the higher the number, the more versatile the robot is in emulating a human operator. A robot with six degrees of freedom is, therefore, considered to be totally flexible, with three arm and body motions and three wristlike rotary motions. Moreover, robots are sometimes mounted on a rail, or *gantry,* to enlarge the working envelope by moving the robot from one position to another. This is usually not a true degree of freedom, although it is very beneficial. Figure 15.11 shows a gantry robot.

Classification by Method of Programming

We have previously discussed the programming method for nonservo-controlled industrial robots in which a programmable controller is used to give a sequence and duration of on-off commands for each axis of motion and where a positive stop determines the final point on each axis. As also discussed, the controller of a servo-controlled robot is very similar to that of an NC machine tool. Therefore, programming methods in both cases are quite similar. Following is a brief description of the three different methods of programming a servo-controlled industrial robot.

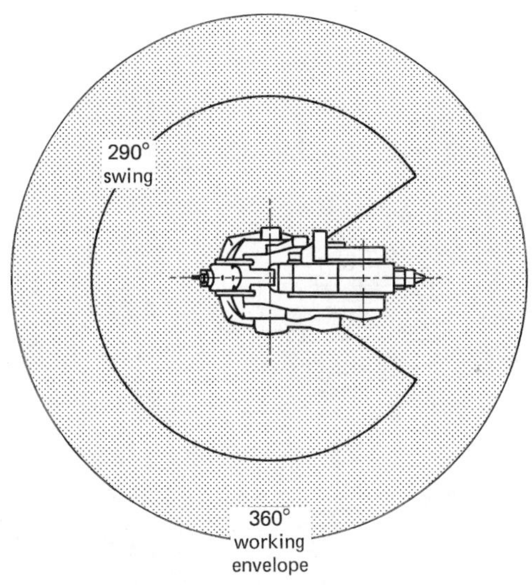

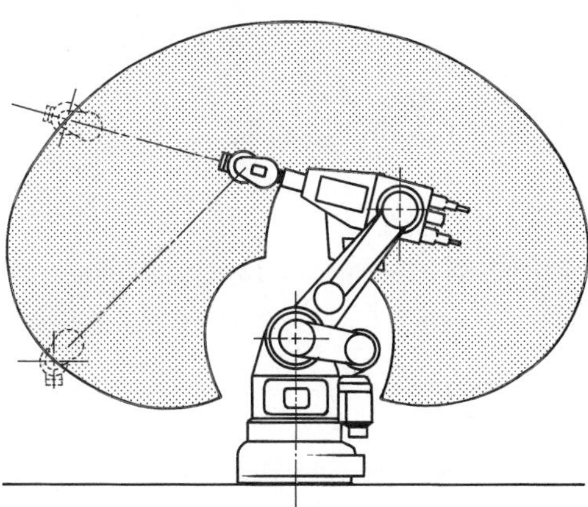

FIGURE 15.8
The working envelope for one model of articulated-arm robot *(Courtesy of Prab Robots, Kalamazoo, Michigan)*

Walk-through programming. In the *walk-through programming* method, the operator manually drives the robot's manipulator by a hand-held pendant through a series of desired positions and "teaches" the robot control unit by pressing a "teach" button after each position is obtained. An auxiliary function command to control the end effectors (grippers) is also programmed corresponding to each position. A playback of what has been stored in the memory of the control unit ensures the desired cycle of events. Walk-through programming is used with most robots in industry today.

Lead-through programming. In the *lead-through programming* method, used with continuous-path advanced robots, the operator physically drives the manipulator through the desired path and sequence of moves, which are stored in the magnetic

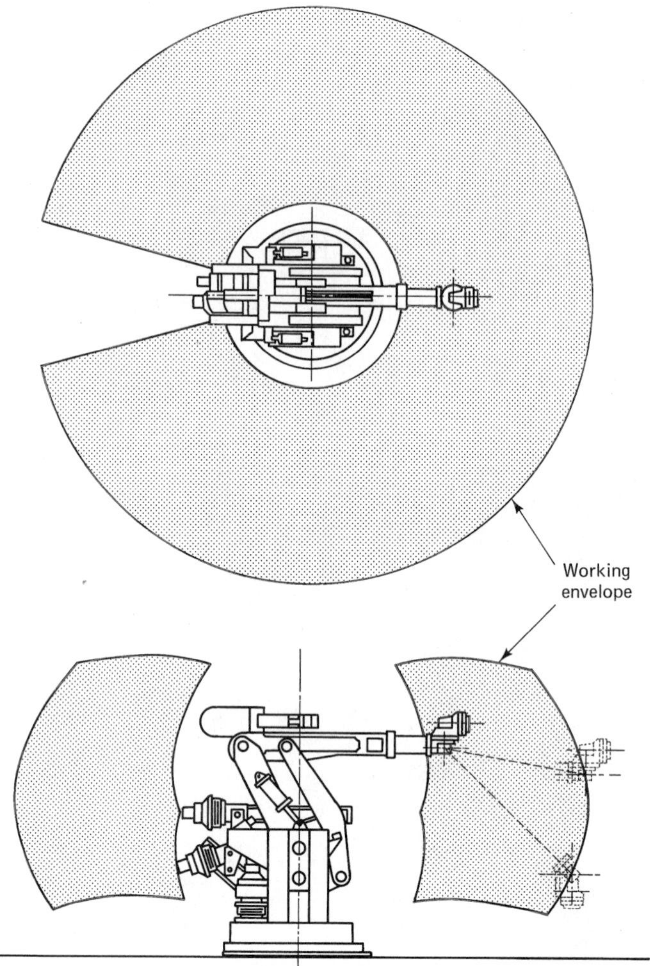

FIGURE 15.9 The working envelope for another model of articulated-arm robot *(Courtesy of Prab Robots, Kalamazoo, Michigan)*

memory of the control unit. When the robot is operated, exactly the same path and sequence of moves can be reproduced (this is similar to digitizing in CNC machine tools). Lead-through programming is very useful for robots employed in spray painting.

Off-line programming. The preceding two programming methods are not suitable for automating operations where the presence of human operators in the work area is prohibited or at least not recommended, such as when handling radioactive billets within nuclear reactors or manufacturing semiconductors in a clean, uncontaminated room. The concept needed for handling such problems is *telerobotics,* or *off-line programming.*

In the off-line programming method, the cycle of events and the tasks to be performed by a robot are accurately defined with the aid of a separate computer without

FIGURE 15.10
A robot mounted overhead *(Courtesy of Prab Robots, Kalamazoo, Michigan)*

involving either the mechanical or the control units of the robot in the programming process (this is similar to computer-assisted part programming and DNC). Detailed information about the path is required. Because a CAD (or CAM) database is usually available, it can be utilized to "teach" the robot. Other databases defining the relationship between the robot and its surroundings in the manufacturing environment (e.g., location, geometry, and operational characteristics of the machine tools) are used. The manipulator's path must be simulated on a CRT and checked, and kinematic analysis must also be carried out before the program is used to operate the robot. This requires powerful and sophisticated software such as that commercially available from GCA Corporation, Cincinnati Milacron, McDonnell Douglas Automation, and others. Although as yet no standardized telerobotics programming language exists, there are signs that such a language will be developed in the future.

Off-line programming is a necessity for the implementation of a computer-integrated manufacturing (CIM) system that includes robots. This emerging role of industrial robots makes full use of their potential and is a result of the fact that robots have very little significance as stand-alone units. Other advantages for off-line programming include increasing the productivity of a robot by not wasting its utilization

FIGURE 15.11
A gantry robot *(Courtesy of Prab Robots, Kalamazoo, Michigan)*

time in programming and refining complicated programs during simulations (before actual operation).

15.3 COMPONENTS OF A ROBOT

Several basic components must be included in any servo-controlled robot of the current generation. They are the *main body* (or base), the *manipulator,* the *feedback devices,* the *control unit,* and the *power supply.* The main body carries the manipulator and the actuators that generate the motions (a separate one for each axis of motion). The manipulator, the obedient slave that performs the desired task, carries the end effectors (or grippers) that hold the workpiece or tool. Positional feedback devices are required for detection and correction of any discrepancy between the desired position (given in the command) and the actual achieved position. As previously mentioned, the brain of the robot is the control unit, which directs the manipulator along the desired sequence of positions. Finally, a power supply is needed for the actuators, and robots have evolved with sensors, computers, and advanced controls. Based on the signals received from the sensors, *adaptive control* can compensate for the effects of unknown system parameters (such as friction in the bearing of the joints) and disturbances in the system (such as changing payloads or misorientation of the workpiece), thus enhancing the performance of the robot. As a result, a robot equipped with adaptive control can adjust to its surroundings in real time. A level beyond adaptive control is *artificial intelligence,* which is discussed in detail in Chapter 16. For now, let us examine end effectors and sensors in some detail before we discuss the various applications of robots.

15.4 END EFFECTORS

End effectors are devices mounted on the end of the manipulator of a robot that pick up, grab, or handle workpieces (or tools) and that, therefore, must be selected to suit the attributes (shape and properties) of the objects to be processed. Consequently, end effectors are, in most cases, custom-made (i.e., tailored to match the nature of the industrial operation as well as the handling requirements). When handling objects for transfer or assembly operations, one of the following types of end effectors is used, depending upon the geometrical shape of the object and its properties (e.g., whether it is fragile or hot): grippers or permanent-magnet, electromagnetic, or vacuum-cup end effectors. If the robot is used for inspection and quality control, the end effector may take the form of a stylus connected to a measuring instrument. End effectors can also be small power tools (like grinding burrs or nut drivers), attachments for welding operations, or guns for spray painting. The discussion here is limited to end effectors for material handling because processing tools and attachments have been previously discussed, each under the appropriate chapter.

Grippers

Grippers are perhaps more commonly used than other end effectors. Generally, the types of grippers include the two-finger, the three-finger, and the soft blank-finger grippers.

Two-finger grippers. Two-finger grippers are suitable for holding parts with simple geometrical shapes (prismatic). Depending upon the design, they can be used either for external or internal gripping. The two-finger type of external gripper resembles the human thumb and the index finger. For this reason, this type of gripper can be used for picking up a tiny object or one that is in a closely packed group. A modified version has two parallel fingers with replaceable inserts that can be machined to suit special shapes. Figure 15.12a and b illustrates the conventional two-finger (angular) gripper and the parallel two-finger type (its applications involve external and internal gripping of special shapes), respectively.

Three-finger grippers. Three-finger grippers are specially suited for grabbing bodies of revolution because they simulate the motions of the thumb, index finger, and middle finger. Figure 15.13 shows a nonservo-controlled robot with this type of gripper.

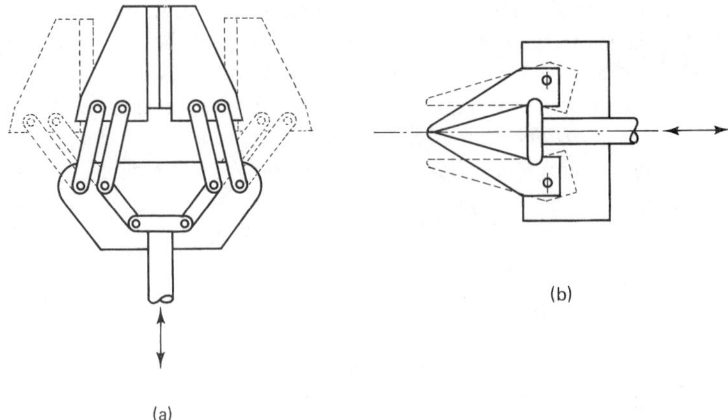

FIGURE 15.12
Two-finger grippers: (a) conventional type; (b) parallel type

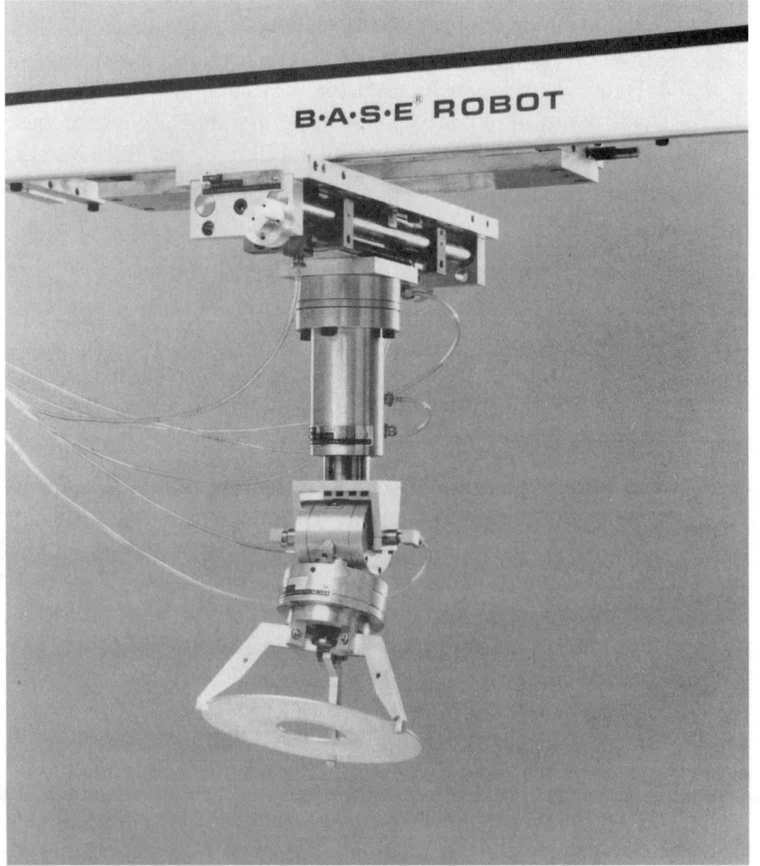

FIGURE 15.13
A nonservo-controlled robot with a three-finger gripper *(Courtesy of MACK Corporation, Flagstaff, Arizona)*

Permanent-Magnet End Effectors

A *permanent-magnet end effector* involves a permanent magnet that travels through and is guided by an aluminum cylinder, as shown in Figure 15.14. When the actuator pushes the magnet to the front end of the aluminum cylinder, it can pick up a ferrous object. When the magnet is withdrawn into the aluminum cylinder, the magnetic field is broken and the object is then released. Although the permanent-magnet end effector is suitable for handling only ferrous objects, it has the advantages of rapid cycle times, the ability to handle parts with irregular shapes, and the ability to handle several parts simultaneously (provided that they are mounted on a special adapter).

Electromagnetic End Effectors

In addition to having the advantages of the permanent-magnet end effector, the *electromagnetic end effector* is compact and easy to operate. Consequently, more than one electromagnetic end effector can be mounted at the end of the robot's arm, thus expanding its capabilities. Moreover, electromagnetic end effectors can still function properly even with some unpredicted disturbance in the system (such as a slight mislocation of the object or changes in its shape, size, or weight). Parts with irregular surfaces, such as rough ferrous castings or contoured parts, can be handled by this type of end effector.

Vacuum-Cup End Effectors

The *vacuum-cup end effector* involves a polymeric cup; when it contacts a flat surface of an object, vacuum is applied within the cup to keep the object attached to it. There must be some kind of control to create and eliminate vacuum exactly at the desired time. This is usually achieved through the use of solenoid valves. There is also a need for a vacuum tank. Vacuum-cup end effectors are most suitable for handling fragile objects, such as CRTs, dishes, light bulbs, and large plastic or sheet metal panels. They are also widely used for picking up and transferring flat stock.

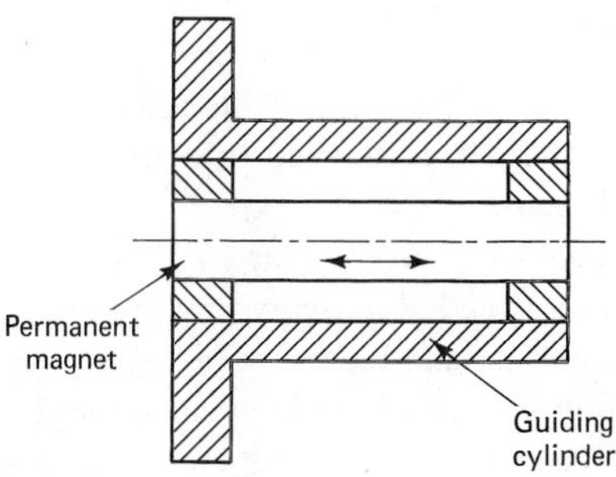

FIGURE 15.14
A permanent-magnet end effector

15.5 SENSORS

As mentioned, an intelligent robot is one that interacts with the industrial environment surrounding it. In order to do this, it must have *artificial* eyes, ears, or fingers that detect, or *sense,* the presence of parts and monitor the handling and processing parameters of the various industrial applications. The electronic devices that provide these artificial senses are called *sensors.* Following are examples of the sensors used.

Tactile Sensors

Tactile sensors are devices capable of defining the features (and/or orientation) of an object through contact combined with continuous measurement of the gripping force. This type of sensor usually involves piezoresistive elements whose electrical resistance increases with increasing contact pressure (or force). Matrix-type compliant tactile sensors have been developed to provide information not only about the force but also about the geometric shape, center of gravity, and area of the object pressed against the sensor. They are basically made of a conductive elastomer with electronic circuitry that provides an electrical signal proportional to the deflection of the elastomer and that conditions and processes that signal to provide the necessary information. Because tactile (and force) sensors are employed to provide feedback when gripping objects, they are important in material-handling and surface-mapping applications.

Distance Sensors

Distance sensors are employed in measuring the distance between the end effector and an object. There are a variety of distance sensors based on different principles. A capacitive sensor, for example, employs the change of capacitance of an electric circuit as a result of the change in the position of an electrode with respect to the metallic object. Ultrasonic waves are also used to detect the location and size of objects when handling workpieces with unknown height. Optical encoders are sometimes used. They convert the position of an object into a digital signal by employing photoconductive elements whose resistance decreases with increasing illumination caused by light reflected from the object.

Arc Welding Sensors

Arc welding sensors detect any discrepancy in the seam by the change in the welding current when weaving. This feedback is used to compensate for that discrepancy in order to obtain the desired seam.

A new system of noncontact seam tracking of welds has been developed by Material Data AB in Sweden. It is called *magnetic wave control* (MWC) seam tracking and involves a sensor transmitting and receiving high-frequency magnetic waves. Any change in the proper location of the joint to be welded yields an unbalanced magnetic field. When the field is processed, it results in a control signal addressed to the servo, which moves the torch. This type of sensor has the advantage of being independent of the welding process parameters (i.e., performance is not affected by any changes in the parameters). It is also particularly suitable for the MIG, TIG, plasma arc welding (PAW), and submerged arc welding (SAW) processes.

Visual Sensors

Visual sensing in industrial robots can be divided into three main categories: *laser gauging, structured light,* and *pattern recognition.* Laser gauging is based on laser interferometry; any slight change in the position of an object results in a change in the pattern of the interference fringes of a laser beam reflected from that object. The method of using structured light employs a two-camera vision system that resembles the binocular parallax of human vision. Any object can, therefore, be reduced to a pair of binocular images or outlines. A computer algorithm is then employed to match the edges and create a picture of the object (in the same way as in descriptive geometry). The method of using pattern recognition is based on using a connectivity analysis program that breaks a binary image into its connected components (units) and then builds a description of each unit when processing the image. Because visual sensors are used to identify and measure the positions and posture of workpieces, they are a key element when using robots for the assembly of parts.

15.6 INDUSTRIAL APPLICATIONS OF ROBOTS

We now know that there are a wide variety of robots having different geometries, constructions, and methods of operation. The task of selecting a robot for a particular application is not, therefore, as easy as it might seem. It is obvious that an advanced robot with sophisticated sensors and supporting software may do any job required; however, we must not forget that its initial capital cost as well as its operating and maintenance costs are far higher than those of a simpler robot that may be capable of adequately performing the same job. In other words, the decision of selecting a robot for a certain application should be based not only on the features and characteristics of a robot but also on economic considerations. We would all agree that using a controlled-path robot for pick-and-place applications is overkill. In this respect, it is worth mentioning that 70 percent of Japan's 50,000 robots (in 1988) were pick-and-place types, and they were meeting the required demands. Following is a survey of some of the important applications of robots.

Machine Loading and Unloading

Machine loading and unloading represents the most promising field of application for industrial robots. It is where their characterizing features are efficiently and fully utilized. Robots have the advantages of accuracy, repeatability, and the ability to be synchronized with the surrounding machine tools. Consequently, they are the unchallengeable candidates for workpiece and tool loading and unloading in manufacturing cells. A single robot can efficiently load and unload two or more CNC machine tools in a properly planned flexible manufacturing cell (see Chapter 16). Industrial robots are also used for this purpose in flexible manufacturing systems, and we can anticipate that intelligent robots will play a key role in the automated factory of the future, where they will completely replace human operators in loading workpieces, setting up machine tools, and changing tools whenever required.

The current generation of industrial robots is also employed for loading and unloading a wide variety of machines in addition to the CNC machining centers. These include sheet-working presses, rubber-forming machines, magnetizers, testing equipment, mold presses, and the like. Generally, these robots have either cylindrical or articulated-arm geometries and can be controlled through four to six degrees of freedom.

Spot Welding

Excellent repeatability of weld location and consistency of weld quality make robots attractive for spot welding, and they are thus widely used by automobile manufacturers all over the world. Reachability and interchangeability are further advantages of robots that cause these manufacturers to favor them over the hard-tooled, dedicated spot-welding line that was previously used in the automotive industry. Generally, articulated-arm robots are recommended for this application.

Arc Welding

The arc welding environment is hazardous to the health of human operators. At the same time, the arc welding process requires highly skilled labor. It is, therefore, rather costly to produce high-quality welds when employing human operators. For this reason, arc welding applications of industrial robots are growing at a very rapid pace. Continuous-path robots of the articulated-arm type with six degrees of freedom are generally employed in arc welding. As can be seen in Figure 15.15, these robots have electric servomotors and actuators to provide smooth motion and eliminate the problems of oil leakage. They are also fitted with adaptive control as well as suitable arc welding sensors for seam tracking. It is clear that the use of industrial robots for arc welding not only has economic advantages (labor-cost saving) but also results in consistently predictable weld quality.

Material Handling

The material-handling tasks performed by robots include packing and unpacking, piling and unpiling, palletizing and depalletizing, plus transferring products to and from conveyors. Almost all types of robots are employed in material-handling applications, ranging from the simple polar-coordinate type to the articulated-arm robot with six degrees of freedom, depending upon the particular application required.

Material Processing

Robots are employed in flame cutting of steel boards and other mechanical components. They are also used to hold and guide the high-pressure water jet for cutting ceramics, composites, and the like. This type of robot is very similar to that used in arc welding applications (i.e., continuous-path type with electric servomotors and actua-

FIGURE 15.15
An arc welding robot *(Courtesy of Prab Robots, Kalamazoo, Michigan)*

tors). In addition to these applications, light machining tasks can be carried out effectively by robots. Examples include deburring, drilling, tapping, grinding, and polishing. Such applications are quite common when the workpiece is very heavy or when it is required to minimize material transfer.

Investment Casting

Repeatability, consistency, and accuracy are real advantages when employing robots in investment-casting applications. These advantages lead to an appreciable reduction in the number of rejects when making the molds. An example of a robot used in investment casting is shown in Figure 15.16.

Painting

Painting is an industrial process carried out in a hostile environment involving toxic fumes and the possibility of explosion. Therefore, it was one of the earliest areas in which industrial robots were applied. The velocity and path of the manipulator are important factors affecting the quality of the painted surface. Consequently, they must be adequately controlled in spray-painting operations. Smoothness of motion is also required, whereas accuracy is not as critical a factor. Only continuous-path robots are

FIGURE 15.16
A robot used in investment casting *(Courtesy of Prab Robots, Kalamazoo, Michigan)*

suitable for this kind of application. The most appropriate method of programming spray-painting robots is the lead-through method, where a skilled painter guides the gun to paint a sample part to create the desired database.

Assembly

The assembly of industrial components represents the future field of application for intelligent robots. Point-to-point servo-controlled robots having a repeatability of about 0.002 inch (0.05 mm) are generally adequate for parts assembly. Nevertheless, sensing devices are required to detect and thus compensate for any discrepancies. Current applications of assembly robots are simple and include insertion of rivets, pins, washers, and rubbers, insertion and tightening of nuts and bolts, insertion of parts in circuit boards, and other assembly work.

Review Questions

1. What is the functional difference between a point-to-point servo-controlled robot and a nonservo-controlled robot?
2. Compare the JIRA definition of *robot* with the RI definition.
3. What are the reasons for using robots? Can a human be a substitute for a robot in all these cases?
4. How can robots be classified?
5. What are the two main units that make up a nonservo-controlled robot?
6. What is the usual power source for nonservo-controlled robots?
7. How can nonservo-controlled robots be programmed? What are the commands? What is the form of a program plan?
8. What are the two main groups of servo-controlled robots?
9. What are the similarities between a servo-controlled robot and an NC machine tool?
10. Why is the point-to-point servo-controlled robot characterized by its high power density and economical performance?
11. How can a path of a robot's manipulator be controlled? What additional features should a robot of this kind possess?
12. How does the load capacity of a continuous-path robot compare with that of a point-to-point robot?
13. Why are electric motors and drives used with continuous-path robots?
14. List the four robot geometries (coordinate systems) and use sketches to illustrate the working envelope for each type.
15. What is meant by the *degrees of freedom* of a robot?
16. Can motions along two parallel linear axes be considered as two degrees of freedom?
17. What is the maximum possible number of degrees of freedom (from the scientific, not the commercial, point of view)?
18. What do we call a robot with six degrees of freedom?
19. How can we extend the mobility and flexibility of a robot?
20. Explain briefly the different methods for programming robots.
21. Which one of these methods renders itself a necessity for the implementation of CIM? What further advantages does that programming method have?
22. What are the main components of a robot? Explain briefly the function of each.
23. What is meant by *adaptive control* as applied to robots?
24. What are *end effectors?* Give examples.
25. What is the difference between end effectors and grippers?
26. For what are two-finger grippers suited?
27. What do two-finger grippers resemble?
28. For what are three-finger grippers specially suited?
29. What do three-finger grippers resemble?
30. What is the working principle behind permanent-magnet end effectors?
31. What are the limitations and advantages of permanent-magnet end effectors?
32. List some advantages of electromagnetic end effectors.
33. For what are vacuum-cup end effectors most suitable?

34. What are *sensors?* For what are they used?
35. What are *tactile sensors?* For what are they used?
36. Explain briefly some different working principles behind distance sensors.
37. Explain briefly the working principles of two types of arc welding sensors.
38. What are the three main types of visual sensors? In what field of application are they mostly used?
39. Do you recommend purchasing an advanced robot with sophisticated functions, no matter what the required application is? Why?
40. Give examples of some loading and unloading applications of industrial robots. What role are these robots going to play in future technology?
41. What makes robots attractive for spot welding applications?
42. Why are robots considered to be better than human operators for arc welding applications?
43. What type of robots do you recommend for arc welding?
44. List some other applications for industrial robots and discuss each briefly.

Chapter 16

Automated Manufacturing Systems

INTRODUCTION

Since the Industrial Revolution, U.S. industry has been undergoing a process of continuous development. During the nineteenth century, new machines expanded the productivity of workers; with the beginning of the twentieth century, new automation technology in the form of mass production through transfer and assembly lines emerged. Factory manufacturing grew to include more and more new sectors of work and finally became a composite of different departments, each separate from the others and dedicated to a certain task. Although automation and computerization of these isolated islands of work have been increasing steadily since the early 1950s, barriers of understanding and methods of working have also been growing. Although the number of blue-collar workers on the shop floor has continually decreased, large numbers of white-collar personnel, ranging from managers to clerks, have been needed to handle paperwork and to transfer information among the different departments in order to tie them together. It is, therefore, obvious that automating isolated tasks in the process of product development, although relatively cost-effective, cannot alone achieve either significant savings in lead time taken to develop a product or gains in productivity. It is also clear that these goals can be accomplished only by automating the flow of information in the business organization and by optimizing the process of product development as a whole through the adoption of a system's approach.

The solution is complete implementation of *computer-integrated manufacturing* (CIM). Figure 16.1a shows how different departments in a corporation

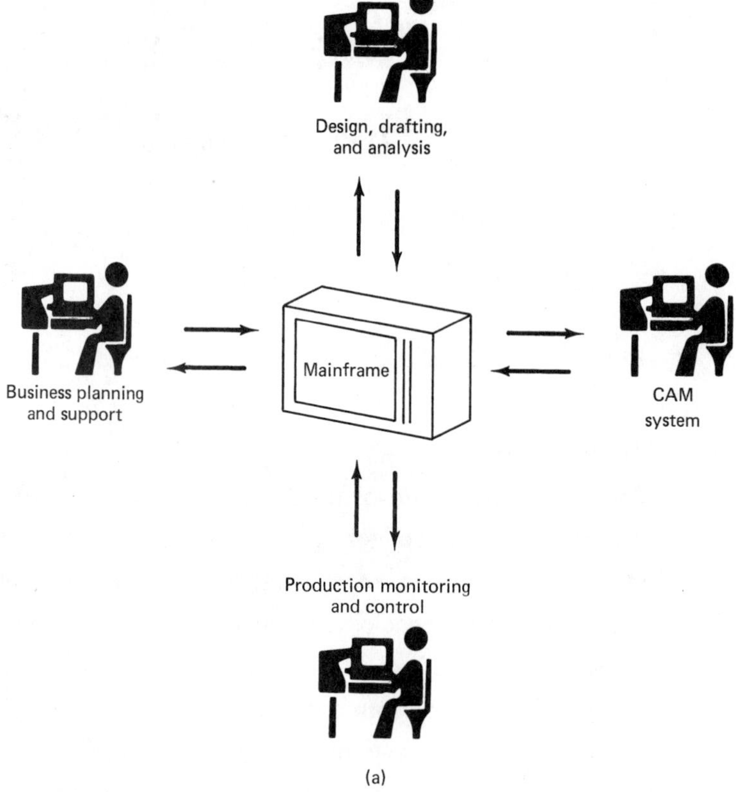

FIGURE 16.1 Types of information transfer between computerized departments of a plant: (a) departments electronically channeled (CIM)

can be electronically channeled so that each department has immediate access to all other departments as well as to the mainframe database. This arrangement ensures efficient control of the corporation and, consequently, optimization of the whole system. On the other hand, Figure 16.1b illustrates how many companies function today, with isolated automated work islands and white-collar workers doing paperwork to pass information from one department to another. A comparison of these figures will reveal to us the anticipated benefits of CIM. But first, let us consider some definitions of CIM.

16.1 COMPUTER-INTEGRATED MANUFACTURING (CIM)

An interesting definition of computer-integrated manufacturing (CIM) was given by Eugene Merchant (the father of the theory of metal cutting):

> [CIM is] a closed-loop feedback system whose prime inputs are product requirements and product concepts and whose prime outputs are finished prod-

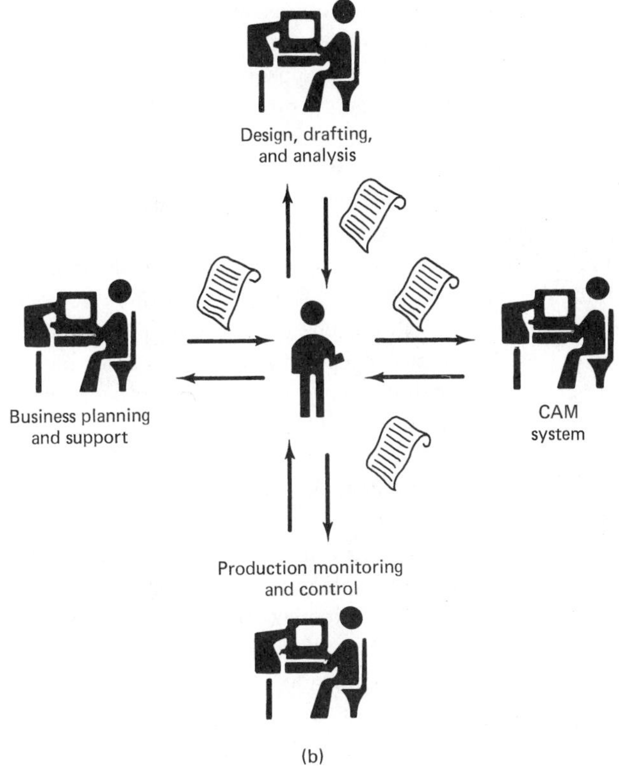

FIGURE 16.1 (CONT.) Types of information transfer between computerized departments of a plant: (b) white-collar workers passing documented information between departments

ucts. It comprises a combination of software and hardware: product design, production planning, production control, production equipment, and production processes.

Another definition, given by Richard G. Abraham in his paper presented at the AUTOFACT III Conference, was adopted and published by the Computer and Automated Systems Association (CASA) of the Society of Manufacturing Engineers (SME). It stated the following:

> A truly integrated CAD/CAM or CIM system provides computer assistance to all business functions from marketing to product shipment. It embraces what historically have been classified as "business systems" applications, including order entry, bill of material processing, inventory control, and material-requirements planning; design automation, including drafting, design, and simulation; manufacturing planning, including process planning, routing and rating, tool design, and parts programming; and shop floor applications such as numerical control, assembly automation, testing, and process automation.

A succinct definition of the business planning and execution functions in a CIM system was also provided. It included economic simulations, long-term business forecasting, customer-order servicing, and finished-goods inventory management. It is important to note that both definitions apply not only to engineering activities but also to management and business activities.

Thus, as a result of CIM, it appears that the role (as well as the quality) of engineers involved in design, manufacturing, and production planning will change. Product design not only will be dictated by the desired functions but also will be affected by manufacturing considerations. It is, therefore, anticipated that the boundaries now existing between the design and the manufacturing phases will disappear, or at least fade. Consequently, the engineers who carry out the process of product design and development will need to be thoroughly knowledgeable about the details of manufacturing. Similarly, an increase in the skill levels of other personnel will also be required. For instance, workers will become responsible for maintenance and initial setup work instead of just the simple, repetitious, and boring tasks involved in transfer or assembly lines. Managers will also need more technical backgrounds in order to be able to make appropriate decisions. In all cases, it is obvious that knowledge of computer systems is an absolute necessity.

In sum, the proper implementation of CIM must be based not only on the mechanization, optimization, and computerization of the various processes but also on achieving these goals in synchronization with the automation of information flow in order to deal with real-time planning and control of the business organization as a whole, from order entry to shipment of the finished product. In addition to a well-structured database, some fundamental methodologies are employed to achieve this goal. Among the methodological tools used for integrating design, manufacturing, and production control are group technology (GT), computer-aided process planning (CAPP), and material-requirement planning (MRP).

Benefits of CIM

In spite of the contributions of computer-aided design drafting (CADD) and computer-aided manufacturing (CAM) in increasing productivity by reducing the total product development costs and time, industrial automation experts predict the greatest gains coming from the implementation of CIM and replacing white-collar personnel and their paperwork with computer control. Following are some of the anticipated benefits of CIM.

Improved product quality. Improved product quality is a result of the CIM concept of developing the product within the computer, thus basing the product development process on rational and profound analysis instead of today's common philosophy of "build and test." Prototypes will still be built, but not to find out how a product performs. Instead, they will be used to verify the results of the analysis carried out by the computer. Other factors that contribute to improving the product quality and consistency include lower probability of human error and ensured uniformity of product attributes because of the use of on-line computer-aided inspection and quality control.

Improved labor productivity. Automating the information flow will result in a decrease in indirect labor, while at the same time increasing the efficiency of information transfer and eliminating redundant data collection. Any decrease in indirect labor will lead to a reasonable decrease in unit cost because the total costs for a typical highly automated production plant (in the United States today) are composed of 0 to 6 percent for direct labor, 18 to 22 percent for overhead and indirect labor, and about 75 percent for materials and machines. (You can see that paperwork costs more than the manufacturing work.) Also, increasing the efficiency of information transfer will result in more effective management.

Improved equipment productivity. Equipment productivity is improved because of the better utilization of machines when CIM is implemented. Factors like programmability of equipment and computerized monitoring and control of the entire manufacturing facility will largely improve the efficiency of machine utilization.

Lower costs of products. Higher labor and equipment productivity will certainly result in lower product cost. This is added to the advantages of designing the product with the required manufacturing processes in mind (i.e., design for manufacturing), which can easily be achieved through the integration of CAD and CAM. The use of design for manufacturing (DFM) will ensure the production of a part through the easiest and cheapest methods, thus reducing its cost. In fact, it has been found that designs that take into account only the product and its functions generally create the need for special manufacturing equipment, leading to a noticeable increase in the production cost.

Increased market share and more profit. By its very nature, CIM increases the flexibility of the manufacturing facility, thus enabling it to react quickly to fast-changing market demands. The reason is that much less lead time is taken to develop a product in a corporation where CIM is implemented (lead time is the time from the moment at which design work begins until the moment the product is shipped out of the factory). Also, less lead time means lower manufacturing cost of the products, which translates into greater manufacturing flexibility.

Implementation of CIM

It is clear that the development of a totally computer-integrated manufacturing system is a difficult task that requires a long-term commitment. However, the implementation of CIM can be accomplished gradually because a computer-integrated business organization consists of computerized modular subsystems that are interconnected through a network. The nature and basic functions of each of these modular subsystems depend primarily upon the products and the activities of business corporations and differ from one company to another. It is, therefore, impossible to purchase a turnkey CIM system from a vendor. Of course, components of a CIM system can be purchased, but the backbone of a CIM system has to be established within the corporation or at least tailored to suit the particular conditions of that corporation.

The first step that management should take in order to implement CIM is to study and review the activities to be performed in each subsystem, the type and amount of information flowing to and from that subsystem, and the interaction between the different subsystems in the business organization. Second, rational development of a long-term plan involving the architecture of the CIM system has to be carried out.

Third, a feasibility study and/or economic justification of each subsystem should be performed to establish the order of designing and implementing the subsystems during the gradual phase-in. Each subsystem is then integrated into the system directly after it is implemented, according to the main plan, with the end result being a fully integrated computer-aided manufacturing system.

The CIM Database

As mentioned, the heart of integration is a well-structured database that enables designers, manufacturing engineers, production managers, marketing and purchasing personnel, and partners and subcontractors to have access to the same set of factual data about products, production, and inventory. The typical architecture of a CIM system should be based upon a group of computer systems (CAD, CAM, and others), a neutral data format, a number of processors linked to a computer system to convert local data into neutral data (free format) and vice versa, and, finally, a database management system (DBMS). The major purpose of a DBMS is to see that all users or elements (such as the terminal or machine controller) automatically receive the updated version of data when any user alters data in his or her local system or element. For example, if, for some reason, a manufacturing engineer in the CAM department changes the design to facilitate the manufacturing phase, all modifications made will be transmitted directly to the database and to the CAD department.

Classes of CIM database. For easier and logical management of a CIM database, it is appropriate to group the data in that database into four classes: product data, manufacturing data, production control and operational data, and business data. Product data involve the attributes and geometries of the objects to be manufactured. Manufacturing data deal with how parts are to be manufactured. Operational data involve lot sizes, schedules, routes, and the like. These three classes are mostly technical, contrary to business data, which deal with resources, people, and money.

Logical and physical databases. The logical database relates to algorithms and programming methodologies or, in other words, what the user is concerned with. On the other hand, the physical database relates to what hardware personnel see. However, these two concepts, though distinct, are interrelated, and, therefore, it is advisable to separate them, making each more flexible regarding its own functions and demands, in order to have a successful CIM database that is responsive to fast-changing technology and users' needs. Because the logical and physical structures are interrelated, it is necessary to define a concept of the CIM database in such a manner that it is completely independent of any specific logical or physical requirement. This is done through the use of *data standards* that control the meanings of data and are defined using a data-modeling tool (like IDEFIX from USAF ICAM). They describe data in terms of entities and attributes of entities and are stored and maintained in a *data dictionary.* A data dictionary defined by Daniel S. Appleton in his original paper entitled "The CIM Database" was adopted and published by CASA as a Rosetta stone for providing access to users into a physical database environment.

Communication Networks

A CIM system is a complex system involving *communication networks* (i.e., a large number of separate but interconnected computers that must communicate with one another through many computer networks). These can be different from the viewpoint of computer system structure or configuration and can incorporate different computers, operating systems, and interfaces. However, the computers must be able to communicate with one another on a real-time basis and must also be able to process one another's data and execute one another's programs. In a conventional network, the distance between the user and his or her data may be several hundred miles. This is not the case with *local-area networks* (LANs) used in CIM, in which the system consists of many processors (computers, machine controllers, etc.) located close together. The individual LANs are connected through internetworking to form the CIM system. The structure of a network can take different forms, each having its own set of advantages and disadvantages. Nevertheless, almost all communication networks and LANs consist of two components: switching elements (specialized computers or IMPs) and transmission lines (circuits).

Network structures. Network structures involve either *point-to-point channels* or *broadcast channels.* In the first type of structure, the network contains numerous channels, each one connecting a pair of nodes or interface message processors (IMPs), which are the switching elements. Figure 16.2 shows some topologies of network structure. The *star* configuration (Figure 16.2a) was used in the early days of networking. But because star-type LANs suffer from the disadvantage of not being decentralized, *ring* topology (Figure 16.2b) became more common because of its decentralized structure and decentralized access techniques. Nevertheless, because this topology necessitates passing messages unidirectionally from node to node until they reach their destinations, any failure at a single node can bring the network down unless a bypass is provided. On the other hand, the second type of network structure (i.e., broadcast channels) involves a single communication channel shared by all IMPs (nodes). In this type of structure, messages sent by any IMP are simultaneously received by all other IMPs (nodes). This type is usually referred to as *bus* topology (Figure 16.2c) when it is used in LANs. Each node knows to whom the message is being sent and can, therefore, ignore any message not intended for itself. This is achieved by providing a piece of information (in the message itself) that specifies who the message is for. In fact, the bus topology has an important advantage in that it has the ability to insert splitters and create branches, thus facilitating network reconfiguration.

Network architectures. Because of the complexity of the system and the different communication needs of nodes, CIM networks are organized as a series of layers or levels. Each layer is built upon its predecessor and is designed to provide certain services to the higher-level layers without involving them in the details of how those services are implemented. According to this architecture, a layer with a certain level (order) on one machine communicates only with a layer having the same order on another machine. The set of rules and conventions stating how these two layers should interact during a conversation is known as the *protocol.* In reality, physical

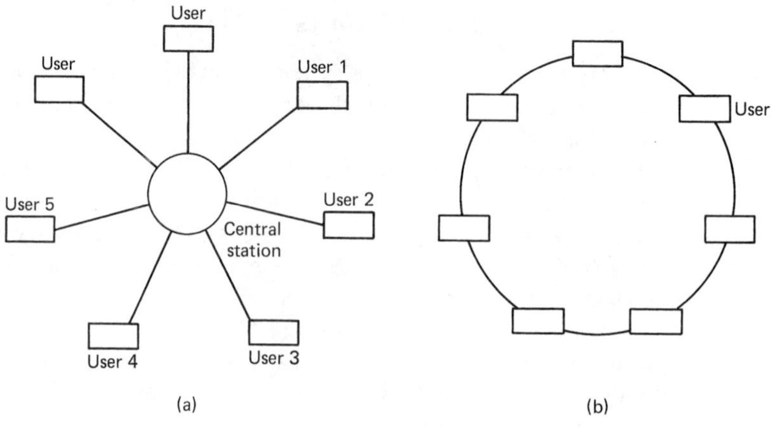

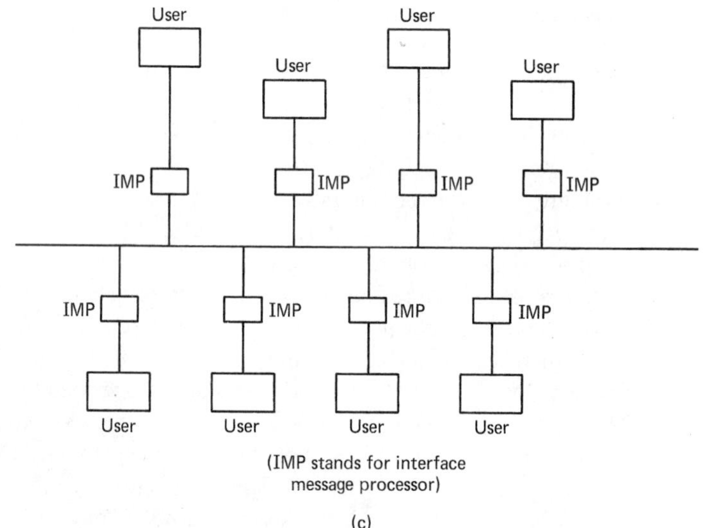

FIGURE 16.2 Some topologies of network structure: (a) star type; (b) ring type; (c) bus type

communication between two machines takes place only at the lowest layer, whereas communication between higher layers on two machines is only virtual. What happens is that each layer passes data to the layer directly below it until it reaches the lowest layer, where physical communication is possible. This idea is illustrated in Figure 16.3, which also indicates the need for an interface between each two successive layers to define the primitive operations and services that are offered by the lower layer to the upper one. It has been agreed to call the set of layers and their protocols the *network architecture.* In the ISO standard, as well as in the Manufacturing Automation Protocol (MAP), the network architecture is comprised of seven layers—physical, data link, internetwork, transport, session, presentation, and applications—as illustrated in Figure 16.4.

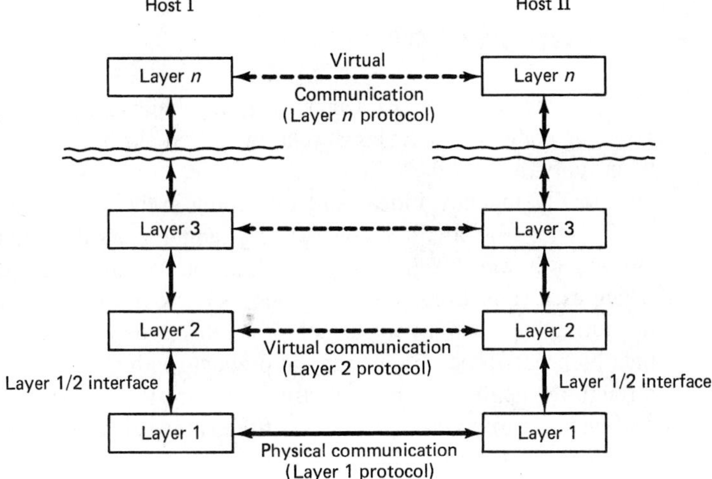

FIGURE 16.3 Layers involved in network architecture

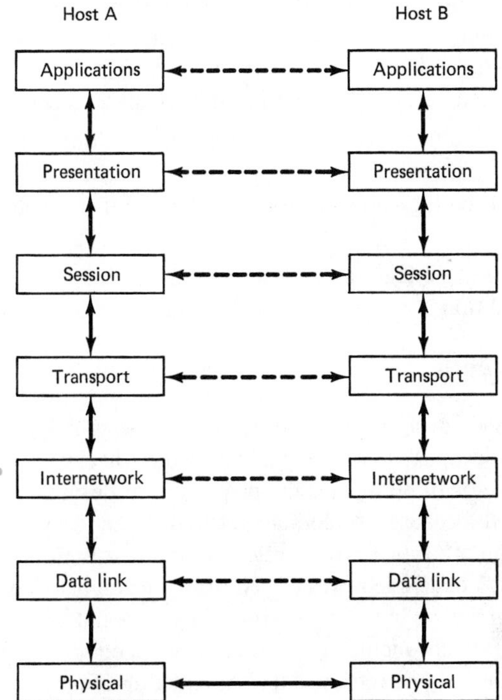

FIGURE 16.4 Network architecture when MAP is used

16.2 GROUP TECHNOLOGY (GT)

Group technology (GT) is a manufacturing philosophy that involves identifying and grouping components having similar or related attributes in order to take advantage of their similarities in the design and/or manufacturing phases of the production cycle. Historically, this novel idea came into being in the United States in 1920, when Frederick Taylor supported the concept of grouping parts that required special operations. He was followed by the Jones and Lamson Machine Company in the early 1920s, which used some crude form of group technology to build machine tools. Their manufacturing approach involved such principles as departmentalization by product rather than by process and minimizing the product routing paths. Today, GT is implemented through the application of well-structured classification and coding systems and supporting software to take advantage of the similarities of components.

Reasons for Adopting GT

Modern manufacturing industries are facing many challenges caused by growing international competition and fast-changing market demands. These challenges, which are exemplified in the following list, have and can be successfully met by group technology:

1. There is an industrial trend toward low-volume production (small lot sizes) of a wide variety of products in order to meet the rising demand for specially ordered products in today's affluent societies. In other words, the share of batch-type production in industry is growing day by day, and it is anticipated that 75 percent of all manufactured parts will be produced in small lot sizes.
2. As a result of the trend toward batch-type production, conventional shop organization (i.e., departmentalization by process) is becoming inefficient and obsolete, with wasteful routing paths of products between the various machine tool departments.
3. There is a need to integrate the design and manufacturing phases in order to cut short the lead time, thus achieving competitiveness in the international market.

Benefits of GT

Benefits in product design. In product design, the principal benefit of GT is that it enables product designers to avoid reinventing the wheel or duplicating engineering efforts. In other words, it eliminates the possibility of designing a product that was previously designed because it facilitates storage and easy retrieval of engineering designs. When an order of a part is released, the part is first coded, and then existing designs that match that code are retrieved from the company's library of designs stored in the memory of a computer, thus saving a large amount of time in design work. If the exact part design is not included in the company's computerized files, a design close to the required one can be retrieved and modified in order to satisfy the requirements. A further advantage of GT is that it promotes standardization of design features, such as corner radii and chamfers, thus leading to the standardization of production tools and work-holding devices.

Standardization of tooling and setup. Because parts are grouped into families, a flexible design for a work-holding device (jig or fixture) can be made for each family in such a manner that it can accommodate every member of that family, thus reducing the cost of fixturing by reducing the number of fixtures required. Also, a machine setup can be made once for the whole family (because of the similarities between the parts of a family) instead of several times for each of the individual parts.

More efficient material handling. When the plant layout is based on GT principles, such as dividing the plant into *cells,* each consisting of a group of different machine tools and wholly devoted to the production of a family of parts, material handling is more efficient because of the minimal routing paths of parts between machine tools. This is in contrast to the "messy" flow lines in the conventional departmentalization-by-process layout. For comparison, both layouts are clearly illustrated in Figure 16.5a and b.

Improved economies of batch-type production. Usually, batch-type production involves a wide variety of nonstandard parts, seemingly with nothing in common. Therefore, grouping parts (and processes) in families allows economies that are obtainable only in mass production to be achieved.

Easier scheduling. Grouping the parts into families facilitates the task of scheduling because this work will be done for each family instead of for each part.

Reduced work-in-process and lead time. Reduced work-in-process (WIP) and lead time result directly from reduced setup and material-handling time. Parts are not repetitively transferred between machining departments because material handling is carried out efficiently within each of the individual cells. This is in contrast to the production in a typical plant with a process-type layout, where a piece that requires only a few minutes of machining may spend days on the shop floor. This situation involves increased WIP, which adversely affects inventory turnover and the cash-flow cycle. Also, lead time for a product manufactured in a plant designed according to GT principles is far shorter than that of a product manufactured in a plant with a process-type layout.

Faster and more rational process planning. Group technology paves the way for automated process planning. This can be achieved through an appropriate parts classification and coding system, where a detailed process plan for each part is stored under its code and thus can be easily retrieved.

Factors Preventing Widespread Application of GT

Problems associated with rearrangement of physical equipment. As mentioned, GT is always associated with the concept of *cellular manufacturing.* The formation of cells necessitates rearrangement of the existing physical equipment (machine tools) and involves costly and cumbersome work that is sometimes difficult to justify.

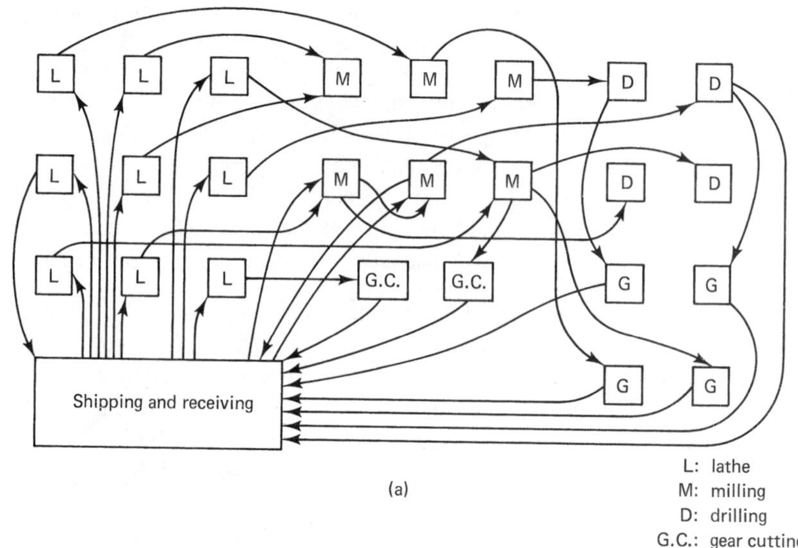

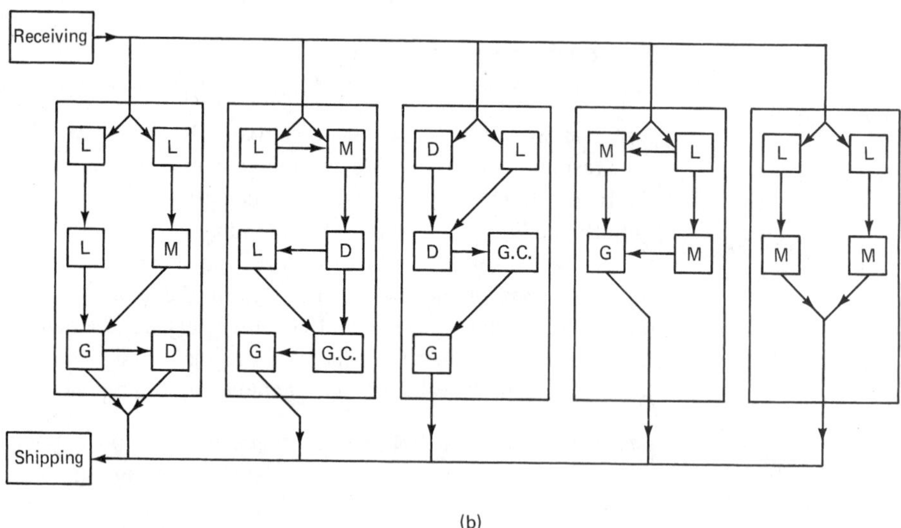

FIGURE 16.5 Flow of parts processed in a plant: (a) when departmentalization is by process; (b) when group technology rules are applied

Need for large amount of up-front work. In order to implement GT, it would seem that every single part in the inventory of the industrial corporation must be coded so that parts families can be established. This appears to be a huge task that in itself would create a barrier to any tendency toward the implementation of GT. However, a manufacturing corporation already deals with ordered and similar groups of parts because of the area of specialty of that corporation and/or its product line (i.e., range of products).

Accordingly, an appropriate approach for solving the problem of the up-front work required is to do it gradually by coding just the blueprints released to the workshop. As a result, the number of truly new (uncoded) parts released to the workshop tends to level out after a short period of time.

Natural resistance to anything new. As human beings, we naturally shy away from anything risky or unknown. For this reason, many managers and administrators avoid the adoption of new concepts and philosophies such as group technology.

Classification and Coding of Parts

Implementation of a classification and coding system. In order to implement a classification and coding system based on GT principles, parts must be classified according to suitable features, and then a meaningful code must be assigned to reflect those features. The process of retrieving or grouping parts with similar features is rather simple. As an example, consider zip codes as representing the basic features of a classification and coding scheme. A zip code indicates a geographic location by progressively classifying it into subdivisions, starting with the state and proceeding to county, city, neighborhood, and street. Codes that are numerically close indicate locations that are, in reality, geographically close. It is this particular feature that similarly enables the formation of a family of parts based on codes, without the need for physically examining the parts or their drawings.

Although many classification and coding systems have been developed all over the world, none of them as yet have become universally standard. The reason is that a system must meet the specific needs of the organization for which it has been developed. The right approach, therefore, is to develop a GT classification and coding system based on the specific needs of the client or to tailor an existing turnkey system to meet those needs.

While there are many benefits for group technology, they fall within two main areas of application: design and manufacturing. Although it is always the ultimate goal to combine the advantages in both areas, it is usually very difficult to do so, and the result is either a design-oriented or a manufacturing-oriented system.

Construction of a coding system. A coding system can be based only on numbers or only on letters, or it can be alphanumeric. When letter codes are used, each position (or digital location) has 26 different alternatives; when number codes are used, position values are limited to 10. Consequently, letters are employed to widen the scope of a coding scheme and make it more flexible.

There are basically two types of code construction: monocodes and polycodes. A *monocode,* also referred to as a *hierarchical* or *tree-structure code,* is based on the approach that each digit amplifies the information given in the preceding digit. It is, therefore, obvious that the meaning of each digit (or what a digit indicates) is dependent upon the digits preceding it. Monocodes tend to be short and are shape-oriented. However, they do not directly indicate the attributes of components because of their hierarchical structure. Consequently, they are usually used for design storage and retrieval and are not successful for manufacturing applications.

In contrast, the meaning of each digit in a *polycode* is completely independent of any other digit and provides information that can be directly recognized from its code. For this reason, a polycode is sometimes referred to as an *attribute code.* Figure 16.6 indicates how a polycode is structured. We can easily see that a polycode is generally manufacturing oriented because the easily identifiable attributes help the manufacturing engineer determine the processing requirements of parts. Moreover, a polycode involves a string of features—a structure that makes it particularly suitable for computer analysis. Nevertheless, polycodes tend to be long, and a digit location must be reserved whether or not that particular feature applies to a part of a family being coded. Common industrial practice, therefore, uses a hybrid construction that combines the advantages of each of the two basic types of codes while eliminating their disadvantages. In a combination code, the first digit divides the whole group of parts into subgroups, where shorter polycodes are employed. Also, in order to eliminate completely the possibility of error when coding a part, an interactive conversational computer program is employed, where the computer asks questions and automatically assigns a code for the part based on answers provided by the user at the computer terminal. An example of this kind of automated coding is the Metal Institute Classification System (MICLASS system), which was developed by the Netherlands Organization for Applied Scientific Research and which has gained industrial application in the United States during the last decade.

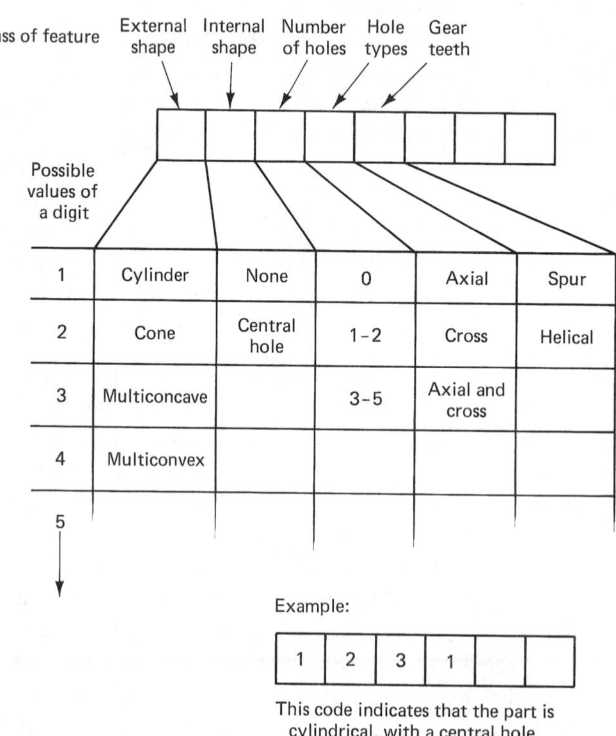

FIGURE 16.6
Structure of a polycode

Design of Production Cells

Before we see how a production cell can be designed, we need to understand a new concept, that of the *composite part.* This is a hypothetical part that has all of the processing attributes possessed by all of the individual parts of a family. Consequently, the processes required to manufacture the parts of a family will all be employed to produce the composite part representing that family. Any part that is a member of that family can then be obtained by deleting, as appropriate, some of the operations required for producing the composite part. Figure 16.7 illustrates the concept of a composite part.

The next step is to design the machining cell to provide all machining capabilities based on the processing attributes of the composite part for the family of parts that is to be manufactured in that machining cell. The number of each kind of machine tool in a manufacturing cell depends upon how frequently that machining operation is needed. In other words, the number of each kind of machine tool in a machining cell is not necessarily the same for all the different kinds of machine tools in the cell. After determining the kinds and numbers of machine tools in the cell, the layout of machines within that cell is planned to achieve efficient flow of workpieces through the cell. Of course, the cells are also arranged to guarantee easy flow of raw stock into the cells and finished products out.

Production-Flow Analysis

Production-flow analysis (PFA) is a method in which parts families are identified and machine tools are grouped based on the analysis of the sequences of operations for the various products manufactured in the plant. Parts that may not be similar in shape but that require identical or similar sequences of operations are grouped together to form

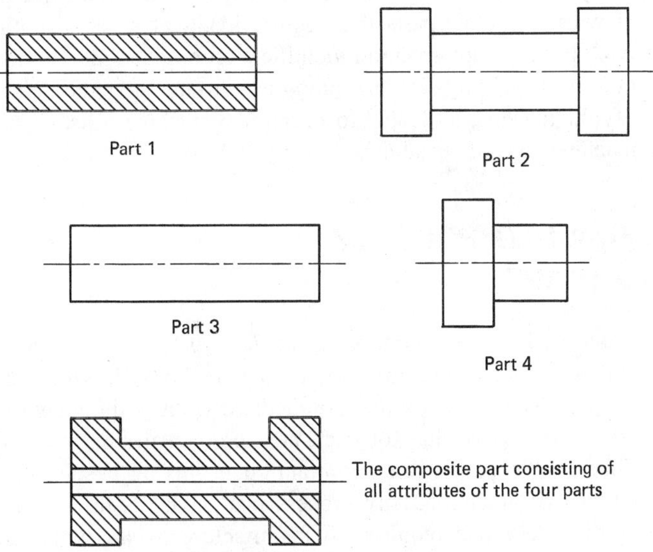

FIGURE 16.7 The concept of a composite part

FIGURE 16.8

Establishing machine cells using PFA: (a) initial PFA chart with codes of machine tools for each pack; (b) PFA chart after arrangement with possible cells of machine tools

Pack identification letter

Code number of machine tools	A	B	C	D	E	F	G	H	I	J	K
1	X	X	X	X	X				X		X
2	X	X	X		X				X	X	
3			X							X	X
4	X				X						
5											
6	X			X		X		X	X		X
7	X	X	X		X	X	X	X		X	
8											
9				X		X			X		X
10	X	X	X		X				X		
11											X
12		X				X	X			X	
13						X	X			X	
14		X					X				

(a)

Pack identification letter

Code number of machine tools	E	B	C	A	I	D	K	F	G	H	J
1	X	X	X	X	X	X	X				
2	X	X	X	X	X						X
3			X				X				X
4	X			X							
5											
6				X	X	X	X	X		X	
7	X	X	X	X				X	X	X	X
8											
9					X	X	X	X			
10	X	X	X	X	X						
11							X				
12		X						X	X	X	X
13								X	X	X	X
14		X								X	

(b)

a family. The resulting families can then be used to design or establish machine cells. Although PFA is simple and readily applicable, it does very little to improve and optimize the machine routing being used. It is, therefore, used just as a first step toward the application of the group technology concept in an industrial firm.

After collecting the required data (i.e., the part number and machine routing for every product), the computer is employed to sort out these products into groups, each of which contains parts that require identical process routings and is called a *pack.* Each pack is then given an identification number, and, by means of graphical illustration, packs having similar routing are grouped together. Next, zoning is used to identify the machine tools that form rational machine cells. Figure 16.8a and b shows how machine cells are established using PFA.

16.3 COMPUTER-AIDED PROCESS PLANNING (CAPP)

Process planning is defined in the *Tool and Manufacturing Engineer's Handbook* as "the systematic determination of the methods by which a product is to be manufactured, economically and competitively." It is the manufacturing engineer's task to set up a process plan for each new product design released from the design office. This process plan is an important stage linking design and manufacturing in an industrial organization. Any process plan involves the sequence as well as the details of the operations required to manufacture a part whose design is given. That sequence of detailed operations is documented on a route sheet. The routing must be optimal to bring the production cost of a part to a minimum, thus making the product economically competitive. Unfortunately, this is not the case today in conventional process planning, where the plan is determined primarily by the experience and the opinion of the planner. In other words, if ten process planners are asked separately to develop a process plan for a given part, they will probably come up with ten different plans. Moreover, it is possible that none of these plans is optimum or close to the optimal plan. In fact, it has been proven by experimental evidence that a conventional process plan usually reflects a stubborn commitment to the personal preference of the planner and his or her industrial and educational backgrounds.

In order to rationally design a process plan for manufacturing a product economically and competitively, a planner has to study all possible alternatives (i.e., different combinations and/or sequences of operations, machine tools available, process parameters, and the like). Obviously, such a task requires thousands of lengthy calculations. Consequently, a computer must be used if that task is to be accomplished within a reasonable period of time. And time is indeed a decisive factor because optimal process plans do not usually remain static but change with changing conditions, such as lot sizes, available equipment, and any emerging new technology. In short, a process plan that is developed for a part today may be different from the plan that would be developed after a year for the same part in the same manufacturing facility. It is, therefore, a major advantage of CAPP that it accounts for any variation in the manufacturing parameters.

Benefits of CAPP

CAPP, together with group technology (which helps pave the way for it), bridges the gap between engineering design and manufacturing and is a key factor in integrating the activities in a manufacturing organization. Some of the benefits of CAPP are as follows.

Improved productivity. Improved productivity is due to the more efficient utilization of resources such as machines, tooling, stock material, and labor, which is based on accurate and lengthy computations, as opposed to conventional process planning, where personal experience, preference, and sometimes even prejudice are the determining factors.

Lower costs of products. Improved productivity leads to cost savings in direct labor, tooling, and material and ultimately results in lower product cost.

Consistency of process plans. Because the plans are based on the same analytical logic, with every planner having access to the same updated database, planners will come up with the same plan for the same part. This also occurs because of the elimination of human error during the lengthy computations performed with a computer.

Reduction in time required to develop a process plan. As a result of computerizing the work, a job that used to take several days can now be done in about 20 minutes.

Consequently, increased volumes of work can be easily handled with CAPP, which is not the case with conventional manual process planning.

Faster response to changes in the production parameters. The fact that the logic is stored in the memory of the computer makes CAPP more responsive to any changes in the production parameters than the manual method of process planning. When CAPP is used, it takes the process planner a few minutes to input updated data, such as lot size, machines available, and cost of raw material, and to obtain a modified optimal version of the process plan.

Less clerical effort and paperwork. The paperwork involved is far less than in manual process planning because its routing through the system (between different specialized planners) is greatly reduced or even eliminated. This is in agreement with and promotes the goals of CIM in reducing clerical white-collar jobs and thus ending up with a paperless factory.

Types of CAPP

The two types of CAPP, *variant* and *generative,* are based on two completely different approaches.

Variant type. The variant approach involves preparing beforehand a group of standard process plans, one for each of the parts families produced by the plant. The parts families should be identified based on GT principles through an appropriate parts classification and coding system. The standard process plans are filed and stored in the memory of a computer, and each one can be retrieved and edited by inputting its associated GT code. When a new part is given to the process planner, he or she can relate it to an appropriate family. Next, the standard plan for the parts family is edited and modified by deletion or addition of some operations to suit the given part (remember our discussion about the composite part concept). The computer is employed in the initial analysis and computations required to develop the standard process plans because it is available in the organization. However, during the actual running of a variant CAPP system, the role of the computer is limited to that of a word processor.

Thus, the main problem with a variant CAPP system would appear to be the large amount of up-front work required to establish the group of standard process plans (which may include hundreds of plans). In addition, these standard process plans need to be revised and modified on occasion to support any changes in the production requirement or parameters. It is primarily for these reasons that the generative approach of CAPP has been developed.

Generative type. In the generative approach, the computer is used to synthesize each individual plan automatically and without reference to any prior plan. What is stored in the memory of the computer are rationales and algorithms that permit the appropriate technological decision to be made. The human role involved in running the system is minimal and includes inputting the GT code of the given part design and monitoring the function; it is the computer that determines the sequences of operations and the manufacturing parameters. It is, therefore, the computer's role to select the processes and the machines, to determine the sequences of operations, and, finally, to sum these up into the optimal process plan.

In spite of the attractive features of a generative CAPP system, it is important to note that the decision logic often varies from company to company and that, therefore, the theoretical process parameters have to be adjusted to conform to the practical manufacturing conditions. For this reason, it is very difficult to establish a truly generative system. Nevertheless, there are some attempts that are quite close to that goal. Generative process planning (GENPLAN), which was developed by Lockheed Corporation, Georgia, is among the successful attempts. The manufacturing logic included in GENPLAN's database was based on a comprehensive analytical study of process plans that were developed over the course of 25 years. As soon as a code is assigned to a part design by the process planner, the software quickly accesses the different alternatives and makes the appropriate decisions to establish an optimal process plan for a generative CAPP.

Another successful attempt is the DCLASS system, which was developed by Brigham Young University of Utah. The DCLASS approach is based on using a hierarchical tree of orderly classified processes as a common reference, to which process parameters and part-attributes information trees are compared and evaluated. These trees take the form of key words and GT codes. In spite of these efforts, much R & D still has to be carried out in this area. Moreover, use has to be made of emerging concepts in computer science, such as artificial intelligence, in order to achieve the goal of a complete and truly generative CAPP system.

16.4 MATERIAL-REQUIREMENT PLANNING (MRP)

Let us now look at what *material-requirement planning* (MRP) is and why it is important. Consider a plant that is well operated by an active, qualified manufacturing staff and that is suddenly faced with the fact that there is no stock material. Production will stop, although machine tools and manufacturing personnel are available, because there is simply nothing that can be manufactured. It is, therefore, of supreme importance to manage inventories to ensure a continuous supply of stock material if production is to run uninterrupted. MRP is a computerized method for assuring this. The function of MRP software is to continually update all information about inventory, to check whether the necessary raw materials and purchased parts will be available when required, and to issue purchase orders whenever necessary.

MRP software packages can be obtained from various vendors, but they have to be tailored to the specific needs of the client. Major computer producers also provide MRP packages as part of the software developed for their machines. Software houses and consulting bureaus offer MRP packages that can be run on different types and makes of computers. It is even possible, in some cases, to obtain MRP on a time-share basis.

A recent trend in MRP development calls for linking inventory planning with the financial system of the company in order to achieve a total business plan. In this case,

MRP is usually referred to as *material-resource planning* (MRP II). Whether MRP or MRP II is employed, the gains have been impressive. Elimination of late orders and delays, increased productivity, and reduced WIP inventories are among the benefits of MRP systems. A further benefit is that MRP promotes the integration of manufacturing systems.

16.5 THE POTENTIAL OF ARTIFICIAL INTELLIGENCE IN MANUFACTURING

The Latin root of the word *intelligence* is "intelligere," which literally means to gather, to assemble, and then to form an impression. The word thus has a meaning that involves understanding and perceiving, followed by the skilled use of reason. If we assume that an artifact gathers knowledge, chooses among facts based on sound reasoning, understands, and perceives, then what we have is indeed an *artificial intelligence* (AI). The desired artifact or reasoning machine must have the special capability of processing knowledge information at very high speeds. In fact, this was the goal of the Japanese who, in 1982, established the Institute for New Generation Computer Technology (ICOT) to guide a ten-year research program aimed at developing hardware and software for the fifth generation of computers, or the knowledge information processing systems (KIPS).

Let us now look at the differences between fifth-generation and fourth-generation computers. The design of all fourth-generation computers is generally the same, and they are called *von Neumann* machines. Each consists of a central processing unit CPU (a memory, an arithmetic unit, and a controller) and input/output devices. A characterizing feature of all von Neumann machines is that they operate in an orderly sequence (serial fashion). The difference is not, therefore, in the working principles but in the building unit of the hardware or central technology. Whereas first-generation computers were composed of vacuum tubes, the second generation was transistorized, the third involved integrated circuits, and the fourth is based on very large scale integrated (VLSI) circuits. Fifth-generation computers, however, involve new computer architectures, new memory organizations, and new programming languages. These computers are designed to handle and process symbols and not just numbers, like the von Neumann machines. Also, programs can be run in any order (i.e., not necessarily in serial fashion). This requires a special computer architecture involving parallel processors (more than one computer working together at the same time). There are special AI programming languages; the most commonly used are PROLOG and LISP. PROLOG has become the favorite in Europe and Japan and was also selected for the Japanese ICOT R & D project. LISP has been widely used in the United States and has become the dominant language.

So, what role can artificial intelligence play in the future of advanced manufacturing systems? In order to answer this question, we should keep in mind that AI programs are mainly concerned with symbolic reasoning and problem solving. It is, therefore, anticipated that fields of AI technology such as expert systems, artificial vision, and intelligent robots will have great potential in manufacturing. Following is a brief discussion of each of these AI technology applications in manufacturing.

Expert Systems

Definition. An *expert system* can be defined as an intelligent computer program that has enough knowledge and capability to allow it to operate at the expert's level (i.e., it can solve extremely difficult problems that can be solved only by employing the expertise of the best practitioners in the field). The person who develops an expert system is called a *knowledge engineer,* and his or her job involves acquiring knowledge and structuring it into an AI program. The program usually consists of a large number of if–then rules, perhaps as many as a few thousand, in addition to a knowledge base.

Applications in manufacturing. It is expected that expert systems will have widespread application in CIM systems. They can be employed in maintaining the systems and quickly detecting any faults or disturbances that may occur. Another emerging application involves intelligent control of machine tools. As previously discussed, CNC, DNC, and computer-assisted part programming are different forms of preplanned computerized control of machine tools. In all cases, the tool path has to be established beforehand through a program. The person who prepares the program employs his or her experience in order to bring the processing time to a minimum without causing any damage or distortion to the workpiece. This is, in many cases, a difficult problem involving many factors, alternatives, and constraints. It is exactly where an expert system is needed.

Artificial Vision

Artificial vision is currently the most exciting emerging AI technology in manufacturing. Although artificial vision is commercially available today, a lot of research is being carried out on more advanced aspects of this technology. Generally, it can be stated that artificial vision is aimed at locating, identifying, and guiding the manipulation of industrial components. As research continues in computer science to develop a superior pattern-recognition methodology, we would expect to see more and more commercial and industrial applications of artificial vision.

Intelligent Robots

The *intelligent robot* has always been the dream of manufacturing engineers in terms of making the fully automated factory of the future attainable. It is artificial intelligence that will make this dream come true. By definition, an intelligent robot is one that is able to think, sense, and perform so as to cope with a changing environment and learn from experience. Because thinking is a brain function, it is obvious that it would fall within the domain of artificial intelligence if it is to be performed by a computer. An integration between reasoning, sensing, and performing would unify artificial intelligence and robotics, with the final outcome being an intelligent robot.

16.6 FLEXIBLE MANUFACTURING SYSTEM (FMS)

Flexible manufacturing is one of the most exciting emerging automation concepts. In order to understand this concept, we must first discuss automation in general. As we know, mechanization started with the Industrial Revolution, when machines performed some of the functions of labor. Automation included mechanization and went beyond it. With the twentieth century came the new concept of mass production. The automation technology then available (i.e., the assembly line) markedly increased the productivity of workers, and the United States grew to lead all other nations both in productivity and prosperity. Both the assembly line and the transfer line belong to the domain of *fixed automation,* in which the sequence of processing operations remains unchanged throughout the production run because the sequence of operations is determined mainly by the nature and arrangement of the physical equipment in the production line, which, once established, cannot be changed (except with great difficulty). It can, therefore, be seen that fixed automation, which is based on integrating and coordinating simple operations, has three characterizing features. First, it requires very high initial capital cost; second, it is suitable only for high production volume (to pay back for its high capital cost); and third, it is inflexible and cannot handle product changeover or even major changes in the product design.

In contrast to fixed automation, *programmable automation* is flexible and can handle a large variety of parts. A good example of this kind of automation is a stand-alone CNC machine tool, where new part designs are accommodated simply by preparing new part programs, without the need for changing the machine. Programmable automation is characterized by its low production volume, its flexibility and ability to tolerate product changeovers, and the fact that its initial capital cost is not as high as that of fixed automation.

An interesting way of illustrating the difference between these two types of automation is to plot the number of different shapes (or designs) of workpieces that the automated system can tolerate versus the annual production volume. This is clearly indicated in Figure 16.9, which also shows the recommended successful domain for each of the two automated systems. Notice that a gap exists between the high-production transfer lines (fixed automation) and the low-production, though flexible, individual CNC machines. This gap involves medium production volume accompanied by a limited variety of part designs. In fact, it is this kind of production that is required most today because of the need for insurance against unexpected circumstances, which calls for flexibility and tailored products demanded by customers in quantities that are not suitable for mass production. A reasonable solution to this problem is to develop a hybrid of fixed and programmable automations that combines the best features of both. This new production system should be responsive to the changing needs of manufacturing, yet highly automated. It is, therefore, appropriately called a *flexible manufacturing system* (FMS).

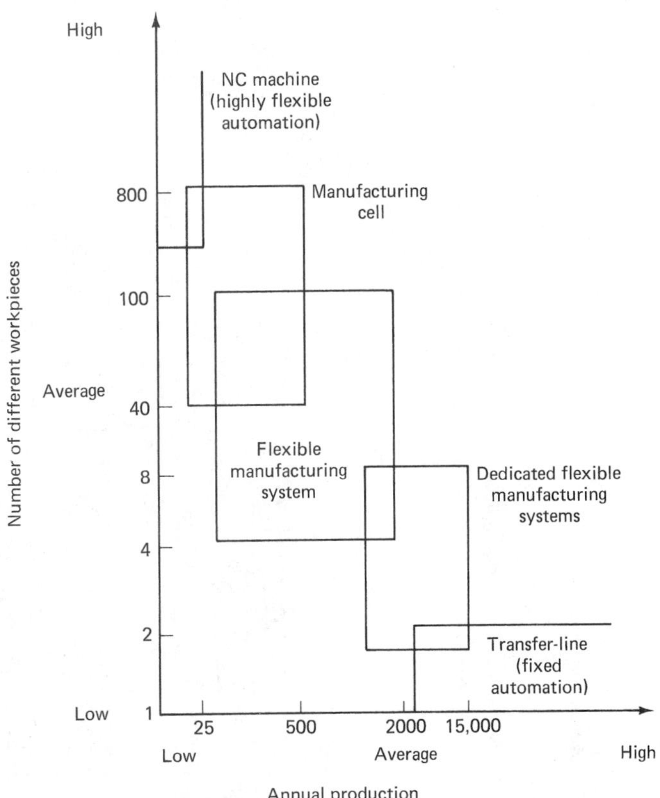

FIGURE 16.9 Areas of application of the different automated manufacturing systems

Elements of a Flexible Manufacturing System

Although FMSs may differ slightly in features, depending upon a system's manufacturer, they are all based on the same idea of incorporating several individual automation technologies into a unified, highly responsive production system. An example of an FMS for the complete machining of gears is shown in Figure 16.10. We can easily see that this new method of automation is actually based on three main elements: machine tools, an automatic material-handling system, and a computer-control system that coordinates the functions of the other two elements in order to achieve flexibility. Figure 16.11 is a chart of these basic FMS elements.

Machine tools. An FMS consists of a mixed range of CNC machine tools. The characterizing features of these machine tools depend upon the configuration of the components to be manufactured as well as the desired degree of flexibility of the system. Therefore, some manufacturing experts tend to divide FMSs into two categories: *dedicated* and *random* FMSs. Dedicated FMSs possess a low degree of flexibility and are built to meet long-term, well-defined manufacturing requirements. Such systems can, therefore, tolerate only a limited variety of processing needs. Special machine tools

FIGURE 16.10
An FMS for the complete machining of gears *(Courtesy of Fritz Werner Machine Tool Company, Carol Stream, Illinois)*

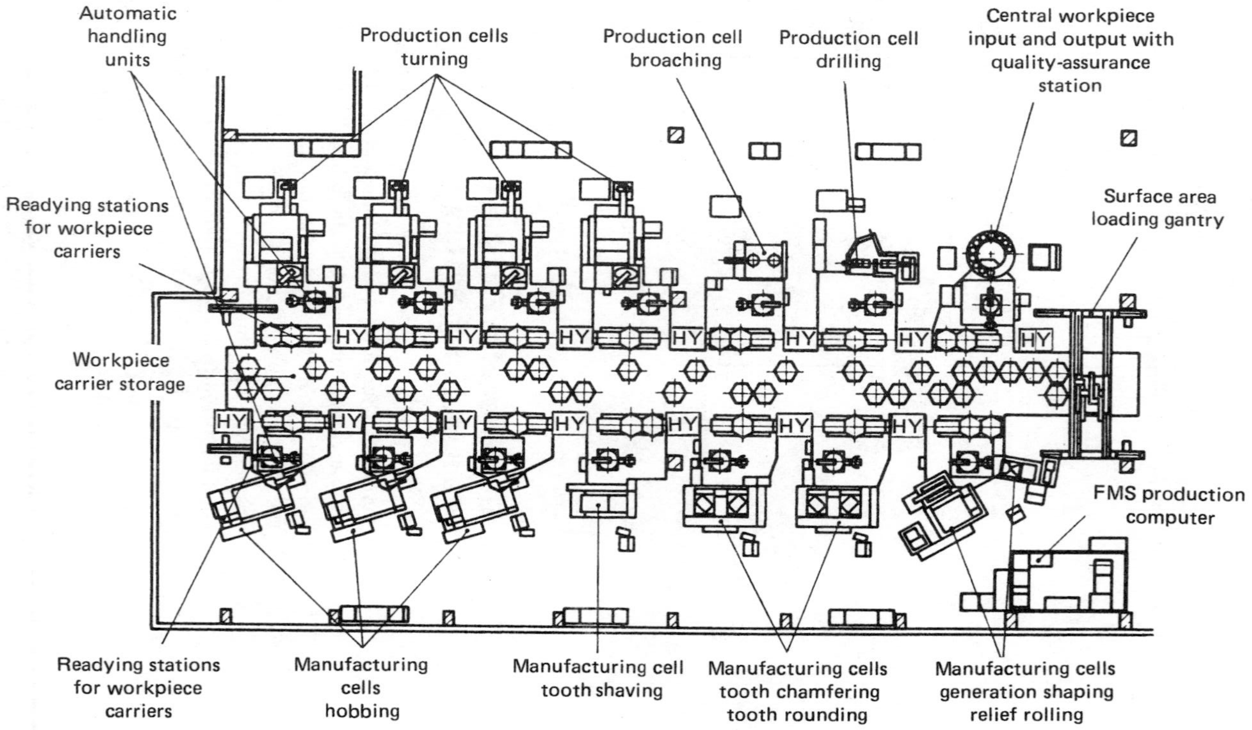

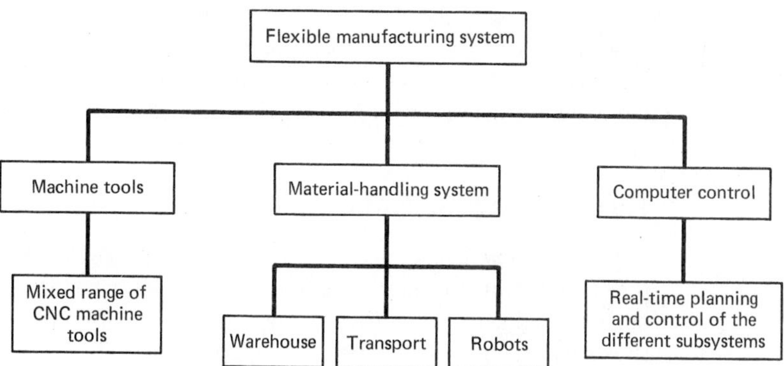

FIGURE 16.11
Elements of an FMS

form the backbone of this type of system. Random FMSs are designed to machine (or process) a wide range of different parts in random sequence. Therefore, the machine tools used must be highly flexible. New part designs can easily be handled by random FMSs. Four- and five-axis CNC machine tools are commonly used with this type of system.

Although the machining processes involved differ from one FMS to another, they often include operations like turning, milling, drilling, tapping, cylindrical grinding, spline hobbing, and broaching. Batches usually fall within the range of 25 to 100 and can be completed, in most cases, in three days or less. The system and the machine tools are designed and planned so that they operate without human supervision during unfavorable work hours.

Material-handling system. The material-handling system involves robots that are used for automatic loading and unloading of machine tools and a means of transportation that links the various machine tools together and moves parts around (e.g., power roller conveyor, shuttle carts on fixed railroads, vehicles that float and move on an air or a fluid cushion, industrial robots that travel on the factory floor). All movements of parts (or pallets) in the system must be randomly independent (i.e., parts can move from one station to another without interference). These movements are directly controlled by the main computer of the FMS. Each pallet is uniquely coded so that it can be tracked by the computer. Codes (which are usually binary) are affixed to a coding tag that is mounted onto a pallet identification strip. This enables the computer to use conveniently located sensors to identify and accurately locate pallets. Industrial robots are then used to load pallets and/or parts in the machine and also return pallets loaded with machined parts to the transport system. Temporary storage is provided at each machining station so that a workpiece can be promptly loaded in the machine as soon as a machined part is removed from it, thus increasing machine utilization by saving time. Another function of the robots is to select components for inspection under computer control.

Computer-control system. An FMS cannot be operated without a digital computer because of the complexity of the system and the need for real-time planning and control of the different subsystems. The function of the computer is not limited to the storage and distribution of the NC part programs, the operation of load–unload robots, and the control of shuttle-cart traffic between stations. It also covers tool monitoring (following each tool throughout the system and monitoring its life) and production control (employing a data entry unit in the load–unload station to act as an interface between operator and computer).

Flexible Manufacturing Cell

A *flexible manufacturing cell* (FMC) usually consists of two CNC machining centers (or one machining center and one or two special machines) that are linked with each other as well as with their tool storage area through an industrial robot. Figure 16.12 shows an example of an FMC. Because of the flexibility inherent in the design of an FMC, its productivity is higher than that of separate stand-alone machines, and it can also handle a larger variety of parts configurations. FMCs are, therefore, used when the production volume does not justify the purchase of an FMS yet cannot be met by stand-alone machines. FMCs are a kind of automation that fills the gap between FMSs and individual stand-alone CNC machine tools. The domain of successful application for FMCs, therefore, involves a production volume between 20 and 500 pieces and a variation in configuration ranging between 30 and 500 different shapes.

The economic advantage of using FMSs and FMCs can easily be realized by comparing each system with individual machine tools in relation to the production

FIGURE 16.12
Example of an FMC *(Courtesy of Fritz Werner Machine Tool Company, Carol Stream, Illinois)*

cost per component. These two comparisons are shown in Figures 16.13 and 16.14. As can be seen in Figure 16.13, the production cost of a piece when employing an FMS is only 72 percent of that cost when the piece is produced in a plant including six machining centers with a pallet changer. Figure 16.14 indicates that the production cost per piece when an FMC is used amounts to only 69 percent of that cost when the workshop consists of three machining centers with a pallet changer. Although these figures indicate particular cases, we must not forget that the production volume plays a very important role in determining the cost per piece. Figure 16.15 shows the cost per piece versus the production volume for each of the different kinds of automated manufacturing systems.

FIGURE 16.13

Characteristics and cost comparison to determine productivity of an FMS *(Courtesy of Fritz Werner Machine Tool Company, Carol Stream, Illinois)*

		6 machining centers with pallet changer	4 machining centers with pallet pool	1 flexible manufacturing system
	System configuration	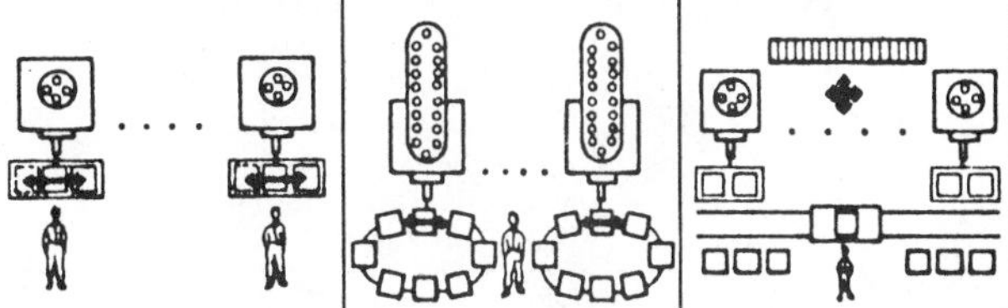		
Technical and organizational characteristics	No. of machines in system	6	4	4
	No. of tools in system	6 X 60	4 X 72	568
	No. of workpiece storage positions	0	40	40
	Machine running time during the 3rd shift (hrs.)	0	6.0	6.6
	Machine utilization time per year (hrs.)	16,205	16,034	17,285
	Degree of utilization (%)	80	82	86
	No. of jobs processed simultaneously	0	8	12
	No. of operators during 1st and 2nd shifts	6	2	2
Economic characteristics	Capital investment (DM)	4,165,980	3,919,240	4,476,000
	Tool costs (DM)	739,000	650,000	501,000
	Fixture and jig costs (DM)	84,000	209,790	139,860
	Total system costs (DM)	5,042,980	4,779,030	5,116,860
	Annual machine costs (DM)	1,118,072	1,305,958	1,445,410
	Annual personnel costs (DM)	921,672	307,224	307,224
	Annual system costs (D (DM)	2,039,744	1,613,182	1,752,634
	Annual hours of utilization (hrs.)	16,205	16,034	17,285
	System costs per hour (DM)	125	101	101
	Average costs per piece (DM)	182	154	132
	Piece-cost relation (%)	100	85	72

FIGURE 16.14
Characteristics and cost comparison to determine productivity of an FMC *(Courtesy of Fritz Werner Machine Tool Company, Carol Stream, Illinois)*

		3 machining centers with pallet changer	2 machining centers with pallet pool	1 duplex cell
	System configuration	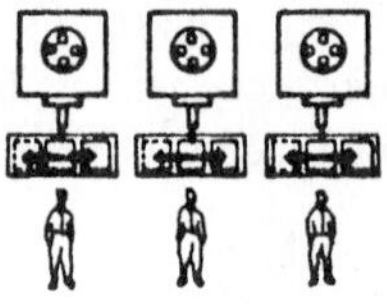	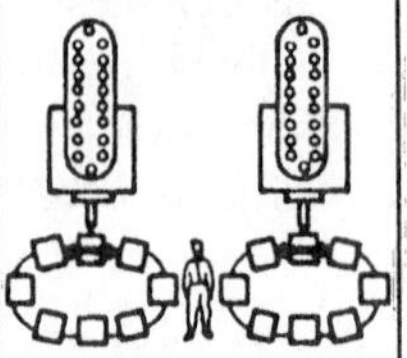	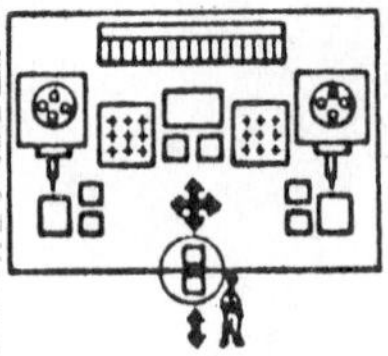
Technical and organization characteristics	No. of machines in system	3	2	2
	No. of tools in system	3 X 30	2 X 60	200
	No. of workpiece storage positions	0	2 X 16	32
	Machine running time during the 3rd shift (hrs.)	0	8	8
	Machine utilization time per year (hrs.)	8,130	8,751	9,173
	Degree of utilization (%)	77	79	83
	No. of jobs processed simultaneously	1	5	8
	No. of operators during 1st and 2nd shifts	3	1	1
Economic characteristics	Capital investment (DM)	1,367,550	1,407,560	1,715,200
	Tool costs (DM)	634,218	516,906	397,620
	Fixture and jig costs (DM)	62,100	168,960	105,600
	Total system costs (DM)	2,063,868	2,093,426	2,218,420
	Annual machine costs (DM)	463,732	592,162	635,065
	Annual personnel costs (DM)	460,836	153,612	153,652
	Annual system costs (DM)	924,568	745,774	788,677
	Annual hours of utilization (hrs.)	8,130	8,751	9,173
	System costs per hour (DM)	114	85	86
	Average costs per piece (DM)	45	36	31
	Piece-cost relation (%)	100	80	69

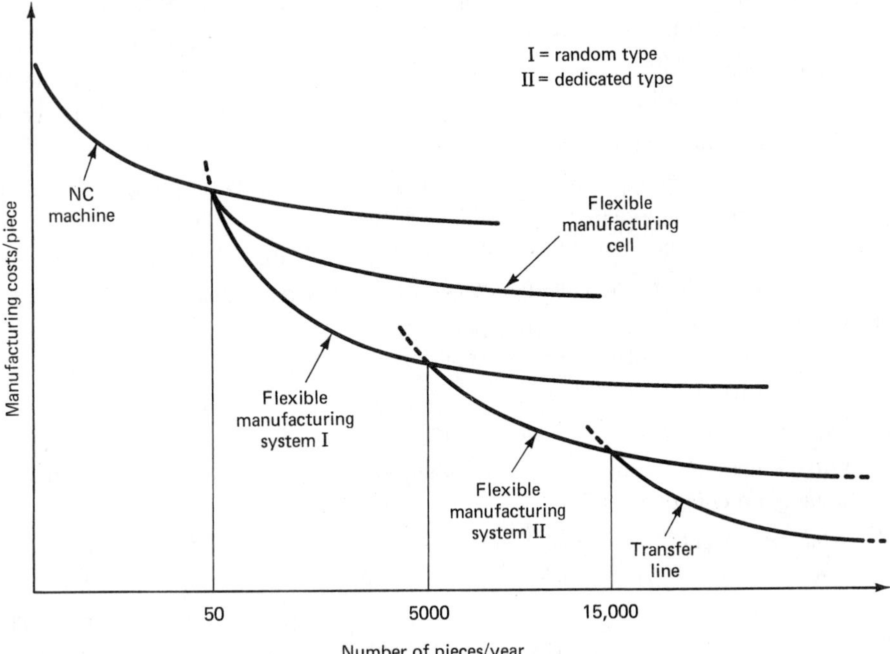

FIGURE 16.15 Cost per piece versus production volume for different automated manufacturing systems

Review Questions

1. What does CIM stand for and what does that mean?
2. Why is the implementation of CIM necessary? Can CAD, CAM, and other advanced technologies be a substitute? Why?
3. Describe the role (and the quality) of the engineers who will operate a CIM system.
4. What steps should be taken in order to implement CIM properly?
5. List the benefits of CIM. Briefly describe each.
6. Describe the typical architecture of a CIM system.
7. What is the major purpose of a DBMS?
8. What are the classes of CIM? Briefly describe each.
9. In a CIM system, what is meant by *communication networks?* What is their function?
10. List and use sketches to illustrate some LAN topologies. What are the advantages and disadvantages of each?
11. What is meant by a *network architecture?*
12. Explain and use sketches to illustrate the layers included in the network architecture involving MAP.
13. What is *group technology?*
14. List the reasons for adopting GT.
15. List the benefits of GT. Briefly describe each.
16. What factors prevent widespread application of GT?

17. Describe the classification and coding of parts. What important feature does this system have?
18. Is there a universally standard classification and coding system? Why?
19. How can a coding system be constructed?
20. What are the basic types of code construction? Explain briefly the characteristics of each, as well as the advantages and disadvantages.
21. What is meant by *cellular manufacturing?*
22. What is the difference between a plant design based on GT principles and another one based on old-fashioned concepts?
23. Explain the meaning of a *composite part.*
24. What is the right approach to take when designing a production cell?
25. What is meant by *computer-aided process planning?* How is it linked with group technology?
26. How do you compare CAPP with the conventional process planning performed today?
27. List the benefits of CAPP. Briefly describe each.
28. What are the main types of CAPP?
29. Give some examples of commercially available CAPP systems.
30. What is meant by *material-requirement planning?* What is its function?
31. What is the recent trend in MRP development? How does this serve the goal of implementing CIM?
32. What is *artificial intelligence?*
33. What are the differences between fifth-generation and fourth-generation computers?
34. Differentiate between the first four generations of computers.
35. What computer languages are suitable for artificial intelligence?
36. What does the term *expert system* mean?
37. What application will expert systems have in manufacturing?
38. What is meant by *artificial vision?* What is its main function?
39. What is the difference between an intelligent robot and a conventional industrial robot? Explain.
40. What are the two basic types of automation? List the main characteristic features of each.
41. Are the two basic types of automation enough to meet today's market demands?
42. Why is there a need for flexible manufacturing systems?
43. What are the main elements of an FMS? Briefly describe each.
44. What are the two kinds of FMSs?
45. Are these two kinds of FMSs enough to fill the gap that exists between the two basic types of automation? If not, how can this gap be filled? Describe a similar automated manufacturing system that can fill this gap.

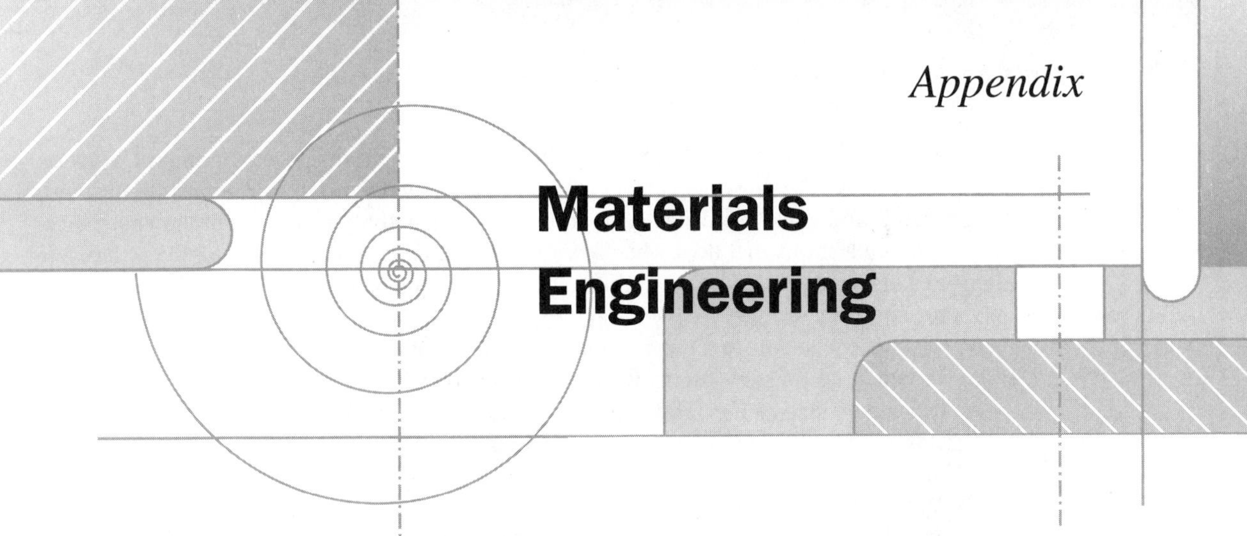

Appendix

Materials Engineering

INTRODUCTION

This appendix is aimed at those students who have not taken any materials science or engineering courses. It provides them with the fundamentals of this important subject so that they can understand the various manufacturing processes and the concept of design for manufacturing. It is not meant as a substitute for a textbook for those who want to study materials engineering more comprehensively. Industrial engineering students will find it helpful to study this appendix thoroughly before reading the current text.

The coverage here is concise and limited to those areas that provide the background necessary to learn material processing.

A.1 TYPES OF MATERIALS

There are three basic classes of engineering materials: metals, ceramics, and polymers. How the atoms of a certain material are bonded together and how they are arranged have the ultimate influence on the nature and the properties of that material. Let us now scratch the surface of each of the three basic classes of materials to obtain a view on the atomic scale or level.

Metals

Metals can be elements or alloys (combinations of elements). As you studied in chemistry, the atoms of a metal are relatively large and heavy. Therefore, the attraction forces that keep the electrons circling in the outer orbits are not strong enough to sustain that dynamic equilibrium. As a consequence, the electrons are free to move throughout the piece of metal, forming an electron mist. On the other hand, the atoms that lose these electrons have positive charges. It is, therefore, clear that a solid metal is composed of atoms (that have positive charges) held together by a matrix of electrons (that have negative charges). This basic nature of metals is responsible for their useful properties. For instance, good electric conductivity is a result of the abundance of free electrons that flow when a wire is subjected to a magnetic field. Other properties of metals include their ability to undergo major permanent deformation and their opacity.

We can see from the preceding discussion that metals take the solid form when a certain bond (positive atoms with positive charges in an electron mist) exists between the component atoms. This type of bonding is referred to as a *metallic bond.* The atoms of a metal are arranged in a repetitive three-dimensional pattern to form tiny (usually) microscopic crystallites or grains. The smallest arrangement of atoms that, when repeated in all six directions in space, produces a grain is referred to as the *unit cell.* Although there are fourteen different types of unit cells, only three types are common, and they are shown in Figure A.1. Each type has a different name: body-centered cubic (BCC), face-centered cubic (FCC), and hexagonal close-packed (HCP). Each metal (element) has one of these types of atomic arrangements; for example, iron and chromium are BCC, copper and aluminum are FCC, and magnesium and zinc are HCP. As we will see later on, the properties of a solid metal are affected by the type of crystal structure it has.

In a few cases, metals have two or more crystal structures (e.g., one at room temperature and one at high, or at low, temperature). A specific example is iron; it is BCC at room temperature but suddenly changes to FCC at 1674°F (912°C). This phenomenon is called *allotropy,* and change from one type of unit cell to another is called *allotropic transformation.* The properties of the metal (such as its ability to dissolve other elements) also change before and after the allotropic transformation.

When metals undergo permanent deformation as a result of applying loads, atoms slide over one another along planes and directions where the atom population density is maximum. This mechanism of deformation is known as *slip.* On the other hand, if the crystal lattice has a defect in the form of impurities or distortion, which is referred

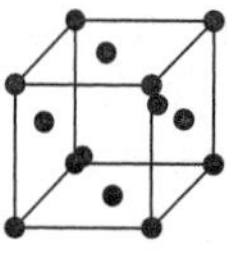

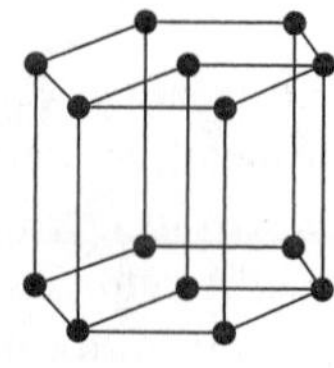

FIGURE A.1 The most common types of unit cells *(From Askeland, D. R.,* The Science and Engineering of Materials, *3rd ed. Boston: PWS Publishing, 1994)*

to as *dislocation,* slip and, therefore, deformation will be more difficult. In fact, the distortion in the crystal lattice that impedes slip can be a result of deformation in the first place. In other words, the more deformation a specimen of metal undergoes, the more difficult it is to produce more deformation. This is usually referred to as *work-hardening* and is encountered when forming metals in their cold state. Work-hardening can, however, be eliminated or inhibited if the temperature of the metal is raised. The reason is that the movements and activity of atoms at elevated temperatures eliminate dislocations and create new grains or crystals that are distortion-free. The process of creating new crystals is known as *recrystallization,* and it always takes place above a certain temperature, called the *recrystallization temperature,* which differs for different metals.

Ceramics

A *ceramic* can be defined as a chemical compound including one or more metals with a nonmetallic element. The atoms comprising a ceramic body are bonded together with either *ionic* or *covalent bonds.* As you studied in chemistry, these types of atomic bonding are very strong and rigid as compared with the metallic bonds that exist in metals. Therefore, the mechanism of failure in ceramics is not slip as in the case of metals, but rather the separation or cleavage failure that characterizes brittle materials. That is to say that tensile loading tends to result in cleavage failure in ceramics and is accompanied by very little or no deformation.

Because of the absence of any ability to undergo plastic deformation and the strong bonds between atoms, ceramics possess high hardness. Also, because the electrons are tied up in that strong type of bonding, ceramics are chemically inert (i.e., do not react with other materials) and are good electrical insulators. Examples of ceramics include aluminum oxide, boron nitride, and silicon carbide.

Polymers

Polymers is the scientific name for the engineering materials commonly known as *plastics.* The word *polymer* stems from two Greek words: "poly," which means many, and "meras," which means parts. Polymers are composed of long chains, each being a giant molecule that, in turn, includes many small molecules linked together. For instance, if many molecules of ethylene (C_2H_2) are linked together, the result will be a long molecular chain of the polymer polyethylene, which has many applications in our everyday life. As can be seen in Figure A.2, the carbon atoms

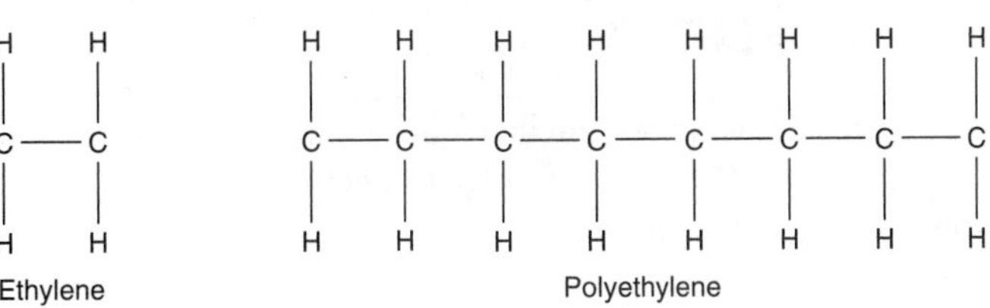

FIGURE A.2 Structural formula of ethylene and polyethylene

in ethylene have unsaturated valence bonds (carbon has a valence of 4, and hydrogen has a valence of 1). Therefore, if other ethylene molecules attach to each side of that molecule, the valence bonds on the carbon atoms in the molecule will be satisfied. In other words, for the carbon-to-carbon bond to satisfy valence requirements, numerous ethylene molecules must attach to one another and form long chains. When the chains grow longer, they get tangled, and a sample of polyethylene is analogous to a bowl of spaghetti.

Polymers are classified into two groups based on their manufacturing properties: *thermoplastic* and *thermosetting* polymers. In thermoplastic polymers such as polyethylene, the only force that binds these molecular chains together is the *van der Waals attraction bond,* which is rather weak. Consequently, the mobility of these molecular chains relative to one another is easily achievable through mechanical loading or thermal activation. For this reason, thermoplastic polymers creep and collapse easily under mechanical loads. They soften, melt, and viscously flow when heated, then solidify when cooled, and then melt again when heated again. In contrast to thermoplastics, the molecular chains of thermosets are cross-linked, forming a three-dimensional network. As a consequence, thermosets can carry mechanical loads higher than those that thermoplastics can withstand. Also, thermosets do not melt and flow, but rather char and burn. Still, in both thermoplastics and thermosets, the backbone of the molecular chains is the element carbon, although it is replaced by silicon in a few cases where high-temperature applications are required.

A.2 PROPERTIES OF MATERIALS

There are hundreds of properties that can be measured accurately in the laboratories. The properties of a certain material are a determining factor when considering the applications and processing methods of that material. During the process of selecting a material for certain applications, the properties play a major role. In order to handle the wide variety of properties, they are usually classified into three main categories: *physical, chemical,* and *mechanical* properties. Physical properties are those that pertain to the science of physics and include, for example, the color, the density, and the magnetic characteristics. Chemical properties involve how a material reacts with the various acids and alkalines, as well as with the aqueous solutions of salts. Mechanical properties are the characteristics of a material that are revealed when that material is subjected to mechanical loading.

A.3 STANDARD TESTS FOR OBTAINING MECHANICAL PROPERTIES

As you may have expected, there are standard tests and procedures that are performed in a materials laboratory in order to obtain the mechanical properties. Following are some of these standard tests.

Tension Test

The simplest way to obtain the important mechanical properties of a material is by a *tension test.* It is carried out by employing a tensile-testing machine that incorporates a means of applying a tensile force and a means for measuring that force. A standard test specimen like that shown in Figure A.3 is used. It has enlarged ends to facilitate gripping in the machine and a uniform cross section in the middle. Gauge marks are scribed on the middle section so that the extension can be measured during loading. The load–extension curve can, therefore, be obtained for any desired material. A typical graph for mild steel is shown in Figure A.4. As can be easily realized, the cross-sectional area of the specimen markedly affects its load-carrying capacity. For this reason, we rely on using the concept of stress or load intensity when making comparisons between the mechanical properties of materials. The *engineering stress* is given by

$$S = \frac{P}{A_o} \tag{A.1}$$

where A_o is the cross-sectional area of the specimen. On the other hand, it is not difficult to see that, for the same level of stress, the extensions of rods are dependent upon the initial length. It is, therefore, appropriate to consider the extension per unit length rather than the total extension. The extension per unit length is referred to as the *engineering strain* and is given by

$$e = \frac{\Delta\ell}{\ell_o} = \frac{\ell - \ell_o}{\ell_o} \tag{A.2}$$

where: ℓ is the current length

ℓ_o is the initial length

Now, it is time to plot the relationship between the engineering stress and the engineering strain. It will look like the load–extension curve; the difference is the scale of the X and Y axes, as shown in Figure A.5. Looking at the graph in Figure A.5, we can easily distinguish three distinct regions, as follows:

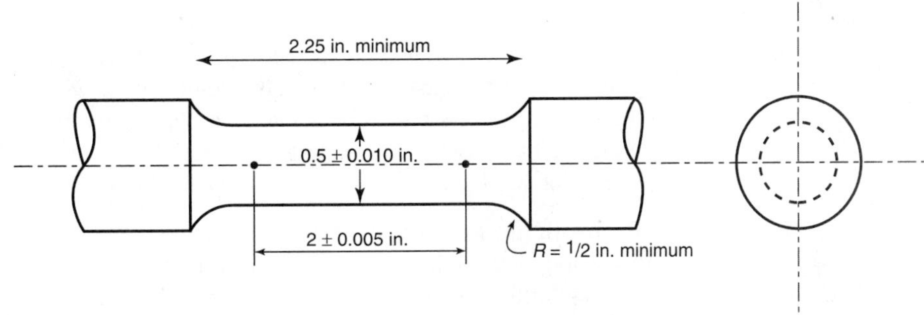

FIGURE A.3
A standard tensile test specimen

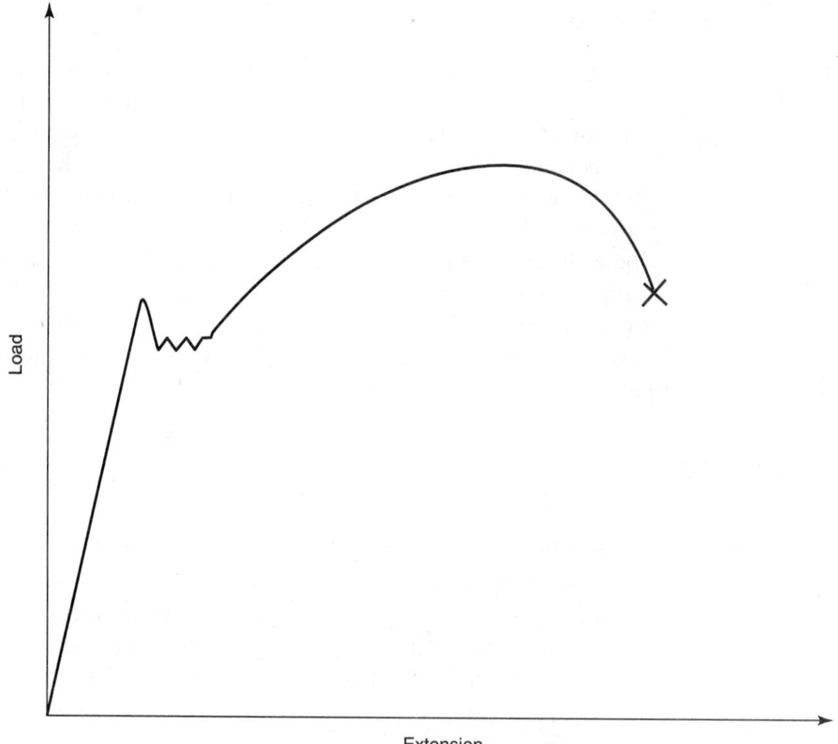

FIGURE A.4
A typical load–extension curve for mild steel

1. Region I, in which the relationship between the stress and the strain is linear, is referred to as the *elastic region.* As is evident, the strain is directly proportional to the stress. Elastic strain is also reversible (i.e., it disappears as soon as the stress is removed). The relationship between the stress and the strain can, therefore, be expressed as follows:

$$\frac{S}{e} = E \tag{A.3}$$

where E is Young's modulus of elasticity and has the same units as the stress. Region I ends at the elastic limit (i.e., the point beyond which the relationship between the stress and the strain is no longer linear). Finally, the bar (specimen) yields, as indicated by constant or dropping stress that is accompanied by appreciable strain. The point at which this process starts is called the *yield stress* (yield strength).

2. Region II, in which major plastic deformation occurs and is accompanied by strain-hardening or work-hardening (i.e., the stress beyond which the material deforms increases with increasing strain), is a result of the piling up of lattice faults or dislocation.

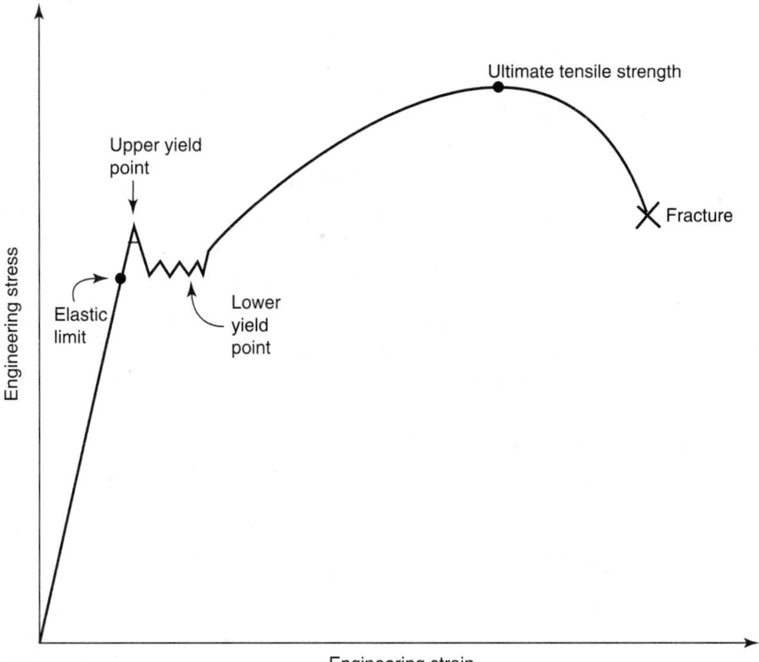

FIGURE A.5
The engineering stress–strain curve

3. Region III is where the specimen being tested collapses at the end. At the beginning of this stage, the middle section of the specimen contracts, resulting in necking. Consequently, the specimen cannot take the high level of stress, stress decreases while the specimen continues to elongate, and, finally, fracture takes place.

Following are some specific tensile properties:

- *Young's modulus of elasticity* is the slope of the linear elastic curve.
- *Yield stress* is the stress at which the material first yields and undergoes plastic deformation.
- *Ultimate tensile strength* (UTS) is the maximum engineering stress that is observed during the test.
- *Ductility* is the ability to undergo large plastic deformation. Ductility has two common measures, as follows:

$$\text{elongation percentage} = \frac{\ell_{\text{fracture}} - \ell_{\text{original}}}{\ell_{\text{original}}} \times 100 \tag{A.4}$$

where: ℓ_{fracture} is the final length of the specimen
ℓ_{original} is the original length

$$\text{reduction in area percentage} = \frac{A_o - A_f}{A_o} \times 100 \qquad \textbf{(A.5)}$$

where: A_o is the original cross-sectional area

A_f is the final area after fracture

- *Resilience* is the ability of the material to store elastic energy. It is more correctly called the *modulus of resilience* and is equal to the area under the linear elastic part of the engineering stress–strain curve. As you can see, this property is very important when selecting materials for springs.
- *Toughness* is the ability of the material to absorb energy until it breaks. A quantitative indication of toughness is the *modulus of toughness,* which is the whole area under the engineering stress–strain curve. The modulus of toughness is a measure of the ability of the material to withstand shock loading. Machine components that are subjected to sudden loading during their service life must be made from materials that possess high toughness.

Hardness Test

Hardness is defined as the ability of a material to resist scratching, abrasion, or indentation. It is probably the easiest criterion or measurement in acceptance tests and quality control of raw stock as well as manufactured products. Although there are numerous ways of measuring hardness, the most commonly used hardness-testing methods are the *Brinell* and the *Rockwell* hardness tests.

Brinell hardness test. The Brinell hardness test involves forcing a hardened-steel ball into a metal specimen under a definite static load and then measuring the size (diameter) of the impression produced by the ball. In this case, the hardness index, which is called the *Brinell Hardness Number,* is the static load acting on the ball divided by the spherical area of the impression (the unit is kilograms force per square millimeter). The hardened-steel ball has a diameter of 10 mm, the applied load is 3000 kg, and its duration is at least 10 seconds, for ferrous metals. For nonferrous metals, the load is 500 kg, and its duration is at least 30 seconds. The diameter of the impression is usually obtained by optical magnification projected on a screen, and the Brinell Hardness Number corresponding to this value can be obtained from tables, thus eliminating the need for calculations. It is not difficult to realize that the Brinell Hardness Number and the ultimate tensile strength for a metal are correlated.

Rockwell hardness test. Similar to the Brinell hardness test, the Rockwell hardness test involves forcing an indentor into a test specimen under static load. However, in the Rockwell testing method, the hardness index is determined by measuring the increment of depth (of the impression) as a result of applying a primary and a secondary load instead of by measuring the diameter. Consequently, there is no need for optical measurements or calculations, and the *Rockwell Hardness Number* is readily shown on a dial indicator. The Rockwell hardness test is, therefore, commonly used in industry because of its simplicity and the ease with which it can be performed.

Two standard indentors are used with two hardness scales to determine the Rockwell Hardness Numbers for nearly all the common metals and alloys. These indentors are a hardened-steel ball having a diameter of 1/16 inch (1.59 mm) and a diamond cone having an apex angle of 120° and a rounded tip 0.2 mm in radius, called the *brale,* and they are used with scales designated as B and C, respectively. The working range for scale B, which is used for nonferrous metals and annealed low-carbon steels, is from R_B 0 to R_B 100. For the sake of measurement accuracy, when the hardness of the material being tested exceeds R_B 100, you must switch to scale C; if the hardness is less than R_B 0, another appropriate Rockwell hardness scale should be used. The useful range of the C scale, which is used for hardened and tempered steels, is from R_C 20 (equivalent to R_B 97) to slightly above R_C 70. Owing to inherent inaccuracies associated with shaping the brale, the C scale should not be used for measuring hardness below R_C 20; instead, the hardened-steel ball and scale B are usually employed.

Figure A.6 shows the procedure for performing a Rockwell hardness test on the C scale. First, the test specimen is placed on the anvil at the upper end of the elevating screw. The capstan wheel is then rotated so as to bring the surface of the test specimen in contact with the indentor. By further rotation of the wheel, the test specimen is forced against the indentor, and a minor load of 10 kg is slowly applied in order to seat the specimen firmly. At this moment, the dial indicator of the apparatus is set to zero. Next, an additional load of 140 kg (90 kg in a test on the B scale) is applied by means of a release handle mounted on the side of the apparatus. The total major load will now be 150 kg, and the duration of its application should be at least 10 seconds. Obviously, the application of this load forces the indentor into the specimen to an additional depth. Still, this depth must not be considered as an indication of hardness because it includes an elastic as well as a plastic deformation. Therefore, the additional load is released without removing the minor load, and the hardness index is then shown on the dial indicator. The reading reflects the permanent or plastic increment of

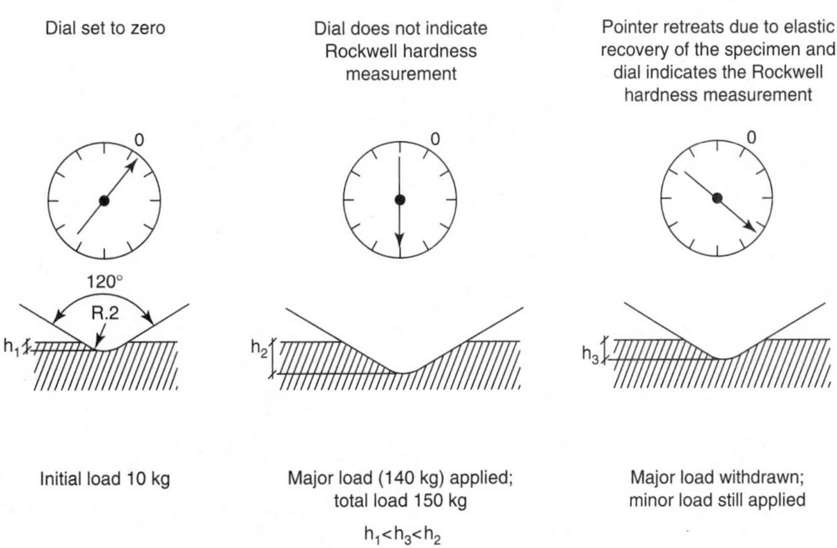

FIGURE A.6 Procedure for performing a Rockwell hardness test on the C scale

penetration depth resulting from the increment of load between the minor and major loads. It does not indicate the total depth of penetration of the indentor. As is the case in the Brinell hardness test, care must be taken to ensure that the surface conditions of the test specimen (its flatness and its thickness) are within the limits specified by the standards.

Impact Test

Impact strength, a measure of a material's ability to withstand shock loading, is defined as the energy required to fracture a given volume of material (the common unit is joules per cubic centimeter). A pendulum-type impacting machine is commonly em-

FIGURE A.7
Commonly used impact tests

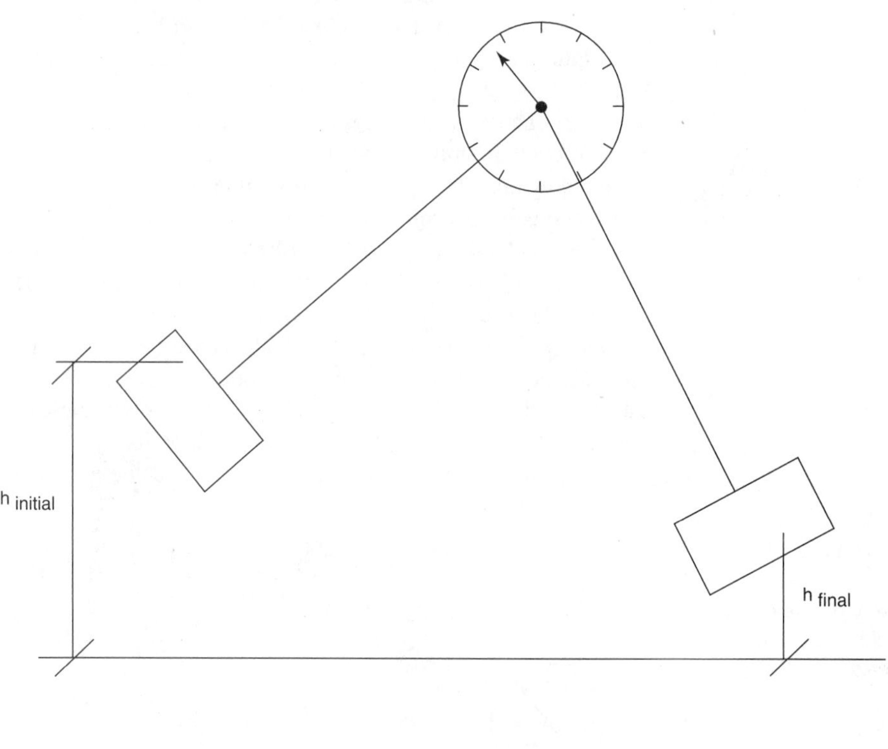

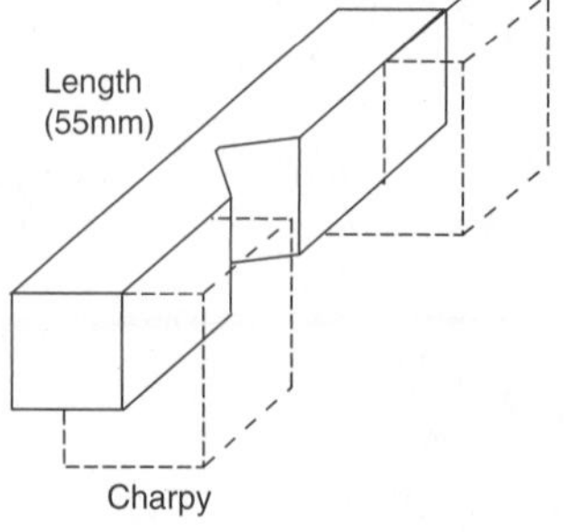

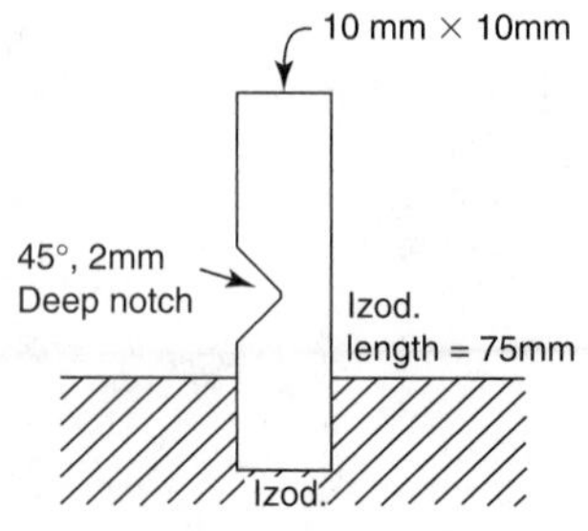

ployed when measuring the impact strength of metals and polymers. There are, however, two types of impact tests that find widespread application: the *Charpy* and the *Izod* impact tests, which are shown in Figure A.7.

Usually, the impact specimen has a notch at the spot where it is desired to promote fracture. In such cases, the impact data are reported as Charpy V or notched Izod. This is not the case when testing brittle materials; there is no need to have a notch because the impact strength of a brittle material is naturally low. As you may have guessed, the notched impact data cannot be compared with the unnotched data.

Note that the impact strength of a material is affected by temperature. Some metals experience a sharp drop in impact strength at low temperature. In fact, it has been recently revealed that this was the main reason behind the sinking of the *Titanic,* as well as many other ships during World War II, in the chilly North Atlantic waters.

Fatigue Test

We can see that the results of a tensile test cannot be used when the part or machine component is to be subjected to dynamic alternating loads. The test used should emulate the service life conditions so as to be a good measure of how well a material will withstand the dynamic loading, such as that shown in Figure A.8. The number of cycles for which a component or a specimen can withstand an alternating load primarily depends upon the magnitude (or amplitude) of that load. The higher the magnitude of the alternating load, the smaller the number of cycles after which it fails. It has also been observed experimentally that the magnitude of the load that causes failure is much less than the yield stress of the material, indicating that the mechanism of failure in this case is different from one in which the specimen is subjected to uniaxial tension. This phenomenon is called *metal fatigue,* and the failure is due to initiation and then propagation of a crack within the cross-sectional area of the part.

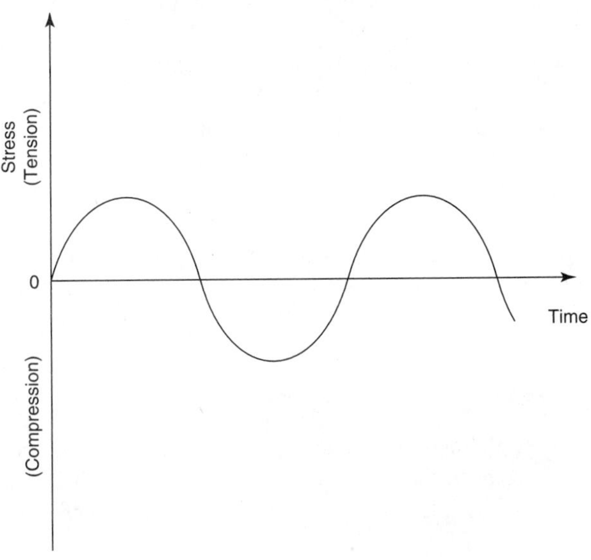

FIGURE A.8 A typical dynamic alternating load

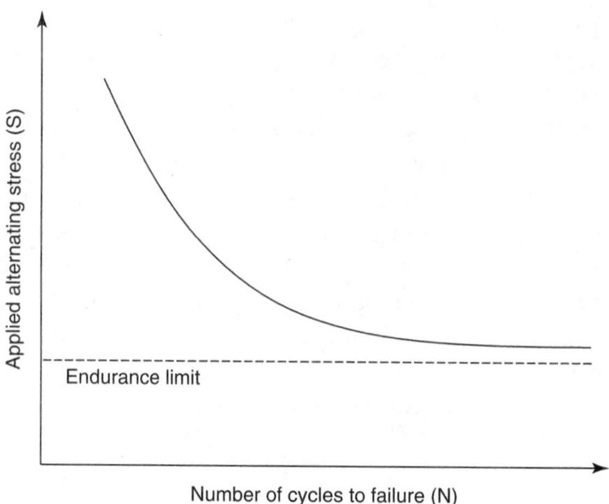

FIGURE A.9
The S–N curve

In order for a machine component to be designed properly to carry an alternating load, it must be capable of withstanding an infinite number of cycles at a particular level of loading. Consequently, it is important to determine the maximum alternating load that the desired material can withstand for an infinite number of cycles. Such a load is referred to as the *endurance limit* of the material. This property can be obtained experimentally by constructing an S–N curve for the desired material. A typical procedure involves preparing a set of identical test specimens, subjecting each to a different magnitude of alternating stress, and recording the number of cycles until failure occurs in each case. Higher magnitudes of load mean a smaller number of cycles, and vice versa. By plotting the magnitude of the stress (S) versus the number of cycles to failure (N), we can obtain the S–N curve, as shown in Figure A.9.

The S–N curve is asymptotic to a certain value of stress, which is clearly the endurance limit of the material. The endurance limit is affected by many factors, such as the surface roughness of the test specimen or the presence of stress raisers that promote the initiation and propagation of fatigue cracks. Sometimes, the S–N curve is plotted on semilogarithmic scale in order to facilitate the determination of the endurance limit. In addition, an increase in the material's temperature will result in a decrease in the endurance limit of the specimen. The frequency with which the load is applied also influences the endurance limit, especially in the case of polymers, where the dissipated energy reappears in the form of heat.

Creep Test

Creep can be defined as plastic deformation at elevated temperature under constant load. It occurs even though the applied stress is less than the yield stress at that temperature. It is also interesting to note that some polymers undergo creep at room tem-

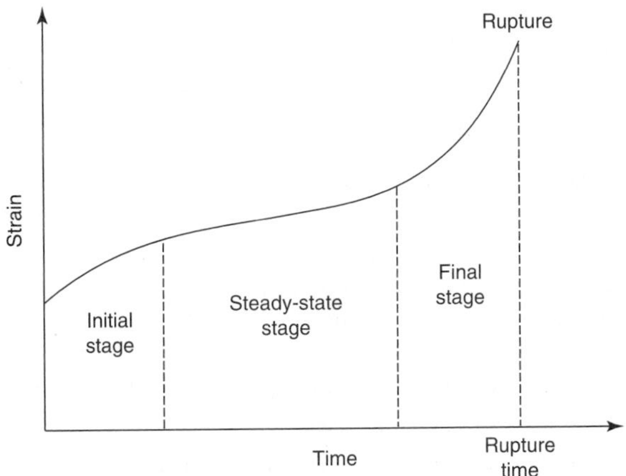

FIGURE A.10
A typical creep curve

perature. The creep characteristics of a material are determined by means of a *creep test,* which involves subjecting the specimen to different constant stresses at elevated temperatures and observing the corresponding strain. For each condition of a constant stress and a constant temperature, the elongation (or strain) is measured as a function of time and plotted to give a creep curve, as shown in Figure A.10. As can be seen, the curve involves three stages: an initial stage, a steady-state stage, and a final stage where deformation takes place at an accelerated rate after necking begins. The final stage continues until failure. For design purposes, the creep data are presented in the form of a family of straight lines on a log-log scale that indicates the relationship between stress and rupture life, with temperature as a parameter that determines the specific straight line to use, as shown in Figure A.11.

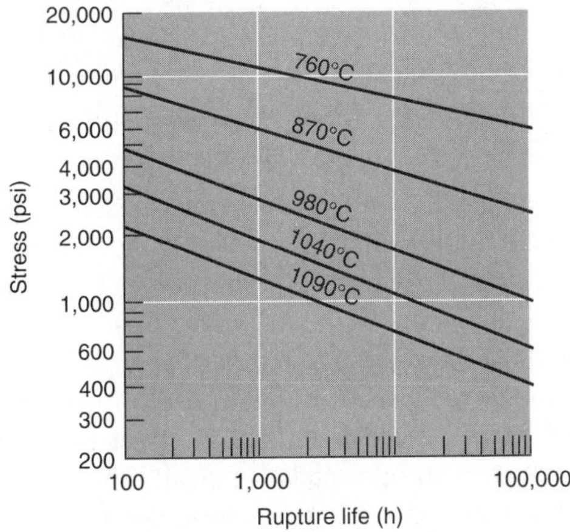

FIGURE A.11
Stress–rupture curves for an iron-chromium-nickel alloy *(From Askeland, 3rd ed., 1994)*

A.4 PHASE DIAGRAMS

Phases and Components

Before we examine phase diagrams, we should fully understand what a phase is. A *phase* can be defined as a portion of material that is chemically and physically homogeneous and is separated from other portions by a well-defined surface (interface). Now, if we have ice and water in an isolated chamber, they must be considered as two phases and not one phase. Although the system is chemically homogeneous (both are H_2O), it is not physically homogeneous (one is a solid and one is a liquid). In fact, what we have is a system under *equilibrium* (i.e., the ratio between ice and water will continue to be the same until an external factor such as pressure or temperature acts to disturb that state of equilibrium). There are different types of systems based upon the *components* (elements or compounds) that form the system. A *unary* system has a single component (like the preceding example), a *binary* system is generated or initiated by two components, and a *ternary* system involves three components.

Binary Phase Diagrams

In order to simplify the discussion of phase diagrams and alloying, only binary phase diagrams will be covered here. Interested readers are advised to consult a specialized text on materials science such as *The Science and Engineering of Materials* by D. R. Askeland (see the reference list at the end of this book).

A *phase diagram* is like a road map in that it is a graphical illustration of the various phases present in equilibrium for different compositions of two alloying elements at different temperatures. By convention, the composition is always plotted horizontally and the temperature vertically. An alloy having a composition of x percent of component A—you can see that it has $(100 - x)$ percent of component B—at a temperature T is represented by a point. Zones or regions in which specific phases exist are bounded by lines or curves. Phase diagrams are useful tools in studying alloying and alloys and in optimizing parameters in casting processes and heat treatment operations. The shape of a binary phase diagram depends upon whether the components are mutually soluble, partly soluble, or totally insoluble in their solid state. Let us discuss each of these cases in the following sections.

Isomorphous phase diagrams. An *isomorphous* phase diagram corresponds to the case of two elements that are completely mutually soluble in their liquid state as well as in their solid state. A typical example of this case is the copper-nickel (Cu-Ni) equilibrium phase diagram shown in Figure A.12. The first vertical line to the left indicates an alloy containing 100% Cu and 0% Ni (i.e., pure copper). The melting point of copper indicates the boundary between the two phases for pure copper: Above it is the liquid copper phase, and below it is the solid copper phase. As can be seen in Figure A.12, the same applies to pure nickel. For all other alloys, the region above the upper-boundary curve represents a liquid solution of nickel in copper. That curve is, therefore, referred to as the *liquidus*. The lower curve is called the *solidus* because the

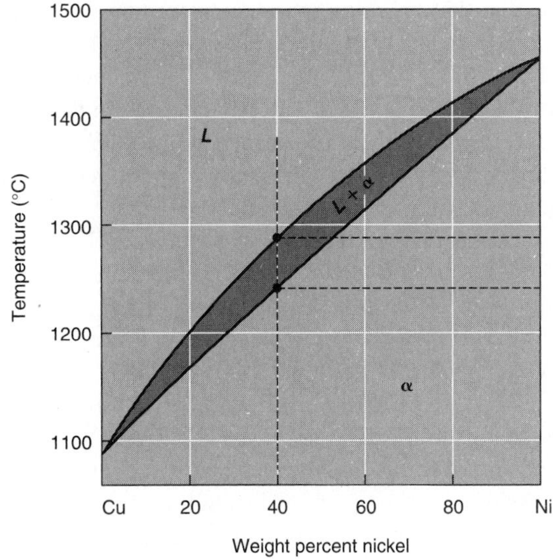

FIGURE A.12 Copper-nickel equilibrium phase diagram *(From Askeland, 3rd ed., 1994)*

area below it represents a solid solution of nickel in copper. The area between the two curves is a transition zone in which both a liquid and a solid phase exist. If we draw a vertical line representing a specific alloy (say, 60% Cu and 40% Ni), it will intersect with the liquidus at a point that indicates the temperature at which solidification starts and will intersect with the solidus at the temperature at which that alloy becomes completely solid. The difference between these two temperatures is the freezing range. Cast alloys must usually have a relatively wide freezing range in order to flow and fill the mold completely before solidification occurs.

Now, let us take a closer look at how various alloys containing different percentages of nickel solidify. This is achieved by plotting the temperature–time relationship when the molten alloy is left to cool down naturally. The graph is referred to as a *cooling curve.* As can be seen in Figure A.13, the cooling curve for a pure metal has a

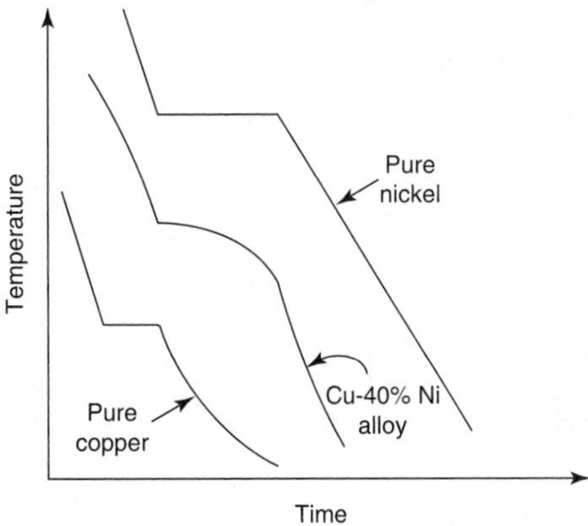

FIGURE A.13 Cooling curves for different alloys

plateau, indicating that solidification takes place at a constant temperature over a period of time during which the molten metal loses the latent heat of fusion. This is not the case with the Cu-40% Ni alloy, where solidification is gradual and there is no plateau. In this case, crystals of a solid solution rich in copper precipitate first, and the remaining liquid phase is, therefore, richer in nickel. With further cooling, more crystals rich in copper precipitate, and the process continues, as explained earlier, until the alloy completely solidifies.

Next, we have to answer the following question: How do the mechanical properties of a copper-nickel alloy differ from those of either copper or nickel? In order to answer this question, let us consider the atomic arrangement. Copper has an FCC lattice, and so does nickel. When they form a solid solution, some nickel atoms replace copper atoms in the lattice. But because a nickel atom is larger than a copper atom, the lattice becomes distorted. As a consequence, it is difficult for atoms to slide or slip over one another when the alloy is subjected to mechanical loading. In other words, the strength of the alloy will be higher than that of pure copper (or nickel), while the ductility will decrease. This type of solid solution is referred to as a *substitutional* solid solution, and it is common when both alloying elements have the same crystal lattice and more or less the same size of atoms. On the other hand, the phenomenon known as *solid solution strengthening* is shown in Figure A.14.

Note that there is another type of solid solution that does not necessarily yield the same type of phase diagrams, and it is known as the *interstitial* solid solution. As the

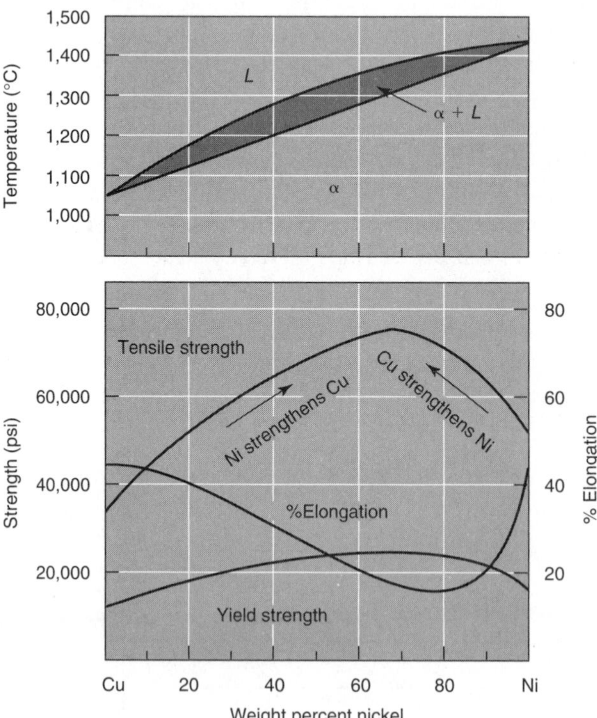

FIGURE A.14
Mechanical properties of copper-nickel alloys *(From Askeland, 3rd ed., 1994)*

name suggests, atoms of the alloying additive occupy the interstitial cavities in the crystal lattice of the base metal. In fact, this is always the case when carbon or nitrogen, which both have smaller atoms, dissolve into metals having FCC lattices, such as iron at elevated temperatures like 1673°F (slightly above 912°C).

Eutectic phase diagram. A *eutectic* phase diagram corresponds to the case of two components that are completely mutually soluble in their liquid state but exhibit negligible or no solubility in their solid state. A typical example of this case is the bismuth-cadmium (Bi-Cd) binary phase diagram shown in Figure A.15. As can be seen in the figure, alloy I at a temperature above the liquidus takes the form of a homogeneous liquid solution and is rich in bismuth. When that alloy is supercooled to a temperature just below the liquidus, solid crystals of bismuth start to precipitate (because the liquid alloy is supersaturated in bismuth). As a consequence, the remaining liquid alloy will have less bismuth. When it is further cooled, more solid bismuth crystals precipitate, and the concentration of cadmium continually increases in the molten fraction that remains after the precipitation of the solid bismuth crystals. On the other hand, if a molten alloy rich in cadmium, such as alloy II, is supercooled, solid crystals of cadmium precipitate first, and the remaining liquid fraction will become richer in bismuth. As you may have guessed, there is a certain alloy with a specific concentration in which solid crystals of both bismuth and cadmium precipitate simultaneously when that molten alloy is supercooled below the liquidus. When solidification is completed, the solid constituents will form a mechanical mixture that can be revealed by microscopic examination.

A better understanding of the solidification of this type of alloy can be gained through examination of the cooling curves. It can be seen in Figure A.16 that alloy III solidifies at a constant temperature that is below the melting points of both bismuth and cadmium. For this reason, both the reaction and the alloy are referred to as *eutectic* (after a Greek word meaning "easily fusible"). After complete solidification, alloy I will consist of bismuth crystals and the eutectic, while alloy II will consist of cadmium crystals and the eutectic.

FIGURE A.15
Bismuth-cadmium binary phase diagram

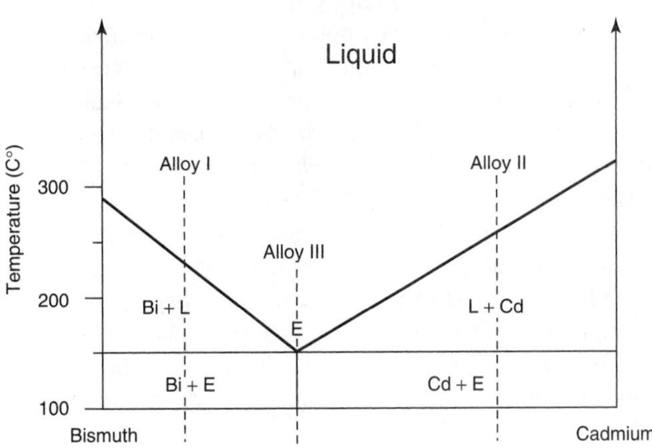

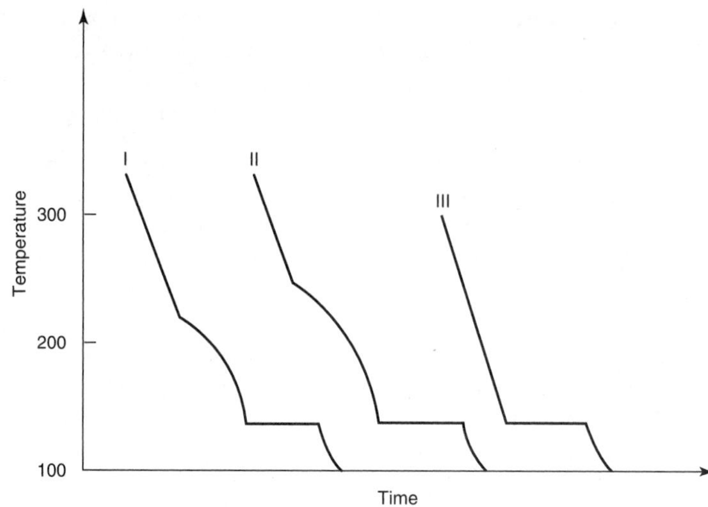

FIGURE A.16
Cooling curves for alloys of a eutectic system

Before leaving our discussion of eutectic systems, we need to see how the composition of an alloy affects the strength of such an alloy. The easiest alloy to consider first is the eutectic alloy. The very nature of the fine mechanical mixture impedes the slip of atoms (each phase forms an obstacle to the next one), thus resulting in an increase in the strength. As a consequence, alloys having a higher percentage of the eutectic possess strength higher than that of alloys containing less of the eutectic.

Alloys showing limited solid solubility in the solid state. A typical example of this case is the lead-tin (Pb-Sn) equilibrium phase diagram shown in Figure A.17. As anticipated, this phase diagram is a hybrid of the preceding two types. Instead of two elemental components forming a mechanical mixture, the eutectic in this case is a mechanical mixture of two solid solutions. As can be seen in Figure A.17, there is a line below the solidus in the limited solid solubility region. This line indicates decreasing solid solubility with decreasing temperatures, and it is known as the *solvus.* At 362°F (183°C), the solubility of tin in lead is 19% and decreases to only 2% at room temperature. As a consequence, if any alloy containing between 2% and 19% tin cools down past the solvus, the solubility limit is exceeded, and the surplus tin precipitates in the form of a solid solution of lead in tin, which is referred to as β. The properties of this type of alloy can be controlled by controlling the amount and characteristics of the β phase. If we can make it take the form of tiny particles that are distributed all over, then slip and deformation will be inhibited, and the result will be an increase in the strength of the alloy. This mechanism for controlling the properties of the alloy is known as *dispersion strengthening.* This phenomenon is very important for the heat treatment of nonferrous alloys. Now, it is not difficult to see that any alloy containing from 19% to 61.9% will have a microstructure consisting of α solid solution surrounded by a solidified eutectic mixture called a *microconstituent.* This type of alloy is known as a *hypoeutectic* alloy, as opposed to one that contains from 61.9% to

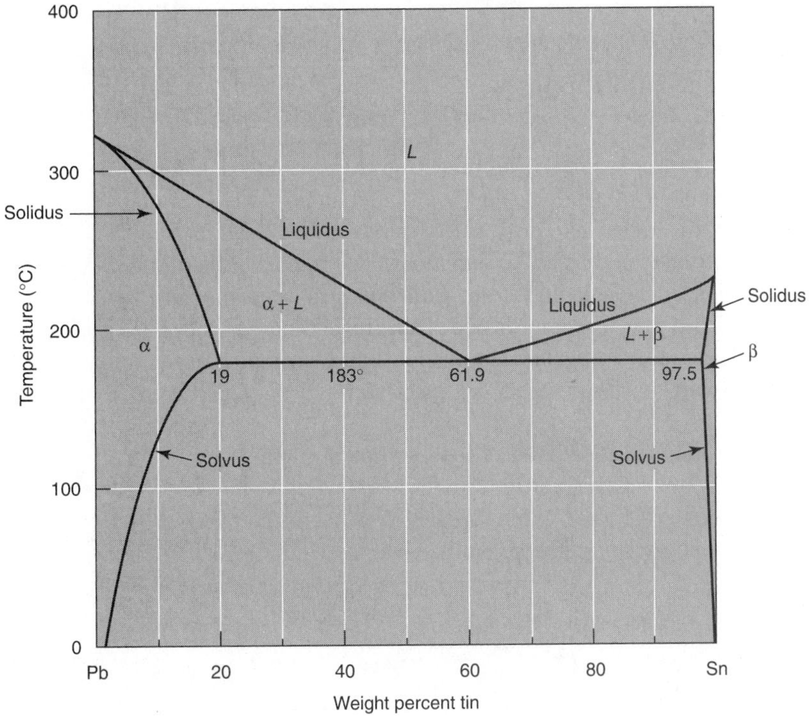

FIGURE A.17 Lead-tin equilibrium phase diagram *(From Askeland, 3rd ed., 1994)*

99% tin, which is called a *hypereutectic* alloy and has a microstructure consisting of β solid solution surrounded by a eutectic.

A.5 FERROUS ALLOYS

Iron-Carbon Phase Diagram

In order to thoroughly understand ferrous alloys, we first must become familiar with the phase diagram. The iron-carbon (Fe-C) phase diagram is slightly more complicated than the ones we have covered so far. Iron and carbon form an intermetallic compound (iron carbide, Fe_3C) that contains 6.7% carbon. The phase diagram has, therefore, two components: iron and iron carbide. It is actually an iron-iron carbide equilibrium phase diagram. Nevertheless, for all practical purposes, it is the percentage of carbon that is indicated and not the percentage of iron carbide. We just trim the phase diagram at 6.7% carbon (i.e., the percentage at which the intermetallic compound is formed).

Figure A.18 shows the iron-carbon equilibrium phase diagram. Alloys containing more than 2.1% carbon are known as *cast irons.* They have a low melting point 2100°F (1148°C) and a wide freezing range because of the eutectic reaction that takes place at 2100°F (1148°C) and 4.3% carbon. Alloys containing less than 2.1% carbon are

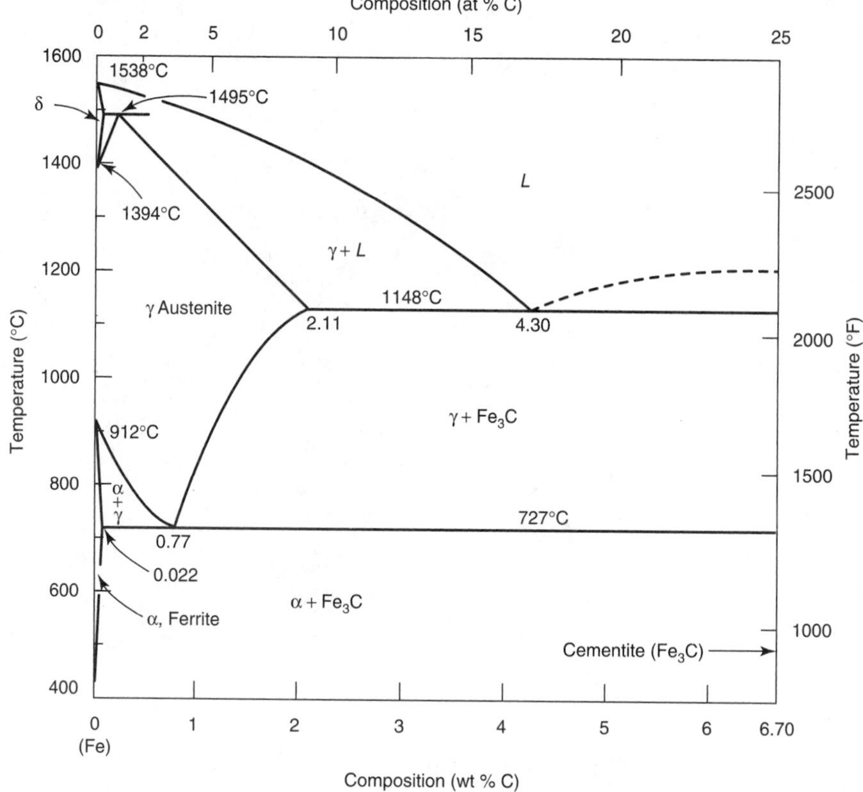

FIGURE A.18 Iron-carbon equilibrium phase diagram *(From Askeland, 3rd ed., 1994)*

referred to as *steels.* They are shown in the lower left-hand section of the phase diagram. (A good engineering student will memorize this section and know it by heart.)

Now, let us discuss each of the phases shown in the iron-carbon phase diagram. The first phase is α-iron, or *ferrite.* It has a BCC lattice and is stable up to 1673°F (912°C), where an allotropic reaction takes place. Ferrite is soft and ductile, is ferromagnetic below 1418°F (770°C), and can dissolve a maximum of 0.02% carbon (because the center of the BCC unit cell is occupied by an iron atom). The other phase is γ-iron, or *austenite.* It has an FCC lattice and can, therefore, dissolve up to 2% carbon (because the center of the unit cell has no atoms). Austenite appears as a result of an allotropic reaction at 1673°F (912°C). It is stable up to 2543°F (1394°C), where another allotropic reaction takes place. Austenite is soft like ferrite, but it is not ferromagnetic. The intermetallic compound Fe_3C (iron carbide) is called *cementite* by metallurgists. It is hard and brittle because of its complex crystal structure. As can be seen from Figure A.18, the microstructure of carbon steels at room temperature is usually a combination of these phases. There is, however, a eutectic-like reaction at a temperature of 1340°F (727°C) and a carbon content of about 0.8%. This reaction involves the decomposition of austenite into a mechanical mixture consisting of lamellar alternate layers of cementite and ferrite. This microconstituent is known as *pearlite* because, under the microscope, it looks like mother-of-pearl. Because this reaction takes place in the solid state, it is referred to as a *eutectoid* reaction (to distinguish it from the eutectic reaction that involves liquid phases). Pearlite has fairly good strength and toughness because the cementite layers impede deformation while the ferrite layers are soft and ductile. In fact, all carbon steels contain pearlite to varying degrees. Steel that contains less than 0.8% carbon is known as *hypoeutectoid* steel and has a microstructure that consists of pearlite surrounded by ferrite. Steel that contains more than 0.8% carbon is called *hypereutectoid* steel and has a microstructure that consists of pearlite surrounded by cementite.

These microstructures can be obtained only when the cooling rate is very slow and diffusion of carbon atoms can take place, thus emulating equilibrium conditions. However, with sudden cooling of austenite, the crystal structure changes from FCC to body-centered tetragonal (distorted cube). Carbon atoms do not have enough time to diffuse and get trapped in the lattice, producing a supersaturated solid solution of carbon in α-iron. Because the lattice is not BCC, but rather tetragonal, the resulting phase is metastable (i.e., does not appear on the equilibrium phase diagram) and is known as *martensite.* In practice, sudden cooling is achieved by *quenching* (i.e., dropping the heated steel part into water or oil at room temperature). Only pearlite undergoes the martensite transformation. If hypoeutectoid steel is quenched, its microstructure will consist of ferrite and martensite; if hypereutectoid steel is quenched, its microstructure will include cementite and martensite.

Heat Treatment of Steel

The main reason for heat treating steels is to control their mechanical properties (such as hardness, strength, and ductility). These properties are adjusted so as to enable the structural component under consideration to withstand the conditions to which it is going to be subjected during its service life. Let us now briefly review the various heat treatment operations.

Hardening. *Hardening* involves heating steel, keeping it at an appropriate temperature until all pearlite is transformed into austenite, and then quenching it rapidly in water (or oil). The temperature at which austenitizing rapidly takes place depends upon the carbon content in the steel used (and can be obtained from the iron-carbon equilibrium phase diagram shown earlier in Figure A.18). The heating time should be increased with increasing component size to ensure that the core will also be fully transformed into austenite. (See Figure A.19.)

As previously mentioned, the microstructure of a hardened-steel part is ferrite and martensite for hypoeutectoid steel, martensite for eutectoid steel, and cementite and martensite for hypereutectoid steel. As a consequence, there is no real gain in hardness when quenching hypoeutectoid steels having a low-carbon content (e.g., 0.1% or 0.2% carbon). Because martensite is a metastable phase, the hardening operation must be followed by another heat treatment in order to control the strength and hardness of the component, as will be explained later.

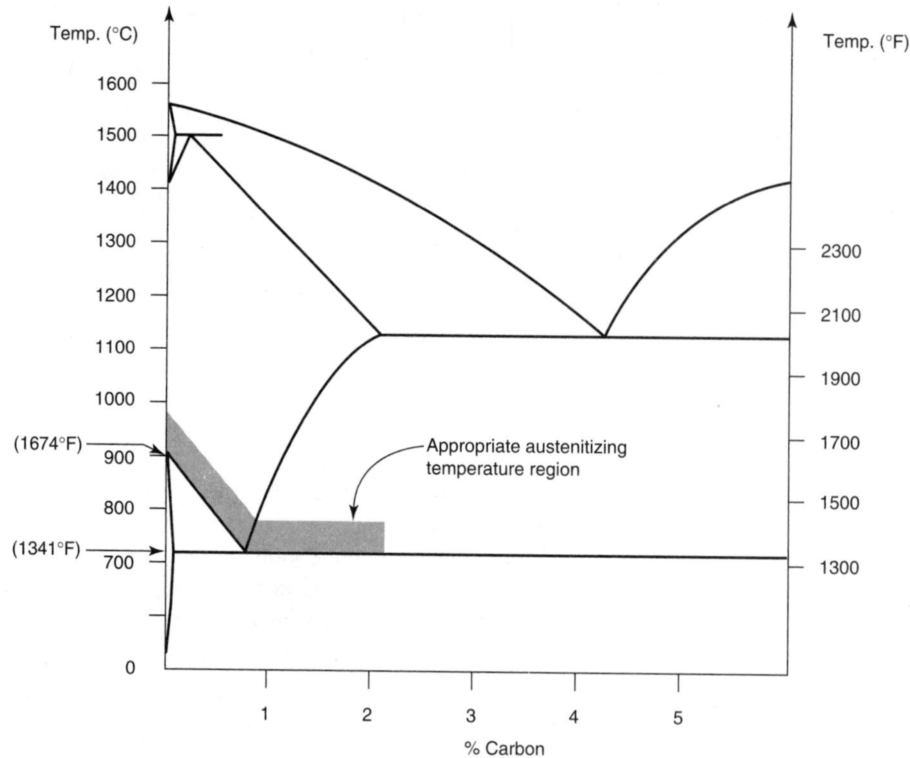

FIGURE A.19 Temperatures appropriate for austenitizing

Tempering. *Tempering* involves heating steel that has been quenched and hardened to a temperature lower than 1340°F (727°C) for an adequate period of time so that the metal can be equilibrated. During the operation, martensite is decomposed, resulting in spheroidal particles of carbide embedded and distributed in a matrix of ferrite. The hardness and strength obtained depend upon the temperature at which tempering is carried out. Higher temperatures, those just below 1340°F (727°C), will result in high ductility but low strength and hardness; low tempering temperatures will produce low ductility but high strength and hardness. In practice, metallurgical engineers make use of an experimentally obtained curve showing hardness versus temperature in order to select the appropriate tempering temperature that will produce the desired level of hardness and strength. This operation is performed on all carbon steels that have been hardened in order to reduce their brittleness so that they can be used.

Annealing. *Annealing* involves treating steel up to a temperature high enough to transform all the pearlite into austenite and then cooling it very slowly to room temperature. The resulting microstructure will be lamellar pearlite possessing high ductility and toughness but low hardness. In industry, annealing is performed by heating a component to the appropriate temperature, soaking it at that temperature, and then shutting off the furnace while the piece is in it. It is always recommended that stock steel be subjected to annealing before being processed by cold forming in order to reduce the load and energy requirements and to enable the metal to undergo large strains without failure.

Normalizing. *Normalizing* involves heating steel to the austenite temperature range, keeping it at that temperature for a period of time, and then cooling it in air. The resulting microstructure is a fine and feathery mechanical mixture of ferrite and cementite and is known as *bianite*. It has higher strength and hardness than pearlite but lower ductility. Normalizing is performed on structures and structural components that will be subjected to machining because it improves the machinability of carbon steels.

Surface Hardening

In many engineering applications, it is necessary to have the surface of the component hard enough to resist wear and erosion, while having the core ductile and tough in order to withstand impact and shock loading. This can be achieved through one of two mechanisms: local austenitizing and quenching for medium- and high-carbon steels and diffusion of hardening elements like carbon or nitrogen into the surface of low-carbon steels. Let us now discuss the various processes that are used in industry.

Flame hardening. *Flame hardening* uses a combustible gas flame as the source of heat for austenitizing the outer layer of the part. This is followed by rapid water quenching. Flame-hardened parts must then be tempered after hardening. The tempering temperatures used depend upon the desired hardness and the carbon content. This process is usually recommended for plain-carbon steels with carbon contents ranging from 0.4% to 0.95% carbon.

Induction hardening. *Induction hardening* is similar to flame hardening; the difference is the source of heat for austenitizing. In this case, a coil in which an alternating current flows induces current in the part to be hardened. This induced current alternates at frequencies of thousands of cycles per second. As a result of the "skin effect" as well as the resistance to current flow, the outer layer of the component heats up very rapidly. Water quenching and tempering produce the desired level of hardness.

Carburizing. *Carburizing* involves the diffusion of carbon into the surface layer of a component of steel that has a low-carbon content (less than 0.4%). This is achieved in industry by placing the component in a carbon-rich medium at elevated temperature. There are three versions of this process: pack, gas, and salt.

Pack carburizing involves packing the component in a steel container and completely surrounding it by a mixture of granular charcoal and barium carbonate. At elevated temperatures, a series of chemical reactions take place and finally produce atomic carbon that penetrates and diffuses into the surface of the low-carbon steel component. As a consequence, the surface layer becomes rich in carbon to a depth that depends upon the carburizing time as well as the temperature at which the part is soaked. Once the carbon content reaches the predetermined level, subsequent quench-hardening is performed.

Gas carburizing is similar to pack carburizing, except that the source of atomic carbon is a hydrocarbon gas such as natural or propane. The carburizing time is more or less the same as in pack carburizing, but this process provides the advantage of allowing quenching to be done directly at the carburizing temperature. It is for this reason that conveyor furnaces are often used for gas carburizing.

Salt carburizing is performed in a heated molten-salt bath. The typical salt used is sodium cyanide (NaCN), which is an extremely poisonous chemical compound and represents the major drawback of this process. On the other hand, its main advantage is the short heating cycle achieved through liquid convection. With today's strict environmental regulations, environmentally safe disposal of cyanide compounds creates a further problem in using salt-bath carburizing. The process is also sometimes referred to as *cyaniding* or *cyanide case-hardening.*

Nitriding. *Nitriding* involves the diffusion of monatomic nitrogen into the surface of the steel being treated. A chemical reaction takes place at elevated temperatures, resulting in iron-nitrogen or iron-alloy-nitrogen chemical compounds. The latter compounds form an extremely hard case, one that is by far harder than tool and carburized steels. A major advantage of this operation is that it is performed at subcritical temperatures and no quenching or heat treatment is required. Because the nitrogen atmosphere surrounding the workpiece is relatively inert, scaling or discoloration does not take place.

The most commonly used source of the nitrogen atmosphere is dissociated ammonia. At the nitriding temperature, which is between 925 and 1050°F (500 and 570°C), ammonia dissociates, resulting in atomic nitrogen that diffuses into steel and hydrogen that is exhausted and pumped out of the system.

The presence of some alloying elements with steel enhances the formation of a continuous, tenaciously hard case of nitrides. Such elements include aluminum, chromium, molybdenum, vanadium, and tungsten. Plain-carbon steels (not having these alloying elements) are, therefore, not well suited to this surface-hardening operation. It can, however, be applied to stainless steels, although it may reduce their corrosion resistance.

Carbonitriding. *Carbonitriding* is a surface-hardening operation that involves the diffusion of both nitrogen and carbon into the steel surface. The atmosphere surrounding the workpiece is a mixture of a carbon-rich gas such as propane or methane and ammonia. Because of the presence of nitrogen, the temperature at which this operation takes place is lower than that used in carburizing operations. In addition, unlike the nitriding operation, the process is suitable for low-carbon steels as well as low-carbon alloy steels.

Steels

Let us now review the different types of steels that are used in industry and discuss the classification system that is used in the United States. International readers are advised to contact their national standards organizations for information regarding the systems used in their countries.

Carbon and alloy steels. *Carbon steels* are alloys of iron and carbon, with up to 2% carbon and only residual amounts of other elements except those added for deoxidization (such as aluminum). There are also limits for various additives, such as silicon 0.6%, copper 0.6%, and manganese 1.65%. Carbon steels are also referred to as *plain-carbon* steels and *low-carbon* steels. These steels comprise the largest fraction of steel production and are available in all forms, such as sheet, bar, pipe, slab, and wire.

Alloy steels can be defined as those having a carbon content up to about 1% and having a total alloy content below 5%. These steels are commonly used for structural components that are required to possess superior wear, strength, and toughness properties. Note that the ability of alloy steels to be hardened (i.e., their hardenability) is far superior to that of plain-carbon steels. Some of these steels can be quenched in air instead of water or oil.

In the United States, the commonly used designation system for carbon and alloy steels is the one adopted by the American Iron and Steel Institute (AISI) and the Society of Automotive Engineers (SAE). Four digits are usually employed for identifying any steel. The first digit indicates the major alloying element, as shown in Table A.1. If the first digit is 1, the major alloying element is carbon, and the alloy is accordingly plain-carbon steel. The second digit indicates the relative percentage of a primary alloying element. If a steel is identified by 23XX, the primary alloying element is nickel, and its percentage in the alloy is about 3. The last two digits indicate the carbon content in hundredths of a percent. In addition to these numbers, letters are used as prefixes and suffixes to provide information about production methods and additives.

Tool steels. *Tool steels* are actually alloy steels having a high content of alloying elements. The amount of impurities in tool steels is much lower than that in ordinary alloy steels because tool steels are always melted in electric furnaces. Tool steels are also subjected to a rigorous quality control and inspection process that is not applied to other kinds of steel, resulting in superior alloy-content control as well as excellent cleanliness.

The AISI classification system for tool steels is based on use or application. There are basically four categories, plus one for special purposes. A prefix (letter) is used to indicate the use category. The prefix is followed by one or two digits that identify the specific alloy within the category. The first category is *shock-resisting* tool steel, and it is identified by the prefix S. The second category is *hot-work* tool steel for use in forging dies and plastic molds; the prefixes H and P identify this group. The third category is *cold-work* tool steel, which is divided into four subgroups depending upon the quenching media. The prefixes W, O, and A are used for water, oil, and air, respectively; the prefix D identifies the fourth subgroup as high-carbon, high-chromium cold-work tool steel. This subgroup contains a very large chromium-carbide content when hardened. As a consequence, it possesses excellent abrasion resistance and is, therefore, the most commonly used steel for cold-work tooling. The last category is *high-speed* tool steel, which derives its name from its intended application (i.e., machining other metals at high cutting speeds). High-speed steels can have either molybdenum or tungsten as the major alloying element, and they are identified as the M series and the T series, respectively. In both cases, they

TABLE A.1
Major groups in the AISI-SAE steel designation system

Class	AISI Series	Major Constituents
Carbon steels	10XX	Carbon steel
	11XX	Resulfrized carbon steel
Alloy steels		
Manganese	13XX	Manganese 1.75%
	15XX	Manganese 1.00%
Nickel	23XX	Nickel 3.50%
	25XX	Nickel 5.00%
Nickel-chromium	31XX	Nickel 1.25%-chromium 0.65 or 0.80%
	33XX	Nickel 3.50%-chromium 1.55%
Molybdenum	40XX	Molybdenum 0.25%
	41XX	Chromium 0.95%-molybdenum 0.20%
	43XX	Nickel 1.80%-chromium 0.50 or 0.80%-molybdenum 0.25%
	46XX	Nickel 1.80%-molybdenum 0.25%
	48XX	Nickel 3.50%-molybdenum 0.25%
Chromium	50XX	Chromium 0.30 or 0.60%
	51XX	Chromium 0.80%, 0.95%, or 1.05%
	5XXXX	Carbon 1.00%-chromium 0.50, 1.00, or 1.45%
Chromium-vanadium	61XX	Chromium 0.80 or 0.95%-vanadium 0.10 or 0.15% min.
Multiple alloy	86XX	Nickel 0.55%-chromium 0.50%-molybdenum 0.20%
	87XX	Nickel 0.55%-chromium 0.50%-molybdenum 0.25%
	92XX	Manganese 0.85%-silicon 2.00%
	93XX	Nickel 3.25%-chromium 1.20%-molybdenum 0.12%
	94XX	Manganese 1.00%-nickel 0.45%-chromium 0.40%-molybdenum 0.12%
	97XX	Nickel 0.55%-chromium 0.17%-molybdenum 0.20%
	98XX	Nickel 1.00%-chromium 0.80%-molybdenum 0.25%

contain the highest percentage of alloying elements of any of the tool steels. Their alloying elements provide resistance to softening at elevated temperatures. High-speed steels can be hardened up to 67 RC and maintain their hardness at temperatures up to 1000°F (540°C).

Stainless steels. *Stainless steels* are steels with at least 10.5% chromium that resist corrosion from oxidizing environments by exhibiting remarkable passivity. There are different types of stainless steels, and each has its own characteristics and applications.

The first type is *ferritic* stainless steel, which has a carbon content of less than 0.2% and a chromium content up to 27%. As the name suggests, the structure of this steel is ferrite at room temperature. Ferritic stainless steels are nonhardenable, are notch sensitive, and exhibit poor weldability. Still, they are used for cutlery and cookware.

The second type is *martensitic* stainless steel, which has a carbon content as high as 1.2% and a chromium content of 12% up to 18%. This steel can be hardened by heat treatment. Sharp tools and knives can be made of martensitic stainless steel.

The third type is *austenitic* stainless steel, which contains nickel as a major alloying element, in addition to iron, chromium, and carbon. Nickel is added as an austenite-stabilizing element that promotes the formation of an austenitic structure (γ-iron) at room temperature. Austenitic stainless steel is hardenable only by cold-working and is used for parts that require good chemical resistance, such as piping and tanks.

A fourth type of stainless steel includes the *precipitation-hardening* (PH) alloys. PH stainless steels can be martensitic, semiaustenitic, and austenitic. These steels possess very high strength and hardness and are, therefore, used in structural components and springs.

Other types of steel. Other types of steel include high-strength low-alloy steels, ultrahigh-strength steels, and austenitic manganese steels.

A.6 ALUMINUM ALLOYS

Aluminum possesses high electrical and thermal conductivity and good ductility and formability. It has poor tensile properties and poor rigidity (low value of E), which limits its use in building bridges or high-rise metallic structures. It has a low density of 2.7 g/cm^3, and its specific strength (i.e., strength-to-weight ratio) is, therefore, excellent, which is why it finds widespread application in the aerospace and automotive industries.

Aluminum alloys can be divided into *wrought* and *cast* alloys, corresponding to the method of fabrication. Further, we can divide each of these groups into *heat-treatable* and *nonheat-treatable* alloys. Each group (i.e., wrought and cast alloys) has a four-digit alloy-designation system that indicates the major alloying element (e.g., 2024 is a wrought aluminum alloy with copper as the major alloying element). For wrought alloys, the alloy designation is followed by a suffix that can have a number after it. The suffix indicates the kind of thermal treatment or degree of work-hardening. This system was developed by the Aluminum Association and is used mainly in the United States. International readers should consult with the national standards organizations in their countries.

The most important alloying elements in aluminum alloys are copper, manganese, silicon, magnesium, and zinc. An aluminum-copper alloy with 4% copper is a typical example of a heat-treatable alloy. The procedure involves heating the alloy to a temperature of 930°F (500°C) in order to form a homogeneous α solid solution. The piece is then quenched, resulting in a metastable phase because copper does not have the time to diffuse out of the solid solution. Next, the piece is subjected to artificial aging by holding it at some low temperature, such as 400°F (200°C), for some time. Copper will be separated and form an intermetallic compound ($CuAl_2$). The presence of tiny particles impedes deformation of the matrix and increases the strength. This phenomenon is, therefore, referred to as *dispersion strengthening,* and the process is known as *precipitation hardening.*

A.7 COPPER ALLOYS

Copper was one of the first metals to be used by humans because it could be found in its metallic form. Now, it is seldom to find copper in its pure form, and it is usually extracted from ores that contain only 5% copper by weight. Copper and its alloys have excellent electrical and thermal conductivity as well as corrosion resistance. In addition, copper (and many of its alloys) has excellent ductility. The preceding properties make copper the ideal metal for electrical conductors. In fact, more than 80% of all copper produced is used in the form of pure copper; the remainder is used in many alloy forms.

The major alloying elements in copper alloys include zinc, tin, and nickel. When zinc is the principal alloying element, the alloy is known as *brass.* Depending upon the percentage of zinc, this alloy can be α brass (70% Cu, 30% Zn) or β brass (60% Cu, 40% Zn). Whereas α brass is ductile and can be formed by various forming processes, β brass is usually produced by casting because of its lack of ductility.

When the principal alloying element is tin, the copper alloy is known as *bronze.* In fact, bronze represents a family of alloys with varying percentages of tin that may also have elements other than tin as the major alloying additive. Bronze products are usually manufactured by casting. They have excellent corrosion resistance and are, therefore, used in marine applications.

Nickel is completely soluble in copper. If it is the only alloying element, a single-phase solid solution will form, no matter what the percentage of nickel is. This family of solid solutions is referred to as *cupronickels.* They possess excellent corrosion resistance and good ductility and can be hardened only by cold-working (because the copper-nickel phase diagram is isomorphous). A common cupronickel alloy is one containing 70% copper and 30% nickel.

Nickel and zinc together are added to copper as alloying elements. With the right combination of nickel and zinc, alloys can be obtained with the appearance of silver. They are, therefore, called *nickel silvers* and are used for silverware, cutlery, and fake jewelry. Some of these alloys have good strength in the cold-worked condition and are, therefore, used in mechanical components.

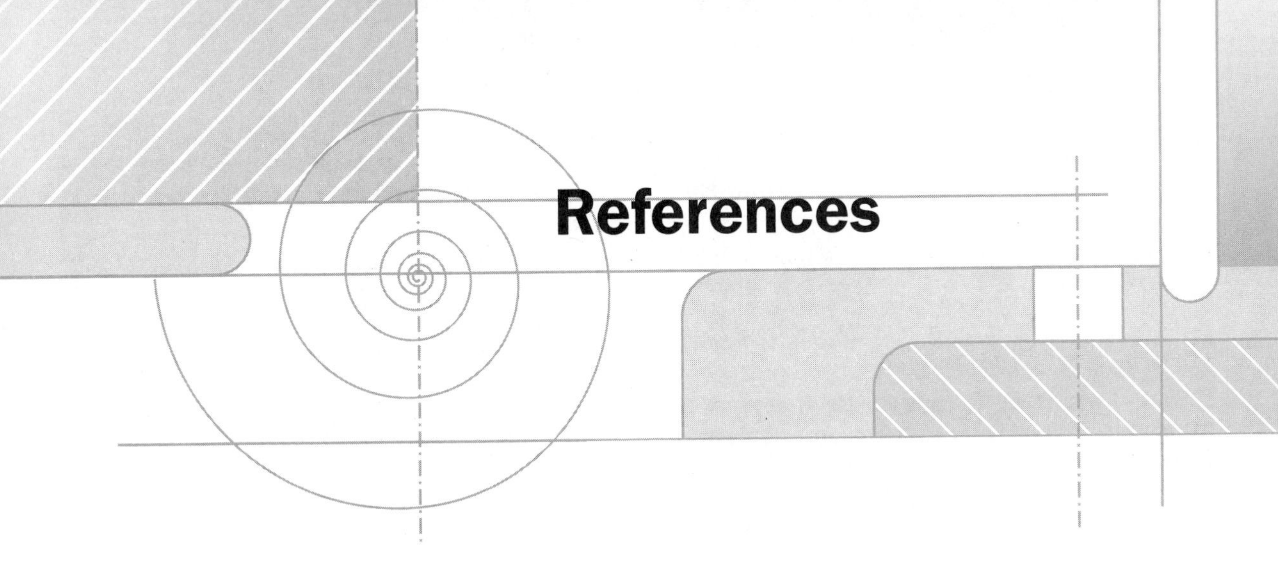

References

A Program Guide for CIM Implementation. Dearborn, Michigan: CASA, Society of Manufacturing Engineers, 1985.

Askeland, D. R. *The Science and Engineering of Materials,* 3rd ed. Boston: PWS Publishing, 1994.

Avitzur, B. *Handbook of Metal-Forming Processes.* New York: John Wiley, 1983.

Avitzur, B. *Metal Forming: The Application of Limit Analysis.* New York: Marcel Dekker, 1980.

Avitzur, B. *Metal Forming: Processes and Analysis.* New York: McGraw-Hill, 1968.

Beck, R. D. *Plastics Products Design.* New York: Van Nostrand Reinhold, 1970.

Blazynski, T. Z. *Metal Forming: Tool Profiles and Flow.* New York: Halsted Press, 1976.

Boothroyd, G. *Fundamentals of Metal Machining and Machine Tools.* New York: McGraw-Hill, 1975.

Boothroyd, G., and P. Dewhurst. *Product Design for Assembly.* Wakefield, Rhode Island: Boothroyd Dewhurst, Inc., 1991.

Boothroyd, G., C. Poli, and L. E. Murch. *Automatic Assembly.* New York: Marcel Dekker, 1982.

Bowman, B. *Teledyne Rodney Metals Handbook: Handbook of Precision Sheet, Strip, and Foil.* Metals Park, Ohio: American Society of Metals, 1980.

Boyer, H. E., and T. L. Gall., eds. *Metals Handbook: Desk Edition.* Metals Park, Ohio: American Society of Metals, 1985.

Brady, G. S. et al., eds. *Materials Handbook,* 14th ed. New York: McGraw-Hill, 1996.

Brown, R. L. E. *Design and Manufacture of Plastic Parts.* New York: John Wiley, 1980.

Budinski, K. *Engineering Materials: Properties and Selection,* 4th ed. Reston, Virginia: Reston-Prentice-Hall, 1992.

Calladine, C. R. *Engineering Plasticity.* New York: Pergamon Press, 1969.

Callister W. D., Jr. *Materials Science and Engineering: An Introduction,* 4th ed. New York: John Wiley, 1997.

Campbell, S. A. *The Science and Engineering of Microelectronics Fabrication.* New York: Oxford University Press, 1996.

Carter, D., and B. S. Baker. *Concurrent Engineering: The Product Development Environment for the 1990s.* Reading, Massachusetts: Addison-Wesley, 1993.

Cary, H. B. *Modern Welding Technology.* Englewood Cliffs, New Jersey: Prentice-Hall, 1979.

Cheremisinoff, N. P. *Product Design and Testing of Polymeric Materials.* New York: Marcel Dekker, 1990.

Computers in Manufacturing, by the editors of *American Machinist.* New York: McGraw-Hill, 1983.

Critchlow, A. J. *Introduction to Robotics.* New York: Macmillan, 1985.

Datsko, J. *Materials Selection for Design and Manufacturing.* New York: Marcel Dekker, 1997.

Davies, R., and E. R. Austin. *Developments in High-Speed Metal Forming.* New York: Industrial Press, 1970.

Degarmo, E. P., J. T. Black, and R. A. Kosher. *Materials and Processes in Manufacturing,* 8th ed. New York: Macmillan, 1996.

Dieter, G. E. *Engineering Design: A Materials and Processing Approach.* New York: McGraw-Hill, 1983.

Dieter, G. E. *Mechanical Metallurgy,* 3rd ed. New York: McGraw-Hill, 1986.

Donaldson, C., G. H. LeCain, and V. C. Goold. *Tool Design,* 3rd ed. New York: McGraw-Hill, 1973.

Doyle, L. E. et al. *Manufacturing Processes and Materials for Engineers,* 3rd ed. Englewood Cliffs, New Jersey: Prentice-Hall, 1985.

Eary, D. F., and E. A. Reed. *Techniques of Pressworking Sheet Metal,* 2nd ed. Englewood Cliffs, New Jersey: Prentice-Hall, 1974.

Farag, M. M. *Materials and Process Selection in Engineering.* London: Applied Science Publishers, Elsevier, 1979.

Flinn, R. A., and P. K. Trojan. *Engineering Materials and Their Applications,* 2nd ed. Boston: Houghton Mifflin, 1981.

Frazelle, E., ed. *Materials Handling Systems and Terminology.* Atlanta, Georgia: Lionheart Publishers, 1992.

Gershwin, S. B. *Manufacturing Systems Engineering.* Englewood Cliffs, New Jersey: Prentice-Hall, 1993.

Grayson, M. *Recycling, Fuel, and Resource Recovery: Economic and Environmental Factors.* New York: John Wiley, 1984.

Greenwood, D. C., ed. *Product Engineering Design Manual.* Malabor, Florida: R. E. Krieger, 1982.

Groover, M. P. *Automation, Production Systems, and Computer-Aided Manufacturing,* 2nd ed. Englewood Cliffs, New Jersey: Prentice-Hall, 1987.

Groover, M. P., M. Weiss, R. N. Nagal, and N. G. Odrey. *Industrial Robotics: Technology, Programming, and Applications.* New York: McGraw-Hill, 1986.

Halevi, G. *The Role of Computers in Manufacturing Processes.* New York: John Wiley, 1980.

Harman, R. C., ed. *Handbook for Welding Design,* London: Pitman, 1967.

Hein, R. W., C. R. Loper, Jr., and C. Rosenthal. *Principles of Metal Casting.* New York: McGraw-Hill, 1967.

Hill, R. *The Mathematical Theory of Plasticity.* New York: Oxford Press, 1950.

Holland, J. R. *Flexible Manufacturing Systems.* Dearborn, Michigan: Society of Manufacturing Engineers, 1984.

Hosford, W. F. *Metal Forming: Mechanics and Metallurgy,* 2nd ed. Englewood Cliffs, New Jersey: Prentice-Hall, 1993.

Housner, H. H., and M. K. Mal. *Handbook of Powder Metallurgy.* New York: Chemical Publishing, 1982.

Iredale, R. "Automatic Assembly—Components and Products," *Metalworking Production,* 8 April 1964.

Jambro, D. *Manufacturing Processes: Plastics.* Englewood Cliffs, New Jersey: Prentice-Hall, 1976.

Johnson, W., and P. B. Mellor. *Engineering Plasticity.* New York: Van Nostrand, 1973.

Jones, W. D. *Fundamental Principles of Powder Metallurgy.* London: Edward Arnold, 1960.

Kalpakjian, S. *Manufacturiing Processes for Engineering Materials,* 3rd ed. Reading, Massachusetts: Addison-Wesley, 1997.

Kim, G. et al. "A Shape Metric for Design-for-Assembly," Proceedings of the 1992 IEEE, International Conference on Robotics and Automation, Nice, France, 1992.

Kissel, T. E. *Understanding and Using Programmable Controllers.* Englewood Cliffs, New Jersey: Prentice-Hall, 1986.

Kobayashi, A. *Machining of Plastics.* New York: McGraw-Hill, 1967.

Koeningsberger, F. *Design Principles of Metal-Cutting Machine Tools.* New York: Pergamon Press, 1964.

Koren, Y. *Computer Control of Manufacturing Systems.* New York: McGraw-Hill, 1993.

Koren, Y. *Robotics for Engineers.* New York: McGraw-Hill, 1985.

Krick, E. V. *An Introduction to Engineering and Engineering Design,* 2nd ed. New York: John Wiley, 1969.

Kronenberg, M. *Machining Science and Application.* New York: Pergamon Press, 1966.

Lancaster, J. F. *Metallurgy of Welding.* London: Chapman and Hall, 1993.

Landau, I. D. *Adaptive Control.* New York: Academic Press, 1979.

Lane, J. D., ed. *Automated Assembly,* 2nd ed. Dearborn, Michigan: Society of Manufacturing Engineers, 1986.

Lenel, F. V. *Powder Metallurgy: Principles and Applications.* New York: American Powder Metallurgy Institute, 1980.

Levy, S., and J. H. Davis. *Plastics Product Design Engineering Handbook.* New York: Van Nostrand Reinhold, 1976.

Lindbeck, J. R., and R. M. Wygant. *Product Design and Manufacture.* Englewood Cliffs, New Jersey: Prentice-Hall, 1997.

Lindberg, R. A. *Processes and Materials of Manufacture,* 4th ed. Englewood Cliffs, New Jersey: Prentice-Hall, 1997.

Machining Data Handbook, 3rd ed. Cincinnati, Ohio: Machinability Data Center, 1980.

McDonald, A. C. *Robot Technology: Theory, Design, and Applications.* Englewood Cliffs, New Jersey: Reston-Prentice-Hall, 1986.

McMahon, C., and J. Browne. *CAD/CAM from Principles to Practice.* Reading, Massachusetts: Addison-Wesley, 1993.

Metal Caster's Reference and Guide, 2nd ed. Des Plaines, Illinois: American Foundrymen Society, 1989.

Metal Casting and Molding Processes. Des Plaines, Illinois: American Foundrymen Society, 1981.

Metals Handbook (multivolume). Metals Park, Ohio: American Society of Metals, 1970.

Miles, B. L., "Design for Assembly—A Key Element within Design for Manufacture," D03388, Proceedings from the Institution of Mechanical Engineering, London, England, 1989.

Miyakawa, S., and T. Ohashi, "The Hitachi Assemblability Evaluation Method (AEM)," Proceedings of the International Conference on Product Design for Assembly, Newport, Rhode Island, April 1986.

Miyazawa, A., "Productivity Evaluation System," *Fujitsu Science Technology Journal,* 29.4, pp. 425–431, (December 1993).

Modern Plastics Encyclopedia. New York: McGraw-Hill, annual.

Nadai, A. *Theory of Flow and Fracture of Solids,* 2nd ed. New York: McGraw-Hill, vol. I, 1950, vol. II, 1963.

Nagey, F., and A. Siegler, eds. *Engineering Foundations of Robotics.* Englewood Cliffs, New Jersey: Prentice-Hall, 1987.

Nanfara, F., T. Uccello, and D. Murphy. *The CNC Workbook: An Introduction to Computer Numerical Control.* Reading, Massachusetts: Addison-Wesley, 1995.

Niebel, B. W., and A. B. Draper. *Product Design and Process Engineering.* New York: McGraw-Hill, 1974.

Ostwald, P. F. *Engineering Cost Estimating.* Englewood Cliffs, New Jersey: Prentice-Hall, 1992.

Ostwald, P. F., and J. Munoz. *Manufacturing Processes and Systems,* 9th ed. New York: John Wiley, 1997.

Pohlandt, K. *Materials Testing for the Metal Forming Industry.* Berlin: Springer-Verlag, 1989.

Ranky, P. G. *Computer Integrated Manufacturing.* Englewood Cliffs, New Jersey: Prentice-Hall, 1986.

Reed-Hill, R. E., and R. Abbaschian. *Physical Metallurgy Principles,* 3rd ed. Boston, Massachusetts: PWS Publishing Company, 1994.

Rembold, V., B. O. Nnaji, and A. Storr. *Computer-Integrated Manufacturing and Engineering.* Reading, Massachusetts: Addison-Wesley, 1993.

Rowe, G. W. *Elements of Metalworking Theory.* London: Edward Arnold, 1979.

Rowe, G. W. *Principles of Industrial Metalworking Processes.* London: Edward Arnold, 1977.

Rowe, G. W., C. E. N. Sturgess, P. Hartley, and I. Pillinger. *Finite-Element Plasticity and Metalforming Analysis.* London: Cambridge University Press, 1991.

Sava, M., and J. Pusztai. *Computer Numerical Control Programming.* Englewood Cliffs, New Jersey: Prentice-Hall, 1990.

Schaffer, J. P. *Materials Engineering.* Homewood, Illinois: Irwin, 1994.

Seames, W. S. *Computer Numerical Control: Concepts and Programming.* Albany, New York: Delmar Publishers, 1986.

Shina, S. G. *Concurrent Engineering and Design for Manufacture of Electronics Products.* New York: Van Nostrand Reinhold, 1991.

Sindo, Kou. *Transport Phenomena and Materials Processing.* New York: John Wiley, 1996.

Singh, N. *Systems Approach to Computer-Integrated Design and Manufacturing.* New York: John Wiley, 1996.

Slocum, A. H. *Precision Machine Design.* Englewood Cliffs, New Jersey: Prentice-Hall, 1992.

Source Book on Brazing and Brazing Technology. Metals Park, Ohio: American Society of Metals, 1980.

Source Book on Electron Beam and Laser Beam Welding. Metals Park, Ohio: American Society of Metals, 1981.

Source Book on Innovative Welding Process. Metals Park, Ohio: American Society of Metals, 1981.

Starrgaard, A. C., Jr. *Robotics and AI: An Introduction to Applied Machine Intelligence.* Englewood Cliffs, New Jersey: Prentice-Hall, 1987.

Steel Castings Handbook, 5th ed. Rocky River, Ohio: Steel Founders Society of America, 1980.

Teicholz, E. *CAD/CAM Handbook.* New York: McGraw-Hill, 1985.

Thomsen, E. G., C. T. Yang, and S. Kobayaski. *Mechanics of Plastic Deformation in Metal Processing.* New York: Macmillan, 1964.

Tipping, W. V. "Component and Product Design for Mechanized Assembly," Conference on Assembly, Fastening, and Joining Techniques and Equipment, P.E.R.A., 1965.

Tobias, S. A. *Machine Tool Vibrations.* Glasgow: Blackie & Son, 1965.

Tool and Manufacturing Engineers Handbook, 4th ed. Dearborn, Michigan: Society of Manufacturing Engineers, 1983 and 1984.

Trucks, H. E. *Designing for Economical Production.* Dearborn, Michigan: Society of Manufacturing Engineers, 1974.

Turning Handbook of High-Efficiency Metal Cutting. Detroit, Michigan: Carboloy Systems, General Electric, 1980.

Ulrich, K. T., and S. D. Eppinger. *Product Design and Development.* New York: McGraw-Hill, 1994.

Ulsoy, A. G., and W. DeVries. *Microcomputer Applications in Manufacturing.* New York: John Wiley, 1989.

Vajpayee, S. K. *Principles of Computer-Integrated Manufacturing.* Englewood Cliffs, New Jersey: Prentice-Hall, 1995.

Valliere, D. *Computer-Aided Design in Manufacturing.* Englewood Cliffs, New Jersey: Prentice-Hall, 1990.

Van Vlack, L. H. *Materials for Engineering: Concepts and Applications.* Reading, Massachusetts: Addison-Wesley, 1997.

Wagoner, R. H., and J. Chenot. *Fundamentals of Metal Forming.* New York: John Wiley, 1997.

Welding and Fabricating Data Book. Cleveland, Ohio: Penton/IPC, 1981.

Yankee, H. W. *Manufacturing Processes.* Englewood Cliffs, New Jersey: Prentice-Hall, 1978.

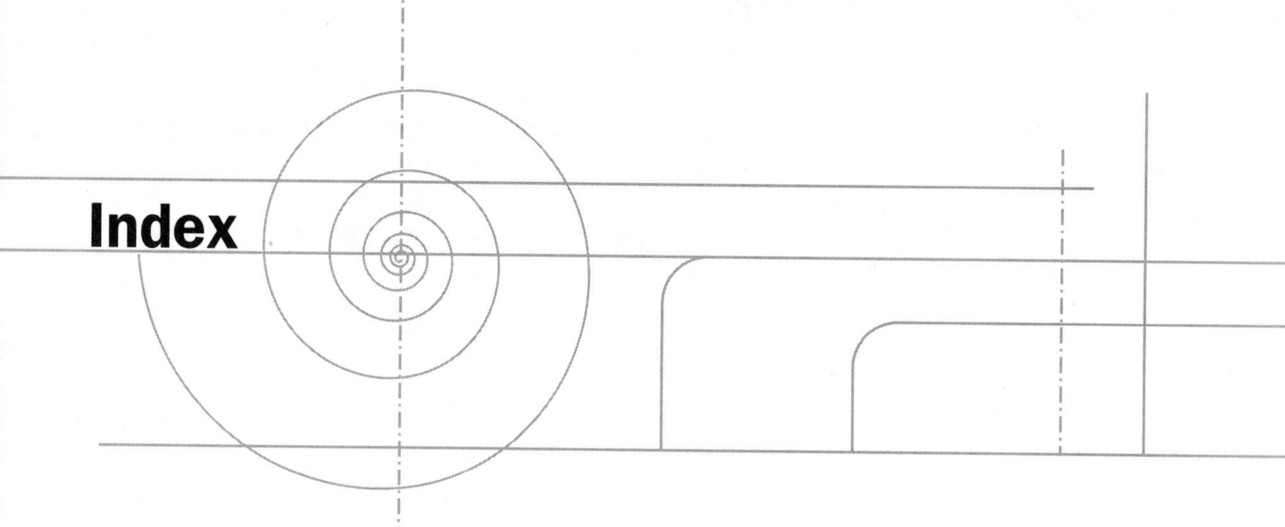

Index